THE INVERTEBRATE WORLD OF AUSTRALIA'S SUBTROPICAL RAINFORESTS

Geoff Williams

CSIRO PUBLISHING

To Alfred Russel Wallace (1823–1913)
for giving us the theory of the origin and divergence of species through natural selection, and the theory of biogeography.
But first and foremost for being an inspiring natural historian.

A catalogue record for this book is available from the National Library of Australia.

ISBN: 9781486312917 (hbk)
ISBN: 9781486312924 (epdf)
ISBN: 9781486312931 (epub)

How to cite:
Williams G (2020) *The Invertebrate World of Australia's Subtropical Rainforests*. CSIRO Publishing, Melbourne.

Published by:

CSIRO Publishing
36 Gardiner Road, Clayton VIC 3168
Private Bag 10, Clayton South VIC 3169
Australia

Telephone: [+613] 9545 8555
Email: csiropublishing@csiro.au
Website: www.publishing.csiro.au

Front cover: (main) Richmond birdwing butterfly, *Ornithoptera richmondia* (Papilionidae) (photo: Carl Bento, Australian Museum). (top, left to right) *Hadronyche* sp. (Hexathelidae), *Scaptia barbara* (Tabanidae), *Paraschizognathus ?ocularis* (Scarabaeidae), *Cabasa ?pulchella* (Asilidae), *Calodema regale* (Buprestidae) (photos: Geoff Williams)
Back cover: (top) *Eborophyllus middletoni* (Cerambycidae), (middle) Regent Skipper, *Euschemon rafflesia rafflesia* (Hesperiidae), (bottom) *Australodon nearnsi* (Cerambycidae) (photos: Kevin Mitchell)

All photographs by the author except *Ornithoptera richmondia* (Carl Bento/Australian Museum), *Euschemon rafflesia rafflesia, Australodon nearnsi, Eborophyllus middletoni* (Kevin Mitchell) and *Thersites mitchellae* (Mark V. Robinson).

Edited by Joy Window (Living Language)
Cover design by Andrew Weatherill
Typeset by Envisage Information Technology
Printed by Ingram Lightning Source

Feb26_RP_ILS

Contents

Acknowledgments

Foremost I have to thank my partner Thusnelda for countless hours of great company and untiring help in the field, my children Beren, Sarid and Arwen for lightening the days in so many ways, and my parents Arthur and Patricia for their encouragement (and cheerful acceptance when in earlier years many a strange creature was brought into their lives). To Eloise Moir-Ford, Lauren Webb, Tracey Kudis (CSIRO Publishing) and Joy Window (copy editor, Living Language) I also express special thanks. Without their diligence and encouragement, the publication of this book would not have been possible.

Over the years of field investigation and research I have benefited from discussions and interactions with numerous friends and colleagues: notably Paul Adam at the University of New South Wales and Dan Bickel at the Australian Museum. Many others have contributed to my knowledge and understanding of rainforests and Australia's plant communities and their biota in general, or have given support in diverse ways: Rebecca Johnson, Alexandra Durham, Mike Gray, Glenn Hunt, David McAlpine, Shane McEvey, Jacquieline Rescei, Russell Cox, Chris Reid, Courtenay Smithers, Derek Smith, Helen Smith, Max Moulds, Dave Britton, John Macdonald, Graham Milledge, Natalie Kees, Timothy Lee, (Australian Museum, Sydney), John Hunter, Terry Evans (NSW NPWS), Steve Nikitin, Clarry Chadwick (NSW Department of Agriculture), Ross Storey (Queensland Department of Primary Industries), Günther Theischinger (Grays Point), Ev. Britton, Phil Carne, John Lawrence, Barry Moore, Laurence Mound, Tom Weir (CSIRO, Canberra), Eric Matthews (South Australian Museum), Shelley Barker (University of Adelaide), Ken Walker, (Museum of Victoria), Allen Sundholm, Roger de Keyser (Sydney), Hugh and Nan Nicholson (Terania Creek), Paul Brock (Natural History Museum, London), Terry Houston (Western Australian Museum, Perth), and Geoff Monteith, Christine Lambkin and John Stanisic (Queensland Museum). Kevin Mitchell (Wherrol Flat), Mark Robinson (Sawtell) and the Australian Museum kindly allowed me to use their photos of *Euschemon rafflesia*, *Australodon nearnsi*, *Eborophyllus middletoni*, *Thersites mitchellae* and *Ornithophtera richmondia* respectively. Eric Matthews kindly allowed the use of Fig. 3 and Jane Hosking (Taree) is thanked for drawing Fig. 2. Not least of all, I acknowledge my colleague and friend, the late Charles 'Chuck' Bellamy, previously with the University of Pretoria (Republic of South Africa) and the California Department of Food and Agriculture (Sacramento, USA).

Too many are now gone, but all are thanked for their diverse free giving of time and thoughts. In conclusion this book is indebted to the often unsung research undertaken by Australian invertebrate taxonomists, systematists and field naturalists.

1 Introduction

The Invertebrate World of Australia's Subtropical Rainforests is about the 'lesser animals': the insects, crustaceans, spiders, harvestmen, mites, scorpions, centipedes, millipedes, true worms, flatworms, nematodes, snails and slugs, and the ancient *Peripatus* or velvet worms. These organisms without backbones, inconspicuous as most are, dominate Earth's terrestrial animal diversity.

The scope of this book takes a broad latitudinal sweep spanning ~23.4–37°S (this to about the southern-most occurrence of littoral rainforest), stretching inland to the dry vine thickets of Queensland and New South Wales, and addressing rainforests as 'subtropical' loosely in a geographical sense (see Fig. 1). Figure 2 shows the locations of the places referred to in this book. Within this geographical range there are several overlapping plant thermal response groups:

- a humid mesotherm dominated (mean annual air temperature >14–<24°C) province centred on coastal southern Queensland and northern New South Wales, but extending to the south coast of New South Wales
- a subtropical – lower montane province with a microtherm–mesotherm humid climate (mean annual air temperature >12–<14°C) whose floristic elements encompass upland *Nothofagus moorei*–dominated rainforest stands
- and a geographically extensive subhumid–humid (mesotherm–megatherm/>20–<24°C) warm subtropical zone that extends from the Northern Tablelands of New South Wales, largely reaching the Queensland coast north of Brisbane and extending to the subhumid tropics inland from Cairns (Adam 1992; Nix 1982).

These capture warmer lowland and lower latitude forests that are subtropical by measure of climate (see Corlett 2013), and cooler upland and higher latitude forests that are 'temperate' (see Adam 1992; Nix 1982; Webb and Tracey 1981). Ultimately, however, altitude and geological history will confound characterisations of rainforest based solely on latitude.

Subtropical rainforest in the context of this definition includes 'closed forest' formations of great floristic and structural diversity, existing under marked differences of altitude, mean air temperature range, rainfall and soil type. A floristically circumscribed characterisation of subtropical rainforest is probably elusive, for they contain combinations of the original Gondwana flora (alluding to plants that evolved in areas of the Gondwana supercontinent, now mainly represented by Antarctica, South Ameria, Africa, Madagascar, Australia, India, New Zealand and New Caledonia) (see 'The formation of Gondwanan landmasses' on p. 5 for a cautionary note on the use of the often conflated term 'Gondwanan') that stretch back to at least the Early Miocene (Adam 1987, 1992; Floyd 1990, 2010; Kitching *et al.* 2010; Appendix 5), autochthonous elements, as well as taxa that represent

Figure 1: Distribution of forests and rainforests in Australia, 2013. This book focuses on the subtropical rainforests of New South Wales and Queensland, shown in black shading within the boxed area. Source: ABARES, Commonwealth of Australia, 2016, CC BY 3.0. http://www.agriculture.gov.au/abares/forestsaustralia/profiles/rainforest

later 'tropical' intrusions. These rainforests have acted as refugia over lengthy geological cycles of expansion and contraction (e.g. Dorrigo and the Nightcap–Border Ranges, in New South Wales) and include those (e.g. Washpool) that are the result of colonisation events (Rossetto *et al.* 2015). Such patterns of retreat to refugia followed by recolonisation and expansion during environmentally conducive periods characterise the history of Australia's diverse rainforests (Hilbert *et al.* 2007; Moritz *et al.* 1997; Price and Sobbe 2005; Shapcott *et al.* 2015; VanDerWal *et al.* 2009), for these rainforests have been subject over long periods of time to the oscillating influence of changing climatic events (i.e. periods of warm–wet and cool–dry), in particular the progressive aridification trend from the Late Miocene (McLaren and Wallace 2010) that has led to the fragmentation of an earlier, more widespread and diverse mesic-adapted biota (that existed during warm–wet conditions 20–10 million years ago (mya)), and the emergence of an extensive xeric-adapted one (Adam 1992; Byrne *et al.* 2008; Martin 2006; Price and Sobbe 2005; Truswell 1993). The distribution and areal extent of subtropical rainforest first encountered by European colonists strongly mirrored that influence; that which survived is set, as floristically and structurally distinct plant communities that reflect varying geological episodes (Benson and Redpath 1997;

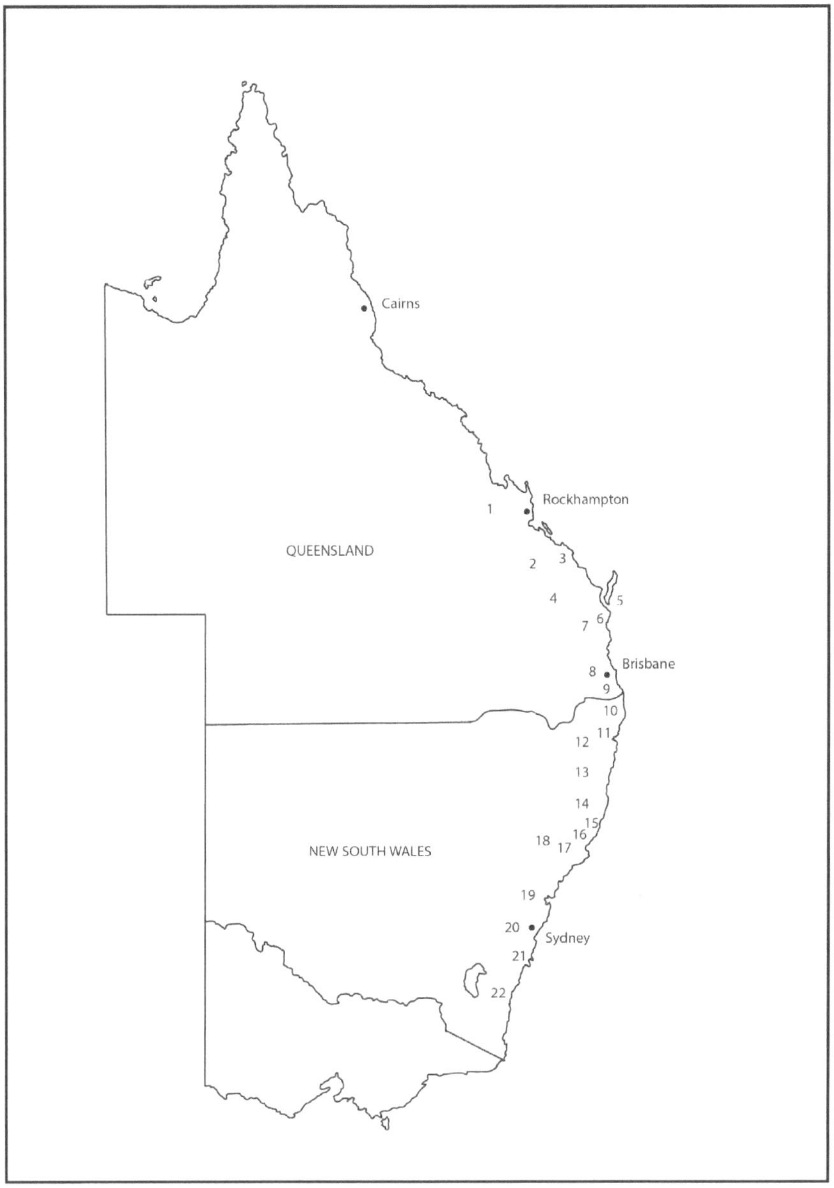

Figure 2: Locations mentioned in text. **Queensland: 1.** Blackdown Tableland, **2.** Kroombit Tops, **3.** Eurimbula NP, **4.** Goodnight Scrub, **5.** Fraser Island, **6.** Cooloola NP, **7.** Conondale NP, Conondale Range, Imbil State Forest, Kenilworth SF, **8.** D'Aguilar NP, D'Aguilar Range, Mt Glennie, Mt Glorious, Mt Mee SF, Mt Nebo, **9.** Cunninghams Gap, Lamington NP, McPherson Range, Main Range, Mount Barney NP, Mt Tamborine, Springbrook NP. **New South Wales: 10.** Acacia Plateau, Boatharbour NR, Border Ranges NP, Brunswick Heads, Cambridge Plateau, Koreelah NP, Lismore, Mallanganee, Moore Park NR, Mount Nothofagus, Mt Warning NP, Murwillumbah, Nightcap NP, Nightcap Range, Richmond Range NP, Stotts Island NR, Terania Creek, Tooloom Scrub, Unumgar SF, Victoria Park NR, Whian Whian SF, Wiangarie SF, Yabbra SF, **11.** Iluka NR, **12.** Forest Lands SF, Gibraltar Range NP, Washpool NP, **13.** Chaelundi NP, Mt Hyland NR, **14.** Bruxner Park, Carrai SF, Dorrigo NP, Dorrigo Plateau, Glenugie Peak, Hastings Valley, Mt Banda Banda, New England NP, League Scrub, Nulla Five-day SF, Styx River State Forest, Werrikimbe NP, Willi Willi NP, **15.** Boonanghi SF, Crescent Head, Kerewong SF, Laurieton, Macleay Valley, Sea Acres NR, Way Way SF, **16.** Black Head, Boorganna NR, Cape Hawke, Comboyne Plateau, Coocumbac Island NR, Coorabakh NP, Dingo Tops, Dooragan NP, Harrington, Kiwarrak SF, Lansdowne SF, Lansdowne NR, Lansdowne-Comboyne Escarpment, Lorien Wildlife Refuge, Tapin Tops NP, Wingham Brush NR, **17.** Allyn River, Barrington Tops, Bulahdelah, Chichester SF, Mt Royal Range, Tuglo Wildlife Refuge, Woko National Park, **18.** Cedar Brush NR, Liverpool Range, **19.** Watagans SF, **20.** Blue Mountains, Kanangra Boyd NP, Mt Irvine, Mt Wilson, Wollemi NP, **21.** Barrengarry Mountain, Cambewarra Mountain, Illawarra Escarpment, Jamberoo Pass, Minnamurra Falls, Moreton NP, Mount Keira, Robertson Plateau, **22.** Brown Mountain, Clyde Mountain, Deua NP, Durras Mountain, Mount Dromedary.

What is 'subtropical' rainforest?

This book considers rainforests and their included biota in a region extending in eastern Australia from about Rockhampton in the north to the south coast of New South Wales. In this sense 'subtropical' is a convenient geographical construct within the latitudinal scope of which there is considerable heterogeneity of rainforest vegetation, stand disjunction and size, climate, soil types, and landscape relief. The included biota is a rich one, reflecting diverse evolutionary histories. Using a structural–physiognomic classification the term 'subtropical' spatially reduces the concept to a suite of notophyll vine forest subformations categorised by complexity of leaf and growth forms, and diversity of strata (see discussions in Adam 1987, 1992; Winter *et al.* 1987). Although the floristic differences in individual stands, owing to changes in soil type, can be striking, rarely can a clear floristic or structural boundary be drawn between what are frequently intergrading rainforest subformations and associations. High numbers of tree species and diverse canopy architecture mosaics define more complex stands, with localised patterns of species recruitment and dominance reflecting episodes of chance dispersal, colonisation and extinction. Altitude alone may allow coarse categorisation of stands, but by itself is not a predictor of species presence/absence or structural complexity.

Although there is a large suite of trees and shrubs that are characteristic of subtropical rainforest, as classically considered in New South Wales (e.g. Adam 1987, 1992; Floyd 1990; Harden *et al.* 2006; McDonald and Hunter 2010; and others), a strict floristic definition is not possible. Variation is the dominating theme and so assigning individual stands to the various floristic alliances that have been proposed (e.g. Floyd 1990) is problematic, and may overlook the dynamic nature of plant composition. Numerous tree species are common to a range of subformations, and are found at greatly varying altitudes and latitudes, such that the occurrence of 'within stand' populations of many cannot always be predicted. Thus the use of floristic criteria to recognise and classify the various alliances that have been proposed, as Adam (1992) notes, may be difficult to apply. However, if 'dry rainforest', 'warm temperate rainforest' and 'littoral rainforest' are considered as floristic derivatives of 'subtropical rainforest', then the distinction can be more narrowly stated as formations that are subtropical by virtue of obvious floristic and structural diversity and upland stands (in which microphyll–nanophyll leaf forms dominate) that are intuitively 'cool temperate'. But regardless of how individual formations are classified or perceived Australia's subtropical rainforests, as a geographical entity, are a rich storehouse of biological diversity.

Bowman 2000; Rossetto *et al.* 2015), in a landscape context in which the intervening vegetation matrix is largely one of eucalypt-dominated 'wet sclerophyll' and 'dry sclerophyll' forests (Adam 1987; McDonald and Hunter 2010) (see Plate 312). However, there is no simple conceptual demarcation of Australian rainforest, for wet sclerophyll forest (a floristically unstable entity) (see Plates 302, 303) has a greater ecological and functional similarity to rainforest than to dry (open) sclerophyll forest (Adam 1987, 1992; Tng *et al.* 2013; Warman and Moles 2009; Warman *et al.* 2013), the nature of the understorey plant community, and edaphic and microclimate conditions, allowing the diffusion of otherwise rainforest-restricted animals and ecological interactions. And, indeed, if wet sclerophyll forest is considered a transient seral stage in the establishment of rainforest (with species such as *Eucalyptus grandis* and *Lophostemon confertus* being relegated to 'early mesic forest colonizer' status), then wetter mixed forest communities with sclerophyll overstoreys may be considered within the category of rainforest (Dale *et al.* 1980; P. Richards 1952; Schimper 1903), albeit, as Adam (1992) notes, one 'with special features'. Ecological processes are not constrained by floristic boundaries, and the invertebrate biota of eucalypt-dominated mesic forests often reflects close ancestral relationships; thus the discussion that follows often considers the interrelated natures of rainforest communities and mesic sclerophyll forest, frequently acknowledging associated ecological and biotic attributes entwined between the two.

Defined solely by latitude, the greatest concentration of Australia's subtropical rainforests occurs mainly from southern Queensland to northern New South Wales (Adam 1987, 1992; Floyd 1990); this is also a region where southern temperate faunas and floras overlap in rainforests occurring along the Great Dividing Range. However, in addition to this core area there are geographically isolated rainforest stands that extend archipelago-like from

The formation of Gondwanan landmasses

The term 'Gondwanan' can imply patterns or processes, such as vicariance events due to the fragmentation of the supercontinent, or distribution in the Southern Hemisphere by more recent processes such as long-distance dispersal of younger age.

The diversification of numerous plants and animals relates to the breakup of the megacontinent Pangaea (Jokat *et al.* 2003; McLoughlin 2001; Taylor *et al.* 2009), which itself formed from the coalescence of earlier landmasses, with the evolution of many living families of insects, and other terrestrial invertebrates and vertebrates, originating during the Mesozoic (collectively the Triassic, Jurassic, Cretaceous periods) (Grimaldi and Engel 2005), an era of geological time extending from ~250 to 65 mya. The breakup of Pangaea, beginning in the Early–Middle Jurassic (~175 mya) and giving rise to the supercontinents Laurasia (North America, Europe, Asia) and Gondwana, shaped the development and subsequent widespread separation (vicariance) of closely related invertebrate lineages (e.g. Borda *et al.* 2008; Martínez-Aquino *et al.* 2017).

Following the rifting of Pangaea, Gondwana further split, initially into East Gondwana (Africa, Madagascar, India) and West Gondwana (Australia, Antarctica, South America, New Zealand, New Caledonia). During the Early Cretaceous (121 mya) Madagascar and India broke away from Africa, rafting to the south-east, with India separating from Madagascar in the Late Cretaceous (88–84 mya). India then drifted northward to eventually collide with Asia in the Early Tertiary (~50 mya). South America began to separate from Africa in the Early Cretaceous (135 mya), moving south-west to join Antarctica, together with Australia and New Zealand, as a single landmass. Subsequently, these separated from each other in the Oligocene (30–28 mya). Australia had finally separated from Antarctica by ~35–33 mya, moving northwards to its present position (Neall and Trewick 2008). There are variations on this time scale, and competing arguments as to the nature of plate/landmass associations and divergences (e.g. Ali and Aitchison 2008; McCarthy *et al.* 2007; Sanmartín and Ronquist 2004; Waters and Craw 2006). In particular, molecular phylogenetic studies have challenged the traditional timetable of the sequential continental breakup, indicating that the connection among major landmasses might have been prolonged until the Late Cretaceous based on the timing of divergence (e.g. Van Bocxlaer *et al.* 2006; Krause *et al.* 1997); this implies an extended opportunity for biotic interchange.

Nevertheless, as Dennis McCarthy and his colleagues reflected (McCarthy *et al.* 2007), Gondwana is a biological and paleontological phenomenon, 'and today remains an elegant symbol of the intricate relationship between geological and biological history'.

about Gladstone and Rockhampton in central eastern Queensland to the western slopes and south coast of New South Wales (Adam 1992; Curran *et al.* 2008; Harden *et al.* 2006; Helman 1987; Mills 1987; Kooyman *et al.* 2012), with floristically depauperate rainforests occurring in eastern Victoria, at altitudes <500 m (which are considered temperate examples [Cameron 1987]). There is a distinct north–south loss of floristic richness, but within this trend there are southern stands of high floristic diversity (e.g. Sea Acres Nature Reserve in Port Macquarie, northern New South Wales).

These closed forest communities possess an invertebrate biota that reflects diverse evolutionary histories (see Williams 1993, 2002, 2003a; Williams and Bickel 2010), upon which, together with my own field records over the last 40 or so years, much of this book is founded. In addition to the impact of climate oscillations over long spans of geological time (see Appendix 5) the evolution of the Australian subtropical biota has been influenced by colonisation from the north as the Australian continental plate rafted towards Malesia; this is the biogeographical region that includes the Malay Peninsula, Indonesia, New Guinea and the Bismark Archipelago. Nevertheless, a high proportion of the fauna appears to be autochthonous (considered to have originated in this region).

The rainforests represent a significant zoogeographical refugium in terms of the evolution of Australian invertebrate taxa (see Bryant and Krosch 2016 for a review of eastern Australian biogeographical barriers to rainforest taxa). This is especially true of the Gondwana Rainforest World Heritage Area (formerly known as the Central Eastern Rainforest Reserves of Australia, or 'CERRA') located from the Barrington Tops in north-eastern New South

Wales to south-eastern Queensland (Adam 1987, 2017; Kitching *et al.* 2010; see Appendix 2). In this book I use 'Gondwana rainforests' (as in Kitching *et al.* 2010) to indicate taxa recorded from the World Heritage Area. It replaces the acronym 'CERRA' used in Williams (2002) and Hunter (2003). 'Gondwana rainforests' is a useful identifier, specifically for World Heritage Area sites, that highlights the biodiversity and biogeographical significance of the World Heritage reserves (Adam 1987, 2017), rather than the alternative of the repetitive and unwieldy 'northern New South Wales–southern Queensland', and allows a cross-reference to the individual invertebrate records previously and more fully cited in Williams (2002, 2003a).

Examples of evolutionarily significant invertebrate taxa include the terrestrial snail family Charopidae, the freshwater snail family Hydrobiidae, the beetle taxa Adeliini, Cyphaelini, Dichotomini, Denticollinae, Rutelinae, Melolonthinae, Lucanidae, flightless beetles in the family Carabidae, the fly family Drosophilidae and dolichopodid subfamily Sciapodinae, parastacid crayfish and their ectosymbiotic flatworms, aradid bugs, the Onychophoran family Peripatopsidae and mygalomorph spiders (Williams 2002). Many taxa are thought to be 'Gondwanan' in origin (see Heatwole 1987; Main 1987), this term taking its name from the indigenous Gond tribe of India. Their extant distributions suggest a Gondwanan origin, but the occurrence of taxonomically related organisms in these southern rainforests does not necessarily establish a Gondwanan ancestry. Some groups are cosmopolitan or have widely distributed rainforest-associated relatives. Others have invaded through periods of geological time from the north or are related to southern Australian temperate rainforest invertebrate faunas, and are restricted to rainforests at higher elevations; these are captured within the 'subtropical' net of this book simply by virtue of the presence of distinctive faunistically overlapping zones in this range of latitudes.

Apparent extant biogeographical patterns are seductive, but may not truthfully reflect past events and lineages. In some instances, as with the conspicuous rainforest ant genus *Leptomyrmex*, a series of selective extinctions have resulted in an apparent Gondwanan affinity simply because organisms have disappeared from elsewhere (Lucky and Ward 2010; Williams and Bickel 2010). This applies equally to the bombyliid fly genus *Comptosia*, for although the extant fauna is restricted to Australia, with close relatives in southern South America, Eocene (56–33.9 mya) fossils from Germany and North America argue against a previously held view that the genus was of Gondwanan origin (Wedmann and Yeates 2008). And sometimes plants and animals display an asymmetrical ancestry, as in the instance of several leafhopper bugs (Membracidae), which occur on the distinctly Gondwanan plant families Myrtaceae and Proteaceae yet the bugs themselves are more geologically recent invaders (Wallace and Deitz 2006). In addition, many northern invaders reach their limits of distribution in the subtropical rainforests of northern New South Wales and southern Queensland. There is often a latitudinal 'sifting' or 'filtering' as particular taxa extend south or north along the east coast (as in the beetle family Carabidae and numerous land snails).

Alfred Russel Wallace, the father of evolutionary biogeography

This book starts with a dedication to a scientist that few may have heard of: Alfred Russel Wallace (1823–1913). He was a self-educated optimistist and Utopian socialist of impoverished means, and ultimately obliging and generous to his detriment. By unanticipated twists of circumstance Wallace declined credit for first establishing the theory of evolution by natural selection, yet founded the science and theory of biogeography (or zoogeography for those who study only animals), the study of the geographical distribution of plants and animals (Wallace 1876). Wallace independently discovered natural selection (Costa 2014), and is notoriously known for writing the letter that prompted Charles Darwin to finally construct a hurried outline of his theory that would be formulated in his book, *On the Origin of Species.* Darwin had digested his thoughts on the subject for many years, but had written nothing concrete or finished. Driven by economic necessity, the unmarried and relatively obscure Wallace had spent years in the Moluccas (what is now largely Indonesia), studying and collecting animals on the many islands there. In February of 1858, alone on the small island of Ternate and suffering from the debilitations of malaria, Wallace wrote *On the Tendency of Varieties to Depart Indefinitely from the Original Type*; the so-called 'Ternate paper'. It was a complete and lucid picture of evolution by natural selection. His paper was 3764 words long. At the time Darwin's thoughts were unformulated, for he was still grappling with the riddle of divergence, and any final resolution seemed beyond his grasp. In early March of that same year, Wallace, for seemingly good personal reasons, sent his paper to Darwin (whom he revered) for comment, and in the hope that Wallace's idea would assist Darwin by supplying the missing factor that Darwin needed to explain the origin of species. The rest is contentious history (Brackman 1980). Wallace returned to Britain to write many things about many subjects (Berry 2002), but until the end he stood in loyal defence of Darwin – not least of all in the pages of *Darwinism*, Wallace's authoritative summary of evolutionary biology published in 1889, several years after Darwin's death. Wallace's Line (also called the Wallace Line), that demarcation (though now perhaps a scientific curiosity) between the particular absence and presence of species, and located between Borneo and Sulawesi and Bali and Lombok, still bears fitting tribute to him, as does the transitional zone known as 'Wallacea' (see Holt *et al.* 2013a, 2013b; Kreft and Jetz 2010; 2013; and Proche and Ramdhani 2012 for a discussion of zones). But the islands in the time of Wallace, then so rich in forests and wildlife, live in their former biological glory now only on the pages of his monumental *The Malay Archipelago* (van Oosterzee 1997).

Nevertheless, there is a significant number of taxa with Gondwanan or 'Old Southern endemic' affiliations (Table 1). This group includes taxa with relatives on other Gondwanan landmasses such as South America, Africa and the Indian subcontinent, and those restricted to Australasia. Some are restricted to Australia but with close relatives in either New Caledonia or New Zealand, but not both. Within known ranges individual taxa can be widespread or relictual. Examples of 'southern' fauna exist across all taxonomic levels – from species to that of subphylum. Notable higher taxonomic rank examples are the megascolecid earthworms, mygalomorph and amphectid spiders, harvestmen in the family Acropsopilionidae, the mite family Pheroliodidae, and the terrestrial snail families Athoracophoridae, Charopidae, Cystopeltidae and Rhytididae. The insects are well represented, and include the beetle tribes Pamborini, Adeliini, Epistomentini and Stigmoderini, the subfamily Spilopyrinae, the family Migadopinae, flies in the family Pelecorhynchidae, the subfamily Arachnocampinae and the tribe Pangoniini, the moth families Hepialidae and Micropterigidae, the birdwing butterfly genus *Ornithoptera*, the hemipteran bug families Idiostolidae and Peloridiidae, the neuropteran lacewing subfamilies Kempyninae and Stenosmylinae, ambrositrine and hyptiogastrine wasps, and the plecopteran families Austroperlidae, Eustheniidae and Griptopterygidae.

The region is also one in which the fauna exhibits a high degree of short-range endemism (Table 1) (notably in stream-, cave- and karst-associated species); this is defined by species with a range <10 000 km^2 and further characterised by poor powers of dispersal and confinement to discontinuous habitats (Harvey 2002), though not always rainforest. Groups in this category include Oligochaeta, Onychophora, Mygalomorphae, Schizomida, Diplopoda and Decapoda. With respect to subtropical rainforest only, much of the terrestrial and freshwater invertebrate fauna is endemic at various hierarchical levels, particularly those of species and genus but also extending to that of family. In addition, many species are currently known only from single localities. Instances of generic and species endemism are particularly high in the snail families Hydrobiidae and Charopidae; the earthworm family Megascolecidae; the crayfish family Parastacidae; the flatworm family Temnocephalidae; the subphylum Onychophora; the mite family Platyameridae; and the spider suborder Mygalomorphae. Examples within the insects are the fly families Dolichopodidae, Platystomatidae, Exeretonevridae, Pelecorhynchidae and superfamily Tipuloidea; the beetle tribes Onthophagini and Canthonini; the beetle families Buprestidae, Lucanidae, Carabidae, Rhinorphidae and Lamingtoniidae, and scarabaeid subfamilies Melolonthinae and Rutelinae; australembiid webspinners; cicadelloid and mezirine Hemiptera; oechophorine moths; and the lacewing family Hemerobiidae and the subfamily Kempyninae (Williams 2002).

However, significant invertebrate biodiversity values in south-eastern mainland Australia are not restricted to taxa inhabiting subtropical rainforest. Although subtropical rainforest, before European occupation, did include notable examples of expansive rainforest stands (e.g. lowland Illawarra, the Comboyne and Dorrigo plateaux, and the Big Scrub rainforests in New South Wales), subtropical rainforest more normally is encountered as patch-like stands of varying size and floristic and

Table 1: Examples of terrestrial invertebrate taxa exhibiting high levels of endemism, Gondwana affiliations or zoogeographical significance within the Gondwana Rainforests region (after Williams and Bickel 2010).

Refugial zoogeographical significance	Gondwanan affiliations	Taxa with high incidence of endemicity at genus and species level
Snail family Charopidae	Snail family Charopidae, earthworm family Megascolecidae	Snail families Charopidae, Camaenidae
Beetle tribes Adeliini, Cyphaleini, Dichotomini, beetle families Denticollinae, Rutelinae, Melolonthinae and family Lucanidae	Mygalomorph and amphectid spiders	Earthworm family Megascolecidae, flatworm family Temnocephalidae
Flightless carabid beetles	Harvestman family Acropsopilionidae	Subphylum Onychophora
Fly family Drosophilidae and subfamily Sciadopinae	Mite family Pheroliodidae	Mygalomorph spiders
Bug family Aradidae	Crayfish family Parastacidae	Crayfish family Parastacidae
Velvet worm family Peripatopsidae	Snail families Athoracophoridae, Charopidae, Cystopeltidae, Rhytidae	Fly families Dolichopodidae, Exeretonevridae, Pelecorhynchidae, Platystomatidae, Tipuloidea
Mygalomorph spiders	Beetle family Phloeostichidae, subfamily Migadopinae, and tribes Adeliini, Epistomentini, Pamborini, Stigmoderini	Beetle families Lucanidae, Carabidae, subfamilies Rutelinae, Melolonthinae, Stigmoderinae and tribes Onthophagini, Canthonini
Stygofauna (taxa associated with groundwater)	Fly family Pelecorhynchidae, subfamily Arachnocampinae and tribe Pangoniini	Australembiid web-spinners
	Scorpion fly family Nannochoristidae	Bug subfamily Mezirinae
	Moth families Hepialidae, Micropterigidae	Moth subfamily Oecophorinae
	Butterfly genus *Ornithoptera*	Lacewing family Hemerobiidae and subfamily Kempyninae
	Bug families Idiostolidae, Peloridiidae	Wasp family Thynnidae
	Lacewing subfamilies Kempyniinae, Stenosmylinae	
	Wasp subfamilies Ambositrinae, Hyptiogastrinae, family Thynnidae	

structural composition in an intervening matrix of sclerophyllous forests, and more rarely woodlands, shrub, heath and swamp-mire complexes, and associated freshwater ecosystems. These habitats sustain distinctive non-rainforest invertebrate faunas that nevertheless are often ecologically interactive (e.g. the recruitment of certain pollinators and decomposers) or share a close evolutionary ancestry. Important higher taxa inhabiting these ecosystems are ruteline Christmas beetles in the tribe Anoplognathini, the family Elateridae, the speciose buprestid genus *Castiarina*, the tenebrionid tribe Heleini, the fly genera *Pelecorhynchus* and *Trichophthalma*, petalurid dragonflies, lestoideid damselflies, and athoracophorid, helicarionid and glacidorbid snails. Limestone outcrops, often set within a rainforest context, are important focal points of terrestrial snail endemism and diversity (Stanisic 1994, 1997; Stanisic *et al.* 2010). Cave systems are important for the conservation of associated, often endemic or localised, arachnid and insect faunas. Boggy seepages and swamp-mires are important habitat for the conservation of several endangered endemic dragonflies (Theischinger and Hawking 2006). Dissecting freshwater streams (Plate 310) are critical habitats for the larval and adult stages of many rainforest invertebrates, and for the functioning of broader ecosystem food webs. However, seasonal changes in flow rates, climate stability, substrate dynamics and individual life histories strongly influence the biomass, trophic structure, functionality and species richness of in-stream invertebrate assemblages (Cummins and Lauff 1969; Griswold *et al.* 2008; Lancaster *et al.* 2009; Pearson 2014).

The natural and evolutionary history of Australia's subtropical rainforests has been previously discussed by Adam (1987, 1992), Barlow (1981), Baur (1957), Connell *et al.* (1984), Floyd (1990, 2010), Hunter (2003), Rossetto *et al.* (2015), Webb (1978), Webb *et al.* (1984), Young and McDonald (1987), and others. These provide an overview and analysis of Australian subtropical rainforest communities. A introductory review of the significance and conservation of invertebrates in New South Wales rainforests was given by Nadolny (1984), and invertebrates of the Gondwana rainforests region have been discussed in Adam (1987), DASET (1992), Williams (1993, 2002, 2003a) and Williams and Bickel (2010).

The invertebrates discussed here constitute an indication of the known fauna. It is not an exhaustive treatment, nor is the structure of presentation a strictly systematic one; and indeed it is somewhat idiosyncratic (though not too dissimilar to the artificial categorising of 'moths' and 'butterflies'). Some groups are absent. Families, and in several instances higher ranks or informal groups (such as Sphaerotheriida, Psylloidea, Tipuloidea, Hydradephaga), are listed alphabetically. This is intended to facilitate access by non-specialist readers and reduce unnecessary complexity in the Index. However, Appendix 4 indicates the position of families or superfamilies that are discussed in the text within higher divisions of Coleoptera, Diptera, Hemiptera, Hymenoptera and Mollusca.

Individual distribution records can be found in Williams (1993, 2002, 2003a, and papers cited therein); however, these represent only a glimpse of the published records. A considerable body of later work has been published, and this I have drawn upon. I have also cited references that may be considered dated, but these have more than historial significance (nor are they necessarily obselete) for they demonstrate the foundations upon which current trends and understandings of the fauna stand. Nevertheless references such as the *Zoological Catalogue of Australia* series and *Insects of Australia* are increasingly out of date (and thus need to be considered cautiously), yet remain gateways to an understanding of the Australian fauna. The discussion also draws heavily on the many years of my own field work in the region, subsequent to the publication of earlier reviews. This personal field work has encomposed a much broader group of forests and included several significant New South Wales north coast rainforests that are not in the World Heritage Area (the focus of Williams 2002, 2003a; Williams and Bickel 2010). These rainforests include endangered floodplain rainforest remnants, isolated pockets in the Liverpool Range, mountain sites in close proximity to the coast (e.g. Glenugie Peak south-east of Grafton, Way Way State Forest near Nambucca Heads, North and Middle Brother mountains at Laurieton), and extensive areas in the New South Wales Lower Hastings, Camden Haven and Manning River Catchments (notably the montane forests of the Dingo–Tapin Tops, Comboyne Plateau, and Lansdowne–Comboyne Escarpment). Other areas include the rainforests in the Myall River at Bulahdelah, and Newcastle regions, small stands such as Mountain Lagoon and Sassafras Gully in the Blue Mountains west of Sydney, forests on the New South Wales central and south coast, and Queensland rainforests extending to the outlying Bunya Mountains and north to Kroombit Tops, Eurimbula and the Blackdown Tableland; as well as Araucarian vine thickets located in central eastern Queensland (such as Goodnight Scrub). Littoral rainforests, a series of distinctive subtropical–dry rainforest subformations now occurring as remnants on headlands and Holocene sands (see Adam *et al.* 1989; Floyd 1990; Williams 1993), have been extensively visited.

The mixed warm temperate and subtropical dominated rainforests of the Lansdowne–Comboyne Escarpment have been investigated over many years (owing to the presence there of my own property, the Lorien Wildlife Refuge and Conservation Area); the long-term study through various seasons, as well as opportunistic encounters, has resulted in the recognition of the escarpment's role as a refugium for numerous species originally thought only to occur on the Comboyne massif itself; recognition of many previously unrecorded plant–animal relationships; new and rare plant discoveries; and the discovery of invertebrate species new or poorly known to science, and species previously believed restricted to rainforest communities at much higher altitudes or at lower latitudes. In this Lorien has served the function of a long-term field studies station by default, a purpose similar to that of privately owned forests at Mt Glorious and Mt Tamborine (Queensland), and Tuglo Wildlife Refuge (north-west of Singleton) (Smithers 1981) in the Mount Royal Range (New South Wales).

Much rudimentary inventory work remains to be done to assess the invertebrate faunas of individual Australian subtropical rainforest stands, but, even with the knowledge at hand, it is evident that even quite small, physically degraded and geographically isolated remnants can possess many ecologically specialised invertebrate species and numerous Australian endemics. These possess high heritage and scientific reference attributes not necessarily suggested by the often disturbed state of the rainforest structure, and the impoverished abundance and diversity of the surviving plant and vertebrate communities. The endangered floodplain rainforests of Wingham Brush and Lansdowne nature reserves, surviving in an urban and agricultural context on the lower north coast of New South Wales, are cases in point. In particular, littoral rainforests, listed as endangered ecological communities under state and federal legislation, retain high invertebrate biodiversity values although now just remnants of a once greater area (following sand-mining and clearing for agriculture or residential development).

Despite being the subject of more than 200 years of scientific investigation there remain significant inventory data gaps in our knowledge of higher taxa (families and above in the taxonomic hierarchy) within Australia's floristically diverse subtropical rainforests. Many taxa have limited dispersal capabilities and, as a consequence, serve as useful tools for better identifying and understanding geological refugia and speciation events (e.g. carabid, tenebrionid and scarabaeine ground beetles, Phasmatodea). Other taxa have specialised and obligate host dependencies, or can function as indicators of environmental impacts (e.g. ants, in-stream fauna), and as such can serve to highlight inadequacies and the need for changes in habitat management strategies. Examples of rainforest-inhabiting taxa with inadequate knowledge bases are hydrobiid freshwater snails, terrestrial leeches, land planarians, flatworms, millipedes, and the cricket families Anostostomatidae and Rhaphidophoridae. Even the composition and distribution of such conspicuous invertebrate groups as bees (Apoidea/Apiformes) in montane rainforests are poorly known. Historically, considerable collecting effort has been targeted towards butterflies, jewel beetles (Buprestidae) and ground beetles generally, but little published data, specific to individual rainforest reserves and stands, are available. This restricts the foundation upon which management plans can be built.

Various authors have used a variety of terms to describe and define biogeographical zones and provinces (e.g. Allsopp 1995; Burbidge 1960; CSIRO 1996; Matthews 1972; Matthews and Bouchard 2008; McAlpine 2001; McMichael and Hiscock 1958; Smith and Kershaw 1979). Frequently these overlapped in range or were synonymous, and often had doubtful validity when applied across major taxonomic boundaries (i.e. utility is restricted to individual taxa). The terms 'Eyrean', 'Bassian' and 'Torresian' can now be viewed as obsolete and clouding perceptions of the development and origins of the Australian fauna, and may be of little use in delineating faunal zones to the level of definition demanded by most modern taxonomists and biogeographers. More recently C. Reid (2016) has used the term 'Australopapua' to define a faunal province that includes Papua, the Australian mainland and its immediate off-shore islands (such as Kangaroo Island and those of Bass Strait), and Tasmania, but excludes Lord Howe Island and other more distant islands that are defined as 'Australian' by virtue of political inclusion. However, rather than attempt to bring the various definitions into a standardised system, with risk of distorting or misinterpreting the original concepts, these have been retained.

The very concept of zoogeoraphic divisions, however, is open to question. This is because the degree of faunal overlap may defy recognition of distinct regions, we lack the benefit of hindsight to accurately relate extinction, radiation and speciation histories, and the high number of undescribed and poorly studied invertebrate taxa may place us in a position premature for understanding taxonomic affiliations and zoogeographical boundaries. Concepts such as Matthew's and Stanisic's 'refugium' areas (see discussion under Coleoptera – Tenebrionidae, and terrestrial Mollusca) may have greater validity for characterising more discrete biogeographical zones important in the evolution of the Australian invertebrate fauna.

As a caveat, it should be highlighted that biogeographical patterns, and the inferences and analyses that can be drawn from them, are only as good as (1) the level of the underlying taxonomy, (2) the adequacy of field sampling at individual sites and landmasses, and (3) known fossil history. Anomalous disjunct distributions, for example, may prove with time to be artefacts of insufficient sampling and deficient taxonomic rigour, or the result of accidental introduction.

2 Australia's subtropical rainforests – the plant context

Rainforests within the region of subtropical Australia's east coast include formations of considerable structural and floristic diversity (Adam 1987, 1992; Baur 1965; Floyd 1990; Harden *et al.* 2006; Helman 1987; Shapcott *et al.* 2015; Williams *et al.* 1984; Young and McDonald 1987). Floristic alliances and suballiances for New South Wales rainforests are detailed in Floyd (1990). Tidally inundated mangrove communities (Plate 315) may also be considered a distinctive rainforest formation (Adam 1992) (see 'Mangroves – a distinctive maritime rainforest formation' for discussion).

Mangroves – a distinctive maritime rainforest formation

Mangroves, and mangrove communities (Plate 31) more inclusively, are generally not thought of as a rainforest ecosystem. Traditionally, mangroves have been considered a distinct vegetation type, but on structural grounds (that of a 'closed canopy' in more developed stands), their placement within the broader conceptual notion of 'rainforest' may be appropriate. This is not a new idea, for their consideration as but one of numerous rainforest formations has previously been discussed and advocated by Paul Adam (1992), and regarded as such by Whitmore (1984). Citing the structural schemes of Specht (1970, 1981), in which 'closed forest' and 'rainforest' are treated as synonyms, Adam opines that mangroves should be regarded as a distinctive salt-tolerant maritime rainforest formation.

Mangroves are usually restricted to the intertidal zone. In Australia they are widely found along the eastern and northern coasts but are essentially tropical and subtropical in distribution (Adam 1992; Specht *et al.* 1995). Only five species reach the subtropical region of the east coast and only two, *Avicennia marina* (Avicenniaceae) and *Aegiceras corniculatum* (Myrsinaceae), are commonly encountered at higher latitudes. They vary structurally from closed canopy forest to open woodland, being floristically diverse at lower latitudes and most extensively developed in protected estuary zones. Regional variation in estuary length and size, temperature, salinity, groundwater, tidal amplitude and patterns of inundation, and possibly rainfall, influence species richness (Reef and Lovelock 2015; Saenger and Moverley 1985; Smith and Duke 1987). Some mangrove stands may represent a transitional community, but exposed shores and hypersalinity can act as a barrier to colonisation (Adam 1992; Mitchell and Adam 1989).

Their tidal-prone location, however, greatly restricts the kinds of terrestrial invertebrates that can occupy the mangrove environment. Various crustaceans and semi-aquatic and semi-arboreal molluscs characterise resident populations, but truly terrestrial forms (such as ants), particularly those species that at some stage of their life cycle are dependent upon the terrestrial substrate, are subject to tidal inundation, and so are limited to incorporating mangrove communities only within ephemeral foraging patterns (e.g. visiting flowers), but not occupation. Those that may be encountered comprise species capable of active flight (e.g. butterflies, cicadas, wasps, bees, flies, certain beetles) and largely sessile taxa, which, once able to reach foliage and branches (e.g. spiderlings that may disperse in air currents), are then able to survive because their immature life stages are not dependent on temporal residency under or on the intertidal-zone ground surface, prey items are dependable, and their habitat remains fully or partly above cyclical tide levels.

Climate, soil type and fertility, underlying geography, altitude and latitude, aspect and gradient, disturbance history (such as fire and wind impacts [Bowman 2000; Lange *et al.* 1981]), plant spatial and temporal heterogeneity, and random variation in conditions affecting reproductive or mortality rates influence both the extent and nature of individual stands (Adam 1992; Connell 1978; Floyd 1990; Williams and Adam 1999b). Rainforests trend to less floristic and structural complexity with increases in altitude and at higher latitudes, with decreased rainfall, and with lower soil fertility. Rainforest at high (temperate) altitudes experience greater seasonal variation in temperature, those at Barrington Tops being subject to snowfall. Individual plant species (and invertebrates [Lambkin *et al.* 2011; Majer *et al.* 2001; McKie *et al.* 2005; Ollerton and Cranmer 2002; Ødegaard and Diserud 2011; Yeates 1985]) respond to gradients in these (e.g. Kooyman *et al.* 2012; Laidlaw *et al.* 2011), such that rainforest plant communities at high latitudes, and at sites of low or irratic rainfall and low soil fertility, are floristically depauperate. Epiphytes respond to microenvironmental variation (Sanger and Kirkpatrick 2017), but in rainforests growing in such environmental extremes epiphytes may be noticeably few, or absent. Rainfall alone is a major factor in determining the complexity of rainforest, with annual rainfall in the region exceeding 1400–2000 mm (though not uniform throughout the region). But patterns of precipitation, not just total annual rainfall, also compound rainforest establishment and survival (and that of individual plant species). Chance events of seed dispersal (or conversely seed theft and predation) can also influence the floristic composition of stands as much as a predisposition to successful foundation, possibly more so. To all of these, individual invertebrate taxa, in addition to plants, respond accordingly.

Rainforest canopies occupy the interface between the biosphere and the atmosphere. They are exposed to impacts of wind, rain, humidity, and seasonal and daily temperature regimes to an extent far greater than the strata below. The frontal canopy of wind-sheared rainforest on exposed headlands (see Plate 305) may not exceed 1 m in height; deformed trees found there (e.g. *Syzygium smithii* – Myrtaceae) are able to grow to greater height in protected areas of the stand (Adam *et al.* 1989). Canopies also experience greater fluctuations in air chemistry, and for littoral rainforests there is the added burden of salt impacts (Parsons and Gill 1968; P. Adam pers. comm.). Plants and animals living at the edges of the canopy must be adapted physiologically and behaviourally to tolerate or avoid these extremes (Ozanne 2013).

Differences, often marked, in tree species dominance and occurrence in separate stands is normal and the intergrading of plant communities is commonplace. There can be considerable differences in structure ranging from complex canopy and subcanopy formations, to simple low vine thickets that, nevertheless, meet Andreas Schimper's (1903) original definition of 'rainforest'(*regenwald* in German). The rainforests of Australia's subtropical region are characterised by floristic alliances (Floyd 1990; Webb *et al.* 1984; and the forestry-based classification of Baur 1965) and physiognomic/structural features (see e.g. Adam 1992; Beadle and Costin 1952; Webb 1959, 1978). McDonald and Hunter (2010) note the close correlation between Baur's classification and the physiognomic/structural classification of Webb (1959, 1978). Historically, the rainforest subforms in New South Wales have been frequently termed 'subtropical', 'dry', 'warm temperate' and 'cool temperate' (Baur 1965; Harden *et al.* 2006; Williams *et al.* 1984, and others). Significant occurrences in Queensland and New South Wales are given in Appendix 3. These four simple subform terms are used throughout this work, but it is important to highlight that rarely are they found as distinct structural and floristic units (dry rainforest–vine thicket patches within woodland and savannah dominated landscapes being examples of exceptions), instead occurring as intergrading communities with fluid boundaries.

Subtropical rainforest is the most structurally and floristically complex of the four types (Adam 1992; Floyd 1990), and altitudinally extends from sea level to >900 m (see Plates 301, 306, 310, 311). It includes rainforests alternatively termed 'complex notophyll vine forest', 'notophyll vine forest' and 'araucarian vine forest' (Webb 1978; Young and McDonald 1987). The canopy is uneven with up to three layers, trees are diverse with 10–>60 species in the canopy, entire (untoothed) leaf margins dominate, there are many species with compound leaves, leaf size is generally large with notophylls and mesophylls dominating (>0.5–25 cm long), trunk buttressing and the presence of palms and figs are common, and vines, ferns and epiphytes are conspicuous and diverse. Subtropical rainforest is favoured by high rainfall (>1300 mm), sheltered aspect, and richer volvanic and alluvial soils (after Harden *et al.* 2006; McDonald and Hunter 2010; Williams *et al.* 1984). Littoral rainforest is a subform predominantly of subtropical rainforest, found on headlands, behind dunes, and in coastal estuaries, but may include trees that characterise dry rainforest, such as the presence of *Backhousia sciadophora* (Myrtaceae) and associates at Black Head on the lower north coast of New South Wales (Williams 1993) (see Plates 304, 305, 313).

Trees that are typical of subtropical rainforest in northern New South Wales and southern Queensland include numerous species of Cunoniaceae, Elaeocarpaceae, Escalloniaceae/Rousseaceae, Lauraceae, Euphorbiaceae/Phyllanthaceae, Icacinaceae/Cardiopteridaceae, Meliaceae, Atherospermataceae/Monimiaceae, Moraceae, Myrtaceae, Proteceae, and Sapindaceae (Floyd 1990; Young and McDonald 1987). In northern New South Wales *Dendrocnide excelsa* (Urticaceae), *Brachychiton acerifolius*, *Argyrodendron actinophyllum* (Sterculiaceae/ Malvaceae), and *Gmelina leichhardtii* (Verbenaceae) are frequently encountered as members of the upper canopy. *Araucaria cunninghamii* (Araucariaceae) sometimes occurs as an emergent, and *Archontophoenix cunninghamiana* palms (Arecaceae) are common, frequently dominating the canopy of sheltered gullies. At higher latitudes (e.g. Durras Mountain, Mount Dromedary) mixed subtropical rainforest–warm temperate rainforest occurs at lower elevations in sites of higher soil fertility. Trees are less diverse with *Syzygium smithii* (Myrtaceae), *Dendrocnide excelsa* (Urticaceae), *Citronella moorei* (Icacinaceae/ Cardiopteridaceae), *Ficus obliqua* (Moraceae), and the palm *Archontophoenix cunninghamiana* (Arecaceae) being characteristic species (Helman 1987).

Endangered subtropical rainforest in New South Wales and Queensland

The total rainforest in Australia is ~2 000 000 ha, this surviving mainly as scattered patches.This is a very small area, less than the total Amazon rainforest cleared annually. Australian rainforest is spread over a latitudinally extensive area, its origin predating that of the continent's now iconic eucalypt forests and woodlands. Of the moister coastal rainforest it is estimated that two-thirds of the original Queensland forest and three-quarters of that present in New South Wales has been cleared; the area of original rainforest at European occupation therefore being ~80 000 km^2 (Webb and Tracey 1981; Winter *et al.* 1987). What remains are fragmented and often highly disturbed and isolated 'habitat islands'(e.g. Bradshaw 2012; P. Rose 2014; Scanlan *et al.* 2018).

About 250 000 ha of rainforest survives in New South Wales (Pople and Cowley 1982). Of the remaining subtropical rainforest subform in New South Wales, ~73 400 ha is structurally intact. This is only equal to the estimated size of 'The Big Scrub' rainforest in 1842, of which only ~300 ha in small remnants now survives (Floyd 1990). Situated near Lismore in the far north-east corner of New South Wales, this was once the largest stand of tall subtropical rainforest in Australia. It took <40 years for European farmers to almost totally clear, its loss testimony to Australia's thoughtless and ongoing destruction of native vegetation. A similar fate befell the luxuriant New South Wales north coast rainforests of the Dorrigo Plateau and Comboyne Plateau, and the lowland rainforests of 'The Illawarra' south of Sydney. Of the estimated original littoral rainforest only 1200 ha remains, this in numerous weed infested fragments, surviving largely from coastal sand mining. Dry rainforest is estimated to have ~76 100 ha remaining, but this includes floristically depauperate vine thickets and remnant scrubs on the Western Slopes, in addition to more complex stands. Warm temperate rainforest totals 46 600 ha; however, about half of this has been heavily logged with consequent structural and floristic changes occurring. The conservation status of cool temperate rainforest has been relatively unaffected by logging, though cool temperate stands on the Comboyne Plateau, in particular, have suffered from clearing. Now only relict stands of *Nothofagus moorei* survive there (Bale and Williams 1994).

Despite their acknowledged rich biodiversity (Adam 1987, 1992; Department of the Environment, Sport and Territories 1994; Kitching *et al.* 2010), however, Australia's subtropical rainforests (and their included biota) remain under threat; directly or indirectly they are ensnared by the impacts of what has been termed the 'Anthropocene human-induced mass extinction event' (see Braje and Erlandson 2013; Ceballos *et al.* 2015). Throughout their range in Queensland and New South Wales, stands continue to be vulnerable to weed invasion, pathogen attack (e.g. *Austropuccinia psidii*, Myrtle Rust) (Williams 2018; Williams and Adam 2019), ongoing loss of connectivity (see Beier and Noss 1998), the deleterious impact of 'edge effects' (e.g. reduction in core area and vegetation buffers, changes to wind, temperature and light regimes, altered drainage patterns and nutrient inflow) (Fox *et al.* 1997), fire damage initiated from adjacent open forest and woodland, climate change and clearing (including underscrubbing) (Saunders *et al.* 1991). Logging operations in New South Wales state forests threaten isolated small stands due to boundary encroachment, and vine thicket communities in Queensland remain at risk from broadscale land clearing (see Plate 314). Feral animals also make significant impacts on the biota, potentially catastrophic ones, and though the threats to native vertebrates are generally understood, or can be predicted with a high degree of accuracy, the impact on the invertebrate fauna (e.g. by exotic rodents and cane toads [Shine 2010]) remains to be rigorously examined.

In response to this ongoing vulnerability state and federal legislation gives a measure of recognition and protection to surviving stands. In New South Wales various rainforest communities (e.g. 'Lowland Rainforest in the New South Wales North Coast and Sydney Basin Bioregions', 'Lowland Rainforest on Floodplain in the New South Wales North Coast Bioregion, 'Illawarra Subtropical Rainforest in the Sydney Basin Bioregion') are considered endangered ecological communities under the *Biodiversity Conservation Act 2016*. In addition to being listed as an endangered ecological community under this Act, threats to lttoral rainforest were further recognised under the NSW State Environmental Planning Policy (S.E.P.P.) *No. 26 – Littoral Rainforest*. The federal *Environment Protection and Biodiversity Conservation Act 1999* also lists several floristically diverse subtropical rainforest communities in New South Wales and Queensland as 'critically endangered' ecological communities (e.g. 'Lowland Rainforest of Subtropical Australia', 'Littoral Rainforest and Coastal Vine Thickets of Eastern Australia', 'Semi-evergreen Vine Thickets of the Brigalow Belt [North and South] and Nandewar Bioregions'). How these communities will be protected in the face of increasing socio-economic pressure remains to be tested.

Biological consequences of rainforest fragmentation

The historical, anthropogenic, fragmentation of subtropical rainforest ecosystems has resulted in the creation of isolated, island-like remnant patches of various sizes existing within agricultural, and sometimes urban, landscapes (Fox *et al.* 1997; Horton 1999; P. Rose 2017; Scanlan *et al.* 2018; Williams and Gerrand 1998b, 1998c). These altered landscapes form matrices, usually with abrupt boundaries, which are hostile to forest organisms (Harris and Silva-Lopez 1992; Zipperer 1993) (Plate 314).

Fragmentation causes significant, and often dramatic, changes to the physical environment of forest remnants (e.g. alteration to fluxes in radiation, wind and water [Nobel 1981; Saunders *et al.* 1991]) and this in combination with the reduction in the area of remaining habitat has important consequences for the survival of individual relict plant and animal populations (Didham 1997; Cagnolo *et al.* 2009; Krosch 2011; Laurance and Bierregaard 1997). These impacts are often subtle and extend over long periods; examples are the breakdown and loss of species mutualisms (Harris and Johnson 2004), invasion by exotic plants and animals (Christian 2001; Taylor *et al.* 2011; Williams and Gerrand 1998a, 1998b, 1998c; see also Crichton *et al.* 2019 for a discussion of pollinator 'super-generalism' potentially facilitating weed invasion), loss of gene flow and genetic drift (Honnay and Jacquenym 2007), and the eventual failure of trees to reproduce (Hobbs and Yates 2003). Although population size might place rare species at a heightened extinction risk in fragments (Thomas 1993), common species may be as, or more, susceptible to the genetic consequences of habitat fragmentation as rare species (Honnay and Jacuenym 2007). Stork *et al.* (2017) highlight the fact that trends in species richness at local scales may not reflect the resilience of ecosystems to disturbance, and functional feeding groups (e.g. herbivores, coprophages, predators, pollinator guilds) may respond differently to individual disturbance events and fragmentation more broadly. And so consideration of the effect of fragmentation on functional composition, rather than just on numbers of species and their abundance within remnants, is also important when assessing disturbance threats.

The longevity of many plants can mask the loss of ecological function due to the extinction of insect pollinators and the subsequent absence of seed set (Williams and Adam 1999b). Surviving patches of subtropical rainforest are frequently characterised by large numbers of vascular plant species often at low population levels with individuals commonly scattered widely about in the stand. Pollen flow between distantly spaced conspecifics is predominantly non-directional (pollinators usually exhibit low fidelity) and so levels of pollen transfer between conspecifics are likely to be low and seasonally unreliable. This threat is magnified when plants are obligate out-crossers (Williams and Adam 1994, 1999b, 2010), or do not flower annually (Williams 1995). Even when a degree of pollen transfer and fertilisation is successful, but with lowered fecundity, levels of seed production can be insufficient to sustain recruitment back into the stand (e.g. in the face of seed predation pressures). The surrounding cleared land experiences alterations to the radiation balance, resulting in increased surface temperatures during the day and the potential for frost at night. This had adverse impacts on remnants, especially along edges, and higher latitudes can impose significantly more solar radiation (and thus higher temperatures) than that in unfragmented landscapes. A consequence of this is that rainforest edges may have a predominance of colonising plant species, and shade-tolerant species are restricted to the interior 'core' or at different distances along a tolerance gradient. Indeed, edge effects may be so intrusive that no area of the remnant remains unaffected. Nutrient cycling; soil microorganism function, abundance and diversity; soil moisture retention; the decomposition of leaf litter; plant growth and plant phenological patterns; and the survival of individual animal and plant species can also be affected directly and indirectly by soil heating and increased atmospheric temperature regimes (e.g. Klein 1989; Norton 1994; Saunders *et al.* 1991). Reduction in the area of rainforest, and its retention in otherwise cleared landscapes, alters the nature of air flow with resultant flow-on impacts involving changes to wind turbulence patterns, exposure and severity. These changes in turn lead to modification of the canopy structure (e.g. crown shearing, lower overall canopy height) and atmospheric gas fluxes above the canopy (Nobel 1981; see also Pincebourde and Casas 2016), as well as canopy destruction and crown damage (e.g. salt burn of littoral rainforest [see Adam *et al.* 1989; Parsons and Gill 1968]), increased vulnerability to isolated tree windthrow and consequent increased light regimes at ground level (with concomitant impacts on leaf chemistry [Frankel and Berenbaum 1999]), increased litter fall and thus alteration of animal microhabitat. Substantial increases in spring and summer temperatures, seasonal episodes of prolonged and repeated drought, and vulnerability to hot dry winds can lead to greater tree mortality and the suspension of invertebrate maturation and emergence cycles. Thermal tolerances in individual animal species may be exceeded (Hemmings and Andrew 2017; Kaspari *et al.* 2015), with localised microclimates acting as amplifiers of macroclimatic conditions, exacerbating even seemingly small fluctuations in temperature (Pincebourde *et al.* 2016). Canopy destruction can also occur where large populations of flying foxes concentrate in small remnants (as at Wingham Brush Nature Reserve and Maclean, northern New South Wales), resulting in loss of herbivore and pollinator resources. Fragmentation also alters the hydrological cycle, changing the rates of rainfall interception and evapotransportation, modifying the penetration of water into the soil profile, and changing decomposition rates and altering seed-bed characteristics. Attendent impacts on the suitability of habitat for the biota follow. The reduction in the area of habitat that fragmentation intrinsically involves results in a loss of carrying capacity in both species abundance and diversity (Howden and Nealis 1975; Recher *et al.* 1987), fewer microhabitat and niche opportunities, and increased vulnerability to penetration by edge effects, both biotic and abiotic (Didham and Lawton 1999; Murcia 1995). Reduction in the populations of individual species can result in diminished 'encounter frequency', the resultant drop in offspring risking a sequential cascade in numbers towards localised extinction events. Impacts may be as seemingly subtle as the loss of nesting and foraging opportunities for native bees (Batley and Hogendoorn 2009) with unpredictable impacts on plant reproduction and floristic complexity. The ability of species to recolonise largely depends on the capacity of plants or animals to disperse across the cleared landscape matrix (Hill 1995), and on the distance between isolated remnants and alternative rainforests from which locally extinct populations might be repopulated over time. Even where secondary forest is allowed to re-establish, this still may act as a barrier to dispersal between old growth rainforest stands (Li *et al.* 2017), though individual species may react differentially to the degree of connectivity between fragments (Beier and Noss 1998; Hill 1997; da Silva and de Cerqueira Rossa-Feres 2017). However, the position of individual rainforest remnants in the landscape, their size and shape, and the presence/absence of food plants and participating mutualism partners,will largely determine the viability of relict populations and the ability of remnants more generally to recruit from rainforest surviving elsewhere (Laurance and Bierregaard 1997; Moritz *et al.* 1997).

Rhodomyrtus psidioides and Myrtle Rust – an example of a possible plant–insect association extinction event?

Rhodomyrtus psidioides (Plate 299), a member of the Myrtaceae, is an endemic large shrub or small tree distributed in rainforest from Gosford on the central coast of New South Wales to Gympie in south-eastern Queensland (Harden 1991). Its flowers are massed and conspicuously presented, are creamish-white in colour, and are pollinated by insects, but may be wind pollinated as well; a mixed biotic-abiotic pollination strategy that has been termed 'ambophily' (Bullock 1994; Williams 2018; Williams and Adam 2010). Its pollinators are predominantly small ecologically unspecialised insects, typically including the beetle families Cerambycidae, Chrysomelidae, Cleridae, Elateridae, Mordellidae and Oedemeridae, as well as flies, wasps and several native bees (Williams 1995, 2018). However, the unforeseen and widespread entry of the South American fungal pathogen Myrtle Rust (*Austropuccinia psidii*) into Australia has resulted in the widespread death of mature *Rhodomyrtus psidioides* trees.

Infection by Myrtle Rust results in crown dieback, branch death and mortality of sensitive members of the Myrtaceae (Carnegie *et al.* 2015; Invasive Species Council 2011; Pegg *et al.* 2017). It was first detected in New South Wales in April 2010, and its impact is such that it has the potential to fundamentally alter the ecology of Australia's vegetation communities (and that of the associated fauna) at a catastrophic landscape scale (Makinson 2018a, 2018b). Myrtle Rust has already proved capable of affecting at least 358 Australian native plant species or subspecies (Makinson 2018b). *Rhodomyrtus psidioides* is potentially able to regenerate from root stock, but root suckers are constantly re-infected by spores, resulting in the deformation and death of young leaves, such that suckers are not able to progress to the re-establishment of reproductively mature plants. And so, although none of the recorded insect visitors constitutes species with known obligate plant dependencies (Williams 2018; Williams and Adam 2019), *Rhodomyrtus psidioides* flower–putative pollinator interactions appear extinct, at least locally. Impacts on associated herbivorous insects are unknown; however, many Chrysomelidae, for example, are known to exhibit a strong host preference for members of the Myrtaceae.

Littoral rainforest on headlands in New South Wales includes numerous figs (*Ficus* spp.), *Euroschinus falcatus* (Anacardiaceae), *Elaeodendron australe* (Celastraceae), *Disopyros pentamera* (Ebeneaceae), *Cryptocarya rigida* (Lauraceae), *Syzygium smithii*, *Lophostemon confertus* (Myrtaceae), *Guioa semiglauca* (Sapindaceae), *Pouteria australis* (Sapotaceae), and *Dendrocnide photinophylla* (Urticaceae). Individual headland stands can be floristically diverse (as at Black Head–Hallidays Point). Stands associated with nutrient-enriched sands growing on and behind beach frontal dunes (and often called coastal scrubs) tend to be more depauperate, with impoverished remnants dominated by *Cupaniopsis anacardioides*, *Arytera divaricata* (Sapindaceae), *Alphitonia excelsa* (Rhamnaceae), *Banksia integrifolia* (Proteaceae) and various species of Rutaceae, Myrtaceae and Euphorbiaceae/Phyllanthaceae and the pine *Podocarpus elatus* (Podocarpaceae) (Floyd 1990; Williams 1993, 1995).

Dry rainforest, including structurally and floristically simple vine thickets, usually occurs in warm areas of lower (630–1100 mm) and less predictable seasonal rainfall, often on shallow soils and steep slopes, and along stream lines (see Plates 300, 316). It includes rainforest forms termed subtropical dry semi-evergreen rainforest, Araucarian notophyll vine forest, araucarian microphyll vine forest, low microphyll vine forest and low microphyll vine thicket classes. Dry rainforests are of a simpler structure than subtropical rainforest, typically of two strata, the lower continuous but the upper mostly of emergent scattered *Araucaria* (see Plate 314) or semideciduous species, with 10–30 species in the lower canopy and with smaller microphyll leaf sizes (2.5–7.5 cm) dominating. Leaf types may be hard and blunt-tipped, plank buttressing is rare and palms are absent. Large vines and prickly-leaved shrubs are generally common, but large epiphytes, mosses and ferns are rare or absent. However, some species of epiphytic orchids may be common (Curran *et al.* 2008; Floyd 1990; Harden *et al.* 2006; McDonald and Hunter 2010; Williams 1993; Williams *et al.* 1984). Dry rainforest species have been found to have higher drought survival than mesic congenors and with seedling mortality generally lower on loamy than sandy soils; this supports the idea that drought resistance of seedlings can shape tree species distribution at regional scales (Curran *et al.* 2013) (conversely for the biota, however, Houston and Melzer [2012] found that the invertebrate fauna of dry rainforests was sensitive to reduced soil moisture).

Emergent trees in northern New South Wales dry rainforest stands include *Araucaria cunninghamii* (Araucariaceae) and *Flindersia australis* (Rutaceae). In mixed dry–subtropical rainforest remnants large *Ficus macrophylla* (Moraceae) can be conspicuous emergents (Williams 1993). Species typical of the canopy and subcanopy layers include *Capparis arborea* (Capparaceae), *Ehretia accuminata* (Boraginaceae), *Croton insularis*, *Drypetes deplanchei*, *Mallotus philippensis* (Euphorbiaceae/Phyllanthaceae), *Neolitsea dealbata* (Lauraceae), *Ficus rubiginosus* (Moraceae), *Austromyrtus bidwillii*, *Backhousia myrtifolia*, *B. sciadophora* (Myrtaceae), *Olea paniculata* (Oleaceae), *Aphananthe philippinensis* (Ulmaceae/Cannabaceae), and numerous Sapindaceae. Some small gully-restricted stands consist of little more than low *Backhousia myrtifolia* (Williams 1993; Williams *et al.* 1984); this species is also dominant in dry rainforest at lower altitudes in southern New South Wales (Helman 1987). In Queensland *Agathis robusta* (Araucariaceae) occurs in rainforest on deep sands from Noosa to Fraser Island (Harden *et al.* 2006) and can be a dominant emergent tree in dry marginal rainforests (Boland *et al.* 1984).

In southern Queensland Araucarian forest Celastraceae, Ebenaceae, Euphorbiaceae/Phyllanthaceae, Rubiaceae, Rutaceae and Sapindaceae predominate (Young and McDonald 1987). Araucarian forest was once widespread in southern Queensland but following European occupation is now relictual in distribution (see Plate 314). In Queensland emergent *Brachychiton australis* and *B. ruprestris* (Sterculiaceae/Malvaceae) bottle trees typify low vine thickets, commonly with *Ehretia membranifolia* (Boraginaceae), *Alectryon conatus* (Sapindaceae), *Erythroxylum australe* (Erythroxylaceae), *Bridelia leichhardtii* (Euphorbiaceae/Phyllanthaceae), and *Flindersia australis* and *F. collina* (Rutaceae) as associates (Young and McDonald 1987).

Warm temperate rainforest is loosely synonymous with Webb's (1978) simple notophyll evergreen vine forest, occurring at moderately high altitudes (450–1200 m), on soils of lower fertility (see Plate 309). It possesses only two tree strata, and has a generally uniform canopy. Warm temperate rainforest frequently intergrades with subtropical rainforest at lower altitudes, this characterised by broad ecotones of mixed composition. Typically it is less diverse than subtropical rainforest with only three to 15 species in the upper canopy, but the diversity of small trees in the understorey can be a rich one. Leaf types are dominated by those with toothed margins, leaf size is in general medium with notophylls and microphylls common, and plank buttresses and palms are rare or absent (though the diminutive palm *Linospadix monstachya* [Arecaceae] may be common in some understorey patches). Tree trunks are relatively slender, many of which support distinctive lichen growth. Ferns are common but large epiphytes, though present, may be uncommon.

Frequently encountered trees in the northern New South Wales–south-eastern Queensland region are *Ceratopetalum apetalum*, *Ackama paniculata* (syn. *Caldcluvia paniculosa*), *Schizomeria ovata* (Cunoniaceae), *Doryphora sassafras* (Atherospermataceae), *Polyosma cunninghamii*, *Quintinia sieberi*, *Q. verdonii* (Escalloniaceae/Rousseaceae), *Orites excelsa* (Proteaceae) and various Lauraceae (Floyd 1990; Williams 1993). In isolated gully rainforests west of Sydney are several floristically simple rainforests dominated by *Ceratopetalum apetalum*, *Doryphora sassafras*, *Syzygium smithii* and the relict pine *Wollemia nobilis* (Araucariaceae) (Benson and Allen 2007). *Ceratopetalum apetalum* reaches its southern-most distribution in the Illawarra. Here there are warm temperate rainforests of low floristic diversity, with some intergrading examples of subtropical rainforest (e.g. on the Robertson Plateau), in which *Cryptocarya glaucescens* (Lauraceae), *Syzygium smithii*, *Polyosma cunninghamii* and *Doryphora sassafras* are prominent (Fuller and Mills 1985; Mills 1987).

Cool temperate rainforest, alternatively termed 'microphyll mossy forest' and 'microphyll fern forest' (Webb 1959, 1978), is usually confined to elevations >900 m (see Plates 307, 308). Annual rainfall often exceeds 1700 mm. In cool temperate rainforest ferns, moss and lichens are conspicuous but the rainforest understorey is frequently open. Forests at higher elevations are seasonally very cold; that at Barrington Tops is subject to snow, and frequently mist-shrouded even during summer days. At higher altitudes from Barrington Tops to the McPherson Range (= Border Ranges, also called the MacPherson Range in some parts of the literature) cool temperate rainforest is frequently dominated by *Nothofagus moorei* (Nothofagaceae); this was placed in the genus *Lophozonia* by Heenan and Smissen (2013), though this change is contentious (see Hill *et al.* 2015) and is not followed here. *Nothofagus moorei* is a large tree that can grow to 45–50 m, is often the only tree in the upper canopy (or may occur as an emergent), and is found at altitudes up to 1550 m between Barrington Tops and the Queensland border. Relictual *Nothofagus moorei* stands, from past land clearing, survive on the Comboyne Plateau at elevations between 570 and 630 m (Bale and Williams 1994; Chisholm 1925).

Additional trees present in cool temperate rainforest from Barrington Tops north include *Elaeocarpus holopetalus* (Elaeocarpaceae), *Doryphora sassafras*, *Ceratopetalum apetalum*, *Orites excelsa*, and the small understorey trees *Vesselowskya rubifolia* and *V. venusta* (Cunoniaceae). These may occur as associates in *Nothofagus*-dominated rainforest at lower canopy levels, occur as co-dominants in the canopy and subcanopy (as in Werrikimbe National Park), or constitute cool temperate rainforest where *Nothofagus* is absent (McDonald and Hunter 2010; Williams 1993). *Nothofagus* is inexplicably absent from Mt Hyland and the Gibraltar Range of northern New South Wales (Adam 1992).

Cool temperate rainforest is absent between the Barrington Tops and the Illawarra, but south from the Barrington Tops are scattered occurrences of *Atherosperma moschatum* (Atherospermataceae), a species more typical of cool temperate rainforests in eastern Victoria and Tasmania. The cool temperate rainforests of southern New South Wales are often dominated by *Eucryphia moorei* (Eucryphiaceae) (Helman 1987), with the species reaching its northern limit in the Illawarra (Mills 1987). Isolated stands in the Kosciusko National Park are dominated by *Atherosperma moschatum*, *Elaeocarpus holopetalus* and *Eucryphia moorei* (Doherty *et al.* 2011).

Plate 1: *Argonemertes ?australiensis* (Prosorhochmidae)

Plate 2: *Didymogaster sylvaticus* (Megascolecidae)

Plate 3: Velvet mite (Trombidiidae)

Plate 4: Uropodid mites on *Pamborus alternans* (Carabidae)

Plate 5: Phoretic mite associated with passalid larva

Plate 6: Mites associated with millipede

Plate 7: *Eriophora* sp. (Araneidae)

Plate 8: *Plebs bradleyi* (Araneidae)

Plate 9: *Arkys lancearius* (Arkyidae)

Plate 10: Ant-mimicking ?*Leichhardteus* sp. (Corinnidae)

Plate 11: *Menneus aussie* (Deinopidae)

Plate 12: *Progradungula carraiensis* (Gradungulidae)

Plate 13: *Tarlina* sp. (Gradungulidae)

Plate 14: *Dimidamus dimidiatus* (Nicodamidae)

Plate 15: *Megadolomedes australianus* (Pisauridae)

Plate 16: *Helpis mintabunda* (Salticidae)

Plate 17: *Myrmarachne* sp. (Salticidae)

Plate 18: *Opisthoncus quadratarius* (Salticidae)

Plate 19: *Opisthoncus* sp. (Salticidae) with *Ectenopsis* (Tabanidae) fly prey

Plate 22: *Tmarus* sp. (Thomisidae)

Plate 20: *Heteropoda* sp. (Sparassidae) with ?sciarid flies and captured *Methana* cockroach

Plate 23: *Missulena bradleyi* (Actinopodidae)

Plate 21: *Theridion* sp. (Theridiidae) with captured *Leptomyrmex* ant

Plate 24: *Paraembolides* sp. (Hexathelidae)

Plate 25: *Hadronyche* sp. (Hexathelidae)

Plate 26: Elevated retreat tube, *Misgolas* sp. (Idiopidae)

Plate 27: ?*Ixamatus* sp. (Nemesiidae)

Plate 28: ?*Namea* sp. (Nemesiidae)

Plate 29: *Equitius* sp. (Trianenonychidae)

Plate 30: ?*Megalopsalis* sp. (Neopilionidae)

Plate 31: ?*Megalopsalis* sp. (Neopilionidae)

Plate 32: *Cercophonius squama* (Bothriuridae)

Plate 33: Glass shrimp (Atyidae)

Plate 34: *Macrobrachium* sp. (Palaemonidae)

Plate 35: *Euastacus spinifer* (Parastacidae) with flatworm ectosymbionts

Plate 36: *Cormocephalus westwoodi* (Scolopendridae)

Plate 37: *Epicyliosoma* sp. (Sphaerotheridae)

Plate 40: *Diphlebia lestoides* (Lestoideidae)

Plate 38: *Walesbolus lobatus* (Spirobolellidae)

Plate 41: *Austroargiolestes amabilis* (Argiolestidae)

Plate 39: ?*Nesomachilis* sp. (Meinertellidae)

Plate 42: *Synlestes weyersii* (Synlestidae)

Plate 43: *Austrophlebia costalis* (Aeshnidae)

Plate 44: *Telephlebia godeffroyi* (Aeshnidae)

Plate 45: ?*Molytria* sp. (Blaberidae)

Plate 46: *Panesthia australis* (Blaberidae)

Plate 47: ?*Rhabdoblatta* sp. (Blaberidae)

Plate 48: *Methana* sp. (Blattidae)

Plate 49: *Calofulcinia* sp. (Mantidae)

Plate 52: ?*Anostostoma* sp. (Anostostomatidae)

Plate 50: *Orthodera* sp. (Mantidae)

Plate 53: ?*Endotaria* sp. (Gyrillidae/Phalangopsidae)

Plate 51: *Titanolabis colossea* (Anisolabidae)

Plate 54: *Ornebius* sp. (Gyrillidae/Mogoplistidae)

Plate 55: ?*Australotettix* sp. (Rhaphidophoridae)

Plate 56: ?*Australotettix* sp. (Rhaphidophoridae) feeding on agaric fungus

Plate 57: ?*Vingselina* sp. (Tetrigidae)

Plate 58: *Anthophiloptera dryas* (Tettigoniidae)

Plate 59: *Austrosalomona falcata* (Tettigoniidae)

Plate 60: ?*Diastella* sp. (Tettigoniidae)

Plate 61: *Mastigaphoides haffneri* (Tettigoniidae)

Plate 62: *Torbia* sp. (Tettigoniidae)

Plate 63: *Paraphisis* sp. (Tettigoniidae)

Plate 64: *Phricta* sp. (Tettigoniidae)

Plate 65: *Torbia* sp. (Tettigoniidae)

Plate 66: *Candovia robinsoni* (Diapheromeridae)

Plate 67: *Pachymorpha* sp. (Diapheromeridae)

Plate 68: *Acrophylla* sp. (Phasmatidae)

Plate 69: *Eurycnema goliath* (Phasmatidae)

Plate 70: *Extatosoma tiaratum* nymph (Phasmatidae)

Plate 71: *Drakiessa hackeri* (Aradidae)

Plate 72: *Chlorocysta vitripennis* (female) (Cicadidae)

Plate 73: *Chlorocysta vitripennis* (male) (Cicadidae)

Plate 74: Dead *Henicopsaltria eydouxii* (Cicadidae) being attacked by fungi

Plate 75: *Ledra* sp. (Cicadellidae)

Plate 76: Immature *Canungrantmictus morindiana* (Coreidae)

Plate 77: *Zoraida* sp. (Derbidae)

Plate 78: *Desudaba psittacus* (Fulgoridae)

Plate 79: *Physopelta gutta* (Largidae)

Plate 80: *Oncopeltus sordidus* (Lygaeidae)

Plate 81: *Vitellus ?rufospinosus* (Pentatomidae)

Plate 82: *Myiophanes* sp. (Reduviidae)

Plate 83: ?*Horcinia* sp. (Reduviidae)

Plate 84: *Ptilocnemus ?femoralis* (Reduviidae)

Plate 85: *Pristhesancus plagipennis* (Reduviidae)

Plate 88: *Oncoscelis australasiae* (Tessaratomidae)

Plate 86: *Leptocoris ?tagalicus* (Rhopalidae) on *Alectryon coriaceus* (Sapindaceae)

Plate 89: *Neohoodiella jennibeardae* (Phlaeothripidae)

Plate 87: *Cantao parentum* (Scutelleridae)

Plate 90: ?*Archichauliodes* sp. (Corydalidae)

Plate 91: ?*Italochrysa* sp. (Chrysopidae)

Plate 92: *Calomantispa venusta* (Mantispidae)

Plate 93: *Porismus* sp. (Osmylidae)

Plate 94: *Psychopsis* ?*mimica* (Psychopsidae)

Plate 95: *Lemodes coccinea* (Anthicidae)

Plate 96: *Euphanteos* ?*ornatus* (Anthribidae)

Plate 97: *Ancylotropis waterhousei* (Anthribidae)

Plate 98: *Rhinotia ?marginella* (Belidae)

Plate 99: *Rhinotia* sp. (Belidae)

Plate 100: *Calodema regale* (Buprestidae) on *Tristaniopsis laurina* (Myrtaceae)

Plate 101: *Castiarina acuminata* (Buprestidae)

Plate 102: *Castiarina insignis* (Buprestidae)

Plate 103: *Castiarina maculipennis* (Buprestidae)

Plate 106: *Metaxymorpha grayii* (Buprestidae) on *Syzygium floribundum* (Myrtaceae)

Plate 104: *Castiarina variegata* (female with male insert) (Buprestidae)

Plate 107: *Nascioides tillyardi* (Buprestidae)

Plate 105: *Chalcotaenia lamberti* (Buprestidae)

Plate 108: *Neobuprestis williamsi* (Buprestidae)

Plate 109: *Prospheres aurantiopictus* (Buprestidae)

Plate 110: *Chauliognathus* sp. (Cantharidae)

Plate 111: *Cratoferonia phylarchus* (Carabidae)

Plate 112: *Mystropomus* sp. (Carabidae)

Plate 113: *Notonomus angustibasis* (Carabidae)

Plate 114: *Pamborus ?alternans* (Carabidae) larva

Plate 115: *Pamborus guerini* (Carabidae)

Plate 116: *Trichosternus vigorsi* (Carabidae)

Plate 117: *Athemistus harrisoni* (Cerambycidae)

Plate 118: *Ceraegidion dorrigoensis* (Cerambycidae)

Plate 119: *Disterna* sp. (Cerambycidae)

Plate 120: *Distichocera kirbyi* (Cerambycidae)

Plate 121: *Distichocera superba* (male) (Cerambycidae)

Plate 124: *Piesarthrius frenchi* (Cerambycidae)

Plate 122: *Hesthesis variegata* (Cerambycidae)

Plate 125: *Rhipidocerus australasiae* (Cerambycidae)

Plate 123: *Homaemota bicolor* (Cerambycidae)

Plate 126: *Temnosternus dissimilis* (Cerambycidae)

Plate 127: *Temnosternus niveoscriptus* (Cerambycidae)

Plate 128: *Thyada barbicornis* (Cerambycidae)

Plate 129: *Tropocalymma dimidiatum* (Cerambycidae) on *Acronychia imperforata* (Rutaceae)

Plate 130: *Zoedia longipes* (Cerambycidae)

Plate 131: *Calomela ?pallida* (Chrysomelidae)

Plate 132: Possibly undescribed *Eucolaspinus* sp. (Chrysomelidae) feeding on *Nothofagus moorei*

Plate 133: *Hoplostines viridipennis* (Chrysomelidae) feeding on *Elatostema reticulatum*

Plate 134: *Liliocerus chamelus* (Chrysomelidae)

Plate 135: *Poneridia australis* (Chrysomelidae)

Plate 136: *Poropteromela epipleuralis* (Chrysomelidae)

Plate 137: *Eunatalis ?planipennis* (Cleridae)

Plate 138: ?*Diethusa* sp. (Curculionidae)

Plate 139: *Elleschodes hamiltoni* (Curculionidae) and *Eupomatia laurina* (Eupomatiaceae) flower

Plate 140: *Enteles vigorsi* (Curculionidae)

Plate 141: *Ilacuris laticollis* (Curculionidae)

Plate 142: *Perissops ocellatus* (Curculionidae)

Plate 143: *Poropterus ellipticus* (Curculionidae)

Plate 144: *Poropterus ?westwoodii* (Curculionidae)

Plate 145: *Tranes* sp. (Curculionidae) on *Lepidozamia peroffskyana* (Zamiaceae) cone

Plate 146: *Aphileus lucanoides* (Elateridae)

Plate 147: *Paracalais ?macleayi* (Elateridae)

Plate 148: ?*Parasaphes* sp. (Elateridae)

Plate 149: *Episcaphula duplopunctata* (Eryotylidae)

Plate 150: *Australobolbus austrinus* (Geotrupidae)

Plate 151: *Liparochrus fossulatus* (Hybosoridae)

Plate 152: *Metriorrhynchus* sp. (Lycidae)

Plate 153: *Cacostomus squamosus* (Lucanidae) (female)

Plate 154: *Cacostomus subvittatus* (Lucanidae)

Plate 155: *Ryssonotus nebulosus* (Lucanidae)

Plate 156: *Safrina grandis* (Lucanidae)

Plate 157: *Safrina polita* (Lucanidae)

Plate 158: *Zonitis* sp. (Meloidae)

Plate 159: Kateretid beetles on *Alocasia brisbanensis* (Araceae)

Plate 160: Nitidulidae, Kateretidae and Curculionidae (genus near *Parimera*) in spathe of *Alocasia brisbanensis* (Araceae)

Plate 161: *Aulacocyclus* sp. (Passalidae)

Plate 162: ?*Macrosiagon* sp. (Ripiphoridae)

Plate 163: *Amphistomus speculifer* (Scarabaeidae)

Plate 164: *Anoplognathus concolor* (Scarabaeidae)

Plate 165: *Anthotocus antennalis* (Scarabaeidae)

Plate 166: *Aulacopris maximus* (Scarabaeidae)

Plate 167: *Aulacopris maximus* (Scarabaeidae) perching at night on tree trunk

Plate 168: *Cheiroplatys latipes* (Scarabaeidae)

Plate 169: *Chlorobapta frontialis* (Scarabaeidae)

Plate 170: *Diorygopyx niger* (Scarabaeidae) perching at night on forest floor leaf litter

Plate 171: *Onthophagus neostenocerus* (Scarabaeidae) perching at night on elevated *Ripogonum* (Ripogonaceae) leaf

Plate 172: *Onthophagus williamsi* (Scarabaeidae)

Plate 173: *Paraschizognathus ?ocularis* (Scarabaeidae)

Plate 174: *Podotenus ?toowoombaensis* (Scarabaeidae)

Plate 175: *Schizorhina atropunctata* (Scarabaeidae)

Plate 176: *Semanopterus subcostatus* (Scarabaeidae)

Plate 177: *Xylonichus eucalypti* (Scarabaeidae)

Plate 178: *Xylotrupes ulysses* (Scarabaeidae) male

Plate 179: *Xylotrupes ulysses* (Scarabaeidae) female

Plate 180: *Diamesus osculans* (Silphidae)

Plate 181: *Ptomophila perlata* (Silphidae)

Plate 184: *Cyphaleus cupreus* (Tenebrionidae)

Plate 182: *Blepegenes equestris* (Tenebrionidae)

Plate 185: *Cyphaleus childreni* (Tenebrionidae)

Plate 183: *Cardiothorax hopsoni* (Tenebrionidae)

Plate 186: *Emcephalus* sp. (Tenebrionidae)

Plate 187: *Hypaulax ovalis* and ?*Leichenum* (Tenebrionidae) feeding on fungi

Plate 188: *Sloanea costata* (Tenebrionidae)

Plate 189: *Cryphaeus punctipenne* (Tenebrionidae)

Plate 190: *Omorgus* ?*squamosus* (Trogidae)

Plate 191: *Harpobittacus* sp. (Bittacidae)

Plate 192: *Mesophysa flavipes* (Acroceridae)

Plate 193: *Blepharotes splendidissimus* (Asilidae)

Plate 194: *Cabasa ?pulchella* (Asilidae)

Plate 195: *Chrysopogon splendidissimus* (Asilidae)

Plate 196: *Orthogonis ornatipennis* (Asilidae)

Plate 197: *Chrysopogon harpaleus* (Asilidae)

Plate 198: *Stomorhina xanthogaster* (Calliphoridae)

Plate 199: *Geron* sp. (Bombyliidae) feeding on nectar of *Trochocarpa laurina* (Ericaceae/Epacridaceae)

Plate 200: *Thevenetimyia tenta* (Bombyliidae)

Plate 201: *Rutilia lepida* (Tachinidae)

Plate 202: Corethrellidae feeding from *Mixophyes* frog host

Plate 203: *Krakatauia macalpinei* (Dolichopodidae)

Plate 204: *Drosophila hibisci* (Drosophilidae) on *Hibiscus splendens* (Malvaceae)

Plate 205: *Tapeigaster* sp. (Heteromyzidae)

Plate 206: *Mimegralla australica* (Micropezidae)

Plate 207: *Metopochetus ?freyi* (Micropezidae)

Plate 208: *Trichophthalma nigrovittata* (Nemestrinidae)

Plate 209: *Neurochaeta inversa* (Neurochaetidae)

Plate 210: *Pelecorhynchus distinctus* (Pelecorhynchidae)

Plate 211: *Pelecorhynchus nigripennis* (Pelecorhynchidae)

Plate 212: *Duomyia ?obscura* (Platystomatidae)

Plate 213: *Euprosopia ?armipes* (Platystomatidae)

Plate 214: *Hysma lacteum* (Platystomatidae)

Plate 215: *Lecomyia ?notha* (Stratiomyidae)

Plate 216: *Xanthoberis* sp. (Stratiomyidae)

Plate 217: *Chalcosyrphus pleuralis* (Syrphidae) on *Rhodosphaera rhodanthema*

Plate 218: *Deineches hackeri* (Syrphidae)

Plate 219: *Dideopsis aegrota* (Syrphidae) on *Avicennia marina* (Avicenniaceae)

Plate 220: *Graptomyza* sp. (Syrphidae) on *Alectryon coriaceus* (Sapindaceae)

Plate 221: *Mesembrius hilaris* (Syrphidae)

Plate 222: *Odneromyia spadix* (Syrphidae)

Plate 223: *Orthoprosopa multicolor* (Syrphidae)

Plate 224: *Xylota flavitarsis* (Syrphidae)

Plate 225: *Caenoprosopon ?trichaerum* (Tabanidae)

Plate 226: *Scaptia auriflua* (Tabanidae)

Plate 227: *Scaptia barbara* (Tabanidae)

Plate 228: *Scaptia guttata* (Tabanidae)

Plate 229: *Scaptia quadrimaculata* (Tabanidae)

Plate 230: *Scapta subcana* (Tabanidae)

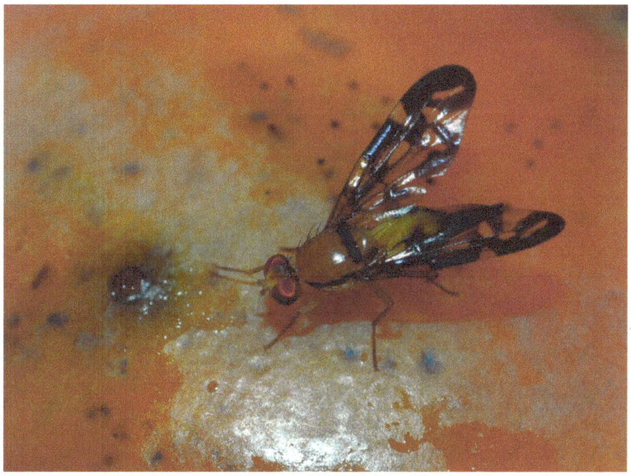
Plate 231: *Austrorioxa acidiomorpha* (Tephritidae)

Plate 232: *Agapophytes* sp. (Therevidae)

Plate 233: *Collessiama narelleae* (Therevidae)

Plate 234: *Dapanoptera richmondiana* (Tipulidae)

Plate 235: Tipulid larva

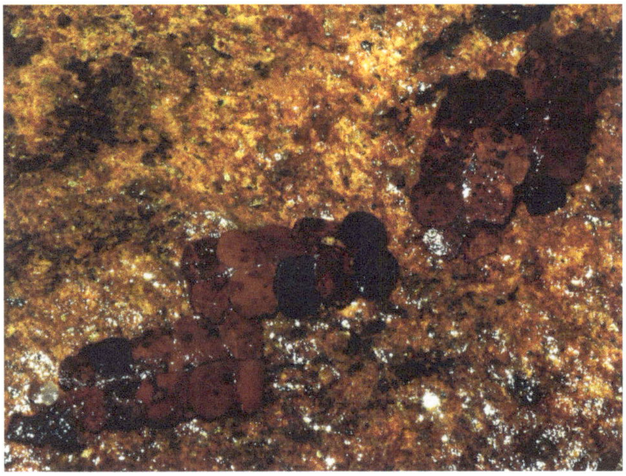

Plate 236: Larval cases, *Caenota* sp. (Calocidae)

Plate 237: *Anisocentropus bicoloratus* (Calamoceratidae)

Plate 238: *Anthela* sp. (Anthelidae)

Plate 239: *Chelepteryx chalepteryx* (Anthelidae)

Plate 240: *Asota iodamia* (Erebidae)

Plate 241: *Niceteria macrocosma* (Geometridae)

Plate 242: *Thalaina* sp. (Geometridae)

Plate 243: *Abantiades labyrinthicus* (Hepialidae)

Plate 244: *Aenetus eximia* (Hepialidae)

Plate 245: *Zelotypia staceyi* (Hepialidae)

Plate 246: *Fodina* sp. (Noctuidae)

Plate 247: *Destolmia* sp. (Notodontidae)

Plate 248: *Snellenia* sp. (Oecophoridae)

Plate 249: *Coequosa australasiae* (Sphingidae)

Plate 250: *Theretra nessus* (Sphingidae)

Plate 251: *Lactura suffusa* (Zygaenidae)

Plate 252: *Doleschallia bisaltide* (Nymphalidae)

Plate 253: *Heteronympha mirifica* (Nymphalidae)

Plate 254: *Delias nigrina* (Pieridae), on *Pittosporum undulatum* (Pittosporaceae) nectar

Plate 255: *Euschemon rafflesia* (Hesperiidae) larva

Plate 256: *Wilkiea huegeliana* (Monimiaceae). Regent skipper, *Euschemon rafflesia,* food plant.

Plate 257: *Erysichton lineata lineata* (Lycaenidae) on *Alectryon coriaceus* (Sapindaceae)

Plate 258: *Graphium macleayanus* (Papilionidae)

Plate 259: *Amblyopone* sp. (Formicidae)

Plate 260: *Leptomyrmex* ants feeding on the nectar of *Acronychia oblongifolia* (Rutaceae)

Plate 261: *Myrmecia nigrocincta* (Formicidae) feeding on *Alphitonia excelsa* (Rhamnaceae) nectar

Plate 262: *Myrmecia tarsata* (Formicidae) with dead dragonfly (*Telephlebia godeffroyi*)

Plate 263: *Orectognathus mjobergi* with larvae

Plate 264: *Polyrhachis pilosa* (Formicidae) and leaf nest

Plate 265: *Enicospilus* sp. (Ichneumonidae)

Plate 266: *Heteropelma scaposum* (Ichneumonidae)

Plate 267: *Aulacus* sp. (Aulacidae)

Plate 268: *Leucospis* sp. (Leucospidae)

Plate 269: *Dimorphothynnus dimidiatus* (Thynnidae) on *Elaeocarpus obovatus* (Elaeocarpaceae)

Plate 270: *Hemithynnus rufiventris* (Thynnidae)

Plate 271: Female *Diamma bicolor* (Thynnidae) on *Alphitonia excelsa* (Rhamnaceae)

Plate 272: *Ferreola handschinii* (Pompilidae)

Plate 273: *Calopompilus raptor* (Pompilidae)

Plate 274: *Heterodontonyx* sp. (Pompilidae)

Plate 275: Pompilid wasp with mygalomorph prey

Plate 276: *Pseudotrielis zonata* (Scoliidae)

Plate 277: *Bidentodynerus bicolor* (Vespidae)

Plate 278: *Deuterodiscoelius ephippium* (Vespidae)

Plate 279: *Pergagrapta polita* (Pergidae)

Plate 280: *Styracotechys dicelysma* (Pergidae)

Plate 281: *Xyloperga* sp.(Pergidae)

Plate 282: *Amegilla bombiformis* (Apidae) and predatory mantid (*Orthodera* sp.) on *Senna acclinis* (Fabaceae)

Plate 283: *Tetragonula carbonaria* foraging for pollen on male *Rhodosphaera rhodanthema* (Anacardiaceae)

Plate 284: *Thyreus lugubris* (Apidae)

Plate 285: *Megachile mystaceana* (Megachildae)

Plate 286: *Nomia aurantifer* (Halictidae)

Plate 287: *Xylocopa aruana* (Apidae)

Plate 288: *Amphylaeus nubilosellus* (Colletidae)

Plate 289: *Triboniophorus graeffei* (Athoracophoridae)

Plate 290: Juvenile *Triboniophorus graeffei* (Athoracophoridae)

Plate 291: *Triboniophorus graeffei* sheltering in shallow tree hollow

Plate 292: *Thersites mitchellae* (Camaenidae) (photo credit Mark Robinson)

Plate 293: *Thersites novaehollandiae* (Camaenidae)

Plate 294: *Hedleyella falconeri* (Caryodidae)

Plate 295: *Pygmypanda atomata* (Caryodidae)

Plate 298: *Mysticarion porrectus* (Helicarionidae)

Plate 296: *Cystopelta* sp. (Cystopeltidae)

Plate 299: *Rhodomyrtus psidioides* (Myrtaceae)

Plate 297: *Fastocarion virens* (Helicarionidae)

Plate 300: Dry rainforest understorey, Kiwarrak State Forest

Plate 301: Montane subtropical rainforest, Tapin Tops National Park

Plate 302: *Lepidozamia peroffskyana* (Zamiaceae) (foreground), Lansdowne Escarpment

Plate 303: Wet sclerophyll forest, Lansdowne Escarpment

Plate 304: Littoral rainforest understorey, Harrington

Plate 305: Littoral rainforest, Redhead

Plate 306: Lowland subtropical rainforest, Wingham Brush Nature Reserve

Plate 307: Forest margin, cool temperate rainforest, Barrington Tops

Plate 308: Cool temperate rainforest understorey, Barrington Tops State Forest

Plate 309: Warm temperate rainforest, Werrikimbe National Park

Plate 310: Subtropical rainforest, Nightcap National Park

Plate 311: *Archontophoenix cunninghamiana* palm stand, Terania Creek

Plate 312: Rainforest pocket within sclerophyll forest matrix, Kroombit Tops

Plate 313: Dissecting track, littoral rainforest near Eurimbula National Park

Plate 315: *Avicennia marina*-dominated mangrove community, Manning River estuary

Plate 314: Araucarian vine forest remnant near Goodnight Scrub demonstrating extent of clearing in an increasingly agricultural landscape

Plate 316: Author netting insects from flowering *Croton insularis* (Euphorbiaceae), dry rainforest, Kiwarrak State Forest

3 The invertebrates – the 'other 99 per cent' of subtropical rainforest biodiversity

Invertebrates are literally 'the other 99 per cent' of the world's animal kingdom, and dominate the fauna of all Australian terrestrial and freshwater ecosystems. Both in numbers of species and abundance of individuals invertebrates are the major elements of biodiversity, and as such can lay claim to being the keystone biotic participants in rainforest functionality.

There are estimated to be ~250 000 species of terrestrial arthropods (Chelicerata, Crustacea, Uniramia) in Australia (world total ~5 000 000 species) (Chapman 2009; Yeates *et al.* 2003). This may be a substantial underestimation of the actual size of the fauna (Raven and Yeates 2007). The majority of the Australian terrestrial arthropod fauna comprises Arachnida (spiders), Acarina (mites), Hymenoptera (wasps, bees, ants), Diptera (flies) and Coleoptera (beetles). For insects alone Chapman (2009, largely citing Yeates *et al.* 2003) gives the estimated number of Australian Coleoptera as ~80 000–100 000 species, followed by Hymenoptera (~44 000 species), Diptera (~30 000), Lepidoptera (~20 000 species) and Hemiptera (~11 500). However, very few groups have been studied in detail, and for many invertebrates, the ecology, particularly that of their immature stages, remains poorly known. For most species the very nature of the form and behaviour of their larval stages, potential dependence on mutualistic associations with other plant or animal organisms, necessary population thresholds so that any given species can reproduce and survive in isolated habitat patches, and basic physiological and environmental constraints acting on reproduction remain a mystery. Only for butterflies do we have anything approaching an expansive knowledge of immature stages, food plant associations, and mutualisms with other organisms.

THE ROLE OF INVERTEBRATES IN SUBTROPICAL RAINFOREST ECOSYSTEMS

Ecological communities are structured by interactions of competition and predation, mutualistic and parasitic relationships between species, and chance events that cannot be predicted. Interactions between species can be negative, as with predation and competition, or positive as in the case of mutualisms (Ellwood *et al.* 2016). Thus communities are structured with regard both to species diversity and functional composition (Stork *et al.* 2017).

Invertebrates play pivotal roles in the function and survival of rainforest communities (Ashford *et al.* 2013; Culliney 2013) with species richness predicted to increase in rainforests of greater floristic and structural complexity, where a broader range of niches are available, providing more microhabitat and microclimate heterogeneity. Invertebrates are predators, herbivores (literally predators of plants), fungivores, hosts to other organisms (such as mites and flatworms), parasites, primary decomposers of plant material, dispersers of nutrients and seeds within soil profiles, and scavengers and decomposers of dead animals.

A Gulliver in the bush

Recent decades have seen an intense interest in the invertebrate fauna of Australia's rainforests. For those studies scattered throughout the eastern subtropics this has been driven partly by the dedication of the Central Eastern Rainforest Reserves World Heritage Area (Adam 1987, 1992; Cavanaugh 2010), now called the Gondwana Rainforests of Australia. However, a long history of fieldwork by natural historians (often not professionally employed in their field) preceded this, and the knowledge they acquired established the recognition of the unique biological and biogeographical attributes of Australia's subtropical rainforests. Awareness of their contributions fades with the passage of time, and indeed their investigations were largely known only to a handful of far-flung co-workers. Many lived and worked in isolation. Although their research papers continue to speak of their contributions, few left written accounts of their expeditions into the field. One exception was Herbert James Carter, born 1858 in Marlborough, Wiltshire, England. From 1881 to 1891 he was the Second Mathematical Master at Sydney Grammar School, Principal of Ascham until 1914, and an Honorary Entomologist to the Australian Museum. His passion was the world of beetles, and particularly Buprestidae. Carter left a unique record of his insect world, for in 1933 he published *Gulliver in the Bush: Wanderings of an Australian Entomologist* (Angus and Robertson). Within its 234 pages he relates expeditions to Western Australia, tropical Queensland, Tasmania, South Australia and extensively New South Wales, by train, coastal steamer, launch, car, buggy, bicycle, by foot, and on horseback. Its pages are filled with the names of the animals he discovered.

Carter travelled with friends and sometimes family, but often alone. He sometimes overnighted in boarding houses but mostly he camped in tents, for their siting gave immediate exposure to habitats of many kinds. Of particular interest here are his observations of the rainforests of New South Wales and southern Queensland. He speaks of Kangaroo Valley, south of Sydney, at the far end of which 'rises Mount Cambewarra, and here the entomologist can let himself go, for it is a bit of rain-forest, full of natural treasures, animal and vegetable'; then of the 'tree-fern forest' at Brown Mountain further south, where his time was cut short by his companion-driver's itch 'to get on', and long hours 'playing at cards' in the Barrington Tops as he and his zoologist fellows waited for the rain to cease, their venture there preceded by a long ascent by horse and cart from the coastal plain below. Closer to his home he travelled west to the Blue Mountains, collecting in the moist forests that still survive at Mount Tomah, Mount Wilson and Mount Irvine – and getting there by bicycle. At Mount Wilson he encountered *Peripatus*, and the carabids *Notonomus arthuri* and *N. lateralis*, and at Mount Irvine the cerambycid *Ceregidion horrens*. At Wallis Lake Carter hints at collecting in what was likely littoral rainforest, finding there the tenebrionid beetle *Adelium helmsi*. On a later trip to the New South Wales north coast he reflects on the ongoing clearing of rainforest, particularly that of the Comboyne and Dorrigo plateaux, and on the needless waste of timber felled, and the loss of habitat and the animals within. He once stopped at the coastal township of Wingham, noting that there had been 'reserved a considerable area of virgin forest as a park' (now Wingham Brush Nature Reserve). He had earlier been sent an undescribed beetle (*Cardiothorax alternatus*) said to have been collected from there, and was keen to get more. He asked his travelling companion to give him 'a quarter of an hour in this reserve, and within the fifteen minutes I was fortunate enough to find another example of the beaste'. Carter did not neglect Queensland, exploring the rich montane rainforests of the Border Ranges, coastal forests further north at Byfield and, understandably, those of the Wet Tropics. He regretted too few visits, and often too short a visit, to them all. Several of the insects Carter described have been found to be synonyms of ones previously named. In this he is not alone. It was a different time, yet he has left a rare record, of creatures, places, people and events, that for current-day naturalists imparts a clear sense of walking already trodden paths.

They maintain soil structure, contribute to the regulation of hydrological processes, and participate in life cycle mutualisms with other invertebrates and even vertebrates. Some invertebrates are ecological specialists, others are generalists, and many are opportunists (Matthews 1976b; for a discussion as to whether specialisation constitutes an evolutionary dead end see Futuyma and Moreno 1988; Kelley and Farrell 1998; Stireman 2005).

In Australia's subtropical rainforests, and the world's rainforests more generally, invertebrates function as the main and frequently the exclusive pollinators of flowering plants (Williams 1995; Williams and Adam 1994, 1999b, 2010). However, many insects function to impact adversely on plant reproduction, for ~50 per cent of insect species are fully or partially herbivorous (Hendrix 1990), their chewing mouthparts consuming flowers and fruit, in addition to leaves, stems and roots. Of the insect orders 99 per cent of Thysanoptera, 98 per cent of Lepidoptera, 90 per cent of Orthoptera, 47 per cent of Coleoptera and 25 per cent of Diptera are phytophagous (Hendrix 1990). By their combined actions assemblages of insect herbivores can drive rainforest plant diversity and composition (Bagchi *et al.* 2014).

In response to such levels of herbivory (and in addition to that of herbivorous mammals) and the plant selection pressure this represents (Bagchi *et al.* 2014; Jurado-Rivera *et al.* 2008), plants have evolved an array of chemical and mechanical defenses (see Coley *et al.* 1985; Marinho and Teixeira 2018; Rosenthal and Janzen 1979; Willmer 2011), including leaf toughness (Aide and Londoño 1989; Lowman 1992) and defensive plant–insect mutualisms (Huxley and Cutler 1991). The guild structure of phytophagous species has been found to respond to leaf size (Nooten and Hughes 2016), yet over time herbivory (including seed theft) can affect the structure of plant communities and how individual plants are dispersed within populations.

But herbivore attack also induces rapid plant responses. These include allocation of defence compounds (Lowman 1992), translocation of resources to different parts of the plant, changes in the balance between vegetative growth and reproduction, impacts on photosynthetic rates, and even alteration to sex expression (e.g. in those species with both male and hermaphroditic flowers on the same individual) (Doust and Doust 1990). The diversity and diet breadth of herbivores can vary with elevation (Rodríguez-Castañeda *et al.* 2010), and dramatically within small spatial scales (Kemp *et al.* 2017) owing to the heterogeneous dynamics of plant species turnover (extinction and recruitment) within individual stands. Rates of herbivory can change with respect to leaves being present in the sunlit canopy or in the shaded understorey. Lowman (1992), for example, found that rates of herbivory on the tree *Doryphora sassafras*, a species common in upland New South Wales rainforests, were consistently higher for shade than sun leaves (leaves in the sun were tougher). Lowman also noted that levels of defoliation were higher in young rather than old leaves and in lower than upper canopy levels. Herbivore communities also respond in abundance and number of species to openings in the rainforest canopy. However, studies by Yves Basset and his colleagues (Basset *et al.* 2001) indicated most species responded to an increase in leaf production by seedlings rather than to canopy openness *per se*.

Invertebrates can be found at all forest levels and occupy a great diversity of microhabitats; however, the presence of adults and their activity patterns are strongly tied to that of season (Bigger 1976; Frith and Frith 1985; Lowman 1982, 1985) as well as daily rhythms and variations of light and dark (Saunders 1977). Examples of well known diel (24 hour period) responses by individual species and higher taxonomic ranks abound, but such patterns are also exhibited by whole assemblages of invertebrates. For example, at the guild level Basset and Springate (1992) noted that on the tree *Argyrodendron actinophyllum*, in southern Queensland, herbivorous arthropods were generally more active at night, but predators and parasitic species were more so during the day.

Insects and other invertebrates interact with habitat at fine scales. For example, distinctive taxon boundaries may occur with respect to foliage and flower inhabiting faunas, as Wardhaugh *et al.* (2013b) found for beetles inhabiting tropical rainforest canopies. Rainforests with high environmental heterogeneity may possess invertebrate communities and guilds (see Hawkins and MacMahon 1989; Peck and Howden 1984) with concomitant levels of species richness and abundance (Schuldt *et al.* 2015; Sousa *et al.* 2016), and can exhibit distinct assemblages with respect to altitude or the prevalence of moister microhabitats (Hall *et al.* 2014; Houston and Melzer 2012). Although some invertebrate assemblages may exhibit clear vertical stratification in rainforest (as has been shown for moths [Ashton *et al.* 2016]), invertebrate biodiversity in general may not necessarily be skewed to favour individual, seemingly pre-eminent, forest strata (e.g. canopy versus forest floor [Stork and Grimbacher 2006]). Their abundance and species richness (often used synonymously here with 'diversity') may decrease in rainforest with increases in altitude (Laidlaw *et al.* 2011; Lambkin *et al.* 2011; Le Cesne *et al.* 2015; Ødegaard and Diserud 2011), or may vary in abundance in response to canopy height (Kitching *et al.* 1993).

Invertebrates are to be found in leaf litter on the forest floor, in standing and fallen rotten timber, in the upper soil profile, under the loose bark of standing trees, on the leaves of plants of the dim understorey and trees of the sunlit canopy, in ephemeral water pools, in and patrolling above dissecting streams, in tree hollows (Plate 291), below stones and the bases of epiphytic plants, at the base of leaf axils, as cohabiters (commensals) in the nests of invertebrates and vertebrates, in and under the corpses of dead animals, and occupying rocky outcrops and caves.

Life in a rainforest canopy

The potentially rich invertebrate fauna to be found in the canopies of subtropical rainforest trees remains poorly studied, not least due to the difficulty of gaining physical access. Nevertheless there is an increasing interest in the taxonomic composition and the ecological features of invertebrates that are found there (Basset 1990, 1991; Hammond *et al.* 1997; Kitching *et al.* 1993, 2004; Lambkin *et al.* 2011; Lowman 1982, 1985, 1992; Tree and Mound 2011; Walter and Behan-Pelletier 1999; Walter *et al.* 1994). The studies at hand, however, do give insight into the nature of the fauna. One tree that serves as an exemplifier is Black Booyong *Argyrodendron actinophyllum* (Sterculiaceae/Malvaceae), a distinctive large-buttressed species found at intermediate altitudes in subtropical and warm temperate rainforest between Gloucester (New South Wales) and Gympie (Queensland). This was the subject of studies by Yves Basset in the Mount Glorious State Forest north-west of Brisbane (Bassett 1990, 1991). Spiders were one of the major predatory groups inhabiting the canopy stratum. They included ambush hunters, and active predators that forage on foliage, flowers and branches. Families and genera included Theridiidae (*Syntaxus, Thwaitesia, Euryopsis, Achaearana*), Clubionidae (*Clubiona, Cheiracanthanium*), Salticidae (*Helpis, Holoplatys*), Araneidae (*Araneus, Dolophones*) and Thomisidae (*Diaea, Sidymella*). Theridiidae were dominant. All families were also present in forest litter at ground level, as were many of the genera. Dominant Coleoptera were Chrysomelidae, Curculionidae (including Scolytidae), Corylophidae, Staphylinidae, Scarabaeidae, Phalacridae, Latridiidae, Mordellidae, Melandryidae, Cerambycidae, Elateridae, Carabidae, Tenebrionidae and Cleridae. Most are phytophagous or xylophagous in habit but some predatory families are present (Staphylinidae, Carabidae). These families are present throughout the different forest strata, though there is considerable replacement or variation in abundance of species and genera. Brachyceran Diptera included Phoridae, Chloropidae, Empididae, Lauxaniidae, Dolichopodidae and Muscidae. Dolichopodids are distinctive long-legged predators commonly found on sunlit shiny leaves. Lauxaniidae were well represented in samples, but Syrphidae were poorly collected though distinctive of some vines (*Cissus*) and ground herbs (*Pollia*) in lowland and littoral rainforests. Psyllidae (Hemiptera) were the dominant foliage-frequenting insect group. Hemiptera (Heteroptera and Homoptera) were diverse and included, Achilidae, Anthocoridae, Aphidiidae, Aradidae, Carsaridae, Cicadellidae, Cydnidae, Derbidae, Enicocephalidae, Flatidae, Fulgoridae, Isometopidae, Issidae, Lygaeidae, and Miridae. Arboreal ants (Formicidae) on *Argyrodendron actinophyllum* were relatively under represented in samples, though often commonly encountered on the flowers and foliage of trees and shrubs in lowland remnants (Williams 1995; Williams and Adam 1994, 2010). However, Hymenoptera more generally were well represented, including predators, parasitoids and nectar/pollen gathering and feeding taxa: Apidae, Bethylidae, Braconidae, Diapriidae, Encyrtidae, Eulophidae, Halictidae, Ichneumonidae, Mymaridae, Platygastridae, Pteromalidae, and Torymidae. Aculeate wasps were under-represented (owing to the collecting methods used) but are known to be diverse, and seasonally abundant, on flowering lowland subtropical rainforest trees (Williams and Adam 1995). Of the other orders, thrips (Thysanoptera), and particularly Phlaeothripidae, were abundant in samples, and Neuroptera were well represented by Coniopterygidae. Psocidae were represented by foliage-dwelling Caecilidae, Stenopsocidae and Ectopsocidae, and Lepidoptera mainly by the moth families Erebidae, Geometridae and Noctuidae. Orthoptera were mainly the families Gyrillidae, Tettigoniidae and Gryllacrididae; individuals representing disproportionate contributions to overall biomass. The composition of the fauna is in part influenced by ecological determinants of rainforest community and host plant microclimate, leaf, flower and fruiting phenology, branch and foliage architecture (e.g. interdigitation of host branches with adjacent plants), bark sculpture, encrusting epiphytes and fungal growth, host biochemistry, and by the composition and abundance of the fauna itself (including prey networks and prey size, predator abundance). Irregular flowering events alone can dramatically alter the composition of attendant taxa. Broader climatic differences (e.g. drought) and recruitment events such as plant species foundation episodes (and the arthropods associated with them) also promote or constrain the nature of the invertebrate assemblage (Kitching and Arthur 1993). Historical events of colonisation and radiation of specific invertebrate taxa (and their plant hosts) over periods of geological time explain some differences in composition and dominance of invertebrate assemblages reported between different tree species and different bioregions.

Many are only active during the day, many only at night, and some are crepuscular in habit, being active at the brief periods of dusk and to a lesser degree at dawn (Saunders 1977). Invertebrates are also important as soil-disturbers, their actions contributing to soil quality by the mixing of nutrients and the enhanced distribution of water through soil profiles (see Eldridge and James 2009 for a discussion of biotic agents in this role). Though this activity in subtropical rainforest is more conspicuously performed by vertebrates (e.g. brush turkeys, lyrebirds), the combined soil perturbation activities performed by invertebrates (such as ants, worms, scarab larvae) can have wide-ranging effects on both biotic and abiotic processes within the rainforest ecosystem.

There is also an array of internal and external invertebrate parasites associated with mammals, such as the nematode worms of the families Cloacinidae and Capillariidae that are found in native marsupials and rodents (Smales 1994, Spratt 2006), and Gordian worms (Nematomorpha), the juveniles of which live within the body cavities of cockroaches and caddis flies (Neboiss 1991; Roth 1991a). Phthiraptera (lice), Siphonaptera (fleas), and Metastigmata (ticks) have obligate dependencies and have coevolved with their mammal and bird hosts. Some, such as the larvae of dragonflies, mayflies and caddisflies, are restricted to water bodies but as winged adults disperse to terrestrial habitats throughout the forest. Freshwater parastacid crayfish occupy cool flowing rainforest streams but may leave these to undertake uncertain journeys in search of new habitats. Various families of water beetles inhabit streams as larvae and adults, but the latter frequently undertake flights elsewhere, always returning to permanent water to lay their eggs. Though many rainforest-dependent invertebrates (certain papilionid butterflies, large aculeate wasps and cetoniid and buprestid beetles, for example) can disperse and forage over relatively large distances, including movements that extend into adjacent non-rainforest plant communities, the movement patterns of the flightless fauna (e.g. millipedes, earthworms, nemerteans, land snails, lucanid and tenebrionid beetles, many carabid beetles and scarabaeine dung beetles) are in general restricted to the closed forest understorey. For elements of the flightless ground fauna, such as land snails, daily foraging movements may encompass a realm less than several square metres in size. But for all flightless rainforest-dependent invertebrates, many of which are endemic, there is little or no capacity to disperse to other rainforest stands, and intervening roads and trails that dissect patches (Plate 313) may act as unassailable barriers to dispersal, or retreat to preferred diurnal or nocturnal refuge microhabitats. Even linear strips of otherwise seemingly suitable rainforest habitat may preclude dispersal (e.g. Hill 1995). And so, in addition to being sensitive to habitat and climate change, the flightless ground-dwelling fauna is especially vulnerable to localised extinction.

Flightless groups may form mosaic-like populations, rather than uniformly dispersed ones, throughout individual rainforest patches. This is primarily influenced by microhabitat and microclimate constraints, such as zones of higher or more constant levels of humidity, the presence of protective rock formations, mulch depth, moisture levels in leaf litter, the size and temporal duration of sunlit gaps (due to canopy disturbance), or the spatially concentrated availability of algal food substrate. These mosaic-like 'within patch' (heterogeneous) distributions are not confined to the fauna, but are also exhibited commonly by subtropical rainforest trees (Connell *et al.* 1984; Williams 1993), although individual tree species can be predictably and seemingly uniformly present throughout stands, as for *Nothofagus moorei* (Nothofagaceae) in rainforest at high elevations, *Syzygium floribundum* and *Tristaniopsis laurina* (Myrtaceae) in riparian zones at intermediate elevations, *Backhousia sciadophora* (Myrtaceae) in many dry rainforest formations, and *Cupaniopsis anacardioides* (Sapindaceae) and *Alphitonia excelsa* (Rhamnaceae) in dune-associated littoral rainforest.

Many invertebrates are herbivores of living plants and respond to seasonal variation in, and patterns of, plant growth (Lowman 1982; Wolda 1978). They include those that eat leaves, are stem or leaf miners, consume floral parts and pollen, and destroy fruits and seeds, and those that attack roots, living trunks and branches. In this they are acting as simple predators. Few display multiple herbivory feeding strategies, though some leaf eaters may also feed on flowers. Leaf miners are considered to be relatively host specific with their host range restricted to one or few congeneric hosts (Maunsell *et al.* 2017). Maunsell *et al.* (2017), for example, in a study of leaf miner host-specificity in Australian subtropical rainforests found that local plant-specific relationships exceeded 60 per cent within their study suite. Differences in gross facial morphology alone can act as a barrier to the plant component that various herbivores can feed upon. And, indeed, species may possess highly specialised and prolonged mouthparts (as in certain adult curculionid beetles and nectar feeding Lepidoptera) that preclude their ability to switch to different kinds of food resources. Those that feed on leaves may be restricted to a single plant species or are associated with a broader suite of plants; these frequently are taxonomically diverse, not necessarily closely related. Species, over their total known geographical range, may feed from a suite of plants, but individual invertebrate populations at individual rainforest remnants may feed off only a single host plant, as is the instance for the herbivorous chrysomelid beetle *Menippus darcyi*, which maintains an endangered population in littoral rainforest near Sydney. Here the species is associated in a food plant relationship with the tree *Celtis paniculata* (Ulmaceae/Cannabaceae). However, both the tree and the beetle have more extensive recorded geographical and rainforest habitat distributions along the Australian east coast (Harden 1990; Reid and Nally 2008). Importantly, McLeish *et al.* (2017) note that 'specialised associations between insects and plants are a function of varying climatic and evolutionary constraints over the geographic range of the host lineage'.

Some invertebrate higher taxa (e.g. family) may have species that as larvae feed on living plant tissue. Others are known only to attack dead or dying timber, for example numerous species in the beetle families Cerambycidae and Buprestidae (see Bellamy *et al.* 2013; Lawrence and Ślipiński 2013; Ślipiński and Escalona 2013, 2016). In the speciose Cerambycidae there are divergent examples of larval host choice. The larvae of the rainforest-inhabiting ant-like *Mesolita lineata* develop in the dying and dead timber of unrelated *Ceratopetalum apetalum* (Cunoniaceae) and *Argyrodendron actinophyllum* (Sterculiaceae/Malvaceae), while those of *Pentacosmia scoparia* have been recorded from a broad suite of tree species including *Pouteria australis* and *Alectryon coriaceus* (Sapotaceae), *Drypetes deplanchei* and *Glochidion ferdinandi* (Euphorbiaceae/Phyllanthaceae), *Doryphora sassafras* (Atherospermataceae) and *Nothofagus moorei* (Nothofagaceae) (Webb *et al.* 1988; G. Williams 1985). However, the suitability of plants for larval development in species that utilise newly dead or dying wood may be predetermined by the decaying state of the timber (and possible chemical cues that the beetles respond to), the suitability of the bark surface for egg deposition (e.g. bark surface textured rather than smooth, and retained rather than easily shed), and the seasonal timing of branch fall or trunk death (needing to be synchronised with the presence of adult insects), rather than a recruitment response predicated by plant phylogeny. Thus the insects are responding in an opportunistic manner to whatever suitable larval substrate is available.

Among rainforest taxa whose larvae utilise dead wood, there are instances of apparent 1:1 relationships. The buprestid beetle *Nascioides tillyardi* (Plate 107), for example, restricted in distribution to the cool temperate rainforests of mountainous northern New South Wales, only utilises dying branches of *Nothofagus moorei* as its larval host (Williams 1987). But this is a closed host relationship seen only from the view of *Nascioides tillyardi*, for other insects, many kinds of beetles among them, also develop in the dying branches of *Nothofagus moorei*; these include other buprestids (*Melobasis hypocrita*, *M. ignipicta*) that inhabit the same rainforest stands. In *Nascioides*, as is the case with many other herbivore ranks above that of species, the phylogenetically circumscribed beetle herbivore–host plant relationship tends to break down considerably at the level of genus. In this case the majority of the known *Nascioides* species are recorded from multiple phylogenetically unrelated larval food plants (Bellamy *et al.* 2013; Williams 1987; Williams and Bellamy 2002). Thus at the level of genus (*Nascioides*) the larval host plant strategies are diverse, indicating multiple lines of response and radiation within the overall plant-habitat landscape.

Though ecological generalist strategies appear to dominate the world of herbivory, where invertebrates use living plants for the development of their young, there are many examples of species that are ecological specialists, their choice of food plants being restricted to either a single species or a suite of related ones; this is seen with certain butterflies (Braby 2000) and moths (Common 1990), fig wasps of the family Agaonidae (Weiblen 2002; Wiebes 1979), and the weevil genus *Elleschodes* (A. Hamilton 1897; Williams and Adam 2010).

Some rainforest insects are specialist seed feeders, a strategy particularly evident among certain curculionoid beetles (Juniper and Britton 2010). In several instances plants sacrifice floral parts (as in the *Eupomatia*–*Elleschodes* plant–pollinator mutualism) or reproductive components (as in fig–fig wasp mutualisms), in exchange for pollination services; however, various herbivores can be very destructive of flowers and leaves; an example is the chrysomelid beetle *Monolepta australis*, which in some years can emerge in massive numbers and defoliate trees such as *Glochidion ferdinandi* (Euphorbiaceae/Phyllanthaceae) and consume the whole flowers of

Batesian and Müllerian mimicry in insects

'Mimicry' (mimesis), as a term, applies when an organism resembles another, and 'crypsis' when an organism resembles a particular background. But their application can be ambiguous, for example if an insect resembles a leaf, twig or bark nodule (Endler 1981).The subtropical rainforest realm of insects is alive with examples of mimicry based on structure (crypsis and procrypsis, this commonly in combination with colour and/or behaviour). Membracid bugs mimic branch thorns, leaf nodes and broken twigs; phasmatids look like plant stems; tettigoniid grasshoppers and katydids resemble leaves of different kinds; and mantids manage to convincingly blend with the lichen-festooned tree trunks that they patrol. By this means, they benefit by concealment, avoid or reduce risks of predation, or enhance their foraging success as predators or consumers of flower resources. But there are other forms of mimicry, based largely upon shared aposematic warning colour schemes, which many insects resort to. Warning colour schemes typically are bright reds and yellows, frequently in lateral bands across the abdomen and thorax, or are just a shared ground plan of red and black. However, the role that colour (or the perception of it) plays in convincingly projecting an appropriate 'mimic' image is dependent on the development of a predator's colour vision. Many animals appear to be colourblind, or have limited trichromatic perception, thus posing the question as to whether organisms are responding to contrasting and conspicuous patterns rather than colour sensations. The most well-known forms of mimesis are termed 'Batesian' and 'Müllerian' mimicry, and involve taxonomically distinct groups that are assumed to be sharing a conspicuous visual image that projects a warning signal to predators, or suggests they are predators. Each group is based around a distasteful 'model', or is a suite of insects that are all chemically defended. The evolutionary pathways, as to how these strategies came about, are uncertain (though usually attributed to processes of natural selection), and the validity of the mimic themes themselves remain partly conceptual with most studies on the evolution of mimicry and warning coloration in insects considering predators as the major selective force (Rashed and Sherratt 2007). Basically they are good ideas that make sense and that have a level of tested foundation, but how to conclusively establish the mechanisms by which such strategies evolved is somewhat elusive. Predators must learn that prey items are distasteful, and conceivably this means that several non-toxic, potentially palatable species will be encountered first, and 'sacrificed' unless they are uncommon relative to the model. So for this strategy to work, individuals of the model must be much more abundant than the mimic. Batesian mimicry is named after the English naturalist Henry W. Bates, who spent the years 1849 to 1860 wandering through the rainforests of Brazil. Bates observed that many butterflies (he called them Heliconiids, though many are now placed elsewhere) when positioned together exhibited inconsistencies. Though sharing similar appearances, some belonged to different families. Heliconiids were the more numerous and were not killed by predatory birds, and so he thought these were unpalatable or toxic. Bates assumed the less abundant butterflies, which shared the conspicuous coloration of the Heliconiids, gained protection by the pretense of appearing to be Heliconiids. And so predators would avoid what were really quite edible 'mimetic' prey. Often-cited Australian examples are red- and black-coloured Lycidae, Meloidae, Cerambycidae, Belidae, Buprestidae, Pyrochroidae, Scarabaeidae (all beetles), Erebidae and Oecophoridae (moths) complexes based around a distasteful and abundant lycid (*Metriorrhynchus*) model, and black and yellow/red complexes variably comprising beetles (*Hesthesis* – Cerambycidae), wasps (*Pseudabispa* – Vespidae, *Williamsita* – Crabronidae, *Sceliphron* – Sphecidae), flies (*Chrysopogon* – Asilidae, *Syndipnomyia* – Stratiomyidae, *Ceriana* – Syrphidae), and others, in which wasps are the presumed models. The guise of deceit is often enhanced by similar appearances of shape, or by subtle shading of colour that gives the impression of the form of the model. Bates had noticed that occasionally inedible but unrelated butterfly species were also similar in appearance. In 1878 the German zoologist Fritz Müller provided an explanation. Müller had also spent time catching butterflies in Brazil, and also assumed that predators learnt to avoid certain prey after bad experiences. However, a certain number of individuals would be killed, though not eaten, before the predator was conditioned to avoid them. He postulated that if different prey species shared the same warning signal, then the predator would more readily learn to recognise any insect similarly coloured. But Batesian mimics can also be part of a Müllerian complex (as Barker 2006a proposed for the *Castiarina* 'producta' group of buprestid beetles), for the more species in the latter results in a greater number of genuine unpalatable models; but most examples of Müllerian mimicry involve different species with the same signal patterns.

various other rainforest angiosperms. These destructive predatory actions can have cascading negative ecosystem effects where the loss of developing fruit impacts sequentially, not only on plant reproduction and population founding events (Shanahan *et al.* 2001), but also on the populations of dependent fruit-eating and nectar-feeding vertebrates as well.

POLLINATION IN AUSTRALIA'S SUBTROPICAL RAINFORESTS AND THE POLLINATOR FUNCTION OF INSECTS

The pollination ecology of Australia's subtropical rainforests, in which flowering plants dominate (as in all the world's rainforests), has been extensively reviewed and discussed by Williams and Adam (1994, 1999b, 2010, 2013, 2019). Rainforest canopy flowers, collectively as a distinct microhabitat, may offer unique opportunities for host specialisation and diversity, equal to or vastly greater than the invertebrate community that inhabits mature leaves (Wardhaugh *et al.* 2013a, 2013b).

There are instances among subtropical rainforest plants of pollination by wind that involves the auto-release of pollen (e.g. *Podocarpus elatus* – Podocarpaceae, a pine), secondary dispersal by air currents following initial ballistic release of pollen grains (e.g. *Trophis scandens*, *Streblus brunonianus* – Moraceae [Williams and Adam 1993]), and structural adaptations of flowers that suggest a mixed pollination strategy (termed 'ambophily') utilising biotic transfer and abiotic dispersal by wind, as in endangered *Rhodomyrtus psidioides* (Myrtaceae) (Williams and Adam 2010, Williams 2018), and other species in which the stigma surface is broad or furcated in some way (thus increasing the area available for pollen deposition) and held above the surrounding anthers and where pollen grains are generally dry. However, the dominating pollination theme in subtropical rainforests (and indeed those in the tropics, and temperate Tasmania) is that flowering plants are primarily pollinated by animals and that most of these are not only insects, but also insects under 5 mm in size, with pollination by insects being an ancestral trait of early fossil angiosperms (Hu *et al.* 2008).

There are relatively few flowering plants in subtropical rainforest that show floral characters (such as flower form, colour, large nectar volumes, and robust stems sufficient to support large feeding animals), indicating they have evolved to depend significantly on vertebrates as agents of pollen transfer. *Castanospermum australe* (Fabaceae) and *Grevillea robusta* (Proteaceae) are examples, though neither extend naturally, south in range, beyond northern New South Wales. Various vertebrates can be important dispersers of rainforest seed, among which flying foxes (Pteropodidae) can play a keystone role, but their role as pollinators at this broadly defined latitude is primarily restricted to xeric-adapted plant communities and plant taxa such as *Corymbia*, *Eucalyptus* and *Melaleuca* (Myrtaceae). Although the range of the Queensland blossom bat, *Syconycteris australis*, extends into the lowlands of northern New South Wales, it appears to utilise rainforest stands only as roost sites, with individuals moving out at night to feed from the blossoms of trees growing in open forest, such as *Melaleuca quinquinervia* (Myrtaceae), *Banksia integrifolia* and *B. serrata* (Proteaceae) (Law 1992, 1993), though senescent *B. integrifolia* trees can be found in regenerating coastal closed forests associated with beach hind-dune zones (Williams 1993). The apparent latitudinal filtering of angiosperms that meet a vertebrate-adapted pollination syndrome alone precludes what kinds of biotic agents will function as pollinators.

The predominant structure encountered among flowering plants in subtropical rainforest is that of ecologically unspecialised flowers of open form in which there is little depth effect in the flower structure (e.g. *Syzygium smithii* – Myrtaceae, *Scolopia braunii* – Flacourtiaceae/Salicaceae) (Williams and Adam 2010). This allows access by a diverse assemblage of flower-visiting insects, most of which can function as pollinators, even if inefficient ones that individually only carry small pollen loads. Some flower-visiting insects compensate for their inefficient transfer (owing to restricted and random foraging patterns, or small pollen loads) by occurring in great abundance, as thrips often do. During the day these generalised entomophilous flowers attract numerous kinds of flies, beetles, wasps, bees and to a lesser degree butterflies. At night these are replaced by a different, less diverse subset of taxa: small beetles, nematoceran flies, coackroaches and moths among them. Many species may be regularly present in low numbers (large cetoniine beetles, eumenine wasps, papilionid butterflies), or seasonally absent (many buprestid beetles), but others may seasonally number as a multitude (e.g. *Phyllotocus* spp., *Antitrogus* spp. – Scarabaeidae). Some are distructive of individual floral parts, and some consume whole flowers.

Large flower-visiting insects, such as many aculeate wasps (e.g. *Abispa splendida* – Vespidae, and numerous

Thynnidae), buprestid (*Calodema regale*, *Metaxymorpha grayii*), cerambycid (*Distichocera superba*) and cetoniine beetles (*Schizorhina atropunctata*, *Eupoecila australasiae*) (Williams and Adam 1995, 1998, 2010) commonly undertake relatively long-distance flights between individual flowering trees, and so can be important agents of pollen transfer. Many insects that visit flowers, however, are not rainforest-dependent (either for adult food resources or larval development) and occur also at the flowers of trees and shrubs in adjacent open eucalypt forest, woodland and heath, and even flowering mangroves. These recruit to plant species essentially on the basis of flower type, the nature and quantity of the floral reward, conspicuousness (including that of the fragrance plume) of the plant when in flower, and the seasonal or daily timing of flower opening. Common examples among the larger insects are the beetles *Eupoecila australasiae* and *Glycyphana brunnipes* (Cetoniinae), numerous *Castiarina* (Buprestidae), *Aridaeus thoracicus* (Cerambycidae), wasps of the genera *Scolia*, *Radumeris*, *Pseudotrielis* (Scoliidae), *Heterodontonyx* (Pompilidae) and *Sphex* (Sphecidae), and the butterflies *Graphium sarpedon* and *Papilio aegeus* (Papilionidae). Swift-flying, nocturnal hawkmoths (Sphingidae), such as *Theretra nessus*, also undertake long-distance flights between plants.

In addition, many smaller flower-visiting insects, which are typical of open habitats, also are attracted to the blossoms of mass-flowering rainforest trees and shrubs, in effect allowing many rainforest plants to recruit potential pollinators from a great range of available insects and habitats. Thus a diversity of pollinators can be relied upon, with the strategy of general entomophily facilitating plant-founding events and the pollination of dispersed plant populations within stands. However, Crichton *et al.* (2019) caution that 'super-generalist' (ecologically unspecialised) pollinators participating in largely generalised entomophilous systems may facilitate weed invasion where these do not require ecologically specialised pollinators; the ubiquitous presence of the introduced honey bee *Apis mellifera* in all rainforest formations is an obvious 'super-generalist' candidate.

The pollination syndrome of generalised entomophily is expressed in all rainforest strata, but is particularly prevalent and conspicuous among mass-flowering trees of the canopy layer (Williams and Adam 1994, 1999b, 2010). Though large-sized anthophilous insects are usually absent from the shaded understorey strata, some (e.g. thynnid wasps, trap-lining bees of the genus *Amegilla*) may be encountered in sunlit canopy gaps and stream openings where herbs, shrubs and straggling vines are in flower.

Generalised entomophilous flowers are frequently coloured white and are often highly fragrant, but individually may only offer small nectar rewards. This limitation of the available resource, and the commensurate energetic cost incurred by vertebrates of foraging at them, by itself, is an obstacle to visits by birds, flying foxes and blossom bats, and can act as a constraint on the coevolution of more specialised plant–pollinator relationships. The timing of fragrance emissions can also discriminate against what kinds of insects (diurnal/day-active versus nocturnal/night-active ones) will visit (Willmer 2011).

Nevertheless, there are rainforest plants in which there are varying degrees of depth effect in their flowers. These include flowers that are radially symmetrical (actinomorphic) and bilaterally symmetrical (zygomorphic), or of a more complex form (as in Araceae): orchids (*Pseudovanilla*, *Dendrobium*), various vines (*Pandorea*, *Cynanchum*), herbaceous shrubs and herbs (*Alpinia*, *Crinum*, *Lobelia*, *Typhonium*), and trees (*Brachychiton*, *Castanospermum*) number among them. Many of these are frequented by a more taxonomically circumscribed suite of pollinators such as trap-lining, long-tongued bees (e.g. *Amegilla*), butterflies and hawkmoths.

In *Alocasia brisbanensis* (Araceae), a large herbaceous shrub-like plant encountered on forest margins and sunlit openings below the canopy (Hay and Wise 1991), the complex 'flower' formed by the enveloping spathe and specialised spadix precludes entry by most organisms, the pollinators being small flattened nitidulid, kateretid and derelomine weevils that are able to gain entry to the depths of the flower (Plates 159, 160). The large understorey herb *Alpinia caerulea* (Zingiberaceae) is found in wet sclerophyll forest, and though visited by several small-sized bees (frequently *Tetragonula*) the pollinators are species of *Amegilla*. Flower spikes open sequentially, generally from top to bottom; however, developing flower buds are sometimes eaten by lepidopteran larvae. The upper filament and cup-like stigma are held tightly between two longitudinally positioned anthers. The stigma faces downwards and slightly forward so that bees, on entering the flower, have any pollen on their upper surface scraped into the concave stigma, from which it would be difficult to dislodge when the bee reverses. Likewise the positioning of the stigma restricts entry of self-pollen due to the placement of the stigma in front of the anthers. The diminutive, mauve-coloured, dioecious herb *Lobelia trigonocaulis* (Lobeliaceae) grows in sunlit openings on the forest floor of rainforest and wet sclerophyll forest. Flowers have yellow nectar guides at the top of the labellum and a paler longitudinal zone running along the length of the labellum, in addition to a darker mauve area at the entrance to the 'throat'. Female flowers have the stigma exerted past the encompassing apical stamens. Flower visitors comprise small syrphid flies, halictid and hylaeine bees, nymphalid and hesperiid butterflies. *Apis mellifera* is also a commonly observed visitor.

The red tubular flowers of of the mast-flowering tree *Brachychiton ascerifolius* (Sterculiacae/Malvaceae) suggest pollination by birds; however, the pollinators are apparently bees, the pollination syndrome in this instance being counter-intuitive. The yellow, tubular flowers of the understorey shrub *Pittosporum revolutum* (Pittosporaceae), usually visited and pollinated by nitidulid beetles, may also be frequented by the Australian Eastern Spinebill, *Acanthorhynchus tenuirostris*; the plant's small flowers and delicate stems do not act as constraints owing to the bird's frequent hover-like behaviour when feeding. When nectar-adapted birds, such as honeyeaters (Meliphagidae), are observed at the flowers of subtropical rainforest trees, they are usually foraging for insects, or even fruit.

Though the 'open-to-all-comers' theme of generalised entomophily dominates subtropical rainforests, nevertheless there are examples of specialised plant–pollinator relationships. The classic example is that of the evolutionarily primitive genus *Eupomatia* (Eupomatiaceae) (Plate 139), comprising small trees or large shrubs that are sometimes found in eucalypt-dominated wet sclerophyll forest adjacent to, or intergrading with, rainforest. Two species, *Eupomatia laurina* and *E. bennetti*, occur in Australia's subtropical rainforest (Harden *et al.* 2006), and an additional species (*E. barbata*) is to be found in the rainforests of tropical Queensland. Their pollinators are exclusively those of the weevil genus *Elleschodes*. The Eupomatiaceae are restricted to rainforest and wet sclerophyll forest of Queensland, New South Wales and Victoria (Harden 1990), and retain flower structures that are thought to characterise the hypothetical condition of early flowering plants (Thien 1980; Williams and Adam 1994, 2010).

Eupomatia laurina is the more widely ranging of the two subtropical species, its distribution being from eastern Victoria to north-eastern Queensland, and from sea level to altitudes >1000 m. Its reproductive ecology was first studied in detail by Alexander Hamilton in the 19th century, his observations being published in 1897. Some earlier brief observations had been made by Robert Brown, a naturalist on board Matthew Flinders' ship, HMS *Investigator. Eupomatia laurina* possesses large white waxy flowers that emit a strong aromatic odour. The flowers are formed by basally united stamens and large fleshy staminodia, which are united in a single structure called the synandrium. These fall off some 14–30 hours after opening. Plants often open synchronously between 3.30 and 6.30 a.m., at which time they are in the female phase. The male phase occurs during the evening. Southern *Eupomatia laurina* populations are pollinated solely by the weevil species *Elleschodes hamiltoni* (Plate 139); these leave the flowers at dusk to enter those opening at dawn the following day. They do not damage the flowers but feed on the staminodia, which serve as food resources, enhance visual display, possess osmophores that emit attracting odours, and produce sticky secretions that facilitate the carriage of large pollen loads (Armstrong and Irvine 1990). The weevils lay their eggs in the fused synandrium, which, when falling away, carry the larvae to the ground, there to develop in the soil profile (Williams and Adam 2010).

Growing on the margins of rainforest and in sheltered gullies is the herb *Typhonium brownii* (Araceae) (Harden 1993). Its inflorescence, as in the related *Alocasia brisbanensis* previously discussed, consists of a central spadix made up of many small basal female flowers, and a male zone separated from the female flowers by a collar of bristles. The spadix is partially enveloped by a sheath-like spathe (a large petal-like bract). At the top of the spadix is a sterile appendage that becomes hot when emitting the odours and carbon dioxide that attract its pollinators, thus mimicking mammalian dung. During odour emission the temperature of the appendage can rise to 15°C above ambient levels. The odour emitted is unpleasant, attracting numerous small flies. These are not the pollinators. Instead pollination is performed by dung-inhabiting beetles (Monteith 1986a). Dung beetle-mediated pollination has also been recorded for zygomorphic *Orchidantha inouei* (Lowiaceae – Zingiberales) in a mixed dipterocarp forest in Sarawak (Sakai and Inoue 1999). *Typhonium brownii* is protogynous, the female flowers first maturing, then later the males. The beetles arrive carrying pollen from other plants, and fall into the lower chamber of the flower, formed by the collared base of the spathe. They are temporarily denied escape owing to the smooth inner surface of the spathe and the encircling bristles on the central spadix column. The following day the flower repositions itself in a manner that allows the beetles to escape, in the process collecting pollen on their bodies from the male flowers that have now come into phase.

Pollination by thrips (Thysanoptera) is understated in rainforests (Mound and Williams 2003; Sakai *et al.* 1999; Suetsugu *et al.* 2018; Williams and Adam 2010); rather, these are more usually accounted as agents of flower damage and destroyers of pollen. However, studies undertaken in the subtropical rainforests of New South Wales (Williams 1995; Williams *et al.* 2001) suggest thrips may be important contributors to the pollination of flowering rainforest plants. Approximately 23 species of thrips were recorded from 26 species of subtropical trees, shrubs and vines (Williams *et al.* 2001) (Table 2). These are mainly trees of various sizes, but they include vines (*Smilax glyciphylla* – Smilacaceae, *Maclura cochinchinensis* – Moraceae), and shrubs and shrub-like small trees (*Linospadix monostachya*, *Diospyros australis*, *Alectryon coriaceus* – Sapindaceae, *Myrsine variabilis* – Myrsinaceae/Primulaceae, *Wilkiea huegeliana* – Monimiaceae, *Anopterus macleayana* – Escalloniaceae/Grossulariaceae). These generally possess flowers with little depth effect. But *Wilkiea huegeliana* (Plate 256), a food plant of the threatened Regent Skipper butterfly *Euschemon rafflesia rafflesia* (Hesperiidae), possesses small pale yellow flowers that are globose in shape. *Wilkiea huegeliana* plants are usually monoecious, having male or female flowers. The carpels develop within the fully enclosed female flowers, but entry inside is only by way of a small opening (ostiole) at the apex. Only thrips can gain entry, and only *Thrips setipennis* (Thripidae) was found to be the pollinator of those *Wilkiea huegeliana* plants investigated (Williams *et al.* 2001). But this plant–pollinator relationship is an asymmetrical one, for *Thrips setipennis* is found in the flowers of numerous other species.

Bees are important pollinators of many flowering plants world-wide (Michener 2007; Roubik 1992; Willmer 2011), and have received particular attention from pollination ecologists. However, bees are not always pollinators of the various flowering rainforest plants from which they have been recorded. Rather, they may consume nectar and pollen, and destroy flowers, without any beneficial return to the plant (Williams and Adam 1994, 2010, and references therein). For many flowering subtropical rainforest trees, native bee diversity and abundance coincide with peak flowering events, and at these times

Table 2. Flowering plants in subtropical rainforest from which thrips (Thysanoptera) have been collected (after Williams *et al.* 2001).

Plant families	Plant species
Anacardiaceae	*Euroschinus falcatus*
Arecaceae	*Linospadix monostachya*
Celastraceae	*Denhamia celastroides*
Cunoniaceae	*Ackama paniculata, Schizomeria ovata*
Ebenaceae	*Diospyros australis, D. pentamera*
Elaeocarpaceae	*Elaeocarpus obovatus, E. reticulatus*
Escalloniaceae/Rousseaceae	*Anopterus macleayanus, Cuttsia viburnea*
Flacourtiaceae/Salicaceae	*Scolopia braunii*
Lauraceae	*Cryptocarya microneura*
Monimiaceae	*Wilkiea huegeliana*
Moraceae	*Maclura cochinchinensis*
Myrtaceae	*Rhodomyrtus psidioides, Syzygium floribundum, S. smithii, Tristaniopsis laurina*
Myrsinaceae/Primulaceae	*Myrsine howittiana, M. variabilis*
Rhamnaceae	*Alphitonia excelsa*
Rutaceae	*Acradenia euodiiformis*
Sapindaceae	*Alectryon coriaceus, Guoia semiglauca*
Smilaceae	*Smilax glyciphylla*

competing exotic honey bees, *Apis mellifera*, are frequently prevalent (Williams and Adam 1997) and commonly dominate the upper size class of insects visiting flowers (Williams 1995).

In lowland subtropical rainforests native bees can be important pollinators of some species (e.g. *Diospyros australis* – Ebenaceae, *Acradenia euodiiformis* – Rutaceae), and may be particularly diverse at the flowers of profusely flowering rainforest Myrtaceae (e.g. *Syzygium floribundum*, *Tristaniopsis laurina*). Though often conspicuous at individual flowering trees and shrubs, in abundance native bees usually represent a small fraction of the otherwise fly–beetle–wasp dominated assemblage (Williams 1995). However, small *Exoneura*, *Lasioglossum* and *Hylaeus* bees are often conspicuous among the depauperate assemblage of flower visitors seen on small-flowered vines (e.g. *Cayratia clematidea*, *Cissus hypoglauca* – Vitaceae, *Celastrus subspicatus* – Celastraceae) (G. Williams pers. obs.).

The endangered shrub *Senna acclinis* (Fabaceae – Caesalpinioideae) (Plate 282) is one of several subtropical rainforest species in which native bees do play an essential pollination role. *Senna acclinis* is a 'gap-phase' species, surviving in sunlit locations, and bears clusters of yellow zygomorphic flowers. It is recorded largely from the central and northern coastal rainforests of New South Wales and Queensland. It is a buzz-pollinated species, its anthers needing to be vibrated (sonicated [see De Luca and Vallejo-Marín 2013]) by bees to release pollen (G. Williams 1998). Bees observed vibrating *Senna acclinis* flowers include *Amphylaeus nubilosellus* (Plate 288), *Hylaeus turgicollaris* (Colletidae – Hylaeinae), *Lasioglossum polygoni* (Halictidae) and *Amegilla bombiformis* (Plate 282) (G. Williams 1998; Williams and Adam 2010; G. Williams pers. obs.).

Bees are also important participants in the pollination of the delicate white-flowered understorey herb *Pollia crispata* (Commelinaceae) (Williams and Walker 2003). *Pollia crispata* can form dense mats along sheltered rainforest margins, and in sunlit openings below damaged and regenerating canopies. Populations below gaps are progressively displaced as trees shade and close the opening. Native bees and syrphid flies form a mixed guild of pollinators, for few other insects occur on flowers. However, species of *Myrmecia* ants are occasional visitors. Various species of *Lasioglossum* (Halictidae), of the subgenera *Austrevylaeus*, *Callalictus*, *Chilalictus*, and *Parasphecodes*, dominate bee diversity; the red/orange-banded *Nomia aurantifer* (Plate 286) is the most visually conspicuous. Other bee taxa include *Amphylaeus nubilosellus* and *Hemirhiza melliceps* (Colletidae – Hylaeinae), *Amegilla* sp., *Tetragonula carbonaria* (= *Trigona carbonaria*), *Exoneura* sp. and the introduced *Apis mellifera* (Apidae). The participating Syrphidae comprise the genera *Betasyrphus*, *Episyrphus* and *Melangyna*, as well as several additional species of the tribe Syrphini. Thus, while *Pollia crispata* exhibits a 'generalised' pollination strategy necessitated by its ephemeral presence at sites, the pollinators that participate in it are drawn from a taxonomically circumscribed pool of available taxa.

Alectryon coriaceus (Sapindaceae) (Plate 257) is a shrub or small tree, and is endemic to Australia. It is distributed from near Newcastle, New South Wales, to southern Queensland (Floyd 1989), and frequently occurs on the exposed margins and internal sunlit canopy gaps of littoral rainforest. The flowers of *Alectryon coriaceus* are visited principally by a taxonomically diverse range of small insects (see discussion below), none of which is an ecological specialist. Most of these undertake very limited movements, generally to adjacent florets or inflorescences. However, *Alectryon coriaceus* is a candidate for pollination partly facilitated by ants. The flowers of *Alectryon coriaceus* do not have an obvious fragrance. The flower structure is shallow and readily accessible, the petals and sepals being reduced to an obsolescent rim. The carpels bear dense, elongate, protective, hair-like trichomes, the stamens are bright yellow, the filaments are reduced, and the pollen is yellow and sticky and is not readily dislodged by wind or other physical movement.

Alectryon coriaceus is commonly encountered in clusters of a few individuals, or as isolated plants some distance from congeners. Plants found along exposed, sunlit, forest margins are typically drooping, with overlapping branches and foliage that often touch the ground, such that non-flying flower visitors can readily move along and across stems and leaves, or by way of the soil surface. Insect visitors (>100 species) observed at littoral rainforest remnants in northern New South Wales (Williams 1995) include beetles (Belidae, Chrysomelidae, Coccinellidae, Corylophidae, Curculionidae, Dermestidae, Latridiidae, Melyridae), flies (Asilidae, Bombyliidae, Calliphoridae, Chloropidae, Drosophilidae, Lauxaniidae, Muscidae, Phoridae, Platystomatidae, Sepsidae, Syrphidae, Tachinidae, Tephritidae), bugs (Lygaeidae, Miridae, Psyllidae, Tingidae), wasps (Agaonidae, Bethylidae, Chalcididae, Crabronidae, Encyrtidae, Eulophidae, Eurytomidae, Figitidae, Gasteruptiidae, Mutillidae, Platygastridae, Pompilidae, Pteromalidae,

Scoliidae, Sphecidae, Thynnidae, Torymidae, Vespidae), bees (Apidae, Colletidae-Colletinae, Colletidae-Hylaeinae, Halictidae), ants (Formicidae), day-flying moths and butterflies (Erebidae-Arctiinae, Lycaenidae, Pieridae), and thrips (Phlaeothripidae). Many individual species of these are incidental visitors (especially microhymenoptera), or herbivores of floral parts, and probably play no role in pollination.

However, a distinctive feature of the attendant flower visitors is the abundance of ants that can often be observed foraging during the day, commonly constituting more than 25 per cent of the individuals at flowering trees at any one time. Species of *Crematogaster* and *Polyrhachis* (e.g. *Polyrhachis clio*, *P. hookeri*, *P. pilosa*), in particular, and sometimes *Myrmecia nigrocincta*, are often abundant on flowers, and at times are the only conspicuous foragers present during excessively hot (>35°C) and humid periods of the day. At some flowering plants *Polyrhachis pilosa* (Plate 264) can be especially common for this ant is a leaf web-nesting species, its nests encountered within bound living foliage of the rainforest margin. Neither *Crematogaster* nor *Polyrhachis* is restricted in their foraging to the flowers of *Alectryon coriaceus*. In subtropical lowland rainforest *Crematogaster* is recorded on flowers of the mass-flowering trees *Syzygium smithii* (Myrtaceae), *Alphitonia excelsa* (Rhamnaceae), *Elaeodendron australe* (Celastraceae), and *Mischocarpus pyriformis*, *Guioa semiglauca* and *Cupaniopsis anacardioides* (Sapindaceae). *Polyrhachis* is recorded from *Alphitonia excelsa*, *Elaeodendron australe*, *Euroschinus falcatus* (Anacardiaceae), *Guioa semiglauca* and *Scolopia braunii* (Flacourtiaceae/Salicaceae). The large and aggressive myrmecine ant, *Myrmecia nigrocincta*, likewise exhibits no foraging fidelity, being also recorded on flowers of *Syzygium smithii*, *Alphitonia excelsa*, *Euroschinus falcatus* and *Scolopia braunii* in littoral rainforest remnants.

Alectryon coriaceus meets several of the criteria thought to characterise the ant pollination syndrome (Williams and Adam 2010): plants grow along forest margins and gaps formed by canopy disturbance, commonly in clusters; they possess low and interdigitating branches and leaves which often touch the ground; individual inflorescences frequently touch and overlap each other; flower structures provide unimpeded access; and carpels are protected against ant attack by a dense cover of trichomes. The structural attributes facilitate ant movements between florets and inflorescences, and the often-overlapping nature of the foliage facilitates short-distance movements between inflorescences.

World-wide, ants are a frequent feature of rainforest foliage and flowers, but are usually considered to have no or little role in pollination. This is due to their restricted and nest-redirected foraging patterns, small size, smooth body integument, and the chemicals secreted by their metapleural glands which may function as biocides that reduce pollen viability and germination. However, *Polyrhachis*, which is often abundant on flowers, either does not possess metapleural glands or, in species where these are present, they are atrophied and non-functional (Hölldobler and Wilson 1990). Metapleural glands are present in the frequently observed *Myrmecia nigrocincta*, but it is not known if they are functional. *Crematogaster* ants possess functional metapleural glands. Peakall *et al.* (1991) suggest that even where ants do produce metapleural gland secretions, isolation of pollen grains by chemical or physical means from the ant integument may serve to protect pollen. Such protection against ant secretions and disruption of function may be provided by the sticky pollenkitt that is a feature of the pollen of *Alectryon coriaceus*, thus removing a potential barrier to pollination.

The pollen grains of subtropical rainforest trees, shrubs, herbs and vines possess a variety of structuring on their outer surface (the exine). Surface sculpture ranges from that which is smooth or indistinctly patterned, to pollen types that are armed with conspicuous spines (echinate). Most trees and shrubs possess only relatively smooth or weak to moderately sculptured exine surfaces, but no single plant growth form group possesses a distinct sculpture class (Williams and Adam 1999a). Of the 130 species (in 58 families) investigated by Williams and Adam (1999a) only a small subset of subtropical rainforest plant species were found to have pollen types that were strongly sculptured or echinate (e.g. *Hibiscus splendens*, *H. heterophyllus* – Malvaceae, *Tasmannia insipida* – Winteraceae). This trend was found in additional pollen samples investigated: namely, indistinctly or weakly sculptured (*Toona ciliata* – Meliaceae, *Senna acclinis* – Caesalpinioideae) and moderately to strongly sculptured (*Pararistolochia praevenosa*, *P. latreyana* – Aristolochiaceae, *Ripogonum discolor* – Ripogonaceae) (G. Williams 1998; G. Williams unpubl. data).

Historically it has been suggested that particular patterns of exine sculpture are related to the means of pollen dispersal and are specialised adaptations for carriage by different kinds of pollinators. The smooth grains of gymnosperms (as in the subtropical rainforest gymnosperm *Podocarpus elatus*) are an adaptation to dispersal in wind currents, but strongly sculptured pollen grains were thought to characterise transport by insects. However, insects collect and carry pollen grains of many types. Bullock (1994) suggested that Neotropical plants, which were assumed to be pollinated by small ecologically generalist insects, might instead also be facultatively wind pollinated. The prevalence of relatively smooth-surfaced pollen types in Australian subtropical rainforest flowering plants suggests Bullock's proposal may have much wider application. Nevertheless, in floristically and structurally complex rainforests the inherently non-targeted movement of wind-transported pollen between conspecific flowering plants must confront the uncertainty of numerous ecological, environmental and physical barriers (Rech *et al.* 2016).

It is important to consider that localised patterns of plant species extinction and recruitment are widely reported in rainforests world-wide and may confound the prediction of a 'climax' plant community. The unpredictable retention over time of individual plant species can have significant implications for the evolution and survival of plant–invertebrate relationships, such as those between particular flowering plants and their pollinators, and herbivores with their obligate larval food plants. Plants which are self-incompatible, including dioecious species that have separate male and female plants (see Williams and Adam 1999b, 2010), are at particular risk. Obligate-outcrossing species survive in many small rainforest stands at low, often single, tree and shrub populations that may already be functionally extinct (this masked by the longevity of individual trees). Loss of pollinators, and a consequent loss of plant species, can bring about cascading events that lead to the extinction of other dependent organisms, as in the example of the *Wilkiea huegeliana*–*Thrips setipennis*-threatened Regent Skipper butterfly complex.

4 Invertebrate taxa of Australian subtropical rainforests

The following treatment and discussion is far from an exhaustive consideration of the invertebrate fauna to be encountered in subtropical rainforest communities. Mites, for example, are a mega-diverse group, and many families and genera known from this region of Australia's east coast are omitted or only casually addressed. In strong part, the choice of groups discussed reflects a personal bias. Family group names, even when used colloquially, are treated as nominative plural throughout (Cranston *et al.* 1991).

PLATYHELMINTHES (FLATWORMS, LAND PLANARIANS)

This phylum includes free-living species, often referred to as flatworms and land planarians, of the forest floor, and symbionts that live on the external surface of freshwater crayfish (Sewell and Cannon 1998; Sewell *et al.* 2006; Steel 1897; Winsor 1997, 1998; Wood 1926).

Temnocephalida The Temnocephalida are all either symbiotic commensals or parasitic in habit, and are associated with freshwater hosts. The order comprises the Palaearctic Scutellarielloidea and the Australian and Neotropical Temnocephaloidea. Patterns of diversification and distribution have been affected by the fragmentation of Gondwana (Martínez- Aquino *et al.* 2017).

Temnocephalidae A diverse Temnocephalidae fauna is associated with freshwater crayfish, and other organisms such as hyriid bivalve snails (Sewell *et al.* 2006). *Temnohaswellia* and *Temnosewellia* (Temnocephalinae) are ectosymbionts on species of *Euastacus* (Plate 35) and are widely distributed along the east coast, being especially diverse extending from southern Queensland to southern New South Wales (Sewell *et al.* 2006) (Table 3). Individual species are host specific (e.g. *Temnohaswellia capricornia*, *T. cestus*, *Temnosewellia minima*) or exhibit wider host ranges (*Temnohaswellia comes*, *Temnosewellia gingrina*) (Table 3), and as a consequence in the latter there is a commensurate broader geographical distribution, the former often very narrow (*Temnohaswellia capricornia* on *Euastacus monteithorum* restricted to Kroombit Tops) (Sewell *et al.* 2006). *Temnohaswellia* has a single species in New Zealand, is confined to *Euastacus*, but generally exhibits low host specificity. *Temnosewellia* is not confined to *Euastacus* (also on *Cherax* and *Engaeus* crayfish, isopods, freshwater crabs and shrimp, freshwater bivalve molluscs) but is characterised by stricter host specificity (Sewell *et al.* 2006).

Craspedellinae are ectosymbionts found in the branchial chambers of Australian crayfish (Parastacidae). The distribution of the subfamily suggests an origin in Australia–New Guinea following the separation of the Australian and South American landmasses from Antarctica ~45 mya (Sewell and Cannon 1998). The biogeographical pattern of Craspedellinae matches that of the crayfish genus *Cherax*, implying coevolution and radiation with the host (Sewell and Cannon 1998). Sewell and Cannon (1998) suggest that Craspedellinae may have

Table 3. Some temnocephaline flatworms associated with *Euastacus* crayfish (Parastacidae) in subtropical eastern Australia (after Sewell *et al.* 2006).

Temnocephalidae	Parastacid host
Temnohaswellia capricornia	*Euastacus monteithorum*
Temnohaswellia comes	*Euastacus australienisis, E. brachythorax, E. clarkae, E. dangadi, E. gumar, E. ?guwinus, E. hirsutus, E. jagara, E. maidae, E. mirangudjin, E. neohirsutus, E. setosus, E. spinichelatus, E. spinifer, E. sulcatus, E. suttoni, E. valentulus, E. yanga*
Temnohaswellia cornu	*Euastacus jagara*
Temnohaswellia munifica	*Euastacus hystricosus*
Temnohaswellia simulator	*Euastacus gumar, E. neohirsutus, E. spinichelatus, E. sulcatus, E. suttoni, E. valentulus*
Temnohaswellia subulata	*Euastacus australiensis*
Temnohaswellia umbella	*Euastacus guwinus*
Temnohaswellia verruca	*Euastacus brachythorax, E. crassus, E. polysetosus, E. reductus, E. spinifer, E. yanga*
Temnosewellia aspinosa	*Euastacus valentulus*
Temnosewellia bacrio	*Euastacus maidae, E. valentulus*
Temnosewellia bacrionculus	*Euastacus maidae, E. neohirsutus, E. setosus, E. sulcatus, E. valentulus*
Temnosewellia batiola	*Euastacus hystricosus, E. urospinosus*
Temnosewellia belone	*Euastacus brachythorax*
Temnosewellia cestus	*Euastacus urospinosus*
Temnosewellia comythus	*Euastacus gumar, E. spinichelatus*
Temnosewellia coughrani	*Euastacus mirangudgin, E. sulcatus*
Temnosewellia cypellum	*Euastacus spinifer*
Temnosewellia fasciata	*Euastacus australiensis, E. clarkae, E. polysetosus, E. spinifer*
Temnosewellia fax	*Euastacus hirsutus, E. yanga*
Temnosewllia flammula	*Euastacus neohirsutus*
Temnosewellia gingrina	*Euastacus dangadi, E. gumar, E. sulcatus, E. suttoni, E. valentulus*
Temnosewellia gracilis	*Euastacus ?guwinus*
Temnosewellia maxima	*Euastacus sulcatus*
Temnosewellia minima	*Euastacus sulcatus*
Temnosewellia possibilitas	*Euastacus* spp.

'host-switched' from *Cherax* to *Euastacus*, which is a speciose genus in the Gondwana rainforests region (Morgan 1997). Sewell and Cannon (1998) note, however, that host-switching does not appear to have occurred in the relic and isolated Queensland *Euastacus* populations (although they do not discount the possibility of later extinction).

Tricladida Land planarians (Sluys *et al.* 2009) are an inconspicuous element of the subtropical rainforest ground fauna. Individuals can be found sheltering in and under rotten wood, and under stones. They are usually nocturnal in habit, though some may be observed in the first hours of daylight seeking refuge. There is considerable generic diversity in southern Queensland and northern New South Wales. The Geoplanidae include *Artioposthia*, *Australoplana*, *Caenoplana*, *Australopacifica*, *Geoplana*, *Parakontika* and *Fletchamia*. The Rhychodemidae are represented by *Rhynchodemus*, *Dolichoplana* and possibly *Platydemus*. *Parakontika* and *Fletchamia* reach the northern limit of their range in southern Queensland (Steel 1897; Williams 2002; Winsor 1997, 1998; Wood 1926).

Geoplanids have been recorded preying on scarabaeine dung beetles (Monteith and Storey 1981).

NEMERTEA (ribbon worms)

This is a phylum of unsegmented worm-like organisms that possess an eversible proboscis contained in a fluid-filled cavity or rhynchoel. The proboscis is used mainly to capture prey, but also in locomotion, the proboscis being rapidly projected and then drawn back within the body cavity. The phylum has a global distribution. Most nemerteans are carnivorous, the majority feeding on live prey; however, some are scavengers. Prey include annelid worms, arthropods and snails. Nemerteans have been considered closely related to flatworms (Platyhelminthes) but this phylogenetic position is controversial.

The world fauna numbers ~1500 species. The majority are primitively and predominantly marine and freshwater species, but a few are terrestrial. Terrestrial nemerteans are recorded from the Indo-Pacific region, Australia, New Zealand, and islands within the Atlantic (e.g. Azores, Canary Islands, Bermuda). Given that the group has a strong marine affiliation, the ability of some terrestrial species to live on mountain tops is of interest. Australia is the only continent that possesses endemic terrestrial Nemertea (Main 1987), the fauna occurring in tropical and subtropical rainforest, dry sclerophyll forest and damp woodland (R. Gibson 1995; McDermott and Roe 1985; J. Moore 1975, 1985; Moore and Gibson 1981, 1988; Turbeville 2002).

In Australia, the nemertean fauna includes four terrestrial species placed in the genus *Argonemertes* (Plate 1), family Prosorhochmidae (R. Gibson 1995). Three are considered rare (Winsor 1985). Australian *Argonemertes* may represent an off-shoot from the New Zealand genus *Antiponemertes* (Moore and Gibson 1981). *Argonemertes* is primarily distributed from south-eastern Queensland to Tasmania, with one species (*Argonemerte dendyi*) in Western Australia. *Argonemertes hillii* is restricted to south-east Queensland and northern New South Wales, and *A. stockii* occurs in New England National Park. *Argonemertes australiensis* is more widespread, occurring from south-east Queensland to Victoria and Tasmania (J. Moore 1975; Williams 1993, 2002).

NEMATODA (round worms)

The phylum Nematoda is often stated to be one of the planet's most speciose and abundant groups of multicellular animals – a title contended only by insects. They occupy marine, freshwater and terrestrial environments, occurring in soil and above-ground habitats, including arboreal ones. They are a dominant component of the soil microfauna and play key roles in ecosystem function through decomposition and mineralisation processes (Ferris 2010). Nematode communities are typically composed of bacterial, fungal and algal feeding species, are predators of small invertebrates, and are parasitic and phoretic associates of plants and animals, including ants, Orthoptera, spiders and harvestmen (Baker and Poinar 1994; Downes 2017; Poinar 1985). Nematodes are highly diverse in temperate and tropical rainforests, comprising unique assemblages of species between and within habitats (Boag and Yeates 1998; Porazinska *et al.* 2012; Powers *et al.* 2009).

Mermithidae These are parasites of numerous invertebrates, including insects such as ants (e.g. *Camponotus*, *Polyrhachis*, *Pheidole*) (Downes 2017). The family has a world-wide distribution. Infection may cause dramatically altered appearance of the host, for example gigantism and gaster enlargement in ants, or abnormal behaviour. Hosts may carry multiple nematodes, frequently located in the gasters of ants.

Thelastomatidae Nematodes in this family are parasites of Australian beetles in the family Passalidae (W. Clark 1978), a family frequently found under decaying timber on the forest floor of all rainforest subformations. Three thelastomatid genera (*Anuronema*, *Phalacronema*, *Sprentia*) are known to occupy the hind gut of their hosts (*Aulacocyclus* – Passalidae) (W. Clark 1978).

Tylenchidae Many characteristics of the Tylenchidae are considered to be primitive or ancestral in origin (Reay 1991). The subfamily Tylodorinae includes the Australian endemic genus *Arboritynchus*, which is known only from subtropical rainforests of the Gondwana rainforests region (Reay 1991). Endoparasitic *Rugopharynx* species have a wide host range and have been recorded from the macropods *Petrogale penicillata* and *Thylogale thetis* in the region (Beveridge and Chilton 1999)

ANNELIDA (earthworms, leeches)

Annelida include the segmented worms (Oligochaeta), and predatory terrestrial and aquatic leeches (Hirundinea), which together with several lesser relatives constitute the class Clitellata (Erséus 2005; Purschke *et al.* 2014; Struck *et al.* 2007). In addition there are many marine and freshwater species (e.g. of Polychaeta) (Beesley *et al.* 2000). The diversification of the Annelida has been attributed to, or occurring before, the Cambrian explosion (Conway Morris 1979; Parry *et al.* 2014; Rousset *et al.* 2007; Valentine *et al.* 1996) with the oldest known fossils dating to at least the Silurian (>400 mya) (Eriksson *et al.* 2004), but unequivocal fossils are not known for leeches (Siddall *et al.* 2006).

Oligochaeta (earthworms)

The subclass Oligochaeta are important terrestrial decomposers that contribute to the recycling of vegetable matter in the soil (Edwards and Bohlen 1996; Lachnicht *et al.* 2002). The native earthworm fauna can be influenced by soil pH, soil moisture, available nitrogen (forest litter) and soil organic matter, and is sensitive to habitat disturbance and modification. They are often displaced by introduced earthworms in areas under agricultural cultivation (Hendrix *et al.* 2006). Several taxa (e.g. *Digaster*) contribute to the breakdown of decaying fallen timber in rainforest, and may be encountered within rotting logs.

Geological and climatic oscillations have dramatically influenced speciation and distribution of the Australian earthworm fauna (Moreau *et al.* 2015). The Australian fauna is largely a Gondwanan one and is dominated by the family Megascolecidae. Two megascolecid subfamilies, the Megascolecinae and Acanthodrilinae, occur in Australian rainforests. The Australian endemic earthworms are considered by Jamieson (1981) likely to be derived from successive Gondwanan invasions which were then followed by prolific speciation. These speciation events took place largely in the family Megascolecidae. In contrast, an almost negligible endemic fauna has apparently been derived from post-Miocene (<10 mya) invasion from the Oriental region (Jamieson 1981). This is in contrast to several other northern-derived higher taxa, such as the terrestrial snail family Camaenidae, where postulated Miocene invasion events have been followed by high levels of speciation (Hugall and Stanisic 2011).

The Macpherson–Macleay Overlap, which broadly encompasses the Gondwana Rainforests of Australia World Heritage Area, is characterised by a diverse earthworm fauna (Blakemore 1994, 1997; Jamieson 1963, 1971, 1972, 1975, 1995; Jamieson and Wampler 1979; Pinder and Brinkhurst 1997; Williams 2002). Complex faunistic patterns of superimposed colonisation episodes are associated with the development of rainforest on Tertiary (70–1 mya) lava flows from the Main Range and Mount Warning/Tweed shield volcano. Genera such as *Digaster* and *Plutellus* exhibit their greatest diversity in the overlap zone (Werren and Kershaw 1991).

G. R. Dyne (in Werren and Kershaw 1991) indicates that ~120 oligochaete species and 17 genera were known from Australian rainforests, as of that date, but that this is likely to represent a small fraction of the actual fauna diversity. Many species appear to be highly localised (e.g. *Cryptodrilus bunyaensis*, known only from the Bunya Mountains; *Propheretima hugalli*, known only from Boatharbour Nature Reserve; *Digaster lingi*, known only from Binna Burra [Jamieson 1995]). The endemic genus *Hiatidrilus* is confined to the Gondwana rainforests region and consists of two species: *H. bunya* (Bunya Mountains) and *H. semicinctus* (Grafton) (Jamieson 1994).

Several endemic megascolecid species from the McPherson Range are weakly to very brightly bioluminescent (e.g. *Spenceriella curtisi*, *S. cormieri*, *Fletcherodrilus unicus*) (Jamieson and Wampler 1979). This bioluminescence originates from exuded fluid.

Threats to native earthworm species include changes to soil moisture regimes, displacement by exotic species, disturbance to detritus profiles, removal of rotting timber on the forest floor, fire and predation by feral animals. In lowland tropical Queensland rainforest pigs have been shown to cause significant decline in soil macroinvertebrate density, including that of earthworms (Taylor *et al.* 2011). In small stands (e.g. Wingham Brush Nature Reserve) foraging by dense populations of native birds (e.g. brush turkeys) may also threaten the populations of native species. Long periods of inundation of lowland rainforests by floodwaters (as frequently occurs at Lansdowne Nature Reserve) can cause the wholsale loss of earthworm species. Where there is no connectivity to other stands this potentially results in the localised extinction of remnant populations.

Megascolecidae – Acanthodrilinae This subfamily dominates the earthworm fauna of southern Africa and southern America, New Zealand, New Caledonia and the subantarctic islands (Werren and Kershaw 1991). However, in Australia, the generic diversity of the subfamily is low – especially in comparison to the New Zealand fauna where the subfamily occupies rainforest niches filled in Australia by the Megascolecinae (Werren and Kershaw 1991).

Megascolecidae – Megascolecinae The majority of the Australian rainforest genera belong to the Megascolecinae. The Megascolecinae represent a Gondwanan element (Heatwole 1987), and appear to have evolved in the Australian, Indian and Antarctic sector. They subsequently spread to South-east Asia, China, New Zealand and the Pacific Islands (Werren and Kershaw 1991).

Three megascolecine tribes occur in the Gondwana rainforests region: Dichogastrini, Megascolecini and Perionychini. *Didymogaster* and *Digaster* largely represent the nucleus of Dichogastrini genera in the region. Several *Digaster* species, for example *D. eastoni* and *D. nothofagi*, which occur in cool temperate rainforest, have restricted distributions (Blakemore 1997; Jamieson 1963, 1972, 1975; Williams 2002, 2003a), while the genus as a whole is confined to the montane coastal province of New South Wales and Queensland (Jamieson 1975). The origin and evolution of *Digaster* and *Didymogaster* suggests a presence 'well back into the Tertiary or earlier' (Jamieson 1981).

The Perionychini are absent from Africa and South America but occur in New Zealand, New Caledonia, India, Asia and North America. Taxa occurring in the Gondwana rainforests region include the genus *Plutellus*, which stretches from southern Queensland to Sydney (New South Wales) and Lord Howe Island, and occurs in *Nothofagus* forest. *Plutellus barringtonia* and *P. heteroporus* are known only from Gondwana rainforest sites in northern New South Wales. *Hiatidrilus* (= *Diporochaeta*) *bunya* is restricted to the Bunya Mountains–Clarence River and the related *H. semicinctus* is known only from its original type locality given as Clarence River. *Hiatidrilus* is an endemic genus and is restricted to south-east Queensland and north-east New South Wales. *Zacharius* (*Zacharius zacharyi*) is an endemic, monotypic genus known only from Maclean (Woodburn Island). *Propheretima* is an endemic genus restricted in distribution to Queensland and northern New South Wales. *Propheretima monsmonitionis* has been recorded from Mt Warning (Blakemore 1997, 2000a, 2000b; Jamieson 1976, 1981, 1994).

The endemic perionychine genus *Heteroporodrilus* is distributed from Victoria to Queensland. *Heteroporodrilus* is considered to be a derived (apomorphic) sibling group of *Plutellus* that reached a climax in rainforest and riverine environments in the subtropical regions of south-east Queensland, where it is particularly diverse. The distribution of *Heteroporodrilus* partially corresponds with various river systems. Several species have very restricted ranges: *Heteroporodrilus doubei* is known only from 10 km north-west of Lismore, *H. jamiesoni* is known only from Mt Glorious, and *H. editus* is recorded from Mt Warning (Blakemore 1994; Williams 2003a). *Anisochaeta* (*c.* 100 species) is restricted to Australia (all states) and New Zealand, and is highly speciose in northern New South Wales and southern Queensland. Numerous species are known only from the Gondwana rainforests (e.g. New England National Park, Richmond Range, Washpool National Park) (Blakemore 2000a; Williams 2003a).

A conspicuous large terrestrial worm encountered in lowland rainforest is the Squirt Worm, *Didymogaster sylvaticus* (Plate 2). Adults are commonly the thickness of a finger, and when disturbed can squirt odorous coelomic fluid to a height of 30 cm (Edwards and Bohlen 1996); the coelom is a cavity that extends through the length of the body, and in which the fluid is stored. Fluid is emitted through dorsal pores in response to physical or chemical irritation, a defence response evoked when handled. Squirt worms are found under and in heavily decomposed timber on the rainforest floor, and are able to persist in rainforest stands that are subject to periodic flooding. They are, however, dependent on suitable ground cover for protection from predators, which potentially include mound-building brush turkeys, *Alectura lathami*. *Didymogaster sylvaticus* actively seeks out suitable microhabitat, for adults will move from rainforest across open areas to live in heavy mulch piles positioned around fruit trees in adjacent orchards and gardens.

Phreodrilidae Australia is the centre of phreodrilid diversity with the number of species (>26 species) being high when compared with other landmasses (Pinder and Brinkhurst 1997). New Zealand, in comparison, has seven species. *Astacopsidrilus jamiesoni* is known only from Lamington National Park.

Phreodrilid worms are Gondwanan, occurring in Africa, Sri Lanka, South America, New Zealand and Australia, and some southern Oceanic islands (e.g. South Georgia, Falklands, Kerguelen, Macquarie and Campbell Islands) (Pinder and Brinkhurst 1997). Because several of the islands are geologically relatively recent, the occurrence of Phreodrilidae there suggests post-Gondwanan dispersal (Pinder and Brinkhurst 1997).

Hirudinea (leeches)

The Hirudinea are a small subclass of highly specialised jawed and non-jawed annelids. They include terrestrial and aquatic-adapted species. Richardson (1975) considered that it was highly probable that the terrestrial leech fauna of eastern Australia dates to the Pliocene (10–1 mya). Terrestrial leeches, or land leeches, are semi-sedentary with limited ability to disperse over large distances (Richardson 1975). Although land leeches are adapted to a terrestrial way of life, they are usually restricted to mesic forests subject to seasonal rainfall (Borda *et al.* 2008), seemingly quiescent during prolonged dry periods, yet responding immediately to the onset of rain. Aquatic species are infrequently encountered in freshwater streams that dissect subtropical rainforests.

Members of the blood-feeding family Haemadipsidae (*sensu* Sawyer 1986) are ectoparasites of vertebrates. Haemadipsids occur in the Oriental, Australian, Oceanic and Malagasian zoogeographical regions (Richardson 1974; Borda *et al.* 2008) and are found in rainforests throughout the Indo-Pacific; however, Gondwanan vicariance does not appear to explain the diversification of the extant fauna (Borda and Siddall 2010). Many Haemadipsidae genera (e.g. *Philaemon*, *Chthonobdella*) were placed in the family Domanibdellidae by Richardson (1975), but these are now generally grouped with the former (see Borda *et al.* 2008; Sawyer 1986). *Jaabdella* (*Jaabdella whitmani*) is an endemic genus known only from south-east Queensland–north-eastern New South Wales (Richardson 1967). *Microbdella* is an endemic genus comprising two described species (Richardson 1974, 1975; Williams 2002). *Microbdella gloriosi* is known only from Mt Glorious. Additional genera occurring in the Gondwana rainforests region include *Quaesitobdella* and *Castrabdella*.

The subtropical rainforest fauna includes the common blood-feeding rainforest species *Chtonobdella limbata* (Haemadipsidae), as well as non-blood-feeding species (Erpobdellidae) that prey on aquatic invertebrates; the latter have abandoned the blood-feeding habits of their ancestors (Siddall 2002). An engorged *Chtonobdella limbata* has been observed attached to the injured tail tip of the Pink-tongued Skink *Hemisphaeriodon gerrardii* (G. Williams pers. obs.). Blood-feeding leeches secrete hirudin and histamine in their salva. Hirudin is responsible for bleeding at the site of leech bites (Seiverling *et al.* 2014).

CHELICERATA (spiders, scorpions, harvestmen, mites, ticks, pseudoscorpions, whip spiders)

Chelicerta are a major group of related arthropods that include predators, parasites and symbionts. Groups such as spiders (Arachnida – Araneomorphae and Mygalomorphae), harvestmen (Opiliones) and scorpions (Scorpionida) are predators; however, many mites (Acarina) are predominantly fungus and plant feeders, and decomposers.

ARACHNIDA (spiders)

Araneomorphae ('true' spiders)

The majority of the infraorder Araneomorphae, 'true' or 'modern' spiders, are solitary predators of insects (Foelix 2010), though feeding on floral nectar is documented for several families (e.g. Clubionidae, Corinnidae, Salticidae, Thomisidae) (Jackson *et al.* 2001). True spiders date from the Carboniferous (Framenau *et al.* 2014) and are found within all rainforest strata, but are major predators of the canopy layer (Basset 1991; Russell-Smith and Stork 1995). Many araneomorphs are active hunters by day and night, but some are web-snare and ambush hunters, the latter waiting motionless at the entrance of their burrow or leaf shelter, or patiently sitting upon a flower, to capture their prey (Whyte and Anderson 2017). Araneomorphs have diverse evolutionary histories (Framenau *et al.* 2014), and some exhibit relictual distribution patterns that may reflect past climate change events. There are clear Gondwanan affinities above the level of genus, with the distribution patterns of several taxa (e.g. in Desidae, Lamponidae, Salticidae [Gray 1992; Platnick 2000; Žabka 1990]) suggesting past direct links between Australia and New Caledonia (Williams and Bickel 2010). Along with other invertebrate groups (e.g. *Astraeus*, *Nascioides*, *Notonomus* – Coleoptera, *Neohoodiella* – Thysanoptera, *Myrmecia*, *Orectognathus* – Formicidae), this supports the proposed Cretaceous connection of these two landmasses (Bellamy 1991; J. Griffiths 1971, 1974; Mound and Williams 2003; Williams and Bellamy 2002)

Amphinectidae These are inconspicuous predators of the forest floor. Some inhabit caves. They do not build capture webs and are usually nocturnal in habit. By day they hide in retreat burrows. The subfamily Metaltellinae occurs in Australia and South America. Endemic Australian genera, recorded from the Gondwana rainforests and adjacent areas, are *Quemusia*, *Magua* and *Cunnawarra* (Todd Davies 1998). *Magua* is a monotypic genus, but *Quemusia* includes many species restricted to the region. The second Australian subfamily, Tasmarubriinae, is restricted to Tasmania and Victoria

Anapidae These are very small, sheet-web-building spiders that rarely exceed 1–5 mm in size, and in rainforests are inhabitants of leaf litter and moss (Framenau *et al.*

2014). The distribution of the Anapidae includes New Zealand, New Caledonia, South Africa and the Old World tropics. Eastern Australia includes several endemic anapid genera. *Risdonius* comprises two known species and is distributed from New South Wales to Tasmania. *Risdonius barrington* is known only from Barrington Tops, while the second species, *R. lind*, is known only from Lind National Park in Victoria (Platnick and Forster 1989). *Queenslanapis* is a monotypic genus, known only from Lamington National Park. *Queenslanapis* closely resembles the New Zealand genus *Zealanapis* and may be more closely related to that genus than to other Australian anapids (Platnick and Forster 1989). *Maxanapis* (e.g. *Maxanapis dorrigo*, *M. burra*, *M. crassifemoralis*) is confined to Queensland and New South Wales, with the Gondwana rainforests region being a centre of diversity for the genus. *Octanapis* is restricted to the region from Queensland to eastern Victoria, and consists of two species (*Octanapis cann* and *O. octocula*). The genus *Chasmocephalon* occurs in Western Australia and the region from Victoria to Queensland, and is particularly diverse in eastern Australia (e.g. *Chasmocephalon iluka*) (Platnick and Forster 1989).

Araneidae This family includes the well known large orb-weaving spiders of the genus *Nephila* (sometimes placed separately in Nephilidae) and the widespread genera *Acroaspis*, *Arachnura*, *Araneus*, *Argiope*, *Celaenia*, *Cyclosa*, *Cyrtophora*, *Dolophones* and *Eriophora* (Plate 7). Some are web-snare builders found at night in forest openings, their strong silk webs suspended above the ground (e.g. *Araneus brisbanae*, *A. dimidiatus* – Araneinae). By day individuals hide among foliage. Cyrtarachninae (e.g. *Poecilopachys*) build reduced webs. Several species exhibit distinctive variation in colour form between populations (*Araneus circulissparsus*), and also between sexes (*Araneus talipedatus*). *Araneus praesignis* (e.g. from the Bunya Mountains) possesses dark eye-like markings on its abdomen, suggesting a defensive mimic strategy (Whyte and Anderson 2017). Two-spined orb-weavers assignable to the *Poecilopachys* 'australasica species group' have colourful abdomens that from the rear resemble the faces of snakes.

The large and fat-bodied spiders of the genus *Eriophora* at night build robust web snares that stretch across wide forest trails, their webs snaring moths and occasionally microbats and tiny passerine birds. The orb-weaving *Plebs bradleyi* (Plate 8) is a large brightly patterned species known from the Barrington Tops. Like other orb-weavers it hides below foliage during the day. *Cyclosa* includes species with very elongated abdomens. Prey remains are commonly incorporated into the capture webs. The genus *Dolophones* is a group of procryptic ambush species whose body plan and colour mimic that of twigs and bark. In repose, their bodies characteristically wrap around the stems and branches upon which they sit motionless in wait. *Argiope* species are large sedentary web-snare hunters that build permanently positioned webs strategically placed between suspended foliage and branchlets. *Argiope keyserlingi* and *A. protensa* are recorded from the Border Ranges (Levi 1983). The related *Telaprocera maudae* is recorded from southern Queensland. The genus builds elongated ladder webs positioned close to the tree trunks of rainforest trees (Whyte and Anderson 2017). *Cyrtophora* includes several species (e.g. *Cyrtophora moluccenis*, *C. exanthematica*) that build loose webs, from which they suspend themselves in wait for prey. Araneidae include spiders of the bizarrely formed and brightly patterned subfamily Gasteracanthinae (e.g. *Austracantha*). Many are distinctive in possessing numerous spines that arise from their abdomens, but the most ornate and colourful species are restricted to northern Australia. The related Mastophorinae include the cryptic and ornately structured *Celaenia* that mimic bird droppings or debris, adults being encountered motionless and camouflaged on foliage. The largest members of the family are those of the genus *Nephila* (Nephilinae), the so-called bird-eating spiders. The large females construct expansive and messy webs between branches, and commonly between different trees. The webs are robust and very sticky, frequently appearing to be poorly maintained, but sufficient to snare small birds and bats. The relatively tiny males are often found in association. *Nephila* are usually encountered on forest margins at lower altitude, and although there are southern species (*Nephila plumipes*), the genus is largely a northern one.

Archaeidae Archaeid spiders are active, elongate-fanged, hunters that do not construct a capture web. Prey include other spiders. Commonly referred to as assassin spiders, the family is known from eastern Australia, Madagascar, South Africa, and Baltic amber (Forster and Platnick 1984). Fossil archaeids are also known from Mesozoic and Tertiary deposits (Framenau *et al.* 2014). Australian members of this small family are included in the endemic genera *Austrarchaea* and *Zephyrarchaea* (Framenau *et al.* 2014; Walton 1985b). Species of

Austrarchaea inhabit moss beds and low vegetation in rainforest and wet sclerophyll forest. *Austrarchaea nodosa* is confined to the McPherson Range (Forster and Platnick 1984).

Arkyidae The Arkyidae were previously placed in the Araneidae (Dimitrov *et al.* 2017) and include both colourful (*Arkys cornutus*, *A. lancearius*) and dull-hued species (*Arkys alatus*, *A. speechleyi*) hunters that wait in ambush on foliage (Plate 9). Their forelegs are armed with stout, inwards-facing spines that are useful in holding prey, and their bodies are often bizarrely and cryptically shaped and coloured (*Arkys dilatus*, *A. furcatus*). *Arkys curtulus*, in appearance, resembles bird droppings. The genus includes the colourful, triangular *Arkys lancearius*, a common east coast species that is found active during the day on the leaves of lowland rainforest and wet sclerophyll forest. The family also occurs in New Caledonia.

Desidae These are nocturnal, ground- and tree-trunk-frequenting, generalist hunters. The genus *Colcarteria* includes three species restricted to northern New South Wales: *Colcarteria carrai*, *C. kempseyi*, *C. yessabah* (Gray 1992; Williams 2002). *Colcarteria*, and the New Caledonian *Canala*, possess related characters (i.e. male palpi and female genital characters) (Gray 1992). *Corasoides* includes *Corasoides motumae* known from cool temperate rainforest and wet sclerophyll forest (Clyde Mountain) and *Corasoides terania* from rainforest in the Border Ranges region (e.g. Cunninghams Gap, Lamington, Border Ranges and Nightcap national parks, Nothofagus Mountain). Species hide in tree crevices, and under logs and rocks. *Corasoides* likely originated in rainforest (probably in New Guinea), the 'older' species being rainforest associates, but with other members (*Corasoides australis*, *C. occidentalis*) adapted to more arid environments (Humphrey 2017).

Gradungulidae The Gradungulidae are a relictual group restricted to Australia and New Zealand (Forster *et al.* 1987). The family includes *Progradungula*, which comprises two species restricted to south-eastern mainland Australia (Gray 1982; Milledge 1997b). *Progradungula otwayensis* occurs in *Nothofagus* forest in the Otway Ranges of south-western Victoria (Milledge 1997b). *Progradungula carraiensis* (Plate 12) was originally known only from the Carrai Bat Cave (Forster *et al.* 1987), but has also been recorded from the ferny understorey of mixed *Nothofagus moorei–Doryphora sassafras–Ceratopetalum apetalum* cool temperate rainforest near the Plateau Beech section of Werrikimbe National Park (G. Williams pers. obs). Individual *Progradungula carraiensis* construct retreat web networks among rocks and rock overhangs; though not necessarily in caves (Framenau *et al.* 2014). Their capture webs consist of a zig-zag ladder-like structure held by the spider in a head-down manner. Two additional genera, *Tarlina* and *Kaiya*, are ground hunters. *Tarlina* is restricted to mid-eastern and north-eastern Australia, with several species (*Tarlina noorundi*, *T. milledgei*, *T. woodwardi*, *T. smithersi*) confined to south-eastern Queensland and northern New South Wales (Forster *et al.* 1987; Williams 2002) (Plate 13). Species of *Kaiya* are distributed from mid-eastern upland regions of New South Wales, south to Victoria (Framenau *et al.* 2014).

Hersiliidae Hersiliidae are distributed in tropical and subtropical areas of the World. They are 'almost certainly part of the younger northern faunal element in Australia', and did not arrive before the Miocene (Baehr and Baehr 1987). The Australian hersiliid fauna is rich and diverse; however, the relationships of the non-Australian fauna are poorly known. South-eastern Queensland–northern New South Wales is a major centre of evolution of the genus *Tamopsis*. Baehr and Baehr (1987) suggest that the history of *Tamopsis* colonisation follows a scenario of occupation of favourable ecological refugia during moist climatic periods, with subsequent geographical isolation during drier periods, followed by episodes of speciation and radiation within the refugia. The fauna includes species that are active diurnal and nocturnal predators that hunt without a capture web (Framenau *et al.* 2014). Species are well camouflaged and are typically found positioned cryptically on tree trunks. Prey include small moths and other soft-bodied insects.

Lamponidae Lamponidae are restricted to the Australasian region, and are recorded from Australia (including Tasmania), New Guinea, New Caledonia, Lord Howe Island, Norfolk Island and the Kermadec Islands (Platnick 2000). With the exception of the Kermadec Islands, the subfamily Lamponinae (*Lampona*, *Lamponicta*, *Lamponella*, *Lamponoides*) parallels this distribution. The Pseudolampinae (*Pseudolampona*, *Paralampona*) are endemic to Australia (Platnick 2000). The Centrothelinae are known from Australia, New Guinea and New Caledonia. Most centrotheline genera occurring in the Gondwana rainforests region are Australian endemics (e.g. *Centrina*, *Centsymphia*, *Queenvic*, *Graycassis*, *Longepi*).

North-eastern New South Wales–southern Queensland is a centre for lamponid diversity. Many species have restricted upland distributions (Williams 2002). Species inhabit a variety of rainforest formations including vine thickets (*Asadipus kunderang*), cool temperate rainforest (*Graycassis barrington*), and subtropical rainforest sub-formations more broadly (*Centrothele gordon*, *Graycassis chichester*).

Lycosidae Commonly called wolf spiders, lycosids (e.g. *Artoria*, *Tasmanicosa*, *Venatrix*) (Framenau 2002; Volker *et al.* 2014; Whyte and Anderson 2017) are active hunters of insects at night, their eyes conspicuously shining blue-white when illuminated by torchlight. They are distinctive in carrying their egg sacs attached to their spinnerets. The speciose genus *Venatrix* occurs in the open forests and rainforest of northern New South Wales and southern Queensland. Members of the *Venatrix 'pictiventris'* species group (*Venatrix allopictiventris*, *V. hickmani*, *V. pictiventris*) exhibit a limited and relictual distribution. Such remnant range distributions among spiders in south-east Australia appear to be common phenomena, possibly due to rising sea levels in the Cretaceous and Miocene and climate change, which resulted in disjunction to once continuous distribution patterns (Framenau and Vink 2001). *Trochosa tristicula* is recorded from many sites centred on the New South Wales–Queensland border; including the Bunya Mountains, Lamington National Park, Border Ranges, Nightcap Range and Mt Warning (McKay 1979).

Nicodamidae Distinctively coloured red and black, the Nicodamidae are restricted to Australia (including Tasmania), New Zealand, and New Guinea (Framenau *et al.* 2014; Harvey 1995). *Ambicodamus* is an Australian endemic, with *A. darlingtoni* known only from Barrington Tops. The genus *Dimidamus* has a disjunct distribution with four species known from the montane rainforests of Papuan New Guinea and West Papua, and two species occurring in the rainforests of eastern Australia. *Dimidamus dimidiatus* (Plate 14) is essentially confined to the Gondwana rainforests region, and the second species (*Dimidamus simoni*) is known from a single locality in Victoria (Harvey 1995). The genus *Oncodamus* comprises two species, and is restricted to closed forests in eastern Australia. *Oncodamus bidens* occurs in southern New South Wales, and *Oncodamus dicipiens* ranges from northern New South Wales to North Queensland. Nicodamids build sheet-webs positioned near the ground on stones and logs, and in low vegetation (Framenau *et al.* 2014). Many species, for example those of *Dimidamus* and *Nicodamus*, are active during daylight.

Orsolobidae Orsolobids inhabit forest leaf litter and prey on soft-bodied arthropods such as Collembola and insect larvae. The family occurs in Australia (including Tasmania), New Zealand, the Campbell Islands, Chile, Argentina, and the Falkland Islands (Forster and Platnick 1985). The genus *Tasmanoonops* is endemic to Australia and includes five species (*Tasmanoonops dorrigo*, *T. mysticus*, *T. parinus*, *T. parvus*, *T. pinus*) that occur in rainforests of the World Heritage Area and adjacent sites (Forster and Platnick 1985). *Tasmanoonops* species build silken purse-web retreats in which females lay egg sacs that they guard until the young hatch (Framenau *et al.* 2014).

Pisauridae The Pisauridae, or fishing spiders, in Australia comprise the genera *Dolomedes*, *Megadolomedes*, *Hygropoda*, *Perenethis*, *Dendrolycosa*, *Tasmomedes*, *Ornodolomedes*, *Mangromedes* and *Inola* (Raven and Hebron 2018). Pisaurids are predominantly large, semi-aquatic, spiders that hunt at the water's edge. Most hunt freely, but some build sheet-webs. Individuals use their front legs to detect the vibrations of potential prey on the water's surface. They eat soft-bodied invertebrates and tadpoles, and will also hunt fish (Nyffeler and Pusey 2014). Species of the widespread genus *Dolomedes* (e.g. *Dolomedes facetus*, *D. albicomus*, *D. briangreenei*) resemble Lycosidae, but maternal pisaurids differ in that they carry their egg sacs with their chelicerae. They are strongly nocturnal though some species exhibit day and night activity patterns (Suter and Benson 2014). *Dendrolycosa* (e.g. *Dendrolycosa icadia*) occurs in northern Australia and along the east coast as far as northern New South Wales. Species of *Dendrolycosa* build a sheet-web but with a retreat placed in vegetation (Framenau *et al.* 2014). Also included in the family is the endemic genus *Megadolomedes* (e.g. *Megadolomedes australianus*, *M. trux*), which bears a superficial resemblance to the Neotropical genus *Trechalea* (Todd Davies and Raven 1980). *Megadolomedes australianus* (Plate 15) is an especially large spider and is recorded from Mt Nebo, Kenilworth State Forest, the Richmond Range, and south to Wollongong. *Ornodolomedes* includes species with restricted ranges: *O. benrevelli* (Mt Glorious) and *O. mickfanningi* (lower slopes of Lamington National Park) (Raven and Hebron 2018).

Salticidae The Salticidae, or jumping spiders (more correctly, dancing spiders), are a speciose family that has been highly successful in occupying numerous mesic- and xeric-adapted vegetation communities. Sexual dimorphism is common, some (*Maratus* spp.) exhibit distinctive, highly animated, dance-like courtship-signalling behaviour, and in general salticids are highly adept at prey capture. Salticids discriminate between prey and conspecifics by means of large, anterior median eyes (Blest *et al.* 1981), allowing them to distinguish between rivals and prey exceeding distances of >30 cm (Harland *et al.* 1999). Some salticids have switched from purely predaceous habits to one that also includes feeding from nectar (Jackson *et al.* 2001).

The Australian salticid fauna includes more than 300 species in some 80 genera; however, many new genera and species await description (Žabka and Gray 2004). Numerous species (e.g. of *Astia*, *Astilodes*, *Adoxotoma*, *Apricia*, *Bianor*, *Cytaea*, *Evarcha*, *Helpis*, *Holoplatys*) can be encountered during daylight on tree trunks, leaves and even rocks, particularly in subtropical rainforests along margins and sunlit gaps, as well as the canopy (Basset 1991) (see Plates 16, 17, 18, 19). *Holoplatys* is restricted to Australia, New Guinea, New Caledonia and New Guinea (Žabka 1991), and although most species inhabit open forest and woodland, some (e.g. *Holoplatys complanata*) are known from vine thickets. *Huntglennia* is an endemic genus, and is distinctive in that the eyes are in four rows, a feature rare in Salticidae (Žabka and Gray 2004). The single species, *Huntglennia williamsi*, is known from two widely separated and vegetationally distinct localities in New South Wales: Lansdowne Nature Reserve in a lowland subtropical rainforest–mangrove ecotone, and Warrumbungle National Park in shrubby eucalypt–*Callitris* pine woodland. This lack of fidelity to any particular habitat is commonly encountered in several invertebrates.

The genus *Ocrisiona* is restricted to Australia and adjacent regions (Žabka 1990). *Paraplatoides* is restricted to New Caledonia and Australia, where it is widespread (Žabka 1992). *Sondra* is an endemic genus restricted to northern Queensland–northern New South Wales (Wanless 1988), where it occurs in rainforest and open forest. *Sondra nepenthicola*, *S. raveni*, *S. bickeli* and *S. brindlei* are known only from south-eastern Queensland and northern New South Wales (Wanless 1988; Žabka 2002). *Sandalodes* is recorded from the Bunya Mountains and Mt Lindsay (*Sandalodes scopifer*), and Kyogle and Brooklana (*Sandalodes superbus*), but the genus has an extralimital distribution in New Guinea (Žabka 2000). The distribution of *Prostheclina* is centred on south-eastern Queensland and southern New South Wales (Richardson and Žabka 2007). *Prostheclina pallida* occurs in upland forests of intermediate and higher altitudes south from the Gibraltar Range to Mt Wilson, Jamberoo Mountain and Mt Keira. *Prostheclina amplior* is restricted to the Illawarra Escarpment, and *P. bulburin* ranges from the Kroombit Tops to North Queensland. *Maratus volans* is a small, brightly coloured, foliage-frequenting species found in littoral rainforest, as well as escarpment rainforests and wet sclerophyll forests. The genus *Simaetha* includes small metallic-hued, beetle-like spiders.

Salticidae also include several convincing ant mimics in the genera *Myrmarachne*, *Ligonipes*, *Judalana*, *Rhombonotus* (Whyte and Anderson 2017) (Plate 17). Ceccarelli (2008) suggests that the waving movements of leg 1 in *Myrmarachne* species mimics the antennal movements of ants, and that this functions to increase the spider's resemblance to ants, so as to avoid predation. Salticids exhibit several spider–ant adaptations (Cushing 2012). Ant-resembling myrmecomorphs are Batesian mimics, but myrmecophages are specialised ant predators. Morphological and behavioural resemblances to ants confers strong survival advantages against visually hunting predators (Cushing 2012).

Sparassidae This family includes species previously placed in Heteropodidae. Sparassids are large, somewhat flattened, solitary spiders that have a world-wide distribution (Moradmand *et al.* 2014). In Australia the endemic lineage Deleninae has diversified to form nearly 100 species (Agnarsson and Rayor 2013).

Commonly called huntsmen spiders, sparassids generally do not build webs, and are nocturnal predators in wet sclerophyll forests, rainforests, and more open vegetation habitats. They are long-legged, and when the legs are splayed out, individuals (e.g. of *Holconia*, *Heteropoda*) are able to leap as a response to disturbance, and effectively paraglide to the ground. In wet sclerophyll forest large and highly alert Holconia species are usually encountered under the loose bark of eucalypt trees. *Holconia* is widespread in Australia, excluding Tasmania, but includes *H. immanis* and *H.* insignis from north-eastern New South Wales (Hirst 1991). On the lower north coast of New South Wales *Heteropoda* is commonly encountered in lowland rainforest, individuals being very agile and holding their bodies high off the ground when moving (Plate 20).

The endemic heteropodid genus *Isopeda* is restricted to eastern and southern Australia and occurs from central eastern Queensland to Victoria and southern South Australia, and southern Western Australia. Within this range *Isopeda* is generally confined to areas receiving >500 mm annual rainfall (Hirst 1992). The distribution of individual species can be localised. *Isopeda queenslandicus* is restricted to the region from Kroombit Tops (southern Queensland) to north-east New South Wales (e.g. Tooloom, Nimbin). *Isopeda binnaburra* is known only from the type locality of Lamington National Park (Hirst 1992). *Typostola* is restricted to Australia and New Guinea. *Typostola heterochroma* has been described from the vicinity of Dorrigo (Hirst 1999).

Stiphidiidae In addition to Australia (including Tasmania), Stiphidiidae occur in Madagascar, Mauritius, New Guinea, New Caledonia and New Zealand, and in Australia are most common in temperate and subtropical rainforests (Framenau *et al.* 2014). *Procambridgea* is an endemic genus with several species (e.g. *Procambridgea carrai, P. hilleri, P. lamington*) apparently restricted to rainforest in the Gondwana rainforests region and adjacent sites (Davies and Lambkin 2001). Individuals inhabit hollows in fallen logs, in which they construct sheet-webs. The genus *Therlinya*, is placed in or near the family Stiphidiidae (Gray and Smith 2002) and is endemic to eastern Australia. Framenau *et al.* (2014) place *Therlinya* in the staphidiid subfamily Borralinae. Species such as *Therlinya bellinger, T. wiangaree*, and *T. ballata* are recorded only from the Gondwana rainforests or adjacent sites (e.g. Levers Plateau, Mt Tamborine, Border Ranges National Park, Richmond Range, Cambridge Plateau (Gray and Smith 2002; Williams 2003a). *Borrala* and *Pillara* are restricted to forests north from the Hunter River to the Border Ranges (Gray and Smith 2004). *Jamberoo* ranges from central eastern New South Wales to Victoria, and *Barahna* are small spiders found in wet sclerophyll and dry sclerophyll forests, as well as vine scrub. *Borrala* is restricted to south-eastern Queensland and northern New South Wales (Framenau *et al.* 2014).

Theridiidae These are a very diverse family of generally small-sized spiders frequently found in association with their delicate webs on foliage and branches (Plate 21). Theridiids use sticky silk to capture prey. They include the silver-abdomened *Argyrodes antipodianus* and diminutive species of *Euryopis, Phycosoma* and *Dipoena*. Species of *Moetipa* and *Moneta* are found on foliage. *Argyrodes* includes invasive kleptoparasites of other spiders' webs (e.g. of Araneidae), where they may prey on the host and its eggs and young (Framenau *et al.* 2014). Species of *Phoroncidia* have bulbous, often irregularly shaped abdomens. Although numerous theridiids (e.g. *Thwaitesia, Theridion, Moneta*) frequent canopy and subcanopy foliage in rainforest, a subsuite are hunters that inhabit the forest floor and tree trunks (e.g. *Steatoda*).

Thomisidae The habitat-diverse thomisid flower or crab spiders are sedentary diurnal ambush hunters (Framenau *et al.* 2014) commonly found motionless on rainforest flowers (e.g. *Sidymella rubrosignata, Lehtinelagia evanida*). They are typically long-legged and often distinctively coloured green and white (e.g. *Boomerangia, Diaea* and *Lehtinelagia* spp.), green with contrasting coloured highlights (*Australomisidia, Cetratus*), simply camouflaged in shades of brown (*Australomisidia, Cymbacha, Sidymella, Stephanopis, Tmarus* [Plate 22]), cryptically coloured species (e.g. *Phrynarachne*) that resemble bird droppings, or almost pure white (*Thomisus spectabilis*). Many species are variable in colour (many are able to adjust their colour to that of their host flower [Framenau *et al.* 2014]), juveniles may be differently coloured to that of adults, and sexual dimorphism is common; this all makes identification at times difficult. The widely distributed *Stephanopis barbipes* has dark hair-like tufts on its forelegs, and *Tharpyna* includes specialised hunters inhabiting tree trunks, and hiding by day under loose pieces of bark.

Uloboridae This is a family of small-sized, venomless spiders that build tangled webs (though some are simple capture strands) suspended in vegetation, in hollow logs and below rock ledges; they are predators of tiny insects (Framenau *et al.* 2014; Whyte and Anderson 2017; Wunderlich 1976). Genera commonly encountered in coastal rainforest stands include *Miagrammopes* and *Uloborus*. Several *Miagrammopes* cohabit in rainforest at Minnamurra Falls south of Sydney. The monotypic genus *Ranguma* (*Ranguma lehtineni*) is known only from the McPherson Range (Opell 1984; Walton 1985b; Wunderlich 1976).

Zodariidae Zodariids are predominantly tropical ant-associated, ant-eating and ant-mimicking spiders. *Habronestes bradleyi* mimics ants not only in body shape and behaviour but also their pheromones (B. Baehr 2003). The majority of Australian genera are endemic. The entire *Asteron* complex is endemic to the continent, and most included species are adapted to semi-arid areas. However, several (e.g. *Asteron zabkai, A. grayi*) are restricted to the east coast, occurring in forested locations (Jocqué and Baehr 2001). The endemic genus *Storena* includes *S. cyanea* recorded from Lamington National Park and the Richmond Range (Baehr and Jocqué 1994, Jocqué and Baehr 1992). *Nostera* occurs in open forest; however, *N. lynx* and *N. spinata* inhabit cool upland rainforests of the Gondwana rainforests region. The related *Nosterella* *nadgee* is distributed from southern Queensland and the mid-north coast of New South Wales in rainforest and wet sclerophyll forest (Baehr and Jocqué 2017). The endemic genus *Habronestes* is mainly distributed in arid zones, but some species inhabit upland cooler mesic forests (e.g. at Barrington Tops, Mt Hyland Nature Reserve, Washpool National Park, Richmond Range [B. Baehr 2003]).

Miscellaneous families Clubionidae are a family of abundant, slender-shaped, active hunters found commonly on foliage, on the forest floor and under bark. They are characterised by their large, forward-facing chelicerae. Many are coloured in tones of brown (*Clubiona* spp.), though some are green and blend readily with rainforest foliage. Corinnidae (e.g. *Battalus*, *Nyssus*, *Poecilipta*) were separated from Clubionidae and are agile day-active hunters of the forest floor and tree trunks (Plate 10). They can be found in leaf litter or hiding under loose bark. Several species are ant mimics (Raven 2014). The Cyatholipidae are composed solely of two monotypic genera (*Tekellatus* and *Teemenaarus*) one of which, *Tekellatus lamingtonensis*, is restricted to south-east Queensland (in Walton 1985b). The elongate-bodied Deinopidae (i.e. *Deinopis* and *Menneus* spp.[Coddington *et al.* 2013]) have extremely large ocelli and are medium- to large-sized sedentary net-casting spiders usually found on forest margins. They constuct rectangular silk nets, held by their legs, which are quickly expanded to snare prey. *Menneus* has normal-sized eyes, but the posterior median eyes of *Deinopis* are much enlarged. *Menneus aussie* (Plate 11) is distributed along the east coast in rainforest and wet sclerophyll forest, but is also known from New Caledonia (Coddington *et al.* 2013). The Eutichuridae possess conspicuous large curved chelicerae and are active nocturnal hunters (*Cheracanthium* spp.) on the forest floor, and also on foliage and flowers. Filistatidae include *Wandella orana*, a species recorded in upland forests west of Kempsey (Gray 1994). Though endemic, *Wandella* is widespread on the mainland. One genus and two species comprise the Theridiosomatidae, a small family restricted to the region between the Queensland border and Sydney. *Theridiosoma brauni* is known from the McPherson Range (Walton 1985b). Lynx spiders (Oxyopidae) are active hunters, and in rainforests are abundant on foliage, tree trunks and flowers. *Oxyopes* is particularly diverse, and includes many species with widespread distributions. *Hamataliwa monroei* extends along coastal Queensland as far south as Rockhampton (Framenau *et al.* 2014). Tetragnathidae (Álvarez-Padilla and Hormiga 2011; Hormiga *et al.* 1995) include web-snaring species (e.g. *Leucauge* spp.) associated with moist forest types. *Mesida argentiopunctata*, as well as extremely long- and narrow-bodied species of *Tetragnatha*, frequent rainforest foliage; the latter is often found motionless in repose along the mid-vein of sunlit leaves. The brown-coloured *Tetragnatha nitens* is an associate of vegetation adjacent streams. The genera *Ozitira* and *Storenosoma* (Amaurobiidae) include free-living, rainforest-inhabiting species that are distributed in upland cool forests between central coastal Queensland and eastern Victoria (e.g. *Oztira kroombit*, *Storenosoma altum*, *S. grayi*, *S. supernum*, *S. terraneum*) (Milledge 2011). The related endemic genus *Malala* includes *Malala lubinae*, which was described from rainforest at Lamington National Park (Todd Davies 1993). Additional regional-occurring genera include *Manjala*, *Midgee* and *Daviesa* (Framenau *et al.* 2014). The swift-moving Selenopidae have a characteristically flattened body and live under bark and in rock crevices. *Karaops raveni* is distributed from southern Queensland to central eastern New South Wales (Whyte and Anderson 2017). Zoropsidae (e.g. *Birrana*, *Kilyana*) are related to lycosids and include several species from upland and lowland rainforests of Queensland and northern New South Wales (Framenau *et al.* 2014). Zoropsids are active hunters, but the biology of the family is poorly known. The Tengellidae (*Austrotengella*), or false water spiders, are also related to lycosids and are mostly found in subtropical rainforest and other moist habitats. They are active hunters found on the forest floor.

Mygalomorphae (trap-door and funnel-web spiders)

The infraorder Mygalomorphae, sometimes referred to as primitive spiders, are an ancient group consisting of large spiders that generally live secretively within burrows, or below rocks and fallen logs on the forest floor (Framenau *et al.* 2014). For much of their life they are sedentary and nocturnal ambush predators feeding on invertebrates and small vertebrates, but will actively seek out new sites of habitation within appropriate habitat types. Mygalomorphae are sensitive to desiccation, generally requiring microhabitats that are moist. They are distinguished from the Araneomorphae by fangs that are largely parallel, functioning to strike downwards on their prey. However, their eyesight is poor. The fangs of araneomorphs point sidewards, and species usually have high visual perception. Mygalomorphs have four lungs set as two pairs, but most Araneomorphae only have two lungs.

The various mygalomorphs consist of circumtemperate and/or tropical families or subfamilies with a large and diverse endemic Australian fauna (in Walton 1985b). They are most abundant and diverse in the Southern Hemisphere (Main 1981). With the exception of Asian liphistiomorphs, mygalomorph spiders are the most ancient and primitive group, and their evolution has been independent to that of the Araneomorphae (Main 1981). The oldest mygalomorph fossils date to the Triassic (225–180 mya) and fossils known from Dominican amber appear quite similar to extant species from that area (Raven 1994).

The rich mygalomorph fauna occurring in the Gondwana rainforests World Heritage Area, and that more broadly in subtropical eastern Australia, reflects overlapping faunal patterns. The region is a major focus of generic diversity within Australia, with only south-west Western Australia possessing more genera (Williams and Bickel 2010). Genera occurring in the region include *Cataxia* (vine forest), *Misgolas* (rainforest, vine forest, *Nothofagus* forest, sclerophyll forest), *Missulena* and *Armadalia* (in sclerophyll forest and woodland), *Arbanitis*, *Evoplos* (rainforest and sclerophyll forest), *Seqocrypta* (rainforest) and *Trittame* (vine thicket) (Main 1981; Raven 1994). The region is a mixing area for migration routes of genera, marking the southern limit of northern genera (*Evoplos*, *Cataxia*, *Armadalia*); contains widespread genera (e.g. *Missulena*, *Arbanitis*); and is at the centre of the range of *Misgolas* (= *Dyarcyops*) (Main 1981).

Three evolutionary phases can be recognised in the Australian mygalomorph fauna: (1) ancient elements, (Mesozoic 225–60 mya) and early Tertiary (e.g. *Missulena*, *Misgolas*); (2) autochthonous groups of ancient origin (e.g. *Arbanitis*); and (3) late Tertiary and Pleistocene (1 million–10 000 years ago) northern immigrants (all absent from New Zealand) (e.g. *Conothele*, *Cataxia*, *Armadalia*) (Main 1981). Representatives of all groups occur in the Gondwana rainforests region. *Missulena* is restricted to Australia except for a single species (*Missulena tussulena*) from Chile (Grismado and Goloboff (2006).

Actinopodidae The Actinopodidae are known from Australia and South America (Framenau *et al.* 2014) and are powerfully built spiders, conspicuous by their massive, curved chelicerae. They include the burrow-building genus *Missulena*, placed previously in the Ctenizidae (see Main 1981, 1983; Framenau *et al.* 2014; Whyte and Anderson 2017 for comments on the Ctenizidae, which as currently considered comprises only the genus *Conothele*). Known as mouse spiders, the genus *Missulena* occurs widely in Australia, and also New Guinea (Main 1981). Several species inhabit urban gardens; however, one of these (*Missulena bradleyi*, Plate 23) is recorded from partly cleared subtropical rainforest at Mt Tamborine (Lake 1990), and from Dooragan National Park (Whyte and Anderson 2017).

Barychelidae Spiders in the family Barychelidae occur in littoral zones, open sclerophyll forests, vine thickets and rainforests, but nevertheless are cryptic species not easily discerned within the environment (Raven 1994). Most species belong to the subfamily Barychelinae, and are medium- to large-sized spiders. They are most diverse in the Pacific but this might be partly due to lack of revision of the Neotropical fauna. Ten genera are known from the Austro-Papuan region. In comparison, 11 genera are recognised from the African-Middle Eastern and west Indian subcontinent region, nine genera occur in South America, and the South-east Asian-Australasian region (including the Austro-Papuan) comprises 21 genera (Raven 1994).

Barychelidae occur throughout much of Australia, but are absent from Tasmania, Victoria and southern coastal New South Wales. Endemic genera occur in northern New South Wales–south-eastern Queensland. *Trittame* is distributed from northern to southern Queensland with one species, *T. ingrami*, known only from semi-evergreen vine thicket. This is the most southerly occurring species. *Seqocrypta* is a predominantly rainforest genus restricted to south-east Queensland and north-east New South Wales, occurring as far south as Kempsey. At least three *Seqocrypta* species (*Seqocrypta mckeowni*, *S. jakara*, *S. hamlynharrisi*, *S. jakara*) occur in the Gondwana rainforests region (Raven 1994). Additional genera include the mostly tropical *Ozicrypta* and the widespread *Idiommata* (Whyte and Anderson 2017).

Dipluridae These are web-building spiders, and are the most diverse family of mygalomorph spiders described from the forests of north-eastern New South Wales and south-eastern Queensland. They are apparently absent from Tasmania (Framenau *et al.* 2014). Raven (1981) recognised three subfamilies, Diplurinae, Ischnothelinae and Masteriinae, as occurring in Australia. Individuals are very agile and are predators of snails (Whyte and Anderson 2017).

Many species (e.g. *Austrotralothele maculata*, *A. montana*, *A. nothofagi*, *Cethegus* spp.) exhibit localised distribution patterns (Walton 1985b; Williams 2002). *Australothele* is known from southern

Queensland and northern New South Wales. The genus is particularly common in mesic forest types and moist microhabits. *Austrotralothele bicuspidata* is recorded from cool temperate rainforest in New England National Park (Raven 1984c). *Austrotralothele maculata* is restricted to the McPherson Range and, though found in *Nothofagus*–cool temperate forest, is more prominent in adjacent subtropical rainforest (Raven 1984c; Walton 1985b). The monotypic genus *Carrai* (*Carrai afoveolata*) is recorded only from the Mt Boss–Mt Banda Banda region (Raven 1984c).

Hexathelidae The funnel web family Hexathelidae (previously a subfamily of Dipluridae) is a Gondwanan relic with related genera in New Zealand and South America (Walton 1985b). The family is diverse in subtropical Australia, and species tend to be large to very large in size. When distubed they will feign death, or assume raised defensive postures. The family includes three subfamilies (Atracinae, Hexathelinae, Plesiothelinae), of which the Atracinae include the highly venonous genera *Atrax*, *Hadronyche* and *Illawarra* (Framenau *et al.* 2014).

Hexathelidae include several *Paraembolides* species (e.g. *Paraembolides boycei*) with localised distributions and/or ranges restricted to the Gondwana rainforests region (Raven 1978; Walton 1985b; Williams 2002). *Paraembolides* (which includes many species previously in *Bymainiella*) (Plate 24) is an endemic genus found in south-eastern Australia. Raven (1978) considered *Bymainiella* to be a sister group of the South American genus *Scotinoecus*. *Paraembolides* occurs in rainforest, wet and dry sclerophyll forest, and adjacent lowland forests. Though several species appear to be restricted to rainforest (e.g. *Paraembolides polesoni*, *P. montisbossi*, *P. monteithi*), a portion of the fauna inhabits more diverse forest types (*Paraembolides tubrabucca*, *P. cannoni*).

The endemic genus *Hadronyche* (Plate 25) ranges from south-eastern Queensland to Victoria and Tasmania (Gray 1987) and includes numerous new species. *Hadronyche* males will actively seek out females, and many species are known only from closed forests and wet sclerophyll forests of the Gondwana rainforests region (e.g. *Hadronyche formidabilis*, *H. infensa*, *H. macquariensis*, *H. valida*). *Hadronyche versuta* occurs further to the south from west of Sydney to the Illawarra region. The related *Atrax* (e.g. *Atrax robustus*) is difficult to distinguish from *Hadronyche*, and is confined to south-eastern Australia along the coast and highlands (Gray 1987). *Illawarra wisharti* is a southern forest-inhabiting species.

Atracine spiders need moist conditions and are restricted to south-east Australia (Gray 1987). Their absence from the temperate closed forests and tall open forests (wet sclerophyll forest) of south-western Victoria and western Tasmania is probably related to palaeoclimatic and ecological factors (Gray 1987). Northern, mainland limits to distribution coincide with the inland divergence of the Great Dividing Range, where drier environmental conditions predominate (Gray 1987).

Idiopidae The Idiopidae have an essentially Gondwanan distribution. Many species close their borrow entrances with a hinged door, though *Misgolas* rarely has a trap door (Framenau *et al.* 2014). In Australia numerous species are confined mainly to rainforest and wet sclerophyll forest (e.g. *Arbanitis robertcollinsi*, *A. variabilis*, *Cataxia pulleinei*, *Misgolas pulchellus*, *M. echo*, *M. bithongabel*), though some are found in rainforest (including vine thickets) and open forest (*Arbanitis longipes*), or are currently recorded only from open forest (*Misgolas monteithi*). *Arbanitis* is known only from south-east Queensland and northern New South Wales (Raven and Wishart 2005). *Misgolas* is widespread in eastern Australia but its current diversity possibly reflects contraction of species ranges to refugial areas, and later expansion and radiation in response to availability of suitable habitat and conditions (Main 1981). Many undescribed and localised *Misgolas* species are known from the Gondwana rainforests region, and rainforests further to the south; the number probably exceeds 100 species (M. Gray pers. comm.; Gray and Cassis 1994; Raven and Wishart 2005; Wishart 1992; Wishart and Rowell 2008). Several are recorded only from small isolated lowland rainforests remnants (e.g. *Misgolas kirstiae*). Others are known only from single upland mesic forests (*Misgolas yorkmainae*, Mt Boss; *M. maxhicksi*, Barrington Tops).

Misgolas includes distinctive tube-building species (e.g. *Misgolas robertsi*, *M. mascordi*) (Plate 26). *Misgolas robertsi* was first recorded from rainforest remnants south of Sydney (Wishart 1992). The related *Misgolas mascordi* was described from subtropical rainforest in Dorrigo National Park. This species builds tubes up to 26 cm in height above the ground, attached to their support only at the lip to the entrance. Tubes are sometimes decorated with moss and other plant material (Wishart 1992). Though rarely encountered, tube-building *Misgolas* (of uncertain affinity) inhabit dry rainforests in the Manning Valley (Kiwarrak State Forest, Camels Hump Nature Reserve) south of Dorrigo. Here spiders build tall leathery aerial tubes adhering to plant stems and rocks in the forest understorey, and extending from the burrow below. The tubes are of varying height, frequently >13 cm in length. However, populations within individual rainforest stands have localised distributions, with spiders clustered in particular sections rather than uniformly distributed, even though seemingly suitable habitat is present throughout the stand.

Migidae Migidae occur in Madagascar, South America, New Caledonia, New Zealand, and Australia, including Tasmania (Main 1981), suggesting a Gondwanan origin (Framenau *et al.* 2014). They are a very ancient family and their occurrence in Australia may date from the Late Cretaceous/Early Tertiary (~<80 mya) (Main 1981). However, the Australian fauna is small.

The genus *Migas* is known from New Caledonia, New Zealand, Chile, Norfolk Island, and a single species each from southern Queensland (*Migas variapalpus*) and Tasmania. *Migas variapalpus* is known only from complex subtropical rainforest on the Lamington Plateau (Raven 1984a). *Migas* provides biogeographical evidence indicating that New Caledonia remained as part of Australia to a period post-dating the separation of New Zealand and Norfolk Island (Raven 1995).

Nemesiidae These are sometimes referred to as wishbone trapdoor spiders (see Plates 27, 28) owing to the y-shaped silk-lined burrow constructed by some species (Framenau *et al.* 2014). Individuals are active nocturnal predators and can be aggressive when disturbed. The predominantly rainforest genus *Namea* (e.g. *Namea bunya*, *N. excavans*, *N. dicalcarea*, *N. flavomaculata*) is confined to the moist coastal upland areas from northern Queensland to the Gibraltar Range of northern New South Wales, but reaches its greatest diversity in the Macpherson–Macleay Overlap (Raven 1984b; Walton 1985b; Williams 2002). *Namirea planipes* inhabits rainforest and dry sclerophyll forest in south-east Queensland and adjacent forests in New South Wales (Raven 1984c).

The endemic genus *Xamiatus* ranges from northern Queensland to northern New South Wales, and has at least seven species recorded from the Gondwana rainforests region. All inhabit rainforest. These include *Xamiatus kia* (Bellingen–Nambucca Heads), *X. rubrifrons* (Conodale and D'Aguilar ranges), *X. musgravei* (New England National Park), and *X. fischeri* (Mt Banda Banda) (Raven 1981, 1982; Walton 1985b; Williams 2002). Additional eastern Australian genera include *Kiama* (e.g. *Kiama lachrymoides*) and *Ixamatus* (Framenau *et al.* 2014).

SCHIZOMIDA (whip spiders)

This small group of ancient predatory soft-bodied arachnids is largely restricted to the world's tropical and subtropical regions (Savory 1977; Harvey 2003). Species are associated with debris, and usually live their lives close to the ground, in a microclimate of great humidity. Individuals are vulnerable to desiccation. In Australia Schizomida are mainly found in the northern parts of the continent but some extend to south-east Queensland and southern New South Wales. Schizomida have eight pairs of legs, but the first pair is modified as sensory organs, thus using only six legs for walking. Species are nocturnal and individuals are vulnerable to dessication.

Two families, the Hubbardiidae (formerly Schizomidae) and Protoschizomidae, are recognised. Protoschizomids are restricted to the United States of America and Mexico. *Apozomus woodwardi* (Hubbardiidae) occurs in closed forest of southern Queensland and Tasmania (Harvey 1992).

SCORPIONIDA (scorpions)

Scorpions give birth to live young and are generally restricted to the warmer parts of the world (Savory 1977). They are largely nocturnal in behaviour, have a long reproductive rate, and in Australia comprise more than 40 described species in four families (Buthidae, Bothriuridae, Hormuridae, Urodacidae) (Luna-Ramirez *et al.* 2017). Australian species range in size from ~2 cm to 12 cm in length. The Scorpionida appeared in the Silurian (428 mya) and are fairly numerous in the Carboniferous (Savory 1977); however, the fossil record indicates their body plan has remained largely unchanged since the Paleozoic (Luna-Ramirez *et al.* 2017). Major differences between early and modern scorpions are largely due to adaptations to the terrestrial environment; book lungs replaced book gills, simple eyes replaced compound eyes, and there was evolution of structures to sense movement of air currents.

The Australian scorpion fauna indicates a history of indirect connection with South America followed by a period of isolation and encroachment on South-east Asia (Koch 1981). Bothriuridae genera are restricted to Australia and South America but other families are related to Indo-Asian taxa or are autochthonous, having evolved since the approach of the Australian landmass towards Asia (Koch 1981). Scorpion distribution is more a product of rainfall, temperature and biotic influences (e.g. competitive exclusion) than vegetation type (Koch 1981). The large flattened rainforest scorpion *Hormurus waigiensis* (Hormuridae) occurs in Queensland and north-eastern New South Wales. The mottled *Lychas marmoreus* (Buthidae) is commonly encountered during daylight sheltering under stones, fallen logs, and loose tree bark in wet sclerophyll forest, but also may be encountered in rainforest. It is an active nocturnal predator of spiders, insects and small vertebrates such as juvenile skinks. The generally larger but somewhat similar live-bearing *Cercophonius squama* (Bothriuridae) (Plate 32) inhabits a diversity of rainforest types ranging from those by the sea to at least intermediate altitudes >700 m. A nocturnal predator of soft-bodied invertebrates, by day it shelters in burrows, beneath leaf litter or under bark.

PSEUDOSCORPIONIDA (pseudoscorpions)

Diminutive scorpion-like pseudoscorpions are found sheltering by day under loose tree bark and in rotten fallen wood. Some species are blind, being almost wholly dependent on touch. However, a feature of particular note is that the elongate articulate chelicerae, in addition to their role as grasping and sensory organs, also possess silk glands; these occupy the same position as the venom glands of many spiders (Savory 1977).

Harvey (1981) gives a checklist of the then-known Australian pseudoscorpion fauna. Pseudoscorpions exist in subtropical rainforest in low densities and are usually encountered as solitary individuals. They have sometimes been found in apparent phoretic associations with Hymenoptera (on fossil Ichneumonidae [Savory 1977], extant Braconidae [A. Bradford pers. comm.]), Diptera (Tabanidae, Culicidae), Coleoptera (Cerambycidae, Passalidae, Curculionidae-Platypodinae), Neuroptera (Chrysopidae), Hemiptera (Reduviidae, Cicadidae), Lepidoptera (Sphingidae) (Aguiar and Bührnheim 1998) and microbats (Finlayson *et al.* 2015). Of the families recorded from Australia the Olpiidae are typical of the forest litter fauna. The Pseudotyrannochthoniidae includes *Pseudotyrannochthonius queenslandicus* (Joalah National Park [Walton 1985b]), as well as the troglobitic *P. eberhardi* and *P. jonesii* (Harms and Harvey 2013). These are short-range endemics with highly limited known distributions. The Tridenchthoniidae are most diverse in South-east Asia and Africa (Walton 1985b); however, two genera and three species have been described from Australia (one is restricted to Lord Howe Island). The monotypic genus *Heterolophus* is known only from Mt Tamborine where it occurs in rainforest. *Anaulacodithella australica* and *Morikawa* (= *Tyrannochthonius*) *queenslandica* are also known from the rainforests of Mt Tamborine, and Lamington National Park (Harvey 1994).

OPILIONES (harvestmen)

Harvestmen are delicately formed spider-like predators encountered under rock overhangs, fallen timber and loose rocks on the rainforest floor (Hunt and Hickman 1993). Occasionally they are found motionless on upper leaf surfaces. They are normally active at night. Many species specialise on spiders as prey, and much of the Australian fauna is endemic. The harvestmen of the Gondwana rainforests are dominated by the genus *Equitius* (Trianenonychidae) (Plate 29).

Acropsopilionidae This family is known from Australia, New Zealand, South Africa and South America (Forster 1955), and contains only nine described species in five genera (Cantrell 1980): *Acropsopilio* (Chile, Australia), *Cadella* (South Africa), *Zeopsopilio* (New Zealand), and *Austropsopilio* and *Tasmanopilio* (Australia) (Cantrell 1980).

The monotypic genus *Austropsopilio* is an 'early Australian fauna' (Forster 1955) restricted to the Gondwana rainforests region. *Austropsopilio altus* is known only from New England National Park, and *A. australicus* is known only from Emu Vale in southern Queensland (Cantrell 1980).

Assamiidae Assamiidae known from northern New South Wales–southern Queensland include the Australian endemic genus *Octobunus* (Forster 1955).

Neopilionidae The subfamily Ballarrinae occurs in South Africa (one genus), southern Australia (three genera), and southern South America (one genus). The endemic monotypic genera *Plesioballarra* (e.g. *Plesioballarra crinis*) and *Ballara* (*Ballara cantrelli*) are restricted to south-east Queensland and northern New South Wales (Hunt and Cokendolpher 1991; G. Hunt pers. comm.). Enantiobuninae are the most speciose of the long-legged harvestmen occurring in Australasia. Males may have greatly enlarged chelicerae though this feature is reduced in 'minor' males or is apparently absent in some species (e.g. *Megalopsalis coronata*, *M. puerilis*) (C. Taylor 2013) (Plates 30, 31).

Triaenonychidae *Equitius* (Plate 29) is an endemic genus restricted to the region between south-eastern Queensland (27°S) and the Shaolhaven River (35°S) in southern coastal New South Wales. It is the dominant harvestman genus found in moist forest types and woodland within this range and is among the commonest arachnids found under logs on the forest floor (Hunt 1985). *Equitius* has undergone considerable speciation in northern New South Wales and southern Queensland with five (*Equitius montanus*, *E. richardsiae*, *E. spinatus*, *E. tambourineus*, *E. altus*) of the seven described species confined to the Gondwana rainforests region (Hunt 1985; Williams 2002). *Equitius doriae* occurs in mountainous rainforest, wet sclerophyll forest and snow gum woodland between the Hunter River and Goulburn to the south of the Gondwana rainforests region. Hunt's (1985) seventh species, *Equitius formabilis*, inhabits patches of subtropical rainforest in lowland coastal areas between Port Hacking (Sydney) and north of the Shoalhaven River. Some species are sympatric; examples are *Equitius altus* with *E. spinatus* at Barrington Tops, and *E. richardsae* with *E. montanus* in a rainforest remnant between Dorrigo and Ebor (Hunt 1985). No hybrid examples are known but individual species can exhibit apparent altitudinal preferences: *Equitius altus* and *E. richardsae*, from the examples above, are more common at lower altitudes, and *E. spinatus* and *E. montanus* occur mainly at higher altitudes. Habitat preferences are also apparent with species occurring more commonly (or exclusively) in rainforest or eucalypt forest (Hunt 1985).

ACARINA (mites, ticks)

The often bizarrely formed Acarina (= Acari, Acarida) occur in vast numbers within most ecosystems (Halliday *et al.* 2001; Saito 2010; Walter and Proctor 2013). Walter *et al.* (1998) suggest that more than 2000 species of mites may occur in one section of Lamington National Park alone. However, though many are of ancient lineage, the origin of much of the world's extant mite fauna is uncertain (Walter 2001)

Along with other members of the arthropod mesofauna (e.g. Collembola), mites are useful bioindicators of soil health. They exhibit numerous coevolutionary relationships with animals and plants, including predatory and phoretic ones (e.g. Atyeo 1992; Binns 1982; Domrow 1979; Gibbs and Stanton 2001; Guerra *et al.* 2010; Perotti and Braig 2009; Tagami and Halliday 2013) (see Plates 4, 5, 6), and frequently have restricted distributions within particular rainforest stands and strata. In rainforests mites occupy diverse habitat niches including leaves, bark and hanging detritus (Beaulieu *et al.* 2006; Walter and O'Dowd 1992, 1995). Many are canopy specialists in subtropical rainforests. Some taxa inhabit the domatia, 'little houses', of rainforest plant leaves, and others are commensals in the nests of vertebrates and invertebrates (e.g. of bees). Others ride upon the bodies of pollinators, which eventually alight on flowers; the mites there consume pollen and nectar (Colwell 1995), or secrete themselves (as larvae and adults) under the elytra of carabid beetles (Constantine and Seeman 2014). Mite diversity in subtropical rainforest ecosystems is related to plant diversity and because mites are so small, typically <500 μm in length and as small as 80 μm, even the smallest habitat (leaf surface, leaf domatia, insect body, moss or lichen patch) can provide a wealth of microhabitat opportunities (Walter and O'Dowd 1992, 1995; Walter and Proctor 2013).

Mites are a ubiquitous invertebrate group with many endemic and highly localised taxa in subtropical eastern Australia (Williams 2002). In addition to having a diverse soil-associated, phoretic and parasitic fauna, Australian rainforests also possess a largely unrecognised and distinctive arboreal fauna associated with forest canopies (R. Kitching pers. comm.; Walter and Behan-Pelletier 1999). The Australian mite fauna stands at ~2700 described species, and about one-third of these are parasites of animals or plants. The former are largely endemic and those associated with plants appear to be mostly exotic taxa. However, the origins of the remaining mite taxa are uncertain and the geographical origins of many species are unknown (Walter 2001).

The mite suborder Mesostigmata (e.g. Ascidae, Ameroseiidae, Uropodidae) is a large and successful group of taxa that have radiated to occupy many habitats (Walter 1999). Most are predators. However, some are scavengers and parasites of mammals, birds, reptiles, and other invertebrates. In rainforests they occur in the soil, in the leaf litter of the forest floor, in rotting wood, as commensals in animal nests, on ferns and in moss clumps. Some are phoretic on beetles (Monteith and Storey 1981). The Astigmata (e.g. Algophagidae) are fungi feeders and scavengers, but include many that are parasites of vertebrates. They include 'feather mites' that are parasites of birds (Atyeo 1992). Species of *Histiostoma* (Anoetidae/Histiostomatidae) and *Caloglyphus* (Acaridae) are associates of the brood masses and nests of the scarabaeine dung beetle *Cephalodesmius armiger* (Monteith and Storey 1981). Histiostomatidae are also associated with the giant earwig *Titanolabis colossea* (Tagami and Halliday 2013).

The Prostigmata (e.g. Erythracidae, Paratydeidae, Raphignathidae, Stigmaeidae, Trombidiidae, Trombiculidae) are the most diverse of the major mite groups in numbers of species, morphology and behaviour. Trombidiidae (Plate 3) include the distinctively bright red-coloured velvet mites, which as adults are active predators. Trombiculidae (chiggers) are parasites of vertebrates (Domrow 1978, 1979, 1991). The suborder exhibits a cosmopolitan distribution, is found in many habitats, and includes terrestrial, freshwater and marine forms. Included in freshwater fauna are the hyporheic water mites (Hydracarina), found beneath and alongside streams, whose larvae are usually parasitic on arthropods but the latter stages are predaceous. In Australia the Hydracarina comprise more than 400 species in 89 genera and 24 families (Smit 2007). The fauna from south-eastern Queensland and northern New South Wales is a rich one. Species recorded from the Bunya Mountains and the Main Range alone include *Anisitsiellides circularis* (Anisitsiellidae), *Axonopsella fragala* (Aturidae), *Australiobates hyporheicus*, *Aspidiobates geometricus*, *Kallimobates australicus* (Hygrobatidae), *Koenikea crinita* (Unionicolidae), *Frontopodopsis obscura* (Frontopodopsidae), *Barwontius lunoka*, *Twarntaturus alphus*, *Davidsia aurita* (Notoaturninae), *Momoniella parva*, *Austromomonia grownsae*, *Partidomomonia tertia* (Momoniidae) and *Arrenurus bunyaensis* (Arrenuridae) (Smit 2007).

Mites in the order Oribatida (or Cryptostigmata) probably evolved in the Lower Devonian, and today are an important and dominant component of the arboreal, floor litter and soil-associated Acarina (Norton *et al.* 1988). Oribatid mites have been found to number in densities of several hundred thousand individuals per square metre in organic soil horizons (Battigelli *et al.* 1994; Peterson and Luxton 1982). They feed mainly on soil microfungi and decaying vegetable material. Some will feed on nematodes. Changes to soil moisture regimes, humidity, rainfall patterns, the types and quantity of forest litter, and differences in soils influence the diversity and abundance of oribatids and their responses to microhabitat availability; such that species may exhibit patchy 'within habitat' distributions, both laterally and vertically within a forest (reviewed in Hunt 1994). Orabatid families occurring in the rainforests of subtropical eastern Australia include (but not exhaustively) Brachychthoniidae, Carabodidae, Eutegaeidae, Galumnidae, Nothriidae, Oppiidae, Otocepheidae, Pedrocortesellidae and Scheloribatidae (Williams 2002).

The Metastigmata mites differ so dramatically in size from their relatives that they are referred to as ticks. Engorged females of species in subtropical rainforest can be in excess of 5 mm in length. Ticks retain their ectoparastic habit and feed on the blood and tissue fluids of vertebrates. Males may not feed. 'Hard ticks' of the family Ixodidae (e.g. *Ixodes* and *Haemaphysalis* spp.) are often specific to certain hosts, or host groups; notably macropod and murid mammals in rainforest and wet sclerophyll forest, but also reptiles (*Amblyomma*, *Aponomma*) and birds. Hosts of the Argasidae ('soft ticks') include small microchiropteran bats and birds (F. Roberts 1970).

Adhaesozetidae This family includes *Adhaesozetes polyphyllos*, known from Lamington National Park, Mt Glorious, Border Ranges and Nightcap national parks, and the Allyn River adjacent the Barrington Tops (Colloff and Halliday 1998). *Adhaesozetes* has been recorded from the leaves of numerous rainforest plants including the trees *Karrabina benthamiana* (= *Geissois benthamii*), *Elaeocarpus angustifolius*, *Doryphora sassafras*, *Nothofagus moorei*, *Pseudoweinmannia lachnocarpa*, *Pennantia cunninghamii* and *Ackama paniculata*, and the vine *Palmeria scandens* (Walter 2004).

Alicorhagidae The Alicorhagidae include species from Lamington National Park assignable to the genus *Alicorhagia* or a possible new genus (Walter 2001). This family is known from Devonian fossils and is placed in the Endeostigmata, a group of mites 'described from some of the earliest terrestrial [fossil] faunas known ...' (Walter 2001). The Australian endeostigmatan fauna is a rich one but the affinities (or 'endemicity') of many included taxa are uncertain. *Alicorhagia* species include *A. usitata* from subtropical and warm temperate rainforest (Bunya Mountains, Lamington National Park–southern Queensland, and northern New South Wales) and *Stigmalychus veretrum* from rainforest in Lamington National Park. The related *Terpnacarus gibbosus* has been collected at Mt Glorious and Lamington National Park (Williams 2003a).

Algophagidae These are algal and detritus feeders. Algophagid mites occur in the Palaearctic, Nearctic, Oriental and Australian regions, and subantarctic islands. The family includes *Lamingtonacarus*, which is endemic to south-east Queensland and is associated with water-filled tree holes (Fashing *et al.* 2000).

Ameridae The Ameridae include two known species (both in the genus *Hymenobelba*) from Australia. *Hymenobelba domahidyi* occurs in the Nightcap Range, and the New England and Lamington national parks. An undescribed species is known from Lansdowne State Forest, near Taree, northern New South Wales (Colloff and Halliday 1998).

Ameroseiidae This family is poorly known, but includes fungi and pollen feeders. *Epicriopsis walteri* is known only from southern Queensland and northern New South Wales (Halliday 1997). *Neocypholaeops rotundus* is recorded from Lamington National Park and Mt Tamborine. *Neocypholaeops* is associated with flowers and the animals that visit them (Halliday 1997).

Anderemaeidae The single known Australian species, *Anderemaeus australiensis*, is associated with *Nothofagus* forest in the Barrington Tops area (Colloff and Halliday 1998).

Arrenuridae *Arrenurus kitchingi*, recorded from Lamington National Park (Smith and Harvey 1989), was the first member of the genus known to occupy water-filled tree hole habitats. The only water mites previously recorded from this habitat were members of the genus *Thyopis* (Hydryphantidae), which occur in Ohio, USA (Smith and Harvey 1989). The genus *Arrenurus* is distributed worldwide, and is found in a wide range of habitats that include temporary pools, deep lakes and seepage areas. Larvae of some species, along with their hosts, are adapted to marginal aquatic habitats (Smith and Harvey 1989). The endemic genus *Lamingtonacarus*, a member of the family Algophagidae, has also been discovered in water-filled tree cavities in south-east Queensland (Fashing *et al.* 2000).

Ascidae Ascidae live in soil and leaf litter, flowers and subcortical situations, and are associated with other animals. The Australian ascid fauna is almost unknown (Halliday *et al.* 1998). *Hoploseius australianus* has been described from south-east Queensland (Walter 1998). This Gondwanan ascid mite genus is otherwise unknown from Australia but possesses morphological characteristics shared with species from India and Central Africa. *Proctolaelaps nesbitti* is restricted to southern Queensland (Halliday *et al.* 1998). *Lasioseius* is known from Australian rainforests but is apparently absent from New Zealand (Walter and Lindquist 1997). *Asca* species are common in eastern Australian forests (Walter *et al.* 1993). *Asca garmani* and *A. macromela* are recorded from Iluka Nature Reserve. Additional ascid taxa recorded from south-east Queensland and northern New South Wales include a species of *Cheirosus* from Lamington National Park (Kitching and Callaghan 1982).

Austrachipteriidae This family includes several *Austrachipteria* species that inhabit subtropical rainforests in the Border Ranges (*Austrachipteria bidactylus*, *A. breviseta*, *A. mariae*), and the endemic genus *Leebates* known only from Lamington National Park (Colloff and Halliday 1998).

Brachychthoniidae An undescribed species of *Liochthonius* is known from Lamington National Park (Colloff and Halliday 1998). *Liochthonius* is recorded from Queensland, South Australia and Argentina.

Carabodidae The genus *Austrocarabodes* occurs in south-east Australia in wet sclerophyll forest and rainforest. A number are only known from Mt Glorious (e.g. *Austrocarabodes gressitti*, *A. agressor*) (Colloff and Halliday 1998).

Eriorhynchidae The endemic Eriorhynchidae are restricted to south-east Australia, and Tasmania, and includes five species in the genus *Eriorhynchus* (e.g. *Eriorhynchus walteri*, *E. womersleyi*) (Qin and Halliday 1997).

Eutegaeidae These are relatively large mites. In the Gondwana rainforests region the eutegaeids include the genera *Atalotegaeus*, *Neoeutegaeus* and *Neseutegaeus*. *Neseutegaeus monteithi* and *N. phyllophorus* are known only from cool temperate rainforests in northern New South Wales (Colloff and Halliday 1998).

Galumnidae These are large-bodied mites. Two genera and seven species, *Allogalumna* (three species) and *Galumna* (four species), occur within the Gondwana rainforests region. All of these are known only from the McPherson–Border Ranges area (Colloff and Halliday 1998).

Hammeriellidae Hammeriellidae include the endemic genus *Labiogena*, which is confined to eastern Australia. *Labiogena walteri* and *Novazelandiella kellyi* are rainforest species confined to the Gondwana rainforests region (Hunt 1996c; Colloff and Halliday 1998).

Malaconothridae A single undescribed *Zeanothrus* is known from rainforest in south-east Queensland (Colloff and Halliday 1998).

Macrochelidae At least seven species of *Macrocheles* are recorded from the Gondwana rainforests region (Williams 2002). *Macrocheles angustus* and *M. fungicolus* are known only from Lamington National Park (Halliday 2000). *Macrocheles spatei* is a phoretic associate of the endemic dung beetle *Aulacopris maximus* collected in a bat cave at Coorabakh National Park, northern New South Wales (Halliday 2000; Williams 2003b; G. Williams unpubl. data). *Macrocheles tesellatus* and *M. tenuirostris* are phoretic associates of the endemic dung beetle *Cephalodesmius armiger* (Halliday 2000; Monteith and Storey 1981; G. Williams unpubl. data).

Mesoplophoridae An undescribed *Apoplophora* species is known from Lamington National Park, south-east Queensland, and Whian Whian State Forest, north-east New South Wales (Colloff and Halliday 1998). The genus is otherwise known in Australia by *Apoplophora pantotrema*, which is widely distributed from India to Australia and Papua New Guinea (Colloff and Halliday 1998).

Nothridae The nothrid genus *Novonothrus* is recorded from south-east Queensland, Tasmania and New Zealand (Colloff and Halliday 1998).

Oppiidae The family Oppiidae is very speciose in the rainforests and wet sclerophyll forests of north-east New South Wales, with Tooloom Scrub near Urbenville being the type locality for many species (Balogh 1982; G. Hunt pers. comm.). Though among the most numerous of the mite fauna, oppiids are also quite small and difficult to identify (Hunt 1994). Genera recorded from the Gondwana rainforests region include *Arcoppia*, *Brachioppiella*, *Ctenoppia*, *Basiloppia*, *Lanceoppia* and *Setoppia* (Colloff and Halliday 1998; Williams 2002). Numerous species, in particular those of *Lanceoppia*, have been described from Mt Glorious (*Lanceoppia microtricha*, *L. lancearia*, *L. microtrichoides*) (Colloff and Halliday 1998). The genus *Brachioppiella* is confined to south-east Australia and includes *B. biseriata*, also from Mt Glorious.

Otocepheidae Members of this family are relatively large bodied and common in moist forests. The genus *Pseudotocepeus* (subfamily is Gibbicepheinae) is distributed in north-east New South Wales, Queensland and South Australia (Balogh 1982; Balogh and Balogh 1983; Colloff and Halliday 1998). Northern New South Wales and southern Queensland are major centres of speciation for the genus. Species inhabit cool temperate rainforest (*Pseudotocepheus bacilliger*, *P. monteithi*), subtropical rainforest (*Pseudotocepheus coarctatus*), and wet sclerophyll forest with a rainforest understorey (*Pseudotocepheus szentivanyorum*, *P. vicarius*) (Williams 2002). Also occurring in the Gondwana rainforests region is the endemic genus *Lophotocepheus*, represented by a single species (*Lophotocepheus simplex*) (Colloff and Halliday 1998).

Pedrocortesellidae This family includes the genera *Pedrocortessella* and *Hexachaetoniella*. *Hexachaetoniella* is confined to eastern Australia, New Zealand and Norfolk Island (Hunt 1996b). *Hexachaetoniella bunya* is known only from the Bunya Mountains National Park (Hunt 1996b, Colloff and Halliday 1998). *Hexachaetoniella dispersa* is more widely distributed (New England National Park to south-east Queensland), inhabiting rainforest, wet sclerophyll forest and dry sclerophyll forest. The speciose endemic genus *Pedrocortesella* has an essentially Gondwanan distribution, and is widely distributed in Australia where it has undergone major radiations, especially in drier habitats (Hunt 1994). Several species are restricted to individual World Heritage Area reserves (e.g. *Pedrocortesella bithongalba* found in *Nothofagus* rainforest in Lamington National Park, *P. conundrum* in subtropical rainforest at Mt Allyn and Taree) (Colloff and Halliday 1998; Hunt 1996a; Williams 2002).

Pheroliodidae This is principally a Gondwanan family with a known distribution comprising South America, New Zealand, South Africa and Australia, and outlier taxa in Florida and the Mediterranean (Hunt 1996d).

The Australian fauna consists of 11 species, nine of which are in the genus *Pherolioides*. *Pherolioides* also occurs in South America, Galapagos Islands and Florida (one species) (Hunt 1996d). Several species exhibit localised distributions. *Pherolioides barringtonensis*, for example, is restricted to the Barrington Tops, and adjacent Chichester State Forest. An additional genus, *Octoliodes*, known from the region, is restricted to Australia and New Zealand. *Octoliodes robustus* inhabits cool temperate and subtropical rainforests and is distributed from northern New South Wales to Tasmania (Hunt 1996a; Williams 2002).

Phytoseiidae The phytoseiine genus *Phytoseius* is cosmopolitan but endemic species are to be found in southern Queensland and northern New South Wales; examples are *Phytoseius oreillyi* (known only from Lamington National Park in subtropical rainforest), and *P. leaki* (from Mt Glorious, Mt Warning and Border Ranges national parks in subtropical rainforest and wet sclerophyll forest) (Walter and Beard 1997; Williams 2002).

Platyameridae Platyameridae are an endemic family confined to the Gondwana rainforests region (Colloff and Halliday 1998). *Platyamerus peculiaris* is the only member of the family and is known only from subtropical rainforest in the Nightcap Range (Balogh and Balogh 1983; G. Hunt pers. comm.).

Scheloribatidae This is a major group within the Australian mite fauna. The possibly endemic genus *Megascheloribates* is recorded from South Australia (two species), and Lamington National Park and the Nightcap Range (one species) (Colloff and Halliday 1998).

Steganacaridae The eastern mainland and Tasmania includes numerous species of the genus *Austrophthiracarus*

(Atropacarinae). Several species with localised distributions are known only from the Gondwana rainforests region (e.g. *Austrophthiracarus michaeli* and *A. nicoleti* – Joalah National Park, *A. multisetosus* and *A. wallworki* – New England National Park). *Notophthiracarus* also includes species with localised distributions (e.g. *Notophthiracarus ramsai* – Joalah National Park) (Colloff and Halliday 1998).

Stigmaeidae This family includes *Neilstigmaeus lamingtoni*, known from Lamington and Mt Warning national parks, and the Allyn River further south, near the Barrington Tops (Gerson and Smith Meyer 1995). *Neilstigmaeus lamingtoni* has been recorded on the leaves of the vines *Parsonsia fulva* (Apocynaceae) and *Cissus antarctica* (Vitaceae), and the trees *Sloanea woollsii* (Elaeocarpaceae) and *Commersonia fraseri* (Sterculiaceae/Malvaceae).

Trombiculidae The Trombiculidae comprise many parasitic species associated with native marsupial and rodent hosts (Domrow 1978). These include *Ascochoengasta rattus*, *Guntheria kallipygos* and *Trombicula quadriensis* (on the possum *Trichosurus caninus*), *G. parameles* and *T. mackayensis* (on the bandicoot *Parameles nasuta*), *G. pannosa* and *G. wongabelensis* (on *Rattus fuscipes*) and *Neotrombicula antechinus* (on *Melomys cervinipes*) (Williams 2002).

Miscellaneous families The subtropical rainforest mite fauna includes many additional families. These have diverse zoogeographical affinities, and microhabitat and host associations. Additional mite families and representative genera include Ameronthribidae (*Capillibates*), Ascouracaridae (*Gallichus*), Astegistidae (*Sulcoribula*), Archeonthridae (*Loftacarus*), Ceratozetidae (*Sagittazetes*), Compactozetidae (*Compactozetes*), Crassoribatulidae (*Crassoribatula*), Parakalummatidae (*Neoribates*), Peloppiidae (*Pseudoceratoppia*), Sellnickiidae (*Sellnickia*), Suctobelbidae (*Suctobelbella*), Tegeocranellidae (*Tegeocranellus*) and Tegoribatidae (*Neophysobales*, *Paraphysobates*) (see Williams 2002), as well as Podapolipidae (*Eutarsopolipus*) (Constantine and Seeman 2014). Many of these are known only from single, or few localities. Their feeding strategies and microhabits are diverse; species of Ameronthribidae include soil mites, Parakalummatidae feed on plant tissues, Astegistidae and Tegeocranellidae are associated with the leaves of rainforest trees (e.g. *Pseudoweinmannia lachnocarpa*, *Pennantia cunninghami*), and various Tegoribatidae inhabit tree trunks. Ascouracaridae include the species *Gallichus jonesi*, which inhabits the quills of the mound-building Australian brush turkey (*Alectura lathami*) and feeds on the spongy medulla inside the quill (Proctor 1999).

Mites and rainforest: an invisible super-abundance

Mites, variously Acarina, Acari and Acarida, numbered some 45 000 described species by the end of the 20th century (Walter and Behan-Pelletier 1999), but are predicted to be a megadiverse fauna that may number 500 000–1 million species. They represent some of the smallest of terrestrial invertebrate size classes in subtropical rainforest, and within the rainforest environment are found in all strata and niches. Mites occupy the stems, leaves and flowers of trees, shrubs, vines and herbs; are found in ephemeral water bodies in tree holes, and in excavated insect galleries; and hide among the fern, moss and lichen clusters that festoon tree trunks, rocks, stream banks and the forest floor (Walter and Proctor 2013). Mites in vast numbers and diversity also inhabit soil, leaf litter and the bodies (in parasitic and phoretic relationships) of the living and the dead (Perotti and Braig 2009). But, ubiquitous and abundant as they are, mites are rarely seen, and except for the occasional sighting of a brightly coloured velvet mite, or the 'scrub itch' skin irritation of chiggers, they are usually ignored.

Mites are free-living arthropods, that through their very minute size (80 μ–16 mm), yet great ecological diversity, are able to take advantage of the myriad microhabitats (e.g. plants with tomentose leaves with consequent higher relative humidity and accumulated biomatter) and niches that rainforests (especially the more complex ones) provide. The greatest abundance and diversity of mites has historically been reported from soil, ground litter and suspended detritus (Battigelli *et al.* 1994; Hunt 1994; Peterson and Luxton 1982). Many tens of thousands, and a multitude of coexisting species, may be found per square metre of the upper 10 cm of the soil profile. Orabatids are usually considered characteristic of the soil mite fauna; however, a rich diversity (distinct from that of the soil fauna) is also found in the canopy (Hunt 1994; Walter and Behan-Pelletier 1999). Mesostigmata and Prostigmata are also inhabitants of the forest canopy, and these include obligate plant parasites, fungivores and predators. Both these groups include a large number of small and minute species that are able to disperse in air currents, and that live in galls or in domatia ('little houses') on the undersurface of leaves (as in numerous Meliaceae, Sapindaceae and Anacardiaceae [Harden *et al.* 2006]) (Walter and O'Dowd 1992). Domatia are the distinctive cavities, hair tufts or swellings found in the angles between the main and lateral raised leaf veins. The mites that are plant parasites tend to be host specific, and the diverse mite fauna recovered from humus suspended in the canopy is generally distinct from the floor fauna (Walter *et al.* 1998). Arboreal oribatids include detritous feeders, but many species feed on lichen, algae and fungi. The texture and microclimate of tree bark can influence the overall nature of the associated mite assemblage (Nicolai 1986). Mites that utilise epiphytes may live in isolated, 'island-like', populations, and in floodplain rainforest (e.g. Stotts Island, Coocumbac Island, and Lansdowne nature reserves) ground faunas are subject to repeated flood events that can periodically inundate the forest floor in its entirety and sweeping away litter cover in the process. Mites (e.g. Mesostigmata spp.) are also associates of bracket fungi (Walter and Behan-Pelletier 1999), and true water inhabitants (e.g. *Arrenurus kitchingi, Lamingtonacarus* sp.) may be found in tree holes (Smith and Harvey 1989; Fashing *et al.* 2000). There is also a distinct mite fauna that specialises on plant stems and branches, as well as others that graze the surfaces of leaves (the phylloplane) (Walter and O'Dowd 1995; Walter *et al.* 1994), some of which may cross to fruit. Phylloplane associates gain protection, fluids and nutrition directly from leaves (Lindquist 1998), but one phoretic group infests flowers, riding upon the backs of pollinators, and on alighting consumes nectar and pollen from their floral hosts (Guerra *et al.* 2010; Williams and Adam 2010). Mites participate in phoretic relationships with ground-frequenting invertebrates (e.g. beetles, earwigs) (Halliday 2000; Tagami and Halliday 2013) and are commonly phoretic on rainforest vertebrates (Proctor 1999).

Phoretic mites, especially Mesostigmata, are also likely to be the most abundant arthropods found on carcasses and corpses (Perotti and Braig 2009) and, like phoretic flower mites, travel upon the backs of scavenger species (e.g. silphid beetles). They are known to be highly specific in the choice of host carriers, and no other. But this is an aspect of mite ecology that, in Australian subtropical rainforests, has been largely neglected.

CRUSTACEA (amphipods, isopods, syncarids, crayfish, shrimp and related groups)

PERACARIDA (amphipods, isopods)

The Pericarida include marine, freshwater, littoral and terrestrial species. The freshwater Peracarida have little changed from their marine ancestors (Williams and Bickel 2010). Their capacity to occupy and survive in freshwater environments is dependent on suitable water chemistry, temperature and flow rates, permanency of the aquatic habitat, and their ability to disperse (Adlem and Timms 2000). They are sensitive to environmental degradation (Bradbury and Williams 1999).

Amphipods are mainly cryptic detritivores and scavengers, but are also one of the major contributors to the diversity and abundance of the stygofauna; that is, the fauna inhabiting groundwater (see Appendix 1). Amphipods recorded from the Gondwana rainforests region include species of *Pseudomoera* (Eusiridae) and *Neoniphargus* (Neoniphargidae), *Austrocrangonyx barringtonensis* and *A. hynsei* (Paramelittidae), and *Arcitalitris sylvaticus* (Talitridae), all from the Barrington Tops and Nundle–Walcha area (Bradbury and Williams 1999; Adlem and Timms 2000; Williams 2002). *Neoniphargus* is a rare relict taxon. *Austrocrangonyx* is restricted to the Barrington Tops–Walcha region. *Agilesta hylaea* (Talitridae) occurs further north in Lamington National Park, Mt Tamborine, Border Ranges National Park, Mallanganee, Bruxner Park near Coffs Harbour, and Dorrigo National Park (Friend 1982, Williams 2002). The genus is endemic.

The Isopoda first made their appearance in the Carboniferous ~300 mya. They include woodlice (Oniscidea). Relatives live in marine, freshwater and terrestrial environments. Some are grazers and filter feeders, a few are predators, or feed on decaying animal and plant material.

SYNCARIDA

The shrimp-like Syncarida are omnivores that feed on detritus; however, a number are probably predators (Thurgate *et al.* 2001). There are two living orders: the Anaspidacea, which have a Gondwanan distribution, and the Bathynellacea. In Australia the Syncarida are mainly restricted to the south-east of the continent andTasmania (Ahyong and Alonso-Zarazaga 2017). The fauna includes several endemic taxa, with localised distributions and

Gondwanan affinities, which occur in cave and interstitial waters (those within spaces between sediment particles) in south-eastern Australia (Appendix 1, Tables 6, 7). 'Two congeneric species of an undescribed anaspid [Syncarida] family occur in caves and interstitial waters of the North Coast Region in New South Wales and they are considered to have close affinities with Koonungidae. … Both species possess specialised raptorial grasping appendages and are probably active predators … in contrast to most other syncarids that are omnivorous detritivores' (P. Serov, and S. Eberhard, cited in Thurgate *et al.* 2001).

DECAPODA (crayfish, shrimp, freshwater prawns)
Freshwater shrimp are a prominent component of the invertebrate fauna found in the streams in many rainforests of lowland and upland subtropical Australia. These include the glass shrimp of the family Atyidae (e.g. *Atya striolata*, *Paratya australiensis*) (Plate 33), an ancient group of mainly small-sized crustaceans widely distributed in Australia. The atyid genus *Australatya* is endemic to eastern Australia (Page *et al.* 2007).

Also found in these water bodies are freshwater prawns, also called long-armed shrimp, of the genus *Macrobrachium* (Palaemonidae) (Riek 1951) (Plate 34). *Macrobrachium* is mainly restricted to northern and eastern Australia; however, several species reach southern latitudes (e.g. *Macrobrachium atactum atactum* – Conondale, Queensland and New South Wales border, *M. australiense australiense* – Rockhampton, Gayndah, Brisbane) (Riek 1951). Though dependent on permanent sources of fresh water, individuals can be found in dissecting subtropical rainforest streams that are subject to extreme seasonal fluctuations in flow rates, standing pool size, and water temperature (e.g. Lansdowne Escarpment). At this location species of Atyidae, *Macrobrachium*, and Parastacidae crustaceans inhabit the same stream pools (G. Williams pers. observ.).

Parastacidae This is the only family of Australian freshwater crayfish, and is restricted to the Southern Hemisphere. All species are endemic, and the fauna is of particular conservation and evolutionary interest (McCormack 2012; Riek 1969; Shull *et al.* 2005). Jell (2004) lists a parastacid (juvenile) compression fossil (*Austrolestidion duaringae*) from Tertiary sediments of central Queensland, noting that it was first erroneously described as a zygopterid dragonfly larva.

Individual species and populations are potentially threatened by localised harvesting, land management practices and aquaculture (Kohen and Merrick 1998; McCormack 2012; Merrick 1991, 1995, 1997); however, such impacts are difficult to monitor (Merrick 1995). Natural aquatic enemies include eels and platypus, but crayfish venturing onto land are vulnerable to predation from various native and exotic vertebrates. Cane toads, and translocated and invasive *Cherax destructor* crayfish, may also pose threats. Large adult *Euastacus* will often strike a formidable defence pose, forward-facing, claws open and outstretched, when threatened. This pose is maintained even during their backwards retreat to shelter. Defence posturing is often accompanied by an audible snake-like hissing sound. This is produced by abdominal stridulatory organs in which projecting bristles and ridges on the surface of abdominal segments rub against each other (e.g. see Sandeman and Wilkens 1982).

The Australian freshwater crayfish fauna comprises more than 130 known species, currently placed in 10 genera (Merrick 1991, 1995; Crandall *et al.* 1999; McCormack 2012). Most of the known world parastacid genera are endemic to Australia. The remaining genera occur in New Zealand (one species), southern Chile and Brazil (three species), and Madagascar (one species) (Crandall *et al.* 1999).

Tenuibranchiurus has a restricted range in southern Queensland (Riek 1969; McCormack 2012). It comprises a single very small species, *Tenuibranchiurus glypticus* (Bells Creek, Caloundra). The genus *Euastacoides* comprises several cold-water, stream-dependent species now placed in *Euastacus*: *E. maidae* (Currumbin Creek), *E. setosus* (Mt Glorious) and *E. urospinosus* (Maleny). Monteith (1986b) recorded an unidentified species of *Euastacoides* from rainforest in the Kroombit Tops; it is now *Euastacus monteithorum*.

The endemic spiny crayfish genus *Euastacus* (Plate 35) is similar in many respects to the Madagascan genus *Astacoides* and has undergone marked diversification in New South Wales and Queensland. Northern New South Wales and southern Queensland encompass numerous endemic species with exceptionally localised distributions and are a major focus of speciation within the genus (Borsboom 1998; E. Clark 1936; Coughran 2002, 2005, 2008; Coughran and Leckie 2007; Coughran and McCormack 2011; Lawler and Crandall 1998; McCormack 2012; Morgan 1997; Williams 2002). *Euastacus* species are scavengers, grazers, opportunistic hunters, and cannabalistic predators (of juveniles), but live in association with infesting temnocephalan flatworms

(Sewell *et al.* 2006) and other invertebrate ectosymbionts. They inhabit flowing, highly oxygenated, cool streams, particularly in montane eastern regions of the Great Dividing Range. Warm eutrophic (nutrient enriched, or oxygen depleted) lowland or plateau streams are unsuitable habitat (Morgan 1997). All construct burrows either directly in stream banks, or deeper in the stream profile between large stones. Crayfish must periodically shed their exoskeletons. Individuals leave their burrows when ready to moult, after which their soft articulated body leaves them especially vulnerable to attack or damage until hardened. Juveniles may occupy stream microhabitats such as banks or minor drainage lines that minimise the risk of predation by animals such as eels and herons, and by predatory adult crayfish. Individuals are nocturnal, though some species are active during the day, and are capable of limited overland movement. However, steep ridges and intervening unsuitable habitat restrict dispersal to such an extent that populations may be regarded as semi-isolated, although specific populations may be distinctive at the end of their ranges (Morgan 1997). The effectiveness of biophysical barriers to movement, however, is difficult to evaluate (Morgan 1997).

There are more than 50 species of *Euastacus*. These range from northern Queensland to southern-most Victoria. More than 30 of the currently recognised species are recorded from New South Wales (McCormack 2012). Distributions closely correspond to river drainages (Lawler and Crandall 1998), and those of individual species may be highly localised within them, for example, *Euastacus girurmulayn* (Nightcap and Whian Whian national parks at elevations >400 m), *E. guruhgi* (Mt Warning and Wollumbin national parks), *E. jagabar* (Border Ranges), and *E. clarkae* (Werrikimbe National Park). Some species, however, do overlap in geographical range (e.g. *Euastacus spinifer* with *E. reductus*), and although numerous species are restricted to upland streams (e.g. *Euastacus hystricosus*, *E. jagara*, *E. maccai*) several are distrubuted widely from coastal lowlands to higher elevations (e.g. *Euastacus dangadi*, 50–550 m.a.s.l., *E. neohirsutus*, 50–>1200 m.a.s.l.) (McCormack 2012). The high vulnerability of species, owing to greatly restricted habit, is exemplified by *Euastacus dalagarbe*, only known from a small sector of the Border Ranges National Park (Coughran 2005). This species occurs in small streams that feed into Brindle Creek, but not in Brindle Creek itself. Its habitat, though seemingly aquatic, consists of drainage lines in small moist gullies that are devoid of surface water for a considerable part of the year.

The related genus *Cherax* is restricted to Australia and New Guinea. *Cherax* reaches its greatest diversity in south-east Queensland; however, the McPherson Range is a biogeographical barrier to coastal *Cherax*, separating the northern New South Wales and southern Queensland fauna from one another (Munasinghe *et al.* 2004; Williams and Bickel 2010). *Cherax* and *Euastacus* generally tend not to co-occur in water bodies, the two genera having differing tolerances to water temperature and water flow. *Cherax* species in subtropical coastal Queensland and New South Wales include *Cherax cairnsensis*, *C. robustus*, *C. dispar*, *C. depressus*, *C. punctatus*, *C. cuspidatus*, and *C.rotundus and C. setosus* (McCormack 2012; Riek 1969).

UNIRAMIA (peripatus, centipedes, millipedes, springtails, insects and related groups)

ONYCHOPHORA (peripatus, velvet worms)

The Onychophora are soft-bodied, predatory, legged worms of ancient lineage that are now wholly terrestrial, living in the leaf litter of the forest floor (De Sena Oliveira and Mayer 2012; Geoffroy and Ruhberg 2006). They may represent the 'sister group' to all the other arthropods; these being all the invertebrates with jointed legs and external skeletons. Onychophorans are characterised by the possession of numerous pairs of stump-like legs, conspicuous long antennae and a ventral mouth with sharp armed jaws. The subphylum has a pantropical distribution, with ~100 known species, and is found on most of the Gondwanan landmasses. Fossils from the Cambrian have been allied with the Onychophora, and fossil taxa related to modern forms have been found in amber dating from the Mid Cretaceous (Grimaldi and Engel 2005; see Appendix 5). Caterpillar-like velvet worms, also commonly called peripatus, hunt and capture prey by expelling a proteinaceous fluid from oral papillae. This sticky fluid snares their victims which include snails and tiny arthropods.

The most archaic living onychophoran taxa occur in south-eastern Australia and Tasmania, and these regions preserve a close relationship with taxa in New Zealand, South America and South Africa (A. Reid 1996). However, of the 30 onychophoran genera known from Australia only *Ooperipatellus* is shared with New Zealand.

The remaining 29 are endemic. The Australian species are Gondwanan relics with the subphylum being more widespread before the breakup of Pangaea.

There has been a marked radiation of species in eastern Australia. Within Australia, Onychophora are confined to the wetter coastal zone extending from northern Queensland to Tasmania with relictual populations in South Australia and south-western Western Australia. Confinement to mesic refugia, especially during Quaternary (1 million–<10 000 years ago) climatic fluctuations, may have led to isolation of populations, and may explain the presence of isolated relic and endemic mountain faunas along the Great Dividing Range (A. Reid 1996).

Onychophora are vulnerable to desiccation and consequently are restricted to moist and humid microhabitats such as rotting logs and leaf litter (A. Reid 1996). This sensitivity to desiccation and confinement to moist habitats, in conjunction with apparently small and spatially localised populations, suggests that they are potentially threatened by a range of land use and environmental impacts (e.g. fire, underscubbing of vegetation, drought conditions).

Peripatopsidae The Gondwana rainforests, and additional forests in the vicinity, possess a rich peripatopsid fauna (eight genera and 10 species being cited in A. Reid 1996). The fauna from the Border Ranges area is particularly diverse. Factors which may contribute to this regional diversity are suitable moist habitats, a variety of vegetation types, diversity of microclimate, and diverse topography and altitudinal range.

Peripatopsidae known from the south-eastern Queensland and north-eastern New South Wales include *Aethrikos setosa* (endemic genus from New England National Park, Styx River), *Centorumis trigona* and *Wambala makrothele* (endemic genera restricted to cool temperate rainforest in the Barrington Tops and adjacent areas), *Regimitra quadricaula* (an endemic genus occurring in open forest and cool temperate rainforest in the Barrington Tops–Chichester State Forest area), and *Nodocapitus inornatus* (an endemic genus, and species known only from subtropical rainforest in the Gibraltar Range National Park). Five species in endemic genera (*Dactylothele habros*, *Hylonomoipos akares*, *H. brookensis*, *Nodocapitus barryi*, *Sphenoparme hobwensis*) are restricted to the upland mesic forests of the New South Wales–Queensland border area (A. Reid 1996). The genus *Planipapillus* is diversely represented in south-eastern Australia, with species (e.g. *Planipapillus taylori*) penetrating north along the coastal ranges (A. Reid 2000).

CHILOPODA (centipedes)

Centipedes are actively mobile hunters that during the day rest below fallen logs and rocks. In subtropical rainforests they are frequently encountered sheltering in heavily decomposed wood on the forest floor.

Genera known from the Gondwana rainforests region are *Ethmostigmus* and *Rhysida* (Otostigmidae), *Paralamyctes* (Henicopidae) and *Cormocephalus* (Scolopendridae) (Chamberlin 1920; Edgecombe 2001). *Paralamyctes* has a Gondwanan distribution and is known from eastern Australia (including Tasmania), southern Africa, Madagascar, India, New Zealand and Chile. *Paralamyctes hornerae*, *P. neverneverensis* and *P. cassisi* inhabit wet forest types in northern New South Wales, *P. grayi* is distributed from the Illawarra region north to the Barrington Tops, and *P. monteithi* has a disjunct distribution extending from the Queensland–New South Wales border to the Wet Tropics (Edgecombe 2001). *Ethmostigmus rubripes* and *Rhysida suvana* are restricted to the Mt Tamborine area. The genus *Cormocephalus* is widely distributed, occurring in South Africa, Madagascar, Sri Lanka, New Guinea and New Zealand (Koch 1983), and includes at least six species from northern New South Wales and south-east Queensland. *Cormocephalus westwoodi* (Plate 36) is widespread in Australia (North Queensland to Tasmania), and in the Gondwana rainforests region is encountered commonly in subtropical and warm temperate rainforests in both upland and lowland sites. It is a habitat generalist that in addition to rainforest and sclerophyll forest readily frequents urban areas (Williams 1993). Additional rainforest-inhabiting *Cormocephalus* species include *C. brachycerus* (Bunya Mountains–Dorrigo), *C. brevispinatus* (Mt Tamborine), *C. monteithi* (Lamington National Park–Dorrigo), *C. spinosior* (Border Ranges) and *C. strigosus* (Dorrigo, Bruxner Park) (Chamberlin 1920; Koch 1983; Williams 2002). Adults exhibit brood care, and remain as companions to clusters of their young.

DIPLOPODA (millipedes)

Millipedes are among the earliest arthropods, and are relatively slow-moving herbivores and scavengers found day and night on the forest floor and sometimes on the trunks of standing trees. They shelter in leaf litter, under loose bark and stones, and in cavities in rotten wood. Millipedes possess chemical defence glands that secrete a range of irritants and repellents (reviewed in Shear 2015), and these may also have an anti-fungal or anti-microbial function (Stanković *et al.* 2016). These glands appear to be present in species recorded in Early Cretaceous fossils. The order Polydesmida, which makes up the largest group of millipedes world-wide (>2700 known species), produce hydrogen cyanide gas that can be fatal to other arthropods, and even potentially to small vertebrates, in a confined environment (Shear 2015).

The Australian millipede fauna is rich but largely unknown (Black 1997). Nine of the 15 orders, and 20 of the ~100 families of the world fauna, occur here. The greatest number of families occurs in eastern and south-eastern Australia, but only the Peterjohnsiidae, which occur in Queensland, New South Wales, Victoria and Tasmania, are endemic (Black 1997). Some species, such as the *Attemsobolus dorsovittatus* (Spirobolellidae) and *Epicyliosoma excavatum*, both recorded from the Upper Richmond River, appear to have very localised distributions (Jeekel 1986). The distinctively blue-green-coloured *Walesbolus lobatus* (Spirobolellidae) (Plate 38) is an indicator species of littoral and dry rainforest (Williams 1993), found in and under decaying wood on the forest floor. In littoral rainforest it is not uncommon to find several individuals clustered motionless together within the sandy detritus below logs.

Dalodesmidae This family comprises the endemic, monotypic genus *Orthorhachis* (*Orthorhachis pallida*), which is known only from the Bunya Mountains in dry rainforest (Jeekel 1985).

Paradoxosomatidae Paradoxosomatid millipedes have an almost world-wide distribution and are the most widespread and speciose family in Australia (Car 2016). The subfamily Australiosomatinae is the dominant paradoxosomatid group in the Australian region (Jeekel 1968) and their entry into this region may date from the Mesozoic. Lower sea levels during the Pleistocene apparently did not lead to an exchange of New Guinean and Australian faunas. Isolation of the New Guinean and Australian faunas may date from the end of the Mesozoic or the beginning of the Tertiary (Jeekel 1968). There is no close relationship between the Australian and South American Paradoxosomatidae (Jeekel 1968). Most Neotropical paradoxosomatid genera are referable to the tribe Catharosomatini. Adults, as with the majority of millipedes encountered in subtropical rainforest, are largely active at night, being encountered on the forests floor and often on tree trunks and plant stems. *Cladethosoma* is confined to the east coast with the eight known species likely to be small-range endemics. Several species exhibit localised distribution patterns in south-eastern Queensland, including *Cladethosoma monticola* from the Main Range and *C. musgravei* associated with the McPherson Range (Mesibov 2018).

The Australiosomatinae contains three tribes; the Antichiropodini (Australia, inclusive of Tasmania, Lord Howe Island and New Caledonia), Australiosomatini (naturally confined to eastern mainland Australia, one species [*Akamptogonus novarae*] introduced to New Zealand and Western Australia), and the Aschistodesmini (Halmahera to Solomon Islands) (Car 2016; Jeekel 1968). Australiosomatini include the endemic genus *Phyllocladosoma*, which consists of four species (e.g. *Phyllocladosoma dorrigense*, *P. annulatipes*) restricted to south-eastern Queensland and northern New South Wales (Jeekel 1987). *Dicladosomella* (Australiosomatini) includes *D. vexillifera* and *D. fagina* (in *Nothofagus* forest) from the Barrington Tops, and *D. mesibovi* from Morton National Park (Car 2016). Most known *Dicladosomella* species are restricted to south-eastern New South Wales.

Sphaerotheriida. The order Sphaerotheriida, or giant pill-millipedes, exhibits a disjunctive range comprising South Africa, Madagascar, the entire Oriental region, eastern Australia and New Zealand (Jeekel 1974). Species are herbivorous.

The family Sphaerotheriidae (Plate 37) occurs in South Africa, Madagascar, India, Sri Lanka, Australia and New Zealand. The Sphaeropoeidae are distributed throughout the Oriental region from Assam to Java, Sulawesi (Celebes) and Halmahera, and the Philippines. The two families are separated by the valley of the Ganges River and the eastern part of the Indo-Australasian archipelago. The tribe Cyliosomatini is restricted to Australia and New Zealand. There is a close relationship between the sphaerotheriids of Australia and New Zealand (Jeekel 1974). The present Sphaeropoeidae distribution existed in the Upper Jurassic (~<160–135 mya), and the separation of the Sphaerotheriidae and Sphaeropoeidae probably occurred in the late Palaeozoic (<600–225 mya) or early Mesozoic (Jeekel 1974).

The endemic sphaerotheriid genus *Epicyliosoma* (e.g. *Epicyliosoma excavatum*, *E. froggatti*, *E. queenslandicum*) is restricted to Queensland and New South Wales (Jeekel 1986; Williams 2002).

COLLEMBOLA (springtails)

Collembola are soft bodied invertebrates closely related to, but distinct from, 'true insects'. Fossil Collembola date from the Early Devonian (Greenslade and Whalley 1986; Grimaldi and Engel 2005), and are considered allied to Protura, Diplura and the centipede-like Symphyla, though exact relationships are uncertain (Greenslade 1991; Hopkin 1997; Kristensen 1991; Luan *et al.* 2005). Along with the Protura and Diplura, collembolans form the Entognatha; this group is characterised by mouthparts that are enclosed in folds that extend from the front of the head, and distinguished from 'true insects' (the Ectognatha) in which the mouthparts are exposed. Entognatha and Ectognatha together constitute the superclass Hexapoda (Kristensen 1991; Kukalová-Peck 1991; New 2011)

Collembolans are mainly found in the upper soil layer and in detritus on the forest floor; however, some are encountered in flowers and under tree bark. There is also a rich and abundant Collembola fauna in the rainforest canopy (Kitching *et al.* 1993). Most collembolans feed on microorganisms. Some feed on pollen. Numerous species inhabit subtropical rainforests but humidity is an important determinant of their distribution and activity patterns. In subtropical rainforest Collembola have been found to use tree trunks as pathways, thus establishing ecological connections between ground and canopy habitats (Rodgers and Kitching 1998).

Entomobryidae *Lepidosira australica tamburinensis* is known only from the type locality (Mt Tamborine, Queensland). *Lepidosira* is recorded from Australia, New Caledonia, New Zealand, Sri Lanka and Rwanda (in W. Houston 1994). *Lepidocyrtoides hopkini* is known only from wet sclerophyll forest associated with the Lansdowne Escarpment (i.e. Lorien Wildlife Refuge) (Cipola *et al.* 2017). Entomobryidae have been recorded from the flowers of trees in New South Wales lowland subtropical rainforest stands: ?*Lepidosira* (on *Alphitonia excelsa*), *Lepidosira*, *Lepidocyrtoides* (on *Cuttsia viburnea*), and *Lepidosira* (on *Tristaniopsis laurina*) (Williams 1995). *Sinella* is an associate of the brood nests of the dung beetle *Cephalodesmius armiger* (Monteith and Storey 1981).

Neanuridae The subfamily Neanurinae includes the endemic genus *Australonura*, two species (*Australonura quarta*, *A. scoparia*) being known only from their type locality in the northern extreme of the Gondwana rainforests region (in W. Houston 1994). Uchidanurinae are an ancient subfamily (Greenslade 1991), are of great conservation importance, being restricted to undisturbed old growth forests and heaths, and are associated with old rotting logs and other timber. Logging, burning and other disturbances seem to eliminate them (P. Greenslade pers. comm.). Undescribed neanurid taxa are known from Lamington, New England and Barrington Tops national parks, and from Dorrigo (P. Greenslade pers. comm.).

Paronellidae Paronellidae include *Paronellides mjobergi* described from Lamington National Park. *Paronellides* is a Gondwanan taxon known from South America, New Zealand and Australia (in W. Houston 1994).

PROTURA

Proturans are minute insect-like invertebrates that inhabit leaf litter, soil and moss beds (Tuxen 1967). They are almost unpigmented, and though distributed across all zoogeographical regions proturans are rarely seen. Their biology is poorly known, but they may feed on mycorrhizal fungi (Imadaté 1991). They are not known to be arboreal, and are absent from the fossil record (Grimaldi and Engel 2005). Although possessing six legs (hexapodous), they are functionally tetrapods for the front legs are held forward, and are not used in locomotion. Rather, they are held aloft from the ground and serve as sensory appendages (Grimaldi and Engel 2005). Unlike other Entognatha (Diplura, Collembola) proturans do not moult after attaining maturity.

The Australian fauna is placed in the families Eosentomidae, Protentomidae and Acerentomidae, but appears distinct from that of other regions, especially that of South-east Asia (Tuxen 1967). Species recorded from southern New South Wales include *Eosentomon australasicum* (Eosentomidae), *Australentulus tillyardi* and *Berberentulus capensis* (Acerentomidae). *Eosentomon australasicum* has been recorded from warm temperate rainforest, *Australentulus tillyardi* inhabits wet and dry sclerophyll forest and *Berberentulus capensis* has been collected from tree fern litter (of *Cyathea* sp.) in wet sclerophyll forest (Tuxen 1967).

DIPLURA (two-pronged bristletails)

These are narrow-bodied hexapods that lack eyes. Some taxa (e.g. robust Heterojapyginae) resemble earwigs (Dermaptera). Diplura are presumed to have originated in the Early Devonian, but their fossil record is poor (Grimaldi and Engel 2005). Diplura include herbivorous and predatory species that in subtropical rainforest are occasionally found under fallen logs and stones, but are particularly found in leaf litter (and are sensitive to its removal). Diplurans also respond to the prevalence of moist microhabits. Houston and Melzer (2012) found that moisture-requiring and shelter-requiring invertebrates, Diplura among them (but also earwigs, pseudoscorpions, gastropod snails, isopods and Symphyla), characterised the composition of invertebrate communities in Queensland dry rainforest stands, in comparison to the fauna of adjacent eucalypt woodland.

In Australia the order is represented by the families Campodeidae, Heterojapygidae, Japygidae, Parajapygidae and Projapygidae (Condé and Pagés 1991). Heterojapyginae (Heterojapygidae) include *Heterojapyx tambourinensis* from Mt Tamborine in south-eastern Queensland. *Heterojapyx* occurs in Australia, New Zealand, Madagascar and the Indian subcontinent (Pamir, Tibet) (W. Houston 1994), and has an apparent Gondwanan distribution. Some *Heterojapyx* are fierce ambush hunters that sit motionless in the soil, ready to capture soft arthropod prey with the strong forcep-like structures (forcipate cerci) at the apex of the abdomen. *Heterojapyx* has also been known to exhibit cannabalism (Condé and Pagés 1991).

INSECTA ('true' insects)

There are lots of insects. Excluding the wingless insect relatives Collembola, Diplura and Protura (the Hexapoda/Entognatha), and the primitive wingless Archaeognatha and Thysanura (Zygentoma), Grimaldi and Engel (2005) estimate the number of extant true insects world-wide at well over 900 000 described species. Gaston (1991) estimates the actual total number of living insects is around 5 million. Others have given different figures, the most contentious possibly being that of Erwin (1982), who hypothesised that there may be as many as 30 million (Majer *et al.* 1994).

The listing of orders generally follows the classification sequence in Naumann (1991a), but families are generally listed alphabetically under order (see also Appendix 4). As with other groups the inclusion of families, and lesser taxonomic ranks within families, is not exhaustive.

ARCHAEOGNATHA (bristletails)

These are frequently placed with 'true' insects (Pterygota), and were previously grouped with silverfish (Thysanura/Zygentoma). The Archaeognatha (bristletails) are free living, commonly nocturnal, silverfish-like invertebrates that by day shelter under leaf litter, loose stones, and tree bark (Watson and Smith 1991). They are herbivores and detritivores. Their bodies are clothed in small scales that easily detach, and which may help these seemingly defenseless animals to escape the grasp of predators. Though flightless, individuals are also able to avoid threats by rapid jumping responses. Some species are active by day and can be found basking on sunlit rocks, but the Archaeognatha encountered in subtropical rainforests are characteristically nocturnal species found on the trunks of trees; the dull-coloured individuals are usually located no higher than a metre or so above the ground (G. Williams pers. observ.). Although Archaeognatha superficially resemble silverfish there is no close relationship, with the two groups evolving 'independently at least since the Carboniferous [350–270 mya] period' (G. Smith 1998).

Meinertellidae All known Australian species belong to this family (Sturm and Smith 1993) (Plate 39). The Meinertellidae are the most primitive living family and, although found in all zoogeographical regions, are essentially southern (Watson and Smith 1991; G. Smith 1998).

The Australian species are poorly known. The genus *Nesomachilis*, which occurs in the Gondwana rainforests, is restricted to the Australia–Indonesia region with species also recorded from New Caledonia, New Zealand, Papua New Guinea and Lord Howe Island (G. Smith 1998; Sturm 1980; Watson and Smith 1991).

EPHEMEROPTERA (mayflies)

Ephemeroptera are the most primitive of extant winged insects and provide a visual window on the appearance of the earliest flying insects. They stem from the Permian, but close relatives are known from the Carboniferous. Numerous extinct families are known from Permian–Jurassic compression fossils (Grimaldi and Engel 2005).

The nymphal stages (naiads) of all mayfly species are aquatic and are abundant in cool, flowing freshwater streams. Nymphs may be commonly found on the undersurface of rocks and stones, or associated with accumulations of leaves and twigs in stream bed depressions. They feed on detritus, algae on rocks, are filter feeders, or are predators. However, adults do not feed (mouthparts are vestigal or absent), live ephemeral lives, and may be seen mating in flight over, or in proximity to, water bodies, and occasionally venturing to fly in quite large sunny rainforest gaps or adjacent to rainforests. Adults may swarm but are weak fliers, and flight patterns are usually over short distances and primarily no higher than several metres above ground. Mayflies are preyed upon by freshwater fish, amphibians, birds, spiders and numerous predaceous insects such as dragonflies.

Faunal assemblages are influenced by the nature of available substrates (e.g. submerged leaf litter, stones and logs), with small-scale differences in hydraulic conditions created by combinations of velocity, depth and substrate roughness (of sand, gravel beds) having an important role in determining spatial distribution of individual species (Brooks *et al.* 2005). Heavy, episodic flow discharges (as are frequent in escarpment streams) can dramatically affect the availability of suitable microhabitat and thus the diversity of species and population abundance. Ephemeroptera are particularly sensitive to water impurities, low dissolved oxygen conditions, changes in water pH, water temperature, water velocity and depth, and seasonal variation in flow rates (Christidas 2003; Connolly *et al.* 2004; Leigh 2013). Consequently they are an important taxon within the suite of in-stream macroinvertebrates that can serve as indicators of stream health (e.g. Turak and Bickel 1994).

The faunas of Australia, New Zealand and South America are very similar (Peters and Campbell 1991), and this is particularly evident in the families Ameletopsidae, Coloburiscidae, Oniscigastridae and Siphlonuridae. The occurrence of these families in Australia is the result of Gondwanan dispersal (Peters and Campbell 1991). The order is widely distributed and is especially abundant in the highlands of south-eastern Australia.

Among the families occurring in the region the Ameletopsidae include *Mirawera megaloprepia*, the genus being widespread from North Queensland to Victoria (W. Houston 1988; Peters and Campbell 1991). Beatidae (*Pseudocloeon*) may largely be of northern origin. *Pseudocloeon* occurs in the mountain streams of coastal Queensland and New South Wales (Peters and Campbell 1991). Caenidae are northern in origin and include the endemic genus *Irpacaenis* (e.g. *Irpacaenis deani*), which comprises three species and occurs along the east coast from northern Queensland to eastern Victoria (Suter 1999). The Leptophlebiidae include *Atalophlebia maculosa* and *Atalophlebioides annulatum*, both known from New England National Park (W. Houston 1988). *Atalophlebioides* nymphs are commonly encountered in upland streams. The genus *Tasmanophlebia* (Oniscigastridae) occurs on mainland Australia from Barrington Tops to South Australia (Peters and Campbell 1991). The probable only Australian species of the Teloganodidae, *Austremerella picta* (= *Ephemerellina picta*), was described from stony streams in southern Queensland (Peters and Campbell 1991). Immature life forms, possibly assignable to this species, have also been collected from the Barrington Tops, and the Clarence, Bellinger, Macleay and Manning river basins (Chessman and Boulton 1999). *Austremerella* is placed in the subfamily Austremerellinae, which has an enigmatic distribution occurring in Australia, Vietnam, and south and south-eastern China. The Austremerellinae are possibly Gondwanan in origin (McCafferty and Wang 1997) and in Australia are restricted to south-eastern Queensland and northern New South Wales (Chessman and Boulton 1999).

ODONATA (dragonflies, damselflies)

The Odonata have a long geological history with ancestral groups (Protanisoptera, Archizygoptera, Triadophlebiomorpha, Tarsophlebioptera) arising in the Early to Late Permian, and further developing in the later Triassic and Jurassic (Grimaldi and Engel 2005; Rehn 2003).

A consensus on classification and diversity of Odonata is now available (Dijkstra *et al.* 2013). Generally this is used for systematics and nomenclature in the following text, and differs significantly from that of Theischinger and Hawking (2006). World-wide there are now ~6000 species of Odonata (dragonflies) in the three suborders Anisoptera (dragonflies proper) (= Epiproctophora), Anisozygoptera and Zygoptera (damselflies). These suborders constitute the extant Odonata (see Dijkstra *et al.* 2013, 2014). Anisoptera and Zygoptera possess ~3000 species each. Only four species are included in Anisozygotera (Dijkstra *et al.* 2014). In the subtropics adults are mainly present in spring and summer. Anisoptera are most conspicuously differentiated in the field by the habit of adults, when at rest, holding the wings in a flat plane outstretched from the body, whereas most Zygoptera hold their wings parallel to the long thin abdomen. In addition, Anisoptera are robust in build, but Zygoptera are more delicate and thinly elongate. Behaviourally, some Anisoptera are territorial, and many dragonflies and damselflies perch on particular items of vegetation from which they sally forth, and then return to.

Odonata are predators, and their association with rainforest and wet sclerophyll forest is dependent on the availability of suitable inclusive water bodies, adjacent habitat such as seepages and boggy swamps, and even artificial dams and ponds built for fire-fighting purposes. Adults hunt on the wing and are mainly diurnal. However, in subtropical rainforests there are several species that display crepuscular flight behaviour (e.g. *Austrophlebia costalis*, *Telephlebia cyclops*, *T. godeffroyi*). Larvae (nymphs, naiads) undergo progressive development (are hemimetabolous) and are stealthy ambush hunters. Larval development is determined by water temperature (Theischinger and Hawking 2006) and, though essentially aquatic, there are exceptional terrestrial species (*Antipodophlebia asthenes*) (Watson and Theischinger 1980). Larvae shelter among debris and protective stones, but survival in intermittently raging escarpment streams may be precarious owing to the sweeping away of sheltering debris and prey, and physical damage owing to the forceful movement and collision of rocks and logs during excessive stream flow.

A high proportion of stream-dwelling Odonata principally occur in the eastern and south-west regions of Australia (Theischinger and Endersby 2009; Theischinger and Hawking 2006; Theischinger and Houston 1988). Watson and O'Farrell (1991) suggest that these taxa can be interpreted as southern, possibly Gondwanan relics. Approximately 40 per cent of the fauna possesses southern (including Gondwanan) origins and most of these are centred on the montane and coastal fringe of eastern Australia (Watson and Theischinger 1984). There is also a high level of endemicity in the Australian Odonata fauna (e.g. in the Aeshnidae, Argiolestidae, Gomphidae, Isostictidae, Synthemistidae) except in the families Coenagrionidae and Libellulidae, which are relatively recent arrivals from the north (Houston and Watson 1988). The Hunter River Valley south of the Northern Tablelands of New South Wales (i.e. Barrington Tops, Mt Royal Range) presents a major disjunction of distribution for Odonata, as well as for Megaloptera and Plecoptera, and this has resulted in isolation of taxa and taxonomic divergence in the fauna (Watson and Theischinger 1984). Alarmingly, modelling studies by Bush *et al.* (2014) suggest that a high proportion of the Australian Odonata fauna (especially the endemic component) is at risk from projected climate change impacts, including the limited ability to disperse to alternative refugia.

Zygoptera (damselflies)

Argiolestidae Argiolestids are small to very large damselflies previously placed in Megapodagrionidae (see Dijkstra *et al.* 2014 for a discussion). Argiolestid distribution encompasses Australasia, Africa, Madagascar and Asia (Kalkman and Theischinger 2013). The presence of Argiolestidae in regions north of Australia is the result of at least three colonisation events emanating from Australia (Kalkman *et al.* 2018). All known Australian species are placed in a single subfamily, the Argiolestinae (Kalkman and Theischinger 2013). *Austroargiolestes* (e.g. *Austroargiolestes amabilis*, *A. brookhousei*) and *Griseargiolestes* (e.g. *Griseargiolestes bucki*, *G. fontanus*) are recorded from the Gondwana rainforests region (Theischinger 1998c; Theischinger and O'Farrell 1986). Both genera are endemic for Australia, and their respective distributions are largely focused on the south-eastern region of Australia (Kalkman and Theischinger 2013). *Griseargiolestes bucki* frequents *Nothofagus*-dominated cool temperate rainforest (Theischinger 1998c). The subtropical rainforest *Austroargiolestes* fauna includes the spectacular black and orange-bodied *A. chrysoides* and *A. amabilis* (Plate 41). Females of the former lay their eggs in moss (Theischinger and Hawking 2006).

Coenagrionidae This family represents the greatest damselfly radiation (Dijkstra *et al.* 2014). Species vary greatly in size and include some of the most colourful of species (e.g. *Pseudagrion aureofrons*) to be encountered on the banks and vegetation of rainforest dissecting freshwater streams and rivers. The most widespread genera include *Pseudagrion*, *Aciagrion*, *Austroagrion*, *Ischnura* and *Agriocnemis* (Theischinger and Hawking 2006). *Austroagrion* occurs in New Guinea, New Caledonia and Australia (Theischinger and Hawking 2006). Many coenagrionid species are vagrants in swamps, seepages and ponds and will enter moist forest types along sunlit drainage lines and large open disturbed areas.

Isostictidae The Isostictidae (e.g. *Neosticta*) are a small family of slender dull-coloured damselflies restricted to Australia, New Caledonia and New Guinea, and adjacent islands (Houston and Watson 1988; Watson and O'Farrell 1991). *Neosticta* is endemic and regionally includes the subspecies *N. canescens dorrigoensis*. The two species of *Rhadinosticta* (*Rhadinosticta simplex*, *R. banksi*) are restricted to eastern Australia (Theischinger and Hawking 2006).

Lestidae These are medium- to large-sized damselflies (Theischinger and Hawking 2006; Watson and Moulds 1979; Watson and O'Farrell 1991) that include *Lestes concinnus*, a species widely distributed in northern Australia, and also from India to New Caledonia. No Australian lestid genera are endemic. The genus *Austrolestes* is regionally diverse (e.g. *Austrolestes psyche*, *A. analis*, *A. cingulatus*) (Theischinger and Hawking 2006). Species frequent ponds, streams and ephemeral pools.

Lestoideidae This family incorporates taxa previously cited as Diphlebiidae (e.g. in Theischinger and Hawking

2006). Included is the bright blue-coloured *Diphlebia lestoides* (Plate 40), a robust damselfly commonly found at higher elevations (e.g. Barrington Tops), where adults rest on streamside rocks in bright sunlight. Two subspecies are recognised: *Diphlebia lestoides lestoides* (found south of the Hunter River), and *D. l. tillyardi* (found north of this) (Theischinger and Hawking 2006). Additional species include *Diphlebia coerulescens* and *D. nymphoides*. *Diphlebia* is essentially restricted to eastern Australia.

Synlestidae These are large damselflies. Three endemic genera are known from Australia (*Synlestes*, *Episynlestes*, *Chorismagrion*). Recorded from the Gondwana rainforests region is the very slender *Episynlestes albicauda*. Only three species occur in this genus, which is restricted to the coastal region between northern Queensland and northern New South Wales (Houston and Watson 1988; Theischinger and Hawking 2006; Theischinger and Watson 1985). Regionally occurring *Synlestes* are *S. weyersii* and *S. selysi* (Plate 42). This genus is distinctive in that adults frequently rest with their wings fully or partly spread, rather than folded along the length of the abdomen. Though encountered in the shady understorey adults rarely venture far from streams.

Anisoptera ('true' dragonflies)

Although populations are often restricted in rainforest by the absence of an aquatic larval habitat, nevertheless many adult dragonflies are to be seen in large numbers patrolling rainforest margins and canopies; their presence relies on the availability of water, or boggy swamps and seepages, close by.

Aeshnidae This group incorporates species previously included in Telephlebiidae (see Theischinger and Hawking 2006). The non-Gondwanan Australian aeshnids are medium- to large-sized dragonflies, all are strong fliers, but none of the genera is endemic (Theischinger and Hawking 2006). Genera present in subtropical eastern Australia are *Adversaeschna*, *Anax*, *Hemianax*, *Austrogynacantha* and *Gynacantha*. Species inhabit still and slow-flowing waters. Some are crepuscular (*Austrogynacantha heterogena*). The Gondwanan section of Aeshnidae (Peters and Theischinger 2007) are all swift and strong fliers whose larvae are found in streams of varying size and flow. *Austroaeschna multipunctata* and *A. sigma* mainly inhabit streams at higher elevations. *Austroaeschna obscura* is found in permanent streams, ephemeral stream lines, and pools. The larvae of *Austroaeschna unicornis* inhabit freshwater streams and their larvae are found in submerged logs and detritus. *Austroaeschna cooloola* is restricted to southern Queensland and inhabits mostly densely vegetated streams, often in dune situations (Theischinger and Hawking 2006), and *A. subapicalis* is a montane species recorded from Werrikimbe National Park, though also more widely distributed. *Acanthaeschna victoria*, the only known member of the genus, is restricted to coastal southern Queensland and coastal New South Wales (Theischinger and Hawking 2006). *Austrophlebia* contains the enormous *A. costalis* (Plate 43). The larvae reach a total length of more than 50 mm, and possess wings that have dark maroon-coloured leading edges. Adults are mainly crepuscular and cruise the surface of rainforest streams and forest margins at dusk in search of prey. They occasionally are active during daylight, but tend to restrict their movements to shaded stream banks and log overhangs. At southern latitudes the endemic genus *Spinaeschna* includes the single large *Spinaeschna tripunctata*, whose larvae are found on the underside of submerged rocks. *Notoaeschna* is restricted to south-eastern Australia, and includes the large *N. sagittata* and *N. geminata*. *Telephlebia* species are large, brown-coloured, usually crepuscular or fully nocturnal dragonflies, whose larvae inhabit small streams and damp litter (Plates 44, 262). The genus is confined to the mainland of eastern Australia (Theischinger 1985). Several regional species are present (*Telephlebia godeffroyi*, *T. brevicauda*, *T. cyclops*, *T. tryoni*) (Theischinger and Hawking 2006). Adult *Antipodophlebia aethenes* are crepuscular and fly close to the ground. The larvae are apparently terrestrial (Theischinger 1977; Watson and Theischinger 1980).

Libellulidae These include several small dragonflies found as vagrants to rainforest margins and canopies. Many are wide-ranging on the mainland (*Nannophlebia risi*, *Agrionoptera insignis*, *Crocothemis nigrifrons*, *Orthetrum villosovittatum*, *Nannophya australis*, *Diplacodes haematodes*, *Neurothemis stigmatizans*, *Tholymis tillarga*), though the range of species such as the blue-bodied *Orthetrum boumiera* is more restricted (Theischinger and Hawking 2006). The family includes *Rhyothemis braganza*, *R. graphiptera*, *R. phyllis* and *Tramea stenoloba*, *T. eurybia* and *T. loewii*; these are large species with beautifully patterned and coloured wings. Adult *Rhyothemis graphiptera* and *R. phyllis* have a distinctive fluttering flight behaviour, and are conspicuous species encountered on the margins of littoral rainforest.

Formerly listed as Urothemistidae are distinctive red-bodied robust species (*Aethriamanta circumsignata*,

Macrodiplax cora, *Urothemis aliena*) that are encountered perching on stream and lagoon margins, and atop emergent pond vegetation.

Libelluloid genera of uncertain status These are small to medium-sized dragonflies, previously placed in the Corduliidae, later in Austrocorduliidae, Cordulephyidae, Gomphomacromiidae (G. Theischinger pers. comm.; for comparison see Theischinger and Hawking 2006), and generally found near streams, including intermittent ones, and pools. A number are crepuscular (*Apocordulia macrops*). The larvae commonly are found below submerged rocks and large boulders, including those of the rainforest-frequenting *Micromidia convergens*. Additional species include *Austrocordulia refracta*, *A. leonardi* and *Micromidia atrifrons*. Other regional species are *Archaeophya adamsi*, the larvae of which are found under rocks or litter along stream margins, *Cordulephya pygmaea* and *C. montana*.

Petaluridae Petalurids were common in the Early Jurassic (~190 mya) (Theischinger 1999a). The only Australian genus is the endemic *Petalura* (Watson and O'Farrell 1991). They are large to very large dragonflies that inhabit swamps and boggy areas. *Petalura gigantea* is found through a broad range of altitudes from the low coastal zone to over 1100 m. *Petalura litorea* is restricted to lowlands. The larvae of both grow to over 45 mm in length. Neither is a true rainforest-inhabiting species, but they are associated with moist closed forests where suitable habitat, including artificial ponds, adjoins.

Synthemistidae These are small to medium-sized dragonflies (Theischinger 1998a, 1998b; Theischinger and Hawking 2006), the adults often metallic-hued and yellow spotted or striped. Regional Synthemistidae include the endemic Australian genera *Tonyosynthemis* (Theischinger 1998d) and *Eusynthemis*. *Eusynthemis* includes species with restricted distributions: *E. ursula*, which is associated with *Nothofagus* forest and is known from Chichester State Forest on the south-eastern margin of the Barrington Tops, and *E. ursa*, which is recorded from Barrington Tops (Theischinger 1998d, 1999b; Theischinger and Hawking 2000). Additional species include *Archaeosynthemis orientalis*, *Synthemis eustalacta*, *Parasynthemis regina* and *Choristhemis flavoterminata*. Larvae of these generally inhabit boggy areas, mud, swamps, seepage lines and detritus.

Miscellaneous families The remaining fauna includes Austropetalidae which are large dragonflies (i.e. *Austropetalia patricia*, *A. tonyana*) with distinctive brown blotches on the leading edges of their fore- and hindwings. *Austropetalia* is an endemic genus. The Corduliidae include the wide-ranging *Hemicordulia australiae*, *H. intermedia*, *H. tau* and *Procordulia affinis*. These are medium-sized, often thick-set, dragonflies found in streams, lakes and ponds. They are generally vagrants to rainforest margins.

Gomphidae are numerous in the south-east, and are inhabitants of streams, rivers, lakes and riverine pools. Genera include *Austrogomphus*, *Austroepigomphus*, *Antipodogomphus* and *Hemigomphus*. Macromidae are represented by *Macromia tillyardi*, a large species that inhabits streams and pools.

PLECOPTERA (stoneflies)

The oldest stonefly fossils date from the Permian, and the extant world fauna (in the suborders Antarctoperlaria and Arctoperlaria) is divided into Gondwanan and Laurasian lineages (Zwick 2000). This biogeographical separation relates to the breakup of Pangaea in the Late Jurassic. Though Plecoptera have been considered primitive, and historically have been placed near Odonata and Ephemeroptera (e.g. Theischinger 1991a); their quite different larval gill types point to independent origins of aquatic lives (Grimaldi and Engel 2005).

Plecopteran nymphs live entirely in fresh water and the few that are terrestrial are restricted to moist microhabitats. Nymphs mainly feed on detritus or are omnivores, though some are carnivorous (Grimaldi and Engel 2005). Adults exhibit a somewhat awkward flight behaviour and are attracted to light.

The majority of Australian species inhabit cooler regions, particularly the south-east, and are included solely in the families Austroperlidae, Eustheniidae, Gripopterygidae and Notonemouridae (Theischinger 1991a). These are restricted to the Southern Hemisphere. The Austroperlidae, Eustheniidae and Gripopterygidae are Gondwanan and occur in Australia, New Zealand and South America. The Notonemouridae are restricted to Australia, New Zealand, South America, South Africa and Madagascar and are a derived subgroup of an exclusively northern evolutionary line (Theischinger 1991a). All species, and all but one of the six Australian genera are endemic; *Notonemoura* is shared with New Zealand (G. Theischinger pers. comm).

Austroperlidae The Austroperlidae are a small family known only from Australia, New Zealand and South America. All five Australian genera are endemic

(Michaelis and Yule, in W. Houston 1988). One *Austroheptura* species (*Austroheptura picta*) ranges from south-east Queensland to north-east New South Wales (e.g. Border Ranges, New England National Park) (Michaelis and Yule, in W. Houston 1988; Monteith 1993; Riek 1973b).

Eustheniidae Two subfamilies are recognised: the Eustheniinae which are confined to Chile and south-east Australia, and the Stenoperlinae which occur in Chile, New Zealand and eastern Australia (Theischinger 1991a). The sister family, Diamphipnoidae, is known only from South America (Michaelis and Yule, in W. Houston 1988). All genera are endemic. Australian species previously placed in *Stenoperla* (which occurs in New Zealand) have been transferred to the endemic genus *Cosmioperla* (McLellan 1996). *Cosmioperla* (e.g. *Cosmioperla australis*, *C. denise*, *C. wongoonoo*) is distributed widely within the Gondwana rainforests region (Theischinger 1983).

Gripopterygidae This is the dominant stonefly family in Australia. All known Australian genera are endemic (e.g. *Dinotoperla*, *Dundundra*, *Illiesoperla*, *Trinotoperla*, *Leptoperla*, *Newmanoperla*, *Odontoperla*, *Riekoperla* [Mynott *et al.* 2017; Theischinger and Cardale 1987]). Gripopterygids are also known from New Zealand and South America (Michaelis and Yule, in W. Houston 1988). The New Zealand fauna shows a closer relationship with the South American fauna than with that of Australia (Winterbourn 1980). *Neboissoperla* has one species occurring in montane zones of south-east Queensland (Theischinger 1991a). The monotypic genus *Dundundra* (*Dundundra wanungra*) is recorded from the Lamington Plateau. *Dinotoperla* is particularly diverse in south-east Queensland and northern New South Wales. The genera *Illiesoperla* and *Trinotoperla* include several species (i.e. *Illiesoperla mayi*, *I. frazieri*, *Trinotoperla yeoi*) confined to the region (Theischinger 1982).

Notonemouridae In contrast to the gripopterygid fauna, the Australian and New Zealand notonemourid genera are more similar to each other than to the South American taxa (Winterbourn 1980). Notonemouridae are confined to the Southern Hemisphere and are the only Australian plecopteran group to be derived from an exclusively northern evolutionary line, the Arctoperlaria. All Australian genera are endemic except for *Notonemoura*, which also inhabits New Zealand (Michaelis and Yule, in W. Houston 1988). In Australia, *Notonemoura* (e.g. *N. maculata*) ranges from south-east Queensland to Victoria. *Austrocercella* ranges from south-east Queensland to Tasmania. *Austrocerella weiri* is known from New England National Park (Michaelis and Yule, in W. Houston 1988; Theischinger and Cardale 1987). *Kimminsoperla* has one species (*Kimminsoperla hystrix*) extending north from eastern New South Wales into south-east Queensland (Theischinger 1991a; Theischinger and Cardale 1987).

BLATTODEA (cockroaches, termites)

Evidence from DNA sequences indicates that termites have evolved from wood-feeding cockroaches (Lo *et al.* 2000). Lo and his colleagues considered that a 'clear picture of termite ancestry is crucial in understanding how these insects evolved eusociality', fully social nesting strategies in other insects being found only in ants and some bees (Hymenoptera). The results of their study supported an earlier hypothesis of the origins of Isoptera (Cleveland *et al.* 1934), which proposed a close relationship between termites and the wood-feeding cockroach *Cryptocercus*; this was on the basis of their switch in nesting and feeding habits, and including the presence in their hind guts of symbiotic cellulose-consuming protozoa, found nowhere else in nature. Consequently, as now considered, the order Blattodea comprises the suborder Blattaria (cockroaches) and the infraorder Isoptera (termites), the latter previously a separate order (Roach and Rentz 1998b; Watson and O'Farrell 1991). Along with the related praying mantids (Mantodea), termites and cockroaches are grouped in the superorder Dictyoptera.

Blattaria (cockroaches)

Blattaria are members of the seemingly disparate superorder Dictyoptera (linked by distinctive structures in male and female reproductive systems, DNA sequencing, and relict transitional species) (Lo *et al.* 2000). Worldwide, cockroaches are diverse. There are ~4500 described species in 460 genera (Rentz 2014), most of which are tropical. Many of the more than 500 Australian species are endemic, particularly the majority of the Blattidae, Epilamprinae and Panesthinae (Blaberidae), and Ectobiidae (previously Blattellidae) (Roth 1991a). There are both diurnal (e.g. *Medastinia*) and nocturnal (*Calolampra*) species, and species that are terrestrial (*Laxta*, *Panesthia*) or wholly or fully arboreal (*Balta*, *Ellipsidion*, *Medastinia*) (Rentz 2014).

With the exception of several genera in the subfamily Panesthiinae, a defining character of cockroaches is the laying of eggs in egg pods, called oothecae (Rugg and Rose 1984; Walker and Rose 1998), a feature shared only with mantises. Nearly all termites have lost this trait, laying their eggs singly. Unlike termites, neither cockroaches nor mantises are eusocial, though some cockroach nymphs aggregate with adults (e.g. as in *Panesthia*).

Although Australian fossil species are known from the Middle Triassic (Jell 2004), fossils of modern cockroaches are no older than the Cretaceous. The known extant blattarean fauna is now small, but during the Palaeozoic era cockroach-like animals were among the most diverse of the insect orders (Roach and Rentz 1998a; Roth 1991a). The Palaeozoic 'roachoids' (Archimylacrididae, Necymylacrididae, Mylacrididae), not true modern cockroaches, were one of the most abundant groups in the extensive coal swamps of the Carboniferous. They differed from extant Blattaria, among other characters, by the possession of a long external ovipositor, a primitive feature that was retained into the Late Jurassic (Grimaldi and Engel 2005). In the Mesozoic the Palaeozoic roachoids were replaced by other roach-like families, one of which (Phylloblattidae) is considered the sister group of all living Dictyoptera. One extinct Mesozoic family, the Late Jurassic to Mid Cretaceous Raphidiomimidae, may have been predatory; this behaviour is suggested in part by their long prognathous face and fully exposed head (not covered by a large shield-like pronotum), long palps and long slender body (Grimaldi and Engel 2005). This body plan, particularly with reference to the prolongation of the head and mouthparts, is reflective of certain modern beetles (e.g. Meloidae, Stigmoderini–Buprestidae) that feed on pollen and nectar (see Williams and Adam 2010).

Cockroaches are generalist feeders, consuming decaying and fresh leaves, rotting fruit, and animal droppings. *Panesthia* are wood feeders. Cockroaches have a nymphal stage, the nymphs bearing a resemblance in miniature to the adults. In subtropical rainforests cockroaches are mainly nocturnal and cryptic members of the forest floor fauna, sheltering during the day under stones, or under or in decaying fallen wood. Several small-sized species, however, are active during the day on foliage and tree trunks.

Records of cockroaches visiting flowers are few, in no small part owing to the nocturnal and terrestrial habits of most of the fauna, and the consequent lesser chance to observe them. The height that most rainforest trees grow to reduces further the opportunity for observation. Although cockroaches have been postulated to be the main pollinators of *Uvaria elmeri* (Annonaceae) in a Sarawak lowland rainforest (Nagamitsu and Inoue 1997), the role of cockroaches as pollinators in Australian subtropical rainforests has been largely overlooked. Williams (1995) provides a single record of *Ectoneura pallidula* (Ectobiidae/Blattellidae) individuals visiting the flowers of *Alphitonia excelsa* (Rhamnaceae) at night in a littoral rainforest remnant at Harrington, northern New South Wales. But as Williams and Adam (2010) note, if cockroach visits to flowers are destructive of floral parts, devouring stigmas and stamens, then the presence of such visitors may be at a negative cost to the plant.

Blaberidae Blaberidae are poorly represented in Australia (Rentz 1996, 2014) (Plates 45, 46, 47). However, six of the 11 subfamilies are recorded (Rentz 2014; Roach and Rentz 1998a). Commonly encountered blaberid genera in subtropical rainforests are *Panesthia* (Panesthiinae), *Neolaxta* and *Laxta* (Perisphaeriinae); *Neolaxta monteithi* is recorded from the Bunya Mountains, Lamington National park, Tooloom and Unumgar State Forest (M. Mackerras 1968c; Williams 2002). *Laxta* (e.g. *Laxta friedmani*, *L. granicollis*) is distributed from Queensland to South Australia. Females are very flattened and flightless, but males may be fully winged. In subtropical rainforest and adjacent wet sclerophyll forest, individuals are common under loose bark and under fallen timber, and sheltering beneath stones. Epilamprinae are widely distributed in South America and Asia (Roth 1991a) and are represented in Australia by four genera, the largest being *Calolampra* (Rentz 2014). Males of *Calolampra* (e.g. *Calolampra irrorata*, *C. mjoebergi*) have distinctively speckled wings, and during the day can be found sheltering below loose bark. Females are flightless and shelter during the day in leaf litter and under bark. Species of the similar looking *Molytria* (*Molytria inquinata*, *M. perplexa*) forage at night on foliage. The endemic Geoscapheinae include Australia's largest-sized cockroaches (e.g. *Macropanesthia*, *Geoscapheus*) (Walker and Rose 1998). These are strongly armoured burrowing cockroaches that produce a hissing sound by expelling air through the hind spiracles (Rentz 1996). They inhabit the soil and feed on dry leaf litter. The Panesthiinae are robust insects with a hard integument, and are mainly Oriental and Palaearctic in distribution with an extension into Australia (Rentz 1996; Roth 1991a). *Panesthia* (e.g. *Panesthia australis*) is the most widespread of the panesthiine genera. They are common under and in decaying timber, adults often found in company with numbers of young. Individuals utilise amoeba that live in their intestines to digest cellulose. *Panesthia tryoni* is the largest species encountered in the rainforests of northern New South Wales (Monteith 1986a; Roach and Rentz 1998a).

Blattidae This is the largest of the world's cockroach families with over 500 species in four subfamilies. The Australian fauna numbers more than 200 species currently placed in 38 genera (Rentz 2014). *Celatoblatta* (Blattinae) is confined to Australia, New Caledonia and New Zealand, and individuals are found under bark and stones, and in crevices in rotten wood. In eastern Australia the Polyzosteriinae (>200 species) include the genera *Cosmozosteria*, *Desmozosteria*, *Melanozosteria*, *Methana*, *Platyzosteria*, *Polyzosteria*, *Scabina* and *Temnelytra* (M. Mackerras 1966, 1967a, 1967b, 1968a, 1968b; Rentz 2014; Roach and Rentz 1998a; Shaw 1925) (Plate 48). *Scabina antipoda* is a rainforest species; however, many species in these genera are restricted to open habitats. Except for *Methana*, all Polyzosteriinae are flightless. *Methana* (e.g. *Methana canae*, *M. curvigera*, *M. marginalis*, *M. parva*) is restricted to the Austro-Malayan subregion. *Tryonicus parvus* (Tryonicinae) occurs on the Lamington Plateau, but its range extends to the Illawarra region of southern New South Wales (M. Mackerras 1968a). The genus is otherwise restricted to New Caledonia and the high mountains of northern Queensland (Monteith 1993; Roth 1991a). The genus *Polyzosteria* occurs widely in Malesia and extends to New Caledonia and New Zealand. *Polyzosteria* is especially diverse in the Gondwana rainforests region (10 species are cited in Williams 2002), including species from the Barrington Tops (*Polyzosteria castanea*, *P. melanaria*), Lamington National Park and adjacent areas (*Polyzosteria nitidella*, *P. stradbrokensis*), and the Bunya Mountains (*Polyzosteria perpolita*, *P. scabrella*). The range of several species overlaps (Williams 2002).

Corydiidae/Polyphagidae This is a widespread family of delicate cockroaches recorded from Asia, North Africa, Australia and Central America. The Australian fauna is a poor one, represented by the endemic genera *Austropolyphaga* (*Austropolyphaga perkinsi*, *A. queenslandicus*) and *Polyphagoides* (*Polyphagoides cantrelli*) (Rentz 2010). Corydiidae are the most primitive of the superfamily Blaberoidea (Roth 1991a) and are thought to have 'diverged early from the main Blattoid stem' (M. Mackerras 1968d).

Ectobiidae/Blattellidae This is largest cockroach family, and occurs world-wide (Roth 1991a). There are ~20 genera in Australia, *Balta* (Pseudophyllodromiinae) and *Ellipsidion* being the most diverse. Most species are nocturnal or crepuscular (e.g. *Balta*, *Mediastinia*), though some are active in daylight (*Ellipsidion*). Many frequent rainforest foliage (e.g. *Mediastinia platycephala*, *M. australica*), and include several taxa (e.g. *Beybienkoa* spp.) that feed at night on pollen and flowers. Many species of *Balta* live in leaf litter (Rentz 2014).

Roth (1991b, 1992) recorded the endemic genera *Robshelfordia*, *Eowilsonia*, *Hensaussurea* and *Choristima* from the Gondwana rainforests region, though other genera (*Beybienkoa*, *Carbrunneria*, *Dyakinodes*, *Ectoneura*, *Johnrehnia*, *Mediastinia*, *Neotemnopteryx*, *Paratemnopteryx*) are present (Rentz 2014). *Hensaussurea newtonorum* inhabits cool temperate rainforest in New England National Park (Roth 1991b). *Hensaussurea* species are flightless and live in leaf litter. The pale green-coloured *Medastinia platycephala* is wide-ranging along the east coast, extending from North Queensland to northern New South Wales.

Isoptera (termites)

Isoptera are now considered an infraorder of Blattodea (Lo *et al.* 2000). Termites are predominantly a tropical and subtropical group of soft-bodied social insects whose species establish subterranean, mound or arboreal nests. Some species may construct mounds in part of their range (as in some *Coptotermes*), but elsewhere may be arboreal (Lee *et al.* 2017). Each termite colony consists of several castes, behaviourally and morphologically highly adapted to undertake specific tasks of reproduction, foraging and nest defence. There is a primary separation of castes into reproductive and non-reproductive ones. The pathways to the development of individual castes are complex, and differ with respect to particular families (Watson and Gay 1991).

The Isoptera are highly evolved and adapted cockroaches, whose digestion of plant cellulose utilising symbiotic gut microorganisms 'has allowed them to invade a vast, largely unexploited niche among insects' (Grimaldi and Engel 2005). But unlike extant true cockroaches (Blattaria), and the closely related Mantodea (Roach and Rentz 1998b), termite eggs are layed singly, rather than as oothecae. Despite their modest diversity (~2900 described species world-wide) termites rank with bees as probably the most ecologically important of the insects. They are a recently evolved group, their emergence in the fossil record dating from the Late Cretaceous (the oldest undisputed fossils being those of Hodotermitidae), having developed from Palaeozoic cockroach-like ancestors

(Kukalová-Peck 1991). Given this late emergence Raymond *et al.* (2000) question how the abundant sources of plant material (and plant-derived matter such as herbivore dung) were processed before the Cretaceous.

In some soils termites are the dominant invertebrates, and can be particularly abundant in terms of biomass. However, temperature, moisture, and vegetation type are factors which strongly influence termite distribution. Moisture plays a key role in the impact termites have on dead wood decay and carbon loss (Cheesman *et al.* 2018), and increasing rainfall with elevation has been found to positively affect termite diversity (Casalla and Korb 2019). Because several species form vast colonies their impact on the soil, and consequently the nature and distribution of plant species and other animals within communities, may be considerable; in effect they are ecosystem engineers. Termites change the nature of soils in several ways. Through the physical disturbance of soil profiles their activities alter the nature of organic matter and the distribution of plant nutrients, which in turn change soil fertility and subsequent microsite opportunities for seed germination and plant establishment. Some species form subterranean galleries, which affect soil water regimes by increasing porosity, aeration and water infiltration (Stork and Eggleton 1992). Several invertebrate taxa have coevolved with termites (and ants) and are found in association in their nests (e.g. some Aphodiinae, Dynastinae and Staphylinidae beetles). Termites are an important food source for various birds, mammals, reptiles and other invertebrates, and in Australia several quite distinct ecologically specialised vertebrate animals have coevolved with them: the monotreme Short-beaked Echidna *Tachyglossus aculeatus* (Tachyglossidae), the dasyuroid Numbat *Myrmecobius fasciatus* (Myrmecobiidae) (now extinct in most of its former known range), and the worm-like blind snakes of the genus *Rhamphotyphlops* (Typhlopidae).

Subtropical south-eastern Australia includes several species of *Coptotermes* (*Coptotermes lacteus*, *C. acinaciformis*, *C. cooloola*) in the family Rhinotermitidae (Lee *et al.* 2015, 2017). Some *Coptotermes* are among the world's most invasive and destructive termites, but the genus also contains sophisticated mound-building species. The genus occurs in Africa, Asia (where it is now most diverse), Central and South America, and Australia – all Gondwanan landmasses. Lee *et al.* (2015) consider that *Coptotermes* likely originated in Africa, and dispersed to, and subsequently radiated within, other regions after the breakup of the Gondwana supercontinent. And that ancestral Australian *Coptotermes* were likely to be tree nesting species from which mound nesting arose multiple times.

By their very size and location termite nests are often conspicuous features in wet sclerophyll forests in south-eastern Australia, their brick-red mounds arising like towers from the forest floor, or their nests standing stark, high up upon a eucalypt tree trunk. Seemingly healthy trees, such as tall Flooded Gum (*Eucalyptus grandis* – Myrtaceae) can actually be 'piped' by the activities of termite colonies (e.g. by *Porotermes* – Termopsidae), the central excavated pipe packed with moist clay-like material. However, in subtropical rainforests the presence of termites is more cryptic. At a glance they appear neither particularly diverse nor abundant, and are often simply just hard to find.

The presence of related genera (*Stolotermes*, *Porotermes*) in Australia, Africa, New Zealand and South America suggests vicariance events during the breakup of Gondwana (Bourguignon *et al.* 2015); however, the termite fauna of Australian rainforests generally appears to be a poor one (*Stolotermes australasicus*, *S. queenslandicus*, *Coptotermes dreghorni* and *Parrhinotermes queenslandicus* are restricted to tropical rainforests [Calaby and Gay 1959]), impoverished in both species and numbers in comparison to the termite fauna that inhabits xeric-adapted, and structurally more open plant communities (e.g. Dawes-Gromadzki 2005; Lee *et al.* 2017). This is in contrast to the rich termite faunas of other tropical rainforests (Constantino 1992; Fittkau and Klinge 1973; Watson and Gay 1991). For example, Fittkau and Klinge (1973) estimated that termites and ants together made up 30 per cent of the total animal biomass in the Central Amazonian rainforest, and in a study of the termite fauna inhabiting rainforest in the Pasoh Forest Reserve, Malaysia, Abe and Matsumoto (1979) found 249 epigeal (close to the ground) and arboreal termite nests per hectare. Although Lee *et al.* (2015) consider that Australian *Coptotermes* most likely evolved and radiated from rainforest-inhabiting ancestors to more open forest and arid-adapted forms, the reason for the current impoverished state of the Australian rainforest termite fauna is puzzling and has been queried (e.g. Adam 1992). Calaby and Gay (1959) and Watson and Gay (1991) suggest that the depauperate fauna is due to the fact that rainforest is now restricted to relatively small and discontinuous stands, and that rainforest soil types, of themselves, may not be conducive to occupation by termites. It is tempting to suggest that a partial answer to this paucity in Australian rainforests may rest with the nature of tree timber itself, for many trees quickly rot away in rainforest environments, relative to the longevity of fallen and dead eucalypt timber (e.g. *Eucalyptus microcorys*) in hardwood forests and the abundance of termites there. But this does not reconcile with the richness of termites in rainforests elsewhere. Termites are potentially massive consumers of timber and plant biomass more generally and, given the relatively rapid decay of dead and fallen timber in rainforest, perhaps there is a patch size threshold afterall, below which termite colonies and their populations are not sustainable – at least not large and abundant ones.

Kalotermitidae Only one of the eight Australian genera are endemic (Watson *et al.* 1998). *Neotermes* occurs in coastal rainforests from Victoria to the Torres Strait. *Ceratokalotermes spoliator* ranges from southern to northern Queensland, in eucalypt-associated forests of the coast and adjacent montane areas (Watson and Gay 1991). *Glyptotermes brevicornis* occurs naturally in Queensland and New South Wales, and has been probably introduced to New Zealand, Lord Howe Island and Norfolk Island (Watson *et al.* 1998).

Termitidae In contrast to the low level of generic endemicity in the Kalotermitidae, 20 of the 25 Australian termitid genera are endemic (Watson *et al.* 1998). Genera known from the region include *Microceratermes* and *Nasutitermes*, both of which exhibit generally similar Indo-Papuan, Oriental, Afrotropical, Madagascan and Neotropical distributions. In addition, *Nasutitermes* occurs in the Nearctic and Pacific, and *Microceratermes* occurs in the Palaearctic (Watson *et al.* 1998).

Termopsidae This family includes two Gondwanan genera: *Stolotermes* with three mainland and one Tasmanian species (elsewhere *Stolotermes* occurs in New Zealand and South Africa), and *Porotermes* has one species (*Porotermes adamsoni*) in south-eastern Australia, one in Chile and one in South Africa (Watson and Gay 1991). *Porotermes adamsoni* is the only representative of the subfamily Porotermitinae. Stolotermitinae occurs in mountainous country from Victoria to southern Queensland, and on the Atherton Tableland (Watson and Gay 1991).

MANTODEA (praying mantids, mantises)

Mantodea are grouped with cockroaches and termites in the superorder Dictyoptera. The origin of the Mantodea is considered to be at the beginning of the Jurassic, with most modern praying mantids originating in Gondwana in the Cretaceous (Svenson and Whiting 2009). Balderson (1991) cites 21 of the then 37 Australian genera as endemic, but there are no endemic families or subfamilies (Balderson *et al.* 1998a). The group is most diverse in the tropics, and includes many species in semi-arid and woodland habitats (Milledge 2005). There is a strong Indo-Malaysian element, with numerous species and genera shared with New Guinea (Balderson *et al.* 1998a).

Praying mantids are strictly predatory insects that include active agile hunters, and cryptically shaped and coloured species that hunt by stealth and ambush. Camouflage also reduces risk of attack by predators, but nymphs suffer heavy mortality. Most mantids seem to be solitary, males are winged, but females of several genera (e.g. *Paraoxypilus* – Amorphoscelidae) are apterous or possess abbreviated wings and thus are unable to fly. But even for winged species, recourse to flight is infrequently observed, flights usually only being of a short distance to the nearest branch or adjacent plant. Eggs are laid in an enclosed ootheca, fashioned from soft material produced by accessory glands, and which becomes hard or leathery when dry. A series of capped exit apertures are arranged along the midline of the ootheca, through which the newly hatched nymphs can emerge (Balderson 1991). Brood care by adults, either of the oothecae or of the young, does not normally occur. Parasitism by small wasps (e.g. Chalcididae [Bouček 1988]) is common, and predation on oothecae by small insects, birds and mammals is recorded.

As far as is known praying mantids are diurnal. In subtropical rainforests mantids frequent canopy foliage and subcanopy sunlit openings and tree trunks, but are not commonly encountered on the forest floor, probably owing to the lack of suitable day-active prey there (e.g. Orthoptera, Diptera and other soft-bodied arthropods, and, potentially, small vertebrates). The much-touted image of praying mantids as cannabalistic may be rare in nature. The availability of prey probably limits mantid occurrence and population abundance; however, dispersed populations appear typical of Mantodea regardless of species or the habitat they occupy.

The subfamily Paraoxypilinae (Amorphoscelidae) comprises small species (e.g. *Paraoxypilus*) that occur on the ground and on the trunks of rainforest trees. Individuals may be territorial. Males are fully winged but females may be incapable of flight. The family Mantidae has a world-wide distribution, and includes most of the

Australian mantid fauna. Size ranges from small, delicate species to quite robust ones. Many, such as the foliage-frequenting species of *Orthodera*, are coloured green (Plate 50). There are several endemic Australian genera, which are recorded from southern Queensland and northern New South Wales (Williams 2002). The Iridopterygine genus *Calofulcinia* (*Calofulcinia paraoxypila*) is restricted to eastern mainland Australia and New Guinea (Plate 49). Females are flightless (Balderson *et al.* 1998a) and may be encountered active by day on tree trunks. Similarly, the mantid tribe Archimantini (e.g. *Archimantis latistyla*, *A. sobrina*) is restricted to Australia (14 species) and New Guinea (one species) (Milledge 1997a). They inhabit shrubs or tall grasses.

DERMAPTERA (earwigs)

Earwigs are omnivorous browsers, sometimes ectoparasitic, rarely strictly herbivorous or carnivorous. They have a long fossil history, the earliest fossils dating to the Late Triassic–Early Jurassic (Grimaldi and Engel 2005). There are several related, now extinct, groups (e.g. families of the Archidermaptera). Modern dermapteran adults are easily recognised by their stout terminal cercal forceps, and short forewings (the latter a feature they notably share with most staphylinid beetles). The abdominal forceps are used to capture prey, in mating, and in wing-folding. Egg batches are usually laid in a burrow or below protective stones, or under or in rotting logs. Females undertake brood care, cleaning and protecting eggs and early instar nymphs against invasive fungi, and may aggressively defend the nest site (Rentz and Kevan 1991).

Dermapterans are nocturnal and have a global distribution (except Antarctica and the extreme Arctic), most living in tropical, subtropical, and warm temperate regions. They are to be found under fallen tree trunks, in tree crevices, under loose bark, and in leaf litter and debris on the forest floor. Adults are usually solitary. More than 60 species are known from Australia and most of these are endemic (Rentz and Kevan 1991). The more primitive superfamily Pygidicranoidea is widely distributed in Australia, but most of the dermapteran species are found in the wetter regions of Australia, with adults sheltering by day under fallen rotting wood or loose bark on standing living or dead trees.

Pygidicranidae are considered the most primitive of living Dermaptera but are rarely encountered, the slender and soft-bodied species of *Cranopygia* occasionally being found under loose bark on standing trees.

Anisolabididae The Anisolabididae are cosmopolitan and include the large *Titanolabis colossea* (Plate 51), which is found under stones and rotten logs in rainforest and wet sclerophyll forest (Monteith 1993). *Titanolabis colossea* has an extralimital distribution and is also recorded from New Caledonia and Vanuatu (Cassis 1998). Phoretic mites (e.g. *Paradiplogynium nahmani*, *Histiostoma australiense*, *H. titanolabi*) are recorded from adults (Seeman 2007; Tagami and Halliday 2013).

Apachyidae The Apachyidae are a highly specialised family comprising two genera and 15 species restricted to the Eastern Hemisphere. Adults are characteristically flattened, and live under bark. The genus *Apachyus* (three species) occurs from northern Queensland to northern New South Wales, on coastal plains and tablelands in rainforest, wet sclerophyll forest and open forest (Cassis 1998; Rentz and Kevan 1991).

Labiduridae This is a cosmopolitan family naturally represented in Australia by 15 species (Balderson 1991). Two additional species have been introduced (Cassis 1998). *Gonolabidura meteor* is described from Dorrigo, and is the only representative of the subfamily Allostethinae in Australia. *Gonolabidura* occurs in the Australian and Oriental regions (Cassis 1998).

Spongiphoridae This family is poorly represented in Australia (15 species), four subfamilies being recorded (Balderson 1991). The subfamily Spongophorinae includes *Spongovostox hackeri*, described from Mt Tamborine. *Spongovostox* is distributed in Oriental, Afrotropical and Neotropical regions (Cassis 1998). The subfamily Labiinae includes *Paraspania*, which is distributed in the South-east Asian–Australasian region, with at least two species (i.e. *Paraspania australiana*, *P. brunnea*) occurring in the Gondwana rainforests region (Sakai 1993).

ORTHOPTERA (grasshoppers, katydids, crickets)

Orthoptera are mainly medium-sized insects, but include very large (Anostostomatidae, Phyllophorinae), and often cryptically formed (Phrictini), species. Traditionally they have been divided into the suborders Ensifera (e.g. Anostosmatidae, Phalangopsidae, Gryllacrididae, Rhaphidophoridae, Tettigoniidae) and Caelifera (e.g. Acrididae, Pyrgomorphidae), which have sometimes been treated as separate orders (e.g. Plates 52, 53, 56, 58). Depending on the classification followed there are ~10 000 species of Ensifera and 11 000 of Caelifera. Numerous families within the Ensifera are known from the Mesozoic, the Tettigoniidae for example, recognised from the Early

Cretaceous (Grimaldi and Engel 2005). The earliest definitive Caelifera date from the Triassic.

Orthopterans are often solitary, though Rhaphidophoridae in subtropical rainforest are commonly gregarious, though usually so in small numbers. Rentz (1991, 2010) notes that stridulatory-produced 'songs' are a characteristic of orthopteran behaviour. Sounds are produced by a variety of mechanisms, including by simple and complex files (e.g. see Bailey and Rentz 1990; Rentz 1985, 1996), and mainly by males. Calling songs, as in the subfamily Tettigoniinae, for example, can be an important taxonomic character in differentiating species that are morpholocally similar (Rentz 1985). The calling songs of most Phaneroptinae are complex, with females answering males (Rentz 2010). Some species do not produce sound by stridulation, but rather sound is produced while in flight (crepitation). Sound production serves as a defence and warning signal, and may be group-orientated or related to territorial behaviour, or in mating and reproduction. However, for non-singing Orthoptera, vision (and associated visual mating cues) and the release of pheromones are important in successful mating (Rentz 1996).

World-wide more than 20 000 species are known, and these include phytophagous, omnivorous and even predatory types. They are found on the forest floor, in tree tops, and in caves. There are nocturnal and diurnal species, and many that inhabit subtropical rainforest are relatively sedentary (e.g. Rhaphidophoridae), remaining during the day under dark protective rock overhangs, or hide beneath loose bark or in deep crevices. Many shelter below leaves on the forest floor. However, large Anostostomatidae are frequently venturesome, and leaf feeders undertake regular short-distance movements within canopies and between food plants. Certain small Grylloidea (e.g. *Oecanthus*, *Ornebius* [Plate 54]) are found openly on leaves during the day. Species of Phalangopsidae (= Gryllidae – Phalangopsinae) are common at night on tree trunks and rocks. Flower-frequenting species are not abundant, though some taxa (Zaprochilinae) possess elongate, forward-facing mouthparts indicative of a nectar or pollen-feeding habit.

There is a high level of generic endemicity in the Australian fauna (Rentz 1991). For example, 44 per cent of the Grylloidea, 100 per cent of the Tettigoniinae and ~90 per cent of the Acridoidea genera are endemic.

Acrididae The acridid subfamily Oxylinae includes ~20 species in the genus *Praxibulus*, occurring in moist grassy habitats of eastern New South Wales, Victoria and south-east Queensland (Key 1989). Several species are recorded from the Gondwana rainforests region; two species (*Praxibulus queenslandicus*, *P. triangularis*) are known only from the region. However, most are vagrants entering from adjacent open vegetation communities. Acrididae are mostly fully-winged, strong fliers capable of dispersal over relatively long distances, and best represented in open forests, woodlands and grasslands.

Anostostomatidae Anostostomatids, formerly Stenopelmatidae, are often known as king crickets and wetas (Johns 1997; Monteith and Field 2001; Rentz 1991, 1996) (Plate 52). Their immature stages, and many adults, are flightless. Adult males are armed with powerful biting jaws, and hind tibiae that possess large, sharp spines. The Australian fauna includes more than 60 species. All occur in the forests of the east coast, except for a single species from Western Australia (Johns 1997; Rentz 1991). Much of the fauna is confined to rainforest, with the majority restricted to the Torresian faunal province, in particular the narrow coastal zone extending from Cape York to northern New South Wales. The Australian fauna has affinities with that of New Zealand and India (G. Monteith pers. comm.). The genus *Onosandrus* shares species with southern Africa (Rentz 1991). However, anostostomatids are absent from Victoria and Tasmania, even though apparently suitable habitat is present, numerous other Gondwanan invertebrates are found there, and the family is well represented in New Zealand.

The Australian Wet Tropics is a centre of Australian anostosmatid diversity; however, south-eastern Queensland–northern New South Wales is a secondary centre of diversity. Three of the five described species of *Anostostoma* occur in south-east Queensland (*Anostostoma australasiae*, *A. opacum*, *A. spinosum*), but south of the border anostosmatid diversity declines sharply. At least one species extends to the lower north coast of New South Wales, where they can be common on the forest floor at night (Williams 1993; G. Williams unpubl. data). *Penalva* occurs from northern Queensland to southern New South Wales in both open forest and rainforest. *Hemiandrus* occurs in Australia and New Zealand (Monteith and Field 2001), but in Australia there is a pronounced faunal disjunction with most species restricted to the Wet Tropics between Cooktown and Ingham, and only a single geographically isolated species occurring on the Lamington and Springbrook Plateaux (G. Monteith pers. comm.).

Gryllacrididae These are mostly fully winged crickets. More than 20 genera in the Gryllacrididae have been described from Australia. This number represents a disproportionate high level of diversity in the Australian gryllacridid fauna (Rentz 1991; Rentz and John 1989). The genus *Wirritina* is endemic and is distributed from northern New SouthWales to eastern Victoria (Rentz and John 1989). It comprises two described and one undescribed flightless species. *Wirritina naumanni* has been described from New England National Park. *Arrolla turramurrae* is common in the Sydney region and inhabits wet forests (Rentz 1996).

Pyrgomorphidae Only the subfamily Pyrgomorphinae occurs in Australia (Rentz 1991). *Monistria discrepans* has been recorded from the Gibraltar Range (Moss and Popple 2000) and is included in the endemic tribe Monistrini.

Rhaphidophoridae Members of this family are harmless, nocturnal, opportunistic scavengers (including plants and animals) and are commonly referred to as cave or camel crickets (Rentz 1996; A. Richards 1987) (Plates 55, 56). Several taxa occur within the region (e.g. *Cavernotettix wyanbenensis, Australotettix* spp.), but only the Macropathinae are known to occur in Australia. The Macropathinae have a circum-Antarctic distribution (Rentz 1991). All Australian species are confined to wet forests, rock outcrops and caves at elevations extending from montane hinterland sites (e.g. Upper Hasings Valley >1100 m) to almost to sea level (Rentz 1991; Williams 1993). Species are flightless and generally terrestrial, rather than arboreal. Rhaphidophorids are extremely sensitive to temperature changes, require high humidity (Rentz 1996), and are heavily dependent on their sense of touch. And although species are often associated with caves and holes, they can be especially common openly foraging on the forest floor during warm nights.

The flightless nature of rhaphidophorids, their restriction to particular microhabitats and their limited dispersal capabilities make individual taxa and populations vulnerable to edge effects, removal of understorey habitat niches, modification of habitat light and moisture regimes (e.g. through tourism development of cave systems and rock overhangs), and predation by feral rodents.

Tetrigidae Members of this family are often called pygmy grasshoppers (Rentz 1996) (Plate 57). They are dull coloured in tones of brown and grey and resemble stones, twigs or pieces of wood. The diminutive species of *Tetrix* and related genera are often encountered on the ground near watercourses; however, little is known about their biology. Species of *Vingselina*, or closely related genera, can be found at night on the lower trunks of rainforest trees, their form easily allowing them to be mistaken for part of the textured bark surface.

Tettigoniidae Australian tettigoniids (Rentz 1991, 1993; Rentz and Balderson 1979) are unrelated to the South African fauna, but exhibit strong relationships with the fauna in South America and the genus *Neduba* from North America (Rentz 1985). In temperate Australia the subfamily Tettigoniinae is most closely associated with heathlands and some subtropical taxa are associated with grasses (Rentz 1985). Disjunct examples of these habitats occur within Gondwana rainforests at higher altitudes.

Several tettigoniids recorded from forested areas in the Gondwana rainforests region belong to the subfamily Zaprochilinae. Zaprochilinae are entirely confined to Australia (Rentz 1993; Rentz and Clyne 1983). Adults are pollen and nectar feeders. Two genera are present: the monotypic genus *Anthophiloptera* (*Anthophiloptera dryas*) (Plate 58), which ranges widely from Batemans Bay in southern New South Wales, north to Cairns (Rentz 1996) (including records from the Bunya Mountains, Lamington National Park, Dorrigo), and two species of *Zaprochilus* (*Zaprochilus australis, Z. mongabarra*). *Zaprochilus mongabarra* is known only from a few specimens collected at three disjunct localities in North Queensland (Windsor Tableland), south-east Queensland (Mt Glorious) and far northern New South Wales (type locality, Minyon Falls near Nightcap National Park) (Rentz 1993). Rentz (1993) attributed the rarity of *Z. mongabarra* in the vicinity of the type locality to successive forestry burning practices, which eliminate necessary understorey habitat.

Possibly the most spectacular tettigoniid inhabiting subtropical rainforests and wet sclerophyll forests is the broadly leaf-winged *Mastigaphoides haffneri* (Pseudophyllinae) (Plate 61). The species extends south to the Lansdowne Escarpment on the lower north coast of New South Wales (Williams 1993, 2002), and is otherwise disjunctly recorded from the Tweed River to far northern Queensland (Rentz 2010). By day adults frequent foliage but at night descend to lower strata where they can be found on tree trunks and large vines. Other leaf-like Pseudophyllinae include *Mastigapha crassicornis*, but species of the spiny katydid genus *Phricta* are ornately patterned so as to blend into the lichen-festooned trunks of trees on which they can be found (Plate 64). Additional subfamilies present in wet sclerophyll forest or rainforest include Listroscelidinae

(*Conocephalomima barameda*, *Paraphisis* spp.), Conocephalinae (*Austrosalomona falcata*) and Phaneropterinae (*Ozphyllum nasreckii*, *Caedicia* spp., *Diastella*, *Torbia* spp.). *Paraphisis*, or spider katydids, are delicate, foliage-frequenting species that have distinctive elongate spines on their fore- and mid-legs (Plate 63). *Austrosalomona falcata* (Plate 59) is a large tettigoniid that inhabits rainforest and wet sclerophyll forest but also commonly occurs in gardens and houses. Phaneropterinae (e.g. Plates 60, 62, 65) include large katydids with elongate wings that resemble eucalypt leaves.

PHASMATODEA (stick insects)

The Phasmatodea are phyllophagous (especially on *Eucalyptus*) (Key 1991) and in form adults and nymphs parallel the cryptic camoufllage that is often seen exhibited by the related Orthoptera (Plates 66, 67, 68, 69). Some are veritable giants of the insect world, with *Pharnacia kirbyi* from Borneo reaching over 550 mm in length. As their name implies, stick insects often resemble twigs and stems, and also leaves. A particular case of crypsis is seen in the early instars of *Extatosoma tiaratum* (Plate 70) that appear to mimic ants of the genus *Leptomyrmex* (Plate 260). Eggs are also cryptic in form, resembling plant seeds, small bark nodules or stones, but may be parasitised by microhymenoptera (Riek 1955b). There is, however, little apparent care in egg-laying; these either being scattered about within leaf litter, or sometimes buried in the soil or cemented to plants. There is a high incidence of pronounced sexual dimorphism, such that the physical nature of some males and their respective consorts are quite dissimilar. Flightlessness (aptery) is common, and asexual reproduction via parthenogenesis is recorded (e.g. *Podacanthus viridiroseus*, *P. wilkinsoni*) (Brock and Hasenpusch 2015; Hadlington 1965). Flightless adults (e.g. female *Extatosoma tiaratum*) and wingless juveniles, confront particular problems of dispersal, and are vulnerable to impacts such as fire.

Fossils bearing phasmatodean traits date from the Permian or Triassic (Rasnitsyn and Quicke 2002; Willmann 2003). There are now some 3000 species; however, the order is most abundant in the tropics. Three families, Diapheromeridae, Phasmatidae and Phylliidae, occur in Australia (Brock and Hasenpusch 2015), the current status of the fauna somewhat different to that of Balderson *et al.* (1998b). Phylliidae are restricted to far northern Queensland. Phasmatidae include the largest-sized insects to be encountered in the rainforests and associated wet sclerophyll forests of subtropical eastern Australia (e.g. *Extatosoma tiaratum*, *Podacanthus* spp. – Tropidoderinae, *Eurycnema goliath* – Phasmatinae). *Extatosoma* is restricted to Australia, Lord Howe Island and New Guinea. A highly disjunct distribution is found in *Parapodacanthus* (Phasmatidae) (Brock and Monteith 2018) with *P. hasenpuschorum* being restricted to the Wet Tropics of North Queensland, and the second known species, *P. ailaketoae*, recorded only from the Border Ranges region (Lamington National Park, Cunninghams Gap); the intervening gap of over 1500 km between the two suggests the genus had emerged before the retreat of rainforest in eastern Australia (Brock and Monteith 2018). A similar geographical disjunction is found in *Onchestus* (Brock and Hasenpusch 2015).

Stick insects are found in all rainforest and wet sclerophyll forest strata: on foliage, branches, trunks and the forest floor. Some, such as the large but harmless *Eurycnema goliath* (Plate 69), can be encountered on broad-leaved *Acacia* growing on the margins of littoral rainforest remnants (such as at Harrington). In addition to those already discussed, regional species include *Austrosipyloidea carterus*, *Candovia* spp., *Sipyloidea whitei*, *Pachymorpha squalida*, *Gibbernecroscia kooymani* (Diapheromeridae), *Acrophylla* spp., *Anchiale austrotessulata*, *Austrocarausius mercurius*, *Ctenomorpha marginipennis*, *Didymuria violescens* (outbreaks of which can severely defoliate trees), *Onchestus gorgus*, *Paronchestus pasimachus*, *Extatosoma tiaratum*, *Podacanthus typhon*, *P. viridoroseus*, *P. wilkinsoni*, *Tropidoderus childrenii* and *Xeroderus kirbii* (Phasmatidae) (Brock and Hasenpusch 2015; Brock *et al.* 2019; Hadlington 1965; Williams 1993, 2002). Many are widely distributed and found in open sclerophyll forest, in rainforest only as vagrants from adjacent non-rainforest communities, or more characteristically in rainforest (as with *Austrocarausius mercurius*). However, a number have limited known distributions, such as the already discussed *Parapodacanthus ailaketoae* and the newly described *Gibbernecroscia kooymani*, which is known only from south-east Queensland and far northern New South Wales. Adults of the latter have been recorded feeding on the leaves of *Neolitsea dealbata* (Lauraceae), and myrtaceous *Backhousia myrtifolia* and *Syzygium* species (Brock *et al.* 2019).

Onchestus gorgus has been reported to stridulate (Brock and Hasenpusch 2015), and will quickly open its wings when threatened. Startle displays are also exhibited by *Acrophylla titan*, *Austrosipyloidea carterus*, *Ctenomorpha marginipennis*, *Eurycnema goliath*, *Tropidoderus childrenii* and *Xeroderus kirbii*, though this may be performed only by specific sexes (e.g. female *Austrosipyloidea carterus*). *Didymuria* may feign death when disturbed. The spiny and large female of *Extatosoma tiaratum* is capable of causing wounds. *Candovia robinsoni* (Plate 66), a species found in small pockets of rainforest in southern New South Wales as well as more temperate rainforest on the mid-north coast (e.g. Tapin National Park), secretes defensive fluids when disturbed (Brock and Hasenpusch 2015).

EMBIOPTERA (web-spinners, embiids)

The Embioptera, also known as Embiodea or Embiidina, are mainly a tropical group of communal insects, with some members extending to temperate regions. They are distantly related to stick insects (Phasmatodea) and earwigs (Dermaptera), and feed on plant material such as dead leaves, lichen and moss (Ross 1991). Females do not have wings and this may limit the dispersal of populations. World-wide there are over 360 species (Grimaldi and Engel 2005). Fossil Embioptera are known from the Cretaceous and Tertiary, but may date to the Permian (Kukalová-Peck 1991). Web-spinners are gregarious, rather than subsocial, and inhabit and live in protective silk galleries; the silk is produced from special glands located on enlarged foretarsi, and extruded through specialised hollow setae. Their elongate, flattened and short-legged bodies are able to navigate the galleries with speed, being adept at rapid backwards movements. Another defining character of the order is specialised wing veins through which haemolymph is pumped, but when withdrawn the wings are able to collapse in on themselves; repumping of fluid occurs when flight is necessary. Females take care of their young, and may inhabit shared gallery structures with other females (Edgerly 1987, 1994).

Australembiidae are endemic to the eastern Australian mainland and Tasmania (Ross 1991). Members of the genus *Metoligotoma* live mostly in leaf litter. The genus ranges from coastal central Queensland to Tasmania, with the majority of mainland species occurring east of the Great Dividing Range. The Oligotomidae include *Aposthonia*, the most widespread Australian genus, adults of which inhabit loose bark and rock crevices (Ross 1991).

PSOCOPTERA (psocids, book lice)

Psocoptera are small (*c.* 1–10 mm) soft-bodied insects (Smithers 1991b). The majority are fully winged as adults, and these are often very agile. Their membranous wings are commonly patterned with dark blotches and stripes, and are found in great diversity on foliage, stems, trunks, under bark and stones, and in leaf litter. They feed on fungal hyphae, algae, lichen, spores, and plant and insect tissue (Smithers 1991b). Aggregations of nymphs may occur, but frequently the adults are solitary. Populations in rainforest may show little seasonal variation (Smithers 1977).

The order is thought to have derived at least from the Permian (Kukalová-Peck 1991; Rasnitsyn and Quicke 2002), but Grimaldi and Engel (2005) note that Palaeozoic fossils (of the Permopsocida) attributed to them possess numerous features that are considerably more primitive than living Psocoptera, and that the oldest fossils attributable to extant Psocoptera date from the Late Jurassic (*c.* 150 mya) and definitively from Cretaceous and Tertiary ambers.

Psocoptera are a rich component of forest arthropod communities in subtropical eastern Australia (Smithers 1977, 1997b; Williams 2002), and are particularly diverse and abundant in the rainforest canopy (Kitching *et al.* 1993). Of the arthropod assemblage sampled by Kitching *et al.* (1993) in subtropical rainforest in southern Queensland >20 per cent were Psocoptera, the second most significant group following the Collembola.

Numerous new or poorly known psocopteran species have been recorded by Smithers (1994a, 1996, 1997a) from the Mount Royal Range on the southern edge of the Barrington Tops Plateau, and even small isolated rainforest remnants within otherwise cleared agricultural and urban landscapes may contain a rich psocopteran fauna. Smithers (1997b), for example, listed 17 species, including a new species (*Ptycta willamsorum*) and the first known males of *Ectopsocus hartleyi*, from Coocumbac Island Nature Reserve, a small rainforest stand surviving on the floodplain of the Manning River, New South Wales.

Amphientomidae include *Hemiseopsis alettae*, which is restricted to the Gondwana rainforests region (Smithers, in Wells 1996a). The genus *Taeniostigma* (Amphipsocidae) includes some of the largest Australian Psocoptera (Smithers 1991b). Species are leaf dwellers. *Taeniostigma trickettae* is restricted to rainforest and occurs in the Barrington Tops, Mt Royal Range and Lansdowne Escarpment. The foliage-frequenting Caeciliidae have a world-wide distribution. Two endemic genera, *Caecilius* and *Paracaecilius*, are recorded from the Gondwana rainforests region. *Caecilius griseus* and *C. tugloensis* are known only from the Mount Royal Range (Smithers, in

Wells 1996a). Additional genera include *Enderleinella* and *Maoripsocus*.

Ectopsocidae occur in drier leaves and leaf litter, and are saprophages associated with *Syzygium* fruit (Juniper and Britton 2010). Numerous species of *Ectopsocus* (*Ectopsocus albiceps, E. aldretei, E. australis, E. briggsi, E. californicus*) inhabit rainforest and are particularly well known from elevated hinterland localities (e.g. Gibraltar Range, Mt Royal Range). Elipsocidae are a world-wide family. Species include *Paedomorpha gayi*, *Pentacladus eucalypti*, *Propsocus pulchripennis* and *Spilopsocus ruidis* (Smithers 1996). The moth-like Lepidopsocidae (e.g. *Echmepteryx brunnea*, *Nepticulomima tridentata*) generally have scaly wings and are found in leaf litter and under bark. Myopsocidae are large, mottled, species found on bark and timber, and eat algae and lichens. They include the rainforest species *Myopsocus australis*, *M. furcatus* and *M. incomptus* (Smithers 1975, 1997a).

The Peripsocidae are mainly found on bark, and are well known from higher elevations (Smithers 1994a). Species recorded from rainforest include *Peripsocus hamiltonae*, *P. milleri*, *P. morulops*, *P. roseus* and *P. tillyardi*. Philotarsidae include the genera *Aaroniella*, *Haplophallus* and the diverse *Latrobiella*, including the rainforest associated species *L. ornata* (Border Ranges, Dorrigo National Park, Barrington Tops) (Smithers 1996).

Pseudocaeciliidae frequent leaves and green twigs. They are diverse in the moist forests of subtropical eastern Australia. Two endemic pseudocaeciliid genera, *Austropsocus* and *Pseudoscottiella*, occur in northern New South Wales–southern Queensland. *Pseudoscottiella medialis* is known only from the type locality of Lamington National Park (Smithers, in Wells 1996a). *Pseudoscottiella alettae* and *Heterocaecilius rotundus* are known only from the Mount Royal Range area (Smithers 1996). *Heterocaecilius rotundus*, and *H. nigricans* from Madagascar, share several features (such as loss of ocelli) and appear to have 'undergone remarkable parallel evolution …' (Smithers 1996).

Psocidae frequent bark, and include the rainforest-restricted *Kaindipsocus emarginatus* and *K. marksae*. In Australia *Kaindipsocus* is known only from the Gondwana rainforests region, and the genus is otherwise known only from one species occurring in New Guinea (C. N. Smithers pers. comm.; Smithers 1997a). Additional genera include *Blaste*, *Lasiopsocus*, *Clematostigma*, *Ptycta*, *Sigmatoneura*, *Spilopsocus* and *Tichadenotecnum*. Stenopsocidae include the rainforest-frequenting *Stenopsocus albipileus* and Trogiidae include the cosmopolitan *Cerobasis guestfalica*, *Lepinotus inquilinus* and *L. stoneae*. Trogiids are often found in domestic situations.

Rarely are records given for plant associations; however, Smithers (1995) listed species collected from the fresh fronds of the cycad *Lepidozamia peroffskyana* (Zamiaceae) at Lorien Wildlife Refuge, north of Taree, New South Wales (Plate 302). *Lepidozamia peroffskyana* has a limited and disjunct distribution (Harden *et al.* 2006) and at this site occurs mainly in a wet sclerophyll forest–subtropical rainforest ecotone. Species listed by Smithers were *Caecilius quercus*, *Paracaecilius lemuris* (Caecilidae), *Taeniostigma trickettae* (Amphipsocidae), *Ectopsocus australis* (Ectopsocidae), *Peripsocus milleri* (Peripsocidae), *Lobocaecilius monicus*, *Pseudoscottiella rotundata* (Pseudocaeciliidae), *Sigmatoneura formosa* and *Clematostigma maculiceps* (Psocidae). Previous known records were that *Caecilius quercus*, *Paracaecilius lemuris*, *Lobocaecilius monicus* and *Pseudoscottiella rotundata* usually are found on fresh foliage; *Taeniostigma trickettae* is more widely known from rainforest elsewhere; *Ectopsocus australis* usually lives in dead leaves; and *Peripsocus milleri*, *Sigmatoneura formosa* and *Clematostigma maculiceps* have been collected from bark (Smithers 1995).

PHTHIRAPTERA (lice)

Phthiraptera are highly modified flightless ectoparasites of mammals (but excluding monotremes and bats), and most groups of birds (Grimaldi and Engel 2005). Their apparent absence from bats is enigmatic, given that bats are hosts to fleas and mites. Lice can be specific to particular parts of the host's body. They spend their whole life on the host and as a consequence the distribution of individual Phthiraptera species is almost completely determined by host movements (Parma and Barker, in Wells 1996a). This apterous and obligate relationship speaks to the evolution of phthirapterans, and hints at the possibility of host associations (e.g. with pterosaurs) that pre-dated the emergence of modern birds and mammals (Rasnitsyn and Zherikhin 1999). However, Phthiraptera are almost absent from the fossil record, a possible peculiarity of the difficulty of host-parasite copreservation, but given that basal mammal lineages extended from the Triassic, and that of birds at least from the Mid Jurassic, then the emergence of phthirapterans probably occurred in the Late Jurassic–Early Cretaceous (140–150 mya) (Grimaldi and Engel 2005).

Species may be host specific such that the endangering and extinction of host populations can lead to extinction of lice species (see Grimaldi and Engel 2005; Smales 1994). Their bodies are characteristically flattened and strong-legged, allowing them to move easily, and concealed, about the host's body. Adults and nymphs have a similar diet, and depending on the host this can consist of the keratin of hair and feathers, skin fragments, blood, mucus or serum (A. Marshall 1981). Adults cement their eggs onto feathers or hairs. Birds and mammals have adopted a range of behavioural, morphological and physiological adaptations to control their lice populations, though infestations can at times be massive, inflicting considerable discomfort, if not stress, on the host.

The majority of the Australian fauna is contained within the families Boopiidae, Menoponidae and Philopteridae. The majority of the Australian fauna are endemic species occurring on endemic hosts (Calaby and Murray 1991); however, none of the families is endemic. Laemobothriidae, Menoponiidae, Philopteridae, and Ricinidae utilise birds as hosts. Menoponidae have adapted to fill a wide variety of ecological niches, feeding on blood, mucus and particular components of feathers of numerous bird species (Palma, in Wells 1996a). Hoplopleuridae mainly infest rodents (e.g. *Hydromys chrysogaster, Rattus fuscipes*). A list of known Phthiraptera-host records is given in Wells (1996a).

Boopidae This family is mainly confined to Australian and Papuan marsupial hosts, and most lice in the suborder Amblycera (i.e. Boopiidae, Laemobothriidae, Menoponidae, Ricinidae), found on mammals, are boopiids (Calaby and Murray 1991; von Keler 1971). South American marsupials are parasitised by Trimenoponidae, but this family is not closely related to Boopiidae (Calaby and Murray 1991).

Macropophila breviarcuata is described from material collected from the macropod *Thylogale stigmatica* at Mt Lindsay in south-eastern Queensland (Palma and Barker, in Wells 1996a). Additional host records for wet forest-inhabiting mammals include *Boopia brevispinosa* (on *Antechinus stuartii, A. swainsonii*), *B. greeni* (on *Antechinus swainsonii*), *B. bettongia* (on *Isoodon macrourus, Parameles nasuta*), *B. spinosa* (on *Phascogale tapoatafa*), *Heterodoxus ualabati* and *Latumcephalum lesouefi* (on *Wallabia bicolor*), *Macropophila biarcuata* (on *Thylogale thetis*), *M. breviarcuata* and *M. forcipata* (on *Thylogale stigmatica*), and *Paraheterodoxus erinaceus* (on *Potorous tridactylus*).

HEMIPTERA (true bugs, leafhoppers, cicadas, mealybugs, lerps)

The Hemiptera, in the broad sense (Sternorrhyncha and Auchenorrhyncha [Homoptera], and Heteroptera) (Carver *et al.* 1991; Cassis and Gross 1995, 2002; Appendix 4) are mainly phytophagous and feed through an elongate rostrum. But notable predatory groups (that usually feed on other arthropods) include the terrestrial Anthocoridae, Aenictopecheidae, Reduviidae and Enicocephalidae, and several aquatic and semi-aquatic families (e.g. Corixidae, Gerridae, Nepidae, Belostomatidae, Gelastocoridae).

Hemiptera are found in all forest strata but distinct assemblages may be present at different levels, and between plant species in just one stratum. This may be due to host-specific hemipterans targetting a particular plant. However, there is a general trend to a reduction in species with increases in latitude, and rare species may be vulnerable to the direct environmental consequences of climate change, and to indirect effects caused by changes to the distribution of their host plants (Andrew and Hughes 2005). Some herbaceous Hemiptera specifically feed on flowers and plant shoots due to the high levels of soluble nitrogen present in those structures (see Verne 2007). Hemiptera associated with flowering trees in lowland rainforest include Alydidae, Anthocoridae, Aphididae, Aphrophoridae, Cicadellidae, Eurymelidae, Flatidae, Issidae, Lygaeidae, Membracidae, Miridae, Nogodinidae, Pentatomidae, Psyllidae, Reduviidae, Scutelleridae, Tessaratomidae and Tingidae (Williams 1995), though some such as Eurymelidae only rarely. Members of the Auchenorrhyncha superfamilies Cercopoidea (froghoppers), Membracoidea (leaf and treehoppers) and Fulgoroidea (planthoppers) include some of the most powerful jumpers in the insect world (Burrows 2014; Donaldson 1983, 1991).

The order arose in the Early Permian and was well established by the Late Permian. Relictual Gondwanan taxa are defined by a combination of primitive morphological features and host relationships, disjunct distributions (especially in montane rainforests), and by being related to taxa in other Gondwanan landmasses, especially South America (e.g. *Kumaressa* – Aradidae) (Carver *et al.* 1991; Monteith 1969).

Taxa originating in Australia (autochthonous elements) consist of those derived from the original Gondwanan stock that diversified in response to climate change and evolution of vegetation (e.g. many Eurymelidae, Cicadellidae, Cicadidae, Coreidae–Amorbini, Psylloidea and Coccoidea) (Carver *et al.* 1991). There is a high level of endemism in the Cicadelloidea (e.g. Cicadellidae, Eurymelidae) with more endemic groups being recorded from Australia than in any other area (F. Evans 1981; J. Evans 1966). J. Evans (1966) suggests that this is related to the long isolation of Australia during the Tertiary period. The dominant cicadelloid family Eurymelidae is confined to Australia, New Guinea and New Caledonia (Carver *et al.* 1991; J. Evans 1966).

Achilidae This is a cosmopolitan family of sap-sucking bugs found in rainforest, wet sclerophyll forest and open forest types. They inhabit rainforest canopies and understoreys (Basset 1990, 1991; Basset and Charles 2000; Casson and Hodkinson 1991). Adults hold their forewings flat and are generally phloem feeders (phloem being the vascular tissue that conducts synthesised food such as sugars), though some are mycetophagous (fungus feeding). Nymphs live under loose tree bark and crevices in rotting wood.

The Achilidae include the endemic genus *Aneipo*, adults of which are usually brightly coloured. *Aneipo ceres* and *A. minerva* are recorded from upland forests (e.g. Lamington National Park, Dorrigo, Barrington Tops, Upper Allyn River) (Lambkin 1978; Williams 2002).

Alydidae This is a widely distributed phytophagous family with most species occurring in the world's tropical and subtropical regions (Cassis and Gross 2002). They were once regarded as a subfamily of the Coreidae. Regionally occurring genera include *Riptortus* and *Noliphus*. In subtropical regions, species are sometimes found in loose aggregations on grasses and low shrubs on forest margins, but individuals also frequent *Acacia* in regrowth vegetation. Monteith (1982) recorded large aggregations during the dry season in monsoon forests of tropical Queensland.

Anthocoridae Anthocorids are similar in appearance to Miridae, though differ markedly in behaviour, and are known from Late Mesozoic compression fossils (Grimaldi and Engel 2005). Most are fully winged. They feed on insect eggs, the haemolymph of other arthropods, pollen and plant material, and are sometimes associated with the nests and burrows of birds and mammals (Carver *et al.* 1991; Kelton 1978). The Anthocorinae are commonly found on flowers, but many anthocorids are known from forest litter (Cassis and Gross 1995). Species respond to increases in altitude, with communities in rainforest in Lamington National Park being most abundant and species-rich at intermediate elevations (Ødegaard and Diserud 2011).

There is a high level of endemicity in the Australian fauna (Cassis and Gross 1995). The tribe Oriini has a world-wide distribution but is apparently poorly represented in Australia (Postle *et al.* 2001; Woodward and Postle 1986). The largest genus is *Orius*, which is diverse in southern and south-eastern Asia but only a few species (e.g. *Orius chadwicki* from Dorrigo, *O. gracilis* from Mt Tamborine and Murwillumbah) are described from Australia. *Orius chadwicki* is recorded on *Ficus coronata* (Moraceae) and *Dendrocnide excelsa* (Urticaceae), and *Orius gracilis* on *Cissus antarctica* (Vitaceae) (Postle *et al.* 2001); the intensely painful leaves of *Dendrocnide excelsa* are apparently no deterrent to habitation. *Australmeida derricki* is associated with leaf mould in rainforest (Cassis and Gross 1995),

Aphrophoridae Members of this spittle bug family are poorly represented in Australia, and are one of several hemipterans commonly referred to as froghoppers. The nymphs live above ground enclosed in spittle. These have a ventral abdominal channel into which air is taken and then expelled through a film of anal fluid; and thus bubbles of spittle are formed (in Carver *et al.* 1991). The resultant frothy mass likely serves as a defensive barrier against parasites and desiccation, for individuals are quite soft and otherwise vulnerable to attack during the nymph stage. The fauna includes *Eoptylelus australis* (Bunya Mountains), *Philagra concolor* (Mt Tamborine) and *P. recurva* (Binna Burra) (J. Evans 1966).

Aradidae Called flat bugs or bark bugs, the Aradidae are a widely distributed family that is well represented in tropical and subtropical rainforests, and wet sclerophyll forest (Monteith 1967; 1997, Williams 2002). They are mycetophagous, feeding on fluids derived from fungal myceliae and fruiting bodies in moist decaying wood. Many species are wingless. The Australian region is the only zoogeographical zone in which all (eight) subfamilies occur (Cassis and Gross 2002). The family is known from Cretaceous compression fossils, and from Burmese amber of the same period.

Adults and nymphs are highly flattened bugs that live under loose bark on standing but especially fallen rotting tree trunks. Species that live under bark are often gregarious (e.g. species of *Chiastoplonia*, *Drakiessa*). Some species, however, can be individually encountered on the outer surface of tree trunks, their form and patterned colouring allowing ready camouflage, even in daylight. A number (e.g. various *Drakiessa*, *Chiastoplonia*) coat themselves with soil and debris (Monteith 1997), such that this encrusting material affords them a high degree visual deception. This coating in some instances may be due to their litter-dwelling behaviour and heavy cuticle to which soil readily adheres (e.g. as in *Glochocoris*).

The Carventinae include the endemic flightless rainforest genus *Glyptoaptera*, which has no close relatives among the Australian taxa (Monteith 1967). Species are restricted to the eastern seaboard of the mainland and several are confined to the Gondwana rainforests region (e.g. *Glyptoaptera montana*, *G. dorrigoensis*, *G. woodwardi*).

The subfamily Chinamyersiinae is a relict group of possible southern origin, but the presence of the chinamyersiine genus *Gnostocoris* in Vanuatu and New Caledonia makes the evolutionary interpretation of the subfamily uncertain (Monteith 1969). The wingless genus *Kumaressa* (*Kumaressa carraiensis*, *K. scutellata*) is endemic to the Gondwana rainforests region, and otherwise is related to the wingless New Zealand genus *Tretocoris*.

The Australian Mezirinae exhibit no readily discernible affinities with the South American or South African fauna (Monteith 1997). There are, however, some relationships with the New Zealand and New Caledonia faunas, suggesting radiation of the group post-dating separation from Africa and South America. In Australia, most mezirines are confined to the high rainfall belt along the east coast. Seventy-five per cent of the Mezirinae are restricted to warm rainforest, and their distribution and evolution is strongly linked to the history of rainforest in Australia (Monteith 1997). The mezirine fauna within the Gondwana rainforests region separates into a Southern Queensland zone and a Northern New South Wales zone. The latter comprises the Ebor–Dorrigo and Carrai Plateau–Barrington Tops areas with 14 species (Monteith 1997). The Southern Queensland zone includes the Mt Tamborine, Main Range, McPherson Range and Mt Warning complex, and the Bunya Mountains, Kroombit Tops and Davies Range to the north. Twenty-five species are known from this zone (Monteith 1997). The Border Ranges has a high diversity of species, many of which are endemics. However, within the Border Ranges 'complex' (*sensu* Monteith 1993), there is considerable variation in the distribution of Mezerinae species between different rainforest massifs and, as with flightless carabid beetles, this lack of congruence between any two sites may reflect low dispersal capabilities of species across intervening lowland landscapes (Monteith 1993).

The eroded remnants of the Mt Warning shield volcano are an evolutionary centre for Aradidae and eight wingless Mezerinae are found on its remnants (Monteith 1997). Rainforest-inhabiting mezirine taxa within the Gondwana rainforests region include the cosmopolitan genus *Neuroctenus* (*Neuroctenus grandis*), *Ctenoneurus* (*Ctenoneurus meridionalis*), which has centres of diversity in the Afro-Malagasy and Indo-Pacific regions; *Brachyrhynchus* (*Brachyrhynchus sulcatus*, *B. wilsoni*), which occurs in Africa, Madagascar and the Indo-Pacific region; and the endemic genera *Drakiessa*, *Mesophloeobia* and *Neophloeobia* (Monteith 1997). The apparently rare *Drakiessa cantrelli*, which is restricted in range to mountainous rainforests associated with the Mt Warning shield volcano, forms a closely related species-pair with *D. glaebula* from north Queensland (Monteith 1997). *Drakiessa minor* inhabits dry rainforest and open eucalypt forest between Eungella, in central coastal Queensland, and Brisbane. It is sympatric with *D. tertia* and *D. consobrina* in the southern part of its range, but is not known from more moist rainforest types inhabited by these two species (Monteith 1997). *Drakiessa hackeri* (Plate 71) is a large species found on logs in open forest and marginal rainforest. *Neophloeobia montrouzieri* is the most southerly distributed member of its genus and, in concert with *Mesophloeobia australica*, is the southern-most apterous mezirine found in Australia. The small-sized *Chiastoplonia minuta* extends from northern Queensland to southern New South Wales (e.g. Minnamurra Falls), occurring particularly in upland rainforests (Monteith 1997). The Indo-Pacific genus *Glochocoris* includes species restricted to rainforest and wet sclerophyll forests (e.g. *Glochocoris monteithi*) as well as normally open forest species that also exhibit seemingly anomalous isolated occurrences in rainforest (e.g. *Glochocoris brisbanensis* at Fraser Island).

Cercopidae Cercopidae are a cosmopolitan family of jumping bugs that are particularly speciose in tropical regions. They are known from Late Eocene compression fossils (Grimaldi and Engel 2005) and, in similar habit to the closely related Aphrophoridae, nymphs secrete a frothy mass in which they live while feeding on plants. The spittle mass usually completely envelops the nymphs such that they remain quite hidden. Species include *Eoscarta vacuola*, *Tonnoiria chinae* and *T. tasmaniana*, all recorded from south-eastern Queensland (J. Evans 1966).

Cicadellidae The cicadellid leafhoppers are one of the largest insect families and world-wide number more than 22 000 known species (Wang *et al.* 2017) (Plate 75). The Australian fauna is particularly associated with open forests and woodlands, and has a high proportion of arboreal species; adults and nymphs are almost exclusively feeders in the xylem and phloem of angiosperms. The family is richly represented in the forests of subtropical south-eastern Australia (Williams 2002), including species that inhabit rainforest (e.g. *Balocha xantho*, *Bharoopa clavostignata*, *Stenalsella testacea*, *Dorrotartessus dorrigensis*, *Austrotartessus monteithi*, *Eutartessus gressitti*) (F. Evans 1981; J. Evans 1966; Webb 1983). *Ishidaella angustata* is sometimes associated with mass-flowering canopy trees (G. Williams 1995).

There are several endemic taxa of subfamily or tribal status (Austroagalloidinae, Reuplemmelini, Stenocotini, Tartessinae, Thymbrini, and Trocnadini), and of these the Stenocotini (e.g. *Smicrocotis*), Tartessinae (e.g. *Pingellus*, *Tartessus*) and Thymbrini (e.g. *Macroceps*, *Stenalsella*) occur in the Gondwana rainforests region (F. Evans 1981; J. Evans 1966). The tribe Ulopini (e.g. *Woodella*) are pre-Tertiary (>70 mya) relics. Tartessinae are Australia-centred and are adept at colonising rainforest (F. Evans 1981).

Of the 16 recognised Ledrinae-Thymbrini genera, 11 are restricted to Australia (Stevens 1994). The genus *Putoniessa* includes *P. minima* (known from the McPherson Range), *P. watsoni* (Dorrigo National Park), and *P. rieki* (restricted to the Gondwana rainforests region and adjacent areas) (Stevens 1994). *Putoniessa* is confined to Australia, New Guinea, New Zealand and the Chatham Islands. Within Australia *Putoniessa* exhibits a disjunct Bassian distribution, with several species extending north into the south-eastern sector of the Torresian zone (Stevens 1994).

The subfamily Idiocerinae is cosmopolitan in distribution but of the 28 genera only six have been recorded outside of the Australian region (M. Webb 1983). Although most Australian genera are recorded from the north-east (which overlaps with the Oriental region), >33 per cent of known idiocerine generic diversity occurs in northern New South Wales and south-eastern Queensland (see M. Webb 1983). Four monotypic genera occur in this region: *Nyndgama* (*Nyndgama arowa*), which is restricted to the McPherson Range; *Bharoopra* (*Bharoopra clavosignata*), which is confined to rainforest between south-east Queensland and southern New South Wales; *Bundabrilla* (*Bundabrilla clovella*) from Lamington National Park; and *Tinderella* (*Tinderella maondica*), which is known only from Mt Tamborine (M. Webb 1983).

Cicadidae Cicadidae (cicadas) are a conspicuous and seasonally noisy component of the subtropical wet sclerophyll forest, and at times the subtropical rainforest, environment. Cicadas are related to the Cercopidae and Cicadellidae, these being placed in the infraorder Cicadomorpha (see Carver *et al.* 1991, Sanborn and Villet 2014).

True cicadas are known from the Early Jurassic (Grimaldi and Engel 2005), and there are now ~1200 species world-wide. Species are phytophagous, adults laying eggs in a series of slits along branches of the host plant, and nymphs feeding subterraneously (often deep underground) by piercing small plant roots and sucking up sap. There is a high level of host specificity, but eucalypts are favoured by many species. Adults may choose living plant hosts (as by *Cystosoma saundersii*) or dying branches (*Henicopsaltria eydouxii*) to lay their eggs (Moulds 1990). Nymphs are fossorial, and adult males ascend to favoured trees, there to call to females. Emergence can be synchronous; however, cicada species may exhibit different emergence patterns (Ewart 2001). Species of *Psaltoda* often exhibit an explosive emergence pattern, whereas others have no clear emergence peak, their numbers rising progressively, or apparently intermittently at low abundance levels, throughout spring and summer. The presence and abundance of species and populations appears linked at least partly to seasonal rainfall, with low cicada abundance and diversity occurring in years of low or infrequent rainfall.

The Australian Cicadidae are highly endemic with the great majority of genera, and almost all species, restricted to Australia (Carver *et al.* 1991; Moulds 1990, 2012). Numerous smaller cicadas were placed in the family Tibicinidae (De Boer 1997); however, these should be retained in the Cicadidae (M. Moulds pers. comm.).

Of the Tibicininae, three small endemic genera, *Chlorocysta* (three species), *Glaucopsaltria* (one species) and *Cystosoma* (two species), commonly referred to as bladder and bottle cicadas, occur in wet forest types of the Gondwana rainforests region (De Boer 1997; Moulds 1990). These three genera are restricted to the east coast from northern Queensland to the central coast of New South Wales. The distribution of *Chlorocysta vitripennis* (Plates 72, 73) is primarily confined to south-east Queensland and northern New South Wales (Moulds 1990). Tibicininae that inhabit subtropical rainforest and wet sclerophyll forest include *Birrima castanea*, *Abricta curvicosta* (also inhabits open environments), *Cicadetta denisoni*, *Cicadetta forresti*, *Chlorocysta vitripennis*, *Glaucopsaltria viridis* and *Pauropsalta rubristrigata* (Moss and Moulds. 2000; Moulds 1990; Williams 2002). The species most closely identified with subtropical rainforests is arguably the large bladder cicada *Cystosoma saundersii*, its low drone call often continuing into the early hours of the evening, well after the onset of night. Its distribution is largely centred on south-east Queensland and northern New South Wales, but isolated populations extend to North Queensland.

Rainforest and wet sclerophyll forest-inhabiting Cicadinae include the genera *Arunta* (*Arunta perulata*, *A. interclusa*), *Psaltoda* (*Psaltoda brachypennis*, *P. moerens*, *P. plaga*), *Henicopsaltria* (*Henicopsaltria eydouxii*), *Cyclophila* (*Cyclophila australasiae*), *Tamasa* (*Tamasa rainbowi*) and *Topha* (*Topha saccata*) (Moss and Moulds 2000; Moss and Popple 2000; Moulds 1990; Williams 2002). These species also frequent open forest; *Arunta perulata* occurs in mangroves but commonly in *Banksia integrifolia* trees where these inhabit regenerating littoral rainforest stands. The related *Arunta interclusa* inhabits a range similar to *A. perulata*, though not extending so far south (see Moulds 1990), but is almost exclusively restricted to mangroves. *Henicopsaltria eydouxii* (Plate 74) and *Topha saccata* frequent lowland wet sclerophyll forest and riparian rainforest, where the massed chorusing of gregarious males in summer project the most discomforting of cicada 'songs'. Most *Myopsalta* occupy open forest, woodland and grassland; however, *M. umbra* occurs in dry rainforest (Expedition Range, Queensland). Populations inhabit remnant patches, being typically found in the middle stratum on woody vegetation (Popple 2017).

The closely related Tettigarctidae, or hairy cicadas, include just two species (*Tettigarcta crinita*-New South Wales–Victoria, *T. tomentosa* – Tasmania). Although unknown from the Gondwana rainforests region, suitable habitat of tall eucalypt forest with a dense fern understorey exists in the Barrington Tops. The most northern known record is Mt Wilson (900 m), for *Tettigarcta crinita* (Moulds 1990). Tettigarctidae are a primitive relict group confined to south-eastern Australia, but *Eotettigarcta* from the Eocene of Scotland suggests the family was once more widespread (Grimaldi and Engel 2005).

Cixiidae This is a cosmopolitan family of phytophagous, membranous-winged leafhoppers (Carver *et al.* 1991; Löcker 2015; Löcker *et al.* 2006, 2007). Species are often a species-rich and abundant component of tropical and subtropical rainforest hemipteran communities, and are mainly generalist sap suckers with low host-specific relationships (Basset 1999; Casson and Hodkinson 1991; Dem 2015; Dem *et al.* 2013; Le Cesne *et al.* 2015). Cixiidae have been shown to exhibit a reduction in species richness in rainforest with increases in altitude (Le Cesne *et al.* 2015). Adult males and females are known to emit vibration signals of short uniform pulses, of a simple temporal pattern. These are aubible to the human ear as separate clicks (Tishechkin 2016).

Species recorded from subtropical eastern Australia include *Ozoliarus umbella*, *O. bumarangoides*, *Oteana lubra* (Pentastirini), *Innobindus licinus*, *I. imarginatus*, *I. robinae* and *I. alternans* (Brixiini) (Löcker *et al.* 2006, 2007). Brixiini are known from Lower Cretaceous fossils (Löcker *et al.* 2006), though the family as such is also recorded from the earlier Jurassic (Grimaldi and Engel 2005).

Coreidae Coreidae, or leaf-footed bugs, are a cosmopolitan, mainly tropical, family with ~1800 described species world-wide, and are known from Early Cretaceous compression fossils (140–120 mya) (Grimaldi and Engel 2005). They are often large in size and produce strong odorous fluids when threatened. Several species have distinctly swollen hind femora and expanded flattened flanges on the hind tibiae. Cassis and Gross (2002) cite 83 Australian species, the fauna being especially diverse in tropical and subtropical regions. Most species are phytophagous, are known from a broad range of hosts (frequently on *Eucalyptus*), and may feed on plant reproductive parts (Schaefer and Mitchell 1983; Schuh and Slater 1995). Regional species include *Amblypelta nitida*, *Amorbus alternatus*, *A. damelus*, *A. obscuricornis*, *A. rubiginosus*, *A. robustus*, *Pachycolpura manca*, *Pomponatus luridus* and *Tambourina kelsalli* (Brailovsky and Monteith 1996; Williams 2002; G. Williams unpubl. records). However, a spectacular species is the tortuously-named *Canungrantmictus morindana* (Brailovsky 2002) (Plate 76). This is a robust coreid in which the anterolateral angles of the pronotum are armed with large stout and serrated horns that often project past the head. The known distribution of the species extends from Carnarvon National Park in central Queensland, south to lowland subtropical rainforest of the Lansdowne Escarpment and littoral rainforest at Camden Head in New South Wales (Brailovsky 2002; G. Williams unpubl. records) and to warm temperate rainforest near Sydney (P. Woodward pers. comm.). It is associated with the vine *Gynochthodes jasminoides* (syn. *Morinda jasminoides*) (G. Monteith pers. comm.), a widely distributed member of the Rubiaceae.

Derbidae Derbidae are a huge family of small fragile-winged insects (M. Fletcher 2013), and are known from

Eocene Baltic amber (Szwedo 2005). They occur in all zoogeographical regions. Nymphs are known to be mycetophagous, especially fungi found in rotting wood. The adults are phytophagous. The genus *Zoraida* (Plate 77) has distinctive long and narrow wings. Species are occasionally encountered on foliage in subtropical rainforest and wet sclerophyll forest. *Lydda elongata* is recorded feeding on *Pandanus* (Pandanaceae) in tropical Queensland, but to date no Derbidae have been recorded from southern populations of that genus; *P. tectorius* reaches its most southern point of distribution along coastline fringes and seaward littoral rainforest margins in the Laurieton–Red Head region of northern New South Wales.

Eurymelidae This leafhopper family is restricted to Australia, New Guinea and New Caledonia (Carver *et al.* 1991; J. Evans 1966). Early Australian eurymelids (*Mesojassus*) are known from the Upper Triassic (Jell 2004). Species are phytophagous and feed on plant phloem. Nymphal stages are gregarious, and this behaviour may extend into the adult stage (J. Evans 1988). Some species are notable for secreting sugar-rich 'honeydew' (a digestion by-product of feeding on phloem), which attracts attending ants (trophobiosis), and upon which the ants feed (Hölldobler and Wilson 1990; Williams and Woinarski 1997). Though several eurymelids are to be met in wet forest types (e.g. undetermined species on *Alphitonia excelsa* [Williams 1995]), they are better known from open forest and woodland where species are often associated with eucalypts, *Acacia*, and she oaks (*Casuarina*, *Allocasuarina* – Casuarinaceae). Basset (1991) notes the absence of Eurymelidae in samples of the arthropod fauna collected from Black Booyong *Argyrodendron actinophyllum* (Sterculiaceae/Malvaceae) in rainforest in south-east Queensland, and McQuillan (1993) does not list the family from Australian *Nothofagus*, though he highlights its association with *Eucalyptus*.

Most published regional records are for species found in upland forests (e.g. *Eurymela rubrolimbata*, *Eurymelessa moruyana*, *Eurymeloides marmorata*, *E. perpusilla*, *Pauroeurymela parva*, *Ipo pellucida*, *Ipoella colmani*, *I. fidelis, Katipo rubrivenosa*) (Day and Fletcher 1994; J. Evans 1966). The distribution of many of these is widespread on mainland Australia.

Flatidae Adults of this plant-feeding family are characterised by wedge-shaped forewings that, when folded, give a laterally compressed profile similar to that of the Ricaniidae, but more steeply angled upwards (Carver *et al.* 1991; M. Fletcher 1985, 1988, 2005). In combination with green or brownish coloration (depending on the species), this triangular shape provides a cryptic plant-like image when individuals are motionless. But when disturbed, adults and free-living nymphs produce rapid, strong jumps, achieved by a catapault-like propulsion mechanism in which the hind legs are powered by a pair of large elastic recoil muscles located in the thorax (Burrows 2014). Nymphs, though, are usually restricted in movements, and produce conspicuous wax-like filaments.

Adults are normally associated with sunlit foliage, either in the canopy or at lower levels of exposed rainforest margins. Species of *Siphanta* (M. Fletcher 1985) are recorded from flowering trees in lowland rainforest remnants (i.e. *Guioa semiglauca*, *Scolopia braunii*, *Syzygium smithii*, *Tristaniopsis laurina*) (Williams 1995). *Massila* is recorded from *Alphitonia excelsa* and *Guioa semiglauca* (Williams 1995; Williams and Adam 2001).

Fulgoridae This is principally a tropically distributed family, popularly called lantern flies, of relatively large plant hoppers, and includes numerous colourful and ornately shaped Neotropical and Asian species (e.g. *Pyrops*, *Phrictus*, *Egregia*). They are phytophagous and feed on phloem. Nymphs and adults are found on leaves, but mostly are encountered on branches and tree trunks. Basset (1999) found them only rarely associated with tree seedlings in a Guyana rainforest. Some species aggregate and ovipositing is thought to take place on tree trunks (Johnson and Foster 1986). In Australia the Fulgoridae (e.g. *Desudaba psittacus*, *D. aulica*, *Eurinopsyche obscurata*) are mainly found along the east coast (Carver *et al.* 1991; Constant 2010). However, they are not common in lowland subtropical rainforest (G. Williams pers. observ.), though they have been collected in various canopy studies in south-east Queensland (e.g. Basset 1990, 1991). Adult *Desudaba psittacus* (Plate 78) from wet sclerophyll forest of the Lansdowne Escarpment have bright red-orange areas at the base of the hindwings. Large *Erilla* from the same site have forewings that in their basal half are dun-coloured, but largely clear in their apical half.

Gelastocoridae Ancestral Gelastocoridae, or toad bugs, originated in the Mid to Upper Jurassic, being abundant and diverse in the fossil record of that time (Grimaldi and Engel 2005). With the exception of the Palaearctic, gelastocorids are now found in all the world's major zoogeographical regions, but the majority of species occur in Australia, Melanesia and the Neotropics. *Nertha* is the only Australian genus and, as their common name suggests, they are flat and squat in appearance. Though closely related to the fully aquatic Belostomatidae and Nepidae, toad bugs are riparian ground-dwelling predators. Both nymphs and adults actively hunt their prey. Most inhabit muddy stream banks, pond shores, shallow ephemeral pools and rocky areas; however, dull-coloured *Nertha* will venture out into the rainforest understorey where they can be encountered in leaf litter or exposed on the surface of the forest floor.

Several *Nertha* species, including *N. alaticollis* and *N. sinuosa*, have been described from World Heritage Area reserves in northern New South Wales (Cassis and Gross 1995). The *alaticollis* species group (e.g. *Nertha alaticollis*, *N. hylaea*) is confined to temperate and subtropical Australia with the majority of known species found in south-eastern Australia (G. Cassis pers. comm.). *Nertha alaticollis* and *N. hylaea* are habitat generalists, inhabiting rainforest and sclerophyll forest types.

Idiostolidae Idiostolidae are a possibly ancient family of ground-dwelling, flattened bugs confined to Australia and southern South America, and were first placed as a subfamily of Lygaeidae (Scudder 1962). Though little is known of their biology it has been suggested they are phytophagous (Schuh and Slater 1995), but this is contentious (Cassis and Gross 2002). Two genera (*Trisecus*, *Monteithocoris*) and three species occur in Australia (Cassis and Gross 2002). All species apparently inhabit moss beds associated with *Nothofagus*-dominated cool temperate rainforest, a habitat similar to that of the Peloridiidae, and one similar to that of the South American idiostolid *Idiostolus insularis*. The genus *Trisecus* is distributed from southern Queensland to Tasmania (Carver *et al.* 1991). *Trisecus armatus* is recorded from the Border Ranges, Barrington Tops, and New England National Park (Cassis and Gross 2002; Monteith 1993).

Largidae Though having a cosmopolitan distribution, the Australian largid fauna is a small one, comprising just two genera (*Delacampius*, *Physopelta*) and four species in a single subfamily (Physopeltinae) (Cassis and Gross 2002; Schuh and Slater 1995). Species are considered to be phytophagous, feeding on seeds or plant sap. Largidae are usually found on foliage; however, *Physopelta gutta* (Plate 79) has been collected at light traps on the margins of lowland rainforest and wet sclerophyll forest, adults sometimes occurring in large numbers (G. Williams unpubl. data).

Lygaeidae Lygaeidae, as established by Henry (1997), have a world-wide distribution and comprise three subfamilies: Lygaeinae, Ischnorhynchinae and Orsillinae. Some authors have elevated several subfamilies (e.g. Orsillinae) to family rank but here I follow Cassis and Gross (2002). Several introduced species are important pests of crops. Lygaeinae are the most diverse and are widely distributed, but Australian generic edemicity is low. The Australian fauna numbers more than 80 species (Cassis and Gross 2002; Slater 1975). Several species occurring in subtropical eastern Australia possess red-black aposematic warning coloration (e.g. *Scopiastella nigricoronata*, *Scopiastes degeeri*, *S. impeditus* – Lygaeinae).

Lygaeids are usually diurnal, feeding on the nutrient-enriched endosperm of seeds, and are recorded from a broad range of plant families. The feeding habits of the Orsillinae (given family status by Sweet 2000) are the most generalised, eating vegetative parts, seeds and flowers (Ashlock 1967; Schuh and Slater 1995). Species of *Arocatus* (Lygaeinae) and *Nysius* (Orsillinae) exhibit broad host associations in lowland rainforests (Williams 1995). Species of *Nysius* are recorded from flowering *Alectryon coriaceus*, *Alphitonia excelsa*, *Euroschinus falcatus*, *Guioa semiglauca*, *Syzygium floribundum*, and *S. smithii*. *Arocatus* is recorded from flowering *Alectryon coriaceus*, *Alphitonia excelsa*, *Cryptocarya microneura*, *Cuttsia viburnea*, *Drypetes deplanchei*, *Elaeocarpus obovatus*, *Euroschinus falcatus*, *Syzygium floribundum* and *S. smithii* (Williams 1995). *Parsonsia straminea* (Apocynaceae), a vine common to wet forest margins in lowland sites, is a host to *Arocatus aenescens* and *A. rusticus* (Lygaeinae). *Oncopeltus sordidus* (Lygaeinae) (Plate 80) also is associated with *Parsonsia straminea* and, in addition, is a pollinator of the mangrove *Avicennia marina*, its ventral surface and legs possessing long setae among which pollen grains are snared (G. Williams unpubl. records). Although *Oncopeltus sordidus* undertakes few movements to adjacent trees, *Avicennia marina* is somewhat self-compatible (Clarke and Myerscough 1991); thus movements between individual floweres in the same cluster, or adjacent clusters on the same tree, can be sufficient to achieve a level of pollination. *Oncopeltus sordidus* has a broad altitudinal and habitat range extending from the immediate coast (e.g. Harrington, littoral rainforest), intermediate zones (Lansdowne Escarpment, subtropical rainforest and wet sclerophyll forest) to higher elevations (Barrington Tops, cool temperate rainforest). *Parsonsia straminea* extends throughout this range.

Additional species, recorded from regional rainforests and wet sclerophyll forests, include *Pylorgus australensis* (Ischnorhynchinae), *Arocatus montanus*, *Scopiastes laticeps* and *Thunbergia augur* (Lygaeinae) (Cassis and Gross 2002).

Membracidae The phytophagous treehopper family Membracidae includes numerous species with bizarre pronotal horns and projections (see Carver *et al.* 1991; Grimaldi and Engel 2005). This conspicuous feature visually characterises the family, giving many species the cryptic appearance of tree twig nodules. Species inhabiting open forest are often gregarious though those inhabiting rainforest are usually encountered as isolated individuals on twigs and leaves. Specimens are known from Miocene Dominican amber; however, the family is considered a geologically recently evolved one.

In general Australian membracid genera are more closely related to one another than to other world faunas (Day 1999). Although no membracid bugs are endemic at the subfamily or tribal level, there are numerous endemic genera (e.g. *Ceraon*, *Eutryonia*, *Lubra*, *Neocanthuchus* [Day 1999]). The subfamily Centrotinae is possibly Gondwanan in origin, and the Australian and South African faunas exhibit close relationships (Day 1999). The Australian mainland rainforest genera comprise taxa (e.g. *Lubra*) possibly widespread in ancient rainforests and elements (e.g. *Sertorius*) that are possibly more adapted to open habitats, and which are only occasionally encountered in rainforest. Day (1999) suggests that the colonisation of rainforest by this second group may have been geologically more recent. Geologically recent membracids associated with the Gondwanan plant families Myrtaceae and Proteaceae are participating in an asymmetrical relationship (Wallace and Deitz 2006), for their apparent origins are quite dissimilar.

Genera recorded from the subtropical south-eastern Australia comprise *Acanthucalis*, *Acanthuchus*, *Alocanthella*, *Alosextius*, *Bucktoniella*, *Bunyella*, *Cebes*, *Ceraon*, *Eufairmairia*, *Eufairmairiella*, *Eufrenchia*, *Eutryonia*, *Lubra*, *Neocanthuchus*, *Otinotoides*, *Pogonella*, *Pogonotypellus*, *Polonius*, *Sarantus*, *Sertorius*, *Strzeleckia* and *Terentius* (Day 1999; Day and Fletcher 1994; J. Evans 1966), but only a subset of these inhabit rainforest and wet sclerophyll forest (e.g. *Eutryonia monstrifer*, *Lubra spinicornis*, *Neocanthuchus barringtonensis* [Williams 2002]). Though a rich group in Australia, Membracidae have been notably absent in some studies of subtropical rainforest canopy faunas (e.g. Basset 1991), and Williams (1995) encountered them only rarely (among >60 000 specimens overall) during the mass-sampling of flowering trees in lowland rainforest remnants (undetermined species on *Alphitonia excelsa*, *Pogonella minutus* on *Scolopia braunii*, *P. bispinus* on *Tristaniopsis laurina*). Membracids are, however, canopy-frequenting, honeydew-producing trophobionts of the weaver ant *Oecophylla smaragdina* in North Queensland tropical rainforest (Blüthgen and Fiedler 2002).

Miridae Miridae are known from Cretaceous amber, and are now the most diverse family of heteropteran Hemiptera (Cassis and Gross 1995; Shuh 1995). They currently number more than 12 000 described species; the Neotropical fauna being particularly rich. The majority of species are plant feeders, but some are predators of small soft-bodied arthropds, such as aphids, and the eggs of arthropods. Phytophagous species are generally specific to particular host plants (Grimaldi and Engel 2005). Ant mimicry is common and has evolved several times within the family (e.g. *Myrmecoridea* – Orthothylinae). Adaptations include shortening and deceptive patterning of the wings, and elongation of the head and constriction of the abdomen (Cassis and Gross 1995). In subtropical rainforests adults are found on foliage, in leaf litter and on flowers. Many are coloured green, are difficult to discern when motionless on leaves, but are very agile and readily take flight when disturbed. When disturbed some quickly fall from whatever leaf they were on, quickly land on lower foliage, and then relocate to other leaves.

Mirids have been recorded as minor pollinators of Asian *Shorea* species (Dipterocarpaceae), and (in concert with Anthocoridae), as primary pollinators of *Macaranga tanarius* (Euphorbiaceae) in Okinawa and the Amami Islands, Japan; adults feed on nectar from extrafloral nectaries and possibly breed in chambers formed by flower bracts (Ishida *et al.* 2009; Williams and Adam 2010). In North Queensland the widely distributed *Corizidolon australiense* (Mirinae) is found abundantly on the flowers of *Ryparosa kurrangii* (Flacourtiaceae/Achariaceae), feeding on nectar, and perhaps pollen, but seemingly doing little apparent damage to floral organs (Webber *et al.* 2008). Most individual bugs were found to be non-dispersive nymphs, and so it is uncertain what pollinator role *Corizidolon australiense* might play. In subtropical rainforests undetermined species of Miridae have been recorded from flowering *Acradenia euodiiformis*, *Alectryon coriaceus*, *Alphitonia excelsa*, *Cuttsia viburnea*, *Euroschinus falcatus*, *Guioa semiglauca*, *Rhodomyrtus psidioides*, *Syzygium floribundum*, *S. smithii* and *Tristaniopsis laurina* (Williams 1995; Williams and Adam 2010). None was observed damaging flowers.

Two monotypic genera, *Porophoroptera* and *Corizidolon*, have been described from the Gondwana rainforests and are largely known only from northern New South Wales and south-eastern Queensland (Cassis and Gross 1995). *Peritropis basseti*, a member of the large and diverse subfamily Cylapinae, was described from specimens collected on the rainforest tree *Argyrodendron actinophyllum* at Mt Glorious (Moulds and Cassis 2006). The Australian *Leucophoroptera* (Phylinae) group fauna (e.g. *Blesingia*, *Porophoroptera*, *Sejanus*) has a Gondwanan association but is intermediate between that of southern Africa and South America. The group is strongly centred on the Australian region, islands to the north, in India, and South Africa (Carvalho and Gross 1982). Three Phylinae species (*Shizopteromiris queenslandicus*, *Blesingia latezonata*, *B. tamborinea*) described from the Gondwana rainforests region are known only from their respective type localities (Cassis and Gross 1995). Additional mirid species recorded from the region include *Trilaccus nigroruber*, *Chaetedus longiceps*, *Dolichomirus linearis*, *Trogonotylus doddi*, *Porophoroptera excellens*, *Sejanus brittoni*, *Austrocapsus annulipes*, *Corizidolon australiense*, *Synthlipsis annulipes* and *Taylorilygus nebulosus* (Carvalho 1978; Carvalho and Gross 1982; Cassis and Gross 1995; Williams 2002).

Myerslopiidae This is a small family with a Gondwanan distribution (Rakitov 2015; Szwedo 2004), and was previously placed in the Cicadellidae (Carver *et al.* 1991). The tribe Myerslopiini has a disjunct 'austral' distribution and occurs in Australia, New Zealand and Chile (K. Hamilton 1999). The Sagmatiini are distributed in Madagascar and Australia, which may represent relict populations of a previously widespread fauna. *Sagmation horridum* is recorded from Lamington National Park (K. Hamilton 1999). The biology of the Australian fauna is little studied; however, the Chilean species *Mapuchea chilensis* has been collected from forest litter in temperate forests and shrubby bogs. The New Zealand fauna has been collected from moss and litter. They may feed on phloem sap. Adults and nymphs are strong jumpers (Rakitov 2015).

Ochteridae This is a small semi-aquatic family of littoral predators distributed from tropical to temperate regions of the world (M. Baehr 1989; Carver *et al.* 1991; Cassis and Gross 1995). They probably originated in the Upper Jurassic (Grimaldi and Engel 2005). Ochterids are reported to feed on fly larvae, Collembola and aphids. Nymphs are found in shallow water or partially buried. Oriental ochterids appear to be closely related to the Australian fauna (M. Baehr 1989). Two genera occur in the Australian region: *Megochterus* and *Ochterus*. *Ochterus atridermis*, *O. secundus secundus* and *O. australicus* are recorded from south-eastern Australia (Cassis and Gross 1995), but both are widely distributed elsewhere in eastern and northern Australia. *Ochterus australicus* is also recorded from New Caledonia, New Hebrides and the Solomon Islands.

Peloridiidae Peloridiids, or moss bugs, are small hemipterans restricted to eastern Australia, Lord Howe Island, New Zealand, New Caledonia, Chile and Patagonia. They are considered Gondwanan relics. Related 'primitive peloridiomorphs' occurred in the Jurassic and Cretaceous (Grimaldi and Engel 2005). Moss bugs are primarily phytophagous, are found in moist, rain- and mist-prone habitats, and are associated with moss and liverworts (Woodward 1956). *Hackeriella veitchi*, recorded from the McPherson Range and New England National Park, occurs in subtropical and cool temperate rainforests and is associated with mosses on dead branches of Antarctic Beech, *Nothofagus moorei* (Nothofagaceae) (Cassis and Gross 1995).

Pentatomidae Pentatomids (Plate 81) are commonly called shield bugs or stink bugs, and are mainly green or brown, and sometimes greyish in colour. They are medium-sized to large, convex bugs, and in diversity are second only to Miridae. The family is most likely no older than the Cretaceous, and is now found in all zoogeographical regions. Pentatomidae number more than 4000 species, and are especially diverse in the subtropics and tropics (Cassis and Gross 2002; Schuh and Slater 1995). Some undertake brood care (Grimaldi and Engel 2005).

They are found during daylight exposed on foliage and tree trunks, or sheltering under loose bark, but are uncommon on flowering rainforest trees or shrubs (e.g. on *Cuttsia viburnea*, *Syzygium floribundum*) (Williams 1995). Adults and nymphs are mainly sap-feeding herbivores, associated with a broad range of plant families, and feed on leaves, stems, immature fruits and seeds (Schuh and Slater 1995). However, Asopinae are predatory (Carver *et al.* 1991), feeding for example on Membracidae (Grimaldi and Engel 2005). When disturbed they may move away, carrying their impaled prey with them. Predatory species include *Oechalia schellenbergii*, *Amyotea hamata* and *Cermatulus nasalis*.

The majority of Australian genera and species are endemic (Cassis and Gross 2002). Genera represented in subtropical eastern Australia include *Anaxarchus*, *Antestiopsis*, *Aspideurus*, *Biprorulus*, *Commius*, *Cuspicona*, *Diaphyta*, *Diemenia*, *Eriobotes*, *Eysarcoris*, *Gilippus*, *Glaucias*, *Hypogomphus*, *Lubentius*, *Monteithiella*, *Morna*, *Nezara*, *Niarius*, *Ocirrhoe*, *Oncocoris*, *Paramenestheus*, *Plantia*, *Platycoris*, *Poecilometis* and *Vitellus* (Amhad and Kamaluddin 1989; Cassis and Gross 2002; Gross 1972; Williams 2002; G. Williams unpubl. data); however, only a subset of these (and included species) inhabit wet forest types. *Cuspicona*, *Oncocoris* and *Poecilometis* have speciated extensively in Australia (Carver *et al.* 1991; Gross 1972).

Pseudococcidae Early, definitively assigned, coccids originated in the Cretaceous (Koteja 1999), with fossil evidence suggesting a post-Eocene radiation (see Grimaldi and Engel 2005). The Australian true mealybug coccoid fauna includes more than 200 species in 66 genera (Ben-Dov 2012; Qin and Gullan 1989; D. Williams 1985). The majority of species are endemic to Australia and are possibly of Gondwanan origin. By comparison, the New Guinea mealybug fauna is poorly represented and comprises mainly South-east Asian elements (D. Williams 1985). The endemic genus *Conulicoccus* is constituted by three species, one of which (*Conulicoccus eucalypti*) is known only from New England National Park, where it occurs under the bark of *Eucalyptus pauciflora* (Myrtaceae) (D. Williams 1985). Rainforest-associated species include *Dysmicoccus laportae* and *Nipaecoccus passlowi*.

Coccids are preyed upon by ladybird beetles (e.g. *Cryptolaemus montrouzieri*, *Diomus*) of the family Coccinellidae (Ślipiński 2007). Female mealybugs are sedentary and lack wings, but the small winged males are gnat-like in appearance. Females attach themselves to their host plant (e.g. in subtropical rainforest *Gymnostachys anceps* – Araceae [G. Williams unpubl. obs.]) and secrete a powdery waxy layer. Many close symbiotic relationships with ants have evolved in this family, for they produce honeydew, which is fed upon by ants, and as ant-attended trophobionts mealybugs are able to draw ants into a closer relationship with the plant; the interrelationship of pseudococcids, plant and honeydew food resource influences the dynamics of interactions among different ant species and patterns of ant distribution and foraging behaviour (Blüthgen *et al.* 2004; Gullan and Cranston 2014). In Malaysia and Java nomadic ants of the genus *Hypoclinea* actively transport their mealybug 'herds' to different optimal feeding sites (Hölldobler and Wilson 1990).

Psylloidea The superfamily Psylloidea includes several families (Aphalaridae, Calophyidae, Homotomidae, Psyllidae, Triozidae) known as lerp insects, plant lice or psyllids, and are small active jumping insects that are phloem feeders (Burckhardt and Ouvrard 2012; White and Hodkinson 1985). They include many lerp-building species (e.g. *Glycaspis* in Psyllidae [K. Moore 1970; Yen 2002; Sharma *et al.* 2015]). Most Australian species are endemic (Carver *et al.* 1991). Psyllids are mainly associated with eucalypt forests, where large populations can be injurious (Carnegie and Angel 2005; Stone 1996), and interact with Bell Miners (*Manorina melanophyrs*) to cause tree dieback (Nageli *et al.* 2016; Whyte *et al.* 2011). Numerous species may frequent the foliage of rainforest trees and shrubs, and can be the most abundant of insect groups inhabiting the canopy stratum (Basset 1991). However, they are less frequently collected at the inflorescences of mass-flowering rainforest trees (Williams 1995).

Protective sugar-rich shelters (lerps) are constructed from excreted honeydew, and can take the shape of simple shell-like structures, or more complex cones and basket-shaped forms. Some Psylloidea, though, are gall formers (Fernandes and Santos 2014; Sharma *et al.* 2015). However, the probability of gall survivorship, and the susceptability of plant species to gall formation, may be dependent on leaf toughness (sclerophylly), which itself can be related to the age of leaves and the position of plants within various rainforest strata (Ribeiro *et al.* 2014). Nymphs are usually associated with particular plant hosts but adults may frequent, and oviposit on, a variety of plants (see Carver *et al.* 1991). *Acizzia pendulae* (Psyllidae) is associated with the mistletoe *Amyema pendulum* (Loranthaceae). Although the genus *Acizzia* has a widespread world distribution (G. Taylor 1999), the parasitic plant family Loranthaceae is highly host specific, with high levels of endemicity, and is considered Gondwanan in origin (G. Taylor 1999). *Trioza eugeniae* (Triozidae) attacks species of rainforest-inhabiting *Syzygium* (e.g. *Syzygium floribundum*, *S. paniculatum*, *S. smithii*) (Young 2003). Nymphs establish in pit galls on the undersurfaces of leaves, shoots and developing fruit. This can result in stunted plant growth. Invertebrate predators include larval coccinelid beetles, syrphid flies, ants, spiders and orthopterans (Young 2003). Homotomidae are associated with *Ficus* (Moraceae) (Carver *et al.* 1991).

Reduviidae Assassin bugs are a family of small to large-sized, predatory hemipterans that world-wide number ~6800 described species (Cassis and Gross 1995;

Hwang and Weirauch 2012) (Plates 82, 83, 84, 85). Based on their diversity of extant species, and their wide distribution within terrestrial ecosystems, reduviids represent one of the most successful groups of predatory animals. The family originated in the Mid Jurassic (178 mya), underwent significant diversification in the Late Cretaceous (97 mya) (Hwang and Weirauch 2012), and now has a cosmopolitan distribution; the Australian fauna comprises more than 200 species placed in 13 subfamilies. Some are ground dwelling, hiding by day under fallen rotting logs and loose tree bark, or on foliage or flowers, motionless in wait of prey (bees and other flower-visiting insects). Certain northern Australian species (e.g. of Harpactorinae) are cave dwellers (Moulds and Bannink 2012).

Adults may be brightly coloured red or pink (*Catasphactes*, *Gminatus*), variously patterned (*Peirates*, *Pristhesancus*), or simple tones of brown (*Dicrotelus*, *Horcinia*, *Ischnolestes*, *Myiophanes*, *Opistoplatys*, *Piestolestes*, *Sphedanocoris*, *Thodelmus*). Many are robust and ovoid in shape. However, *Ischnolestes* and the feather-like *Myiophanes* are elongate and seemingly delicate in appearance. Species of *Opistoplatys* are somewhat flattened, suggesting they hide under loose bark. *Opistoplatys australasiae* has been collected at light traps. World-wide the Reduviidae include generalist predators (e.g. Harpactorinae) that feed on many kinds of prey species (mainly other arthropods), and specialists that have evolved restricted prey choices, for example the millipede-feeding Ectrichodiinae, ant-feeding Holoptilinae and termite-specialist Salyavatinae (Cassis and Gross 1995; Hwang and Weirauch 2012). Holoptilinae (e.g. *Ptilocnemus*) have very conspicuous plumose hind tibiae, and glandular areas which secrete chemicals that apparently attract and paralyse ants upon which they prey (Gross and Malipatil, in Naumann 1991a; Weirauch and Cassis 2006).

Individual species exhibit daily patterns of hunting activity, regional species such as *Nebriscoides nitens*, *Nebriscus pupus* (Ectrichodiinae) and *Gardena brevicollis* (Emesinae) being nocturnal hunters and *Gminatus australis* and *Pristhesancus plagipennis* (Harpactorinae) hunting by day. The subtropical rainforest fauna includes arboreal ambush hunters that make limited movements, and agile species that actively seek out their prey on the forest floor. *Pristhesancus plagipennis* (Plate 85) is the most common species encountered on flowers (e.g. *Guioa semiglauca*) in lowland rainforests. However, *Pristhesancus plagipennis* is not a rainforest specialist, being also encountered in wet sclerophyll forest and tidally inundated mangrove communities. *Ischnolestes* (Harpactorinae) is also an occasional ambush hunter found on flowers.

Reduviidae listed from the Gondwana rainforests world heritage area include *Ploiaria musgravei*, *Pseudobargylia involucrata*, *Coanus trabeatus*, *Gminatus australis*, *Ectomocoris ornatus*, *E. patricius*, *Sastrapada australica*, *Thodelmus impicticornis*, *Archilestidium ornatulum*, *Armstrongocoris singularis*, *Mendola puellula*, *Myiophanes tipulina*, *Nebriscoides nitens* and *Nyllius australicus* (Cassis and Gross 1995; Williams 2002).

Rhopalidae Known as scentless plantbugs, the Rhopalidae are a cosmopolitan family in Australia comprising only the genera *Liorhyssus* and *Leptocoris*. Species feed on seeds. *Leptocoris mitellatus* feeds on the ripe seeds of *Geijera parvifolius* (Rutaceae) and *Alectryon tomentosus* (Sapindaceae) (Kumar 1966). *Leptocoris ?tagalicus* (Plate 86) feeds on the seeds of *Alectryon coriaceus* in littoral rainforest (G. Williams unpubl. data).

Rhyparochromidae The Rhyparochromidae comprise more than 1800 species world-wide, with more than 180 species recorded from Australia. Numerous species have extralimital distributions (Cassis and Gross 2002). The family was earlier placed as a subfamily of Lygaeidae (see Carver *et al.* 1991; Cassis and Gross 2002; Henry 1997), where it had been the dominant subfamily. Flightlessness is common in the Rhyparochromidae (e.g. *Tomocoris australiensis*, *T. scutellaris*), and some species are considered ant mimics. Species are mainly ground dwellers that feed on fallen mature seeds (e.g. *Grossander major*, *Henicorthaea yeoi*, *Australotarma keirae*). Several are blood feeders (e.g. *Laticlerada monteithi*, *L. naumanni*) on vertebrates and are recorded from the nests of birds, possums and rodents.

The family includes the genus *Megadrymus* (Cassis and Symonds 2012), which in subtropical eastern Australia includes the rainforest-inhabiting species *M. aurorae* (Border Ranges, Mt Keira), *M. brunneus* (Victoria Park Nature Reserve, Mt Keira), *M. tenuicornis* (Mt Tamborine), and *M. terraereginae* (Wingham Brush Nature Reserve, Kenilworth State Forest).

Schizopteridae These are minute, large-eyed bugs, no bigger than ~2 mm in size, found in forest litter, soil, moss beds, and also are attracted to light (Cassis and Gross 1995). They are thought to be predators of small invertebrates. Schizopterids are found in rainforest (e.g. *Duonota decoricaudata*, *D. fusca*), including *Nothofagus* forest and wet sclerophyll forest, but also in open vegetation types

(*Rectilamina torquata*). Additional species recorded from subtropical eastern Australia include *Cryptomannus furvus* (Upper Yarraman State Park, Barrington Tops), *Duonota bicamaca* (Mt Clunie), *D. bimaculata* (Gibraltar Range National Park), *Hypselosoma oncerochilotum* (Bulburin State Forest), *Ogeria biprojecta* (New England National Park), *Pateena elimatula* (Moreton National Park), and *Rectilamina spinosisura* (Lamington National Park) (Cassis and Gross 1995; Williams 2002).

Scutelleridae Jewel bugs are medium- to large-sized bugs that are often brightly coloured, or with a metallic iridescence (Plate 87). The scutellum is particularly enlarged in this family, covering the abdomen like the elytra of beetles. World-wide there are more than 450 species (Cassis and Gross 2002). Cassis and Vanags (2006) list 25 Australian species placed in 13 genera; most species are geographically broadly distributed. The family is most diverse in the tropics and subtropics. Many species exhibit aggregation behaviour and several undertake brood care with females protecting eggs from predators. Female *Tectocoris diophthalmus* guard their own egg clusters, as well as unrelated eggs, until hatching (Giffney and Kemp 2016).

The family is exclusively sap-feeding herbivores, but unlike Miridae where host specificity is high, scutellerids tend to be found on a broad range of plants. Recorded plant assocations for rainforest-frequenting Scutelleridae include *Coleotichus artensis* (on *Alphitonia excelsa*); *Cantao parentum* (on *Mallotus philippensis*, *M. claoxyloides*, *Araucaria cunninghamii*); *Tectocoris diophthalmus* (on *Brachychiton acerifolius*); and *Lampromicra senator* (syn. *Philia aerea*) (on *Breynia oblongifolia*) [Cassis and Gross 2002; Cassis and Vanags 2006]). *Lampromicra senator* is a common species occurring in littoral rainforest (e.g. Iluka, Harrington) and lowland escarpment rainforest where it can be observed resting on the foliage of numerous plants. *Tectocoris diophthalmus* has become established in urban gardens in eastern Australia, utilising ornamental members of the Malvaceae as hosts.

Tessaratomidae This is mainly a tropical family, best represented in the Eastern Hemisphere (Carver *et al.* 1991; Cassis and Gross 2002; Kumar 1969; Leston and Scudder 1957; D. Sinclair 2002a, 2002b), and in Australia is mainly restricted in distribution to the east coast of the mainland. Commonly referred to as stink bugs, tessaratomids are able to spray defensive, highly irritant chemical fluids (containing alkanes, cimicine and aldehydes) from glands in the thorax (Monteith 2011). The Oncomerinae are the only subfamily occurring in Australia (Cassis and Gross 2002). In Australia the injurious Bronze Orange Bug *Musgraveia sulciventris* is the best known of the family. Species are foliage-frequenting and phytophagous, adults are large in size, and both adults and nymphs are typically broadly ovoid-elongate in shape. Several genera (e.g. *Lyramorpha*) undertake brood care (Monteith 2006, 2011), the nymphs aggregating in association with adults.

Species occurring in rainforest and wet sclerophyll forest of subtropical eastern Australia include *Erga longitudinalis*, *Lyramorpha rosea*, *Musgraveia sulciventris*, *Oncoscelis australasiae*, *Peltocopta crassiventris* and *Stilida indecora*. *Lyramorpha rosea* is common in littoral rainforests, and is often found on the foliage of *Cupaniopsis anacardioides* and *Arytera divaricata* (Sapindaceae). Numerous Sapindaceae are recorded hosts for *Stilida indecora* in subtropical rainforest (Monteith 2006). The tree *Mallotus discolor* (Euphorbiaceae) and the vine *Austrosteenisia blackii* (Fabaceae) are recorded hosts for *Peltocopta crassiventris*. *Oncoscelis australasiae* is recorded from *Melicope micrococca* and *Acronychia imperforata* (Rutaceae) (Cassis and Gross 2002; G. Williams pers. observ.). *Peltocopta crassiventris* reaches its southernmost known distribution at Iluka (Monteith 2006). At over 30 mm in length male *Oncoscelis australasiae* (Plate 88) are the largest members of Heteroptera occurring in south-eastern Australia, the species common in littoral rainforest but also occasionally in rainforests at intermediate altitudes of the hinterland (e.g. Lansdowne Escarpment).

Thaumastocoridae The family Thaumastocoridae is a Gondwanan relic and has an anomalous distribution, occurring in Australia and India, and also in the Neotropics and Florida (Cassis and Gross 1995). Individuals are sap-feeding herbivores, are small in size, and are strongly dorsoventrally flattened in similar fashion to Aradidae. Species from subtropical eastern Australia include *Baclozygum depressum*, *Onymocoris barberi*, *O. hackeri*, *Thaumastocoris australicus*, *T. hackeri* and *T. macqueeni* (Cassis and Gross 1995). *Elaeocarpus obovatus* (Elaeocarpaceae) is recorded as a host of *Thaumastocoris hackeri* and *Banksia integrifolia* (Proteaceae) is recorded as a host of *Onymocoris hackeri* (Drake and Slater 1957; H. Rose 1965).

Tingidae Tingidae are moistly small-sized bugs that world-wide number more than 2000 species (Cassis and Gross 1995). They are commonly known as lace bugs owing to the characteristic raised mesh-like appearance of their pronotum and forewings. Nearly all are phytophagous, are diurnally active, and in rainforest are commonly found on leaf surfaces, where they feed on the epidermis. Species are sometimes encountered on the inflorescences of mass-flowering trees: *Teleonemia scrupulosa* (*Alectryon coriaceus, Euroschinus falcatus, Guioa semiglauca, Schizomeria ovata, Scolopia braunii*), and an undetermined species (on *Alphitonia excelsa, Cuttsia viburnea, Tristaniopsis laurina*) (e.g. Williams 1995).

Species recorded from subtropical eastern Australia include *Allocader cordatus, Bunotingus camelins, Callithrincus serratus, Eritingis trivirgata, Inoma angusta, Lepturga magnifica, Nethersia nigritarsis, N. setosa, Parada popla, Tanybyrsa secunda, Tingis hackeri* and *T. perkensi* (Cassis and Gross 1995; Williams 2002). Although some have restricted known ranges (e.g. *Allocader cordatus, Inoma angusta*), many have latitudinal distributions that extend beyond this region.

Neotropical *Leptobyrsa decora* and several species of *Teleonemia* (*Teleonemia scrupulosa, T. elata*) were released in Australia in an attempt to control the exotic invasive weed *Lantana camara* (Verbenaceae). Individual species either have not established or have made little impact, with vast areas of northern New South Wales remaining subject to infestation. The South African *Carvalhotingis visenda* has been released as a potential biocontrol agent of Cat's Claw Creeper (*Macfadyena unguis-cati* – Bignoniaceae), a rampant invasive vine of rainforests in Queensland and northern New South Wales (Dhileepan *et al.* 2006, 2007).

Veliidae Known as small water striders and riffle bugs, this is a small family of predatory and scavenging Hemiptera that feed on aquatic animals such as mosquito larvae and ostracods. They are found in a variety of water bodies, including quiet running freshwater streams, and pools remaining in otherwise intermittently dry drainage lines. Several species are semi-aquatic (e.g. *Microvelia fluvialis*). Some aggregate in large numbers (Cassis and Gross 1995).

In the Gondwana rainforests region the Veliidae include the eastern Australian endemic genus *Drepanovelia*, and the monotypic genus *Lacertovelia* (*Lacertovelia hirsuta*) endemic to northern New South Wales (Andersen and Weir 2001).

THYSANOPTERA (thrips)

Thysanoptera are tiny fringe-winged insects that may have originated in the Early Triassic, and are now best known as plant pests. The structure of their mouthparts suggests a close relationship to Hemiptera; however, these are less specialised for sucking plant fluids (Grimaldi and Engel 2005). Thrips are generally phytophagous as adults and larvae, some are fungal spore feeders (e.g. *Herathrips nativus*), and a number are probably predatory (e.g. *Haplothrips bituberculatus, Xylaplothrips clavipes*). Leaf-frequenting species include gall formers (e.g. *Leeuwenia* spp.) and leaf rollers (e.g. *Teuchothrips* spp., *Hoodiella convergens, Tolmetothrips smilacis*). Some leaf rollers are known to coexist.

Though commonly encountered in the upper strata of rainforests thrips are also highly diverse and abundant in the understorey (Mound and Williams 2003; Williams 1995; Williams *et al.* 2001; Tree and Mound 2011). Tree and Mound (2011), for example, collected 81 species of thrips in litter samples and intercept traps from the understorey of subtropical rainforest in Lamington National Park, Queensland. However, the occurrence of thrips in rainforest is capricious. In a collection of 27 000 insects caught in total in flight intercept traps set in the understorey of rainforest at Lorien Wildlife Refuge (Lansdowne Escarpment), Lansdowne Nature Reserve and Kiwarrak State Forest (Manning Valley), only three thrips were collected (Williams 1995; Williams *et al.* 2001).

The Aeolothripidae are flower thrips and faculative predators on small arthropods (Marullo and Mound 1993; Mound 1972; Mound and Heming 1991). The family has traditionally been considered to be the most basal of the order (Grimaldi and Engel 2005). Aeolothripidae include the genus *Franklinothrips*. This is essentially a tropical genus with six species known from the Neotropics, two from Africa, one from Taiwan, and two from Queensland. The two Australian species, *Franklinothrips variegatus* and *F. basseti*, are confined to south-east Queensland (Mound and Marullo 1998). The monotypic *Erythridothrips* is confined to south-east Queensland and northern New South Wales (Marullo and Mound 1993) and occurs in subtropical and littoral rainforest (Williams 1995). *Asprothrips* (Dendrothripinae) comprises three species and is known only from India, Japan and South-east Australia (Mound 1999). *Cranothrips ibisca* has been recorded breeding in the flowers of *Pentaceras australis* (Rutaceae), and is the only eastern species of an otherwise western and arid zone genus (Tree and Mound 2011).

Thripidae (along with Phlaeothripidae) comprise more than 95 per cent of of thrips species (Grimaldi and Engel 2005), and are usually leaf feeders. The family includes the endemic genera *Kladothrips* (*Kladothrips maslini*) and *Pseudanaphothrips* (*Pseudanaphothrips araucariae*), as well as genera that are more widely distributed (*Asprothrips, Anaphothrips, Bolacothrips, Heliothrips, Neohydatothrips, Parabaliothrips, Thrips*) (Wells 1996a). *Pseudanaphothrips araucariae* was described from the male cones of *Araucaria bidwilli* in southern Queensland (Mound 2002). *Ensiferothrips primus* occurs on Casuarinaceae in Australia and New Caledonia (Mound and Heming 1991) as well as on a variety of subtropical rainforest trees, shrubs and vines (Mound and Williams 2003).

Phlaeothripidae feed on green leaves, fungal spores, pollen, and mosses (Mound and Heming 1991). They are regionally diverse (e.g. species of *Akainothrips, Antillothrips, Ethirothrips, Euoplothrips, Haplothrips, Heligmothrips, Horistothrips, Leeuwenia, Liothrips, Phaulothrips, Podothrips, Teuchothrips, Tolmetothrips*) (Mound 2004; Pitkin 1973; Tree and Mound 2011; Wells 1996a; Williams 2002). These include endemic genera (*Akainothrips, Heligmothrips, Teuchothrips*) and those with much broader distributions (e.g. *Ethirothrips, Haplothrips, Liothrips, Podothrips*). *Euoplothrips bagnalli* is likely a cleptoparasite within galls of other thrips, including those of *Liothrips* on *Piper novae-hollandiae* (Piperaceae) (Tree and Mound 2011). Gall formers include *Leeuwenia diospyri* (on *Diospyros pentamera* – Ebenaceae), *L. polyosmae* (on *Polyosma cunninghamii* – Escalloniaceae/Grossulariaceae), *L. sclopiae* (on *Scolopia braunii* – Flacourtiaceae) and *L. tetrastigmae* (on *Tetrastigma nitens* – Vitaceae) – all members of the tribe Leeuweiini, which usually induce galls on the leaves of rainforest trees (Mound 2004). Phlaeothripidae also include *Neohoodiella jennibeardae* (Mound and Williams 2003) (Plate 89), the larvae and pupae of which are associated with the leaves of *Ficus coronata* (Moraceae) and *Ripogonum elseyanum* (Ripogoneaceae), two quite unrelated families. The only other representative of the genus is *Neohoodiella grandisetis* from New Caledonia. *Neohoodiella jennibeardae* larvae are usually restricted to the heavily-veined undersurface of *Ficus coronata*, and unlike common flower-living phaeothripids are sluggish in behaviour.

In addition to *Neohoodiella* numerous other thrips are associated with leaves (Mound and Williams 2003), though such records do not necessarily indicate a true host plant relationship. Individual species may utilise (or at least be found on) a wide range of plants, including different growth forms; as does *Heliothrips haemorrhoidalis* (e.g. on *Eupomatia laurina, Palmeria scandens, Rhodomyrtus psidioides, Backhousia sciadophora, Citriobatus pauciflorus, Ripogonum album, Mischocarpus pyriformis, Pouteria australis, Tetrastigma nitens*), and *Ensiferothrips primus* (on *Hibiscus heterophyllus, Bridelia exaltata, Ficus rubiginosus, Trophis scandens, Streblus brunonianus, Notelaea longifolia, Citriobatus pauciflorus, Aphananthe philippinensis*). Many thrips recorded by Mound and Williams (2003) were more restricted in the range of plants upon which they were found; however, further collecting may increase the number of plant species.

Some thrips are pollen feeders, and these include an unrecognised suite of pollinators of Australian subtropical rainforest plants (Williams 1995; Williams and Adam 1994, 2010; Williams *et al.* 2001; Table 2). Flower-frequenting thrips can be seasonally found in considerable numbers. Plants in this pollinator relationship include vines (*Smilax australis, Maclura cochinchinensis*), shrubs (*Breynia oblongifolia, Wilkiea huegeliana*), and a diversity of small and large trees (Williams 1995; Williams *et al.* 2001). Thrips are also associated with *Ripogonum discolor* (Ripogonaceae), a vigorous-growing, creamish-white-flowered and highly fragrant vine common in the understorey of subtropical rainforests at intermediate elevations. Flowers do not possess obvious nectar rewards, but the pollen is dry and readily dusts onto the bodies of visiting thrips (G. Williams pers. obs.).

Flowers are often used as egg deposition sites in which larvae will later emerge, a requirement for this use being that flowers need to be structured in such a way as to conceal and protect the visiting adults and developing young (Williams and Adam 2010). This is especially demonstrated by the flowers of *Wilkiea*, in which entry to the urceolate flower is only gained by way of a tiny osteole (Williams *et al.* 2001). However, floral structures like those of the small understorey tree *Cryptocarya rigida* (Lauraceae), and even tiny florets packed closely in the massed inflorescences of canopy trees, can provide ample shelter.

MEGALOPTERA (alder flies)

Megaloptera, or alder flies, are also known as dobson flies and fish flies, and in flight can be mistaken for lacewings (Neuroptera). The earliest fossils, considered to be Megaloptera (assigned to the extinct Parasialidae), date from the Late Permian (Grimaldi and Engel 2005). Compression fossils are known from the Paleocene of Australia (Lambkin 1992).

The world fauna is best represented in temperate regions and is placed in two families, Corydalidae and Sialidae (Riek 1954b; Theischinger 1991b, 2000; Theischinger and Houston 1988). However, the greatest diversity of Corydalidae in Australia is in tropical Queensland, and none is present in Tasmania. Wingspans of some non-Australian species range up to 175 mm (G. Theischinger pers. comm.), but their flight is generally awkward, again similar to certain Neuroptera. Species are mainly associated with hinterland streams of the Great Dividing Range (Theischinger and Houston 1988). Most adults are crepuscular or fully nocturnal in behaviour and are attracted to lights set in rainforest. Larvae are aquatic, possess strong biting mouthparts, and live under rocks and rotting organic material in stream beds.

The 26 known Australian species (22 Corydalidae, four Sialidae) are endemic and constitute ~10 per cent of the world fauna (Theischinger 1991b). Fauna known from the region include the genera *Archichauliodes*, *Protochauliodes* (Corydalidae) and *Stenosialis* (Sialidae).

Corydalidae All Australian species are placed in the Chauliodinae, or fish flies (Theischinger 1991b), the earliest fossils (*Eochauliodes striolatus*, *Jurochauliodes ponomarenkoi*) of which are known from the Mid Jurassic of Inner Mongolia, China (Liu *et al.* 2012). *Protochauliodes* (with a few species also in western North America), *Taeriochauliodes* and *Nothochauliodes*, distributed in Australia, southern Africa and Chile respectively, represent Southern Hemisphere relics or living fossils of Gondwana (Liu *et al.* 2012). *Archichauliodes*, which also occurs in New Zealand and Chile, includes species found in south-east Queensland and northern New South Wales (*Archichauliodes guttiferus*, *A. neoguttiferus*, *A. pictus*, *A. plomleyi*, *A. polypastus*) (Smithers 1993; Theischinger 2000) (Plate 90). *Protochauliodes*, which also occurs in Chile and western North America, is known from southern Queensland and eastern New South Wales (*Protochauliodes biconius*) (Smithers 1993; Theischinger 1991b; Theischinger and Houston 1988).

Sialidae Both genera of Australian sialids, *Austrosialis* and *Stenosialis*, are apparently endemic although similar to some South American, South African and Indian genera (Theischinger 1991b). The Australian fauna equals ~5–10 per cent of the World fauna (Theischinger and Houston 1988). Larvae have been found in slow-flowing coastal streams (Theischinger and Houston 1988). The endemic genus *Stenosialis* (*Stenosialis australiensis*) is distributed along the coastal ranges of eastern Australia from North Queensland to southern Victoria (Riek 1954a; Theischinger 2000).

NEUROPTERA (lacewings, ant-lions)

This is an ancient group characterised by their delicate and intricately veined wings (Grimaldi and Engel 2005). Definitive fossil evidence for the emergence of the order dates from the Late Permian (represented by the family Permithonidae), with the origin of some extant families (Hemerobiidae) occurring around the end of the Triassic (Garźon-Ordũna *et al.* 2016). Fossil species dated to the Triassic are recorded from south-east Queensland (Lambkin 2014). Athough the Neuroptera are less diverse now than they once apparently were, nevertheless they are found in all zoogeographical regions, though some groups are limited in distribution (Lambkin 1986; New 1988a, 1991; New and Smithers 1994; Smithers 1991a, 1993; Wells 1996b).

Larval habits determine the distribution of species within forested landscapes (New 1991). For example, subterranean pit-trapping species may exhibit a preference for the location of their conical pits, not just based on protection from adverse environmental conditions, but on the particle size of the substrate (Devetak and Arnett 2015). This can lead to additional patchiness in distributions (e.g. in fine sediments accumulated below rock overhangs) of species populations within individual rainforest stands. Adults are encountered on foliage by day, or actively flying at night. Many species are attracted to light. When at rest some (e.g. Ascalaphidae) can be confused as dragonflies, but the flight behaviour of many is awkward. Others are moth-like in appearance (e.g. Ithonidae) or resemble praying mantids (Mantispidae). Adults are predatory, phytophagous or omnivorous, and some Chrysopidae are specialised honeydew feeders (New 1991). However, neuropterans are generally not encountered on the flowers of rainforest trees. Some adults emit defensive odours when handled. Larvae have conspicuous protruding jaws and are either free-living predators on twigs and leaves, or excavate conical pits that snare prey. Many Chrysopidae have tangling body filaments that snare small fragments of leaves and other debris, and this lends a degree of camouflage.

The Australian Neuroptera have a very high level of endemism and include many of the world's more archaic groups (New 1991; Wells 1996b). More than 90 per cent of the Australian species are endemic (New 1991). Nymphidae, Ithonidae, Myrmeleontidae-Stilbopteryginae, and some osmylid taxa are completely (or almost) confined to the region.

Chrysopidae These are the green lacewings (Naumann 1991; Winterton 1995), delicate adults of which may be attracted to light at night, or found blending with rainforest foliage by day. They are one of the largest families of Neuroptera. Though their bodies are commonly green the wings are usually transparent. However, there are also dull-coloured species (e.g. of *Italochrysa*) (Plate 91).

The primitive subfamily Nothochrysinae includes the endemic genera *Triplochrysa* and *Dictochrysa* (New 1991). *Triplochrysa pallida* has been described from the Bunya Mountains to the immediate north-west of the Gondwana rainforests World Heritage Area (in Wells 1996b). *Dictochrysa peterseni* has been recorded from the Mount Royal Range (Smithers 1993), an area contiguous with the southern boundary of the Barrington Tops. Adult Chrysopinae are predators (e.g. *Calochrysa extranea*, *Chrysopa innotata*, *Nathancycla verreauxii*, *Italochrysa* spp.). *Calochrysa* and *Nothancyla* are endemic (New 1991).

Coniopterygidae These are small, mainly arboreal, lacewings that are usually <6 mm in length. In this family the larval mouthparts are short and needle-like. Coniopterygidae include the subfamilies Aleuropteryginae (*Heteroconis ornata*, *H. pulchra*) and the more diverse Conipteryinae (*Coniopteryx maculithorax*, *C. orientalis*, *Neosemidalis acuta*, *N. farinosa*, *N. nervalis*). *Conipteryx* occurs world-wide but *Neosemidalis* is endemic (New 1991; Smithers 1993).

Hemerobiidae Hemerobiidae, or brown lacewings, are predatory in both the larva and adult stage and, though widespread, endemism in the Australian fauna is high. The Australian hemerobiids are strongly related to the New Guinea fauna. Although this family is the most frequently encountered (New 1988a) most Australian species are apparently uncommon (Wells 1996b), rare or localised (New 1988a). Several species (e.g. *Carobius trifurcatus*, *Notherobius nebulosus*) are known only from the Gondwana rainforests region (New 1988a). Additional species include *Carobius pulchellus*, *C. trifurcatus*, *Drepanara binocula*, *Hemerobius australis*, *Megalomina bridwelli*, *Micromus tasmaniae*, *M. timidus*, *Notherobius nebulosus*, *Notiobella viridis*, *Psectra nakahari*, *Psychobiella sordida* and *Zachobiella submarginata* (New 1988a, 1991; Smithers 1993; Wells 1996b). *Notherobius* and *Carobius* are endemic, *Psectra* is known from the Western Pacific, and *Notiobella* also occurs in New Caledonia, Indonesia and New Guinea (New 1988a). *Micromus* and *Zachobiella* have extralimital distributions.

Ithonidae The Ithonidae, or moth lacewings, are virtually restricted to Australia (Riek 1974). Elsewhere the family is represented by *Oliarces* from North America (Riek 1974; Wells 1996b). The endemic genus *Megalithone* (*Megalithone tillyardi*) occurs widely in the montane areas of southern Queensland and New South Wales (New 1991).

Mantispidae Adults are diurnal or nocturnal, possess raptorial forelegs, and convincingly resemble praying mantids. Adults frequent foliage and may also be found on flowers (Williams 1995). A number are brightly coloured (e.g. *Calomantispa venusta*, Plate 92) and may mimic certain flower-frequenting Coleoptera (e.g. Cantharidae – *Chauliognathus*, Cerambycidae – *Nenenia*) (see Plate 110). Mantispinae are specialised parasites of spiders, feeding on eggs (New 1991).

The family is diverse in subtropical eastern Australia and, in addition to *Calomantispa venusta* mentioned above, the regional fauna includes *Campion australasiae*, *C. callosus*, *C. rubellus*, *Ditaxis meridiei*, *Spaminta minjerribae*, *Therestria stigma* and *Toolida infrequens* (Lambkin 1986; Smithers 1993; Wells 1996b). *Therestria* is endemic (New 1991).

Myrmeleontidae This is the largest family of Neuroptera in Australia, with adults ranging in size from 10–>70 mm (New 1991). The wings can be distinctively patterned (e.g. *Glenoleon pulchellus*). The larvae are ground dwelling, inhabit trees or live underground. Some construct conical pits in which they await unfortunate prey that have fallen within. The fauna includes *Callistoleon erythrocephalus*, *Dendroleon dumigani*, *Glenleon pulchellus*, *Mossega reticulata*, *Myrmeleon acer* and *Myrmeleon pictifrons* (New 1991; New and Smithers 1994; Smithers 1993). Of these, the genus *Myrmeleon* includes the true ant-lions whose larvae live in pits (New 1991).

Neurorthidae This is a small family, and only the genus *Austroneurothrus* (*Austroneurothrus brunneipennis*) is known from Australia. Its larvae are mainly aquatic but can also be found in damp leaf litter. Closely related taxa occur in Europe and Japan (New 1991; Wells 1996b). Adults are usually found in damp places, including stream-side vegetation and rainforest.

Nymphidae Nymphidae are predominantly restricted to Australia, with the majority of species (e.g. *Myiodactylus osmyloides*, *Nymphes aperta*, *N. myrmeleonides*, *Osmylops sejunctus*) occurring in eastern Australia

(Oswald 1997; Smithers 1988b, 1993; Wells 1996b). Extralimital distribution of the family occurs in Lord Howe Island (New 1991), and Papua New Guinea and adjacent islands (Oswald 1997). The larvae of *Myiodactylus* and *Osmylops* are arboreal. The white filament-stalked eggs of *Nymphes myrmelenoides* are commonly encountered on the undersurface of fallen tree trunks. This is one of the largest and most commonly encountered neuropteran. Its larvae live in leaf litter and are covered with debris.

Osmylidae This predatory, insect-feeding family is diverse in Australia. Regional species include *Eidoporismus pulchellus* (Eidoporisminae), *Kempynus acutus*, *K. kimminsi*, *K. thecatus*, (Kempyninae), *Porismus strigatus* (Porisminae), *Euporismus albotrox*, *Oedosmylus brevis*, *O. latipennis*, *O. montanus*, *O. nebulosus* and *O. tasmaniensis* (Stenosmylinae) (New 1983, 1986, 1989, 1991; Smithers 1990, 1993; Wells 1996b; G. Williams pers. observ.) (Plate 93). *Porismus strigatus* is a large species with distinctive black-patterned wings that inhabits wet sclerophyll forest (e.g. Lansdowne Escarpment), as well as more open eucalypt-dominated forest.

Eidoporisminae are endemic, and Kempyninae and Stenosmylinae (the two largest subfamilies), are of Gondwanan origin. Kempyninae are known from Australia, New Zealand and temperate South America (New 1983; Wells 1996b). The Kempyninae include some of the most conspicuous osmylid lacewings and, with the exception of *Kempynus incisus*, all species are endemic (New 1983). Few kempynines are common, with most species being known from single or few localities (e.g. *Kempynus acutus* known only from New England National Park [New 1986]). The diverse subfamily Stenosmylinae also occurs in South America (New 1991). *Oedosmylus* is an endemic genus restricted to the eastern Bassian province. *Oedosmylus brevis* and *O. nebulosus* are confined to the Gondwana rainforests region (New 1989).

Psychopsidae Adult Psychopsidae are medium to large in size (up to 30 mm), broad-winged, cryptic and mainly nocturnal (Plate 94). However, the wings of some species are attractively patterned with subtle markings. Psychopsids are distributed in Australia, South Africa and the Oriental region (New 1991), though the family is best represented in Australia (being most diverse in eastern Queensland and New South Wales) and South Africa. *Psychopsis gracilis* has been described from a threatened floodplain rainforest remnant at Booyong in far northern New South Wales. The spectacular *Psychopsis illidgei* (= *Megapsychops illidgei*) was described from Mt Tamborine (New 1988b; Wells 1996b). *Psychopsis* is endemic.

Additional Neuroptera include the Ascalaphidae and Nemopteridae. Ascalaphidae are known as owl flies, and are medium-sized to large lacewings. The larvae initially aggregate but later disperse within foliage or leaf litter. Nemopteridae are a small family whose adults can be found feeding on flowers (though the larvae are predators) (New 1991).

COLEOPTERA (beetles)

It is difficult to think of the order Coleoptera without bringing immediately to mind the comment reputedly made by the British scientist J. B. S. Haldone, who when asked what conclusions might be deduced about the work of the Creator from the study of his works, was said to have answered 'an inordinate fondness for beetles'. There is some uncertainty as to what Haldone actually did say, but even if popular myth, it is a nice one. For the sheer number of living beetle species speaks to the reality, though this is beginning to be questioned since in Europe, for example, it appears that there are more Diptera and Hymenoptera species than Coleoptera (D. Bickel pers. comm.). Approximately 40 per cent of insects, and ~30 per cent of all living animals, are beetles. This number exceeds that of fungi and vascular plants, and is much greater than that of all vertebrate animals (Lawrence and Britton 1991). Beetles are diverse, as Grimaldi and Engel (2005) posit, 'probably because they are the only flying insects besides ants to have thoroughly invaded cryptic niches on the ground'. The great Early Cretaceous–Tertiary radiation of Angiosperms, the flowering plants (of which there are now more than 350 000 extant species [Friis *et al.* 2011]), is often cited as a driver leading to the present diversity of the Coleoptera, but this does not account for the independent development of Adephaga (e.g. Carabidae, Gyrinidae, Dysticidae) and Staphylinoidea (e.g. Staphylinidae), well before angiosperms had emerged as the dominant terrestrial plant group (Grimaldi and Engel 2005).

In addition to colonising diverse terrestrial environments, beetles have also been successful in occupying aquatic habitats (Short 2018), and there are now ~420 000 described beetle species in the world (Lawrence and Ślipiński 2013, Ślipiński *et al.* 2011). This is a substantial increase from the figure of more than 300 000 named species given by Lawrence and Britton in 1991; this increase reflects the intensity of field work and taxonomic studies undertaken over the last decades by Coleopterists (those who study beetles).

Presently there are more than 23 000 described species of beetles, in 117 families (listed in Lawrence and Ślipiński 2013), recorded from Australia; the fauna of south-eastern Australia is especially rich (e.g. see Farrow 2016; B. Moore 1980–96). Many more species are known but are yet to be formally named (see Chapman 2009). By far the largest family is that of the weevils, the Curculionidae, with >6000 species. Carabidae, Chrysomelidae, Scarabaeidae and Staphylinidae are less numerous but still represent faunas that individually number in the thousands.

Coleoptera are one of the three major orders, with Diptera and Hymenoptera, dominating the pollinator assemblages on mass-flowering canopy trees in lowland rainforest (Williams 1995; Williams and Adam 1994, 2010). Major families comprising the flower-frequenting fauna include Aderidae, Anthicidae, Buprestidae, Cantharidae, Cerambycidae, Chrysomelidae, Cleridae, Coccinellidae, Corylophidae, Curculionidae, Dermestidae, Elateridae, Latridiidae, Lycidae, Melyridae, Mordellidae, Nitidulidae, Oedemeridae, Phalacridae, Pyrochroidae, Scarabaeidae, Scirtidae, Staphylinidae and Tenebrionidae (Williams 1995). Some are highly adapted ecological specialists, but many are pollen predators that may additionally function unintentionally as minor pollinators. The presence and abundance of each varies with plant species (the resource), geographical and altitudinal location, season and time of day. The nocturnal fauna is poorly studied, though Scirtidae are commonly observed at night on canopy blossoms.

The canopy fauna in general is a diverse and abundant one (Basset 1990, 1991; Kitching *et al.* 1993; Lowman 1992, 1985; Stork and Grimbacher 2006) but fluctuates with regard to season, with peak activity, diversity and abundance corresponding to late spring–summer and, for phytophagous species, episodes of leaf flushing (Lowman 1982). Using 'knock down' fogging techniques Hammond *et al.* (1997) recorded 454 species of canopy-associated Coleoptera in a single rainforest study plot in Lammington National Park. Major families in cascading order included Curculionidae, Staphylinidae, Chrysomelidae, Corylophidae, Anthribidae, Pselaphinae (as Pselaphidae), Ptiliidae, Coccinellidae, Tenebrionidae, Melandryidae, Latridiidae, Cryptophagidae, Cantharidae, Scraptiidae, Elateridae, Aderidae, Mordellidae, Cerambycidae, Ptilodactylidae, Carabidae and Phalacridae. These comprise plant-feeding herbivores, species that feed primarily on wood (xylophages), species that feed on decaying plant and animal matter (saprophages), fungivores and predators. Of these more than 50 per cent were stratum generalists that occuppied more than one forest zone, and ~20 per cent were considered likely to be 'tourists', simply passing through and using the forest canopy as a way station of varying ecological significance to each one. In general, beetle assemblages in both the canopy stratum, and that occupying the forest floor, may contribute equally to the biodiversity of rainforest Coleoptera (Stork and Grimbacher 2006); however, their feeding guild structure is a function of the diversity, and the temporal and spatial availability, of microhabitat and trophic resources found in each habitat (e.g. Wardhaugh *et al.* 2013a).

In addition to the flower-frequenting and general canopy faunas in subtropical rainforests, there are also distinctive beetle assemblages associated with tree bark (Menzel *et al.* 2004). Menzel *et al.* (2004) found that rainforest trees with rougher bark had a greater diversity (species richness) of beetle species and families, had a higher proportion of rare species than did trees with smooth-barked trunks, and harboured significantly larger-sized beetles; this owing to the greater physical complexity of the rough-barked tree habitat and the more extensive opportunites for shelter and feeding specialisation, as well as opportunities for species aggregation.

Many beetle groups had their origin during the Jurassic (180–135 mya) (e.g. Carabidae, Chrysomelidae, Curculionidae, Elateridae, Ripiphoridae, Scarabaeidae, Staphylinidae) (Crowson 1981; Hsiao *et al.* 2017; Lawrence and Britton 1994) and these are generally well represented and diverse in south-eastern mainland Australia. Three families occurring in subtropical Australia are endemics – Lamingtoniidae (*Lamingtonium*), Rhinorhipidae (*Rhinorhipis*) and Acanthocnemidae (*Acanthocnemus*) – but only the Acanthocnemidae are widely distributed. The level of endemism in the Australian beetle fauna is high (e.g. Scarabaeidae) and many of the larger Australian families have radiated extensively during the Tertiary in association with *Eucalyptus* (Myrtaceae)- and *Acacia* (Mimosoideae)-dominated plant communities. The geologically oldest families are represented by the relictual Cupedidae and Ommatidae (Howden 1981) of the primitive suborder Archostemata, which were common in the Mesozoic. The cupedid *Distocupes varians* occurs from southern Queensland to Tasmania. The genus *Adinolepis* also occurs in eastern Australia. At least one species of ommatid (*Omma stanleyi*) inhabits south-east Australia (Lawrence and Britton 1994; Lawrence and Ślipiński 2013).

Numerous taxa recorded from this region have Gondwanan affinities. Several have representatives in New Zealand, and sometimes New Caledonia (e.g. Pamborini-Carabidae). Examples of groups with representatives in New Zealand and South America are Migadopini (Carabidae), *Ceratognathus* (Lucaninae-Lucanidae) and Lampriminae (Lucanidae), Cavognathidae, Adeliini (Tenebrionidae) and Belidae. The closest relatives of some taxa occur in South America but without representatives in New Zealand (e.g. *Syndesus* – Lucanidae), while others have representatives in southern Africa and Madagascar (e.g. Byrrhidae) (Lawrence and Britton 1991, 1994).

The families (listed alphabetically) and species mentioned below are only a portion of the diversity of the beetle fauna known from subtropical rainforests, and in some instances eucalypt-dominated sclerophyll forest with a rainforest understorey, of mainland Australia. More extensive species compilations can be found in Williams (1993, 1995, 2002, 2003a) and the papers cited therein. Information on the morphology and systematics of the Australian fauna is detailed in Lawrence and Britton (1991, 1994) and Lawrence and Ślipiński (2013). The suborder placement of individual families discussed is given in Appendix 4.

Anthicidae The family has world-wide distribution, and has been recorded from Mesozoic Burmese amber (Grimaldi and Engel 2005; Rasnitsyn and Ross 2000). Their larvae are found in decaying vegetation, leaf litter and rotting logs. The adults of distinctively red-coloured *Lemodes coccinea* frequent fungus-hosting rotting logs in lowland rainforest and wet sclerophyll forest (Plate 95). The lowland subtropical rainforest fauna also includes several day-active species that inhabit flowering trees (e.g. *Egestriomima albolineata* on *Syzygium smithii*).

The flowers of lowland rainforest trees frequented by various species of *Anthicus* are floristically diverse (*Acradenia euodiiformis, Alphitonia excelsa, Ackama paniculsa, Cuttsia viburnea, Drypetes deplanchei, Euroschinus falcatus, Guioa semiglauca, Scolopia braunii, Syzygium floribundum, S. smithii, Tristaniopsis laurina*). Anthicids (e.g. *Egestrina*) are also to be found in adjacent wet sclerophyll forest where they occur on various *Eucalyptus* species and related genera, as well as rainforest trees of the wet sclerophyll forest understorey (e.g. *Lagrioida*, or a closely related genus, on *Schizomeria ovata* flowers [Williams 1995]). Flower-frequenting Anthicidae are ecological generalists that feed on nectar and/or pollen but exhibit no fidelity to a particular flowering plant.

Anthribidae The cosmopolitan Anthribidae, or fungus weevils, are related to true weevils (Curculionidae) and are widely distributed, including South America. The Indo-Austromalayan fauna is diverse, but the Australian and New Zealand faunas do not appear, at least superficially, to be similar (Zimmerman 1994a). The Australian fauna is dominated by geologically recent invaders from Indonesia, but remains poorly known (Zimmerman 1994a). Many new species and endemic genera have been collected from the region between south-east Queensland and northern New South Wales (Williams 2002); this region being dominated by the subfamily Anthribinae. Most anthribid larvae live in dead wood or are associated with fungi. Adults of some (*Basitropis*) are attracted to light, or will be found on the flowers of lowland subtropical rainforest trees (e.g. on *Diospyros australis* (Williams 1995).

Taxa present include several species of the endemic genera *Erichsonocis* (from rainforest at Barrington Tops, the Tooloom Plateau, Mt Tamborine, New England National Park, and the Lansdowne Escarpment north of Taree), *Euphanteos* (Mt Nothofagus, Richmond River, Lower Barrington Tops), *Taburnus* (Lamington National Park, and north-west of Wauchope), *Telata* (south-east Queensland), and *Teratanthribus* (Bunya Mountains, Lamington National Park, Border Ranges, Tooloom Scrub, Barrington Tops) (Williams 2002; Zimmerman 1994a). Several species occur in wet sclerophyll forest, in addition to inhabiting rainforest (e.g. *Ancylotropis waterhousei, Euphanteos ornatus, Tellala* sp., *Basitropis* sp., *Phaulimia* sp.) (Plates 96, 97). The resplendent red-coloured *Eupanteos ornatus* (Allyn River, Lansdowne Escarpment) (Plate 96) exhibits a striking resemblance in colour to the similarly sized anthicid *Lemodes coccinea* (Plate 95).

Apionidae Apionidae are small weevil-like beetles that feed on seeds, fruits and living plant stems. They are a minor component of the beetle fauna in lowland subtropical rainforest and wet sclerophyll forest. Lawrence and Ślipiński (2013) place the Apionidae as a subfamily (Apioninae) of Brentidae, but the family status of Apionidae (following Zimmerman 1994b) is retained here. They occur in the Indo-Pacific, Indo-Malayan and Austro-Malayan regions. Few species are known from New

Zealand, but New Caledonia has distinctive genera and species. Apionids are considered by Zimmerman to be geologically recent invaders of Australia from the north, New Guinea being the probable source of the Australian fauna. In addition to several species historically placed in the genus *Apion* two endemic genera, *Lissapion* (*Lissapion varistriatum*) and *Notapion* (*Notapion variorostre*), occur in the Gondwana rainforests region (Lea 1926; Williams 2002; Zimmerman 1994b). Both are confined to Queensland and New South Wales.

Archeocrypticidae The larvae and adults of Archeocrypticidae inhabit leaf litter and the softer fruiting bodies of some polypore fungi. The family is widely distributed, occurring in the southern United States of America, South America, Africa, Asia, Australia (including Tasmania) and New Zealand. The Archeocrypticidae comprise the genera *Enneboeus* (southern United States of America–South America, eastern and southern Australia), *Pseudenneboeus* (Brazil), *Sivacrypticus* (throughout Africa and Asia; one species in South Australia) and *Archeocrypticus* (South America, New Zealand, Tasmania). The remaining six genera (*Australenneboeus*, *Enneboeopis*, *Falsoplatydema*, *Gondwanenneboeus*, *Nothenneboeus*, *Wattianus*) are restricted to Australia (Lawrence 1994b). *Enneboeopis ovalis*, *E. tarsalis* and *Wattianus queenslandicus* occur in northern New South Wales.

Attelabidae The herbivorous attelabid weevils (*sensu* Zimmerman 1994a) exhibit an almost world-wide distribution but are absent from New Zealand and Polynesia. As considered by Zimmerman (1994a) the Australian Attelabidae include a single genus, *Euops*, which also occurs in Asia, Africa and Madagascar. These are leaf rollers, the females laying their eggs in cut portions of leaves (Lawrence and Ślipiński 2013). Most of the world's 144 known species of *Euops* are confined to the Wet Tropics (Riedel 2001). Numbers of species within the *quadrifasciculatus* species group are roughly equal between Australia and New Guinea, but relative to that of New Guinea, the broader Australian *Euops* fauna is impoverished (Riedel 2001). Two species of the *quadrifasciculatus* group (*Euops bakewelli*, *E. oberprieleri*) appear to be confined to *Eucalyptus* (Myrtaceae), but Riedel (2001) considers that 'ancestors of the *quadrifasciculatus*-group were originally living on *Nothofagus*', and may then have shifted host when *Nothofagus* became restricted in the changing Australian floristic landscape. *Euops* is absent from New Zealand and New Caledonia and the genus may have arrived from Asia, adapting to *Nothofagus* subsequent to the separation of those two areas from the Australian plate (Riedel 2001).

Several *Euops* are recorded from the Gondwana rainforests region: *E. coxalis*, *E. strigiventris*, *E. suturalis*, *E. tuberculata*, *E. bakewelli*, *E. corrugatus*, *E. hoppla* and *E. oberprieleri* (Riedel 2001; Williams 2002; Zimmerman 1994a). *Euops tuberculatus* inhabits littoral rainforest. However, the genus is usually an uncommon one in lowland subtropical rainforest, though adults are occasionally collected on flowering trees (e.g. *Syzygium smithii* [Williams 1995]).

Belidae These are herbaceous beetles that are very elongate and weevil-like in appearance. Belidae and their lineages may date at least from the Late Jurassic, and along with other families of Curculionoidea (see R. Anderson 1995; Kuschel 1995; Lawrence and Ślipiński 2013; Pullen *et al.* 2014), they are present in Cretaceous deposits (Grimaldi and Engel 2005). Zimmerman (1994a) considers the family a relic 'of Orthoceri, most closely allied to the Oxycorynidae and Aglycyderidae, groups of ancient weevils that … were abundant and widely distributed in Cretaceous [135–70 mya] and Tertiary times …'. The Oxycorynidae are now represented by a few relict taxa in South American and South Africa. Aglycyderids are currently known only from the Canary Islands (and adjacent coast of Africa), Polynesia, New Caledonia and New Zealand (Zimmerman 1994a), and the Belidae are now confined to Australia, New Guinea, Soloman Islands, the Moluccas, New Zealand and South America. None is known from New Caledonia or Africa. They are most diverse in Australia but are poorly represented in tropical Queensland rainforests.

The Pachyurinae (*sensu* Zimmerman 1994a) include many endemic genera (*Agnesiotes*, *Apagobelus*, *Basiliobelus*, *Brachybelus*, *Pachybelus*, *Sphintobelus*) that occur within northern New South Wales and southern Queensland (Williams 2002; Zimmerman 1994a). Two species, *Apagobelus brevirostris* and *Brachybelus undulatus*, are associated with Araucariaceae. Belinae, in this region, are dominated by the diverse genus *Rhinotia*, but the regional fauna also includes *Isacantha*, *Isacanthodes*, *Rhinotoides* and the endemic *Araiobelus*. *Rhinotia* (Plates 98, 99) is most diverse in Australia, but outlier taxa also occur in the Solomon Islands, New Guinea and Lord Howe Island (Zimmerman 1994a), and include several species that mimic Lycidae beetles (Plate 152). The genus is most commonly encountered in open eucalypt forest and woodland (often on *Acacia*); however, a few species inhabit warm temperate and subtropical rainforest (*Rhinotia bidentata*) and lowland wet sclerophyll forest (*Rhinotia spinipennis*). There are few recorded rainforest host plant records. *Rhinotia ruficornis* is recorded from Mt Tamborine associated with *Argyrodendron actinophyllum*. *Rhinotia semipunctulata* has been collected from flowering *Alectryon coriaceus*, and *R. semipunctulata* and *R. tenuis* from flowering *Guioa semipunctata*, in littoral rainforest. *Rhinotia tenuis* has additionally been collected from subtropical floodplain rainforest on flowering *Tristaniopsis laurina* (Williams 1995).

Boganiidae These are small-sized beetles, no greater than 4 mm in length. The family is restricted to South Africa, and eastern and south-western Australia (Lawrence and Britton 1994; Lawrence and Ślipiński 2013). There are few species, but several of *Boganium* occur in dry and wet forests of eastern Australia. *Athertonium* is recorded from lowland rainforests in northern New South Wales. Adults have been collected in floodplain rainforest remnants (Lansdowne Nature Reserve, Wingham Brush Nature Reserve) on the flowers of *Syzygium floribundum*, *S. smithii*, *Rhodomyrtus psidioides* and *Elaeocarpus obovatus*, and from lowland wet sclerophyll forest (Lorien Wildlife Refuge) on flowering *Ackama paniculata*, *E. reticulatus*, *R. psidioides* and *Cryptocarya microneura* (Williams 1995).

Boridae Borids are small to moderate-sized ground beetles that resemble some Tenebrionidae, and like many of them are associated with rotting wood on the forest floor. Boridae include the monotypic subfamily Synercticinae, which is restricted to Australia and Papua New Guinea (Lawrence and Pollock 1994). *Synerctinus heteromerus* is the only Australian species, and ranges from the central coast of New South Wales to north Queensland and New Guinea. In subtropical eastern Australia it is recorded from Mt Tamborine, Boonanghi State Forest (west of Kempsey) and the Wilson River Flora Reserve (Werrikimbe National Park), in subtropical and dry rainforest (Williams 2002).

Bothrideridae Bothrideridae occur in all zoogeographical regions. The genus *Xylariophilus* consists of three species: a single species from India and one each from north and southern Queensland (*Xylariophilus bicoloripennis*). This disjunct distribution suggests a relictual Gondwanan affinity (Pal and Lawrence 1986). Some bothriderids are predaceous on the immature stages of wood-boring Coleoptera and Hymenoptera, but several are unusual in being ectoparasitoids (Grimaldi and Engel 2005). *Xylariophilus* feed on the fruiting bodies of pyrenomycetous fungi. *Xylariophilus bicoloripennis* is known only from the type locality of Lamington National Park.

Brentidae The weevil-like Brentidae (*sensu* Zimmerman 1994a, and thus excluding Apionidae) are mainly tropical in distribution. They are known from Cretaceous fossil beds (Grimaldi and Engel 2005), but do not generally represent part of the ancient Australian weevil fauna. Zimmerman (1994a) recognised two subfamilies: Brentinae and Cylandinae. Most Brenthinae are borers in decaying wood. At least eight species are recorded from upland forests in southern Queensland and northern New South Wales: *Araiorrhinus howittii*, *Catagogus diorymerus*, *Euschizus dictatorius*, *Hormocerus reticulatus*, *Mesetia amoena*, *Microtrachelizus* sp., *Tracheloschizus dichrous*, and an undescribed genus and species known from Lammington National Park and Mt Glorious (Williams 2002, Zimmerman 1994a). *Cordus* sp., *Catagogus diorymerus* and *Euschizus* sp. are attracted to ultraviolet light in lowland subtropical rainforest and wet sclerophyll forest of northern New South Wales (G. Williams unpubl. data). The genera *Catagogus* and *Mesetia* are endemic, and *Euschizus* is confined to Queensland, New South Wales and Norfolk Island. *Hormocerus* and *Araiorrhinus* have an Oriental and Indo-Australasian distribution, *Tracheloschizus* occurs in Melanesia and Australia (Zimmerman 1994b).

Buprestidae The earliest fossil bodies of Buprestidae date from the Late Jurassic (Grimaldi and Engel 2005). Commonly known as jewel beetles, the family is the eighth largest family of beetles, world-wide comprising ~14 700 described species and occurring in all biogeographical regions. The Australian fauna is largely endemic and numbers ~1200 species; however, several genera have extralimital distributions elsewhere in Australasia (Bellamy 2002). Buprestids, as larvae and adults, have an obligatory association with vascular plants. All are herbivorous. Depending upon the species involved, the larvae bore and develop in living or dying stems, branches and roots, or mine leaves. Adults feed on leaves or nectar, or may not feed at all. Bellamy *et al.* (2013) review known larval and adult food plants for the Australian fauna. This includes several species occurring in Australian subtropical rainforests. Some are pollinators of rainforest trees and shrubs (Williams 1993, 1995; Williams and Adam 1994, 1998, 1999b, 2010).

Despite the popularity of this family there remain few species lists from individual regions or reserves. Most distribution records are to be found in research papers dealing separately with generic revisions and species descriptions. Buprestids known from the Gondwana rainforests World Heritage Area are listed in Williams (2002). Species recorded from forests in the Manning Valley, northern New South Wales, are listed in Williams (1993, 1995).

The genus *Chalcotaenia* (Chalcophorinae) is confined to Australia and New Guinea. The endemic rainforest species *Chalcotaenia lamberti* (Plate 105) is restricted to southern Queensland and northern New South Wales, and is poorly known from reserves. Most recent *Chalcotaenia lamberti* records are from threatened littoral rainforest remnants and lowland subtropical rainforest (G. Williams unpubl. data). The monotypic genus *Araucariana* (*Araucariana queenslandica*) (Chalcophorinae – Epistomentini) is known only from vine thickets in southern Queensland (e.g. Ban Ban Range). The Epistomentini are a relict tribe comprising the genera *Cyrioxus* (New Caledonia), *Cyrioides*, *Diadoxus* and *Araucariana* (Australia), and *Epistomentis* (Chile, Argentina) (Levey 1978a). The Australian genera are associated with *Araucaria* – Araucariaceae (*Araucariana*), *Callitris* – Cupressaceae (*Diadoxus*), and *Banksia* – Proteaceae (*Cyrioides*), which probably originated in the Southern Hemisphere (Levey 1978a).

An additional relict taxon recorded from the subtropical eastern Australia is the genus *Prospheres* (Polycestinae) (Bellamy 2002; Carter 1915a; Levey 1978b). *Prospheres* is a numerically small genus represented by single species each from Norfolk Island, New Caledonia, New Guinea and Australia (*Prospheres aurantiopictus*, Plate 109). *Prospheres aurantiopictus* and *P. alternecostus* (from New Guinea) breed in the timber of *Araucaria*, which is thought to have entered Australia from New Guinea sometime between the Oligocene (40–25 mya) and Pliocene (Raven and Axelrod 1972). Adult *Prospheres aurantiopictus* may be found ovipositing on the trunks of fallen *Araucaria cunninghamii*. The distribution, host associations and distinctiveness of *Prospheres* suggest that it originated in the Mid Cretaceous or earlier (Levey 1978b). The closely related *Euleptodema* (also associated with Araucariaceae) occurs in New Caledonia (two species), Malaysia (one species) and the western Pacific (two species) and with *Prospheres* represents an apparently independent Australian lineage of the Polycestinae (Bily 2000).

The polycestine genus *Astraeus* is restricted to Australia (absent from Tasmania) and New Caledonia (Barker 1975; Bellamy 2002). *Astraeus* contains many species found on the foliage of *Casuarina*, *Allocasuarina*, *Callitris* and occasionally *Banksia* (Barker 1975; Bellamy 2002; Bellamy *et al.* 2013), but rarely is the genus found in mesic vegetation types (*Astraeus navarchis* in wet sclerophyll forest, Manning Valley [G. Williams unpubl. data] being one exception). Several species of *Callitris* (*Callitris baileyi*, *C. macleayana*, *C. glaucophylla*, *C. rhomboidea*) and *Casuarina cunninghamiana* are associated with dry rainforest, vine thickets, and other rainforest types (Harden *et al.* 2006); however, any association of *Astraeus* with them is unknown.

The Polycestinae include several very small, easily overlooked, species and genera from wet sclerophyll forest and lowland subtropical rainforests, most of which are rarely collected. All are herbivores. *Paratrachys* consists of one described species (*Paratrachys australia*) from littoral rainforest and subtropical floodplain rainforests (Bellamy and Williams 1995) Adults are found on the foliage of the vine *Trophis scandens* (Moraceae). Minute-sized species of the genus *Helferella* (*Helferella miyal*, *H. manningensis*), previously placed in the subfamily Mastogeniinae, inhabit littoral rainforest remnants (Williams and Weir 1987), but nothing is known about their adult or larval food plants. Equally small-sized beetles of the genus *Habroloma* (Bellamy 2002; Carter 1928; Lea 1895a) frequent the leaves of the tree *Elaeocarpus obovatus*, and occasionally the gymnosperm *Podocarpus elatus*, in littoral rainforest (e.g. Harrington, Broken Head Nature Reserve), and also in lowland wet sclerophyll forest (Manning Valley). In open forest adult *Habroloma* feed on the upper leaf surfaces of *Kennedia* and *Desmodium* growing on the forest floor, but their food plant associations in rainforest are generally unknown.

The subfamily Agrilinae includes the Australian endemic genus *Synechocera* whose larvae are associated with *Gahnia* (Cyperaceae) and *Xanthorrhoea* (Xanthorrhoeaceae). Though not rainforest plants as such, *Xanthorrhoea* is sometimes associated with dry rainforest margins (e.g. as in Williams 1993), and *Gahnia* is encountered on the margins and sunny gap openings of cool temperate rainforest, and commonly on the margins of wet sclerophyll forest. Two species are recorded from the region: *Synechocera burnsi* from the Gibraltar Range, and *S. tasmanica*, which is distributed in upland (Mt Hyland, Werrikimbe National Park, Barrington Tops) and coastal forest and woodland from south-east Queensland to Tasmania (Bellamy 1987).The general morphology of

Synechocera suggests an ancient relationship with several African coraebine genera but has no close relatives in Notogaea (= Australian, Polynesian and Hawaiian regions) or Neogaea (= Western Hemisphere or New World), and therefore might represent an independently derived regional fauna (Bellamy 1987). The association of some species with the endemic plant family Xanthorrhoeaceae supports Bellamy's postulated independent regional derivation of *Synechocera*.

Numerous species of the cosmopolitan and speciose genus *Agrilus* (Bellamy 2002; Curletti 2001; Obenberger 1959) occur in lowland subtropical and littoral rainforests, and dry and subtropical rainforests at intermediate elevations, where their larvae bore and mature in fallen tree braches. Species include *Agrilus walesicus* recorded from vine thickets (e.g. Glenugie Peak), *A. deauratus* from subtropical rainforest (Bunya Mountains, Upper Allyn), *A. carterellus* and *A. decupratus* from littoral rainforest s on headlands and sand masses (Manning Point, Hallidays Point), and *A. frenchi* from dry rainforest (Kiwarrak State Forest) (Curletti 2001; Williams 2002; G. Williams unpubl. data). Larval food plants include dead branches of the trees *Celtis paniculata*, *Argyrodendron actinophyllum*, *Claoxylon australe* and *Drypetes deplanchei*. The minute and cryptic genus *Aphanisticus* includes at least two species that occur in littoral rainforest and lowland mixed wet sclerophyll forest–subtropical rainforest communities (G. Williams unpubl. data). Elsewhere *Aphanisticus* larvae are known to bore in the stems of small rushes, but their rainforest food plants are unknown.

Associated with *Rubus moluccanus* (syn. *Rubus hillii*) (Rosaceae), a sprawling vine common on rainforest margins and wet sclerophyll forest at intermediate elevations, is the genus *Aaaaba* (previously called *Alcinous* and *Aaaba*) (Bellamy 2002, 2013). Adults of the two known species, *Aaaaba fossicollis* and *A. nodosa*, feed on *Rubus moluccanus* leaves and are commonly encountered resting on the upper leaf surfaces in sunlight, but will quickly roll or fly away when disturbed. The two species partition their feeding niches. Adult *Aaaaba fossicollis* feed mainly in the central section of leaves, leaving characteristic patterns of feeding blotches on the upper surface. Adult *Aaaaba nodosa* feed mainly on the leaf margins, leaving excisions indistinguishable from other herbivorous insects. *Aaaaba nodosa* females oviposit on *Rubus moluccanus* canes, which their larvae progressively mine and mature to adults within (Williams 1983). Two additional Agrilinae genera that are widely distributed in Australia in open forest and woodland are *Ethonion* (previously *Ethon*) and *Diphucrania* (previously *Cisseis*). *Ethonion* is generally absent from rainforest but *Diphucrania* is occasionally encountered. *Diphucrania williamsi*, in dry rainforest south of Taree, utilises fallen branches of *Elattostachys nervosa* (Sapindaceae) as a larval host (Barker 2006b), and adults of *D. aurocyanea* have been collected on *Pomaderris* (Rhamnaceae) leaves in upland wet sclerophyll forest – warm temperate rainforest regrowth (Tapin Tops National Park) (G. Williams unpubl. data).

The herbivorous genus *Nascioides* (Buprestinae) (Plate 107) is primarily confined to the east coast of Australia with single species occurring in New Zealand and New Caledonia suggesting a Gondwanan or 'Old Southern endemic' affinity (Williams 1987; Williams and Bellamy 2002). Along with a solitary species of *Maoraxia* (*Maoraxia eremita*), the single New Zealand species (*Nascioides enysi*) constitutes the total buprestid fauna of that country. *Nascioides* resembles the New Guinea genus *Phrixia* and the Central American buprestine genus *Mixochlorus* but is most closely related to the Australian *Nascio*, a genus more typical of open forests and woodlands. *Nascioides*, *Nascio*, and the Chilean genus *Pterobothris* together constitute the Gondwanan subtribe Nascionina (Holynski 1988). A similar Gondwanan relationship occurs between the Australian genus *Selagis* (previously *Curis*) and the Chilean *Ctenoderus*, which collectively comprise the subtribe Curidina (Holynski 1988)

The Gondwana rainforests and Wet Tropics World Heritage Areas are major centres of diversity for *Nascioides* (Williams 1987; Williams and Bellamy 2002). In subtropical eastern Australia *Nascioides* occurs in upland warm temperate, subtropical, dry, and cool temperate rainforests, and extends inland to isolated depauperate rainforest stands in the Liverpool Range (*Nascioides pulcher*). The larvae of four species, *Nascioides quadrinotatus* (Tasmania, Victoria), *N. enysi* (New Zealand), and *N. tillyardi* (northern New South Wales) and *N. nulgarra* (northern New South Wales, south-east Queensland) are obligate feeders in the timber of *Nothofagus* species. Adult *Nascioides macalpinei* are also associated with *Nothofagus* in the Barrington Tops and New England National Park but this species also extends into wet sclerophyll forest in the western Blue Mountains where *Nothofagus* is absent. *Nascioides falsomultesimus* is possibly the most commonly encountered rainforest-dependent species, being recorded from numerous upland and intermediate elevations between the Illawarra and the Queensland border; its larvae have been recorded from *Schizomeria ovata*, *Ceratopetalum apetalum*, *Doryphora sassafras* and *Sloanea woollsii*. *Nascioides costatus* occurs at a similar altitudinal and latitudinal range in the same forest types. At Tapin Tops, north-west of Wingham (New South Wales), the larvae of *Nascioides falsomultesimus* and *N. costatus* mine the fallen branches of the same host tree (*Schizomeria ovata*) in warm temperate rainforest, but at other sites they are apparently restricted to individual hosts, or simply do not cohabit the same site. *Nascioides viridis* occurs in dry rainforest, where its larvae are associated with *Argyrodendron actinophyllum*. *Nascioides pulcher* inhabits subtropical and dry rainforest (e.g. Lansdowne Escarpment, Camels Hump Nature Reserve, Boonanghi State Forest), as well as open forest and has been reared from branches of *Syzygium floribundum*, *Backhousia sciadophora* and *Backhousia myrtifolia*. However, the choice of ovipositing sites appears more related to bark texture rather than plant taxon, for the genus as a whole (including open forest species such as *Nascioides carissimus*, *N. olliffi*) utilises a very eclectic choice of larval host plants (Bellamy *et al.* 2013; Williams 1987; Williams and Bellamy 2002)

Numerous other Buprestinae genera frequent subtropical rainforest formations. *Torresita cuprifera* is a species that inhabits a variety of forest types, but in lowland subtropical rainforests and littoral adults frequent the flowers of *Syzygium floribundum* and *Alphitonia excelsa*, and have been reared from dead branches of *Rhodomyrtus psidioides* (Bellamy *et al.* 2013; G. Williams 1985). The species is sometimes attracted to light (Williams 1982). *Chysobothris* is widely distributed in the African, Nearctic, Neotropical, Oriental and Palaearctic regions (Bellamy 2002; Carter 1925). A single species, *Chrysobothis simplicifons*, is recorded from lowland subtropical rainforest, dry rainforest, and littoral rainforest (including Iluka Nature Reserve). Larval food plants include dead branches of *Myrsine variabilis*, *Alphitonia excelsa*, *Drypetes deplanchei*, *Backhousia myrtifolia* and *Backhousia sciadophora* (Bellamy *et al.* 2013; G. Williams 1985). *Neocuris* includes several species occurring in littoral rainforest (e.g. Harrington) and lowland subtropical rainforest (e.g. Wingham Brush Nature Reserve, Woko National Park) on foliage and flowers (*Alphitonia excelsa*, *Syzygium floribundum*, *S. smithii* var. 'minor') (Williams 1995). *Melobasis* is speciose and widely distributed in Australia, with extralimital species recorded from Lord Howe Island, Norfolk Island, New Guinea, Indonesia, Malaysia, Solomon Islands, New Caledonia, and a single species from Laos (Carter 1923, 1929b; Hawkeswood 1990c; Levey 2012; Obenberger 1930; Williams 2002). *Melobasis* is particularly well represented in open forest and woodland, in particular frequenting *Acacia* foliage. Several species inhabit rainforest and wet sclerophyll forest. Examples include *Melobasis ignipicta* and *M. hypocrita* from cool temperate rainforest (Barrington Tops, Werrikimbe and Border Ranges national parks); their larvae mature in fallen *Nothofagus moorei* branches. The larvae of *Melobasis conica* develop in fallen *Argyrodendron actinophyllum* branches in dry rainforest (Kiwarrak State Forest), and *M. splendida* inhabits a variety of vegetation communities, including littoral rainforest, where it utilises a floristically diverse range of fallen tree branches and dead vine stems as larval food plants (e.g. *Claoxylon australe*, *Elaeodendron australe*, *Myrsine variabilis*, *Maclura cochinchinensis*). Other rainforest inhabiting *Melobasis* include *M. pretiosa*, *M. vittata*, *M. difficilis* (lowland wet sclerophyll forest and subtropical rainforest, Manning Valley), and *M. elderi* (Araucarian vine thicket, Goodnight Scrub). More characteristically associated with dying eucalypt leaves in open forest is *Anilara*, a genus which is largely Australian in distribution (Bellamy 2002; Carter 1926b). *Anilara* is a taxonomically chaotic genus whose members include a small number of species found in rainforest. These include *Anilara obscura* and *A. olivia*, both recorded from rainforests in south-eastern Queensland (Mt Mee) and north-eastern New South Wales (Wilson Nature Reserve). The related *Notographus* (*Notographus sulcipennis*) inhabits dry and subtropical rainforest (e.g. Bunya Mountains, Tooloom Scrub, Woko National Park, Kiwarrak State Forest) where its larvae develop in a variety of dead and fallen tree branches (Carter 1933; Hawkeswood 1988; Obenberger 1930; G. Williams unpubl. data).

Two additional buprestine species that have been reared from the dead branches of numerous larval host trees are *Pseudanilara cupripes* (Bellamy 2002; Bellamy *et al.* 2013; Carter 1923, 1929b), recorded commonly from littoral, dry and lowland subtropical rainforests in northern New South Wales (and also recorded attracted to light [Williams 1982]), and *Maoraxia auroimpressa*, which is widely distributed in littoral rainforest and lowland subtropical rainforest from Eurimbula National Park in Queensland, south to at least Hallidays Point in northern

New South Wales. Two Australian species of *Maoraxia* are described, the second, *M. storeyi*, occurring in northern Queensland (Bellamy and Williams 1985). Both are restricted to rainforest. A possible third, undescribed, species has been collected in vine scrub near Calliope, Queensland (G. Williams unpubl. data). *Maoraxia* has an extralimital distribution extending to the Philippines, Fiji, New Caledonia, New Zealand, Tonga and Lord Howe Island (Bellamy 1991, 2002; Bellamy and Williams 1985). The known distribution of the genus supports the postulated Late Cretaceous (100–65 mya) interconnection of eastern Australia, New Caledonia and New Zealand. However, the exact timing of the loss of a contiguous land connection between Australia and New Zealand is uncertain because of the variation in ocean regression in the Late Cretaceous (see discussion of this in Yeates and Wiegmann 2005).

The endemic buprestine genus *Neobuprestis* (Bellamy 1994; Carter 1933; Levey and Bellamy 2013) includes a single species (*Neobuprestis williamsi*, Plate 108) known from mixed rainforest–wet sclerophyll forest at Barrington Tops, the Border Ranges, and Mt Banda Banda (where adults 'hill-top' on *Acacia* growing on the summit). Species of the genus *Selagis* frequent open forest and woodland, and are pollinators of flowering Myrtaceae (e.g. *Leptospermum*) (Bellamy 2002; Carter 1926b). *Selagis aurifera* and *S. splendens* also occur in rainforest, the former in lowland subtropical rainforest (on flowering *Syzygium floribundum*), and *S. splendens* in Araucarian vine thicket (Goodnight Scrub) and lowland subtropical rainforest (on *Syzygium floribundum*, *S. smithii*).

The numerically dominant Australian tribe Stigmoderini (i.e. *Castiarina*, *Temognatha*, *Stigmodera*, *Calodema*, *Metaxymorpha*), in the subfamily Stigmoderinae, have their closest relatives in the South American genus *Conognatha* (Howden 1981, Gardner 1989) and to a lesser extent *Hiperantha* and *Dactylozodes* (Gardner 1989), and represent a Gondwanan group that has undergone considerable speciation in Australia following the breakup of the Australian and South American plates but post-dating the separation of New Zealand.

The Stigmoderini are very speciose and are largely restricted to mainland Australia and Tasmania, with a few extralimital records from New Guinea. The tribe, and in particular adults of the genera *Castiarina* and *Temognatha*, have a strong food plant association with mass-flowering Myrtaceae (which have also undergone marked levels of speciation in Australia) suggesting coevolution between the two taxa and an important role as agents of pollen transfer (Bellamy *et al.* 2013; Williams and Adam 2010). *Stigmodera* (*sensu strictu*) occurs on both sides of the continent, *Temognatha* is confined to the mainland and Tasmania, *Castiarina* is confined to mainland Australia, Tasmania and New Guinea, and *Calodema* and *Metaxymorpha* are confined to north-east Australia and New Guinea (Barker 1979, 2006a; Gardner 1989; Nylander 2008).

Castiarina (Plates 101, 102, 103) is the largest genus of Australian buprestids, and is distributed throughout the continent, including Tasmania, at all altitudes. The genus includes a subset of species with elongate faces adapted to feed on nectar (discussed in Williams and Adam 2010), and a large number of species which are Batesian and Müllerian mimics (e.g. of the lycid beetle genus *Metriorrhynchus*, Plate 152) (Barker 2006a; Moore and Brown 1989). Numerous *Castiarina* characterise the beetle fauna seasonally found on the flowers of *Kunzea*, *Leptospermum*, *Angophora*, *Eucalyptus* (Myrtaceae) and related genera in open sclerophyll forest, woodland, heaths, and shrubland. Several species are endemic to the Gondwana rainforests region (Williams 2002); however, the fauna known from subtropical eastern Australia is a large one (see Barker 1979, 1980, 1983, 1986, 1987, 1988, 1993, 2006a; Carter 1913a, 1913b, 1924, 1925, 1931, 1933; Deuquet 1938, 1956, 1963; Williams 1993, 1995).Several species once thought restricted to higher altitudes in the Barrington Tops (e.g. *Castiarina variegata*, *C. harrisoni*), or not extending south from far north-eastern New South Wales and south-eastern Queensland (e.g. *Castiarina warningensis*, *C. sulfurea*), extend to lowland subtropical rainforest near Taree, and in the case of *C. variegata* (syn. *Hypostigmodera varigata*) (Plate 104), unexpectedly in littoral rainforest at Iluka Nature Reserve. The so-called 'producta species group' is very diverse in lowland subtropical rainforest and wet sclerophyll forest (e.g. *Castiarina obliqua*, *C. acuminata*, *C. insignis*, *C. producta*, *C. beatrix*, *C. delta*, *C. deliciosa*, *C. harrisoni*, *C. spectabilis* – at Lorien Wildlife Refuge alone) (Plates 101, 102). This group is diverse on the east coast of Australia and the included species are considered classic examples of Müllerian mimicry, possibly of predatory asilid flies (Barker 2006a). A rich *Castiarina* fauna is associated with mass-flowering *Syzygium* species, in particular *S. floribundum*, and the riparian variety 'minor' of *S. smithii*, in lowland subtropical rainforests (Williams 1993, 1995). However, *Castiarina* is often poorly represented on mass-flowering trees in littoral rainforests, as for example *Syzygium smithii*, *Alphitonia excelsa*, *Euroschinus falcatus*, *Scolopia braunii* and *Acronychia imperforata* (all with few or just single species), even though these and others possess open and easily accessible flower structures and ample floral resource rewards (Williams 1995).

Two species of *Metaxymorpha* (*Metaxymorpha grayii*, *M. imitator*) (Plate 106) appear to be confined to moist forests of south-east Queensland and northern New South Wales, with both species extending to the lower north coast of New South Wales (Nylander 2008; Sainval and Lander 1994; Williams 1993; G. Williams unpubl. data). The large black-and-red-coloured *Metaxymorpha grayii* has been recorded from Mt Glorious in south-eastern Queensland, and Acacia Plateau and the Manning Valley in New South Wales; at the latter locality being found on the flowers of *Tristaniopsis laurina* (Myrtaceae) in lowland subtropical rainforest and wet sclerophyll forest. The similarly coloured, mimic-like, *Metaxymorpha imitator*, is recorded from the Manning Valley, with adults frequenting lowland riparian rainforest stands and at higher elevations on stunted flowering vegetation on the otherwise exposed rock faces of volcanic plugs.

The large and resplendent yellow-and-metallic-green-coloured *Calodema regale* (Plate 100) is distributed southwards from North Queensland to about Bulahdelah on the New South Wales lower north coast. This is the most southern-occurring species of a genus whose distribution is otherwise centred on New Guinea (Nylander 2008). *Calodema regale* is recorded from flowering *Cuttsia viburnea* in lowland subtropical rainforest (Williams 1995); however, adults are often numerous during summer on low myrtaceous shrubs (*Kunzea*, *Leptospermum*) growing at the summit of volcanic plugs, these surrounded at their base by rainforest and eucalypt-dominated wet sclerophyll forest.

Callirhipidae This is a small terrestrial family with only a few species found in Australia (Lawrence and Britton 1991, 1994; Lawrence and Ślipiński 2013). The larvae live in rotten logs. The genera *Callirhipis* and *Ennometes* occur in rainforest and wet sclerophyll forests of northern New South Wales and Queensland. *Ennometes ruficornis* occurs in lowland subtropical rainforest and is attracted to ultraviolet light. Crowson (1981), citing the overlooked work of Andrew Murray (1870) on the distribution of Coleoptera, considers the family to be a predominantly warm-climate group. Other Coleoptera within Crowson's warm-climate group, occurring in the region include Passalidae, Dynastinae, Sagrinae, Brentidae and Paussinae.

Cantharidae Cantharidae are relatively young in the fossil record, the oldest found in Eocene Baltic amber and Miocene Dominican amber (Grimaldi and Engel 2005).

Sometimes called soldier beetles, cantharids are small to medium-sized (up to 15 mm) beetles, whose larvae are found in soil and leaf litter. Adults are active during daylight and are often encountered on flowers. Their heads are elongated, facilitating feeding on nectar, and possibly pollen, a morphological trait suggesting a coevolutionary response to the emergence of angiosperms in the Cretaceous, as is also seen in flower-frequenting Oedemeridae, Lycidae, Meloidae, certain Buprestidae, and others. However, some cantharids are also predaceous, a possible pre-adaptation to a later flower feeding habit.

The subfamilies Chauliognathinae, Dysmorphocerinae and Silinae are represented in rainforest and adjacent wet sclerophyll forest. The subfamily Dysmorphocerinae is mainly a Southern Hemisphere group (Calder 1998), which in the Gondwana rainforests region includes numerous species of *Heteromastix* (e.g. *Heteromastix bispinicornis*, *H. longicornis*, *H. scutellaris*) (Lea 1921; Williams 2002). *Heteromastix* are small-sized beetles, and are minor pollinators frequently encountered in small numbers on flowering trees (e.g. *Syzygium smithii*, *S. floribundum*, *Tristaniopsis laurina*, *Cryptocarya microneura*, *Elaeocarpus reticulatus*, *Euroschinus falcatus*, *Schizomeria ovata*) in lowland rainforests, including littoral rainforest (Williams 1995). The Silinae (e.g. *Sphaerarthrum mastersii*, *S. atricornis*, *S. incenstans*) largely occur in the Palaearctic region but extend to the Oriental and south-eastern Australian regions (Britton and Stanbury 1981; Calder 1998). Chauliognathinae occur in the Americas, Australia and New Guinea (Calder 1998). *Chauliognathus* (Plate 110) is a widely distributed chauliognathine genus. Adults produce defensive compounds (Meinwald *et al.* 1968) and participate in mimicry complexes that include lycid, oedemerid, cerambycid and buprestid beetles. *Chauliognathus* can seasonally occur in vast numbers on flowering trees and shrubs in open forest and woodland, their massed presence disturbing and displacing other flower-frequenting animals. In lowland and coastal rainforest remnants in New South Wales *Chauliognathus flavipennis*, *C. lugubris* and *C. sydneyensis* have been recorded from a floristically diverse number of flowering canopy and subcanopy trees (i.e. *Guioa semiglauca*, *Syzygium smithii*,

S. floribundum, *Tristaniopsis laurina*, *Alphitonia excelsa*, *Cuttsia viburnea*, *Euroschinus falcatus*, *Diospyros australis*, *Scolopia braunii*) (Williams 1995), but never in large numbers. Species of *Chauliognathus* are potentially minor pollinators of rainforest trees that possess open floral structures (all of the above except for *Diospyros australis*), though very likely also pollen thieves; pollen lost to their feeding is compensated for by the pollen adhering to their bodies and transferred to other plants. However, none of the *Chauliognathus* species cited above is a rainforest-dependent species. All occur commonly in other forests and woodland communities, and recruit to rainforests in response to plants in flower there.

Carabidae Carabids are a well-defined family of predatory beetles widely encountered in many xeric and mesic-adapted plant communities throughout Australia, and represent useful 'tools' for testing biogeographical concepts (Noonan 1979). They are the most numerous of the families assigned to the Adephaga (see Appendix 4). World-wide there are more than 20 000 known species. In size, carabids found in subtropical rainforest and wet sclerophyll forest environments range from just several millimetres (*Perileptus* – Broscinae, *Euthenarus* – Harpalinae) to more than 6 cm (*Hyperion schroetteri* – Harpalinae, which feeds on scarabaeoid larvae in rotten wood [Lawrence and Britton 1994]). Most of the ground-frequenting species are coloured black, but numerous tree-living species are coloured brown or possess wing covers that are patterned.

Carabidae have invaded most terrestrial habitats, and world-wide include taxa restricted to the forest floor (geophiles), as well as a large number of predominantly smaller-sized species that are arboreal (arboricoles). Others, such as the bizarrely elongate *Eudalia atrata* (Odacanthinae) from the Barrington Tops region, are specialised hydrophiles that live alongside streams (M. Baehr 2005), sheltering by day under stones and emerging at night to hunt prey. Some geophiles have adopted a subterranean, cave-dwelling, existence (species of *Notospeophonus* – Harpalinae, *Anomotarus* – Lebiinae/Harpalinae [B. Moore 1963, 1966]). Approximately half of the Australian carabid fauna are ground dwellers, one-quarter to one-third are arboreal, and the remainder (e.g. Perileptini-Trechinae) are associated with water bodies such as streams, ponds and swamps (Darlington 1961a; Lawrence and Ślipiński 2013). Many genera and species are rainforest specialists, and do not occur in open forest, though a number inhabit the rainforest understorey of wet sclerophyll forest; this also serves as an intezone between otherwise open 'dry' forest and closed 'wet' forest faunas (e.g. *Chlaenius*, *Craspedophorus*, *Notonomus*, *Trichosternus*, *Pamborus*). A large proportion are flightless, the evolution of this state being prominent in the fauna of upland forests. A number (i.e. *Cratoferonia phylarchus* [Plate 111], *Notonomus angustibasis* [Plate 113], *Trichosternus subvirens*) are apparent hosts to mites in the genus *Eutarsopolipus* (Podapolipidae), which as larvae and adults shelter below the elytra (Constantine and Seeman 2014) or live exposed and conspicuous upon the body surface (e.g. Uropodidae on *Pamborus alternans* [G. Williams pers. obs.], Plate 4).

Carabids appear in the fossil record during the Triassic and Late Jurassic (~155–160 mya), and by the Cretaceous there were diverse subfamilies (Crowson 1981; Grimaldi and Engel 2005). One feature that has contributed to the success of the family is the development of chemical-producing pygidial glands, and the evolution and deployment of a vast array of defensive chemical compounds, which are widespread throughout the family (Moore 1979; Moore and Wallbank 1968). Their method of discharge ranges from the directed expression of pungent irritants, such as hydrocarbons and formic acid, to the explosive emission of hot caustic secretions. Most carabids store the compounds in the form in which they are to be expelled, and the orifices through which this occurs are extrusible and mobile and so a degree of targeted accuracy in delivery is possible. The chemicals involved include phenols, organic acids, quinones and aldehydes, depending on the carabid lineage (Moore and Wallbank 1968, reviewed in Moore 1979). Formic acid and methacrylic acid are the two most commonly isolated compounds deployed by carabids inhabiting subtropical eastern Australian rainforests; as examples, the phylogenetically diverse genera *Amblytelus*, *Mecyclothorax* (Psydrinae/Moriomorphinae), *Notonomus* (Pterostichinae/Harpalinae), *Eudalia* (Odacanthinae/Harpalinae), *Dicrochile*, *Siagonyx* (Callistinae/Harpalinae), *Sphallomorpha* (Pseudomorphinae), *Drypta* and *Philophloeus* (Lebiinae/Harpallinae) utilise formic acid (often exclusively), and *Carenum* (Scaritinae), *Castelnaudia*, *Cratoferonia*, *Nurus*, *Prosopogmus*, *Trichosternus* (Pterostichinae/Harpalinae), *Eurylychnus* (Bruschinae) and *Pamborus* (Carabinae) (Plates 114, 115) deploy methacrylic acid. In both groups this is usually in combination with lesser proportions of other compounds (e.g. tiglic acid, undecane). Most pterostichine genera utilise methacrylic acid; however, *Loxandrus* deploys salicylaldehyde, and as already cited *Notonomus* deploys formic acid, as the main compounds (Moore and Wallbank 1968). The main components isolated, as percentages, in the Paussinae are toluquinone (60), ethylquinone (10) and benzoquinone (30) from *Mystropomus regularis*, and toluquinone (90) and ethylquinone (10) from *Arthropterus* (Moore and Wallbank 1968).

Chemical defenses in carabid beetles

Predatory Carabidae are renowned for the chemically eloborate array of irritant compounds that they can secrete in their defence. To do this they employ widely different groups of compounds associated with twin pygidial reservoirs (Forsyth 1972; Moore 1979, Moore and Wallbank 1968). Carabidae are not alone among terrestrial invertebrates in the use of chemical defenses. Other beetles, as well as certain bugs, ants, sawfly larvae, lepidopterans and millipedes also are chemically defended, but as a diverse and abundant family, carabids do it exceedingly well – not least of all the explosive gaseous emissions manufactured by bombardier beetles (*Pheropsophus verticollis*), these being distinctively yellow spotted members of the subfamily Brachininae. These have a two-chamber system in which an aqueous solution of quinols and hydrogen peroxide is released from a storage reservoir into a reaction chamber. There the solution contacts a catalysing enzyme; the ensuing spontaneous hot reaction forms quinones and gaseous oxygen, which are then violently released under pressure via an exit duct (Moore and Wallbank 1968). A burning sensation follows contact, but for many other carabids the defence discharge results in an acrid odour or staining of the skin, of varying repellant impact.

In reviewing the chemicals involved Moore (1979) grouped them as hydrocarbons, aliphalic ketones, saturated esters, formic acid, higher saturated acids, unsaturated acids, phenols, aromatic aldehydes and quinones (see Windholz *et al.* 1978). Cyanide and simple alkaloids have apparently not been exploited (Moore 1979). Nevertheless these are highly reactive chemicals that alone, or in combination, are irritant and toxic, or, as with the hydrocarbon and saturated esters, may function as auxiliary adjuvants to facilitate penetration by the principal compounds. Carabid beetles tend to store their defensive compounds in the chemical form in which they are to be employed, the beetles utilising strong muscular pressure to expel the fluids as a directed spray from abdominal ducts. But some secretions are simply slowly released, oozing out over the abdominal segments; this method was possibly the ancestral means of delivery, with spraying or gaseous expulsion being later developments, and with formic acid frequently associated with a spraying mechanism.

Moore (1979) opined that, given the widespread occurrence of similar pygidial glands in carabid beetles, chemical defence originated early in the evolution of the family, even though the array of chemicals employed by carabid lineages is diverse and methacrylic acid occurs in a variety of families (Moore and Wallbank 1968). As Moore (1979) noted, quite morphologically different carabid groups share virtually identical defensive systems and this suggests convergent evolution, though within groups (e.g. Pterostichinae/Harpalinae) individual taxa can possess different defensive chemistry.

Although carabid beetles are best known as opportunistic predatory ground beetles, as exemplified by their prominent mandibles and fast agile movements, they have evolved extremely diverse ecologies and behaviours (e.g. Erwin *et al.* 1979; Moore 1964). The Paussinae (e.g. *Mystropomus* [Plate 112], *Arthropterus*) and Pseudomorphinae (*Sphallomorpha*) are specialised inquilines in, or associated with, ant nests. Paussinae possess glandular trichomes (hair-like structures) that secret substances that please their hosts. Pseudomorphinae are ambush hunters. Some Carabidae (i.e. *Amblytelus* – Psydrinae/Moriomorphinae, *Trigonothops*, but particularly *Sarothrocrepis* – Harpalinae) are found in the flowers of lowland rainforest trees, presumably there as hunters, or to feed opportunistically on nectar (Williams 1995). The large adults of *Pamborus alternans* (Carabinae/Pamborini) and *Trichosternus vigorsi* (Pterostichinae/Harpalinae) (Plate 116), and their larvae (Plate 114), are aggressive hunters. At night they patrol the forest floor of lowland rainforests, actively seeking invertebrate prey such as snails, worms and spiders (e.g. *Hadronyche*, Plate 25); their numbers are often so great that at least one or two adults may be encountered foraging within several metres of each other (as in Woko National Park) (Williams 1993), and this density is bound to influence the presence, micro-distribution and abundance of prey species. Similar densities of foraging *Pamborus alternans* and *Cratoferonia phylarchus* are to be encountered in the cool temperate rainforests of the upper Hasting River Valley (e.g. Willi Willi and Werrikimbe national parks) in northern New South Wales. Adults of the speciose ground genus *Notonomus* (Pterostichinae/Harpalinae) undertake brood care and the Lebiini (Lebiinae/Harpalinae) have ectoparasitic larvae. The greatly extended mandibles of *Meonis* (Psydrinae/Moriomorphinae) suggest a highly evolved dedication to preying on snails, and the larvae of Rhysodinae are modified to feed on the vegetative stage of slime moulds (Myxomycetes), which are themselves one of the most enthralling of organisms to encounter in subtropical rainforests (Wellman 2017).

The carabid fauna known from subtropical Australia is a massive one, with a large number recorded from the Gondwana rainforests region alone (see lists in Williams 1993, 2002). Within the Border Ranges 'complex' (*sensu* Monteith 1993) there is considerable variation in the distribution of individual carabid species between different rainforest massifs and this may reflect the inability of some species to cross intervening lowland non-rainforest matrices (Monteith 1993). Darlington (1961b, 1979) postulated that, historically, climate was a major influence on carabid dispersal and isolation. Staunton *et al.* (2014) consider lowland flightless carabid beetles in Australia's Wet Tropics to be potentially at risk from climate change impacts. These impacts may be applicable to the flightless lowland carabid beetle fauna of subtropical New South Wales and Queensland.

South-eastern Queensland and north-eastern New South Wales represent a major transition or overlap zone for Carabidae possessing either Australian southern or Papuan-Oriental affinities. There is no distinct latitudinal, altitudinal or plant community boundary between the two (Darlington 1961a); rather, there are transitions that trend north and south, as well as east and west with regard to increases in elevation and the nature of the plant community.

Transitional filtering of carabid fauna

In the late 1950s the American entomologist P. J. Darlington, accompanied by his wife and young son, travelled extensively down the east and south-east coast of Australia, from Cape York Peninsula to Tasmania (Darlington 1960). Reaching Sydney on 2 December 1956, they bought an old Volkswagon pickup truck, which they travelled and lived in thereafter. Darlington's purpose was to collect carabid beetles in the wetter parts of the complete length of eastern Australia. What particularly interested him was the transition of different taxa. He found that that there were considerable differences in the fauna of Australia and New Guinea, and that a high proportion of the Australian fauna was flightless. *Notonomus* (>100 species) was a dominant genus in rainforests south from North Queensland, being well represented in south-eastern Queensland, New South Wales and Victoria. But only an impoverished *Notonomus* fauna reached Tasmania. *Trichosternus* (>25 species), as he considered it, reached its northern limit in North Queensland, was well represented in the subtropical rainforests of south-eastern Queensland and northern New South Wales, but only extended to near Sydney. *Leiradira*, an almost rainforest-obligate genus, ranged from North Queensland to about Dorrigo but with a big gap south of the Atherton Tableland. These are flightless ground-dwelling genera from the tribe Pterostichini. Other flightless ground-dwelling Pterostichini were found to have localised, but isolated, distributions between northern Queensland and Tasmania: *Mecynognathus* (tip of peninsula forests), *Loxogenius* (main tropical forests), *Nurus* (most species), *Liopasa*, *Cratoferonia*, *Zeodera* and *Notolestes* (subtropical rainforest system), *Loxodactylus* (southern Victoria), and *Rhabdotus* (wet forests, Tasmania). *Pamborus* (Pamborini) and *Mystropomus* (Ozaenini) reach their northern limit near Cooktown and extend to near Sydney. Along the east coast the distribution of *Notonomus*, *Leiradera*, *Trichosternus*, *Pamborus* and *Mystropomus* is broken by the large dry and open forest faunal gap extending north of Rockhampton to about Townsville, the rainforest on the Eungella Range west of Mackay being an 'island' in which all five flightless genera occur. In addition, in the north of their range these five are restricted to rainforest but further south they extend also into open forest. Southward through New South Wales both the extent of rainforest, and that of the associated carabid fauna, dimishes. The transition of ground-dwelling flightless temperate carabid groups, numerous in Tasmanian wet forests, northward into New South Wales, sees the progressive filtering out of flightless Trechinae, *Eurylynchus*, *Promecoderus*, *Percosoma*, *Chylnus* (Broscinae – Broscini), *Stichonotus* and related taxa, *Calyptogonia* (Migapodini/Migapodinae), and *Pseudogonica* and *Agonica* (Agonicini/Peleciini). Of these temperate groups only some Trechinae and *Eurylynchus* reach the New South Wales–Queensland border; Migapodinae reach to Comboyne–Ebor, and *Promecoderus, Percosoma, Pseudogonica*, and *Agonica* to southern New South Wales. *Chylnus* and *Calyptogonia* are restricted to Tasmania. Darlington (his foundation work predating the wide acceptance and uptake of the theories and concepts of 'plate tectonics' and 'continental drift') found, however, that rather than the faunal transition of carabid species in wet forests along the east coast being one of smooth regular progression, the pattern for flightless groups was very irregular, more a 'stepcline', overlain by a pattern of more regular small transitional steps of the winged groups. Darlington concluded that the patterns of distribution and faunal isolation in flightless Carabidae he observed were consistent with moderate fluctuation of climate and forest cover, with dispersal and radiation facilitated by land bridges that historically connected the Australian mainland with Tasmania and New Guinea – the former without apparent ecological barriers to potential movement, but that which connected New Guinea being beset by barriers (to movement by ground-dwelling Carabidae) owing to rainforest not being continuous (Darlington 1961a, 1961b). Darlington took a 'dispersalist' view of patterns of distribution (Darlington 1965, 1979; Thiele 1977). In dispersal biogeography organisms originate in one locality, radiate from this origin until a barrier is encountered, and then by some means expand (jump, hop) beyond the barrier to subsequently speciate in isolation, so forming apparent faunal disjunctions. Dispersalist theory has been criticised (e.g. Myers and Giller 1988) for not taking into account the movement of landmasses over time (continental drift); it may inadequately relate phylogeny to patterns of distribution, and needs to address the ecological dynamics of the newly colonised regions into which organisms have supposedly dispersed (not least of all the impact of sea level fluctuations on landmass area and climate on habitat). Nevertheless, there are numerous examples of landscapes that have been re-colonised (by animals and plants) over periods of recent and geologically more extensive time, and in which faunas have differentiated – notably Northern Hemisphere post-glacial landscapes, volcanic islands and oceanic island chains (Beck *et al.* 2006; Simberloff and Wilson 1969). Island-like subtropical rainforests mirror the latter. Some species concepts have changed since Darlington's fieldwork (e.g. Will and Monteith 2018); nevertheless, as to whether current patterns of distribution reflect historical dispersal and founding events, autochthonous diversification and radiation, or past extinction episodes, their respective origins may remain forever uncertain.

The southern cool temperate carabid fauna includes isolated *Eurylychnus* (Broscinae) taxa (*Eurylynchus cylindricus, E. ovipennis, E. regularis*) on the Mt Royal Range, Barrington Tops, Dorrigo Plateau and montane wet forest areas of the eastern Manning and Upper Hastings River valley (Darlington 1961a; Sloane 1911, 1916; Walton 1987; Williams 1993, 2002). *Eurylynchus cylindricus*, and the smaller *E. ovipennis* cohabit *Nothofagus* forest in the Barrington Tops, the latter extending in range in upland mesic forests of various kinds, north to Mt Hyland. *Eurylynchus regularis* is restricted to the Mt Boss–Dorrigo area. These represent the northern limit of wet forest Broscinae in Australia and have their closest relatives in New Zealand (B. P. Moore pers. comm.). Saturated and unsaturated aliphatic acids have been isolated from the chemical defence compounds deployed by species of *Eurylynchus* (Moore and Wallbank 1968). The related flightless broscine *Promecoderus*, utilises saturated aliphatic acids in defence.

The Migadopinae (previously the tribe Migapodini) are restricted to the Southern Hemisphere with localised genera in Tasmania, southern mainland Australia, New Zealand, Falkland Islands and the tip of South America (Darlington 1961a; Howden 1981). Thiele (1977) suggested that the ancestral migadopines originated in the Northern Hemisphere, migrated across the tropics, and later became extinct in the north. The endemic migadopine genus *Decogmus* (*Decogmus chalybeus*) occurs in the Gondwana rainforests region (Ebor, Comboyne and Carrai Plateau), including populations in subtropical rainforest <100 m on the Lansdowne Escarpment. An undescribed species is known from North Queensland (B. P. Moore pers. comm.). Such significant latitudinal disjunctions in the distribution of beetle genera along the eastern coast of Australia are known from numerous beetle groups (e.g. *Aulacopris* – Scarabaeidae), and are only partly explained by parallel disjunctions in suitable habitat.

The primitive pan-tropical Ozaenini (Paussinae) are represented by the genus *Mystropomus*, which is confined to eastern Australia and occurs in montane rainforests and wet sclerophyll forests of coastal New South Wales and Queensland (Darlington 1961a; Lawrence and Britton 1991, 1994; Monteith 1993, Monteith and Turco 2010).

Mystropomus is a nocturnal geophile, beetles sheltering during the day under fallen rotting wood and stones. *Arthropterus* (Paussinae – Paussini), characterised by large flattened antennae and general body form, can be found under loose bark and in rotting wood in subtropical rainforest and wet sclerophyll forest.

The Trechinae are very small beetles that are distributed throughout temperate and tropical regions of the world. *Perileptus* is associated with riparian zones (M. Baehr 1987; B. Moore 1966), the adults during the day being found sheltering under rocks and sandy gravel. One species, *Perileptus constricticeps*, occurs at Lamington National Park. The majority of Australian *Perileptus* species belong to the subgenus *Pyrrhotachys* which has no relatives in South-east Asia or New Guinea (M. Baehr 1987). Isolated endemic Trechinae occur between the Barrington Tops and McPherson Range. *Tachys*, *Trechiella*, endemic *Tasmanitachoides* and the flightless endemic *Eutrechus* are represented in forests of intermediate and upland elevation (B. Moore 1972; Sloane 1916, 1921; Britton and Stanbury 1981; Williams 2002). The endemic *Trechimorphus diemenensis* occurs in cool temperate rainforests of the Barrington Tops.

The winged *Sloaneana lamingtonensis* and *S. similis*, known only from montane rainforests of the Lamington and Springbrook Plateaux, represent major northern distribution extensions for the typically Bassian Merizodiinae (M. Baehr 2002b) (*Sloaneana*, a replacement name for *Brachydema*, is alternatively placed in Trechinae – Zolini by Lawrence and Ślipiński 2013). *Sloaneana tasmaniae*, the third species in the genus, occurs in Tasmania, eastern Victoria and southern New South Wales. Merizodiinae are a cold-adapted subfamily with most Australian species occurring in Tasmania, and with a broader circumpolar Gondwanan distribution being known only from south-east Australia, New Zealand and southern South America. Although the distribution of *Sloaneana lamingtonensis* and *S. similis* is subtropical, the two species only occur in sites >1000 m in cool temperate rainforest. Baehr (2002b) considers that the externally similar appearance of *Sloaneana lamingtonensis* and *S. similis* indicates sympatric evolution from within their present range, rather than colonisation of the Springbrook–Lamington Plateau by two different stocks.

The Pterostichini are the dominant members of the subfamily Pterostichinae (placed in Harpalinae by Lawrence and Ślipiński 2013) in Australia and New Zealand. Genera in subtropical eastern Australia include *Leiradira*, *Cratoferonia*, *Castelnaudia*, *Trichosternus*, *Notonomus*, *Notolestus*, *Zeodera*, *Nurus*, *Setalis*, *Prosopogmus*, *Megadromus*, *Liopasa* and *Loxandrus*. Numerous pterostichine species occur in New South Wales and southern Queensland (B. Moore 1965; Sloane 1911, 1916; Williams 2002). In New Guinea the Pterostichini are replaced by the Anchomenini (B. Moore 1965). The genus *Megadromus* is predominantly New Zealand in distribution but two primitive and relict species (*Megadromus eborensis*, *M. australicus*) occur in Australia, where they are confined to rainforest (Williams 2002). *Megadromus eborensis* was described from New England National Park ~30 years ago and *M. australicus* is known only from the brushes of Barrington Tops but has not been reported for some 80 years (B. P. Moore pers. comm.). *Notolestus sulcipennis* is an isolated species (and monotypic genus) with New Zealand linkages, and *Zeodera atra* and *Liopasa crepera* both represent monotypic genera of obscure relationships (B. P. Moore pers. comm.; Williams 2002).

Many Pterostichini are flightless (e.g. *Trichosternus*, *Notonomus*, *Nurus*, *Leirodera*, *Cratoferonia*, *Castelnaudia*), as is the conspicuous *Pamborus* (Carabinae). *Cratoferonia* includes *C. regalis* that extends from the Illawarra region of New South Wales, north to about the Comboyne Plateau and *C. phylarchus* that ranges north from the Upper Hastings and Dorrigo regions (B. Moore 1965; G. Williams unpubl. records). Neither appear to enter rainforest at low altitudes. However, the flightless carabid fauna also includes less obvious taxa (e.g. *Mystropomus* spp., *Habutarus crassiceps*) that hide in leaf litter, and under logs and stones. *Habutarus* (Lebiinae/Harpalinae) has a mainly Papua–North Queensland distribution; *Habutarus crassicep*, and the open forest *H. laticeps* from Kroombit Tops, being southern outliers. Its species possess unspecialised mouthparts, as do many carabids, suggesting they are generalist predators of tiny insects, spiders and worms (M. Baehr 2008).

Darlington (1961a) considered that *Mystropomus*, *Pamborus*, *Notonomus*, *Trichosternus* and *Leirodira* dominated the flightless Australian rainforest carabid fauna. These are widely distributed genera but reach their northern limit in the Wet Tropics World Heritage Area of North Queensland. *Trichosternus*, *Mystropomus*, *Pamborus* and *Notonomus* have evolved endemic species in temperate rainforests on the Dorrigo Plateau, and the Mt Royal Range on the south-west rim of the Barrington Tops. Several of these have invaded cool temperate *Nothofagus*-dominated rainforest (Darlington 1961a).

Leiradira is principally rainforest-restricted, and its distribution may not extend south of Dorrigo (Darlington 1961a). *Pamborus* is confined to eastern Australia, and most of the species recorded from southern Queensland and New South Wales are restricted to rainforest and wet sclerophyll forest. *Pamborus* is one of two known genera in the tribe Pamborini (Lawrence and Ślipiński 2013; Sloane 1905); the other (monotypic) genus, *Maoripamborus*, occurs in New Zealand. Three species, *Pamborus brisbanensis*, *P. subtropicus* and *P. monteithi*, inhabit forest on the elevated Kroombit Tops Plateau in central Queensland (Monteith 1986b; Takami and Sota 2006). At higher latitudes the large (3 cm) *Pamborus alternans* is a common species in rainforest at low (e.g. Iluka, Harrington, Sea Acres), intermediate (Watagans and Kerewong State Forests, Lansdowne Escarpment, Comboyne Plateau), and high altitudes (e.g. Border Ranges, Washpool, Werrikimbe national parks), as well as isolated inland dry rainforests (Cedar Brush Nature Reserve). The similarly sized but less frequently encountered *Pamborus macleayi* inhabits elevated subtropical rainforest (e.g. Cambridge Plateau). *Pamborus pradieri* coexists with *P. alternans* through much the same range of altitudes, though apparently not in littoral rainforest yet common in cool temperate rainforest in the Barrington Tops. The altitudinal range of the relatively small (<2 cm) *Pamborus guerini* (Plate 115) extends from coastal littoral rainforests, hinterland remnants at intermediate elevations (Victoria Park and Moore Park nature reserves) and upland sites (Dorrigo, Comboyne Plateau). It is sometimes found in open dry sclerophyll forest (Yarratt State Forest). *Pamborus alternans*, *P. pradieri* and *P.guerini* coexist in wet sclerophyll forest and subtropical rainforest in the Lansdowne Escarpment (G. Williams unpubl. records).

The altitudinal range of *Mystropomus* extends (coastwise) to headland and dune-associated littoral rainforests (e.g. Harrington, Black Head), floodplain rainforests (Wingham Brush, Lansdowne nature reserves), and near-coast mountains (North Brother Mountain/Dooragan National Park). At Kroombit Tops *Mystropomus* inhabits subtropical rainforest at ~800 m (G. Williams unpublished data). With the exception of a single species from Western Australia and New Caledonia (Darlington 1961b), the genus *Trichosternus* is restricted to eastern Australia and extends from North Queensland to the central region of coastal New South Wales. The regional fauna is diverse, including *Trichosternus angulosus*, *T. perater*, *T. renardi*, *T. simpliceps*, *T. subvirens* and *T. vigorsi* (Carter 1933; Darlington 1961b; Fricke 1965; Monteith 1993; Sloane 1911, 1916, 1923; Walton 1987). At least 28 species of the flightless genus *Notonomus* are known from the Gondwana rainforests region alone (Britton and Stanbury 1981; Carter 1933; Sloane 1902, 1911, 1913, 1915, 1916, 1923; Walton 1987; Williams 2002), with several new species awaiting formal description (B. P. Moore pers. comm.). This is a genus commonly encountered below fallen rotten timber on the forest floor of rainforests generally. *Notonomus* ranges from high cool temperate rainforests to littoral rainforests at sea level. Several exhibit localised distributions: *Notonomus tessalatus* (Mt Tamborine), *N. hedleyi* (Barrington Tops region) and *N. dives* (Comboyne) (Williams 2002). Many species have more expanded, though still geographically circumscribed, distributions: *Notonumus crenulatus* (Dorrigo, New England National Park, Ebor) and *N. johnstoni* (Werrikimbe National Park–Barrington Tops, Bulahdelah State Forest). Several species inhabit *Nothofagus*-dominated cool temperate rainforest: *Notonomus truncatus*, *N. angustibasis*, *N. leai* and *N. hedleyi*. The range of many extends over marked changes in altitude; for example, *Notonomus amabilis* occurs in high temperate forests (e.g. Barrington Tops, Mt Hyland), rainforests at intermediate elevations (Comboyne Plateau, Watagans and Way Way state forests) and coastal littoral rainforests (Elizabeth Beach, Harrington). Numerous species characterise elevated mesic forests at higher latitudes: *Notonomus variicollis* (Jamberoo Pass, Mt Keira) and *N. lateralis* (Mt Wilson). *Notonomus australis* inhabits rainforests through a range of elevations and is often encountered under stones and fallen wood in association with its egg clusters.

The genus *Nurus* consists of large nocturnal predators with highly restricted altitudinal and latitudinal distributions, and with limited dispersal potential. Adults wait in or near their burrows in ambush of invertebrate prey, but immatures live in leaf litter and are more active than adults. In the Kroombit Tops Plateau, south-west of Gladstone, two species occur: *Nurus nox* (also known from Bulburin and the Blackall Ranges) in rainforest and *N. curtus* in open forest (Monteith 1986b). The largest species, *Nurus rex*, represents an extreme northern outlier of the genus, occurring in rainforest on the summit of Mt Elliot near Townsville. However, most of the known species are restricted to rainforest and wet sclerophyll forest in northern New South Wales and south-eastern Queensland. Locality records for *Nurus atlas* include Lumley

Park (Alstonville) and Victoria Park Nature Reserve, both in far northern New South Wales, and records for *Nurus brevis* include Rotary Park (Lismore), Yabbra Scrub in Yabbra State Forest, Cambridge Plateau and Mallanganee (Greenslade 1994; B. P. Moore pers. comm.; Will and Monteith 2018). Although additional populations of *Nurus atlas* have recently been found, all are within a 10 km radius of Alstonville and connection between populations is unlikely (Charley and Andren 2018). *Nurus latipennis* is distributed from the Queensland–New South Wales border to the Tooloom Plateau and Nightcap Range, and that of *N. perator* includes Mt Cougal National Park, the Upper Tallebudgera Valley, Springbrook and Nightcap National Park. *Nurus moorei* is known from Upper Tallebudgera Creek, Lamington National Park, the Tweed River, and Lower Minyon Valley; however, *N. imperialis* has an isolated distribution, being restricted to the Mt Tamborine area (B. P. Moore pers. comm.; Will and Monteith 2018). *Nurus baehri* represents the southern-most species, being found in the Mt Hyland, New England National Park, Cunnawarra National Park and Styx River State Forest region of northern New South Wales. *Nurus popplei* is known only from the Torrington area south-west of Tenterfield, northern New South Wales, and is distinctive in that it inhabits open forest at altitudes >1000 m rather than mesic formations (Will and Monteith 2018). *Nurus atlas* and *N. brevis* are listed as endangered species under the New South Wales *Biodiversity Conservation Act, 2016.* Both are considered to have been more common before forest clearance following European occupation (Charley and Andren 2018). Habitat requirements are closed forest with dense leaf litter and fallen timber under which adults hide and/or construct breeding and ambush burrows. Their fecundity is uncertain; however, adults probably lay fewer than 10 eggs. Threats to their surviving populations include habitat clearing and removal of forest floor groundcover (including underscrubbing), and potentially cane toads (*Bufo marinus*) and foraging by brush turkeys. The impact of exotic weeds, including their broadscale removal during forest regeneration works, is uncertain, but in some cases invasive weeds may serve to retain suitable microhabitat and soil moisture regimes. Broadscale removal of weedy groundcover may result in soil desiccation to the detriment of relict *Nurus atlas* and *N. brevis* populations. Potential habitat enhancement actions for relict populations include the restoration of gross vegetation cover and groundcover (e.g. additional placement of logs/timber), enlargement of core areas, the establishment of vegetated buffers around the margins of rainforest remnants, and where possible re-establishing connectivity between isolated habitat fragments.

Arboreal taxa include the genus *Philipis* (Bembidiinae), with the two most southern species recorded from Springbrook and Lamington National Park in Queensland (*Philipis subtropica*) and Gibraltar Range National Park in New South Wales (M. Baehr 1995, 2002a). The genus is restricted to the eastern mainland, from the McPherson Range to north Queensland. *Philipis* occurs in rainforest on tree trunks, and almost all species occur in montane rainforests. Many species are highly localised and are known from only single mountain tops. *Philipis* may represent part of the Old Gondwanan element of Australia, with its closest affinities possibly being with the South American genus *Xytosomus* (M. Baehr 1995). The present high species diversity exhibited by *Philipis* may be relatively recent, possibly due to Pliocene/Pleistocene uplift events, erosion and isolation of the Great Dividing Range, climatic fluctuation, and fluctuation in the extent and availability of rainforest (M. Baehr 1995).

Additional arboreal carabid genera include the endemic *Dystrichothorax* and *Amblytelus* (Psydrinae/Moriomorphinae) (M. Baehr 2016; Lawrence and Ślipiński 2013; Sloane 1911, 1916; Walton 1987; Williams 2002). The former is distributed along the east coast of Australia from North Queensland to eastern Victoria, and to Tasmania. *Amblytelus* has a similar eastern distribution, but also occurs in South Australia and south-west Western Australia. Both genera live on tree trunks in open forest and rainforest. In particular the rainforest-inhabiting species possess quite restricted ranges on single or just a few mountain tops; *Dystrichothorax rufinus* and *Amblytelus suturalis* inhabit *Nothofagus*-dominated cool temperate rainforest in the New England National Park (M. Baehr 2016). Numerous *Amblytelus* occur in open forests and woodland (see Williams 2002). However, several *Amblytelus* species are commonly encountered in cool temperate rainforest (Barrington Tops, Border Ranges), and subtropical rainforest at low elevations, including littoral rainforest (*Amblytelus amplipennis*); where they may be found on foliage and small branches. Psydrinae/Moriomorphinae also include the genera *Meonis* (*Meonis niger*), *Mecyclothorax* (*Mecyclothorax punctipennis*), *Laccocenus* (*Laccocenus ambiguous*), *Teraphis* (*Teraphis helmsi*) and *Raphetis* (*Raphetis curta*) (M. Baehr 2003; though alternatively Lawrence and Ślipiński [2014] place only *Laccocenus* in Psydrinae). *Raphetis curta* has been recorded active on tree trunks at night (M. Baehr 2003), but otherwise these genera are usually associated with the forest floor. *Scopodes* are small, but prominently large-eyed beetles, currently placed in the Harpalinae – Pentagonicini. These highly alert and quick-moving carabids can be found during daylight on fallen tree trunks. The genus also occurs in Indonesia, New Guinea, New Caledonia and New Zealand. At least three species (*Scopodes angulicollis*, *S. sigillatus*, *S. tasmanicus*) are recorded from northern New South Wales in *Nothofagus* forest, upland mixed warm temperate and subtropical rainforest, and wet sclerophyll forest (Williams 2002). The closely related, and minute, *Pentagonica vittipennis* occurs in lowland subtropical rainforest, including remnant rainforests on coastal floodplains (i.e. Coocumbac Island Nature Reserve). Species of the distinctively elongate and flattened *Demetrida* (Lebiinae/Harpalinae) forage on foliage and tree trunks in lowland rainforest, and the related *Drypta australis* has been collected at ultraviolet light in subtropical rainforest and wet sclerophyll forest at intermediate elevations (Allyn River, Lansdowne Escarpment), and floodplain rainforest (Lansdowne Nature Reserve). The distinctively, blue-hued *Catascopus chaudoiri* (Lebiinae/Harpalinae) is a semi-arboreal carabid found active on fallen tree trunks at night and sheltering under loose bark by day. The species ranges from littoral rainforest (Harrington), floodplain rainforest (Wingham Brush), hinterland rainforest at intermediate elevations (Lansdowne Escarpment), and montane rainforest (Nightcap Range, Mt Warning) (Williams 2002; G. Williams unpubl. data).

Agile taxa within the Cicindelinae (sometimes cited as a distinct family Cicindelidae) include the autochthonous, but essentially Torresian, genus *Distipsidera*, which is distributed along the eastern coasts of northern New South Wales and Queensland (McCairns *et al.* 1997). *Distipsidera* can be considered part of the 'Younger Northern Element' of I. Mackerras (1970) (McCairns *et al.* 1997). Four species (*Distipsidera hackeri*, *D. papuana*, *D. grutii*, *D. parva*) appear to have invaded New Guinea and New Caledonia in the recent past. *Cicindela* is often found in association with pool margins and ephemeral silty puddles. *Cicindela semicincta* is distributed widely along the Australian east coast occurring in lowland and upland rainforests and adjacent wet sclerophyll forests (Freitag 1979; McCairns *et al.* 1997; G. Williams unpubl. data); however, the genus as such is more an associate of open environments, including beach littoral zones.

Cerambycidae The Cerambycidae are known from the Early Cretaceous (Yu *et al.* 2015), though a cerambycid-like fossil (*Cerambycomima*) dates from the Late Jurassic (*c.* 152 mya) (Grimaldi and Engel 2005). Fossil cerambycids are also recorded from Eocene Baltic amber and Eocene–Oligocene shales, and it is from the Tertiary that their fossil history is better known (Grimaldi and Engel 2005). However, the Cretaceous was a period that saw the great radiation of flowering plants, upon which most cerambycids, as larvae or adults, feed, on the Earth's surface (Friis *et al.* 2011; Taylor *et al.* 2009) . Today there are ~35 000 known species world-wide, most included in three subfamilies (Prioninae, Lamiinae, Cerambycinae), which, along with the leaf-eating Chyrsomelidae and Megalopodidae, are placed in the superfamily Chrysomeloidea (Haddad and McKenna 2016; Haddad *et al.* 2018).

In Australia there are more than 1400 species in ~300 genera, a high proportion of which are endemic (McKeown 1947; Ślipiński and Escalona 2013, 2016). Extensive species lists are given in Williams (1993, 1995, 2002). All cerambycids are phytophagous as larvae and adults (e.g. see Duffy 1963; McKeown 1945, 1947; Ślipiński and Escalona 2013, 2016; Webb 1987; Webb *et al.* 1988; G. Williams 1985, 1995), the family representing the largest radiation of plant-eating beetles.

Adult cerambycids, commonly called longicorn or longhorn beetles, are characteristically elongate in form and possess long sensory antennae that frequently extend past the length of their body, as in *Acolepta*, *Batocera*, *Didymocantha*, *Disterna* (Plate 119), *Eboraphyllus* and *Ancita*. In some species the males possess antennae that are profoundly flabellate, fan-shaped (*Distichocera kirbyi*, *D. superba*, *Eborophyllus middletoni*, *Rhipidocerus australasiae*, *Piesarthrius frenchi*) (Plates 120, 121, 124, 125). In *Thyada barbicornis* the antennae possess distinct tufts (Plate 128). The wing covers (elytra) are normally developed to fully cover the soft abdomen, but are sometimes much shortened so that the membranous flight wings are exposed (as in *Hestheis*, Plate 122), or very narrow (*Enchoptera nigricornis*, *Macrones besti*). A number resemble ants (*Mesolita*, *Iphra*) and wasps (*Hesthesis*, *Macrones*, *Tragocerus*, *Aridaeus*), though most convincingly so when in flight or moving on leaf surfaces. Species of *Iphra* resemble ants of the genus *Polyrhachis* (e.g. arboreal nest-building *Polyrhachis pilosa*, Plate 264), occurring on the same foliage. Adult *Eroschema* and *Chaodolis* resemble red and black lycid beetles in colour and are presumed to participate in mimicry complexes that include similarly coloured Buprestidae. A particular group of cerambycid beetles, *Homaemota tricolor*, *Zoedia longipes* (Plates 123, 130) and especially the bizarre ant-like *Formicomimus mirabilis* (records include Mt Warning, and Cape Hawke near Forster), in colour mimic ants of the genus *Leptomyrmex*. *Homaemota tricolor* and *Formicomimus mirabilis* are associated with foliage and stems; however, *Zoedia longipes* is sometimes encountered feeding at plant nectar (e.g. *Claoxylon australe* – Euphorbiaceae).

Species inhabiting rainforest in subtropical Australia range in size from the diminutive *Mesolita albofasciatus* and *Wahn zonulitis* (<5 mm) to *Batocera boisduvali*, a giant with a body length >5 cm, but measuring the antennae, a length twice that. Larvae are primarily borers in dead and living timber of trees and shrubs, but some mine the dry stems of vines (*e.g. Didymocentrotus denticollis*, *Notoceresium setistriatus*, *Tessaromma nanum* in *Cissus antarctica* – Vitaceae) (Duffy 1963; Webb *et al.* 1988; G. Williams 1985). Several native species (e.g. *Athemistus howitti*, *A. rugosulus*, *Sybra centurio*, *Ceresium australe*) are recorded from the stems of *Lantana camara* (Verbenaceae) (Day and Nahrung 1998; Webb 1987; G. Williams 1985), a serious invasive shrub in lowland subtropical rainforests.

Larval associations with angiosperm hosts are numerous and floristically diverse, but cerambycids also utilise gymnosperms.The larvae of *Sybra centurio* and *Temnosternus planiusculus* have been reared from dead fronds of the cycad *Lepidozamia peroffskyana* (G. Williams 1985), a species sometimes associated with lowland subtropical rainforest and wet sclerophyll forest in coastal New South Wales (Plate 302). The larvae of *Phlyctaenodes ?truncatispinus* and *Temnosternus dissimilis* (Plate 126) have been reared from the dead branches of *Araucaria cunninghamii* growing in vine thicket (Goodnight Scrub, Ban Ban Range respectively), the latter also from fallen *Araucaria cunninghamii* branches in rainforest remnants near Murwillumbah (G. Williams unpubl. records). Duffy (1963) records *Araucaria cunninghamii* as the larval host for *Archetypus frenchi*, *Coptocarpus decoratus* (= *Strongylurus decoratus*), *Diotimana undata*, *Dysthaeta anomala*, *Rhiphidocerus australasiae*, *Syllitus araucariae* and *Temnosternus dissimilis*. *Temnosternus dissimilis* and *T. niveoscriptus* (Plate 127) have been reared from the same dead branches of *Araucaria cunninghamii* collected in rainforest south-east of Murwillumbah (G. and B. Williams unpubl. records).

The adults of many Cerambycidae are nectar feeders and frequent mass-flowering trees in lowland subtropical rainforest remnants, their seemingly smooth body integument able to carry numerous pollen grains (Williams 1995; Williams and Adam 1994, 1998, 2010). Many have pronounced elongate heads to facilitate imbibing nectar. Encountered on flowers (frequently of rainforest Myrtaceae) are species of *Stenoderus* and *Syllitus* (Cerambycinae). When disturbed, these exude odorous defence secretions (toluene and *o*-cresol) from secretory organs that have openings near their mandibles (Moore and Brown 1971). Secretory glands that emit defence compounds are found in other flower-visiting cerambycine genera (e.g. *Tropocalymma*), though absent in most others (e.g. *Eroschema*, *Distichocera*, *Chaodalis*, *Hesthesis*, *Psilomorpha*, *Demonisis*, *Tragocerus*) (Ślipiński and Escalona 2016).

Flower-visiting cerambycids are not usually numerous in subtropical rainforests. Most species are present as scattered individuals. Nor do they exhibit true host fidelity; rather, species are often found on a floristically diverse range of flowering trees and shrubs, their attraction to them most likely dictated by the conspicuous presentation and open structure of the flowers, floral fragrance plumes, and nectar rewards. For example, the spectacular blue-and-yellow-coloured *Tropocalymma dimidatum* (Plate 129) can occur seasonally in large numbers (yet absent in many years) in littoral rainforest on the small tree *Acronychia imperforata* (Rutaceae) when this is the only late summer flowering tree available, but will also feed from the flowers of *Alphitonia excelsa* (Rhamnaceae) when co-flowering, and flowers of *Syzygium smithii* (Myrtaceae) when flowering earlier in the season (Williams 1995). Neither is *Tropocalymma dimidatum* constrained to rainforest for adults will also feed on non-rainforest species (e.g. *Bursaria* – Pittosporaceae) in adjacent open habitats.

Nevertheless, several large day-active species (e.g. *Distichocera superba*, *Tragocerus spencei*) in subtropical rainforests may be important in transporting pollen over relatively long distances to other flowering plants of the same species. This is potentially significant in complex rainforests where individual conspecific flowering plants can be separated by large distances and numerous intervening other plants (Williams and Adam 1994, 1998, 1999b, 2010).

Although numerous species that are found in subtropical rainforest are to be found also in other vegetation communities (e.g. *Ancita crocogaster*, *Agrianome spinicillis*, *Aridaeus thoracicus*), nevertheless many exemplify the fauna. This includes one of the largest Australian beetles, *Batocera boisduvali* (Lamiinae), the range of which extends from New Guinea and the Solomon Islands, southwards from North Queensland to near Sydney (K. Mitchell and A. Sundholm pers. comm.; B. Moore 1980–96). This is a Torresian species with relatives in the Old World tropics, Oriental and Australian regions (Ślipiński and Escalona 2013). Larval hosts include the figs *Ficus macrophylla*, *F. rubiginosa* and *F. watkinsiana* (Duffy 1963; Ślipiński and Escalona 2013) – trees that grow in subtropical, dry and littoral rainforests.

Other characterising Lamiinae include *Eboraphyllus middletoni*, which is associated with cool temperate rainforest at New England National Park (McKeown 1945, 1947; G. Williams pers. observ.), the genus *Athemistus* (e.g. *Athemistus approximatus*, *A. barretti*, *A. harrisoni*, *A, punctipennis*), and the species *Temnosternus flavolineata*, *Trigonopterus gracilis*, *Oricopis umbrosis*, *O. maculiventris*, *Meton digglesii*, *Bucynthia spiloptera*, *Didymocentrotus denticollis*, *D. foveatus*, numerous species of *Disterna* (*Disterna pumila*, *D. canosa*, *D. bifasciata*, *D, cuneata*, *D. plumifera*), and the peculiar *Australodon nearnsi*, a species with enlarged mandibles and patterned elytra that remind one of the splotched lichen growth on the trunks of rainforest trees. Although records for the species range from Cairns, North Queensland, to Cunninghams Gap, Mt Tamborine, Lamington National Park, Dorrigo (Escalona and Ślipiński 2011), Yarras and the Lansdowne Escarpment (M. Thompson and G. Williams unpubl. records), in south-eastern Australia the species has been rarely collected. Specimens from the Lansdowne Escarpment were collected during the day on tree trunks. Found in warm temperate rainforest and subtropical rainforest at intermediate elevations is *Ceraegidion dorrigoensis* (from Dorrigo, Tapin Tops) (Plate 118), a species with distinctive spines on the elytra. The only other member of the genus (*Ceraegidion horrens*) occurs in southern New South Wales, both in rainforest (Fitzroy Falls) and open forest (Sydney). *Ceraegidion horrens* has been reared from *Eucryphia moorei* (Eucryphiaceae/Cunoniaceae). Adults come to lights, but also can be found at night active on fallen trees. The flightless and endemic genus *Athemistus* is endemic to eastern Australia (North Queensland–Victoria) and is a terrestrial genus found usually on the forest floor in leaf litter and among fallen rotten logs (Carter 1926c). Species such as *Athemistus harrisoni* (Plate 117) and *A. punctipennis* are found at high elevations (e.g. Barrington Tops, Doyles River, Ebor), but others such as *A. pubescens* occur in rainforest at lower elevations (Lansdowne Escarpment, Terania Creek).

The subfamily Cerambycinae includes four species of the genus *Psilomorpha* (*Psilomorpha apicalis*, *P. divisis*, *P. marginalis*, *P. pulchra*) recorded from the Gondwana rainforests region (Williams 2002). *Psilomorpha* is restricted to Australia and is predominantly subtropical and tropical in distribution. Representatives of the genus appear to be quite rare, though different species may be found at the same locality (*Psilomorpha apicalis*, *P. divisis*, *P.marginalis* recorded from the Bunya Mountains) (Scambler 1989). *Psilomorpha* has been collected on flowering *Syzygium smithii 'minor'* (*Psilomorpha divisus*, Woko National Park) and *S. floribundum* (*Psilomorpha ?apicalis*, *P. tenuipes*, Wingham) (Williams 1995) but otherwise published records from flowering subtropical rainforest trees are few. Adult *Chlorophorus curtisi* can be commonly encountered active on freshly fallen tree trunks during the day (e.g. Eurimbula, Rotary Park/Lismore, Glenugie Peak). The similarly shaped *Xylotrechus australis* has been reared from fallen *Drypetes deplanchei* and *Argyrodendron actinophyllum* branches, and adults sometimes are found at the flowers of *Syzygium floribundum*. The striking *Uracanthus cryptophagus* (Fricke 1964) has been reared from *Acradenia euodiiformis* (Rutaceae) in subtropical rainforest (Coorabakh National Park). *Uracanthus* is a speciose and widespread genus in Australia. The larvae of the green-brown-coloured *Brachytria pulcherrima* have been reared from *Syzygium smithii* in littoral rainforest (Iluka Nature Reserve) and *Backhousia sciadophora* from dry rainforest (Camels Hump Nature Reserve, Mt Hoad), though adults are found on foliage and sometimes at flowers of non-rainforest trees and shrubs.

Phacodes is distinctive among the Cerambycinae in having multi-facetted eyes. Larval hosts include *Argyrodendron trifoliolatum* (for *Phacodes mirabilis*, *P. longicollis* in southern Queensland) (Duffy 1963). The genus occurs widely in subtropical rainforest and occasionally wet sclerophyl forest through a range of intermediate elevations (e.g. *Phacodes marmoratus*, Numinbah, Nightcap Range, Victoria Park, Lansdowne Escarpment; *P. longicollis*, Tooloom Scrub, Nightcap Range, Woko National Park, Tapin Tops National Park; *P. persomatus*, Mallanganee, Copeland Tops; *P. mirabilis*, Mt Nardi; *P. obscurus*,

Carrai State Forest, Barrengarry Mountain). Adults are attracted to ultraviolet light. *Phoracantha* includes many species that inhabit open forest, where their larvae develop in *Eucalyptus* and other genera, but the subtropical rainforest fauna is apparently an impoverished one. *Coptocercus guttatus*, however, is common in some rainforests (e.g. Mt Warning, Kiwarrak and Boonanghi State Forests), including dry rainforest formations.

Prioninae include numerous large-sized (e.g. *Agrianome spinicollis*, *Eurynassa* sp.), and otherwise distinctive (metallic-hued *Phaolus metallicus*) species, that though particularly associated with eucalypt forest include species that are also encountered in rainforest. Several have sexually dimorphic antennae (e.g. *Rhipidocerus australasiae*, *Eboraphyllus middletoni*, *Sclerocantha gigas*). Adults of the green-coloured *Rhipidocerus australasiae* can be found on freshly fallen trees (e.g. *Schizomeria ovata*) as well as small *Linospadix monostachya* palms in the understorey. Noteable is the large *Eboraphyllus middletoni* (McKeown 1945) that inhabits cool temperate rainforest in New England National Park at altitudes over 1000 m (G. Williams unpubl. records). Adult males are attracted to ultraviolet light. Males of *Sclerocantha gigas* (syn. *Tillyardia mirabilis*) have the maxillary palpi expanded into basket-like structures (Carter 1913a; Escalona and Ślipiński 2011). The species is known from upland forests of the New South Wales hinterland.

Chrysomelidae Chrysomelidae, or leaf beetles, are phytophagous and, with the Cerambycidae and Megalopodidae, constitute the superfamily Chrysomeloidea (Lawrence and Ślipiński 2013; C. Reid 1995). There are estimated to be over 3000 species of chrysomelids in Australia, the majority of these in the subfamilies Chrysomelinae (775 species), Eumolpinae (750 species), Galerucinae (700 species) and Cryptocephalinae (550 species) (C. Reid 2016). A preliminary species list of the chrysomelid fauna known from the Gondwana rainforests region is given in Williams (2002).

Adult Chrysomelidae feed externally, and larvae feed externally or internally, on a floristically diverse number of plant taxa, and a variety of living plant material such as leaves, seeds, stems, flowers and pollen (Lawrence and Ślipiński 2013). Pollen feeding appears to have been central in the early feeding strategies of Chrysomeloidea (Samuelson 1994). Many chrysomelids defend themselves with an array of toxic secretions (e.g. Paropsinae deploy hydrogen cyanide) (Dettner 1987), a strategy utilised by relatively few of the related Cerambycidae within Chrysomeloidea. Chrysomelids (e.g. Galerucinae) may sequester chemicals injested from plant hosts, making them distasteful and toxic to predators (Ferguson and Metcalf 1985).

The Chrysomelidae probably originated in the Late Jurassic to Early Cretaceous (Samuelson 1994), their diversity strongly reflecting a response to the emergence of flowering plants and with many species restricted to individual, or taxonomically related, larval and adult food plants (Jurado-Rivera *et al.* 2008; C. Reid 2016). However, Late Jurassic fossils are difficult to assess and there are few records for Cretaceous Chrysomelidae (Grimaldi and Engel 2005). It is in the Tertiary that chrysomelid fossils become diverse. That the Chrysomelidae (in partnership with Cerambycidae) have undergone a considerable radiation with, and transition to angiosperms as hosts, is evident in the great number of species world-wide and the myriad plant associations they participate in; the superfamily Chrysomeloidea (at ~58 000 species) is one of the most important groups of beetles in terms of their relationship with plants.

Several species are recorded from flowering subtropical rainforest canopy trees with open, readily accessible floral structures (Williams 1995): *Pedethma* sp., *Ditropidus* sp., *Monolepta* spp. (on *Rhodomyrtus psidioides* – Myrtaceae); *Pedethma* sp., *Ditropidus* sp., *Edusella* sp., *Pyrgoides* sp., *Rhyparida* sp., *Monolepta* spp. (on *Syzygium smithii* – Myrtaceae, *S. floribundum* – Myrtaceae); *Aporocera iridipennis*, *Ditropidus* sp., *Edusella* spp., *Monolepta* spp. (on *Tristaniopsis laurina* – Myrtaceae); *Aporocera* sp., *Ditropidus* sp., *Semelvillea* sp., *Monolepta* spp. (on *Acradenia euodiiformis* – Rutaceae); ?*Bruchidius* sp., ?*Ditropidus* sp., *Monolepta* sp. (on *Alectryon coriaceus* – Sapindaceae); *Pedethma* sp., *Ditropidus* spp., *Monolepta* spp. (on *Guioa semiglauca* – Sapindaceae); *Aporocera iridipennis*, *Ditropidus* spp., *Monolepta* spp. (on *Alphitonia excelsa* – Rhamnaceae); *Pedethma* sp., *Ditropidus* sp., *Monolepta* sp. (on *Ackama paniculata* – Cunoniaceae); *Arsipoda* sp., *Pedethma* spp., *Edusella* spp., *Monolepta* spp. (on *Schizomeria ovata* – Cunoniaceae); *Monolepta* spp. (on *Cryptocarya microneura* – Lauraceae); *Arsipoda* sp., *Monolepta* spp. (on *Cuttsia viburnea* – Escalloniaceae/Rousseaceae); *Monolepta australis* (on *Elaeocarpus obovatus* – Elaeocarpaceae); ?*Aporocera* spp., *Monolepta* spp. (on *Elaeocarpus reticulatus* – Elaeocarpaceae); *Ditropidus* sp., ?*Edusella* spp., *Rhyparida* sp., *Monolepta* spp. (on *Euroschinus falcatus* – Anacardiaceae); *Monolepta* spp.

(on *Scolopia braunii* – Flacourtiaceae/Salicaceae) (Williams 1995). The most speciose chrysomelid cohort, however, was that recorded from flowering *Syzygium floribundum* (Myrtaceae) in two riparian subtropical rainforest sites (Wingham Brush Nature Reserve, Lorien Wildlife Refuge) (Williams 1995). This included *Pedethma* sp., *Aporocera ?iridipennis*, *Ditropidus* spp. (three), *Monolepta* spp. (nine), *Pepila* sp. and *Poneridia mastersi*.

This diversity of species suggests that anthophilous chrysomelids may function as pollinators within the diverse and abundant guild of tiny pollinating insects recorded from flowering subtropical rainforest trees (Williams 1995; Williams and Adam 2010). However, flower-associated Chrysomelidae are more often consumers of floral parts (e.g. sepals, petals, stamens) and of pollen grains, and so are likely to contribute only a minor pollination benefit; and this at a reproductive cost to plants. An extreme example is *Monolepta australis* (Galurucinae) whose adults seasonally emerge in vast numbers, and who are capable of destroying whole inflorescences of certain orchard trees (e.g. mangoes) such that no fruit can be produced. Rather, in functioning mainly as primary consumers of living plant tissue, Chrysomelidae are major evolutionary drivers of plant defence strategies against herbivory.

The Spilopyrinae exhibit a Gondwanan distribution pattern, with three species known from southern South America, nine in New Caledonia, and 28 species from Eastern Australia and the highlands of New Guinea (C. Reid 2000, 2016) The endemic spilopyrine genus *Cheiloxena* (previously placed in the Eumolpinae) is confined to mainland Australia between south-east Queensland and eastern Victoria (C. Reid 1992a, 2000). All species are relatively rare, with the rainforest-inhabiting *Cheiloxena tuberosa* and the more widely distributed *C. westwoodi* occurring in the Gondwana rainforests region; *C. tuberosa* ranging from southern Queensland to Dorrigo in New South Wales; and *C. westwoodi* extending from southern Queensland to southern New South Wales and Victoria. The Spilopyrinae also includes spectacular metallic-coloured species in the genus *Spilopyra*, a genus that is restricted to the eastern coast of Australia, from Lismore and Broken Head Nature Reserve, north to Cape York Peninsula and New Guinea (Reid and Beatson 2010). *Spilopyra sumptuosa* occurs in southern Queensland and far northern New South Wales, where in the latter it can be found in littoral rainforest. Known food plants are *Cupaniopsis anacardioides* and *Guioa semiglauca* (Hawkeswood 1991a, 1992; C. Reid 2000) and it has also been collected by beating the leaves of *Mischocarpus pyriformis* (Reid and Beatson 2010). These records are all from Sapindaceae. Eumolpinae include *Eucolaspinus*, the adults of one possible undescribed species attacking the leaves of *Nothofagus moorei* (Plate 132).

The Australian Galerucinae fauna is a rich one. It includes *Menippus*, of which at least three species (e.g. *Menippus darcyi*) feed on *Celtis* (Ulmaceae/Cannabaceae) (Reid and Nally 2008). The genus *Menippus* occurs in South-east Asia, Lord Howe Island and New Guinea. Additional galerucine genera recorded from subtropical rainforest include *Oides*, *Aulacophora*, *Monolepta*, *Nisotra*, *Pedethma*, *Poneridia*, *Ellopidia* and *Hoplostines*. *Monolepta* is widespread, *Oides* (e.g. *Oides dorsosignata*, *O. fryi*, *O. laetabilis*) is widespread in the Old World tropics, *Poneridia* (*Poneridia australis*, Plate 135) is distributed in Australia and South-east Asia, and *Ellopidia* and *Hoplostines* (*Hoplostines viridipennis*) are Australian endemics (C. Reid pers. comm.); the latter is associated with stinging trees (e.g. *Dendrocnide excelsa* – Urticaceae) and the prostrate, stream- and dripline-associated herb *Elatostema reticulatum* (Urticaceae) (Plate 133). Adult *Poneridia* are herbivorus on figs (e.g. *Poneridia australis* on *Ficus rubiginosa*). *Pedethma*, often cited as *Crepidodera* in earlier papers (e.g. Matthews and Reid 2002; Williams 1995, 2018), includes at least two rainforest-associated species (*Pedethma australiensis*, *P. kirejtshuki*) from the region (Lingafelter and Konstantinov 2000). The distinctively shaped tortoise beetles (e.g. *Aspidimorpha*), Cassidinae, include the Australian endemic *Aproida balyi* (Monteith 1970), atypical of the subfamily in that its body plan is elongate rather than rounded. *Eucryphia moorei* is the only known host of *Microdonacia pilosa*, a species from cool temperate rainforest in southern New South Wales (Clyde Mountain) (C. Reid 1992b). *Microdonacia* is confined to cool temperate rainforest and subalpine heath. Adult *Nisotra bicolorata* are associates of native *Hibiscus* (e.g. *H. splendens*) in wet sclerophyll forest and various rainforest types, being commonly found on foliage.

The Chrysomelinae are most speciose in temperate regions and under-represented in the tropics (C. Reid 2006). They exhibit a strong host preference for Myrtaceae (Jurado-Rivera *et al.* 2008). Endemic genera confined to the McPherson–Macleay bioregion (essentially the Gondwana rainforests region) include *Asernoides*, *Ateratocerus*, *Grammicomela*, *Novacastria* and *Poropteromela*.

The distinctive apple green-coloured *Poropteromela epipleuralis* (Plate 136) feeds on certain rainforest Myrtaceae (C. Reid pers. comm.). *Alfius heiroglyphicus* is associated with rainforest at Mt Tamborine and Lamington National Park. *Alfius* has been found on *Drymophila* (Alstroemeriaceae), but this may be a casual instance (C. Reid 2006). The distinctive metallic green *Johannica gemmellata* widely occurs in rainforests of south-eastern Queensland and northern New South Wales (Williams 2002). The vine *Pandorea pandorana* (Bignoniaceae) is a known host (Hawkeswood 1990a). *Johannica* is endemic. The bright red-and-black-patterned *Phyllocaris cyanicornis*, and the related *P. gracilis* and *P. leoparda*, inhabit rainforests at intermediate altitudes (Williams 2002). Adult food plants of *Phyllocaris cyanicornis* include *Clerodendrum tomentosum* (Verbenaceae/Lamiaceae); the adults chewing broad concave incisions from the leaf edge (G. Williams pers. obs.). *Phyllocaris* occurs in Australia and South-east Asia, with five species recorded from the Gondwana rainforests region. *Novacastria* is a monotypic genus (*Novacastria nothofagi*) endemic to the Gondwana rainforests region and feeds on the leaves of *Nothofagus moorei* (Selman and Lowman 1983). *Calomela* (Plate 131) occurs in Australia and New Guinea, and includes at least six species recorded from the Gondwana rainforests region. *Calomela relicta* occurs in rainforest and wet sclerophyll forest at Barrington Tops (C. Reid 1989b). The related *Calomela fugitiva* is associated with *Acacia* (e.g. *Acacia melanoxylon*) growing on rainforest margins. *Callidenum prasinum* is a large apple green-coloured species occurring in lowland subtropical rainforest and wet sclerophyll forest as well as in rainforest at higher elevations (e.g. Nightcap Range).

The Cryptocephalinae include the endemic genus *Semelvillea*, which shares many taxonomic features with the endemic New Zealand genus *Arnomus* (C. Reid 1991). *Semelvillea nothofagi* is associated with *Nothofagus moorei* at Barrington Tops, and *S. parva* has been collected on *Acacia melanoxylon* in rainforest near Brisbane. It is thought that the more derived (less ancestral) *Semelvillea* switched from *Nothofagus* associations to *Eucalyptus* and *Acacia* species as hosts as these diversified and radiated in response to increasing aridity in the Miocene (C. Reid 1991).

The endemic cryptocephaline genus *Platycolaspis* is known from five species, all of which are restricted to south-east Australia, from Tasmania north along the Great Dividing Range to Lamington National Park, and west to Mt Gambier in South Australia. *Platycolaspis lamingtonensis* was collected by fogging the small trees *Dysoxylum rufum* and *Synoum glandulosum* (Meliaceae) in subtropical rainforest (C. Reid 1994). The endemic *Cadmus* is predominantly a genus of open forest, but includes *C. lawrencei* from rainforest at Cunninghams Gap, south-eastern Queensland (C. Reid 1999b), as well as several species from subtropical rainforest in the Manning Valley.

The Criocerinae are a small subfamily that includes the genus *Lema*, and also the bizarre narrow-necked and green-coloured *Liliocerus chamelus* (Plate 134), a species that inhabits littoral rainforest, and inland subtropical rainforest.

Clambidae Clambids are very small, oval-shaped beetles <2 mm in length. Adults are capable of rolling into a ball, the head being able to fit tightly against the prothorax. They are found in leaf litter and feed on fungal spores and slime moulds, and require humid microclimates best provided by moist forest types such as rainforest. The Australian Clambidae can be grouped into three biogeographical elements (Endrödy-Younga 1990): an endemic group (e.g. *Sphaerothorax*, *Clambus* in part), Gondwanan relics with affiliated taxa on other landmasses (*Clambus* in part), and taxa derived from post-Gondwanan immigrations, also with affiliated taxa on other landmasses (*Clambus* in part).

The McPherson Range represents an interzone for autochthonous lineages extending north from cool temperate zones to the south, and groups with tropical affiliations (Endrödy-Younga 1990). Taxa occurring in the Gondwana rainforests region include several species of *Sphaerothorax* (this genus is confined to mainland Australia, Tasmania and New Zealand) and *Clambus* (Williams 2002), including *Clambus sphaericus*, which is a southern wingless species that inhabits cool temperate rainforests. Species locality records in south-eastern Australia include *Sphaerothorax tasmani* (McPherson Range, Tooloom Plateau, New England National Park), *S. rufocastaneus* (New England National Park), *S. tierensis* (Barrington Tops, New England National Park), *Clambus simsoni* (Fraser Island, Mt Tamborine, Barrington Tops, Bulahdelah State Forest, Jamberoo Pass), *C. domesticus* (Fraser Island, Boatharbour Nature Reserve, Dorrigo), *C. terraereginae* (southern Queensland), *C. sphaericus* (Mt Glennie, Richmond Range, Barrington Tops, Clyde Mountain), *C. alienus* (Barrington Tops) and *C. stewarti* (Roseberry State Forest, Dorrigo).

Cleridae Clerids are known from Mesozoic Baltic amber, with the oldest known fossils dating from the Middle Jurassic of China (Kolibáč and Huang 2016). They are predatory, often metallic-coloured, beetles found on foliage, standing tree trunks and freshly fallen trees, flowers, and in leaf litter. Adults are normally active during daylight.

Species that inhabit the rainforests and adjacent wet sclerophyll forests of subtropical Australia include *Omadius prasinus*, *O. olivaceus*, *Eunatalis floccosus*, *Olesterus australis*, *O. rufiventris*, *Lemidia subaenea*, *L. virgata*, *Metademius floccosus* and *Tarsostenodes simulator* (Gerstmeier 1990; Williams 2002). *Lemidia* species may be predators of Psyllidae (see New 1978). Black-coloured *Lemidia* mimic leaf-nesting *Polyrhachis pilosa* ants in littoral rainforest and floodplain rainforest, and are part of a broader mimicry complex that also includes cerambycid beetles (*Iphra*) and salticid spiders (*Myrmarachne*) (Williams 1993). Larvae of *Zenthicola* in tropical Queensland are recorded feeding on termites, and have been collected from the nests of *Mastotermes* (B. Moore 1973). Large dark-coloured *Eunatalis* (Plate 137) can be found at night active on dead standing trees in wet sclerophyll forest.

The clerid fauna that is associated with flowering lowland subtropical rainforest trees can be a rich one, in particular that which is recorded from riparian *Syzgyium floribundum* and *S. smithhi* var. *'minor'* (Myrtaceae) (Williams 1995); adult host plant records include *Cleromorpha novemguttatus*, *Zenthicola funestus*, *Z. australis*, *Phlogistus conspiciendus*, *P. erimita*, *P. plutus*, *Scrobiger splendidus* (on *Syzygium smithii*); *Elaele viridicollis*, *Lemidia concinna* (on *Syzygium floribundum*); *Lemidia pictipes* (on *Alphitonia excelsa*, *Cryptocarya microneura*, *Syzygium smithii*); *Lemidia* sp. near *purpurpea* (on *Schizomeria ovata*, *Syzygium floribundum*); and *Tarsostenodes armiger* (on *Guioa semiglauca*). These are ecological generalists that are also found on many flowering trees and shrubs (particularly those of Myrtaceae) in open forest and woodland.

Coccinellidae Ladybird beetles, or ladybugs, are among the most well-known insects of the invertebrate world, but the rainforest fauna is less understood. Of the 500 or so known Australian species, only about half have been described (Ślipiński 2007).

Fossil coccinellids occur in Tertiary amber, and are mostly predatory as larvae and adults (e.g. on aphids, scale insects, mealybugs, ants, mites), and are usually encountered during daylight on leaves and stems, but also under bark. Some are phytophagous and are pests of crops. There is also a suite of species that frequent the flowers of rainforest trees and shrubs, active there as predators but also commonly observed imbibing nectar or feeding on pollen (Williams 1995). In the Northern Hemisphere coccinellids are known to switch to pollen as a food source when insect prey is scarce, and some species are known to feed on mildew, sap and honeydew as supplementary food sources (Ślipiński 2007).

Coccinellidae are prey to a broad spectrum of predators, and in consequence they deploy a range of defence responses when threatened. These include secreting repellant pungent fluids (alkypyrazines) from the joints between the tibia and femur, but only rarely with toxic alkaloids (coccinellines, in their haemolymph). Beetles may also feign death. Many adult ladybirds also display aposematic (warning) coloration – commonly patterns of red, yellow and black, though colours can also be drab (some *Scymnus*) or the beetle only partly coloured (as in *Cryptolaemus montrouzieri*, *Diomus ementitor*). The large *Australoneda bourgeoisi* is an attractive and conspicuous species coloured yellow-orange contrasted with black markings. Coccinellidae are also hosts to numerous parasitic species, these including Ischneumonidae and Braconidae wasps, and Phoridae flies (Quicke 2015; Ślipiński 2007).

The coccinellid fauna recorded from northern New South Wales and southern Queensland is a diverse one (Williams 2002), including species that inhabit open forest, and species recorded from rainforest (e.g. *Australoneda bourgeoisi*). In Australia the subfamily Coccinellinae comprises more than 30 species, and part of this fauna is shared with New Guinea, New Caledonia and the Solomon Islands (Pope 1988). Of the genera recorded from the Gondwana rainforests region, *Erithionyx*, *Cleobora* and *Archegleis* are Australian endemics; *Australoneda* is shared with New Guinea; *Illeis*, *Coelophora*, *Phrynocaria* and *Micraspis* are variously distributed within the Indian, Asian, New Guinean and Pacific regions; and *Coccinella* and *Harmonia* are cosmopolitan (Pope 1988).

Numerous Coccinellidae are associated with the canopy stratum of subtropical rainforest, some of whom constitute part of the 'minor' pollinating cohort of ecologically unadapted small-sized Coleoptera (e.g. including Chrysomelidae, Corylophidae, Phalacridae, Dermestidae, Mordellidae, Aderidae, Scirtidae) that visit mass-flowering trees (Williams 1995; Williams and

Adam 1994, 2010). Taxa recorded include *Rhyzobius* sp., *Micraspis frenata*, ?*Scymnus* sp., *Archegleis edwardsi*, *Orcus chalybeus* (on *Syzygium smithii*); *Coccinella repanda*, *Diomus notescens*, *Epilachna* spp. (two), *O. chalybeus*, ?*Rhyzobius* sp. (on *Syzygium floribundum*); *Archegleis testudinaria*, *Rhyzobius* sp., *Scymnus* sp. (on *Rhodomyrtus psidioides*); *Rhyzobius* sp. near *nigrovarius* (on *Acradenia euodiiformis*); *Rhyzobius* spp., *C. repanda*, *D. notescens*, ?*Scymnus* sp. (on *Alectryon coriaceus*); *Rhyzobius* sp. (on *Elaeodendron australe*); *D. notescens*, *Rhyzobius* sp., *Scymnus* sp. (on *Cuttsia viburnea*); *Rhyzobius* spp. (two), *Rodalia koebelei*, *Scymnodes* sp. near *terminalis* (on *Drypetes deplanchei*); and *Rhyzobius* sp. (on *Scolopia braunii*). *Erithionyx albatus* has been recorded from flowering *Cryptocarya microneura* (Lauraceae) blossom (Tomaszewska and Ślipiński 2013).

The coccinellid fauna is particularly speciose on several tree species growing in littoral rainforest. On *Euroschinus falcatus*, recorded Coccinellidae are *Diomus notescens*, *Rhyzobius* spp. (two), *Coccinella repanda*, *Cryptolaemus montrouzieri*, *Epilachna* sp., *Micraspis* ?*frenata* and *Rhyzobius* spp. (two). Species recorded on the flowers of *Alphitonia excelsa* were *Amidellus ementitor*, *Cryptolaemus montrouzieri*, *Ephilachna* sp., *Rhyzobius* spp.or a closely related genus (six), and *Rodolia koebeli* (Williams 1995).

But inland the fauna found on *Alphitonia excelsa* in wet sclerophyll forest appears impoverished (*Amidellus ementitor*, *Archegleis testudinaria*, *Orcus* sp.). This seems to be the situation for other rainforest trees in the subcanopy stratum of lowland eucalypt-dominated wet sclerophyll forests (e.g. *Schizomeria ovata*, *Ackama paniculata*, *Cryptocarya microneura*, *Diospyros australis*) where only single species (often only of *Rhyzobius*) may be observed (Williams 1995). This is unrelated to flower structure.

Corylophidae These are small inconspicuous beetles, mostly little more than 2 mm in length (Ślipiński *et al.* 2009). Several genera are associated with leaf litter and fungal spores. Genera recorded from northern New South Wales include *Orthoperus*, *Clypeaster* and *Sericoderus* (Lea 1895b; Williams 1995, 2002). Although Corylophidae are generally considered part of the Coleoptera fauna associated primarily with detritus and leaf litter on the forest floor (Crowson 1981), adult *Sericoderus* and *Orthoperus* also occur on flowering trees and shrubs in the canopy and subcanopy strata of lowland subtropical rainforest and wet sclerophyll forest (Williams 1995). *Sericoderus* has been collected at the flowers of *Syzygium smithii*, *S. floribundum*, *Rhodomyrtus psidioides*, *Tristaniopsis laurina*, *Guioa semiglauca*, *Alectryon coriaceus*, *Alphitonia excelsa*, *Ackama paniculata*, *Schizomeria ovata*, *Cryptocarya microneura*, *Cuttsia viburnea*, *Diospyros australis*, *Scolopia braunii*, *Elaeocarpus obovatus*, and *Euroschinus falcatus*. *Orthoperus* has been collected from *Acradenia euodiiformis* and *Syzygium floribundum*.

Curculionidae Curculionidae are one of several families within the superfamily Curculionoidea that are commonly referred to as weevils, many with an elongate rostrum of various lengths and curvatures (e.g. as in ?*Diethusa*, Plate 138). Additional weevil-like families in Australia are Anthribidae, Apionidae, Attelabidae, Belidae, Brentidae, Caridae and Nemonychidae, but the Curculionidae vastly outnumber all other families combined (Lawrence and Ślipiński 2013). Curculionids, or true weevils, are phytophagous as larvae and adults. They utilise a wide range of plant material as larval hosts, including wood, fruits, seeds, pollen and flowers. The great majority of terrestrial and freshwater plant taxa are eaten by Curculionidae (R. Anderson 1993).

Curculionidae represent the largest family of terrestrial organisms, comprising more than 50 000 described species world-wide (~8000 in Australia [Pullen *et al.* 2014]), and were originally associated with non-angiosperm plants (R. Anderson 1995). True weevils may not have evolved until the Late Jurassic–Early Cretaceous (Grimaldi and Engel 2005; Hall *et al.* 2004; Oberprieler 1995) and though curculionids (and other phytophagous insects) have clearly radiated in response to the evolution and diversification of angiosperms (O'Brien and Askevold 1992), their ancestors fed on gymnosperms (conifers, cycads), as some continue to be associated with today. Often overlooked is the probable role of now extinct Lycophyta, Bennettitales and other early seed plants (Taylor *et al.* 2009) in providing pre-angiosperm pathways for the development of phytophagous insect groups.

R. Anderson (1995) postulates that the great diversity within the family is derived from the contributing factors of (1) use of the highly developed rostrum, by females, in oviposition site preparation and (2) 'a timely association with living structures of the plant lineage considered directly ancestral to angiosperms'. Curculionids were able to exploit nutritionally-rich living plant structures and 'track these resources as the radiation of angiosperms took place throughout the Tertiary' (R. Anderson 1995).

The curculionid weevil fauna of Australia's subtropical rainforests is potentially a massive, yet largely an undocumented one. Many species are known only from a limited number of localities, or await description. One of Australia's largest weevils, the rare *Eurhamphus fasciculatus* (Molytinae), occurs within the Gondwana rainforests region (Mt Glorious, Bunya Mountains, Canungra, Brooklana, Dorrigo) (Anderson *et al.* 2018). Although this species breeds in Hoop Pine, *Araucaria cunninghamii* (Araucariaceae), its range is significantly less than that of its host tree (Monteith 1993), which extends from Queensland to the Nambucca River in northern New South Wales (Harden 1990). Other regional molytine species associated with *Araucaria cunninghamii* are *Ilacuris laticollis* (e.g. from Yarraman, Mt Tamborine, Imbil State Forest), *Notopissodes pictus* (Dorrigo to southern Queensland with an isolated population near Paluma, North Queensland) and *Imbilius araucariae* (Imbil State Forest) (Anderson *et al.* 2018). *Ilacuris laticollis* (Plate 141) is sometimes considered a pest of timber. Species of the related genus *Tranes* are pollinators of cycads (*Macrozamia, Lepidozamia*) (Chadwick 1993; Ornduff 1989; Terry 2001), several of which are associated with subtropical rainforest and wet sclerophyll forest (e.g. *Lepidozamia peroffskyana*, *Cycas megacarpa*, *Macrozamia lucida*, *M. douglasii*, *M. macleayi* [Harden *et al.* 2006]). In *Lepidozamia peroffskyana* the weevils lay their eggs in the mature, spirally splitting, male cones (Plate 145). Although a few larvae develop in the small amount of decaying cone residue sometimes left on the apex of the trunk, or in the central cone core itself, the greater pulpy mass of the cone falls away to the ground carrying many hundreds, if not thousands, of *Tranes* larvae with it. Most larvae enter the soil profile, there to mature and later emerge (upwards of 5 weeks later; G. Williams pers. obs.). Some adults also enter the microsporophylls but larvae are generally not found in these structures. Ornduff (1989) cites *Tranes lyterioides* as the species associated with the cones of *Lepidozamia peroffskyana* in the Dorrigo–Coffs Harbour region; however, Rolf Oberprieler (pers. comm.) considers that the species associated with *L. peroffskyana* is different.

The tribe Amycterini (Cyclominae) (e.g. *Phalidura hopsoni* – Barrington Tops, *Talaurinus apicihirtus* – Bunya Mountains) are flightless ground dwellers, species of which utilise herbaceous *Lomandra* (Lomandraceae) as food plants. Amycterini are restricted to Australia and are of Gondwanan origin. There is no closely related group in South America. Amycterini do, however, have a close but ancient relationship with the African Somatodini (Zimmerman 1993). The endemic genus *Mythites* is confined to Queensland and New South Wales; *Mythites arboricola* is described from Mt Tamborine (Carter 1933; Zimmerman 1993).

Cryptorhynchinae (= Molytinae – Cryptorhynchini) are a large subfamily of weevils frequently found under fallen logs or active on tree trunks. Adults strike a characteristic death feign when distubed. Genera recorded from northern New South Wales and southern Queensland include *Blepiarda*, *Brachyporopterus*, *Decilaus*, *Enteles*, *Eutyrhinus*, *Imalithus*, *Isax*, *Glochinoorrhinus*, *Methidrysis*, *Microporopterus*, *Mitrastethus*, *Nechyrus*, *Neoporopterus*, *Orthoporopterus*, *Pachyporopterus*, *Paleticus*, *Paletonidistus*, *Perissops*, *Poropterus*, *Protopalus*, *Pseudometyrus*, *Scolyphrus*, *Sympiezoscelus*, *Tapinosomus*, *Tychreus*, *Tyraeosus* and *Wiburda* (Hawkeswood 1990b, 1991b; Williams 2002). *Enteles vigorsii* (Plate 140) is an attractive blue-black patterned species found in subtropical rainforests, and utilises the cycad *Lepidozamia peroffskyana* as one of its larval hosts. When viewed from behind (caudal view) the pattern of setae on the apical third of the elytra (when closed) resembles the face of a salticid spider. Prominent apical 'eye spots' are also present on the related *Metrameniomorpha entima*. *Perissops ocellatus* (Plate 142), from the rainforests of the Comboyne Plateau and adjacent Lansdowne–Comboyne Escarpment, possesses conspicuous black 'eye spots' at the base of the elytra. Many species of *Poropterus* (e.g. *Poropterus angustatus*, *P. angustus*, *P. bisignatus*, *P.corvus*) (Plates 143, 144) possess numerous tubercles on their elytra, which provide cryptic bark-like camouflage, allowing them, when motionless, to blend in with the surfaces of coarse-barked tree trunks and rotting timber. Other Cryptorhynchinae are relatively ovoid (e.g. species of *Perissops*, *Tropidotasia*, *Tyrtaeosus*) and resemble the fallen seeds and fruit of Lauraceae, Vitaceae and Oleaceae.

The Cossinae are small elongate and flattened weevils that develop in rotten wood (e.g. *Cossonus praeustus*, *C. platynotus*) and the dead rhachises of tree ferns (e.g. *Cyathea* spp.–Cyatheaceae) (*Cossonidius*, or related genera) that persist attached for lengthy periods of time to the trunk. Adult *Curculio* (Curculioninae) possess long, curved rostrums and are associated with figs (*Ficus* spp.). Adult *Curculio amoenus* occur on fruiting *Ficus obliqua* in littoral rainforest. *Leptopius gladiator* (Leptopinae/Entiminae) is a distinctive large black weevil, with prominent spiny tubercles on its elytra, that inhabits subtropical rainforest at intermediate and higher altitudes (e.g. League Scrub, Ulong, Barrington Tops). *Phaenomerus*, *Mecopus* (Zygopinae/Conoderinae) and *Baris* (Baridinae/Conoderinae) are examples of modern Indo-Malayan (Torresian) elements in the Australian curculionid fauna (Lawrence and Britton 1994, Thompson 1996). Species of *Mecopus* are fast-moving weevils found on the trunks of freshly fallen trees in rainforests at low to intermediate altitudes (e.g. Lansdowne Escarpment, Woko National Park, Lismore). *Mecopus* males are more long-legged than females and participate in territory defence and competion for females against other males.

Baris includes species that frequent the flowers of *Syzygium floribundum*, *Tristaniopsis laurina* and *Alphitonia excelsa* in lowland riparian rainforest and wet sclerophyll forest (Williams 1995). Numerous other weevils are associated with a diverse range of flowering canopy trees and subcanopy shrubs in subtropical rainforests (Williams 1995); however, these are dominated by species of *Meriphus* and *Storeus* (Curculioninae). Records include *Meriphus* sp. (on *Abrophyllum ornans*); *Cytallia sydneyensis*, *Storeus* sp. (on *Cuttsia viburnea*); *Cytallia* sp. (on *Diospyros australis*); *Meriphus* sp., *Storeus* spp. (on *Guioa semiglauca*); *C. sydneyensis* (on *Rhodomyrtus psidioides*); *Meriphus* spp., *Storeus* spp., *C. sydneyensis*, *Neomelanterius carinicollis*, *Orthorhinus* sp. (on *Syzygium smithii*); *C. sydneyensis*, *Eristinus eucalypti*, *Haplonyx* spp., ?*Ipsichroa* (on *Syzygium floribundum*); *Meriphus* spp., *Storeus* spp., *Araecerus* sp. (on *Alectryon coriaceus*); *C. sydneyensis* (on *Acradenia euodiiformis*); *Cytallia* sp. (on *Alphitonia excelsa*); *Storeus* sp. (on *Drypetes deplanchei*); *Meriphus* spp., *C. sydneyensis*, *Cytallia* sp., *Neomelanterius carinicollis* (on *Euroschinus falcatus*); and *Meriphus* sp., *Storeus* sp. (on *Scolopia braunii*). However, the pollination function of Curculionidae in general on flowers is poorly understood. Some are potentially consumers of pollen and possibly nectar, or utilise developing seeds as oviposition sites and food sources for developing larvae. Weevils in the genus *Elleschodes* (Plate 139) are obligate pollinators of the primitive angiosperm family Eupomatiaceae (*Eupomatia laurina*, *E. bennettii*) in southern Australia, and utilise the fused floral synandrium of *Eupomatia* as an oviposition and larval development substrate (A. Hamilton 1897; Williams and Adam 1994, 2010).

Dermestidae Dermestid larvae are preserved in 100 million-year-old amber from Burma, these being the oldest examples of the family known (Grimaldi and Engel 2005). Eastern Australia is a centre of origin for the Dermestidae and the Australian region in general is characterised by a large number of endemic genera (e.g. *Neoanthrenus*) (Mroczkowski 1968). The Australian fauna possesses affinities with the Neotropics, and both regions are centres of diversity for the genus *Trogoderma* (Mroczkowski 1968). Species recorded from subtropical upland forests include *Anthrenocerus blackburni*, *Neoanthrenus parallelus*, *Trogoderma laevipenne* and *Trogoderma silvicolum* (Armstrong 1942, 1949; Roach 2000; Williams 2002).

Adult Dermestidae possess a vestiture of fine hairs or scale-like setae on the integument, and are often notoriously referred to as carpet beetles, owing to the preference some species have for feeding on keratin-rich fur, skin and feathers. Dermestidae often feed on the proteinaceous remains of animal carcasses, particulary during the active 'wet' and 'skeletal' stages of decay (Ndueze *et al.* 2013), but may also be associated with the nests of different vertebrates, as well as those of social insects. In lowland subtropical rainforests, and rainforest trees growing in the understorey of wet sclerophyll forest, adults of *Anthrenus*, *Trogoderma*, *Attagenus*, *Anthrenus*, *Orphenus* and *Thaumaglossa* (*Thaumaglossa nigricans*) are occasionally found on flowers of canopy and subcanopy trees and shrubs, where they feed on nectar and possibly pollen (Williams 1995). Except for *Attagenus* (Attageniinae), these are all members of the subfamily Megatominae. Though the presence of flower-frequenting Dermestidae might result in the loss of pollen, some residual pollen is carried on their bodies, such that there is potential for pollen transport.

Elateridae Elateridae are a large family of phytophagous, saprophagous and predatory beetles well represented in the rainforests and wet sclerophyll forests of south-eastern mainland Australia (Calder 1986, 1996, 1998; Gullan 1977; Neboiss 1957; Williams 1993, 2002). Larvae, where known, are associated with rotting wood and soil. Elateridae probably originated in the Early to Mid Jurassic, with a diverse fauna being known from the Late Jurasicc (160–150 mya) and Cretaceous, and also richly preserved in Eocene Baltic amber. Recent genera were established by the Early Tertiary (Grimaldi and Engel 2005). Commonly called click beetles, adult elaterids possess a clicking mechanism, in which the interaction of a long tongue on the prosternum, with a cavity in the mesosternum, produces a powerful flexation of the body that enables them to jump (Lawrence and Britton

1991, 1994); this acts as a means of escape from threats, but also allows beetles to right themselves after landing on their backs. Two related families that possess a similar facility are Eucnemidae and the small-sized Throscidae.

The adults of many species are found resting on foliage by day (e.g. Denticollinae [= Dendrometrinae], Parablacinae), some are attracted to light (*Aphileus*, *Paracalais*, *Conoderus*, *Lingana*), and several genera frequent flowers (*Anilicus*, *Agonischus*, *Drymelator*, *Ophidius*, *Melanoxanthus*, *Microdesmes*) (Williams 1995). Elateridae that feed on flowering rainforest trees form a minor element of the overall abundance and diversity of visiting insects. In addition, their activities entail only modest inter-plant movements, such that they are unlikely to play any significant role in the long-distance transfer of pollen between individuals of highly dispersed tree populations. Of those encountered, only species of *Microdesmes* (= *Limonius*) may occur in abundance (usually on Myrtaceae, and particularly on *Syzygium smithii* var. 'minor'). Records include *Microdesmes angulatus*, *M. collaris*, *M. pubescens*, *Agonischus basilis*, *A. aulacoderus*, *A. australis*, *Megapenthes futilis*, *Drymelator* sp., *Paracardiophorus bicolor* (on *Syzygium smithii*); *Agonischus* spp., *Microdesmes pubescens*, *Megapenthes futilis*, *Melanoxanthus lativittis*, *Drymelator* sp. (on *Syzygium floribundum*); *Agonischus ?aulacoderus*, *A. australis*, *Anilicus* sp, *Megapenthes futilis*, *Ophidius histrio* (on *Tristaniopsis laurina*); *Agonischus* sp., *Microdesmes collaris*, *Megapenthes futilis*, *Drymelator* sp. (on *Rhodomyrtus psidioides*); *Ophidius histrio*, *Megapenthes futilis* (on *Alphitonia excelsa*); *Agonischus* sp., *Microdesmes angulatus* (on *Ackama paniculata*); *Agonischus* spp., *Melanoxanthus lativittis* (on *Cuttsia viburnea*); *Microdesmes collaris*, *M. ?angulatus* (on *Elaeocarpus obovatus*); *Agonischus ?australis* (on *Elaeocarpus reticulatus*); *Microdesmes collaris*, *M. angulatus* (on *Euroschinus falcatus*); *Microdesmes collaris*, *Megapenthes futilis* (on *Guioa semiglauca*); *Agonischus* spp., *Microdesmes ?collaris*, *Megapenthes futilis* (on *Schizomeria ovata*); and *Megapenthes futilis* (on *Scolopia braunii*) (Williams 1995). As with nearly all flower-frequenting Coleoptera on mass-flowering canopy trees, elaterids show no host plant fidelity; rather, they exhibit foraging constancy to whatever trees with open floral structures are available.

The subfamily Agrypninae is one of the major groups of Australian Elateridae. The genus *Agrypnus* is distributed world-wide but has not been recorded from South America. Centres of species diversity are Madagascar (111 species) and Australia (116 species) (Calder 1996). *Aphileus* (Plate 146) is an endemic genus recorded from Western Australia, Northern Territory, Queensland and New South Wales. Numerous species of *Conoderus* are recorded from the subtropical region of eastern Queensland and New South Wales. The genus is found in most regions of the World, but the greatest number occurs in Australia, South and Central America. Larvae of *Agrypnus* and *Conoderus* live in the soil. *Drasterius* is widespread, but in Australia the genus is only known from coastal New South Wales (three species) and the south-west corner of Western Australia. *Paracalais* (Plate 147), which includes quite large-sized species, occurs in Australia, Norfolk Island and New Guinea. Species of *Paracalais* prey on wood-boring insect larvae. *Pseudaeolus* is endemic to Australia and occurs widely in south-east Australia (Calder 1996). Examples of Agrypninae from subtropical rainforest are *Aphileus lucanoides*, *Conoderus subflavus*, *Paracalais gibboni* and *P. gigas* (>5 cm in length), which is the largest species known from subtropical rainforest (e.g. Murray Scrub, Toonumbar State Forest).

The Cardiophorinae include two genera (*Cardiotarsus*, *Paracardiophorus*) from subtropical eastern Australia that are more widely distributed. *Cardiotarsus* occurs throughout the Oriental, Afrotropical and Australian regions, but in Australia is confined to southern and northern Queensland, and southern New South Wales and the Australian Capital Territory (Calder 1996). *Paracardiophorus* occurs in the Palaearctic, Oriental and Australian regions and is widespread within Australia. *Cardiotarsus* occurs in rainforest, dry sclerophyll forest and heathland, and *Paracardiophorus* has been recorded from a variety of vegetation communities including rainforest, wet and dry sclerophyll forests, and *Melaleuca* forest (Calder 1996). Larvae of *Paracardiophorus* are predators in the soil.

Denticollinae (= Dendrometrinae) include the genus *Hapatesus*, which is restricted in distribution to eastern Australia, Papua New Guinea and New Britain (Neboiss 1957) and occurs in rainforest, and wet sclerophyll forest. All other Denticollinae are apparently endemic (Calder 1998). Species of the genus *Toorongus* are confined to south-east Australia, with the genus reaching its northern-most distribution in south-east Queensland (Neboiss 1957). *Corystelater* is confined to the McPherson Range subregion. The adults of some *Drymelater* species are associated with *Nothofagus* rainforest (Calder 1996).

Elatichrosis species occur in rainforest, wet sclerophyll forest and woodland. Species of *Liteolater*, *Toorongus*, *Microdesmus* and *Glypheus* have been recorded from rainforest, and wet and dry sclerophyll forest. The small rainforest genus *Rousia* contains two known species and is restricted to the region from northern New South Wales to North Queensland (Calder 1996). Adult *Rousia* superficially resemble lycid beetles. Examples of Denticollinae recorded from subtropical rainforest are *Corystelater kyoglensis*, *Hapatesus pervulgatus*, *Glypheus crucifer*, *Crepidomenus fuscogalbus*, *Rousia dumbrellium* and *Crepidomenus frazieri*. *Crepidomenus vitticollis* inhabits cool temperate rainforest. Adults of *Corystelater kyoglensis* have been collected at light traps and the larvae have been found in rotten wood (Calder 1996). The Denticollinae have a predominantly south-eastern coastal distribution and the geographical ranges of species conform to the pattern of Matthews (1972) for onthophagine dung beetles (Scarabaeinae-Onthophagini). There are two centres of denticolline distribution in Australia focused on the McPherson Range–New England Tablelands region, and the Blue Mountains (west of Sydney)–Victorian highlands (Calder 1986). Calder (1986) suggests that south-east Australia is a possible refuge area for Denticollinae (as Crepidomeninae); this incorporates part, or all, of the Gondwana rainforests region depending on how Calder defines the refugium boundaries. Present distribution patterns are believed largely to be the result of climatic and sea-level changes associated with the last glacial period (Calder 1986).

Elaterinae comprise more than 20 Australian genera, most of which are endemic (Calder 1998). Of the endemic genera *Anilicus*, *Ascesis*, *Augenotus*, *Drymelator*, *Dicteniophorus*, *Glyphochilus*, *Lingana*, *Paranilicus* and *Yalganus* occur in the Gondwana rainforests region (Calder 1996). *Anilicus* occurs in central and eastern Australia from northern Queensland to Victoria, southern South Australia and south-western Western Australia, but is absent from Tasmania. *Ascesis* is restricted to eastern Australia and adults have been recorded from rainforest and wet sclerophyll forest (Calder 1996). *Glypholchilus* is probably associated with wet and dry sclerophyll forest types and rainforest, and *Augenotus*, *Lingana*, *Paranilicus* and *Yalganus* are recorded from several rainforest formations, and wet sclerophyll forest (Calder 1996). The non-endemic genera occurring in the Gondwana rainforests region include *Megapenthes* (Palaearctic, Oriental, Afrotropical, Nearctic, northern Neotropical and Australian regions) and *Melanoxanthus* (eastern Palaearctic, Oriental, Afrotropical and Australian regions), which occur in rainforest, and dry and wet sclerophyll forest. *Anchastus* is widespread, but in Australia is confined between eastern Victoria and the northern tablelands of New South Wales. *Anchastus* inhabits *Eucalyptus* forest (Calder 1996). *Drymelater* occurs in eastern Australia (except Tasmania), south-west Western Australia and the Northern Territory. Adults have been collected in rotting logs in rainforests and in leaf litter; however, they may also been found on flowers (Calder 1996). *Dicteniophorus* is distributed from North Queensland to south-eastern Australia. Adults come to light, occur in leaf litter, and also can be found under loose bark.

Lissominae are a small subfamily related to the Throscidae (Calder 1996). The genus *Austrelator* is endemic to Australia and comprises five species. *Austrelator* is distributed from northern Queensland to south-eastern New South Wales, and Lord Howe Island. The endemic genus *Osslimus* consists of one described species (*Osslimus freyi*), which is recorded from the Bunya Mountains, Mount Glorious and the Manning Valley in rainforest and wet sclerophyll forest, and one undescribed species. The small subfamily Negastriinae includes the endemic genus *Rivulicola*. *Rivulicola* is distributed in a wide mainland arc from southern Victoria, through New South Wales and Queensland to the north-western regions of Western Australia (Calder 1996).

The Australian genera previously placed in the Pityobiinae (e.g. *Wynarka*, *Parasaphes*, *Parablax*) are now in the Parablacinae (e.g. Plate 148) (Kundrata *et al.* 2016), and are restricted to south-east Australia, Tasmania and coastal Queensland (Calder 1998). As currently considered the Pityobiinae now only comprise the North American genus *Pityobius*, and tentatively the South American *Tibionema* (Kundrata *et al.* 2016). The monotypic genus *Xuthelater* (*Xuthelater moppiensis*) has been described from *Nothofagus*-dominated cool temperate rainforest at Barrington Tops, and is known only from the Barrington Tops subregion (Calder 1996). *Wynarka* is monotypic (*Wynarka sylvestre*) and is recorded from *Nothofagus* and *Eucalyptus* forests. *Parasaphes* species have been collected in *Nothofagus* and subtropical rainforest, and wet sclerophyll forest (Calder 1996). Adults have been collected under bark and beaten from foliage. *Parablax* (e.g. *Parablax moorda*) is recorded for temperate rainforest and wet sclerophyll forest (Calder 1996). *Ophidius* (previously in the Elaterinae) extends from northern Queensland to southern New South Wales, and includes the flower-frequenting *Ophidius histrio*, a common species found in rainforest and open vegetation types.

Elmidae These are known as riffle beetles because of their life spent within and adjacent to flowing water bodies. Elmids are small to minute beetles that occur worldwide, and as adults are mainly aquatic. All larvae are aquatic. Other aquatic and semi-aquatic beetles are represented by the Psephenidae, Heteroceridae, Limnichidae, Hydrophilidae, Hydraenidae and Hydradephaga. Xylophagous (wood-feeding) elmid larvae are considered important modifiers of submerged wood, their tunnelling activities increasing surface area and consequently facilitating entry by other wood-decaying organisms (McKie and Cranston 1998). In northern New South Wales and south-eastern Queensland Elmidae are represented by several *Austrolimnius* species (e.g. *Austrolimnius atriceps*, *A. menopon*, *A. thyas*), and *Kingolus heroni* (Williams 2002), the latter one of many insects named after W. Heron who collected and discovered new species of Coleoptera in the Dorrigo area in the early 20th century. *Austrolimnius* is the dominant elmid genus in Australia. Restricted to fresh water, it is also found in South and Central America (Hinton 1965). Hinton (1965) lists five species described from the Allyn River on the eastern margin of the Barrington Tops Plateau.

Erotylidae Beetles of this family are known as pleasing fungus beetles. Adults are brightly patterned in red, yellow and black and feed on fungal spores and fruiting bodies, including bracket fungi (particularly Polyporaceae) (Webb and Simpson 1991, Hawkeswood 1986). By day they are found in and under rotting wood on the forest floor, and at night venture across the surfaces of standing and fallen tree trunks. Species are to be found most commonly in rainforests at intermediate elevations. Erotylidae recorded from the Gondwana rainforests region include *Episcaphula flavofasciata*, *E. rufolineata*, *E. duplopunctata* (Plate 149), *Thallis atricornis*, *T. melancholica* and *Euxestus bivulneratus* (Hawkeswood 1986; Lea 1921; Williams 2002; F. Wilson 1923). *Protoloberus singularis* from Queensland and New South Wales is endemic (Lawrence and Ślipiński 2013).

Eucnemidae The cosmopolitan Eucnemidae are known from Eocene Baltic amber. As larvae they are found in rotting wood, and adults occasionally come to lights. Like Elateridae, eucnemids have a clicking mechanism, which allows them to forcefully move their body. The eucnemid fauna of subtropical eastern Australia includes species of *Dyscharachthis*, *Euryptychus*, *Fornax*, *Hemiopsida and Microrhagus* (Calder 1998; Muona 1993). These are all genera with extralimital distributions beyond Australia.

Eurhynchidae The large weevil-like eurhynchids are placed as a subfamily (Eurhychinae) of Brentidae by Lawrence and Ślipiński (2013). Unlike the Brentidae, the eurhynchids are part of the ancient Australian weevil fauna (Zimmerman 1994b). Most are rare or scarce insects that in subtropical Australia (*Eurhynchus* spp.) are more typically associated with the trunks of *Eucalyptus* trees and related genera. The Australian fauna is largely confined to eastern and south-eastern Australia; however, three species of *Ctenaphides* occur in south-west Western Australia. Of the Australian genera only *Aporina* is extralimital in distribution, occurring also in Papua New Guinea (Zimmerman 1994b).

Geotrupidae The Geotrupidae are members of the superfamily Scarabaeoidea (in Australia including the Hybosoridae, Scarabaeidae, Passalidae, Lucanidae, Trogidae) and are sometimes cited as a subfamily (Geotrupinae) of the Scarabaeidae (Howden 1979, 1985; Howden and Cooper 1977). Scarabaeoids that resemble modern Geotrupidae are known from Lower Cretaceous deposits (Crowson 1981). All native Australian species are currently placed in a single subfamily Bolboceratinae (Lawrence and Ślipiński 2013), a group that is best represented in arid and semi-arid zones (Howden 1979, 1985; Howden *et al.* 2007); males of many of these have conspicuous cephalic and pronotal horns. Adults are nocturnal and lay their eggs deep in the soil, their activities noticeable by excavated piles of soil on the forest floor. Food items may include fungal fruiting bodies; however, life histories are poorly known. Several geotrupids (e.g. *Elaphastomus proboscideus*, *Blackburnium angulicorne*) participate as hosts in a phoretic association with mites (Costa and Allsopp 1979).

Although many geotrupids exhibit Eyrean and temperate Bassian distribution patterns, several *Australobolbus* and *Gilletinus* are associated with rainforest and wet sclerophyll forest along Australia's eastern seaboard (Howden 1992). *Australobolbus* is endemic to Australia and New Guinea. Adults are occasionally attracted to light, but more usually will be seen flying at night in the rainforest understorey or on forest margins. Five species of *Gilletinus* are described from Australia, one of which ranges into New Guinea. *Gilletinus williamsi* occurs in subtropical rainforest and wet sclerophyll forest (Manning Valley, Joalah National Park). *Australobolbus austrinus* (Plate 150), *A. lineacollis*, *A. loboscapus* and *Australobolbus gayndalensis* are recorded from southern Queensland and northern New South Wales, individual species mainly being known from upland sites (e.g. Blackdown Tableland, Kroombit Tops, the Bunya Mountains, Tooloom Plateau, Mt Warning, Gibraltar Range). *Australobolbus masneri* inhabits rainforest on Fraser Island and in Cooloola National Park (Howden 1992). The endemic genus *Elephastomus* (Carne 1965) is represented by *E. gellarus* and *E. proboscideus* – species that inhabit open forests but are sometimes encountered in rainforest, including upland and lowland localities (e.g. Bunya Mountains, Acacia Plateau, Dorrigo, Manning Valley). The genus *Blackburnium* is well represented in western and northern Australia (Howden 1979); however, *B. angulicorne* is commonly found in littoral rainforest on nutrient-rich sandy soils (e.g. Iluka, Harrington); the fresh excavations of adults are conspicuous in late summer following rain.

Histeridae Adult and larval Histeridae are predaceous on other insects (Lackner and Leschen 2017), and are often associated with animal dung and carrion (e.g. species of *Saprinus*), and are found under bark and in rotting wood where they may pursue the larvae of wood-boring insects. Some are myrmecophilous. Although histerids are seemingly an inconspicuous component of the rainforest beetle fauna, they have been shown to exhibit high abundance and species richness in Australian tropical rainforests, both in the canopy stratum, and on the forest floor (Stork and Grimbacher 2006).

A bizarre group of histerids inhabiting subtropical rainforests and wet sclerophyll forests is the subfamily Chlamydopsinae (Caterino 2003). The distribution of the Chlamydopsinae is centred upon Australia (also recorded from New Caledonia) and species are inquilines that live in association with the colonies of social insects, especially ants. Individuals exhibit unusual morphological and chemical appeasement specialisations for that lifestyle. Chlamydopsines are occasionally encountered on rainforest foliage, their appearance resembling bird droppings or cryptic thomisid spiders. Several species of *Chlamydopsis* are recorded from subtropical rainforests: *C. striatipennis* (Mt Keira, Lansdowne Escarpment), *C. compressipes* (Mt Tamborine, Mt Glorious), *C. strigicollis* (Allyn River), *C. variolosa* (Mt Glorious) and *C. coronis* (Lansdowne Escarpment) (Caterino 2003).

Hybosoridae Hybosorid-like species are known from the Lower Cretaceous (Crowson 1981; Yan *et al.* 2013). This relatively small family in Australia is placed within the superfamily Scarabaeoidea, and is sometimes cited as a subfamily of the Scarabaeidae. Adults, and particularly those of the subfamily Ceratocanthinae (Paulian 1977), are able to articulate their body into a sphere-like defensive shape. The only other beetles capable of doing this are Clambidae and Leiodidae. A single ceratocanthine, *Cyphopisthes monteithi*, extends to southern Queensland where it inhabits dry rainforest (Ballerio 2013). *Cyphopisthes* is distributed in India, New Guinea and Queensland.

Liparochrus is the most frequently encountered hybosorid genus (Plate 151). Though also found in more xeric habitats, it especially occurs in the rainforests of eastern and northern Australia, and extends to New Guinea, New Caledonia, the Loyalty Islands and Lord Howe Island (Allsopp 1984; Paulian 1980). The species found in southern Queensland and New South Wales are principally inhabitants of rainforest and wet sclerophyll forest. On the basis of its south-west Pacific distribution *Liparochrus* can be considered autochthonous in origin. Species are saprophages and coprophages. *Liparochrus silphoides* has been recorded commonly feeding on the rotting carcasses of flying foxes (Pteropodidae) following massive climatic heatwave kills at Wingham Brush Nature Reserve. In subtropical eastern Australia *Liparochrus* is recorded in rainforest from Eurimbula (*Liparochrus nanus*, *L. bimaculatus*) and Goodnight Scrub (*Liparochrus bimaculatus*) in Queensland, south to at least to the vicinity of Nowra (*Liparochrus fossulatus*, *L. sculptilis*). In New South Wales species occur in littoral rainforest (e.g. Harrington, Cape Hawke, Sea Acres Nature Reserve) (*Liparochrus fossulatus*, *L. silphoides*), island rainforest subject to complete and repeated flooding (Coocumbac Island Nature Reserve) (*Liparochrus silphoides*), rainforests at intermediate altitudes (e.g. Coorabakh, Dooragan and Dorrigo national parks, League Scrub, Kiwarrak State Forest) (*Liparochrus laevis*, *L. sculptis*, *L. silphoides*, *L. fossulatus*, *L. bimaculatus*), and high altitude sites (e.g. Border Ranges, Werrikimbe National Park) (*Liparochrus bimaculatus*, *L. silphoides*), including *Nothofagus*-dominated cool temperate rainforest in the Barrington Tops (*Liparochrus silphoides*) (W. Houston 1992; Paulian 1980; Williams 2002; G. Williams unpubl. observ.; Williams and Williams 1982, 1983a, 1983b, 1983c, 1984).

Hydradephaga Hydradephaga are the aquatic forms of the Adephaga (see Lawrence and Ślipiński 2013; Short 2018). The remaining Australian Adephagan fauna is constituted by the terrestrial Carabidae. Now extinct Hydraphagan families occur in the Late Triassic of Eurasia (*c*. 230–220 mya), but by ~150 mya most modern lineages had appeared, and were diverse by the Tertiary (Crowson 1981; Grimaldi and Engel 2005). There is a pronounced disjunction in hydradephagan distribution in the Queensland–New South Wales border area.

In Australia the Hydradephaga (Lawrence and Britton 1991, 1994; Lawrence and Ślipiński 2013; Watts 1978, 1985, 1989, 1995, 1997, 1998) comprise the aquatic families Dystiscidae, Noteridae, Gyrinidae, Haliphidae and Hygrobiidae. The dystiscids, or diving beetles, are the most diverse (*c*. 265 species). Larval and adult Haliphagidae feed on algae. Noteridae are thought to possibly be detritus feeders, and hygrobiids are bottom-feeding predators that are often associated with stagnant water. Dysticids prey on insects and other invertebrates, but larger species may also prey on small fish and amphibians.

The Hygrobiidae are a small family, consisting of one species in northern Africa, one species in western China, two species in south-eastern Australia and one species in northern Australia (Lawrence and Britton 1994). These are placed in a single genus *Hygrobia* (Britton 1981). No hygrobiids are known from southern Africa, or North and South America. Their current distribution suggests that they are relics of a much wider fauna (Britton 1981).

Dysticidae are abundant predators in a variety of water bodies, and adults frequently come to light, even when these are placed in rainforest sectors some distance from water. The greatest diversity of taxa occurs in the south-east Bassian region of Australia (Lawrence and Britton 1991; Watts 1978). Fifty-five per cent of genera and ~80 per cent of species found there are not found anywhere else in Australia. Overall, the southern region possesses a high level of dytiscid endemicity but with low levels elsewhere in Australia (Watts 1978). This high incidence of endemism suggests a more ancient origin for Dytiscidae, with the other major Australian aquatic families (e.g. Gyrinidae, Hydrophilidae, Hydraeinidae) being more recently derived from the north (Watts 1985). Members of the dytiscid subfamily Hydroporine, such as *Chostonectes gigas* and *Sternopriscus hansardi*, occur in the Gondwana rainforests region. Although relatively few members of the tribe Hydroporini occur in the Southern Hemisphere, Australia is an exception and possesses 59 endemic species placed in eight genera endemic to the Australian faunal region (Balke 1995). The Hydroporini provide an example of intrusion by southern faunas into the Indo-Papuan region: in contrast to much of the New Guinean fauna being derived from Oriental stock, all New Guinea Hydroporini (Hydroporinae) are Australian, or Australian in origin (Balke 1995).

Gyrinids, or whirligig beetles, occur in most regions of the world (Lawrence and Ślipiński 2013), continue to retain very ancient lineages dating from the Late Triassic to Early Jurassic (Gustafson *et al.* 2017), and are possibly the most conspicuous of the hydradephagans to be seen in rainforests; the gregarious adults are active in daylight and occupying even very small pools in otherwise fast-flowing streams during times of drought. In lowland rainforests adults may continue to be present, though in lesser numbers, throughout winter. Gyrinids locate prey using the Johnston's organ, a sensory organ located on the antennae. This is sensitive to water surface disturbances. Species occurring in subtropical eastern Australia include *Aulonogyrus strigosus*, *Macrogyrus striolatus*, *Gyrinus convexiusculus* and *Dineutus australis*; these represent all the Australian genera.

Hydrophilidae The larvae and adults of Hydrophilidae are found at the edges of streams, in water bodies, in damp soil and in decaying plant matter, or are associates of animal dung. Larvae are predaceous; however, adults are phytophagous and saprophagous (Lawrence and Britton 1991, 1994; Lawrence and Ślipiński 2013). Most are found in still water bodies, but some (e.g. *Sternolophus*) are able to live in swiftly flowing streams (J. Anderson 1976). Species include *Helochares marreensis* and *H. percyi* (the genus widespread in the New and Old Worlds), *Dactylosternum marginale* (genus restricted to the Australian region), *Sternolophus marginicollis* (genus distributed in Australasia, tropical Asia, Africa), and several species of *Enochrus* (world-wide genus) (J. Anderson 1976; Newton 1989; Watts 1989, 1995, 1998; Williams 2002).

Lamingtoniidae The Lamingtoniidae are endemic to Australia. There are three species. *Lamingtonium loebli* and *L. thayerae* are found in cool temperate rainforest in Victoria and Tasmania. The third species, *Lamingtonium binnaburrense*, is restricted to the subtropical rainforests of south-eastern New South Wales and southern Queensland (e.g. Lamington National Park). Larvae and adults feed on soft polypore fungi (Lawrence and Britton 1994; Lawrence and Ślipiński 2013; Sen Gupta and Crowson 1969).

Lampyridae Lampyridae, or fireflies, are associated with leaf litter and rotting wood on the forest floor. The family is recorded from Miocene compression fossils and amber (Grimaldi and Engel 2005), and is well represented in the tropics but poorly so in Australia. There are no endemic Australian genera. Adults possess luminous (photic) organs on the abdominal ventrites, and the controlled light flashes from these are used in courtship. In lowland subtropical rainforest and wet sclerophyll forest adults emerge early during warm still nights in September and October. They are rarely observed there after mid-October. Adults are not known to feed but larvae prey on land snails. Larvae paralyse their prey by injecting secretions produced by a pair of glands at the end of the alimentary canal. The prey is digested extraorally, and the resultant liquified tissues are imbibed (Lawrence and Ślipiński 2013).

The Australian lampyrid fauna is exclusively Luciolinae, a subfamily found throughout the warmer parts of the world with the exception of South America (Ballantyne and Lambkin 2000, 2009). The largest Australian genus is *Atyphella*, which is restricted to Australia and New Guinea (Calder 1998). *Atyphella similis* and *A. atra* occur in the Gondwana rainforests region (Ballantyne and Lambkin 2000; Lea 1921).

Lucanidae Commonly known as stag beetles, these are largely robust beetles found under and in decaying wood (Mizunuma and Nagai 1994). Seven of the 17 known Australian genera are endemic, but the Australian lucanid fauna is a relatively impoverished one, being most diverse in the temperate upland rainforests of the south-east (Reid and Beatson 2016). Mandibular dimorphism can be pronounced in males of several regional genera (*Lissapterus, Cacostomus, Ceratognathus*). This may be related to combat and courtship (Crowson 1981; Grimaldi and Engel 2005).

The Lucanidae are placed in the superfamily Scarabaeoidea, along with the closely related Passalidae, Scarabaeidae, Geotrupidae, Trogidae and Hybosoridae (Lawrence and Ślipiński 2013). The family includes newer (Cenozoic; 70 mya–present) taxa (e.g. *Dorculus, Prosopocoilus*) with strong Oriental affinities concentrated in the Torresian subregion (Howden 1981), with older autochthonous and Gondwanan-related elements (e.g. *Ryssonotus, Lissapterus, Lissotes*) focused in the south and south-east of the continent.

Ceratognathus is shared with New Zealand and the Neotropics, and in Australia is mainly confined to the eastern mainland and Tasmania (Hangay and De Keyser 2017; W. Houston 1992). The genus has undergone considerable speciation in New Zealand, derived at least in part from Australian ancestors (B. Holloway 1963). Although there are numerous species the genus is infrequently collected; however, adults and larvae are found under and in decaying timber, but may also occur in the fibrous outer bark of living trees. Some species are attracted to artificial light. Species recorded from subtropical latitudes include *Ceratognathus bitumulatus* from rotting *Nothofagus* wood in the Barrington Tops; *C. flabellatus* and *C. macrognathus* from upland forests in southern New South Wales; *C. niger*, widely distributed but also inhabiting lowland subtropical rainforest; and *C. froggatti* and *C. minutus* from the Illawarra Range and adjacent mountains. The crepuscular species, *C. ocularis*, is widely recorded (e.g. Carrai State Forest, Clyde Mountain, Mount Wilson, Barrington Tops, Dorrigo, Ebor, New England National Park, and Victoria) and is associated with cool temperate rainforest, warm temperate rainforest and open eucalypt-dominated forest.

The widely distributed and common *Syndesus cornutus* is the sole representative of the genus in eastern Australia; an additional species occurs in Western Australia. *Syndesus* also occurs in New Caledonia and South America (Lawrence and Britton 1994). In Australia *Syndesus* inhabits lowland and upland mesic forests, including dry, cool temperate and subtropical rainforests, and subtropical rainforest on floodplains (e.g. Jamberoo Pass, Kiwarrak State Forest, Barrington Tops, Lansdowne Nature Reserve, Fitzroy Falls, Clyde Mountain, New England National Park) (G. Williams unpubl. data). Adults come to light. *Figulus* is a small genus, whose species are found in open forest, lowland subtropical rainforest (e.g. Upper Williams River, Manning Valley) and vine thickets (Goodnight Scrub).

The endemic genus *Safrina* (Plates 156, 157) is distributed from about Rockhampton in central coastal Queensland, to Victoria. *Safrina*, and *Ryssonotus* (now comprising only the single species, *R. nebulosus* [Plate 155]) from which it was derived (see Carter 1921, 1929a; Macleay 1885; Reid and Beatson 2016), is most closely related to *Australognathus* (previously *Sphaenognathus*) from northern Queensland and South America, and *Chiasognathus* from South America. *Safrina* includes winged and flightless species, largely with localised populations (Reid and Beatson 2016). All are relatively uncommon and most inhabit upland montane wet forests of different kinds. Several species are associated with rainforest, but may also inhabit adjacent eucalypt forest, or be restricted to the latter (e.g. *Safrina laticeps* from southern Queensland). *Safrina jaedoni* is the most northern-known species and is endemic to Kroombit Tops, an area of isolated rainforest and eucalypt forest rising to ~800 m in altitude, south-west of Gladstone. *Safrina jugularis* is not a true rainforest-inhabiting species, occurring in wet sclerophyll forest (with a rainforest understorey) and subalpine forest, in the Australian Capital Territory, southern New South Wales and Victoria. *Safrina moorei* is known only from New England National Park and Mount Hyland Nature Reserve, occurring in cool temperate and warm temperate rainforest at elevations >1100 m. *Safrina polita* is restricted to *Nothofagus*-dominated cool temperate rainforest in the Barrington Tops. *Safrina grandis* is sympatric with *S. polita* in the Barrington Tops but also occurs in adjacent subalpine open forest. *Safrina grandis* inhabits mixed subtropical–warm temperate rainforests, and is recorded from Tapin Tops (Dingo Tops) National Park, Chichester State Forest and the Comboyne Plateau north to Dorrigo. The species also occurs in an isolated elevated mixed rainforest–eucalypt coastal forest at North Brother Mountain overlooking the seaside town of Laurieton. The distinctively mottled *Ryssonotus nebulosus* (= *Rhyssonotus nebulosus*) is an active flier, and is widely distributed from northern Queensland to Victoria, being found in a varity of forest types, including commonly in lowland subtropical rainforest near Taree. The species does not express any host specificity, larvae and adults being found in the decaying wood of many tree species, though preferring decaying timber infected by white-rot fungi (Reid and Beatson, 2016; Wood *et al.* 1996). The large *Australognathus munchowae* (= *Sphaenognathus munchowae*) is one of two large-sized beetles in the genus (Moore and Monteith 2004). The species is restricted to the Blackdown Tablelands and Carnarvon national parks west of Rockhampton.

The enigmatic genus *Eucarteria* (comprising *Eucarteria florialis* and *E. subvittata*) is restricted to the Gondwana rainforests region (B. Moore 1994; Williams 2002). However, *Eucarteria* has recently been considered a junior synonym of the endemic genus *Cacostomus* (C. Reid 1999a), which occurs from northern Queensland to northern New South Wales. Three species now comprise this genus: *Cacostomus florialis*, *C. subvittatus* and *C. squamosus* (Plates 153, 154). All are active during the day; *Cacostomus squamosus* is also attracted to light. *Cacostomus* is most closely related to the South American genus *Casignetus*, and both genera are placed within the Gondwanan tribe Casignetini (C. Reid 1999a). Although the adults of *Cacostomus florialis* and *C. subvittatus* can be associated with flowers, they may also be encountered in large numbers on foliage (Hangay and De Keyser 2017; G. Williams unpubl. data). *Cacostomus subvittatus* is associated with cool temperate, warm temperate and subtropical rainforests, its altitudinal range extending from >1000 m in the Barrington Tops to <600 m at the Comboyne Plateau and Coorabakh National Park. Near the Plateau Beech Reserve (now within Werrikimbe National Park) in the early 1980s, many hundreds of adult *Cacostomus subvittatus* adults were observed resting on young coppice foliage following logging operations (G. Williams pers. observ.). Further north *Cacostomus subvittatus* is replaced by *C. florialis*; this species extends in upland subtropical-warm temperate rainforests from Dorrigo to the Gibraltar Range. The larger *Cacostomus squamosus* has a more extensive distribution extending from the Manning Valley (A. Sundholm pers. comm.) and Dorrigo in the south to northern Queensland.

The endemic flightless genus *Lissapterus* has no known close relatives on any other continent. The genus ranges from Victoria to the rainforests of tropical Queensland and occurs in cool temperate, warm temperate and upland dry rainforest. Five species are restricted to the Gondwana rainforests region: *L. tetrops* and *L. hopsoni* from Barrington Tops, *L. notestinei* from New England National Park, *L. pelorides* from the eastern Border Ranges and Nightcap National Park, and *L. obesus* from Acacia Plateau (New South Wales) and the Main Range (Queensland) (Hangay and De Keyser 2017; B. P. Moore pers. comm., Williams 2002). The most widespread species, *Lissapterus grammicus*, occurs in open forest, cool temperate and mixed warm temperate–subtropical rainforest at upland sites from the Blue Mountains east of Sydney, to southern New South Wales.

The genus *Lamprima* is confined to Australia, Norfolk Island, Lord Howe Island and New Guinea (Houston 1992), and is closely related to the Chilean *Streptocerus* (Howden 1981) and the New Zealand genus *Dendroblax* (B. Holloway 1963). The rare *Lamprima imberbis* is recorded from Dorrigo but nothing is known about its biology (Hangay and De Keyser 2017). The day-flying *Lamprima latreillei* and *L. aurata* are difficult to distinguish and have been considered synonymous by Mizunuma and Nagai (1994). In addition to occurring in open forest types adults inhabit mesic upland forest (Ebor) and lowland subtropical rainforest (Lansdowne Escarpment) and littoral rainforest (Harrington). The related *Homolamprima* is represented by a single rare species, *H. crenulata*. It is restricted to southern Queensland and northern New South Wales (Hangay and De Keyser 2017; W. Houston 1992).

Lycidae Lycidae are known from Eocene amber (Grimaldi and Engel 2005); however, their evolutionary history dates to the Mid Jurassic–Lower Cretaceous (176–25 mya) (Masek *et al.* 2018). The tribe Metriorrhynchini (Lawrence and Ślipiński 2013) is of Gondwanan lineage (Masek *et al.* 2018). Lycid larvae live in leaf litter and soil, and are saprophagous, feeding on decaying plant matter. Adults possess elongate heads that are adapted for feeding on nectar and pollen. Lycidae are chemically defended (Crowson 1981; Eisner *et al.* 2008; Moore and Brown 1981, 1989; Moore *et al.* 1990), the common Australian *Metriorrynchus rhipidius* (Lycinae), when threatened or handled, releasing whitish-coloured fluid that contains complex mixtures of acetylenic acids. Adult Lycidae, typically coloured red and black, are often proposed as distasteful aposematic models for mimic complexes that include beetles and other insects (e.g. Buprestidae, Cantharidae, Cerambycidae, Pyrochroidae, Meloidae, oecophorid moths, pompilid wasps), several which also possess compounds that are unpalatable to predators (Moore and Brown 1989).

Although the family is widespread, reaching its greatest diversity in the tropical regions of Asia and the Americas, lycid generic diversity in subtropical eastern Australia is low. *Metriorrhynchus* (Plate 152) is a genus commonly encountered on myrtaceous flowers (e.g. *Leptospermum*, *Kunzea*) in open forest and woodland, but also frequents subtropical rainforest formations and rainforest trees growing in the understorey of wet sclerophyll forest. In these, adults feed on the nectar (and possibly pollen) of numerous flowering trees (i.e. *Syzygium smithii*, *S. floribundum*, *Tristaniopsis laurina*, *Alphitonia excelsa*, *Cuttsia viburnea*, *Diospyros australis*, *Euroshinus falcatus*, *Schizomeria ovata*, *Scolopia braunii*) (Williams 1995). *Metriorrhynchus lateralis* is a pollinator of the vine *Marsdenia fraseri* (Asclepiadaceae) (P. Forster 1989). However, *Metriorrhynchus* undertakes only infrequent interplant movements, adults staying for relatively long feeding bouts at individual inflorescences.

Additional members of the subfamily Lycinae, *Porrostoma* (14 species), *Synchonnus* (one species) and *Xylobanus* (four species), are recorded from the Gondwana rainforests region (Calder 1998; Lea 1921; Williams 2002). *Porrostoma* is widely distributed between India and Australasia; *Xylobanus* has an Afrotropical, Oriental and Australasian distribution; and *Synchonnus* is known from Australia and New Guinea.

Lymexylidae The Lymexylidae are very elongate and narrow beetles that are possibly Lower Jurassic (~180–>160 mya) in origin, suggesting that the development of some higher taxa may have occurred before the breakup of Gondwana (Crowson 1981). However, Grimaldi and Engel (2005) cite the earliest Lymexylidae (*Atractocerus*) from younger 100 million-year-old Burmese amber 'indicating that by the Mid Cretaceous at least some major lineages of lymexylids had appeared'.

Larvae of Lymexylidae bore into dead trees and feed on fungi that grow on the walls of their tunnels. The family contains two genera (*Australymexylon*, *Melittomma*) distributed along the east coast of Australia (Lawrence and Ślipiński 2013). *Australymexylon* is an endemic genus, comprising two species, ranging from North Queensland to Victoria (Q. Wheeler 1986). Adults are attracted to ultraviolet light.

Megalopodidae Megalopodids occur in the Old World tropics, Holarctic, pantropical and Australasian regions, and are phytophagous species placed within the Chrysomeloidea (Lawrence and Ślipiński 2013; C. Reid 1995). *Palophagus* (two species) is restricted to the Gondwana rainforests region, and is associated with *Araucaria* (Araucariaceae) (Kuschel and May 1989, 1996a, 1996b). *Palophagus bunyae* from southern Queensland has been reared from *Araucaria bidwillii*. *Araucaria cunninghamia* is a probable host for the northern New South Wales species *Palophagus australiensis* (Kuschel and May 1989). Related Palophaginae have been reared from *Araucaria araucana* in Chile and Argentina (Kuschel and May 1996b). Members of the genus *Zeugophora* (Zeugophorinae) are specialised leaf miners (C. Reid pers. comm.) with *Zeugophora vitinea* and *Z. williamsi* occurring in a range of subtropical rainforest formations; the altitudinal distribution of the genus extends from lowland littoral rainforest to montane wet forests such as the Bunya Mountains, Border Ranges and Tooloom Plateau of south-eastern Queensland and north-eastern New South Wales (C. Reid 1989a, 1995; Williams 2002). Although C. Reid (2016) has grouped Australia (including Tasmania) and New Guinea in a biogeographical region he named 'Australopapua', the Australian *Zeugophora* species apparently bear little relationship to those recorded from New Guinea (C. Reid 1989a).

Meloidae These are known as blister beetles, renowned for producing the toxin cantharidin, and notoriously the aphrodisiac Spanish Fly, which is lethal in quantity. During mating cantharidin is transferred from males to females, which then use it in the defence of their eggs (Eisner *et al.* 1996). Adult meloidids are elongate in shape (Plate 158), are relatively soft-bodied, and those occurring in the subtropical south-east are primarily reddish and black in colour. This is suggestive of their participation in Batesian or Müllerian mimic complexes (e.g. with Lycidae, Belidae, Buprestidae, Cerambycidae). The adults have distinctive narrow necks and elongate mouthparts, feed on nectar and lay their eggs on flowers (Lawrence and Britton 1991). Meloidae are known to be parasites of bees (e.g. Apidae-Xylocopinae), the flightless larvae (triungulins) relying on carriage by their host so as to gain entry to the host's nest (Crowson 1981; Lawrence and Ślipiński 2013). The later instars eat the developing bee larvae and their provisioned honey.

Species are only occasionally found on flowering rainforest trees or shrubs, though some (e.g. *Palaestra rufipennis*) will exhibit 'hill-topping' behaviour, seasonally occurring on flowering *Leptospermum* atop mountain peaks (e.g. Mt Killiekrankie, Rowley's Peak) and volcanic plugs (often surrounded at their base by rainforest and wet sclerophyll forest). When disturbed adults may feign death, retract their legs, and tumble from flowers upon which they were feeding. *Zonitis dichroa* is rarely attracted to light.

Melyridae Melyridae have a world-wide distribution. They are generally small beetles that in subtropical rainforest are found as adults on foliage, but more commonly on flowers of canopy trees (Williams 1995) and occasionally flowers of understorey shrubs. They are generally found in small numbers, often just single individuals. Adults feed on nectar and possibly pollen, but the larvae are usually predaceous. Adult food plants in subtropical rainforests include *Syzygium smithii* (*Helcogaster*, *Balanophorus*, *Dicranolauis*, *Neocarphurus*), *S. floribundum* (*Balanophorus*, *Helcogaster*), *Tristaniopsis laurina* (*Carphurus*, *Dastyes*, *Helcogaster*, *Neocarphurus*), *Acradenia euodiiformis* (*Dastyes*, *Carphurus*, *Dicranolauis*), *Alphitonia excelsa* (*Carphurus*, *Helcogaster*), *Cuttsia viburnea* (*Dastyes*, *Helcogaster*), *Euroschinus falcatus* (*Dicranolauis*, *Helcogaster*, *Carphurus*, *Neocarphurus*), *Guioa semiglauca* (*Carphurus*, *Neocarphurus*), and *Scolopia braunii* (*Helcogaster*). These are all trees with open, readily accessible flowers in a broad spectrum of families.

Mordellidae This family has a world-wide distribution and comprises ~1500 species. Mordellids have distinctive drop-shaped, laterally compressed and hump-backed bodies, and are further characterised by a long apical tergite that extends beyond the elytra. The family is known from Mid Cretaceous amber, and fossils plausibly assigned to Mordellidae are known from Late Jurassic rocks (Grimaldi and Engel 2005). The larvae are borers of dead wood or stem borers. Adults are commonly found on flowers and when disturbed they quickly fall away. Though commonly found on flowers the adults exhibit no prolongation of the mouthparts that typify numerous other anthophilous beetles.

Adults are often clothed with short hairs and are members of the minor pollinating suite of beetles associated with entomophilous rainforest flowers. Host records in lowland rainforest remnants include *Mordella* (on *Acradenia euodiiformis*, *Alphitonia excelsa*, *Cuttsia viburnea*, *Drypetes deplanchei*, *Elaeocarpus obovatus*, *Euroschinus falcatus*, *Guioa semiglauca*, *Rhodomyrtus psidioides*, *Syzygium floribundum*, *S. smithii*, *Tristaniopsis laurina*); *Mordellistena* (on *Alphitonia excelsa*, *Cuttsia viburnea*, *Diospyros australis*, *Euroschinus falcatus*, *Rhodomyrtus psidioides*, *Scolopia braunii*, *Syzygium floribundum*, *S. smithii*, *Tristaniopsis laurina*); *Tomoxia* (on *Alphitonia excelsa*, *Syzygium smithii*); and *Hoshihanoanomia* (on *Alphitonia excelsa*). Species of *Mordella* are recorded from *Ackama paniculata*, *Elaeocarpus reticulatus*, *Schizomeria ovata* and *Cryptocarya microneura* in wet sclerophyll forest at intermediate elevations (Williams 1995).

Nemonychidae Nemonychid weevils occur in Eurasia, the Americas, Australia, New Zealand, New Caledonia and New Guinea. They are known from Late Jurassic fossils (R. Anderson 1995), and are considered to be the most primitive of the superfamily Curculionoidea. Zimmerman (1994a) aptly terms them relics of bygone ages. Larvae and adults are pollen feeders of the gymnosperms *Araucaria* and *Agathis* (Araucariaceae) (Lawrence and Ślipiński 2013).

Australia is one of three major centres of nemonychid diversity, the other two being North America and South America. Their absence from Africa is considered to be of major biogeographical significance (Zimmerman 1994a). Two tribes, Mecomacerini and Rhinorhynchini (of Rhinorhynchinae), occur in subtropical south-eastern Australia (Kuschel and Leschen 2010). Rhinorhynchini are recorded from New Zealand, South America and North

America, and Mecomacerini also occur in South America, New Caledonia and New Guinea. The mecomacerine genus *Bunyaeus* (e.g. *Bunyaeus eutactae*, *B. monteithi*) and the rhinorhinchine genus *Basiliorhinus* (*Basiliorhinus araucariae*) are endemic to south-east Queensland. Many Australian species are associated with the plant genera *Agathis* and *Araucaria* (Araucariaceae). *Notomacer* (*Notomacer reginae*, the Bunya Mountains) occurs in Australia and New Caledonia, while *Basiliogeus* is known from Australia and New Guinea. *Araucaria bidwillii* is a host plant of *Basiliogeus prasinus* (Zimmerman 1994a).

Nitidulidae The Nitidulidae are known from Late Jurassic fossils and Cretaceous amber (Crowson 1981; Grimaldi and Engel 2005). They are of diverse form, and have a cosmopolitan distribution and are well represented in subtropical eastern Australia (Lea 1921; Kirejtshuk and Lawrence 1990, 1992a; Williams 1995, 2002). Species are found in leaf litter, at flowers, in the cones of gymnosperms, at fungal fruiting bodies and in fermenting fruits. Some are predaceous. *Gymnocychramus bicolor* has been recorded at the mature odorous spadices of exotic *Amorphophallus bulbifer* grown ornamentally adjacent to lowland subtropical rainforest (Lansdowne Escarpment) (G. Williams pers. obs.). *Carpophilus*, a genus often attracted to rotting fruit, is associated with the mature seed capsules of *Cupaniopsis anacardiodes* (Sapindaceae) in littoral rainforest.

The subgenus *Idaethina* (e.g. *I. pilistriata*) of *Aethina* is endemic to eastern Australia, and ranges from south-east South Australia to the Atherton Tableland, northern Queensland (Kirejtshuk and Lawrence 1990). The endemic tribe Cychramptodini contains three genera (*Cylindroramus*, *Miskoramus* and *Cychramptodes*). *Cylindroramus* is restricted to the Australian east coast, in closed and open forest types. *Cylindroramus accretus* is known only from New England National Park in *Nothofagus* forest (Kirejtshuk and Lawrence 1992a). The endemic genus *Thalycrodes* (e.g. *T. australe*, *T. pulchrum*) occurs in most parts of the continent (Kirejtshuk and Lawrence 1992b). *Rixerodes* (*Rixerodes cunninghami* – Nitidulinae) is an endemic genus that is widespread in eastern Australia but known from few localities.

Nitidulids frequent numerous shrubs and canopy trees in subtropical rainforests at lower elevations. They are co-pollinators with thrips of the understorey shrub *Pittosporum revolutum* (Pittosporaceae), with drosophilid flies of *Hibiscus splendens* and *H. heterophyllus* (Malvaceae), and with derelomine weevils of the understorey palm *Linospadix monostachya* (Arecaceae) and the large fleshy herb *Alocasia brisbanensis* (Araceae) (Williams and Adam 2010) (Plates 159, 160). Nitidulids recorded from flowering rainforest trees (e.g. *Syzygium smithii*, *Acradenia euodiiformis*, *Cuttsia viburnea*, *Euroschinus falcatus*) are species mainly of *Epuraea* and *Aethina* (and the nitidulid related kateretid genus *Notobrachypterus*) (Williams 1995). However, this pollinator role is at a reproductive cost to plants as nitidulids and their Kateretidae kin eat a proportion of available pollen. Nevertheless, some foreign rainforest-associated nitidulids are important pollinators of tropical fruit crops, for example custard apple (*Annona* hybrids) (Blanche and Cunningham 2005).

Oedemeridae The Australian oedemerid, or false blister beetle, fauna is placed in the subfamilies Polypriinae and Oedemerinae (including Nacerdinae) (Lawrence and Ślipiński 2013). Oedemerinae are the most diverse. Larval Oedemeridae feed on rotting wood. Adult oedemerids are long and parallel in shape and are often found on flowers, and only rarely come to light. Species that visit flowers usually have the head elongate, this, like in other fully adapted anthophilous Coleoptera, facilitating feeding on nectar or pollen sometimes held deep within flowers. Some oedemerids synthesise the potent toxic irritant cantharidin, and are considered participants in defensive aposematic mimicry complexes (e.g. *Pseudolycus*) that include similarly coloured insects (e.g. beetle families Meloidae, Lycidae and Buprestidae). Cantharidin is a toxic monoterpene anhydride that occurs in haemolymph and is secreted when beetles are harmed (Grimaldi and Engel 2005).

The fauna known from upland sites (e.g. Bunya Mountains, Mt Warning, Barrington Tops) in southern Queensland and northern New South Wales includes *Agasma semicrudum*, *Asclera atkinsoni*, *Copidita hilaris*, *C. languida*, *C. sloanei* and *Sessenia viticollis* (Oedemerinae) (Williams 2002). *Agasma* is endemic and has its closest relatives in the Holarctic genus *Ditylus* (Lawrence and Britton 1994). Several species are recorded from flowering trees in lowland rainforest and wet sclerophyll forest: *Pseudolycus haemorrhoidalis* (on *Syzygium smithii*), *Pseudolycus* sp. (on *Rhodomyrtus psidioides*, *Syzygium floribundum*, *Tristaniopsis laurina*); *Ischnomera* sp. (on *Acradenia euodiiformis*, *Elaeocarpus reticulatus*, *Euroschinus falcatus*, *Rhodomyrtus psidioides*, *Schizomeria ovata*, *Syzygium floribundum*); *Copidita* ?*pachymera* (on

Syzygium floribundum); and *Sessenia nigronotata* (on *Syzygium floribundum*). Of the plants studied by Williams (1995) Oedemeridae were most diverse on *Syzygium floribundum* (Myrtaceae), but individuals were not abundant on any.

Passalidae The terrestrial scarabaeoid family Passalidae, or bess beetles, mainly occurs in tropical and subtropical regions. Passalids are a conspicuous element of the macroinvertebrate fauna encountered in and under rotting wood on the forest floor of subtropical rainforests at intermediate and lowland elevations. Adults are relatively large, flattened, elongate and parallel-sided, and uniformly black in colour (though often brown before hardening of the body). They are found only rarely in standing dead tree trunks, even when these are quite decayed. Adults undertake subsocial brood care (Schuster and Schuster 1985), often being found in assocation with multiple larvae, which feed on wood pulp (Matthews 1984; B. Moore 1980–96). Adults communicate by stridulation, achieved by friction of the hind wings against the upper surface of the abdomen (Lawrence and Ślipiński 2013). Stridulatory organs are also present on larvae. Passalids are hosts to endoparasitic nematodes (W. Clark 1978), and externally to phoretic mites that cluster about the forebody (Plate 5). Species of endosymbiont fungi are also known to parasitise the hind gut of certain Passalidae (Crowson 1981).

Genera inhabiting subtropical rainforests and wet sclerophyll forests are *Aulacocyclus* (e.g. *Aulacocyclus edentulus*, *A. tambourensis*) (Plate 161), which is also distributed in the Oriental Region, New Guinea and New Caledonia, *Mastochilus* (e.g. *Mastochilus quaestionis*), also occurring in New Guinea, *Analaches* (*Analaches australiensis*) and *Pharochilus* (*Pharochilus dilatatus*) (Carter 1933; Dibb 1938; W. Houston 1992; B. Moore 1980–96).

Phalacridae These are minute to small ovate beetles (<5 mm) that in Australia comprise ~100 species (Lawrence and Ślipiński 2013). The family has a worldwide distribution (Gimmel 2013). They feed on fungal spores. However, adults are also commonly associated with flowering trees in subtropical rainforest, the suite of tree species being largely that cited here for other minor pollinating Coleoptera (e.g. Mordellidae). Anthophilous phalacrids recorded from lowland rainforests are species of *Litochrus*, *Olibroporus*, *Phalacrinus* and *Phalacrus* (Williams 1995); *Litochrus* is the most speciose and frequently encountered. *Litochrus* (Phalacrinae) and *Philacrinus* (Phaenocephalinae) are common and widespread genera.

Phloeostichidae Phloeostichidae are members of the litter fauna in forests, and are a relict family with limited world distribution (Crowson 1981). *Rhopalobrachium* is a Gondwanan genus occurring in Argentina, Chile and eastern Australia. The only known Australian species is *Rhopalobrachium crowsoni*, recorded from Lamington National Park in Queensland, and Richmond River in northern New South Wales (Lawrence 1995; E. G. Matthews pers. comm.; Pollock and Lawrence 1995). Larval and adult Phloeostichidae are thought to feed on rotting cambium and other plant material (Lawrence and Ślipiński 2013).

Psephenidae This aquatic family includes the single Australian genus *Sclerocyphon* (Eubrinae), which is most closely related to the Chilean genus *Tychepsephus* (Lawrence and Britton 1994; Lawrence and Ślipiński 2013). The flattened larvae are commonly called water pennies, and adhere to rocks in streams, where they feed on algae. *Sclerocyphon* exhibits distinctive polymorphic larval types characterised by discrete geographical ranges (Davis 1986). Larval type 'B', for example, is restricted to Queensland, along the coastal strip from Lamington National Park, north to Townsville. Larval type 'D' is recorded only from Lamington National Park, and larval type 'E' is known only from the New England Tableland, at Barrington Tops and near Tamworth (Davis 1986). *Sclerocyphon maculatus* reaches its northern most known distribution at Dorrigo (Davis 1986).

Pyrochroidae In Australia this small but widespread family includes the single subfamily Pilipalpinae. Pyrochroidae are able to synthesise cantharidin, and bicoloured adults, in combinations of red, orange and black, are considered participants in mimicy complexes involving unpalatable Lycidae and Cantharidae beetles. Larvae are found in rotting wood and under loose bark. The current distribution of the Pyrochroidae includes amphitropical and disjunct southern faunas (Pollock 1995). Ancestral Pilipalpinae were widespread on Gondwana, and the persistence of relict genera in Madagascar, Australia, New Zealand and South America suggests the subfamily is Gondwanan in origin (Pollock 1995).

Several endemic genera are encountered in the region of northern New South Wales and Southern Queensland: *Morpholycus* (restricted to eastern Australia), *Binnburrum* (east coast of Australia, including Tasmania, east of the Great Dividing Range), *Paromarteon* (Queensland-Victoria, but absent from Tasmania), and the relatively primitive *Temnopalpus* (east coast from Queensland to Tasmania, and south-west Western Australia).

Adults are occasionally encountered feeding on the nectar of mass-flowering subtropical rainforest trees: *Syzygium smithii* (*Paromarteon constans*), *Alphitonia excelsa* (*Parometeon* sp.), *Cuttsia viburnea* (*Parometeon mutible*), *Elaeocarpus obovatus* (*Parometeon constans*, *Parometeon mutabile*), *Elaeocarpus reticulatus* (*Techmessa* sp.) (Williams 1995). Pyrochroids known from higher elevation rainforests (e.g. Bunya Mountains, Border Ranges, Richmond Ranges, Dorrigo) include *Binburrum angusticollis*, *B. bifoveicollis*, *B. ephippiatum*, *Morpholocus apicalis*, *M. costipennis* and *Temnopalpus bicolor* (Williams 2002).

Pythidae Pythids are a significant relict taxon in northern New South Wales and southern Queensland and are probably an ancient Pangaean group because the confirmed members have an amphitropical distribution (E. G. Matthews pers. comm.). The only southern genus in the family is the endemic *Anoplopus*, which has two species: one in the Atherton refugium of tropical North Queensland and one (*Anoplopus tuberculatus*) in the Macpherson refugium (in rainforest) from the Bunya Mountains, Mt Glorious, Mt Tamborine, Lamington and Border Ranges national parks, and Richmond River (Abdullah 1966; E. G. Matthews pers. comm.; Pollock and Lawrence 1995). *Anoplopus tuberculatus* has been collected from *Argyrodendron actinophyllum* in rainforest at Mt Glorious. Larvae are usually found under the loose bark of rotting logs (Pollock and Lawrence 1995).

Rhinorhipidae This problematic and endemic elateroid family is represented by a single species, *Rhinorhipus tamborinensis*, which is known from three high elevation localities (Mt Tamborine, Lamington National Park, Mt Glorious) in rainforest of south-eastern Queensland (Calder 1998; Lawrence 1988; Lawrence and Britton 1994). The larvae are unknown but adults have been recorded in large numbers on introduced vegetation on rainforest margins (Lawrence and Ślipiński 2013).

Rhynchitidae The phytophagous Rhynchitidae (*sensu* Zimmerman 1994a) are small long-snouted, weevil-like beetles. They are treated by Lawrence and Ślipiński (2013) as a subfamily of the Attelabidae. The rhynchitid subfamily Auletinae is cosmopolitan but the fauna is most diverse in Australia. Auletinae, however, are absent from New Zealand (Zimmerman 1994a). Rhynchitidae include numerous species of the genus *Auletobius*, which lay their eggs in flowers and leaf buds, the larvae of some being leaf miners (Lawrence and Ślipiński 2013). *Auletobius* is occasionally encountered on flowering trees in lowland rainforest (e.g. at *Alphitonia excelsa*, *Elaeocarpus obovatus*, *Syzygium floribundum*) and adjacent wet sclerophyll forest (e.g. on *Schizomeria ovata*) (Williams 1995; G. Williams unpubl. records). Several species are recorded from Mt Tamborine (*Auletobius monticola*, *A. pollux*, *A. sulcibasis* [Lea 1926; Zimmerman 1994a]). Zimmerman (1994a) records *Ficus* as a host plant of *Auletobius minor* in northern New South Wales.

Ripiphoridae The oldest Ripiphoridae (= Rhipiphoridae) are recorded from Mid Jurassic and Cretaceous ambers (Grimaldi and Engel 2005; Hsiao *et al.* 2017). Modern genera are recorded from Tertiary amber (Grimaldi and Engel 2005). The extant fauna has a worldwide distribution and is normally a minor element of the beetle assemblage found in the rainforest canopy (e.g. Stork and Grimbacher 2006). However, ripiphorids are infrequently encountered in subtropical rainforest, the species usually collected or observed being those of the subfamilies Ptilophorinae (e.g. *Trigonodera*), Ripiphorinae (*Macrosiagon*) and Ripidiinae (*Rhipidioides*) (Plate 162). Species of *Trigonodera* are often attracted to light. Owing to their tapering shape ripiphorids resemble Mordellidae, but they do not have the distinctive abdominal spine of the latter. In addition, some have a constriction behind the eyes that gives the semblance of a neck, and several have the elytra foreshortened. The males of Ripiphorinae have conspicuous biflabellate antennae (both sides of the antennal joints having fan-like processes), those of females being more simple, but the females of some Ripiphorinae are larviform (resembling a larva), lack wings and have reduced mouthparts. Ripiphorinae lay their eggs in flowers (Lawrence and Ślipiński 2013) and their active triungulin larvae are endo- and ectoparasites on wasps (e.g. Scoliidae, Thynnidae, Vespidae). Triungulins (mirroring the habits of larval Meloidae) attach themselves to a mature wasp, upon which they are carried back to the host's nest. There they attack the host's larvae. Ripidiine triungulins are parasites of cockroaches. Ptilophorinae are predators or parasites of wood-boring beetles.

Adult *Macrosiagon* have been collected feeding on nectar of flowering *Euroschinus falcatus* and *Alphitonia excelsa* trees in littoral rainforest remnants (Williams 1995). *Macrosiagon* and *Rhipidioides* have been observed feeding on the nectar of *Baeckea* shrubs artificially planted on the margin of lowland wet sclerophyll forest (G. Williams unpubl. data).

Scarabaeidae These are mainly robust beetles of small (*Microvalgus* – Valginae/Cetoniinae, numerous Melolonthinae and Scarabaeinae) to large size (Dynastinae, Rutelinae, Cetoniinae). Individual groups are commonly referred to as Christmas beetles (*Anoplognathus* – Rutelinae), rhinoceros beetles (Dynastinae), pasture chafers (Melolonthinae), dung beetles (Scarabaeinae, Aphodiinae), and rosechafers (Cetoniinae). They are often characterised as the symbolic scarabs (*Scarabaeus*) of ancient Egyptian mythology, jewellery, and architectural embellishment, symbolising the eternal cycle of day and night, birth, death and reincarnation.

Scarabaeidae – Aphodiinae Most subtropical rainforest species of Aphodiinae are coprophagous or sometimes saprophagous in habit, and are small, generally dull-coloured beetles commonly termed dung beetles, a name shared with the subfamily Scarabaeinae. Their mouthparts are reduced and modified to feed on a largely fluid diet. Aphodiines are distributed in numerous subtropical rainforest formations at all elevations, but are often absent or poorly represented in littoral rainforest (Williams 2002; Williams and Williams 1982, 1983a, 1983b, 1983c, 1984). Aphodiinae are restricted to the forest floor, and adults are especially found in macropod dung, but also at bird and reptile droppings. Some species come to ultraviolet light (e.g. *Ataenius imparilis*, *A. tweedensis*).

The subfamily includes several tribal groups whose ranges encompass subtropical Queensland and New South Wales. The Odontolochini (i.e. *Odontolochus*) are primarily a pan-tropical tribe. The Eupariini (i.e. *Airapus*, *Saprosites*) occur world-wide, but the greatest number of genera and species is known from the Western Hemisphere and Australia (Stebnicka and Howden 1996). Many eupariines occur in association with, or a confined to, the nests of ants and termites, and their range is coexistent with that of the host (Stebnicka and Howden 1996). The Proctophanini (e.g. *Proctophanes*) are mainly Australian–African in distribution (Stebnicka and Howden 1995). The Aegialini are represented in Australia by *Saprus*, which is restricted to the Border Ranges south to Tasmania. The genus *Podotenus* (Aphodiini) is concentrated in eastern Australia with northern New South Wales and south-eastern Queensland, and northern Queensland, being centres of speciation (Stebnicka and Howden 1994). Species of *Podotenus* are usually attracted to animal dung, but are sometimes encountered under the carcasses of small animals (e.g. *Antechinus*, flying foxes). Individual species are frequently found in abundance feeding within dung masses.

Aphodiine species recorded from subtropical rainforests include *Podotenus allynensis*, *P. channonensis* and *P. toowoombaensis*. *Podotenus wauchopensis* and *P. williamsi* inhabit *Nothofagus*-dominated cool temperate rainforest (Stebnicka and Howden 1994; Williams 2002). Further south *Podotenus pseudovictoriae* inhabits warm temperate rainforest at Mt Wilson west of Sydney. In New South Wales most *Podotenus* are restricted to rainforest at intermediate and high altitudes; however, *Podotenus ?toowoombaensis* (Plate 174) occurs in remnant floodplain subtropical rainforest at Wingham Brush Nature Reserve, New South Wales (G. Williams unpubl. data). Yet *Podotenus* is not a rainforest specialist and can be found in high altitude open eucalypt forest (*Podotenus coffensis*, *P. barringtonensis*). Additional rainforest-inhabiting aphodiines include *Saprus lawrencei* (e.g. Border Ranges, Werrikimbe National Park), *Airapus bruxnerensis* (Bruxner Park Flora Reserve), *A. obscurus* (Lamington and Dorrigo national parks), *A. picinus* (Allyn River), *Saprosites bunyaensis* (Bunya Mountains), *Proctophanes metasternalis* (Mt Glorious), and *Odontolochus weiri* (Border Ranges, Nightcap National Park) (Blackburn 1904; W. Houston 1992; Stebnicka and Howden, 1995, 1996; Williams 2002). *Ataenius strigifrons* inhabits wet sclerophyll forest at the base of Mt Warning.

Scarabaeidae – Cetoniinae Commonly called rosechafers, the often highly coloured and patterned Cetoniinae are distinct among flying beetles in that the hindwings in flight extend sidewards without the hardened forewings (the elytra) needing to be raised. Cetoniinae are somewhat flattened in shape and are swift flyers, the adults feeding on nectar and the larvae occurring in decaying wood and detritus (Bacchus 1974; W. Houston 1992; Hasenpusch and Moeseneder 2010; Lawrence and Ślipiński 2013; Lea 1914; Williams 1995; Williams and Adam 1994, 2010). Large *Chondropyga dorsalis* are commonly encountered swiftly flying above rainforest streamlines and margins. Numerous species also extend over a wide range of altitudes. The large, pale green-yellow hued *Schizorhina atropunctata* (Plate 175), for example, is encountered at sea level in littoral rainforest (e.g. Racecourse Headland near Crescent Head), and inland to upland forests (e.g. Comboyne Plateau). The species presents a bizarre spider-like form as it hovers with outstretched legs before landing on blossoms. Many species are habitat generalists (e.g. *Chlorobapta bestii*, *C. frontialis* [Plate 169], *Chondropyga dorsalis*, *Eupoecila australasiae*, *Neorrhina punctata*) and may be encountered in a variety of plant communities other than rainforest; the attraction is an availability of nectar sources for adults or microhabitats (such as detritus at the base of tree trunks) suitable as ovipositing sites and the development of larvae.

The Australian cetoniine fauna are important pollinators of many mass-flowering subtropical rainforest trees, and undertake relatively long-distance flights between dispersed flowering plants, thus being important for 'outcrossing' (Williams and Adam 1994, 2010). Species commonly frequenting lowland subtropical rainforest remnants (e.g. *Bisallardiana gymnopleura*, *Chondropyga dorsalis*, *Eupoecila australasiae*, *Glycyphana brunnipes*, *Neorrhina punctata*) have been found to carry significant pollen loads, the pollen grains being especially snared on the head, and lateral and ventral surfaces of the prothorax (Williams and Adam 1998). Flowering trees from which these species have been recorded include *Syzygium floribundum*, *Tristaniopsis laurina* (Myrtaceae), *Euroschinus falcatus* (Anacardiaceae), and *Alphitonia excelsa* (Rhamnaceae). Although Cetoniinae generally lack the prominent integumental punctuation of the flower-frequenting Buprestidae they do possess dense setal vestiture (as in *Lenosoma*) that readily captures and holds pollen grains (Williams and Adam 2010).

Most Australian genera (e.g. *Bisallardiana*, *Chlorobapta*, *Chondropyga*, *Dilochrosis*, *Lenosoma*, *Lyraphora*, *Storeyus*) are placed in the tribe Schizorhinini. The Schizorhinini are endemic to the Australasian region. *Glycyphana* (Cetoniini) is a predominantly Oriental genus (in W. Houston 1992). *Storeyus fasciculatus* is distributed from Rockhampton to Ourimbah in central coastal New South Wales. It mainly occurs in rainforest (e.g. at Bunya Mountains, Mt Glorious, Acacia Plateau, Border Ranges, Iluka, Dorrigo, Taree) and has been recorded feeding at flowering rainforest trees that include *Backhousia myrtifolia*, *Syzygium floribundum* (Myrtaceae) and *Pennantia cunninghamii* (Icacinaceae/Pennantiaceae) (Hasenpusch and Moeseneder 2010; G. Williams unpubl. data). The closely related *Lenosoma fulgens* ranges from Eungella and Rockhampton to Many Peaks, Mt Nebo and Mt Glorious in southern Queensland. Adults have been recorded feeding on the flowers of *Melicope micrococca* (Rutaceae), a small tree of subtropical, dry and warm temperate rainforest, and vine thickets, and *Barkyla syringifolia* (Fabaceae/Caesalpiniaceae), a small tree in dry rainforest and vine scrub north from Beenleigh (Queensland) (Harden *et al.* 2006; Hasenpusch and Moeseneder 2010).

Minute beetles of the genus *Microvalgus* (previously placed in the Valginae) are occasional associates of flowering canopy trees (Williams 1995). More characteristically they are found on myrtaceous shrubs (e.g. *Leptospermum*, *Baeckea*) in open forest and woodland. Larvae may be associated with termites (Lawrence and Ślipiński 2013). *Microvalgus* also occurs in the Oriental and Ethiopian regions (in W. Houston 1992).

Scarabaeidae – Dynastinae Commonly called rhinoceros beetles, the Dynastinae include some of the largest insects found in the forests of southern Queensland and northern New South Wales (i.e. *Xylotrupes ulysses*, *Haploscapanes australasicus*). Species are often sexually dimorphic (Plates 178, 179), the males being armed with horns and tubercles which function in competion for mates and breeding sites (Crowson 1981; Eberhard 1980; Emlen 2000; Grimaldi and Engel 2005). Larvae are found under and in rotten logs, and feed on decaying vegetable matter. Adults are mainly nocturnal and many species come to ultraviolet light, though several species fly during the day (e.g. *Corynophyllus haroldi*, *C. macdougalli*). Dynastinae are not diverse in subtropical rainforests (Williams 2002). The more frequently encountered species include *Anomalomorpha anthracina*, *Metanastes picipes*, *Cheiroplatys latipes*, *Cheiroplatys bifossus*, *Dasygnathus trituberculatus*, *D. dejeani*, *Neodasygnathus davidsoni*, *Semanopterus subcostatus*, *Cryptodus passaloides* and *C. tasmannianus*. Some of these also occur in open forest (e.g. *Dasygnathus trituberculatus*) and occupy rainforest through altitudinal gradients from near sea level to upland sites (e.g. *Metanastes picipes*, *Cheiroplatys latipes* [Plate 168]). The fauna inhabiting *Nothofagus* forest (e.g. *Semanopterus subcostatus*, Plate 176) is depauperate. However, two introduced soil pest species are frequently encountered: *Heteronychus arator* introduced from South Africa, and *Cyclocephala signaticollis* from the Americas. Both can be commonly encountered attracted to light at lower altititudes, though *Cyclocephala signaticollis* is also found at high altitudes (e.g. Barrington Tops, Werrikimbe National Park), associated there with mesic forest types.

Approximately 11 per cent of the world's dynastine species are thought to occur in Australia, but this is probably an underestimate owing to inadequate collecting (Carne 1957a). Of the (in 1957) 169 species, 158 or 93 per cent were known only from the Australian mainland. Carne (1957a) records 26 endemic genera (of the 40 genera currently listed as occurring in Australia; Lawrence and Ślipiński 2013), nine of which occur in the Gondwana rainforests region (Table 4). These can be broken up into three faunal units (after Carne 1957a).

According to Carne (1957a) the Eyrean can be equated with the pre-Jurassic (>180 mya) fauna of East Gondwana, and the post-Jurassic Bassian developed in Gondwana after the Jurassic fragmentation but while Patagonia was still connected. The Torresian is considered to have radiated through the north during the Pliocene, rarely extending south of Queensland and not reaching Tasmania. The post-Jurassic element is the dominant dynastine fauna in Australia and this is reflected in those genera known from the Gondwana rainforests region (see Table 4). The Bassian genus *Cheiroplatys* shares species with Mexico and the United States of America, and such a distribution suggests the persistence of the genus since the Miocene (Carne 1957a). The Bassian genus *Anomalomorpha* includes the regionally occurring *A. anthracina*, a species that is widely distributed from central coastal Queensland, south to Victoria, and recorded from Lamington National Park (Queensland) and Ebor (northern New South Wales) (Allsopp 2000). The endemic inquiline Eyrean genus *Cryptodus*, the largest Australian dynastine genus, is extremely specialised (morphologically adapted for an inquiline habit) with its greatest affinity being with the Madagascan and South African genus *Rhizoplatys*. In all *Cryptodus* the mouthparts are distinctively enclosed by a broad mentum, this being a protective adaptation for inhabiting termite or ant nests though the exact nature of the association is poorly understood. There are more than 20 species (most occurring in the north and west of Australia), but also numerous synonyms, and some species can only be recognised with great difficulty. Carne (1957a) considered that their 'restricted habitat prevented the expression of strikingly divergent specific characters, resulting in an assemblage of species which are unusually constant in general structure and all proportions'.

Scarabaeidae – Melolonthinae Melolonthines are phytophagous, and numerous genera and species are recorded from rainforests and wet sclerophyll forests of subtropical Queensland and New South Wales (Britton 1957, 1970, 1978, 1980, 1986, 1987, 1990, 1995, 2000; Britton and Stanbury 1981; W. Houston 1992; Macleay 1886; Williams 1993, 1995, 2002). Many genera are endemic: *Ancylonyx*, *Anthotocus*, *Antitrogus*, *Automolius*, *Byrrhomorpha*, *Cheiragra*, *Cheirodontus*, *Colobostoma*, *Colpochila*, *Diphucephala*, *Epholcis*, *Homolotropus*, *Liparetrus*, *Metatrogus*, *Nitorellus*, *Nosphisthis*, *Ophropyx*, *Opsitomus*, *Phyllotocidium*, *Phyllotocus*, *Rhopaea*, *Scitala*, *Telura*, *Watkinsia*, *Xylonichus* and *Xyrine*.

The Australian Melolonthinae are remarkably isolated taxonomically and, of the genera recognised (>75, in Britton 1957), more than 90 per cent are endemic. The evolution of nearly all tribes, and most genera, must have occurred after the separation of the Australian landmass from other continents (Britton 1957). *Maechidius* appears to have originated in Australia and has numerous species in Papua New Guinea and a single species described from the Moluccas (Britton 1957). Species of *Maechidius* are active at night and are termitophilous, possessing pronotal hypomera that are produced forward so as to form a pocket into which the antennae can be protectively withdrawn against attack. This feature is absent in the putatively termite-associated and related genus *Epholcis*. *Maechidius* is encountered in rainforests at all elevations (*e.g. Maechidius hopeanus*, *M. longitarsus*, *M. macleayanus*, *M. tumidifrons*); however, the genus is also a common melolonthine component of open forests. *Epholcis*

Table 4: Endemic Dynastinae genera occurring in the Gondwana Rainforests region (after Carne 1957a).

Pre-Jurassic or Eyrean	Post-Jurassic or Bassian	Malayan or Torresian
Metanastes	*Cheiroplatys*	*Liteupatorus*
Semanotropus	*Dasygnathus*	
Cryptodus	*Neodasygnathus*	
	Anomalomorpha	
	Corynophyllus	

- Bassian fauna characteristic of NSW, Vic., Tas. and southern Qld
- Torresian found mostly in Qld, NT and north-west Australia
- Eyrean concentrated mostly in the south-west and centre of the continent

Horns and sexual dimorphism in Scarabaeoidea

Although adult head and thoracic armature can be found in several groups of beetles (e.g. Passalidae, Tenebrionidae, and even some Cetoniinae), the phenomenon is particularly pronounced in many male scarab beetles of the subfamilies Dynastinae and Scarabaeinae, and equally so in the scarabaeoid family Geotrupidae. Though present, horns are poorly developed in Passalidae, and in the related Lucanidae dimorphism is expressed in the mandibles, with females possessing small mandibles, but those of males greatly lengthened, toothed and widened (as in *Lissapterus*, *Cacostomus*, *Lamprima*); they function also for mate competition or domination of choice breeding sites.

Single and multiple horns may be formed on the head (including the clypeus) and the pronotum, and are spectacularly developed in the elongate horns of certain exotic dynastids as well as the regionally occurring *Xylotrupes* and *Haploscapanes*. Where females have horns these are poorly developed. That these are the preserve essentially of males indicates that such ornate structures serve in mating competition and sexual selection, for large horns can function as powerful pincers dislodging and overturning opponents in combative contests for breeding sites and females. Significant increases in female fecundity have also been related to increased horn size in mates (D. Cook 1988). Male dimorphism is expressed in relation to size, positive allometry, for large males trend to possessing very large and well-developed armature, and small males have much reduced horns, and in the case of dung beetles (e.g. *Onthophagus*) minor males may little depart from the outward appearance of females. But commonly there are grades between extremes within species. The size to which adults grow and the scale of armature development relate to the size of respective larvae, for such structures find developmental foundation just before the commencement of pupation. But very large horns can be awkward and so there are practical upper limits to development. In addition individuals with large head or pronotal horns tend to have smaller eyes or less developed wings, such that pronounced armament may come at a considerable functional disadvantage (Crowson 1981; Eberhard 1980; Emlen 2000), serving to compromise visualisation and movement through the rainforest landscape or, in the case of Scarabaeinae, disadvantage the ability of males to excavate burrows.

gracilis is widely distributed in littoral rainforests (e.g. Eurimbula, Fraser Island, Iluka, Camden Head).

The speciose genus *Heteronyx* (>300 species) is represented in Australia and South America, suggesting its development before their separation. *Heteronyx* includes numerous species occurring in diverse forest communities in south-east Queensland and northern New South Wales, but also species from rainforest (*Heteronyx incognitus*, *H. rugosipennis*) (Britton 2000; Williams 2002). *Homolotropus metallicus* is a rainforest specialist, often attracted to light, occurring in lowland (Lansdowne Escarpment) to upland rainforests (e.g. Bunya Mountains, McPherson Range, Cambridge Plateau, Upper Allyn River).

The genus *Liparetrus* (tribe Liparetrini) is endemic to Australia but most species are xerophilic, occurring inland in woodland, mallee, grassland and semi-desert (Britton 1980). However, at least three species (*Liparetrus convexus*, *L. erythropygus*, *L. ferrugineus*) are known from within the Gondwana rainforests region. The tribe Scitalini comprises 15 endemic genera out of a total of 16 within that tribe (Britton 1987), and the greatest proportion of Scitalini occurs in regions of high to moderate rainfall in eastern Australia. The scitaline genus *Telura* is an Antarctic taxon closely related to the Chilean genus *Sericoides* (Britton 1987); adults commonly come to light traps set on the edge of cool temperate rainforest and adjacent open forest in the Barrington Tops.

The Gondwanan tribe Xylonychini is represented in the Gondwana rainforests region by *Nitorellus* (*Nitorellus splendidus*) and *Xylonichus* (*Xylonichus eucalypti*, Plate 177); the latter is attracted to light and is associated with wet sclerophyll forest and sclerophyll forest–cool temperate rainforest ecotones (e.g. New England National Park). *Nitorellus splendidus* occurs from coastal regions (e.g. Cape Hawke) to elevated sites (Styx River State Forest), being found in woodland, eucalypt forest and the margins of cool temperate rainforest. Xylonychini possesses five genera in Australia, four in New Zealand and two in South America; this distribution suggests an origin between the Jurassic and Miocene (Britton 1957). In Australia the related Melolonthini have spread from northern immigrants into northern Queensland since the Pliocene. Although the distribution of the tribe is largely tropical, the genera *Antitrogus* and *Rhopaea* have a predominantly southern Australian distribution (Britton 1978).

The Diphucephalini include numerous species (many undescribed) of *Diphucephala* with localised distributions in rainforest and wet sclerophyll forest at various altitudes. Adults are often found in daylight on rainforest margins feeding on foliage, and during mating events numerous individuals may congregate combatitively in the rainforest understorey (e.g. on foliage of young *Archontophoenix cunninghamiana* palms). *Diphucephala* species are occasionally encountered imbibing nectar from flowering canopy trees (e.g. *Diphucephala rufipes* on *Syzygium floribundum*, *D. pygmaea* on *Backhousia myrtifolia*). *Diphucephala lineata*, *D. dicksoniae*, *D. nitidicollis*, *D. parviceps*, *D. richmondia*, *D. rufipes* and *D. sericea* are rainforest-associated species (but may also be found in open forest). Numerous additional species inhabit adjacent eucalypt-dominated forests, but generally do not enter rainforest. The related day-active genus *Watkinsia* (Britton 1995) consists of eight species, all restricted to the area between northern New South Wales and Eungella National Park in central-eastern Queensland. Most are associated with rainforest. Adults of *Watkinsia williamsi* and *W. vicina* are found on foliage and are known only from their type localities; Border Ranges National Park and Mt Mistake Plateau respectively (Britton 1995). Additional regional species are *Watkinsia tenebrosa* (Mary Cairncross National Park and Mt Glorious), *W. bella* (Tapin Tops National Park), and *W. monteithi* (Kroombit Tops) (Britton 1995).

Phyllotocus, *Cheiragra* and, largely, *Anthotocus* are flower-frequenting genera, species of which inhabit open forest (where they can be prolific on flowering Myrtaceae), but are also found feeding from the nectar of several mass-flowering rainforest trees with open floral structures: *Syzygium floribundum* and *S. smithii* (*Phyllotocus australis*, *P. marginipennis*, *P. navicularis*, *P. scutellaris*, *Cheiragra ruficollis*); *Alphitonia excelsa* (*Cheiragra ruficollis*); *Cuttsia viburnea* (*Anthotocus ?fugitvus*), *Euroschinus falcatus*, *Guioa semiglauca*, and *Scolopia braunii* (*Phyllotocus australis*, *Cheiragra ruficollis*); *Rhodomyrtus psidioides* (*Phyllotocus scutellaris*); and *Tristaniopsis laurina* (*Phyllotocus ?marginipennis*, *P. navicularis*, *P. scutellaris*, *Anthotocus ?fugivitus*, *A. luridus*) (Williams 1995). *Phyllotocus bimaculatus*, *P. macleayi* and *P. basalis* are attracted to ultraviolet light traps set in rainforest. *Phyllotocus bimaculatus* is largely restricted to mesic forests at higher elevations (e.g. Barrengarry Mountain, Mt Wilson, Barrington Tops), including cool temperate rainforest. The distribution of *Phyllotocus basalis* includes littoral rainforest remnants (e.g. Harrington). *Anthotocus antennalis* (Plate 165) and *A. perissus*, rather than frequenting flowers, are usually found low in the rainforest understorey on the foliage of ferns, herbs and shrubs.

Scarabaeidae – Rutelinae The main Australian distribution of the Rutelinae is on the east coast with the subfamily being most diverse in New South Wales and Queensland (Allsopp 1989; Carne 1958). The majority of taxa are endemic and the tribe Anoplognathini is fully indigenous (Howden 1981). Anoplognathine genera (e.g. *Anoplognathus*, *Schizognathus*, *Parashizognathus*) radiated extensively during the later part of the Tertiary (Lawrence and Britton 1994). The endemic genus *Paraschizognathus* (Plate 173) has undergone considerable diversification, within sclerophyllous vegetation types, particularly on the New South Wales north coast (Carne 1958, 1974). Although the genus *Anoplognathus* (Christmas beetles) (Carne 1957b, 1981) includes species that are confined to open eucalypt forest and woodland, several anoplognathine taxa occur only in subtropical rainforest and adjacent wet sclerophyll forest (e.g. *Anoplognathus prasinus*, *A. concolor*, *Paraschizognathus ocularis*, *P. olivaceus*, *P. prasinus*, *P. pinarus*, *Schizognathus compressicornis*, *S. macleayi*), *Anoplognathus concolor* being frequently encountered (Plate 164). *Paraschizognathus pinarus* flies at dusk on the margins of cool temperate rainforest and open forest, but is also attracted to ultraviolet light. The enigmatic and rarely collected endemic genus *Trioplognathus*, represented soley by *T. griseopilosus*, is known only from northern and central coastal New South Wales (Carne 1958). Somewhat resembling *Anoplognathus velutinus* in appearance, the species has been recorded attracted to mercury vapour lights on the edge of lowland subtropical rainforest at Victoria Park Nature Reserve, south-east of Lismore (Carne and Monteith 1971; McAlpine 1972a).

Scarabaeidae – Scarabaeinae The Australian Scarabaeinae, or true dung beetles, comprise the tribes Onthophagini, Canthonini and Dichotomini (Gunter and Weir 2017; Matthews 1972, 1974, 1976a; Philips 2016; Tarasov and Dimitrov 2016). Most adults are coloured black. Larvae and adults are generally coprophagous (Halffter and Edmonds 1982; Hanski and Cambefort 1991), with mouthparts modified for feeding on a largely fluid diet. Adult beetles excavate brood tunnels and chambers, usually constructed below food sources (*Onthophagus*, Plates 171, 172), or fashion balls of faecal or synthesised plant material (numerous Canthonini) and roll these to preferred nest sites.

There can be intense competition between species for spatially limited food resources, which in the absence of larger marsupials in small rainforest remnants may be limited in volume. Adults, particularly of *Onthophagus*, have modified mouthparts, shovel-like heads and robust forelegs, which are highly adapted for their feeding and nest burrowing lifestyle. Scarabaeinae are thought to have originated in the Lower Cretaceous, and fossils, possibly of dung balls of the type constructed by modern Scarabaeinae, are known from the Lower Oligocene (Halffter and Matthews 1966). Extant scarabaeine feeding strategies reflect heavy associations with Recent mammalian hosts, but these evolved (or were re-established following extinction events) from precursory relationships with earlier animals (such as herbivorous dinosaurs) and non-angiosperm plants that dominated Mesozoic terrestrial ecosytems (e.g. Chin and Gill 1996; Gunter *et al.* 2016). The effect on Scarabaeinae populations and diversity, owing to the extinction of medium-sized to large terrestrial understorey browsers (e.g. diprotodontids, which would have deposited relatively large quantities of dung) in Australian rainforests and open forests (Adam 1992; Archer *et al.* 1989; Rule *et al.* 2012) is unknown.

Several tropical (e.g. *Monoplistes curvipes*, *Temnoplectron aeneopiceum*, *Onthophagus gulmarri*) and subtropical species (e.g. *Aulacopris maximus*, *Diorygopyx incrassatus*, *Onthophagus neostenocerus*) exhibit perching behaviour at night on shining rainforest foliage or tree trunks (Plates 167, 170, 171) (Howden and Howden 1991; Williams 2003b; G. Williams unpubl. obs.); this may be a strategy that enhances a species' ability to locate dung resources (Peck and Howden 1984). Scarabaeine dung beetles are preyed upon by reptiles, amphibians, birds and mammals. Invertebrate predators include planarian worms and possibly tenebrionid beetles (Monteith and Storey 1981).

The subfamily is speciose in subtropical rainforests, and many species are restricted to rainforest and wet sclerophyll forest (Gunter and Weir 2017; Matthews 1972, 1974, 1976a; Monteith 1986b; Storey 1977; Storey and Weir 1988, 1989; Williams 2002, 2003b; Williams and Williams 1982, 1983a, 1983b, 1983c, 1984). The Onthophagini (Matthews 1972; Philips 2016) are the most recent of the Australian fauna and include a single genus *Onthophagus*, the origin of which probably dates from the Mesozoic (Howden 1981). The Onthophagini constitute ~50 per cent of the Australian scarabaeine fauna and the number of known species exceeds 200. This represents ~12 per cent of the world's estimated fauna of more than 1500 species. In comparison to the diversity of *Onthophagus* species described from Australia, ~40 species are described from the United States of America and ~90 species are described from Europe. No species are known from Chile or New Zealand (Howden 1981).

Onthophagus is a massive genus found world-wide. The Australian *Onthophagus* fauna has major relationships with more recent Oriental elements (Howden 1981). However, the genus may have originated in Africa, which has some 800 species, but extensive autochthonous development in Australia suggests a long history of occupation here (Matthews 1972). Matthews (1972) postulates that, based upon species groups (if natural), each species group may represent a separate invasion, suggesting 34 original invasions. Except for five species that are recorded from New Guinea, all Australian species are endemic. The high levels of endemicity and probable archaic history (although less than that of the Canthonini and Dichotomini) 'gives this fauna a strongly insular aspect, unlike that of any other region of comparable size but reminiscent of that of some continental islands' (Matthews 1972).

Species are frequently sexually dimorphic with males having prominent tubercles, pronotal excavations, and horns (e.g. *Onthophagus leanus*, *O. mamillatus*, *O. neosternus*). These may be reduced in so-called 'minor' males (Matthews 1972). Many species are attracted to ultraviolet light traps. *Onthophagus* adults are more usually associated with animal dung, in Australia especially that of macropods. *Onthophagus dunningi*, however, a species found in wet sclerophyll forest (as well as more xeric-adapted forests), departs from the prevailing coprophagous habit of feeding and brood provisioning, and feeds on *Amanita*, *Agaricus*, *Boletellus* and *Boletus* fungi (mycetophagy) to the apparent exclusion of mammal dung (Bornemissza 1971). Polyphagous habits are uncommon in the Scarabaeinae, and Bornemissza (1971) posits that this switching in feeding behaviour could have been due to changes in the nature and abundance of Australia's marsupial fauna that accompanied the progressive drying of the continent in the Miocene/Pliocene. Australian subtropical rainforests (unlike the Wet Tropics) possess no specialist marsupial species, and no large understorey species in abundance (Adam 1987, 1992).

In Australia *Onthophagus* is most diverse along the east coast and tablelands of the mainland with almost no records from arid areas (i.e. the fauna is predominantly Bassian and Torresian in distribution) (Matthews 1972). Matthews (1972) divided the fauna into 11 distributional patterns. Areas included in the Gondwana rainforests region are his 'south-eastern coastal and montane', 'eastern coastal', 'rainforest pattern', 'Queensland savannah–woodland' and 'northern savannah–woodland' patterns. There is overlap in distribution of each type. These divisions 'represent very broad distribution zones which could be subdivided further along ecological lines' (Matthews 1972). Patterns most relevant to the Gondwana rainforests region, although they generally extend north and south of it, are the south-eastern coastal and montane, eastern coastal and rainforest patterns. These encapsulate the montane mixed sclerophyll forest types, and diverse rainforest subformations occurring in southern Queensland and northern New South Wales. Species restricted fully or substantially to rainforest and wet sclerophyll forests at upland elevations include *Onthophagus arilla*, *O. bornemisszai*, *O. fuliginosus*, *O. kiambram*, *O. leanus*, *O. longipes*, *O. macrocephalus*, *O. mamillatus*, *O. neostenocerus*, *O. pugnax*, *O. squalidus*, *O. sydneyensis*, *O. turrbal* and *O. yourula* (Matthews 1972; Storey and Weir 1988, 1989; Williams 2002). *Onthophagus williamsi* is known only from cool temperate forests in the Barrington Tops (Storey and Weir 1989), but the altitudinal range of several species, such as *O. ouratita* and *O. yarrumba*, can extend from upland rainforest (Mt Lindsay, Mary Cairncross Park) to coastal littoral rainforest (Iluka). *Onthophagus tabellifer* is restricted to coastal littoral rainforests, mostly on sands (Lake Cathie, Camden Head, Harrington, Seal Rocks, Batemans Bay).

The Australian Canthonini (the Scarabaeini–Canthonina of Matthews 1974; though see Gunter and Weir [2017] for a discussion of the Deltochilini or deltochiline-like putative status of the tribe) are Gondwanan in origin. Matthews (1974) noted that ~45 per cent of the then known Australian Canthonini were flightless, though this figure may have changed owing to discoveries subsequent to his revision. The principal behavioural characteristic of the Canthonini is the ability to fashion and roll a ball of food (usually of mammal dung) (Matthews 1974). This is done at the food source by carving out a food mass, and then progressively shaping and rotating this until a roughly spherical ball is formed. Ball masses are then rolled away (by pushing or pulling movements) and buried, and used either as food for the adult or the developing larvae. Some combative behaviour between beetles may be associated with ball construction (Halffter and Matthews 1966).

Australian canthonines exhibit an unusual degree of radiation in comparison with the World fauna (Howden 1981), and constitute ~20 per cent of the world's generic diversity and ~14 per cent of the species diversity. Only South America has greater generic diversity (Howden 1981; Matthews 1974). Although the Canthonini have a Gondwanan distribution (including species in New Zealand and New Caledonia), the cold–temperate areas are nearly devoid of species. In southern Argentina only two species occur as far south as Santa Cruz Territory, and the subtribe is absent from Chile, Tasmania, South Australia and most of Victoria (Matthews 1974). However, New Zealand possesses two endemic genera and the rainforests of northern New South Wales and southern Queensland are rich in endemic taxa (e.g. *Diorygopyx*). Of those genera recorded from the Gondwana rainforests, *Amphistomus* and *Lepanus* have extralimital occurrences in New Guinea and Indonesia. All other Australian genera are endemics. Due to its high level of endemism the Australian Canthonini 'must have been derived from ancestors that invaded a long time ago, perhaps together with early marsupials' (Matthews 1974). However, at present there is insufficient evidence to determine whether the canthonines arrived in Australia from South America via East Asia, or via Antarctica.

Australian canthonines are primarily active at night (some *Lepanus* are active during daylight; e.g. *L. australis*), and in general are restricted to areas with greater rainfall and higher average temperatures than the Onthophagini. Central Australia is largely devoid of the group but subtropical New South Wales–southern Queensland is a centre of diversity for the genera *Aulacopris*, *Diorygopyx*, *Aptenocanthon*, *Matthewsius* and *Cephalodesmius*; and shares with the Wet Tropics of Queensland a high diversity of species within *Amphistomus* and *Lepanus* (see Williams 2002). Most of the known *Diorygopyx* species occur in the Gondwana rainforests region (Matthews 1974) with *Diorygopyx cuspidatus* being an isolated northern outlier. The Canthonini include two genera, *Aulacopris* (Williams 1993) and *Cephalodesmius* (Monteith and Storey 1981), whose adults synthesise fallen leaves and flower segments into dung-like material from which they fashion brood balls, within which their young feed and develop.

Aulacopris includes two southern species (*Aulacopris reichei*, *A. maximus*) (Matthews 1974) and a single much smaller-sized species (*Aulacopris matthewsi*) from North Queensland (Storey 1986). *Aulacopris maximus* (Plate 166) inhabits subtropical rainforest and wet sclerophyll forest at low altitudes (Lansdowne Escarpment, Way Way State Forest), but also upland warm temperate rainforest (Barrington Tops, Dorrigo) and mixed cool temperate rainforest >1100 m (Werrikimbe National Park) (Williams 2002, 2003b; G. Williams unpubl. obs.). Adults have been observed perching at night on rainforest tree trunks (e.g. Banda Banda Beech Reserve) (Williams 2003) (Plate 167). The species is also associated with bat guano in caves (Lea 1923; Williams 2003), and may be a specialist on bat guano (Matthews 1974; G. Monteith pers. observ.). However, adults are commonly associated with garden compost heaps positioned near rainforest margins (G. Williams pers. observ.), and fashion synthesised brood balls from leaf material and the remains of invertebrates (Williams 1993; G. Williams pers. observ). *Aulacopris maximus* is host to the phoretic mite *Macrocheles spatei* (Macrochelidae) (Halliday 2000).

Aptenocanthon has one southern species (*Aptenocanthon hopsoni*) (Carter 1936; Williams 2002; Williams and Williams 1982, 1983a), and additional species in northern Queensland (Storey 1984; Storey and Monteith 2000), this zoogeographical pattern (and that of *Aulacopris*) suggesting extinction rather than expansion events. *Aptenocanthon hopsoni* is restricted to upland rainforests extending from Barrington Tops, Tapin Tops National Park, the Comboyne Plateau and Werrikimbe National Park, with an unusual outlying population at North Brother Mountain, adjacent to the Pacific Ocean at Laurieton (Williams 2002, 2003b; G. Williams unpubl. obs.). In *Nothofagus* forest adult *Aptenocanthon hopsoni* can be very numerous at baited pit-fall traps, but in nature are found feeding both in, but generally under, the food source. Regional *Matthewsius* comprise *M. rossi* (previously *Aptenocanthon rossi*) with a restricted distribution in upland wet forest in the Blue Mountains (Mt Irvine), and Newnes State Forest, west of Sydney, *M. illawarrensis* (previously *Lepanus illawarrensis*) from the Blue Mountains (Mt Wilson), Kanangra–Boyd and Morton national parks, and south to Batemans Bay, and *M. stricklandensis* known only from open forest near Gosford on the New South Wales Central Coast (Gunter and Weir 2017; Matthews 1974; Williams and Williams 1982). A single species of *Boletoscapter* (*Boletoscapter furcatus*) occurs in southern Queensland. Found in open forest and rainforest, adults have been collected at decaying fungi and mammal dung (Matthews 1974; G. Monteith pers. comm.).

Adult *Diorygopyx* are able to fashion and roll food balls. The genus comprises a suite of flightless species that extend north from the lower north coast of New South Wales, with individual species tending to occupy circumscribed geographical distributions without sympatric cohabitation of sites (Matthews 1974; Williams 2002, 2003b). *Diorygopyx asciculifer* is the most southerly occurring species, its range extending from coastal littoral rainforests (Williams 2002, 2003b; Williams and Williams 1984), progressively to flood-prone lowland rainforest (Coocumbac Island and Wingham Brush nature reserves), intermediate altitudes (Allyn River, Copeland Tops, Tapin Tops), inland to the Liverpool Ranges (Williams 2002, 2003b); but it is absent from rainforest at high elevations in the Barrington Tops. On the Comboyne Plateau and the Lansdowne Escarpment, and north along the immediate coast from Laurieton to Lake Cathie, *Diorygopyx asciculifer* is replaced by *D. incrassatus*. Similar replacements, of different *Diorygopyx* species, progress sequentially into southern Queensland (Matthews 1974; Williams 2002, 2003b). *Diorygopyx incrassatus* has been observed at night perching on low foliage in subtropical rainforest (Wilson River Primitive Reserve [Williams 2003b]). Numerous *Diorygopyx niger* have been observed perching at night on leaf litter (e.g. at League Scrub, G. Williams pers. obs.) (Plate 170). Superficially resembling *Diorygopyx* is the genus *Monoplistes*, a single representative of which (*Monoplistes leai*) occurs in southern Queensland and northern New South Wales (Matthews 1974; Williams 2002). Initially thought restricted to upland forests of the McPherson Range and nearby sites, *Monoplistes leai* additionally occurs in dense coastal *Livistona australis* palm–eucalypt forest (Cape Hawke), wet sclerophyll forest (Lansdowne Escarpment, Bulahdelah State Forest), and depauperate lowland riparian rainforest–wet sclerophyll forest (Yarratt State Forest) (Williams 2002; Williams and Williams 1983b, 1983c, 1984).

Amphistomus is also diverse in the rainforests of New South Wales and southern Queensland, with that of *A. speculifer* (Plate 163) extending altitudinally from the coast inland to the depauperate rainforests of the Liverpool Ranges (Williams 2003b), and distributed to the far south coast of New South Wales (the most southerly occurring of the genus) (Matthews 1974). Adults commonly feed directly from below mammal dung, rather than constructing and rolling balls. Most species (e.g. *Amphistomus cunninghamensis*, *A. macphersonensis*, *A. primonactus*, *A. trispiculatus*) occur in upland forests

(Matthews 1974; Williams 2002). Species are commonly collected covered in adhering soil or clay, this coating making them inconspicuous to the eye.

The tiny beetles of the genus *Lepanus* are also well represented in the wet forests of subtropical Queensland and New South Wales (i.e. *Lepanus australis*, *L. bidentatus*, *L. gelasinus*, *L. politus*, *L. ustulatus*, and undescribed species) (Matthews 1974; T. A. Weir pers. comm.; Williams 2002; Williams and Williams 1982, 1983a, 1983b, 1983c, 1984). *Lepanus* adults are capable of fashioning and rolling balls (Matthews 1974). Some species (e.g. *Lepanus australis*) are able to survive in very small and structurally disturbed rainforest remnants (e.g. Crowdy Head, Saltwater Reserve) (Williams and Williams 1984). *Lepanus ustulatus* inhabits a variety of rainforest types at intermediate and moderate altitudes (e.g. Bulburin and Kenilworth State Forests, Woko National Park, Kerewong State Forest, Lansdowne Escarpment) and extends to seaside littoral rainforest on sandy soil at Eurimbula National Park in central coastal Queensland (G. Williams unpubl. data). *Lepanus* may be a bird-dropping specialist (adults being frequently found at that food source), and may also feed on the droppings of small skinks and land snails; these feeding strategies would allow them to survive in tiny relict rainforest habitats devoid of large terrestrial vertebrates. However, on one occasion several *Lepanus politus* were observed feeding on liquids from 'wet phase' decomposing fish (G. Williams unpubl. data). *Lepanus glaber* has been recently transferred to the endemic genus *Monteithocanthon* and is known from Lamington and Springbrook national parks at altitudes of ~1000 m (Gunter and Weir 2017).

Three species of the endemic genus *Cephalodesmius* occur in New South Wales and southern Queensland (*Cephalodesmius armiger*, *C. laticollis*, *C. quadridens*) (Matthews 1974; Williams 2002). The latter two are restricted to rainforest, generally at higher elevations (e.g. Bunya Mountains, Cunninghams Gap, Mt Warning), but *Cephalodesmius armiger* also inhabits open forest. *Cephalodesmius armiger* has an aberrant biology (Dalgleish and Elgar 2005; Monteith and Storey 1981). Males and females are permanently bonded and inhabit subterranean nests on the rainforest floor (Fricke 1967). Males gather fallen leaves and flowers and these are processed by the females into dung-like material, which is then used to construct brood balls in which the larvae feed and develop. During the growth of the young the adults cooperate in progressive provision of the balls. *Cephalodesmius laticollis* exhibits similar brood ball construction and, though Monteith and Storey (1981) did not observe brood ball construction by *C. quadridens*, they noted that nests contained only leaves and other plant material.

Temnoplectron has a predominantly northern Australian and New Guinean distribution, but *T. boucomonti* and *T. major* extend along the eastern coast into subtropical southern Queensland (Matthews 1974; Reid and Storey 2000). Individuals resemble *Diorygopyx* and *Monoplistes*. *Diorygopyx* is flightless, the wings are fully developed in *Monoplistes*, and vestigal or variously developed in *Temnoplectron* (Matthews 1974). *Temnoplectron boucomonti* and *T. major* inhabit woodland but also are found in vine thickets. Matthews (1974) observed *Temnoplectron boucomonti* (as *Temnoplectron laevigatum*) beetles at night constructing dung balls, the process taking 12–30 minutes. Individuals fought over the possession of completed balls. Using the forelegs, balls were rolled away up to distances of 30 m, in the process being coated with sand. But burial was not observed.

Neither a Gondwanan nor Oriental-Papuan biogeographical pattern explains the derivation and evolution of the Australian Dichotomini (the Coprini of Matthews 1976a). The Australian Dichotomini are, in general, distinctive from other dichotomine faunas and this suggests a long period of isolation (Matthews 1976a). They seem more closely related to African-Holarctic and African-Lemuria genera; this points to a 'direct connection with Africa at a remote time' (Matthews 1976a). The relationships of the three Australian genera (*Demarziella*, *Copris*, *Thyregis*) infer that only two invading ancestral species would have been necessary for the evolution of the current taxa.

The genus *Demarziella* (Matthews 1976a; Matthews and Stebnicka 1986) appears to be an isolated representative of very small dichotomine dung beetles, which include *Pedaria* and *Paraphytus* from the Old World, and *Pedaridium* and *Trichillum* from the New World (Matthews 1976a), and may represent an Older Northern Element (Mesozoic) but of uncertain derivation (Howden 1981). The centre of distribution for *Demarziella* (*Demarziella interrupta*, *D. metallica*, *D. scarpensis*, *D. sylvestris*) approximates the south-east Queensland–New South Wales border, with a second suite of species in northern Australia. In northern New South Wales species occur in littoral rainforest on headlands and sand masses (Matthews and Stebnicka 1986; Williams and Williams 1984), and inland to upland rainforests (including

Nothofagus-dominated cool temperate rainforest). The relictual endemic genus *Thyregis* is related to the Holarctic and Ethiopian genus *Copris* and has a southern or Bassian distribution with one species in south-west Western Australia, and three species in south-east Australia. Two species, *Thyregis relictus* and *T. monteithi*, are known only from the Gondwana rainforests region (e.g. Gibraltar Range, Chaelundi State Forest, Dorrigo), and in addition to rainforest also inhabit open forest. *Thyregis* and *Demarziella* are attracted to mammal dung; however, *Demarziella* may also be cleptoparasitic on other scarabaeines (G. Monteith pers. comm.).

Silphidae Silphids, or burying beetles, are large, somewhat flattened, terrestrial carrion feeders (particularly during the 'wet' decay phase) and predators of fly maggots. The family is poorly represented in Australia, where the ecological position of silphids in subtropical rainforests appears at least partly filled by carrion-feeding Hybosoridae and Staphylinidae. The Australian fauna comprises two genera: the endemic *Ptomaphila* (*Ptomaphila lacrymosa*, *P. perlata*) and the more widely distributed *Diamesus* (*Diamesus osculans*) (Plates 180, 181). All occur in the mesic forests of New South Wales and southern Queensland, and both genera are attracted to ultraviolet light. *Ptomaphila* is a Gondwanan genus restricted to Australia and New Guinea, and is most closely related to the Neotropical genus *Oxelytrum* (Lawrence and Ślipiński 2013; Peck 2001); these groups represent ancient lineages from the Southern Hemisphere (King *et al.* 2015). The natural enemies of *Ptomaphila perlata* include reduviid bugs (G. Williams pers. observ.). *Diamesus* consists of two species. *Diamesus osculans* is distributed from Sri Lanka, southern India, through Vietnam and Indonesia to eastern and western Australia, where it occurs widely in sclerophyll forest and rainforest (e.g. Boorganna Nature Reserve, Bruxner Park, Iluka, Lansdowne Escarpment [Williams 1981]). The second species occurs in Taiwan. When in flight adult *Diamesus osculans* make a high-pitched sound of short duration, akin to that of some Diptera, Hymenoptera and Orthoptera (Williams 1981). Adults are able to articulate the tip of their abdomen so as to direct the expulsion of odorous body fluid; this is a defence response when handled or threatened.

Staphylinidae Staphylinidae, or rove beetles, date to the Triassic, and were diverse by the Late Jurassic (Grimaldi and Engel 2005). Staphylinids are currently the largest family of beetles world-wide, comprising >61 000 species in 32 subfamilies (Parker 2016). Aleocharinae and Pselaphinae are the largest subfamilies. Most Aleocharinae possess a tergal gland complex on the abdomen between tergites 6 and 7, which secretes defensive irritants such as quinones (Parker 2016). Staphylinids are characteristically elongate beetles, mostly with short elytra that leave the soft and flexible abdomen largely exposed, and are a major predator and scavenger group that inhabits the rainforest floor and canopy. Some are relatively large and are found in and under rotting animal carcasses (*Creophilus*, *Antimerus*). This is a spatially and temporallyephemeral habitat niche occupied by smaller staphylinids as well (e.g. *Anotylus*). Thus the corpse-adapted fauna needs to be agile and capable of rapid dispersal. It is within the Staphylinidae that the greatest number of myrmecophilous beetle lineages can be found, particularly within the Aleocharinae and Pselaphinae. These groups are highly modified.

Subfamilies associated with rainforest include Aleocharinae, Euaesthetinae, Omaliinae, Oxytelinae, Paederinae, Pselaphinae (= Pselaphidae), Scaphidiinae (= Scaphidiidae), Scydmaeninae (= Scydmaenidae), Staphylininae and Tachyporinae. In lowland subtropical rainforests species of Aleocharinae, Omaliinae, Oxytelinae, Paederinae and Steninae are occasionally associated with the flowers of canopy trees (Williams 1995). In upland forest (e.g. in Lamington National Park) the family is a major foliage-frequenting element of the canopy (Hammond *et al.* 1997).

The Scydmaeninae were once considered a separate family, and are distinct among staphylinids in being the only group in which the elytra fully cover the abdomen. They are minute beetles found in leaf litter, rotten logs, moss beds, soil and ant nests, and are predators of small arthropods. Scaphadiinae feed on spores and fungal hyphae. Pselaphinae (D. Chandler 2001) are common in leaf litter, and also are to be found in rotting logs, moss beds and caves. They are predators of tiny arthropods such as mites and Collembola. Some are specialised species that dwell in the nests of ants, the expression of myrmecophory being well developed in the subfamily (Parker 2016). Paederinae (e.g. *Pinophilina*) are found in leaf litter and on foliage. Several are brightly coloured and produce a potent skin irritant: paederin, the most complex beetle toxin known (Dettner 1987). Oxytelinae are common in leaf litter, but also at animal dung. *Anotylus* (Oxytelinae) is known to breed in the nests of the dung beetle *Cephalodesmius armiger*. Adults and larvae burrow through the

Feeding behaviour and nest building in scarabaeine dung beetles

Scarabaeinae are essentially coprophagous in habit, feeding mainly on dung, and using it to provision for the developmental needs of their young. A few depart from this ecological groundplan to artificially synthesise 'dung' from plant material (*Cephalodesmius* spp., *Aulacopris maximus*), feed on fungi (*Onthophagus dunningi*), or occasionally resort to imbibing the juice of fermenting fruit. Dung is sourced from the droppings of mammals, birds, reptiles, amphibians and even land snails. Scarabaeine dung beetles employ a diversity of feeding and nest building (nidification) strategies that range from feeding within or immediately underneath the dung mass (e.g. many small *Onthophagus, Aptenocanthon hopsoni*), excavating burrows below and into which material is drawn (*Onthophagus pugnax, O. tabellifer*), or fashioning balls of various sizes, which are then rolled some distance by pulling or pushing movements and buried underground (various Canthonini) (Halffter and Matthews 1966; Matthews 1972, 1974). Tiny species of *Lepanus* often feed exposed on the surface of freshly deposited bird droppings, and certain Australian *Onthophagus* (e.g. *Onthophagus muticus, O. parvus, O. symbioticus*) have evolved prehensile claws, stout bristles and an enlargened last tarsal segment to allow them to cling to the hair of their marsupial hosts and quickly grasp faecal pellets as these emerge from the host's anus (Matthews 1972).

Once supplies have been collected and buried, eggs are deposited in or alongside the food mass, which is then fed upon by the developing larvae. In other instances single eggs are laid first and dung is progressively fashioned in a discretely formed mass around each one. Scarabaeine dung beetles are eloquently structured to facilitate provisioning and feeding, for their forelegs and heads are formed so as to function as efficient shovels and rakes. The clypeus, the forepart of the face, is curved and flattened like a broad shovel, and the foretibiae are widened at their end, the apical tarsi being much reduced and almost redundant, so that the front legs serve as powerfull excavators. However, there are problems associated with using dung as a food source, for it is subject to drying out, is competed for by other invertebrates, needs to be available in reliable and sufficient quantity, and may need to be of a certain texture before it can be effectively utilised. Dung size of itself can serve as a niche that influences the size range and abundance of dung beetle species attracted to it (Peck and Howden 1984), and to find dung the beetles are equipped with heightened olfactory senses, for the movements of potential hosts cannot be forecast, and flightless species potentially face long journeys across uncertain terrain to locate it. Species dependent on large mammals, or surviving in small remnant rainforests with an impoverished or itinerent host fauna, may face particular problems of supply. Where hosts have become locally extinct, then dependent dung beetles may confront the same fate. Nevertheless dung beetles can be present, even in the smallest or most isolated rainforest patch, in large and diverse numbers.

The architecture of nest burrows varies and nest building can be undertaken by both or one sex. Equally so with ball rolling, for both may participate as bonded pairs in the process of translocation and burial. Burrows may take the form of single shaft tunnels into which dung is progressively packed, single tunnels with multiple side brood chambers in which individual larvae feed and mature, multiple brood chambers with multiple larvae in each, multiple feeding entrances contacting the undersurface of the dung mass but interconnecting at depth below it, compound superficial nests in which multiple brood balls are deposited and then tended by both sexes, single large subterranean chambers where females tend multiple or single brood balls, or just shallow excavations into which brood balls are deposited (Halffter 1977; Halffter and Edmonds 1982). Individual brood supplies may be fashioned from segments excised from the dung mass, or from quantities of dung provisioned in bulk within each tunnel or chamber, and which then are apportioned to single larvae. The shape of brood masses may differ, ranging from spherical to pear-shaped to broadly elongate. However, the spatial distribution of excrement may determine the food behaviour patterns of individual species (Halffter 1977). Although there have been numerous papers world-wide that address the nature and evolution of nidification and coprophagy (Hanski and Cambefort 1991), the feeding behaviour and nest-building strategies of the rich Australian subtropical scarabaeine fauna remain poorly known, that of the canthonine *Cephalodesmius armiger* being a rare and peculiar exception. *Cephalodesmius armiger* was the subject of a detailed study in southern Queensland by Geoff Monteith and Ross Storey (Monteith and Storey 1981), the species being found to have an unusual provisioning strategy that involved the cooperative gathering and processing of fallen plant material, literally synthesising dung, upon which their young fed and developed to maturity. Such cooperative bisexual nidification behaviour hints at the development of subsocial life.

brood mass. The related genus *Oxytelus* is recorded from the nests of *Cephalodesmius quadridens* (Monteith and Storey 1981).

The Staphylininae include large robust beetles (13–20 mm) of the endemic genus *Antimerus* (Solodovnikov and Newton 2010). These are found on the forest floor and on foliage, and are among the largest staphylinids that inhabit subtropical rainforest and wet sclerophyll forest. Species include *Antimerus auricomus* (e.g. Mt Glorious, Dorrigo, Barrington Tops), *A. smaragdinus* (Brown Mountain), *A. punctipennis* (Conondale Range, Border Ranges, Mt Hyland, Bulga State Forest, Lansdowne Escarpment), *A. jamesrodmani* (Mt Glorious), and *A. monteithi* (Lamington National Park, Acacia Plateau). *Creophilus erythrocephalus* (Staphylininae) is a large black, red-headed, predatory species that is occasionally found in rainforest, though more typically a species of open forests. Although a habitat generalist, it is an ecological specialist which frequents carcasses.

Adult *Stenus* (Steninae) are small but prominently eyed beetles that forage during daylight for prey on the leaf surfaces of rainforest trees, vines and shrubs. *Stenus caviceps* has been recorded from North Queensland to southern New South Wales (e.g. McPherson Range, Lansdowne Escarpment, Barrington Tops, Barrengarry Mountain, Clyde River) in rainforest and wet sclerophyll forest; *S. puncticollis* is widely distributed in cool temperate rainforest, rainforests at lower elevations, and wet sclerophyll forest (C. Reid 1997); and *S. postulifer* is recorded from Brown Mountain, Robertson Nature Reserve and Barrington Tops. The metallic blue-green coloured *Stenus coeruleus* has been collected on the flowers of the tree *Acradenia euodiiformis* in lowland subtropical rainforest (Williams 1995), and will sometimes be found in large numbers on the leaves of *Toona ciliata* (Meliaceae).

Tenebrionidae Tenebrionids in subtropical rainforests are found in and under rotting tree trunks lying on the forest floor, beneath loose bark, active in sunlight on tree trunks (Cyphaleina), on foliage, and on flowers high in the canopy. Most are nocturnal, and often flightless. Larvae and adults generally feed on fungi associated with rotting wood and other vegetable matter (e.g. lichens), though rarely predatory (Bouchard 2002; Bouchard and Yeates 2001; Bouchard *et al.* 2005; Matthews and Bouchard 2008). Some are nectar feeders (Lagriini, Alleculinae), these often being brightly coloured (e.g. *Aethyssius*). Ground-dwelling species are black or dull coloured, or cryptically patterned. The adults of many Tenebrionidae expel or smear defensive chemicals from eversible abdominal glands (constituents tabled in Brown *et al.* 1992).

Although a high proportion of Tenebrionidae are flightless ground dwellers, populations of numerous taxa are able to survive in lowland rainforest remnants completely, or partially, inundated by regular flood events (e.g. Coocumbac, Lansdowne, Stotts Island and Wingham Brush nature reserves), much of the woody shelter of the forest floor regularly being swept away downstream. Examples include *Adelium helmsi*, *A. geminatum*, *A. reticulatum*, *Bassianus rectibasis*, *Ceropria valga*, *Platydema fuligineum*, and species of *Achthosus*, *Cardiothorax*, *Nolicima* and *Promethis*.

The Macpherson refugium Matthews and Bouchard (2008) identify the Macpherson refugium (approximating the Macpherson–Macleay Overlap) as one of 10 significant focal biogeographical centres for tenebrionid genera in Australia (see Fig. 3) (Carter 1905, 1906, 1911, 1913a, 1913c, 1914, 1915b, 1916, 1920, 1922, 1924, 1925, 1926a, 1928, 1930, 1933; Matthews 1998; Matthews and Bouchard 2008; Matthews and Doyen 1989; Sloane 1911; Watt 1974; Williams 2002). These refugia are believed to have been in existence since the Jurassic and have acted as core areas progressively acquiring biotic diversity, partly unique to each, in the face of climatic, vegetation and oceanic fluctuations. Collectively, these 10 refugia account for the whole of the generic diversity (and probably most of the species diversity) within the family (E. G. Matthews pers. comm.) and can be divided into two biogeographical provinces (see Matthews and Bouchard 2008).

The Macpherson refugium contains five genera (i.e. *Nototrintus*, *Dorrigonum*, *Asphalus*, *Sloanea*, *Styrus*) endemic to the region, and the Barrington Tops and Dorrigo subregion appear to be particularly significant (E. G. Matthews pers. comm.). The probable rainforest genus *Nototrintus* (Adeliini) includes five described species, and ranges from Barrington Tops in northern New South Wales to Kroombit Tops in southern Queensland. *Asphalus ebeninus* (Tenebrionini) occurs in moist forest from northern New South Wales to south-east Queensland. The only other species in the genus, *Asphalus striatus*, is known only from Gympie, southern Queensland. *Dorrigonum* (Adeliini) is a further probable rainforest genus of three species (one undescribed) extending from southern Queensland to northern New South Wales. *Sloanea* (Tenebrionini) is a monotypic genus (*Sloanea costata*, Plate 188) restricted to *Nothofagus* forest in the Barrington

Figure 3: Tenebrionidae: approximate areas of endemism at generic level (historic refugia) (modified from Matthews and Bouchard 2008). Names indicate individual refugia.

Tops. *Styrus* (Cyphaleini) is a probable dry sclerophyll forest-inhabiting genus, containing three valid species (*Styrus batesi*, *S. elongatus*, *S. latior*), which occurs on the western slopes of the Great Dividing Range. Although *Styrus* is not a rainforest genus it serves to indicate that the different refugia should be thought of as isolated areas (even islands at one time), each with varying habitats (E. G. Matthews pers. comm.; see also Matthews 1992, 1998; Matthews and Doyen 1989).

Tenebrionidae – Alleculinae This diverse group has a world-wide distribution and is considered by Matthews and Bouchard (2008) to be a northern invader, but with many endemic genera (e.g. *Atoichus*). Adults are capable of flight. Genera recorded from southern Queensland and New South Wales include *Aethyssius*, *Atoichus*, *Dimorphochilus*, *Euomma*, *Homotrysis*, *Metistete*, *Nocar*, *Nypsius*, *Ommatophorus* and *Tanychilus*. *Nypsius foveatus* occurs in cool temperate rainforests (New England National Park). *Homotrysis* and metallic-hued *Aethyssius* are especially well represented in the Gondwana rainforests region (Williams 2002). Larval tenebrionids are often found in rotting wood; however, the adult alleculines frequent foliage and flowers (usually in daylight) and have brush-like adaptations of the maxillary galeae that suggest pollen-feeding habits as well as feeding on nectar. Species of *Aethyssius*, *Homotrysis*, *Ommatophorus* and the diminutive *Atoichus* and *Nocar* are to be found on the flowers of numerous canopy trees in subtropical rainforests as well as subcanopy trees in wet sclerophyll forests (e.g. *Ackama paniculata*, *Acradenia euodiiformis*, *Alphitonia excelsa*, *Euroschinus falcatus*, *Guioa semiglauca*, *Schizomeria ovata*, *Scolopia braunii*, *Syzygium floribundum*, *S. smithii*, *Tristaniopsis laurina*); these are usually species with open floral structures, but also plants with more enclosed corollas (*Atoichus* and *Aethyssius* on *Elaeocarpus reticulatus*) (Williams 1995).

Tenebrionidae – Lagriinae Most rainforest Lagriinae of the tribe Adeliini are found during the day sheltering under or in rotten wood, sometimes under loose stones. Adults produce defensive sounds by stridulation, and when handled evert abdominal glands which contain defensive chemicals. Adults are almost exclusively venturesome only at night. Larvae and adults feed on dead plant matter. The Australian Adeliini are primarily concentrated from southern Queensland to Tasmania, and in the Wet Tropics. These two areas are centres of diversity and endemism (Matthews 1998; Matthews and Bouchard 2008). Twenty-four genera are endemic to Australia, although only one genus (*Diemenoma*) is endemic to Tasmania. The number of Australian genera represents more than 50 per cent of the world's generic diversity within the tribe. (Matthews 1998).

The Adeliini have a classic Gondwanan distribution with species known from Australia, New Zealand, New Caledonia and Chile (Matthews and Bouchard 2008). The genus *Licinoma* occurs in Australia and Chile. The adeliine rainforest fauna of north Queensland has affinities with the New Caledonian fauna and the south-eastern Australian adeliines have several close affinities with the Chilean fauna. Adeliini have persisted since Gondwanan times without diversifying past tribal limits, and are closely associated with, and diverse within, forest communities (especially rainforests) that have largely retained their ancient Gondwanan character. In comparison to rainforests few genera occur in open forests (Matthews 1998). The Adeliini are related to the tribe Laenini (which may have evolved from a common ancestor with the Adeliini), which occur in Madagascar, southern Africa, and the Indian subcontinent at least (these areas broke away from Gondwana 90 mya), and together their distribution forms a complete Gondwanan pattern (Matthews 1998). Laeini probably dispersed subsequently to South-east Asia and the southern Palaearctic (Matthews 1998).

At least nine genera of Adeliini occur within the Gondwana rainforests region (*Adelium*, *Blepegenes*, *Cardiothorax*, *Coripera*, *Dorrigonum*, *Leptogastrus*, *Nolicima*, *Nototrintus* and *Seirotrana*) (Williams 2002). *Nolicima* (*Nolicima angusticollis*, *N. sylvicola*) occurs in rainforest (including *Nothofagus* rainforest), and dry and wet sclerophyll forest, and is mainly distributed from south-east Queensland to Victoria, but with disjunct occurrences in northern Queensland and south-west Western Australia. *Nototrintus asperatus* (along with *Cardiothorax hopsoni* and *C. aeneus*) characterises the tenebrionid fauna found under rotting wood in *Nothofagus*-dominated cool temperate rainforest in the Barrington Tops. *Coripera* (*Coripera morleyana*) may be a rainforest or wet sclerophyll forest genus occurring from southern Queensland to Tasmania, with an isolated occurrence in North Queensland. *Seirotrana* mainly inhabits sclerophyll forest and is distributed in an arc from North Queensland through eastern New South Wales, Victoria, Tasmania and South Australia. *Leptogastrus* occupies a diversity of vegetation types (including rainforest) and is distributed along the east coast, excluding Tasmania, and south-west Western Australia. *Cardiothorax* inhabits rainforest and dry and wet sclerophyll forest and is widely distributed in eastern Australia from Victoria to north Queensland, and is diverse in northern New South Wales and southern Queensland. *Cardiothorax hopsoni* (Plate 183) and *C. aeneus* inhabit cool temperate rainforest in the Barrington Tops, and *C. cordicollis* occurs in several rainforest types including remnant inland riparian remants (e.g. Moore Park Nature Reserve) and littoral rainforest (Iluka Nature Reserve). An additional species surviving in littoral rainforest remnants is *Cardiothorax haagi* (Racecouse Headland, north of Port Macquarie). Populations of the distinctively shaped *Cardiothorax egerius* survive in the small rainforest remnant at Rotary Park, Lismore.

Adelium, with the apparent exception of arid regions, occurs widely in Australia, and is also diverse in subtropical eastern Australia, with at least 15 species occurring in southern Queensland and northern New South Wales (Williams 2002), though not all these are rainforest inhabitants. Many are found in various rainforest formations at upland, intermediate and lowland elevations (*Adelium geminatum*, *A.aboricola*, *A. helmsi*), but may also inhabit wet sclerophyll forest (*Adelium helmsi*). *Adelium ellipticum* inhabits cool temperate rainforest (Barrington Tops) and warm temperate rainforests (Nightcap Range). The large and distinctive *Adelium striatum* occurs in coastal littoral rainforest (Iluka) as well as upland rainforest in the Kroombit Tops. One of the smallest species, *Adelium reticulatum*, inhabits littoral rainforest on headlands, floodplain rainforests, and lowland dry rainforest remnants further inland. The endemic *Seirotrana* occurs in cool temperate rainforest (Barrington Tops, Border Ranges) and other upland forests (Bunya Mountains, Kroombit Tops), as well as lowland subtropical rainforest (Mt Hoad, Lansdowne Escarpment). The genus *Blepegenes* (*Blepegenes equestris*, Plate 182) inhabits montane rainforest and is centred on the area from eastern Victoria to south-east Queensland, with two species recorded from Eungella in central eastern Queensland and one species from the Wet Tropics (Matthews 1998). As discussed previously, *Nototrintus* (*Nototrintus acaciensis*, *N. hackeri*) and *Dorrigonum* (*Dorrigonum umblicatum*) are restricted to southern Queensland and northern New South Wales.

In subtropical rainforests and adjacent wet sclerophyll forest the tribe Lagriini is characterised by the genus *Ecnolagria* (e.g. *Ecnolagria grandis*, *E. tomentosa*), the bronze-brown adults being found on foliage and sometimes flowers (e.g. *Alphitonia excelsa*). The genus extends in altitude from lowland rainforest and wet sclerophyll forest (Lansdowne Escarpment) to elevated sites such as the Bunya Mountains, Nightcap Range, Werrikimbe National Park and the Comboyne Plateau. Also recorded is *Metriolagria affinis* (Williams 2002).

Tenebrionidae – Tenebrioninae The Cyphaleini (*sensu* Matthews 1992; Heleini – Cyphaleina of Matthews and Bouchard 2008) are a Gondwanan or Old Southern faunal element and are confined to the Australian region where they probably originated. They possibly evolved from a primitive *Paraphanes*-like ancestor. The tribe is mainly distributed in southern and eastern Australia, with south-east and northern Queensland possessing the highest diversity of genera. Seven genera are known from each region (Matthews 1992). About half of the genera follow the 'eastern distributional pattern' – this is an autochthonous forest fauna which is best developed in the rainforests of northern New South Wales and southern Queensland (i.e. *Atoreuma*, *Bolbophanes*, *Byallius*, *Phanechloros* [= *Chlorophanes*], most *Cyphaleus*, *Hemicyclus*, *Paraphanes*, *Prophanes*, *Styrus*) (Matthews 1992). Matthews (1992) postulated that the present cyphalaeine distribution represents a retreat of the fauna to the south-east owing to increasing aridity and insufficient time to re-invade northern regions through dispersal and adaptation. In addition, these northern regions are now occupied by competitive Modern Elements of northern origin.

Cyphaleini are diverse in the Gondwana rainforests region (Williams 2002). *Paraphanes* (e.g. *Paraphanes nitidus*) occurs in rainforest and is known from three areas: Cairns–Mossman, Eungella–Mackay, and mountains south of Brisbane. *Platyphanes* (*Platyphanes oblongus*, *P. frenchi*) inhabits rainforest and sclerophyll forest and is distributed from western Victoria to the Daintree area of North Queensland, but the concentration of species is in south-eastern New South Wales and southern Queensland. The genus *Olisthaena* principally inhabits sclerophyll forest and ranges in a wide arc from Tasmania through south-east South Australia, Victoria, New South Wales and North Queensland, but has an apparent truncated distribution in southern Queensland. *Phanechloros* (*Phanechloros punctipennis*) occurs in sclerophyll forest from eastern Victoria north into southern Queensland. *Atoreuma* includes several widespread species in northern New South Wales and southern Queensland, including *Atoreuma rufoaeneum* which is recorded from mixed cool temperate–subtropical rainforest in the Border Ranges. *Mithippia* (*Mithippia aurita*) has two centres of distribution based on south-west Western Australia and south-east Australia north to the McPherson Range. It occupies montane habitats and probably eucalypt forest (Matthews 1992). The endemic *Microthorax* includes two species: *M. convexicollis* from southern Queensland–New South Wales, and *M. breweri* from south-west Western Australia (Matthews 1992). *Cyphaleus* includes several species (*Cyphaleus cupreus*, *C. brevispinosus*, *C. rugosus*, and the bizarre *C. childreni*, Plates 184, 185) that are very agile and active on tree trunks in daylight – a behaviour exhibited also by individuals of the related genus *Hemicyclus* (*Hemicyclus splendens*, *H. variegatus*).

Whereas rainforest was the probable ancestral habitat of the Cyphaleini, the tenebrionine tribe Heleini (*sensu* Matthews 1993) is characteristic of open forests and semi-arid zones. Like the Cyphaleini, the Heleini are absent from South America (but a possible sister group occurs there – E. G. Matthews pers. comm.). Only one genus (*Mimopeus*) occurs in New Zealand. Their extensive radiation in Australia suggests an origin shortly after the Early Tertiary separation of Australia (Matthews 1993). Genera occurring in this region are *Pterohelaeus* (widespread in Australia), *Ospidus* (extreme north-east New South Wales and eastern Queensland, northern Northern Territory) and *Emcephalus* (= *Encara*) (Seram, New Guinea, New Britain, Northern Territory, and disjunct occurrences in eastern mainland Australia) (Matthews 1993) (Plate 186).

Several genera and species enter subtropical rainforest, or are characteristic of it. The day-active *Lepispilus rotundicollis* (a habitat generalist) is found on foliage and flowers in cool temperate, dry and subtropical rainforests (e.g. Border Ranges, Mt Hyland, Nightcap Range, Barrington Tops, Kiwarrak State Forest) as well as open forest and mangroves (*Avicennia marina* – Avicenniaceae). However, most genera occur in and under rotten wood.

Sloanea costata is an inhabitant of the ground stratum in cool temperate rainforest (e.g. Barrington Tops), *Meneristes* (e.g. *Meneristes australis*) is commonly encountered in lowland and upland rainforests (Harrington, Wingham Brush Nature Reserve, Lansdowne Escarpment, Barrington Tops, League Scrub, Dorrigo, Border Ranges), *Asphalus* (*Asphalus ebeninus*) inhabits lowland and upland rainforests (Coocumbac Island Nature Reserve, Lansdowne Escarpment, Moore Park Nature Reserve, Nightcap Range), and *Bassianus* (*Bassianus sydneyensis*) is found at similar altitudes and localities. Arboreal *Emcephalus* and *Pterohelaeus*, neither genera characteristic Heleini inhabitants of rainforest, are sometimes attracted to light in lowland rainforest.

Among the remaining tenebrionine tribes, the Toxicini include the endemic genera *Mychestes* (*Mychestes congestus, M. lignarius*) and *Orcopagia* (*Orcopagia monstrosa*), as well as the more widespread *Cryphaeus* (= *Toxicum*) (Plate 189) (Matthews and Bouchard 2008; Grimm 2013). Adult *Mychestes* and *Orcopagia* are cryptically shaped, in appearance resembling dirt nodules, or fragments of wood. Toxicini represent an invasive 'northern element' in Australia. Their larvae occur in the fruiting bodies of higher fungi and in decaying wood (Matthews and Bouchard 2008). Bolitophagini (*Byrsax macleayi, B. saccaratus*) may be a recent Indo-Malayan element, and are specialised feeders on higher fungi. Adults blend with lichen patterns on tree trunks. Titaenini (*Titaena minor, T. tyrrhena, T. columbina*) display an Austral Gondwanan distribution (Matthews and Bouchard 2008). Their larvae may be lichen feeders. The tropical Amarygmini include several taxa (*Amarygmus, Chalcopteroides*) whose larvae are largely found in rotting wood, but whose adults are encountered under loose bark or active at night on tree trunks. Larval *Amarygmus* may feed on lichen, and some *Chalcopteroides* larvae inhabit the soil (Matthews and Bouchard 2008). Ulomini are considered Indo-Malayan invaders, the robust *Achthosus westwoodi* being a common inhabitant of rotting wood in lowland dry and subtropical rainforests.

Tenebrionidae – miscellaneous subfamilies The Diaperinae are small tenebrionids that include *Platydema* (*Platydema heroni*) and *Ceropria* (*Ceropria peregrina, C. valga*), two widely distributed genera, examples of which are to be found by day under loose bark in lowland subtropical, dry and upland warm temperate rainforests; at night adults venture onto tree trunks and the outer surfaces of fallen rotting wood. *Menimus* (*Menimus triclavata, M. bicolor*) is an endemic genus whose larvae and adults inhabit forest ground litter and rotting wood, tree trunks, moss, and plypore fungi (Matthews and Bouchard 2008; Watt 1974).

The Stenochiinae are a pantropical family best represented in rainforests at intermediate to low altitudes. Represented genera to be found in subtropical rainforests include *Strongylium, Apterotheca, Kaszaba, Promethis, Zophophilus* and the distinctive large-sized *Hypaulax*. Larvae develop in rotting wood, and adults shelter by day under loose bark, and in hollows in fallen tree trunks and branches, venturing out from these at night. Adult *Hypaulax ovalis* feed on fungi (Plate 187).

Thanerocleridae Thanerocleridae are a small family that are found in bracket fungi where they prey on tenebrionid and tiny ciid beetles (Lawrence and Ślipiński 2013). Thaneroclerids were originally placed as a subfamily (i.e. Thaneroclerinae) of the Cleridae. The native species in Australia consist of four species of *Isoclerus*, an endemic genus restricted to the east coast of New South Wales and Queensland. *Isoclerus* is included in the subtribe Isoclerina, which is considered to have arisen in Africa, or the eastern part of North America, Greenland or Europe (Kolibáč 1998). *Isoclerus gerstmeieri* is recorded from cool temperate rainforest and dry rainforest (Bunya Mountains, Lamington National Park, Dorrigo). *Isoclerus cipsek* is recorded from the Border Ranges.

Trogidae Trogids are strongly sclerotised beetles (see Plate 190). All are facultative necrophages found in and under cadavers and also dried hides and bones, and are usually the last in a successional of insects attracted to carcasses. Several species (e.g. *Omorgus costatus, O. brucki*) have been collected from bat guano in caves and hollow trees (Scholtz 1986). Adults occasionally feed on animal dung, the regurgitated pellets of birds, and possibly ants and termites, and are attracted to light. They are often hosts to large numbers of phoretic mites. Larvae live in burrows below carcasses.

Trogidae are related to Lucanidae, Passalidae, Scarabaeidae, Geotrupidae and Hybosoridae (Scarabaeoidea) and, like some other species of Scarabaeoidea, adults stridulate by rubbing the abdomen against the elyta. All the native species are placed in the genus *Omorgus*, though previously *Trox* (Scholtz 1986). *Omorgus* occurs widely in Australia and also in the Nearctic and Neotropical Regions (W. Houston 1992). The Australasian fauna is most closely related to that of the New World, and probably can be linked to western Gondwanan ancestral groups

(Scholtz 1986). The New Guinea trogid fauna is Australian in origin, rather than of Oriental derivation. Trogidae are best represented in open forest and woodland, and in arid and western regions of the mainland. Species recorded from rainforest or wet sclerophyll forest include *Omorgus semicostatus*, *O. brucki*, *O. candidus*, *O. parvicollis*, *O.squamosus* and *O. suberosus* (Scholtz 1986; G. Williams unpubl. data); but individuals are never abundant.

Zopheridae The Zopheridae (considered a subfamily of Tenebrionidae by some authors, and now including the Colydiidae [Carter and Zeck 1937]) are known from the Holarctic, South America, South Africa and Australia (Lawrence 1982). They are usually associated with rotting wood and fungi (Lawrence 1994a; Lawrence and Britton 1994). The family includes the large Ironbark Beetle *Zopherosis georgei* (Zopherinae), which is encountered largely in higher elevation rainforest and wet sclerophyll forest of northern New South Wales and southern Queensland. Adults feed on fungal bodies (e.g. *Ganoderma applanatum*, Ganodermataceae/Polyporales), rotten wood, and vegetable material (Hawkeswood 1990d; Williams 1993). *Cicablabus* (Colydiinae) is an endemic genus, which includes *C. micros* from Queensland and New South Wales rainforests (Ślipiński and Lawrence 1997).

The Colydiinae (Lawrence 1980; Lawrence and Ślipiński 2013), cylindrical bark beetles, include most of the Zopheridae genera recorded from Australia, several of which are also found throughout the warmer zones of the Old World. The small genus *Munaria* comprises two species restricted to the East Indies and Australia. *Munaria tmetus* is restricted to south-east Queensland–north-east New South Wales, and Lord Howe Island (type locality) (Lawrence 1980).

Miscellaneous families The Coleoptera fauna treated above is far from complete, and excludes numerous families that, in a functional context at least, are significant participants in the interconnected ecological processes within the rainforest environment. Many are found in abundance primarily on the forest floor (e.g. Leiodidae), but others inhabit tree bark and rotten logs, or are secreted deep withing moss beds, emerging at night to feed upon the ephemeral fruiting bodies of fungi, inhabit clusters of flowers, or frequent leaf surfaces. They can be abundant, but many also appear to live in low numbers as members of widely dispersed populations.

The minute Ptilidae consume decaying organic matter, fungal spores and hyphae, species of Leiodidae have similar habits but are also general scavengers. *Prionocyphon niger* (Scirtidae) has been recorded from water-filled tree holes (Kitching and Allsopp 1987). Scirtidae can also be found at night feeding on the inflorescences of mass-flowering rainforest trees, Byrrhidae feed on mosses or are found among plant roots, and the Ptilodactylidae feed on rotting plant material; adults of the latter conspicuously fly during the day in early spring on the margins of lowland rainforest and wet sclerophyll forest. The Throscidae resemble small-sized Elateridae and like them are able to jump by means of a clicking mechanism; however, the biology of the Australian fauna is unknown. The flea-like Ptinidae feed on dry animal waste and fungi and a number are associated with ants. The ptinid subfamily Anobiinae notoriously attacks wood. Scraptiidae live in leaf litter and in rotten wood and Salpingidae are to be found on foliage, flowers, bark, and twigs. They feed on bark and phloem sap. Trogossitidae sometime come to light; most adults are predaceous, though some feed on fungal fruiting bodies. The small family Hobartiidae is represented by *Hobartius*, and also by *Hydnobioides*, the larvae and adults of which feed on the softer fruiting bodies of Basidiomycetes fungi (Lawrence and Ślipiński 2013). Cryptophagidae feed on fungal spores and moulds, inhabit leaf litter under loose bark but are also commensals in nests. The distinctly flattened and elongate-bodied Cucujidae are found under bark. The larvae of Passandridae are known to be ectoparasites of wood-boring insects and the broadly ovate to elongate Endomychidae feed on fungi. However, their feeding habits are otherwise poorly known. The larvae of Melandryidae inhabit dead wood or the fruiting bodies of fungi. The adults are active at night. The Ulodidae are mostly confined to south-eastern Australia and feed on fungi (Lawrence 1994a). Adults can be found under bark (Lawrence and Ślipiński 2013). Latridiidae are small beetles (<0.5 mm) that feed on fungal spores and moulds, inhabit leaf litter but also are encountered occasionally on the flowers of mass-flowering canopy trees in lowland rainforests (e.g. *Acradenia euodiiformis*, *Alectryon coriaceus*, *Alphitonia excelsa*, *Elaeocarpus obovatus*, *Euroschinus falcatus*, *Rhodomyrtus psidioides*, *Syzygium smithii*); the participating genera are mainly *Cortinicara* and *Aridius* (Williams 1995; Williams and Adam 2010).

MECOPTERA (scorpion flies)

The Mecoptera are a small insect order, active during spring to autumn (Byers 1991; Riek 1954c, 1973a; Smithers

1973, 1974, 1993; Wells 1996b). Adults are mainly scavengers, though those of Bittacidae are predaceous. Mecoptera arose from an array of mecopteroid forms (e.g. Kaltanidae, Permopanorpidae, Permocentropidae, Permotipulidae, Robinjohniidae) that lived during the Permian and Early Mesozoic, and from which ancestral Diptera diverged in the Permian (Grimaldi and Engel 2005; Krzemiński *et al.* 2015; Rasnitsyn and Quicke 2002; Willmann 1978). Mecoptera have been proposed as an early pollinator group (feeding on gymnosperms and related plants) that predated the rise of angiosperms (Ren *et al.* 2009).

Some Mecoptera can be mistaken for Neuroptera (lacewings) of the family Hemerobiidae. In adults the head is conspicuously elongated below the eyes, the mouth projecting downwards. Most adults inhabit moist microhabitats such as well-vegetated stream banks, but larvae are mainly terrestrial (except aquatic Nannochoristidae). The Australian mecopteran fauna is largely endemic, and there is no connection between the Australian, and South-east Asian or Indonesian fauna (Byers 1991).

Bittacidae This is the most commonly observed mecopteran family, adults often encountered in flight-carrying prey (e.g. flies, wasps). Adults are occasionally encountered attracted to light. Bittacid larvae are saprophagous and caterpillar-like in appearance. Adults are predaceous, capturing prey with their specially modified legs. They are encountered at rest hanging by their forelegs on low vegetation, thus giving rise to their popular name, hanging flies. Prey includes soft-bodied insects and spiders. Adult *Harpobittacus* (Plate 191) may feed on nectar, giving some support to an ancestral pollen-feeding habit (Ren *et al.* 2009).

The family is widespread and has a tropical and temperate distribution (Byers 1991). All the Australian species are endemic (Wells 1996b). *Austrobittacus* and *Edriobittacus* occur southwards from central coastal Queensland (Byers 1991). The endemic genus *Harpobittacus* (*Harpobittacus limnaeus*, *H. scheibeli*, *H. tillyardi*) is restricted to eastern, south-eastern and south-western Australia, including Tasmania, and occupies low open sclerophyllous vegetation communities. *Harpobittacus* occurs in relatively undisturbed habitats and their presence is a good indicator of the preservation of original habitat (K. Lambkin 1994). The endemic genus *Tytthobittacus* (*Tytthobittacus macalpinei*) is recorded from central coastal New South Wales, and is similar in some aspects of wing venation to the Chilean genus *Anabittacus* (Smithers 1973).

Choristidae This is an endemic Australian family (Byers 1991; Riek 1973a). Larvae are saprophagous, inhabit leaf litter, and like Bittacidae are caterpillar-like in appearance. The genus *Taeniochorista* (e.g. *Taeniochorista bifurcata*, *T. nigrita*, *T. pallida*, *T. similis*) occurs from south-east Queensland to near Sydney. *Chorista* and *Neochorista* range from Victoria to eastern New South Wales (Byers 1991).

Nannochoristidae This is a small ancient family known from south-east Australia (including Tasmania), the south island of New Zealand, southern Chile and the south-east of Argentina (Byers 1991; in Wells 1996b). The larvae are aquatic and feed on larval Diptera (Pilgrim 1972). The endemic genus *Nannochorista* (*Nannochorista eboraca*) is known from the Dorrigo Plateau.

SIPHONAPTERA (fleas)

Fleas are wingless, blood-feeding ectoparasites of birds, but mostly of mammals. There are now ~2500 named species. Siphonaptera are the only group of insect ectoparasites in which the body is strongly laterally flattened. Only ~5 per cent of the world fauna are associated with marsupials, the greater proportion feeding on rodents (Grimaldi and Engel 2005). As with Phthiraptera the origins of fleas must be inferred from the relationships of recent species, owing to the paucity of the fossil record. These are known from Eocene and Miocene amber, indicating that by the Mid Tertiary modern species groups had evolved, suggesting an age at least within the Cretaceous. In fact there are flea-like fossils (*Tarwinia*) recorded from the Early Cretaceous Koonwara sediments of Victoria (Jell and Duncan 1986), though *Tarwinia* does not possess all the features of true fleas (discussed in Grimaldi and Engel 2005).

The Australian Siphonaptera fauna is highly endemic, with more than 80 per cent of the known species falling into this category. This parallels the level of endemicity exhibited by their mammalian hosts (Dunnet and Mardon 1991). The Australian fauna is distinct, but with affinities to that of Papua New Guinea. The dominant family is the Pygiopsyllidae with more than 40 species and subspecies. Pygiopsyllidae include *Uropsylla* (a parasite of Dasyuridae), *Bradiopsylla* (on echidnas), *Choristopsylla* (on arboreal marsupials), and *Pygiopsylla* and the endemic *Bibikovana* (both parasites mainly of rodents and marsupials). Endoparasitic larval *Uropsylla tasmanica* infestations are known to cause extensive body trauma (e.g. severe facial lesions) to Spotted-tailed Quolls (*Dasyurus maculatus*) in south-east Queensland (Nelson *et al.* 2018).

Stephanocircidae are known only from Australia and South America, and the subfamily Stephanocircine is completely endemic. *Stephanocircus* is recorded from marsupials and rodents. The Macropsyllidae (with two monotypic genera) are confined to Australia. Hystrichopsyllidae, Ischnopsyllidae (bat-fleas) and Pulcidae are totally or largely endemic (Dunnet and Mardon 1991). Ischnopsyllidae include the endemic genus *Coorilla* (*Coorilla longictena* on microchiropterans) (Dunnet and Mardon 1973). *Xenopsylla* (Pulcidae) occurs on native rodents, and *Acedestia* and *Idilla* (Hystrichopsyllidae) are recorded from small marsupials.

DIPTERA (flies, midges, mosquitoes)

In 1999 David Grimaldi, Curator of Invertebrate Zoology at New York's American Museum of Natural History, gave the number of extant described Diptera as at ~120 000 species. This estimate has grown substantially in the ensuing years as new species are continually being discovered and named (see below).

The Australian Diptera (the true or two-winged flies), represent ~5 per cent of the estimated world fly fauna. Nearly every nematoceran (Lower Diptera; see Appendix 4) and orthorrhaphan family (the more primitive Brachycera; see Colless and McAlpine 1991) includes examples of taxa with Southern Hemisphere distributions (Colless and McAlpine 1991). In comparison, the Cyclorrhapha section of Diptera has few Southern Hemisphere examples. Diptera are particularly rich in the Gondwana rainforests region, and more broadly in subtropical eastern Australia (Kitching *et al.* 2004; Williams 2002, 2003a; and numerous papers by David McAlpine and Dan Bickel), with many species known only from single localities or spatially restricted suites of sites, as with certain bat-associated species (McAlpine 1966, 1993a).

Fossil Diptera are known from the mid-Triassic (Krzemiński and Krzemińska 2003), with close relatives occurring in the Late Permian, such that it is likely the origin of true two-winged Diptera dates to the early Triassic (*c.* 245 mya), with the emergence of most major infraorders taking place by the Late Triassic. All basal lineages of the infraorder Brachycera appeared by the Mid to Late Jurassic, this being some 30–40 million years before the appearance of angiosperms (Grimaldi and Engel 2005), the subsequent radiation of Brachyceran lineages partly coinciding with the origin of flowering plants (Shin *et al.* 2018).

Today dipterans are highly diverse (species-rich) and abundant world-wide, with more than 154 000 described species; this represents ~10–12 per cent of all animal species (Lambkin *et al.* 2013). They include midges, gnats, crane flies and mosquitoes (Nematocera), and are a principal component of the arthropod fauna that frequents the foliage of trees, shrubs, vines and herbaceous plants in all strata, particularly the sunlit canopy (Basset 1990, 1991; Kitching *et al.* 1993, 2004; Leathers 1922). There is a rich fauna (e.g. Dolichopodidae) that as adults frequent tree trunks (D. Bickel pers. comm.). Diptera are also one of the three major groups (with Coleoptera and Hymenoptera), most of the species <5 mm in length, that constitute the primary pollinators (in taxonomically and ecologically varied guild assemblages) of mass-flowering subtropical rainforest trees, understorey shrubs, vines and ground herbs (Williams 1995; Williams and Adam 1994, 1995, 2010). Important flower-frequenting (pollinating) brachyceran groups that are capable of relatively long-distance movements between co-flowering plants in subtropical rainforest include long-tongued Syrphidae, Nemestrinidae and Bombyliidae. Additional brachyceran families with significant flower-visiting habits include Stratiomyidae, Tabanidae, Conopidae, Calliphoridae and Tachinidae (Willmer 2011). Many of the anthophilous Brachycera (e.g. Phoridae, Sepsidae, Chloropidae, Drosophilidae, Lauxaniidae) are short-tongued and this generally limits their feeding to shallow, readily accessible flowers. Pollen and nectar of flowers were probably the ancestral diet of Brachycera (Grimaldi and Engel 2005), so the prevalence of so many of this Diptera infraorder on rainforest flowers is perhaps not surprising. For Diptera (and other insects) that attend to flowering plants of the rainforest (largely trees flowering at canopy level), their distribution is dependent on the seasonal availability of floral resources. In lowland subtropical rainforests most trees flower during spring and mid-summer, and some normally flower each year (e.g. *Alphitonia excelsa*, *Schizomeria ovata*, *Syzygium smithii*), but for the great majority of trees the flowering patterns are not predictable (Williams 1995). In addition, the flowering phenology of individual species is not constant between individual stands, but instead is quite heterogeneous with many tree species not flowering year after year; thus food resources are neither temporally nor spatially dependable at either local or landscape scales (Williams and Adam 2010).

Adult Diptera possess mouthparts variously adapted to imbibe fluids, pierce skin or masticate pollen (the hard external exine layer otherwise being a barrier to access). Some larvae, such as those of Agromyzidae, are leaf and stem miners or gall makers (Spencer 1989). Adults are principally detritus feeders, though some are predators, parasites or herbivores (Colless and McAlpine 1991; Marshall 2012; Williams and Bickel 2010). Though many do not feed at all, numerous groups can be observed feeding on nectar, or associated with fungi (Heteromyzidae) and rotting or fallen fruit (Drosophilidae, Tephritidae, Neriidae); others are associated with animal dung and carcasses (Sepsidae, Phoridae, Calliphoridae, Tachinidae, Sarcophagidae, Heteromyzidae, Platystomatidae), or are predators (Asilidae, Dolichopodidae). Some Syrphidae and Empididae eat pollen. Several families have evolved in association with vertebrate hosts (e.g. the blood-feeding Ceratopogonidae, Corethrellidae, Culicidae, Simulidae, Hippoboscidae, Nycteribiidae, Streblidae). Mandibulate mouthparts adapted to penetrate skin and ingest the body fluids of the host are recognised in Blepharoceridae, Ceratopogonidae, Culicidae, Psychodidae, Rhagionidae, Simulidae and Tabanidae but functionally lost in Tanyderidae and Chironomidae (Downes and Colless 1967); Ceratopogonidae and Tabanidae include species that feed on nectar, instead of blood.

Most families are day-active, though the majority of nematocerans (thread-horned Lower Diptera, as opposed to the short-horned Brachycera) (see Colless and McAlpine 1991) are crepuscular or fully nocturnal (e.g. Tipuloidea, Chironomidae, Culicidae, Ceratopogonidae, Corethrellidae, Sciaridae); and, as such, individual species may function at night as cryptic pollinators of understorey plants (e.g. Cecidomyiidae). The larvae of certain Diptera are aquatic or moisture loving (e.g. Blephariceridae, Dixidae, Culicidae, Chaoboridae, Corethrellidae, Sciomyzidae, Neurochaetidae), even occupying tiny ephemeral water bodies at the base of plant leaf sheaths, in tree hollows, rock pools, and even vehicle trail ruts. However, as adults these are free living.

Acroceridae The Acroceridae are a small family, the fast-flying adults distinguished by their contiguous (holotypic) small eyes and inflated thorax and abdomen. They are parasitoids of spiders (Lambkin *et al.* 2014) and possess exceptionally long probosces, but in repose the proboscis is held back below the body and so is hidden from view. The family is poorly known, and, although well represented in South America, the Australian fauna is relatively impoverished (Paramonov 1957). In the Australasian region, the genus *Ogcodes* (= *Oncodes*) (e.g. *Ogcodesarmstrongi*, *O. basilis*, *O. victoriensis*) occurs in mainland Australia, Tasmania, New Zealand and New Guinea (Schlinger and Jefferies 1989). They are rarely encountered in rainforest, usually being found at rest on foliage. However, the relatively large *Mesophysa flavipes* (Plate 192) is sometimes observed feeding on nectar of flowering trees (e.g. *Rhodosphaera rhodanthema*) in lowland subtropical rainforest (Williams and Adam 2010).

Asilidae The earliest known fossil asilids are from the Late Jurassic and Early Cretaceous. The family has now evolved to a world fauna of more than 5000 extant species (Grimaldi and Engel 2005). Adult Asilidae, or robber flies, are characteristically elongate in shape, and are a significant group of medium- to large-sized flies occurring in subtropical rainforests and wet sclerophyll forests (Williams 2002). Larvae live in rotting wood or soil. Adults are commonly observed at rest during the day on sunlit foliage or branches. Some (e.g. *Orthogonis ornatipennis*, Plate 196) conspicuously bask on tree trunks and stones for long periods of time.

Asilidae include the largest flies that frequent lowland rainforest and wet sclerophyll forests, such as the metallic blue-green, and sometimes territorial, *Blepharotes splendidissimus* (Plate 193), and smaller ones (e.g. *Austrosaropogon*, *Apothechyla*, *Cabasa*) that are occasionally found on flowering rainforest trees and shrubs (e.g. *Alectryon coriaceus*); *Cabasa ?pulchella* (Plate 194) atypically feeds on nectar. Several (e.g. *Neodioctria*) mimic ichneumonid wasps. Black and yellow winged *Orthogonis ornatipennis* resemble pompilid wasps (e.g. *Calopompilus*), and species of *Chrysopogon* (e.g. *Chrysopogon splendidissimus*, *C. harpaleus*, Plates 195, 197) possess aposematic warning coloration suggestive of Eumeninae (Vespidae). Asilids are predatory (exclusively as adults and almost so as larvae), being the only family of the Lower Brachycera that are so (Fisher 2009). During the day adults aggressively patrol the canopy, sunlit gaps in the understorey, and forest margins; their prey includes wasps, bees, beetles, dragonflies, and other flies. Although many species are fast fliers, individual *Heteropelma scaposum* characteristically undertake unhurried searches low down along the edges of vegetation. Asilids usually catch their prey in flight, grasping it with strong legs and claws, piercing it with the hard spear-like proboscis, and injecting it with enzymes; the body juices of the prey are then sucked out.

About 80 per cent of the Australian genera are endemic (Colless and McAlpine 1991; Daniels 1987, 1989d; Hardy

Flies on flowers: pollination by Diptera (myophily and sapromyophily)

Flies are not normally thought of as pollinators. Rather they are popularly deemed creatures that at best are bothersome, or have habits that bring health risks, actual or perceived. Yet of all the myriad insects it is in the Diptera that the earliest fossil evidence for an anthophilous, or flower loving, adaptation is found (Mazzarolla and Amorin 2000; Ren 1998a, 1998b). Evidence of an anthophilous lineage within Diptera extends to the Middle Jurassic (Grimaldi 1999), this being for a bombyllid ancestor, but it is not until the later Eocene and Oligocene that bombyliid fossils show features (elongate probosces) strongly suggestive of pollinator behaviour (Friis *et al.* 2011). Proto-anthophilous acrocerids and empidids also appear in the Late Jurassic. However, it is in the Eocene–Ologocene for the Acroceridae (Plate 192) (Schlinger and Jefferies 1989), and the Cenozoic for the Empididae, that forms with long probosces are known. For other anthophilous groups the Mesozoic fossil record is sparse, but among those of the Nemestrinidae there is one fossil (*Protonemestrius*) that does seem to have a long proboscis, with Grimaldi (1999) consequently suggesting that Nemestrinidae may be the earliest candidates for obligate pollinators of angiosperms.

In subtropical rainforests Diptera is one of the three major arthropod orders (along with Coleoptera and Hymenoptera) that are responsible for the pollination of most flowering plants, be these trees of the canopy, or shrubs, vines and herbs in the understorey (Williams 1995; Williams and Adam 1994, 2010). Most that participate are <5 mm in length. They make up for their diminutive size by their sheer diversity of species and their cumulative abundance. The fly families that are found on the flowers of subtropical rainforest angiosperms include not only larger, and thus more conspicuous, Syrphidae, Nemestrinidae, Tabanidae and Bombyliidae, but also numerous small day-active Bibionidae, Calliphoridae, Chloropidae, Clusiidae, Conopidae, Drosophilidae, Lauxaniidae, Phoridae, Sepsidae, Sciaridae, Stratiomyidae, Tachinidae, and Tephritidae (these predominantly Brachycera), as well as the mostly crepuscular or night-active visitors such as Chironomidae, Ceratopogonidae, Mycetophilidae, Cecidomyiidae, Culicidae, Psychodidae and Tipuloidea (all Nematocera); nematoceran Anisopodidae and Empididae are occasional diurnal flower visitors. The list is not exhaustive. Few possess specialised mouthparts (long-tongued bombyliids and tabanads being particular exceptions), other than the basic Dipteran mouthpart ground plan for imbibing fluids, which indicates a morphological adaptation to feeding on nectar, or structures (e.g. profuse and dense setae) that suggest a potential efficiency in transporting pollen from one flower to another. Small Diptera, vulnerable to desiccation, are better represented in mesic habitats or more moist microhabitats than in sclerophyllous communities and so Nematocera, in terms of diversity, are probably better represented among the anthophilous fauna of rainforest understoreys. The true pollination contribution made by nematocerans to mass-flowering rainforest trees is probably understated as their presence is generally low during daylight hours when many field observations are undertaken (Williams and Adam 2010).

The floral syndromes that refer to plant adaptations for fly pollination are termed 'myophily' and 'sapromyophily'. Flowers so adapted are characterised by unspecialised (actinomorphic) flowers with no depth effect, in which colours are generally light or dull and nectar guides are frequently present (Bertin 1989; Faegri and van der Pijl 1979; Willmer 2011). The greatest diversity of anthophilous Diptera occurs on accessible unspecialised blossoms. However, the syndrome of myophily is also well developed in more morphologically complex flowers such as those of the Asclepiadaceae (Ollerton and Liede 1997), which possess varying degrees of depth effect in their blossoms, and in the zygomorphic Orchidaceae (Pederson 1995). Myophily may have arisen independently several times in angiosperms, not only because of the morphologically divergent flowers that attract flies, but also because of the diverse phylogenetic relationships of the participating plant taxa. In unspecialised myophilous flowers, odours are generally absent or imperceptible and nectar is easily obtainable. Sexual organs are well exposed, though some fly-pollinated blossoms can be short and tubular (Armstrong 1979) (e.g. *Trochocarpa laurina* [Ericaceae/Epacridaceae], though this is also pollinated by small nymphalid butterflies and bees) with relatively inaccessible nectaries and concealed stamens. There is a correlation between short proboscis length and unspecialised shallow blossoms, and increasingly longer proboscis length and specialised tubular blossoms (Armstrong 1979; Williams and Adam 2010).

In addition to the specialised nectar- and pollen-feeding Diptera, there is a second, sapromyophilous, group of flies that are attracted to flowers by foetid odours, which stimulate feeding or oviposition instincts. Such flies, typically Sepsidae, Calliphoridae, Sarcophagidae, Tachinidae and sometimes Drosophilidae, are ecologically poorly adapted (allotropic) species that derive part of a mixed diet from blossoms (e.g. Ollerton and Liede 1997).

In Australian subtropical rainforest, flies of these families are common on the foetid-sweet flowers of the tree *Alphitonia excelsa* (Williams and Adam 2001).

A separate sapromyophilous flower syndrome has been distinguished from that of general myophily. Sapromyophilous flowers are often greenish, purple or brown with no nectar guides, but frequently with dark spots and motile hairs, capable of movement, or appendages. There is generally no nectar attractant, but sapromyophilous flowers produce strong foetid odours and the sexual organs are generally concealed within the blossom (discussion expanded in Williams and Adam 2010).

1920, 1926, 1929a, 1929b, 1930a; Paramonov 1958a) and there is a large number of undescribed species (Daniels 1989d). The subfamily Dasypogoninae is common and widely distributed in Australia, and includes several genera that are dominant in the south-eastern region of the mainland. Elsewhere in Australasia this subfamily is depauperate (Daniels 1989d). *Brachyrhopala* (Dasypoginini) is distributed in Australia and New Guinea (Clements 2000) and is well represented in subtropical eastern Australia (e.g. *Brachyrhopala atra*, *B. bicolor*, *B. meleana*, *B. ochracea*, *B. quadricincta*), occurring from lowland forest (Lansdowne Escarpment) to upland sites (Mt Warning, Mt Tamborine).

The genus *Colepia* (Asilini) occurs in Australia and New Guinea, and includes numerous species occurring in regional rainforest and wet sclerophyll forest (e.g. *Colepia comatacauda*, *C. horrida*, *C. cultripes*, *C. ingloria*, *C. lanata*, *C. malleola*) (Daniels 1987). The Gondwanan rainforest Asilini also include the endemic genus *Dolopus* and the regionally diverse genus *Zosteria*, which occurs in Australia, New Guinea and New Zealand. Several *Zosteria* species are restricted to the Gondwana rainforests region: *Z. caesariata* (confined to Barrington Tops), *Z. hispida* (Barrington Tops and New England areas), and *Z. nigrifemorata* (known only from Mt Tamborine) (Daniels 1987). *Laphria* (Lapriini) occurs throughout subtropical south-eastern Australia and, within Oceana and Australasia, extends from Malaysia to New Guinea and Lord Howe Island but is apparently absent from New Caledonia and New Zealand (see Daniels 1989d).

With the exception of *Chrysopogon papuensis* from Papua New Guinea, all members of the tribe Chrysopogonini are restricted to Australia. Queensland and Western Australia are particularly rich in species. Three species are known only from the Gondwana rainforests region: *Chrysopogon bellus* (restricted to south-east Queensland), *C. catachrysus* (south-east Queensland to northern New South Wales) and *C. megalius* (Dorrigo-Deervale areas of New South Wales) (Clements 1985). *Laphria* (e.g. *Laphria commata*, *L. rififemorata* [Laphriini]) is widely distributed, occurring in South-east Asia, Indonesia, Australia and Lord Howe Island (Daniels 1989d; Paramonov 1958a).

Athericidae Athericidae are closely related to Tabanidae and Rhagionidae (Stuckenberg 1973). Though widely distributed (Stuckenberg 2000), this small family includes several genera restricted to the Southern Hemisphere. *Dasyomma* (e.g. *Dasyomma infernalis*) exhibits a Gondwanan distribution with species occurring in Australia, Chile and Argentina (Nagatomi and Evenhuis 1989a). Larvae are aquatic and predaceous, feeding on small invertebrates (e.g. chironomid larvae), and can be found in rocky riffle areas of moving streams. Adults often frequent vegetation near water. Some may feed on blood.

Axiniidae The Axiniidae, or axe-head flies, are restricted to Australia and New Guinea and are closely related to the Tachinidae (Colless 1994). The family has more recently been placed by Pape and Arnaud (2001) in the Rhinophoridae, but their distinction is retained here. Adults are mainly known from intercept trap captures (e.g. in wet sclerophyll forest). The family is unique in that adult males possess a distinctive axe-head-like, or fissicorn, third antennal segment (Colless 1994). Four species in the genus *Axinia* are known from the Gondwana rainforests region (*Axinia bicolor*, *A. cantrelli*, *A. lucaris*, *A. mutabilis*). *Axinia cantrelli* is mainly known from central coastal Queensland to the McPherson Range (e.g. Rockhampton, Eurimbula, Bunya Mountains, Mt Tamborine), and with a single New South Wales record (Lansdowne Escarpment). The distribution of many Australian taxa coincides with zones distinguished by Burbidge (1960) (e.g. the Macpherson–Macleay Overlap). Colless (1994) postulates that axiniid history begins in the mid-Tertiary with the arrival of an ancestral muscoid fly, derived from the northern supercontinent Laurasia, and with increased aridity the subgenera *Dixicera* and *Axinia* differentiated respectively in the west and the east of Australia.

Bibionidae The nematoceran family Bibionidae comprises flies of small to moderate size, and is well represented in Australia but almost absent from the Oriental region. Though rare in Mesozoic deposits, bibionids are well known from the Tertiary. The family includes diurnal and nocturnal species. Larvae occur in decomposing vegetation and in the soil. In subtropical rainforest adults frequent sunlit canopy and margin foliage and flowers. They are not normally associated with shaded understoreys. The Australian fauna is placed within two subfamilies, Pleciinae and Bibioninae (e.g. *Dilophus*). The world pleciine fauna is particularly abundant in tropical regions and species of Bibioninae are more numerous in temperate areas. Several species of *Dilophus* (*Dilophus atripennis*, *D. partitus*, *D. parvus*), with very localised and disjunct distributions, occur in the Gondwana rainforests region (e.g. Canungra, Dorrigo, Upper Allyn River) (Williams 2002). However, the genus is widely distributed elsewhere, with extralimital distributions in the Afrotropical, Nearctic, Neotropical, Oriental and Palaearctic regions, as well as New Guinea and New Zealand (Bugledich 1999). In Australian lowland subtropical rainforest, large numbers of diurnally active, but sluggishly flying, *Bibio imitator* (Bibionidae) flies can sometimes be found on the flowers of dioecious *Claoxylon australe* (Euphorbiaceae) (Williams 1995), where they feed on floral nectar and possibly pollen. However, their role as pollinators is arguable owing to their preference for visiting male flowers of this species only. *Bibio imitator* is also found on the flowers of *Scolopia braunii* (Flacourtiaceae) and *Syzygium floribundum* (Myrtaceae).

Blephariceridae These are soft-bodied nematoceran flies commonly called net-winged midges. In Australia blephariceridss are mainly limited to the east coast, in particular the south-east and Tasmania. Blepharicerid larvae are aquatic filter feeders, and have suckers on their abdominal tergites used to adhere to the rocks in the streams in which they live; the streams also include those that dissect rainforest. Adult females feed on other insects such as mayflies, grasping their prey with large claws and then feeding on body fluids. The males do not have functional mandibles.

The Australian fauna is represented by the subfamilies Edwardsininae and Blepharicerinae (Zwick 1998). Edwardsininae are a relictual Gondwanan group largely restricted to southern temperate zones but with an outlying northern species of *Edwardsina* (*Edwardsina confusa*) found in cool streams on the Barrington Tops (Zwick 1981, 1998). The closest relatives of *Edwardsina* occur in South America (Zwick 1989). The Blepharicerinae (e.g. *Parapistomyia*) are represented in Australia solely by the tribe Apistomyiini, which is absent from Tasmania, and these reach their greatest diversity in subtropical and tropical areas of the east coast. *Parapistomyia bulbifera* occurs in upland forests.The Neotropical-Afrotropical tribe Paltostomatini is the sister group of the Apistomyiini (Zwick 1989). With the exception of the genus *Apistomyia* all apistomyiine taxa are restricted to Australasia and South-east Asia (Zwick 1998).

Bombyliidae Bombyliidae are known from Eocene, Oligocene and Miocene compression deposits, and Oligocene amber (Evenhuis 2002; Hull 1973; Wedmann and Yeates 2008), with fossil records suggesting a major radiation in the Late Cretaceous–Early Cenozoic. Commonly called bee flies because of their flight behaviour, long probosces and hairy appearance, Bombyliidae are stout-bodied flies that include small delicate species (*Zaclava*, *Geron*) and some of the largest flies to be found in Australian forests (e.g. species of *Comptosia*) (Colless and McAlpine 1991; Evenhuis 1979, 1989a; Hardy 1921a; F. Roberts 1928a, 1928b, 1929; Yeates 1994). In open forests large *Comptosia* species can be seen feeding on the floral nectar of *Leptospermum* and *Kunzea* bushes (Myrtaceae), and the high-pitched sounds of *Sisyromyia* and *Meomyia* (Bombyliinae), when in flight, make known the otherwise unseen presence of bee-like flies somewhere at ground level. Many have distinct hovering flight habits, or undertake very short-distance flight patterns before again landing close by on soil surfaces. Adult bee flies have long probosces, which allow them to feed from floral nectaries deep within tubular flower corollas. Females are obligate pollen feeders as a necessary requirement for the nourishment of developing ova (Gilbert and Jervis 1998; see also Neff *et al.* 2003). World-wide, bee flies are known to parasitise Hymenoptera (Eumeninae, Apidae), Lepidoptera (e.g. Noctuidae, Pyralidae, Psychidae, Tortricidae), other flies (e.g. Therevidae), Orthoptera, and certain beetles (e.g. Scarabaeoidea, Tenebrionidae, Cicindelinae, Cerambycidae) (Hull 1973; Irwin and Yeates 1995), with the ancestral state one of a generalist predator that evolved towards specialisation on particular soil-inhabiting hosts, then to parasitism (Greathead *et al.* 2009). But in general, life histories of the Australian fauna are poorly known (Colless and McAlpine 1991).

The subfamily Bombyliinae is the largest group in Australia, and is a northern element. Almost all

Australian Bombyliinae genera (e.g. *Sisyromyia*) are endemic (Colless and McAlpine 1991). The exception is *Thevenetimyia* (Xuankan *et al.* 2018), of which a number (e.g. *Thevenetimyia tenta*, *T. longipalpis*) are recorded from the region. The endemic genus *Aleucosia* (Lomatiinae) is found in the southern half of Australia, but the majority of species occur in the south-west of the continent. Two species reach Tasmania (Yeates 1991a). *Comptosia* (Lomatiinae) and its relatives are confined to Australia and South America (Wedmann and Yeates 2008; Yeates 1991b). *Geron* (Gerontinae) is widespread, and *Zaclava* (Systropodinae) in the south-west Pacific is also known from New Caledonia and the Loyalty Islands (Evenhuis 1979, 1989a). The Anthracinae include endemic (e.g. *Thraxan*) and more widespread genera (e.g. *Anthrax*). *Anthrax* occurs in the Nearctic, Neotropics, Palaearctic, Ethiopian, Oriental and Australian regions (Hull 1973) and is distributed widely throughout Indonesia and New Guinea, reaching Vanuatu and Tasmania. However, it is apparently absent from New Caledonia and New Zealand (Evenhuis 1989a). *Ligyra* is considered a northern genus and has been placed in the Exoprosopinae (Colless and McAlpine 1991). *Ligyra* is also widespread and has a similar Australasian and Oceanic range to that of *Anthrax*, but is not recorded from Tasmania (Evenhuis 1989a).

Bombyliidae are well represented in upland forests of south-eastern Australia (e.g. *Anthrax maculatus*, *Ligyra cingulata*, *L. inquinita*, *Sisyromyia albavittata*, *S. aurata*, *Neosardus lepidus*, *Aleucosia costalis*, *A. directa*, *Comptosia praeargentata*, *C. apicalis*, *C. moretonii*, *C. neoapicalis*, *C. tutela*) (Bugledich 1999; Evenhuis 1989a; Paramonov 1967; F. Roberts 1928a; Williams 2002; Yeates 1991a, 1991b, 1996; Yeates and Lambkin 1998). The bombyliid fauna in lowland rainforests and wet sclerophyll forest is also diverse (e.g. *Pseudopenthes fenestrata*, *Villa fuscicostata*, *Comptosia brunnea*, *C. walkeri*, *Aleucosia directa*, *Anthrax dolabratus*, *A. maculatus*, *Sisyromyia aurata*, *Thraxon* sp., *Ligyra bombyliformis*, *Geron* spp., *Zaclava* sp., *Thevenetimyia tenta* [Plate 200]) (G. Williams unpubl. data; Yeates 1991b). Most, if not all, are habitat generalists (showing no fidelity to particular flowering plants, or mesic forest), being also found in open forest and woodland (e.g. *Ligyra bombyliformis*, *Comptosia brunnea*, *Geron* spp.), or are only found associated with rainforest and wet sclerophyll forest when there are large sunny openings within otherwise shaded forest understoreys (*Anthrax* spp., *Comptosia heliophila*, *Sisyromyia aurata*, *Villa* spp.). Diminutive species of *Geron* (Evenhuis 1979; F. Roberts 1929) frequent flowering herbs (*Pollia crispata*, *Pratia* sp., *Lobelia trigonocaulis*) in sunlit understorey gaps and forest margins (G. Williams pers. obs), but also attend to the flowers of trees and larger shrubs such as *Trochocarpa laurina* (Ericaceae/Epacridaceae) (Plate 199). The larvae of *Geron* are known to be parasitic on Lepidoptera (i.e. Tortricidae, Pyralidae, Psychidae, and possibly Noctuidae [Evenhuis 1979]); however, adult *Geron* in turn are prey to *Bembix* wasps (Crabronidae) (Evans and Matthews 1973).

Though Bombyliidae are not abundant at rainforest canopies, several species do feed on the nectar of flowering trees there, for example *Ligyra bombyliformis*, *Pseudopenthes fenestrata* (on *Alphitonia excelsa*). However, *Geron* is the most commonly encountered genus (e.g. on *Rhodomyrtus psidioides*, *Syzygium floribundum*, *S. smithii*, *Tristaniopsis laurina*, *Alphitonia excelsa*, *Cuttsia viburnea*, *Diospyros australis*, *Euroschinus falcatus*, *Guioa semiglauca*, *Trochocarpa laurina*), these food plants mostly having open, shallow flowers, but also flowers with deeper corollas (as in *Diospyros australis*, *Trochocarpa laurina*) (Williams 1995; G. Williams pers. obs.).

Calliphoridae Calliphorids are active during the day and are often associated with products of decay, notably carrion, with all species referred to as blowflies (Colless and McAlpine 1991; Crosskey 1965; Hardy 1932b, 1940). However, as with all necrophagous insects attracted to decomposing animals, calliphorid species may recruit differentially to various kinds and sizes of cadavers, and utilise the carrion resource with various levels of success (e.g. for breeding) depending on competition and disturbance pressure from other carrion-feeding and carrion-breeding animals (Kneidel 1984).

Several subfamilies are recognised. Calliphorinae (*Calliphora*) and Chysomyinae (*Chrysomya*) breed in carrion. The Rhiniinae (now considered a separate family, Rhiniidae) have a distinctive prolonged head, and some genera (*Stomorhina*) are known to develop and live in ant and termite nests (Colless and McAlpine 1991; Dear 1977). Several species are common on flowering rainforest trees (including mangrove species), shrubs and vines. These include *Amenia chrysame* (on *Guioa semiglauca*, *Syzygium floribundum*); *Stomorhina xanthogaster* (on *Alectryon coriaceus*, *Alphitonia excelsa*, *Avicennia marina*) (Plate 198); *Stomorhina* spp. (on *Drypetes deplanchei*, *Syzygium floribundum*, *Alphitonia excelsa*, *Clematis glycinoides*); *Chlororhinia exempta* (on *Acradenia euodiiformis*); *Chrysomya megacephala* (on *Alphitonia excelsa*); and *Calliphora hillii* (on *Guioa semiglauca*) (Williams 1995; G. Williams unpubl. data; Williams and Adam 2001, 2010). These are considered sapromyophilous flies, being attracted to flowers that have a foetid odour, and while this is the case with *Alphitonia excelsa* (Williams and Adam 2001),the other species cited either have a pleasant fragrance or no perceptable fragrance at all.

The metallic blue-green-coloured flies of the subfamily Ameniinae are parasites of land snails and comprise six genera in Australia. It is probable that the ancestral Ameniinae migrated to Australia from islands to the north (Colless 1998). Australian ameniines (e.g. *Paramenia*, *Silbomyella*, *Amenia*) 'may very well have originated in a vicariance event that yielded *Catapicephala* in the Moluccas and *Paramenia* in New Guinea' (Colless 1998). *Paramenia* is confined to Australia and New Guinea and closely resembles *Catapicephala* from the Oriental region, this being a genus that does not occur in Australia. Of the species recorded from subtropical rainforests, *Paramenia angustifrons* is distributed more widely, from North Queensland to northern New South Wales, and *Paramenia semiauriceps* is similarly distributed but in addition reaches Victoria. *Silbomyella* occurs in Australia (two species), and from the Moluccas to New Britain. *Silbomyella crosskeyi* is recorded from Lamington National Park and the vicinity of Mt Warning (Colless 1998). It is unknown if *Silbomyella* is an autochthonous genus which moved north, or whether it arrived in Australia from the north (Colless 1998). *Amenia* is endemic to Australia, including Tasmania and the Torres Strait Islands, and is widespread in the southern half of the continent south of southern Queensland and west to Western Australia (Colless 1998; Kurahashi 1989). *Amenia albomaculata*, *A. chrysame* and *A. imperialis* are recorded from upland forests (Mt Tamborine, New England National Park, Barrington Tops), *Amenia chrysame* also commonly from lowland rainforest and wet sclerophyll forest.

Cecidomyiidae Known as gall midges, cecidomyiids are minute and delicately winged nematoceran flies that in tropical forests are known to swarm in great numbers, particularly at dusk and dawn, high in the rainforest canopy (Stork 1991). The family is closely related to Mycetophilidae, Sciaridae and Bibionidae, and there are currently ~5000 (mostly Palearctic, Nearctic) described species (Bugledich 1999; Gagné 1989, 1994; Skuse 1888–1890), but the family is immense and hardly known (D. Bickel pers. comm.). Cecidomyiids are common and diverse in Cretaceous and Tertiary amber. In subtropical rainforests Cecidomyiidae (e.g. *Cecidomyia albulipennis*, *C. helmsi*) are sometimes encountered in the understorey gathered in numbers near spiderwebs low down on tree trunks, but do not seem to be snared by them. Larvae are scavengers in decomposing plant material and in or on fungi in humus or decaying wood, or live in galls which they can induce (Kolesik and Butterill 2015); for example, *Psephodiplosis rubi* induces galls in the subtropical rainforest vines *Rubus moorei* and *R. nebulosus* (Rosaceae) (Kolesik *et al.* 2018), and *Ashondylia daphnandrae* induces galls in *Daphnandra johnsonii* (Atherospermataceae), an endangered small tree endemic to the Illawarra region (Kolesik *et al.* 2019).

Cecidomyiid larvae induce gall formation by applying salivary secretions to leaf tissues, and these then cause the tissues to proliferate and grow around the larva. Many are known to feed on living plants, some are inquilines, and some are predators (on other small insects and mites). Adults will feed on floral nectar; they have been identified as pollinators of tropical angiosperms (Feil 1992; Sakai *et al.* 2000) and there is some evidence that they are pollinators of *Piper hederaceum* var. *hederaceum* (= *Piper novae-hollandiae*) (Piperaceae) (Ollerton 1996), a vine common in subtropical rainforests north from Mt Dromedary in southern New South Wales (Harden *et al.* 2007). They may also be implicated in the pollination of the understorey tree *Daphnandra micrantha* (Atherospermaceae) (Williams and Adam 2010; Williams *et al.* 2001), a species restricted to the Manning and Hastings Valleys of New South Wales (Harden *et al.* 2006). The flowers of *Daphnandra micrantha* are enigmatic, for they possess stigmatic papillae (but no obvious nectar-producing glands) and produce a clear gelatinous exudate on anthesis, the function of which is unknown (Williams 1995). However, in Australia generally, the role of cecidomyiids as pollinators remains poorly studied.

Ceratopogonidae Many adult ceratopogonids (biting midges or sand flies) are minute, usually predatory, ectoparasitic feeders (e.g. *Leptoconops*, *Culicoides* spp., some *Forcipomyia*) on the blood of vertebrates. Worldwide there are now ~5500 species (Borkent and Wirth 1997; Debenham 1989). Fossil Ceratopogonidae probably originated in the Late Jurassic and are well represented in Cretaceous and Tertiary amber, indicating they have very likely plagued land vertebrates for millions of years. Predatory midges are normally active at dusk or late in the afternoon, and can be prolific in littoral rainforest (presumably entering from adjacent tidal areas) and lowland wet sclerophyll forest.

Species of *Forcipomyia* are blood feeders, but some are attracted to flowers where they feed on nectar. The Australian subtropical rainforest vine *Pararistolochia praevenosa* (Aristolochiaceae) (a larval food plant of the Richmond Birdwing Butterfly, *Ornithoptera richmondia*) is pollinated possibly by a species of *Forcipomyia* (Sands *et al.* 1997 citing G. Monteith pers. comm.), though the pollinators remain uncertain (Sands and New 2013) and may include (or be) phorid flies (Grimshaw *et al.* 2013). Ceratopogonids have short mouthparts, and so to undertake pollen transfer adult flies are first obliged to fully enter the deep *Pararistolochia* flowers. In Pakistan the trap flowers of the related *Aristolochia bracteolata* are also pollinated by species of *Forcipomyia* (Razzak *et al.* 1992). The distribution of the Australian *Forcipomyia* fauna conforms in part to species with Bassian or southern affinities (e.g. *Forcipomyia gandangara*), species with mainly tropical distributions (e.g. *Forcipomyia marksae*), in addition to a Younger Northern Element (Debenham 1987a, 1987b). *Forcipomyia* (*Thyridomyia*) *litoraurea* is recorded from Lamington National Park but also occurs more widely in Australia. It is also recorded from Africa, Estonia and probably Japan and the Caroline Islands, and its occurrence in Australia is likely an accidental introduction via shipping (Debenham 1987a).

Brachypogon has a world-wide distribution. Two subgenera, *Brachypogon* (*Brachypogon* [*B.*] *bryanae*) and *Isohelea* (*Brachypogon* [*I.*] *hadrosaurus*), occur in the Gondwana rainforests region. The subgenus *Brachypogon* is known mainly from Africa, but also occurs in the Americas, Europe, New Guinea and the Pacific. *Isohelea* occurs mainly in the Palaearctic region (Debenham 1991). *Monohelea* (*Monohelea umbrosipennis*) has a cosmopolitan distribution (Bugledich 1999; Debenham 1972). *Macrurohelea dycei* is known only from Bruxner Park, near Coffs Harbour, and the Bellenden Ker Range in North Queensland (Bugledich 1999).

Chaoboridae The ghost midges are mosquito-like nematoceran flies that are closely related to true mosquitoes (Culicidae), but they do not possess a long biting proboscis (Colless and McAlpine 1991). Genera occurring in subtropical south-eastern Australia include *Australomochlonyx*, *Promochlonyx* and *Chaoborus* (Borkent 1989b). Chaoborid larvae are aquatic, some being predators of other aquatic organisms, or filter feeders (*Australomochlonyx*) that strain microorganisms using large mandibular fans. Although the fauna is mostly endemic, the Australian Chaoboridae appear to consist of relict species, and outliers of species groups that are mainly distributed in Asia and Africa (Colless 1986).

Chironomidae This is a large family of small and fragile midges (Freeman 1961; Glover 1973; Cranston and Martin 1989) that are common near water bodies, and primarily active at night. Species diversity and abundance changes with respect to altitude (Lambkin *et al.* 2011; McKie *et al.* 2005). Virtually all adult chironomids have only vestigal mouthparts or none at all and thus do not bite, the exceptions being the primitive genera *Archaeochlus* and *Austrochlus* which have biting mandibles (Cranston *et al.* 2002). Adults are attracted to lights and the larvae are largely aquatic, these living either buried in debris or freely living on vegetation. Some species feed on nectar. Flowering host plants recorded in subtropical rainforests include *Acradenia euodiiformis*, *Cuttsia viburnea*, *Guioa semiglauca*, *Syzygium floribundum* and *S. smithii* (Williams 1995; Williams and Adam 2010).

Species recorded from the Gondwana rainforests region include *Macropelopia* sp. (Lamington National Park), *Rheotanytarsus flabellatus* (Chaelundi State Forest) and *Tanytarsus annulatus* and *T. belairensis* (Bruxner Park) (Bugledich 1999; Cranston 1997; Cranston and Martin 1989; Glover 1973; Kitching and Callaghan 1982; Malloch 1927). The endemic *Echinocladius martini* is distributed along the east coast and predominantly inhabits mesic forest and upland streams. Krosch (2011) suggests that dispersal and gene flow among *Echinocladius martini* populations, isolated in closed forest refugia, have been highly restricted historically, with populations exhibiting divergent lineages and the taxon being vulnerable to habitat fragmentation.

Chloropidae Chloropids are small flies, generally no larger than 5 mm in size (Sabrosky 1989). Their larval food habits vary, but include saprophagous and phytophagous species, as well as gall formers and parasites of the egg masses of orthopterans, mantids, sawflies and spiders. In turn chloropids are parasitised by minute wasps of the family Megaspilidae. Spencer (1986) considered that the Australian chloropodine fauna is exclusively Oriental in origin with most ancestral and modern species reaching Australia via New Guinea, probably during land connections in the Pleistocene and earlier in the Tertiary.

In rainforest Chloropidae frequent the upper canopy, where they are abundant, but are also encountered in lower strata (e.g. Kitching *et al.* 2004; Lambkin *et al.* 2011), often in sunlit gaps. They are active not only during the warmer months of spring and summer but remain active during winter (Lambkin *et al.* 2011). Basset and Springate (1992) noted that on the subtropical rainforest tree *Argyrodendron actinophyllum* chloropids were active mainly at night. Chloropidae (*Cadrema* sp., *Conioscinella* spp.) are pollinators of the ground orchid *Genoplesium littorale* (Bower *et al.* 2015), a critically endangered species from coastal northern New South Wales. In combination with Sciaridae, chloropids are considered pollinators of the vine *Marsdenia cymulosa* (Asclepiadaceae) in tropical Queensland (P. Forster 1992). In subtropical rainforest chloropids are recorded from flowering *Syzygium smithii*, *Alectryon coriaceus*, *Alphitonia excelsa*, *Guioa semiglauca* (*Apotropina*), *Alphitonia excelsa* (*Chlorops*), *Diospyros australis* (*Gaurax*) and *Scolopia braunii* (*Deltastoma unipunctatum*) (Williams 1995). However, the biology of the family, more generally, remains poorly known (Ismay 1993).

Genera occurring in south-eastern Australia include *Apotropina* (Siphonellopsinae), *Batrachomyia*, *Cadrema*, *Gaurax*, *Tricimba* (Oscinellinae), *Chloromerus*, *Chloropella*, *Chlorops*, *Diplotoxa*, *Eutropha*, *Melanum*, *Parectocephala* and *Pemphigonotus* (Chloropinae). (Ismay 1993; Malloch 1927; Sabrosky 1989; Spencer 1986; Williams 2002). Regional records for Chloropinae include *Chloropella bipartita* (Upper Allyn River), *Chlorops federatus* (Gibraltar Range National Park, Bunya Mountains), *C. mackerrasi* (Clyde Mountain), *C. nubilipalpis* (Mt Dromedary, Clyde Mountain), *Diplotaxa tertia* and *Parectecephala hardyi* (Mt Wilson), *D. quarta* (Minnamurra Falls, Dorrigo), *Parectecephala elongata* (New England National Park) and *Pemphigonotus macalpinei* (Brunswick Heads) (Spencer 1986).

The genus *Tricimba* (Oscinellinae) has a world-wide distribution. However, approximately half of the described species occur in Australia and the Papuan region. Most of the Australian species occur on the east coast from northern Queensland to Tasmania. Numerous species are recorded from southern Queensland–northern New South Wales (Ismay 1993; Williams 2002). A particularly distinctive attribute is the isolation of the Australian fauna, from the Oriental fauna, at the species level. Within the Australian fauna there is a single example of a trans-Tasman species-pair distribution: *Tricimba biseta* (mainland Australia) and *T. tasmanensis* (Tasmania) (Ismay 1993).

The Australian Tricimban fauna has been assigned to species group assemblages (see Ismay 1993):

- *Tricimba pallidesta* (Gondwana rainforests example: *Tricimba pallidesta*), *T. tibialis* (Gondwana rainforests example: *Tricimba flavoscutellata*), and *T. scutellata* (Gondwana rainforests example: *Tricimba uniseta*) species groups are distributed in Australia and New Guinea;
- *Tricimba lineella* species group (Gondwana rainforests example: *Tricimba facialis*) occurs in all regions, but does not appear to have a centre of endemism;
- *Tricimba selochopina* (Gondwana rainforests examples: *Tricimba carinifacies*, *T. major*), *Tricimba similata* (Gondwana rainforests examples: *Tricimba exvittata*, *T. nitens*, *T. tuberoscula*), *T. convexa* (Gondwana rainforests examples: *Tricimba convexa*, *T. tenuis*) and *T. longigena* species groups (Gondwana rainforests example: *Tricimba excavata*) are restricted to Australia;
- *Tricimba biseta* species group (Gondwana rainforests example: *Tricimba biseta*) occurs in Australia and New Zealand.

Clusiidae Clusiids are small brachyceran flies found mainly in mesic forests, and though inconspicuous in the field they are frequently collected in intercept traps during field surveys. Their larvae breed in rotting wood and the galleries of other insects such as termites, but little else is known about their habits (Colless and McAlpine 1991; Pitkin and Evenhuis 1989). In the Neotropics Clusiidae are found in mossy, humid environments, often attracted to animal dung, but also feeding on rotting vegetation, nectar and sap (Lonsdale and Marshall 2010, and references therein). Males may exhibit agonistic, competing behaviour (S. Marshall 2000).

Heteromeringia is widely distributed in south-eastern Australia with numerous species recorded from open forest and rainforest (Malloch 1926). Some, such as *Heteromeringia imitans* and *H. laticornis*, have distributions that range from North Queensland to central coastal New South Wales. Populations extend from coastal altitudes to elevated localities within the Great Dividing Range (Lonsdale 2009). *Heteromeringia helina*, *H. limaceus* and *H. patula* have more circumscribed distributions, being recorded only from the Werrikimbe–Carrai region. *Heteromeringia spinulosa* is distributed in rainforest at a wide range of elevations that include riverine floodplain rainforest (Lansdowne Nature Reserve), littoral rainforest (Camden Head), and upland mixed subtropical–temperate complexes (Mt Glorious, Bunya Mountains, New England National Park, Sassafras Gully in the Blue Mountains). *Heteromeringia macropa* has been recorded from the flowers of *Cuttsia viburnea* (Escalloniaceae/Rousseaceae) in lowland subtropical rainforest (Lonsdale 2009; Williams 1995). The occurrence of clusiids on the flowers of subtropical rainforest trees is likely to be more widespread.

Conopidae Conopids resemble some Syrphidae flies and share with many of them a flower-frequenting habit. The adults of conopid flies (e.g. *Microconops*, *Australoconops*) feed on nectar and are likely to be minor pollinators of subtropical rainforest trees (e.g. *Syzygium smithii*) (Williams 1995). Species of *Microconops* (Conopinae), a genus also recorded from Chile, are also found on flowering *Leptospermum* and *Baeckea* (Myrtaceae) shrubs adjacent to lowland subtropical rainforest and wet sclerophyll forest. However, conopid larvae are known to be internal parasites that live in the abdomen of their host (e.g. aculeate Hymenoptera, Blattaria, calyptrate flies – Calliphoridae, Tachinidae, etc.) (Camras 1961; Grimaldi and Engel 2005).

Conopidae occur in all biogeographical regions and more than 75 per cent (14 of 19) of the Australian genera are endemic. Four subfamilies of conopid flies occur in Australia: Conopinae, Myopinae, Stylopinae and Notoconopinae. The Conopinae comprise ~90 per cent of the estimated 150 species recorded from Australia (Schneider 2010). World-wide the subfamily includes many species that in colour mimic aculeate wasps and often accompany the insects they mimic. In contrast, however, the Australian species are mainly blackish in colour, and Schneider (2010) suggests these dark forms mimic and parasitise small native bees and wasps.

Although conopids appear to have radiated in more arid areas, and are better represented in open forest and woodland, several taxa can be encountered in mesic forests (e.g. *Microconops* species from subtropical rainforest in northern New South Wales, *Notoconops alexandrei* from Clyde Mountain in southern New South Wales, *Australoconops vespoides* ranging from Victoria to the Blue Mountains and Tapin National Park [Schneider 2010; G. Williams unpubl. records]). Many exhibit wide latitudinal and altitudinal ranges, as with *Smartiomyia arena*, which is distributed from southern Queensland to northern New South Wales, and is recorded from the uplands of the Barrington Tops to adjacent lowland coastal localities of <10 m above sea level.

Corethrellidae These small nocturnal and parasitic nematoceran flies are commonly called frog-biting midges (Borkent 2008, 2014; Borkent and Grafe 2012). The family has previously been placed in the Chaoboridae. Fossil species are known from Lower Cretaceous Burmese amber (Poinar and Szadziewski 2007); however, the extant fauna consists of a single genus, *Corethrella*, which essentially is pantropical and subtropical in distribution. Larvae are associated with water bodies and probably prey on small plankton (Borkent 1989a). Adult females have biting mouthparts and feed on the blood of frogs (Borkent 1989a, 2008; Borkent and Grafe 2012). They are attracted by the calls of males, and in this they differ from other nematocerans in which the attraction is initiated by carbon dioxide emanating from the host (Borkent and Grafe 2012). Corethrellids are vectors of parasitic trypanosomes and their relationship between these and their frog hosts originated at least in the Early Cretaceous (Borkent 2008). Known regional hosts include *Litoria lesueurii* (Hylidae) and *Mixophyes* sp. (Leptodactylidae) (G. Williams pers. obs.) (Plate 202).

Culicidae The Culicidae (mosquitoes) may have originated in the Early Cretaceous, and are now a large family found world-wide (Grimaldi and Engel 2005; Harbach and Kitching 1998, Webb *et al.* 2016). Traditionally, three subfamilies have been recognised: Anophelinae (*Anopheles*), Culicinae (*Culex*, *Aedes*, *Aedeomyia*) and the monogeneric Toxorhynchitinae (*Toxorhynchites*) (Evenhuis and Gon 1989). The Australian culicid fauna is predominantly derived from northern elements. No genera are endemic and there are affiliations between the Australian and Papua New Guinean faunas (Bugledich 1999).

Most female culicids feed on blood, using their long proboscis to pierce their host; in rainforests hosts include mammals, reptiles, birds and amphibians. Adult *Uranotaenia*, for example, attack frogs. Larvae develop in standing water bodies, even temporary pools in tree hollows, wheel ruts, and the base of leaf axils and fallen fronds (e.g. of palms), offering sufficient habitat for maturation to proceed. Culicidae (*Culex*) are frequently encountered at dusk in lowland rainforests but they can be even more prevalent in adjacent eucalypt-dominated forests. Adults usually rest by day and become active before sunset. The fauna includes *Aedes*, an important vector of disease (e.g. Dengue fever by *Aedes aegypti*). *Aedes* species recorded from the Gondwana rainforests region include *Aedes alocasiola*, *A. candidoscutellum* and *A. subauridorsum* (Bugledich 1999; Evenhuis and Gon 1989; Kitching and Callaghan 1982).

Adult culicids have been collected at the flowers of canopy trees and shrubs in lowland rainforests (*Alphitonia excelsa*, *Cuttsia viburnea*, *Syzygium floribundum*), among these the strikingly large *Toxorhynchites speciosus* (on *Guioa semiglauca* – Sapindaceae) in littoral rainforest (Williams 1995); this is the only southern Australian member of the Toxorhynchitinae (Colledge 1911; Evenhuis and Gon 1989, though possibly a highly evolved culicine [Grimaldi and Engel 2005]). Although larval *Toxorhynchites* are predaceous (e.g. northern Queensland *Toxorhynchites splendens* on larvae of *Aedes aegypti*), adults do not bite.

Dolichopodidae This distinctively long-legged family of predatory, small to medium-sized brachyceran flies is a conspicuous component of the day-active arthropod fauna encountered on sunlit glossy rainforest foliage and on the trunks of trees. Some species frequent rocky streams and still water pools, and though adults can be abundant on foliage both in the upper canopy and at lower strata (Lambkin *et al.* 2011), dolichopodids are only infrequently encountered on the flowers of mass-flowering canopy trees in lowland rainforests (Williams 1995). The larvae are predatory or scavengers, and inhabit leaf litter, rotting wood and tree holes. Adult flies possess a highly developed labellum, and are large-eyed, elongate in form and metallic-coloured (Bickel 2009). Agile adults prey on small, soft-bodied invertebrates such as nematoceran flies, thrips, collembolans and mites. These are crushed between the opposed labellae and the prey's body fluids absorbed.

Dolichopodids are well represented in Cretaceous amber, and probably trace their ancestry to earlier Jurassic Empidoidea (Grimaldi and Engel 2005; see also Moulton and Wiegmann 2007; Sinclair and Cumming 2006). World-wide, today the Dolichopodidae may comprise some 15 000 species, of which ~7200 species have already been described. Approximately 960 species in 70 genera are known from the Australasian-Pacific region (Bickel 1983, 1986, 1987a, 1987b, 1992, 1994, 1996a, 1999a, 1999b, 2002, 2008, 2014, 2019; Bickel and Dyte 1989). This family is likely to be the best documented and understood of the subtropical eastern Australian Diptera fauna, and forms a richly documented group within the Gondwana Rainforests World Heritage Area (Williams 2002; Williams and Bickel 2010). Subfamilies represented there are Babindellinae (*Babindella*), Diaphorinae (*Diaphorus*), Dolichopdinae (*Hercostomus*), Medeterinae (*Atlatlia*, *Corindia*, *Medetera*), Neurogoninae (*Arachnomyia*, *Neurigona*, *Antyx*), Sciapodinae (*Abbemyia*, *Amblypsilopus*, *Austrosciapus*, *Heteropsilopus*, *Krakatauia*, *Mesorhaga*, *Negrobovia*, *Parentia*, *Plagiozopelma*) and Sympycninae (*Nothorhaphium*, *Sympycnus*, *Syntormon*, *Teuchophorus*, *Yumbera*). Greatest abundance and diversity of dolichopodids in lowland rainforest occur from mid spring to late summer, with adults virtually absent during winter. The recently described monotypic genus *Minjerribah* (*Minjerribah litoura*) is a member of the poorly known winter fauna (Bickel 2019).

Of these, the Babindellinae include the endemic *Babindella*, with two species, one from the Australian Wet Tropics and the other, *B. whianensis*, from the Border Ranges. *Kowmungia* species occurring in rainforest include *Kowmungia angustifrons* (Mt Dromedary), *K. crassitarsus* (Ku-ring-gai Chase National Park) and *K. flaviseta* (Bunya Mountains) (Bickel 1987b). The Dolichopodinae are depauperate in Australia, but include *Hercostomus crassivenus*, described from floodplain rainforest (Lansdowne Nature Reserve) but also associated with mangroves and extensively distributed along the east coast north of the Manning Valley (Bickel 2008). The species is also recorded from New Caledonia (Bickel 2008), and Bickel notes that several species and genera of Dolichopodidae are known from both Australia and New Caledonia (Bickel 1999b). The distinctive genus *Pindaia* (Bickel 2014) shows a similar distribution pattern with one species in New Caledonia, and *P. bellangrensis* from the Hastings Valley rainforests and *P. ennogerah* from Brisbane Forest Park. The pattern of taxa exclusive to Australia and New Caledonia as either vicariance events subsequent to the fragmentation of the Gondwana landmass, or later episodes of dispersal, perhaps across the Coral–Tasman Sea (Bickel 2008).

Dolichopodidae – Medeterinae In Australia, the subfamily Medeterinae comprises a diverse but poorly known fauna, including many taxa restricted to rainforests in south-east Queensland and northern New South Wales. *Corindia* is a genus with several species occurring in upland mesic forests: *C. capricornis* (Washpool National Park), *C. major* (Werrikimbe National Park, Barrington Tops) and *C. minor* (Barrington Tops, Upper Williams River). *Medetera* is a cosmopolitan genus of more than 300 described species and ranges from the western Indian

Ocean to Hawaii and southern New South Wales. *Medetera* probably had its origin in Early Tertiary Northern Hemisphere moist forests. Although unknown from New Zealand, the Australian region is an important secondary centre of diversity for the genus (Bickel 1987a). The Australian fauna is distinctive, but appears to be derived from the Oriental region. Within Australia, *Medetera* reaches its greatest diversity in the coastal and montane rainforests of New South Wales and Queensland, including *M. austroapicalis* in littoral rainforest (Iluka), and *M. grisescens* from *Nothofagus*-dominated cool temperate rainforest (Barrington Tops) (Bickel 1987a). *Atlatlia* (*Atlatlia grisea*) is a distinctive genus confined to south-eastern and south-western Australia, and New Caledonia (Bickel 1986).

Dolichopodidae - Neurigoninae The genus *Antyx* (Bickel 1999b) is possibly assignable to this subfamily, although its status remains uncertain. *Antyx* is known from New Caledonia, and from south-eastern New South Wales (e.g. *Antyx fagina*, *A. werrikimbe*) to northern Queensland, where it is confined to submontane rainforest (Bickel 1992). In northern New South Wales, *Antyx* occurs in upland cool temperate, warm temperate and subtropical rainforest formations, and in wet sclerophyll forest (e.g. Border Ranges, Tooloom Scrub, Mt Hyland, New England and Werrikimbe national parks, Gloucester Tops). The existing distribution of the genus 'suggests a vicariant Gondwanan distribution, with confinement to mostly upland tropical and subtropical rainforests' (Bickel 1999b). *Arachnomyia* (*Arachnomyia cuprea*, *A. ornatipes*) is an endemic genus found widely in rainforest and wet sclerophyll forests at a range of elevations, but particularly at higher altitudes (e.g. Doyles River, Chichester State Forest, Barrington Tops). *Neurigona* (*Neurigona signata*) is a cosmopolitan genus occurring in wet sclerophyll forest, and in lowland floodplain, escarpment and upland rainforests.

Dolichopodidae - Sciapodinae These generally prefer moist habitats such as rainforests and streams. Approximately 35 per cent of the Australian Sciapodinae are known only from 'type' localities (D. Bickel pers. comm.). The Australian fauna comprises a Gondwanan element, a northern Palaeotropical element, and taxa of uncertain affinity (Bickel 1994). Many taxa are distributed across the Australasian and Oriental zoogeographical regions. Sciadopinae evolution occurred mainly on the continents derived from Gondwana, probably under moist and humid conditions, with their origin set at least in the Lower Cretaceous. In comparison with the Sciapodinae, species of Dolichopodinae are most diverse in Laurasia.

The coastal zone and adjacent ranges, from south-eastern Queensland to south of Sydney, generally forms a single zoogeographical unit with a large common fauna (Bickel 1994). The barrier to less mobile taxa, formed by the Hunter River Gap, evidently is not so significant to sciapodines as three-quarters of the Sydney species also occur on the New South Wales mid north coast (Bickel 1994). Fifty-nine species, of which 12 are endemic to the region, are recorded between Bundaberg in southern Queensland and the Richmond Ranges in northern New South Wales. Sixty-three species (nine endemics) are recorded from the Hastings–Manning region (Bickel 1994).

The southern distributional limit of Sciapodinae with Torresian affinities appears to parallel the limits to distribution of subtropical rainforest in New South Wales. Cool temperate rainforests are almost devoid of Sciapodinae (Bickel 1994). *Parentia* (*Parentia cardaleae*, *P. vulgaris*) is a Gondwanan/Bassian genus with ties to New Zealand and New Caledonia. *Parentia* is absent in Australian cool temperate rainforest but is common in New Zealand *Nothofagus* forest; this may be due to ecophysiological differences (discussed in Bickel 1996a). The Australian and New Zealand *Parentia* faunas are closely related, perhaps reflecting interchange before separation of the two landmasses in the Late Cretaceous 80 mya (Bickel 1994).

Krakatauia (*Krakatauia alanae*, *K. funeralis*, *K. macalpinei* [Plate 203]) is of possible Papuan origin, with *K. macalpinei* being a distinctive indicator species of littoral rainforests and commonly observed on shiny leaf surfaces. *Heteropsilopus* (*Hypsilopus khooi*, *H. meensis*, *H. tweedensis*) is a Gondwanan/Bassian genus found in southern Australia, southern India and montane areas of Sri Lanka. Its vicariant distribution suggests a Lower Cretaceous origin for the subfamily. *Austrosciapus* is almost confined in its entirety to Australia (two species are found on several Pacific Islands and New Zealand [Bickel 1994]), and represents an eastern forest taxa of probable Gondwanan origin. The genus is highly speciose in rainforests of south-eastern Queensland and coastal New South Wales (*Austrosciapus dekeyzeri*, *A. gwynnae*, *A. janae*, *A. minnamurra*, *A. zentae*). *Plagiozopelma* (*Plagiozopelma ashbyi*) represents an Asian-Tertiary (Old Northern) element (Bickel 1994). The genus *Amblypsilopus* (*Amblypsilopus nambourensis*, *A. bereni*, *A. loriensis*, *A. basseti*, *A.*

canungra, *A. wokoensis*) includes some Asian-Tertiary elements (Bickel 1994) and is of Oriental-Papuan origin (Bickel 1996a). Species are diverse in rainforest and wet sclerophyll forest. *Abbemyia* is restricted to the western Pacific, *Abbemyia nigrofasciata* being known from littoral rainforest (Iluka). *Mesorhaga* is a cosmopolitan genus that includes *Mesorhaga canberrensis* in warm temperate rainforest (Tuglo Wildlife Refuge, north-west of Singleton).

Dolichopodidae – Sympycninae Sympycninae are a diverse and cosmopolitan subfamily, but the Australian fauna is little known (D. Bickel pers. comm.). In general, the subfamily has a particularly diverse Neotropical fauna (Bickel 1992). *Sympycnus* (*Sympycnus bandensis*, *S. bulga*, *S. tooloom*) is a cosmopolitan genus that has an amphitropical distribution, but achieves its greatest diversity in temperate South America, New Zealand and Australia and can be considered Gondwanan. *Sympycnus biplagus* exhibits a classic Gondwanan distribution, being recorded from Barrington Tops, Gloucester Tops and Tasmania (D. Bickel pers. comm.). *Syntorum* is a widely distributed genus, in the South Pacific known also from New Zealand, Tonga, New Caledonia and French Polynesia (Bickel 1999a). It includes *Syntorum janelithae* and *S. lucare* from floodplain rainforest (Lansdowne and Coocumbac Island nature reserves) in the Manning Valley. In rainforest the endemic genus *Nothorhaphium* is represented by *Nothorhaphium aemulans*, a widely distributed species in littoral, floodplain, warm temperate and cool temperate rainforest, as well as xeric-adapted woodland, open forest and subalpine communities (Bickel 1999a). In addition to two species from Australia (one each from Queensland and New South Wales), the distribution of *Teuchophorus* includes the Palaearctic, Oriental and Nearctic regions (Bickel 1983). The genus is well represented in Papua New Guinea (Bickel and Dyte 1989); however, the only southern species is *Teuchophorus longifrons* described from Dorrigo National Park (Bickel 1983). Males exhibit courtship behaviour. *Yumbera* is endemic to eastern Australia and is confined to coastal rainforest and wet sclerophyll forest from northern Queensland, south to Tasmania. Four species (*Yumbera callida*, *Y. nudicornis*, *Y. conica*, *Y. signata*) occur in the Gondwana rainforests region (Bickel 1992), with *Y. signata* including lowland riverine to elevated *Nothofagus* rainforests in its habitat and altitude range.

Drosophilidae Drosophilids, or ferment flies, are small acalyptrate flies restricted mainly to rainforest. Some, such as *Tambourella endiandrae*, have beautifully patterned wings. Of the taxonomically well-known insects in Australian rainforests, Drosophilidae are among the most diverse of all the animal groups. Australia possesses ~10 per cent of the world's fauna and 31 of the world's 59 drosophilid genera (Bock 1982), though the Australian fauna is little related to other Gondwanan drosophilid faunas (Bock 1982). Several species have extralimital distributions (Evenhuis and Okada 1989).

Drosophilidae are most abundant in humid and warm wet forests of northern and eastern Australia (McEvey 1992, 1994), but many species are also present in adjacent eucalypt-dominated forest and other less humid forests (van Klinken 1997). The rainforest communities situated between Cooktown and Ingham in North Queensland, and to a lesser extent the McPherson Range region of south-east Queensland and north-eastern New South Wales, contain the most diverse Australian drosophilid faunas (149 species and 108 known species respectively) (McEvey 1994). The McPherson Range fauna primarily has affiliations with more southern regions. Less than 40 per cent of the number of species is shared between the McPherson Range region and North Queensland, suggesting that there is a significant autochthonous endemic element in south-east Queensland and northern New South Wales (McEvey 1994). *Drosophila*, *Leucophenga* and *Mycodrosophila* are the largest genera occurring in Australia; however, drosophilid diversity in general reduces with increases in latitude, south from the main centre of diversity in North Queensland (McEvey 1992). Drosophilidae also exhibit height distributions within rainforest, the vertical distribution of some species matching that of their host resource or, as in the example of the fungivore *Leucophenga scutellata*, most adults during surveys being trapped well above the forest floor 'suggesting behavioural mode can influence vertical distributions' (van Klinken and Walter 2001) independent of the forest strata in which food resources or breeding site hosts are positioned.

Adults frequent moist microhabitats and are associated with fermenting fruit, rotting vegetation, yeasts, ferns and sometimes flowers (Atkinson 1985; Bock and Parsons 1978, 1979; Colless and McAlpine 1991; Cook *et al.* 1977; Malloch 1924; Mather 1955; McEvey 1992, 2009; Parsons and Bock 1977; Williams 1995; Williams and Adam 2001), but the majority of species, as adults, have unknown resource requirements (McEvey 1992), although species breeding in fallen fruit may share resources (Atkinson 1985). *Leucophenga scutellata*,

Scaptodrosophila fungi, *S. ornata* and *S. sydneyensis* are associated with rotting *Boletus* fungi (G. Williams pers. obs.). Larval drosophilids are known to be predators of spider eggs and inquilines in bee nests (Ashburner 1981).

Some Drosophilidae are pollinators, or are part of pollinator assemblages, in subtropical rainforests (Williams 1995; Williams and Adam 2010). Individuals are attracted by colour and/or fragrance cues, the latter including foetid odours to which drosophilids exhibit a sapromyophilous stimulus response. But drosophilids are also attracted to rainforest flowers with sweet fragrances, or no obviously perceptable fragrance at all (at least to humans). *Drosophila hibisci* participates with small nitidulid beetles in the pollination of the rainforest edge shrubs *Hibiscus splendens* and *H. heterophyllus* (Plate 204). Adult flies gather at the base of the fused corolla, where the large and distinctly spined (echinate) pollen grains adhere in small numbers to their bodies. Adults then move to other sequentially opening flowers. The *Hibiscus* flowers are quite large and unfurl for a single day and then drop away, presumably carrying fly eggs with them, to mature in and emerge from the soil profile below. Such cryptic breeding and egg deposition site use has been recorded by drosophilids in *Hibiscus* species elsewhere in Australia (McEvey and Barker 2001). Drosophilds also frequent the small actinomorphic flowers of mass-flowering canopy trees, there feeding on nectar (*Drosophila* spp. on *Syzygium smithii*; *Rhodomyrtus psidioides*, *Tristaniopsis laurina*, *Drosophila* sp., *Scaptomyza australis* on *Syzygium floribundum*; *Drosophila* sp., *Drosophila flavohirta*, *Leucophenga cooperensis* on *Alphitonia excelsa*; *Drosophila* sp., *Scaptomyza australis* on *Euroschinus falcatus*), and also shrubs and trees of the subcanopy (*Drosophila* sp. on *Alectryon coriaceus*; *Scaptomyza australis* on *Acradenia euodiiformis*, *Cuttsia viburnea*; *Leucophenga flavohalterata* on *Diosporos australis*) (Williams 1995). Drosophilids are also attracted to flowering *Archontophoenix cunninghamiana* palms in lowland rainforest (Williams and Adam 2010). However, it is unlikely that small open-structured flowers, which all listed above except *Diospyros australis* have, can function as breeding sites, unless clustered tightly within larger inflorescences.

The upland drosophilid fauna of southern Queensland and northern New South Wales is diverse (McEvey 1992; Williams 2002). The geographical range of many species extends beyond this region, with some inhabiting open forest in addition to rainforest (or may enter rainforest as 'tourists' from adjacent open forest). For example, *Drosophila setifemur* is distributed from the heights of Mount Bellenden Ker in tropical North Queensland, south through the Bunya Mountains, Cunninghams Gap in southern Queensland, Tapin Tops, Mungo Brush (littoral rainforest near Forster), Mt Wilson and Mt Keira in New South Wales, to Victoria (McEvey 2009); these, to cite but a few, also represent considerable variation in altitude (and climate) from lowland sites adjacent the ocean, to high mountains of the hinterland.

Genera and representative species recorded from the rainforests of south-eastern Queensland and New South Wales include, of Drosophilinae, *Balara* (*Balara poecilithorax*), *Chymomyza* (*Chymomyza eungellae*), *Drosophila* (*Drosophila dispar*, *D. immigrans*, *D.pseudotakahashii*, *D. serrata*), *Hirtodrosophila* (*Hirtodrosophila allynensis*, *H. durantae*, *H. lamingtoni*), *Liodrosophila* (*Liodrosophila vitrea*), *Microdrosophila* (*Microdrosophila discrepantia*, *M. joalahae*), *Mycodrosophila* (*Mycodrosophila marksae*, *M. rosemaryae*), *Neotanygastrella* (*Neotanygastrella janeae*), *Nesiodrosophila* (*Nesiodrosophila plana*), *Scaptodrosophila* (*Scaptodrosophila cancellata*, *S. claytoni*, *S. fungi*), *Scaptomyza* (*Scaptomyza australis*), *Sphaerogastrella* (*Sphaerogastrella novoguinensis*) and *Tambourella* (*Tambourella endiandrae*). The Steganinae include *Amiota* (*Amiota albomaculata*) and *Leucophaga* (*Leucophaga flavohalterata*, *L. patternella*, *L. violae*) (Bock 1982; Evenhuis and Okada 1989; Jenkins and Hoffmann 2001; McEvey 1994; van Klinken 1997; Williams 2002). *Scaptodrosophila* is believed to have originated in Southeast Asia. High diversity (six of nine species) within the Australian *Scaptodrosophila* 'coracina species group' occurs in south-east Queensland, and this zone appears to be the centre of diversity for the group (van Klinken 1997). *Scaptodrosophila* diversity falls rapidly south of the McPherson region.

Empididae Empididae, or dance flies, evolved in the Late Jurassic–Cretaceous and are closely related to Dolichopodidae, together constituting the superfamily Empoidea (Colless and McAlpine 1991; Hardy 1930b; Sinclair and Cumming 2006). Both families are a significant component of the arboreal arthropod fauna in subtropical rainforests (Kitching *et al.* 2004; Lambkin *et al.* 2011). Empidids include a large Antarctic element (e.g. Ceratomerinae – *Ceratomerus*) and are most numerous in montane zones. In the Australasian region the family is well represented in New Zealand and several genera are known from New South Wales and Tasmania (K. Smith

1989); however, the Australian and New Zealand faunas are poorly known (B. Sinclair 2000a). There is a rich fauna in south-eastern Australia, especially within northern highland sites, but this remains poorly studied (D. Bickel pers. comm.). Adults and larvae are predators of other insects, in particular Diptera (e.g. Cecidomyidae, Tipulidae), with the larvae found in soil, leaf litter and rotting wood, or aquatic (Clinocerinae) (K. Smith 1989). Adults are occasionally found on flowers. Males participate in complex mating behaviour, including mating swarms and nuptial feeding where the male offers captured prey to the female as part of the mating ritual (Colless and McAlpine 1991; Grimaldi and Engel 2005).

The Clinocerinae include the genus *Clinocera*. Australian *Clinocera* appear to be related to East Asian species, since New Zealand and Chilean taxa are assigned to a different species group (B. Sinclair pers. comm.). B. Sinclair (2000a) postulates that the 'Australian species possibly originated from dispersal of ancestral populations southwards from New Guinea'. Seven species of *Clinocera* occur in eastern Australia, with the greatest diversity again in the south-east. The New South Wales species (e.g. *Clinocera australiana*, *C. irrorata*, *C. rubiventris*) are widespread and primarily found in small, cool, rocky streams and seepages. *Clinocera australiana* and *C. irrorata* inhabit cool temperate rainforest, although the latter is also recorded from wet sclerophyll forest. *Clinocera rubiventris* is known from subtropical rainforest at higher elevations (e.g. Lamington, New England, Barrington Tops national parks). *Hemerodromia* (Clinocerinae/Hemerodromiinae [Colless and McAlpine 1991]) has been observed feeding on nectar of flowering *Syzygium floribundum* trees in lowland rainforest (Wingham Brush Nature Reserve) (Williams 1995).

Ceratomerinae are distinctly Gondwanan. Australian species of the genus *Ceratomerus* are related to species in New Zealand and southern Chile (B. Sinclair pers. comm.). In Australia, *Ceratomerus* has its highest diversity in the south-east (Tasmania has four endemic species) and is confined to cascading rocky streams and rivers, flowing through a variety of rainforest habitats (e.g. temperate, subtropical and riverine rainforest) (B. Sinclair 2003). Nearly all the species known from northern New South Wales and southern Queensland are widespread (*Ceratomerus albistylus*, *C. maculatus*). However, there is one endemic species (*Ceratomerus barringtonensis*) known only from the subtropical rainforest zone of the Williams River at Barrington Tops National Park (B. Sinclair pers. comm.).

Of the Empidinae, the tribe Hilarini is well represented in Australia, with many undescribed species (D. Bickel pers. comm.). Hilarini include the rainforest and wet sclerophyll-associated endemic genus *Bandella*, with five species described from the Gondwana rainforests region (Bickel 2002). *Bandella* has a largely temperate Bassian distribution, occurring in Tasmania, south-eastern and south-western Australia, with a single outlying species (*Bandella maxi*) in tropical Queensland. Bickel (2002) considers that, relative to other hilarine genera, *Bandella* is morphologically isolated but is possibly closest to the Tasmanian genus *Cunomyia*. *Hilara* is a cosmopolitan genus with several species (*Hilara banearia*, *H. kippara*) occurring in hinterland rainforests. In summer adults are commonly to be seen flying back and forth above the surface of pools, including ephemeral puddles. Additional genera include *Apalocnemis* (*Apalocnemis sanguinea*), *Hilarempis* (*Hilarempis pallidifurca*), and *Notiohilara* (*Notiohilara coombadjah*, *N. minnamurra*, *N. numinbah*), these all occurring in upland forests (e.g. Washpool, Dorrigo, Nightcap national parks, Cockerwombeeba Flora Reserve) (D. Bickel pers. comm.; Williams 2002). Of these genera *Apalocnemis* occurs in Australia and New Zealand (K. Smith 1989), *Notiohilara* is endemic, and *Hilarempis* is distributed in the Southern Hemisphere and possibly the Northern Hemisphere (D. Bickel pers. comm.). Adult *Empis*, or related genera, are sometimes observed imbibing floral nectar from the flowers of rainforest trees and shrubs (*Syzygium smithii*, *S. floribundum*, *Acradenia euodiiformis*, *Alectryon coriaceus*, *Alphitonia excelsa*, *Cuttsia viburnea*, *Euroschinus falcatus*, *Guioa semiglauca*, *Rhodomyrtus psidioides*, *Tristaniopsis laurina*) in subtropical rainforest (Williams 1995).

Additional subfamilies occurring in the region are Hybotinae (*Hybos*), Ocydromiinae (*Leptopeza*), Oreogetoninae (*Anthepiscopus*), Tachydromiinae (*Elaphropeza*, *Micrempis*) and Microphorinae (*Microphorella*) (K. Smith 1989).

Ephydridae These are small, diversely adapted predaceous brachyceran flies closely related to Drosophilidae and Chloropidae (Bock 1987; Colless and McAlpine 1991; Cresson 1948; Mathis 1989). The larvae of most known species are aquatic or semi-aquatic, feeding as browsers or filter feeders, or are terrestrial leaf miners or parasitoids of spider and frog egg masses (Mathis 1989). The adults of some species have raptorial forelegs used to capture prey. Adults use their paired labella to masticate captured prey and then imbibe the body fluids.

Ephydridae are often found near rainforest streams, on foliage and occasionally on flowering trees: *Stratiothyrea* sp. on *Tristaniopsis laurina* in a lowland subtropical rainforest–wet sclerophyll forest ecotone, and *Rhynchopsilopa* sp. on *Alphitonia excelsa* in littoral rainforest (Williams 1995). Adult *Rhynchopsilopa* attack and suck juices from the bodies of *Crematogaster* ants without necessarily killing their victims (Colless and McAlpine 1991; Freidberg and Mathis 1985).

The subtropical rainforest fauna is a diverse one and includes the genera *Setacera* (Ephydrinae), and *Hydrellia*, *Paralimna* and *Notiphila* (Hydrelliinae). The Australian *Hydrellia* fauna is primarily related to that of South-east Asia and New Guinea, but has some affinities with the New Zealand fauna (Bock 1990). *Setacera*, *Paralimna* and *Notiphila* have a world-wide distribution (Bock 1987, 1988, 1990).

Exeretonevridae The Exeretonevridae are a monogeneric family (Nagatomi 1977) restricted to Australia, and occur in highland areas from northern New South Wales to Tasmania (Daniels 1989c; Paramonov 1953). Colless and McAlpine (1991) place Exeretonevridae in the Xylophagidae. Adults are weak fliers and can be found clinging to fern fronds and grass stems, often near streams. They are not inhabitants of rainforest in the strick sense, rather occurring on margins. Four species (*Exeretonevra angustifrons*, *E. maculipennis*, *E. tertia*, *E. zentae*) comprise the family and all occur in New South Wales (Daniels 1989c; see also I. Mackerras 1925; Nagatomi 1977; Paramonov 1953). *Exeretonevra angustifrons* is recorded from Ebor, and *E. maculipennis* is recorded from Barrington Tops.

Heleosciomyzidae This family occurs mainly in Australia and New Zealand (Barnes 1989) and regionally includes genera such as *Luta* and *Helosciomyza* that are found in cooler upland mountainous areas. *Luta luteipennis* is recorded from temperate rainforest, and occurs widely on the Northern and Central Tablelands of New South Wales (McAlpine 2012). *Helosciomyza bickeli* is recorded along the east coast and nearby ranges from Eurimbula and Paluma south to Kenilworth State Forest, the Border Ranges, Mount Wilson and as far as Tasmania. *Helosciomyza obliqua* is distributed from New England National Park to Victoria, and *H. macalpinei* is recorded from New England National Park, Mount Wilson and Clyde Mountain; these various records suggest a widespread but still poorly known fauna. Little is known of the biology of Heleosciomyzidae; however, the Australian species *Eurotocus australis* has been collected from agaric fungi (Barnes 1989).

Heteromyzidae Sometimes referred to as wombat flies, this is a mainly temperate family of minute to small-bodied flies (previously placed in Heleomyzidae and Sphaeroceridae) (Colless and McAlpine 1991; McAlpine 1985, 2007a; McAlpine and Kent 1982). Genera recorded from northern New South Wales and southern Queensland include *Tapeigaster*, *Pentachaeta*, *Austroleria*, *Diplogeomyza*, *Borboroides*, *Biroina*, *Poecilosomella*, *Pterogramma* and *Leptocera* (McAlpine 2007a; Williams 2002). Many species are widely distributed, and some are associated with animal dung (Colless and McAlpine 1991). *Leptocera myrmecophila* is phoretic on adult *Cephalodesmius armiger* (Scarabaeinae) beetles (Monteith and Storey 1981). Adults of the stout-bodied genus *Tapeigaster* (e.g. *Tapeigaster luteipennis*), previously placed in the Heleomyzidae, are often associated with the fruiting bodies of agaric fungi on the edges of lowland mesic forest (Plate 205). *Tapeigaster* is restricted largely in its distribution to temperate Australia (McAlpine and Kent 1982), with at least nine species occurring in the Gondwana rainforests region and adjacent areas (Williams 2002). McAlpine and Kent (1982) suggest that the evolution of the genus occurred in isolation on the Australian continent during much of the Tertiary, or that *Tapeigaster* may be a relict group surviving from the extinction of its nearest relatives. *Borboroides* widely occurs in open forests of south-eastern New South Wales and eastern Victoria; however, several species inhabit mesic upland cooler forests (e.g. *Borboroides stewarti*, *B. perkinsi*, *B. atra*) (McAlpine 2007a).

Pentachaeta is placed with the Chilean genus *Dioche* in the tribe Pentachaetini, subfamily Trixoscelinae (McAlpine 2014), and is distributed in the higher rainfall areas of eastern Australia (McAlpine 1985) from North Queensland to Tasmania. Many species are recorded from higher subtropical latitudes (e.g. *Pentachaeta kirkspriggsi*, *P. pinguis*, *P. bickeli*). Several are commonly encountered in rainforests, for example *Pentachaeta impar* known from numerous sites from near Kenilworth in south-eastern Queensland to Taree in New South Wales. Adults have been collected on the flowers of *Schizomeria ovata* (Cunoniaceae) in lowland mixed subtropical rainforest–wet scleophyll forest (Williams 1995).

Of the species previously placed in Sphaeroceridae (Limosininae), seven species were cited by Williams (2002) as occurring in the Gondwana rainforests region:

Biroina dorrigonis, *B. myrmecophila*, *B. trivittata*, *B. vernalis*, *Poecilosomella pectiniterga*, *Pterogramma aestivale* and *Leptocera fenestrata* (Daniels 1978; S. Marshall 1989; Monteith and Storey 1981; O. Richards 1973). However, the fauna is likely to be far more diverse. Sphaerocerids (*sensu* S. Marshall 1989) are associated with decaying organic matter, fungi, carrion and dung. Species have also been recorded from caves.

Hippoboscidae Hippoboscid flies, commonly known as louse flies, wallaby flies and keds, are ectoparasitic blood feeders of mammals (including marsupials) and birds, and are predominantly found in tropical and subtropical regions (Maa 1989a). They have characteristically flattened bodies, and are capable of agile and inconspicuous movements across the host's body. Host specificity is thought to be generally high, though some subtropical species readily translocate to humans, there to feed. Ornithomyiinae (*Icosta*, *Olfersia*, *Ortholfersia*, *Austrolfersia*) are mostly restricted to birds. Larvae are retained internally until ready to pupate (prepupal stage), at which time the adult female deposits the prepupa on the ground, after which the pupa hardens and darkens.

The fauna of the Gondwana rainforests region includes *Icosta* (*Icosta australica*) which, in the Australo-Oriental region, is distributed from Indonesia to Western Samoa, Fiji and Australia; *Ortholfersia* (*Ortholfersia macleayi*, *O. minuta*), which is restricted to eastern Australia; and *Ornithomya* (*Ornithomya* sp.), which also occurs in New Zealand and is widespread in the Oriental region (Maa 1989a; Paramonov 1954).

Hybotidae This poorly known family was originally placed as a subfamily of Empididae (Colless and McAlpine 1991; K. Smith 1989), with which they share similar behaviour and aquatic habitat. However, the Hybotidae form a lineage distinct from empidids, and within the superfamily Empidoidea represent a lineage more basal than the main radiation of Empididae and Dolichopodidae (Moulton and Wiegmann 2007; Sinclair and Cumming 2006). The Southern Hemisphere genus *Leptopezella* includes *L. anatolica* from the Blue Mountains and Brown Mountain (Sinclair and Cumming 2007). The widespread genus *Tachydromia* in Australia is confined to the eastern coast, and possibly originated by dispersal of ancestral populations from South-east Asia (Grootaert and Shamshev 2011). Species recorded from mesic forests of subtropical eastern Australia include *Trachydromia australiensis* (Blue Mountains), *T. bickeli* (Blue Mountains, Doyles River State Forest, Styx River State Forest, Washpool and Barrington Tops national parks) and *T. corticola* (Conondale Range, Carrai State Forest, Blue Mountains). A species of uncertain status is known from Sea Acres Nature Reserve, and Dooragan, Mount Warning and Lamington national parks (Grootaert and Shamshev 2011).

Keroplatidae The Keroplatidae, or glow-worms, have been placed previously with the Mycetophilidae, or fungus gnats (e.g. Colless 1970; Colless and McAlpine 1991). Keroplatidae first appeared in the Late Jurassic (Blagoderov 1993) and many species are known from Cenozoic amber (Matile 1989a). Predaceous keroplatid larvae inhabit hidden, moist places where they spin hygroscopic webs used to snare small invertebrate prey (such as small gnats) (Baker and Merritt 2003). Prey are killed with acid fluid (mostly oxalic acid) secreted from the labial glands and deposited on the web threads (Matile 1989a). However, some larval keroplatids (e.g. *Heteropterna*) graze on fungi.

The subfamily Arachnocampinae comprises a single genus, *Arachnocampa* (Baker 2010; Baker and Merritt 2003; Baker *et al.* 2008; Harrison 1966; Matile 1981, 1989a, 1990), which has a Gondwanan distribution. The bioluminescent and predatory larvae of *Arachnocampa* inhabit cool moist stream banks and rock overhangs, disused mines, narrow ravines, road cuttings and caves. Individual cave populations in national parks are vulnerable to visitor impacts. The larvae trap their prey by constructing hanging mucilaginous threads. Adults may not feed. Larvae have limited dispersal ability and are reliant on high humidity environments for survival (Baker 2010). Luminescence is produced by swollen ends of the Malpighian tubes, the light being reflected through the translucent cuticle (Grimaldi and Engel 2005). Larvae initiate glowing soon after dark, but cease with the approach of dawn. Bioluminescence has been found to increase substantially during nights in which rainfall has been recorded (Merritt and Patterson 2018). Adult *Arachnocampa flava* are short lived, males surviving for ~6 days and that of females ~2.5 days. Mating of *Arachnocampa flava* takes place immediately after the emergence of females, males mate with numerous females, and reduction in the life span of adult males may be related to copulation activity (Baker and Merritt 2003).

Seven Australian *Arachnocampa* species have been described, including *A. tasmaniensis* from Tasmania (Baker 2010; Baker *et al.* 2008). An additional species (*Arachnocampa luminosa*) occurs in New Zealand (Matile 1989a).

The genus has a patchy but widespread distribution. All published *Arachnocampa* sites in Australia are from rainforest or wet sclerophyll forest, or in limestone or granite boulder caves (Baker 2010).Three species occur in southern Queensland and New South Wales: *Arachnocampa richardsiae* (Wollemi, Blue Mountains and Morton national parks, but also Victoria), *A. flava* (Numinbah, Springbrook, Mt Warning and Nightcap national parks, and possibly Kroombit Tops, Mt Tamborine, Lamington National Park) and *A. girraweenensis* (Girraween, Washpool and Barrington Tops national parks) (Baker 2010). *Arachnocampa richardsiae* and *A. flava* are broadly sympatric in the Gondwana rainforests region, and although *Arachnocampa* is commonly observed in many Gondwana rainforest sites, few collections have been made and identified (D. Bickel pers. comm.). Glow-worms of an unknown species are abundant in wet sclerophyll forest and rainforests of the Lansdowne Escarpment, some exposed clay banks there seemingly populated by countless thousands such that the light projected by the massed larvae vividly exposes the immediate forest surrounds (M. Richardson pers. observ.; G. Williams pers. observ.).

Keroplatinae include the endemic monotypic genus *Tamborinea* (*Tamborinea commoni*, which is restricted to the Gondwana rainforests region), *Euroceroplatus cantrelli* (known only from south-east Queensland and near Kuranda, North Queensland), and mycophagous *Heteropterna affinis* (from northern New South Wales) (Bugledich 1999; Matile 1981).

Lauxaniidae Lauxaniidae are a large cosmopolitan family of small-sized brachyceran flies, adults often with conspicuously patterned wings (e.g. *Trypetisoma*). The family is very diverse in tropical regions (Kim 1994), Australasia and Oceania (Evenhuis and Okadome 1989). The Australian fauna is complex and includes a number of undescribed genera and species (D. K. McAlpine pers. comm). Adults are abundant in the canopy stratum of subtropical rainforest (Basset 1991), but can be found on sunlit foliage in all strata. Activity patterns and abundance are influenced by season and altitude (Lambkin *et al.* 2011). Some species feed on fungal hyphae and spores by grazing the microepiphyte cover of foliage (Basset 1991), the labellum being modified so as to cut and rake in food material (Broadhead 1984). Adults may also feed on nectar (Williams 1995). Larvae are saprophagous and breed in rotting detritus and leaf litter.

Genera occurring in subtropical Queensland and New South Wales include *Ceratolauxania*, *Depressa*, *Melanina*, *Minettia*, *Paranomina*, *Poecilohetaerus*, *Sapromyza*, *Steganopsis*, *Trigonometopsis* (Lauxaniinae), *Homoneura*, *Noeetomima* and *Trypetisoma* (Homoneurinae). *Sapromyza* and *Homoneura* are the largest genera (Evenhuis and Okadome 1989). Of these *Poecilohetaerus* is confined to Australia and New Zealand, with an unconfirmed record from Lombok, Indonesia. The Australian distribution pattern is essentially Bassian with the genus being confined to south-west Western Australia, south-east South Australia, northern coastal Queensland and south-east Australia, including Tasmania. The majority of known species are restricted to southern rainforest and wet sclerophyll forest (Schneider 1991). *Poecilohetaerus aquilus*, *P. schineri*, *P. albolineatus*, *P. xanthopus* and *P. pinnatus* have been recorded from the Gondwana rainforests region (Schneider 1991). The endemic genus *Paranomina* includes *P. dayi* (Gibraltar Range, Bunya Mountains) and *P. unicolor* (Gibraltar Range, Tuglo Wildlife Refuge) (McAlpine 2019). *Homoneura* occurs in all faunal regions except for New Zealand and the Neotropics (Kim 1994). The Australasian and Oceanian regions possess the greatest number of recorded species for any region. The genus is diverse in northern New South Wales and south-eastern Queensland with more than 20 described species (Williams 2002). Several species (e.g. *Homoneura centrimella*, *H. tamborinensis*, *H. canungrae*) are confined to that region (Kim 1994). The high level of species endemicity (91 per cent) in the Australian *Homoneura* fauna suggests either that it developed autochthonously from ancestral stock or reflects very early immigration from the Oriental region, possibly just after the Miocene (Kim 1994). In Australia, *Noeetomima* is restricted to eastern coastal regions, but elsewhere *Noeetomima* is known from Nepal, Thailand, Manchuria and eastern Siberia. *Noeetomima parva* is widespread in Australia (North Queensland–Australian Capital Territory), but has only been collected at four localities over this range. In the Gondwana rainforests region it has been collected from the Boyd River, northern New South Wales. *Trypetisoma* is known from North and South America, South-east Asia, Seychelles, Oceania, Papua New Guinea and Lord Howe Island. In Australia, *Trypetisoma* reaches its greatest diversity on the east coast (*Trypetisoma ilukae*, *T. macalpinei*, *T. sentipeniculus*, *T. uptoni*), the genus ranging in altitude from littoral rainforests (Iluka) to upland rainforests (Bunya Mountains, Barrington Tops).

Lauxaniids are a significant element of the flower-frequenting Diptera fauna in lowland rainforest. Adult *Depressa*, *Homoneura*, *Melanina*, *Noeetomima*, *Sapromyza*, *Steganopsis* and *Trypetisoma* have been recorded at flowering trees and shrubs growing in the canopy and on lower margins (e.g. *Ackama paniculata*, *Acradenia euodiiformis*, *Alectryon coriaceus*, *Alphitonia excelsa*, *Cryptocarya microneura*, *Cuttsia viburnea*, *Drypetes deplanchei*, *Elaeodendron australe*, *Euroschinus falcatus*, *Guioa semiglauca*, *Rhodomyrtus psidioides*, *Schizomeria ovata*, *Scolopia braunii*, *Syzygium floribundum*, *S. smithii*, *Tristaniopsis laurina*) (Williams 1995). These include species with strong, sweet smelling fragrances (as in *Syzygium floribundum*, *Rhodomyrtus psidioides*), and species with foetid (*Alphitonia excelsa*), musty (*Alectryon coriaceus*) or less perceptable odours (*Cryptocarya microneura*). House (1985) also records Lauxaniidae at flowering *Diospyros pentamera* in tropical Queensland rainforest, a tree that also occurs in subtropical rainforests.

Lygistorrhinidae The lygistorrhinids (= Lygistorrhininae ex Mycetophilidae of Colless and McAlpine 1991) are a small family of gracile nematoceran fungus gnats that first appeared during the Cretaceous (Blagoderov and Grimaldi 2004). Adults possess a distinctive long, thin proboscis. The family now occurs widely in the warmer zones of the Nearctic, Neotropical, Palaearctic, Oriental, Afrotropical, Oceanic and Australasian regions (F. Thompson 1989). They are a putative sister group to the Mycetophilidae (Bugledich 1999). Lygistorrhinids are thought to comprise a single genus, *Lygistorrhina*, which is known from the Afrotropical, Palaearctic and Oriental regions, New Caledonia and Australia. *Lygistorrhina insignis* is the only recorded Australian species, and the only additional recorded species from the Australasian region is *L. carayoni* from New Caledonia (F. Thompson 1989).

Micropezidae Micropezids are medium-sized, distinctively long-legged, brachyceran flies. They are principally cosmopolitan and better represented in the Tropics, but are apparently absent from New Zealand (McAlpine 1998). However, there are more temperate than tropical species in Australia. Tropical North Queensland and the south-eastern mainland are centres of Australian micropezid diversity. Adults are predaceous on small insects, but may be attracted to dung and rotting fruit. Many species have more or less banded wings. Larvae may be phytophagous or saprophagous, but little is known of their habits. Groups of temperate origin usually live in *Eucalyptus*-dominated forest. Those with tropical origins are normally inhabitants of rainforest. *Metopochetus* (Plate 207) is known from the Gondwana rainforests region and these species are apparently restricted to rainforest and wet sclerophyll forest. Mainland species of *Cothornobata* are confined to rainforest. Adults of many micropezids are apparently Batesian mimics of ants (e.g. of *Leptomyrmex*) and possibly Hymenoptera (McAlpine 1998). Adults (males?) may exhibit antagonistic behaviour when approached by other individuals, or when alone wave their forelegs in front of the head in a signalling fashion. In rainforest environments adults do not apparently frequent flowers, instead seeming to prefer resting on foliage at ground level. Micropezids are often found, both in lowland forest and cool temperate forest, displaying on sunlit *Lomandra* stems, this plant often forming dense mats along stream banks and forest margins. McAlpine (1998) records *Crepidochetus ios* ovipositing in the borer holes made by Coleoptera in a fallen log.

Metopochetus is the largest Australian genus and includes several subgenera (*Crus*, *Seva*, *Metopochetus sensu strictu*) recorded from the Gondwana rainforests region. *Seva* (*Metopochetus bivittatus*, *M. regius*) is restricted to south-eastern and south-western Australia and Tasmania. *Crus* (*Metopochetus compressus*, *M. freyi*) is restricted to southern Australia, but appears absent from Tasmania. *Metopochetus* (*Metopochetus equalis*) is known from tropical and subtropical eastern Australia and extends to New Guinea and the islands of the Bismark Archipelago (McAlpine 1998). Individual species are distributed through diverse altitudal and latitudinal ranges (*Metopochetus freyi*, *M. bivittatus*, *M. aequalis*), including coastal lowland rainforest and cool temperate forests. Several currently are known from restricted localities (*Metopochetus regius*, Mt Royal). *Cothornobata* has a mainly tropical distribution, with a single species (*Cothornobata aczeli*) extending from tropical North Queensland, south to Stotts Island and Victoria Park nature reserves, Mt Warning and the Border Ranges national parks (McAlpine 1998), and the Lansdowne Escarpment (G. Williams unpubl. data) in New South Wales. *Mimegralla* has a single species (*Mimegralla australica*, Plate 206) that extends into subtropical eastern Australia; its southern-most recorded locality is Port Macquarie (McAlpine 1998). Adults have been recorded in lowland subtropical rainforest with *Araucaria* emergents near Murwillumbah, north-eastern New South Wales (B. Williams pers. record).

Muscidae This family includes the ubiquitous and noisome house fly *Musca domestica* and bush fly *M. vetustissima*. The latter may be common in open habitats adjacent lowland rainforest remnants, but adults do not enter the rainforest understorey in any abundance unless the remnants are narrow or heavily structurally degraded with minimal, if any, core area surviving. Larval muscids breed in animal waste (Hughes *et al.* 1972) and rotting vegetation, and some species are associated with bird nests, there either sucking the blood of nestlings or as scavengers (Colless and McAlpine 1991). Adults imbibe waste fluids (including products of animal decomposition), but are generally uncommon on the flowers of subtropical rainforest trees (occasionally at *Alectryon coriaceus*, *Guioa semiglauca*, *Cuttsia viburnea* and others) (Williams 1995; G. Williams pers. observ.). Crome and Irvine (1986), however, cite Muscidae as participating in the pollination of *Syzygium cormiflorum* (Myrtaceae) in tropical North Queensland rainforest. Adult Muscidae may exhibit height preferences within rainforests, this being a response to host presence. Mavoungou *et al.* (2013) found in a Gabon rainforest that the vertical distribution of blood-feeding species of *Stomoxys* (a genus also present in Australia) responded differentially to height above ground and the availability of feeding hosts.

Muscid flies occur in all zoogeographical regions (Pont 1973, 1989) but are not well represented in subtropical rainforests. Regionally, the subfamily Phaoniinae includes *Dichaetomyia parimpar*, *Helina hirtibasis*, *H. hypopleuralis*, *Neohelina semivittata*, *Phaonia fergusoni* and species of *Atherigona* (Daniels 1978; Malloch 1925; Pont 1986, 1989). The *Atherigona* fauna recorded from the region includes species with restricted ranges (e.g. *Atherigona collessi* – known only from Australia), those with regionally circumscribed distributions (e.g. *Aricimba oryzae* – Australasian and western Pacific regions) and widely distributed species (e.g. *Aricimba atripalpis* – ranging from the Indian subcontinent, Indonesia to Australia) (Pont 1986). *Atherigona apicemaculata* is a little known species but has been recordly widely in the Indo-Australasian region (i.e. China, Sarawak, Flores, Papua New Guinea, Australia). The distributions of *Atherigona hennigi* and *A. ferrari* are more limited, with both species being recorded from Indonesia, Sarawak, Queensland and New South Wales (Pont 1986).

The subfamily Muscinae includes *Neomyia* (= *Orthellia*) and *Hydrotaea*. *Neomyia* occurs in the Indo-Australasian regions, but is absent from New Zealand (Pont 1973, 1989). Species recorded from the Gondwana rainforests region also have extralimital distributions (e.g. *Neomyia lauta* occurs from Iran to Australia). *Hydrotaea rostrata* (= *Australophyra rostrata*) has an extralimital distribution comprising Australia, New Zealand, Norfolk Island and Lord Howe Island (Pont 1973, 1989). *Pyrella tasmaniae* is restricted to eastern Australia and is recorded from the Bunya Mountains and the Barrington Tops (Pont 1973). The cosmopolitan genus *Lispe* (Coenosiinae) includes *L. nigrimana*, *L. setigena* and *L. sydneyensis*. Adult *Lispe* are found in damp places. Their subaquatic larvae feed on a variety of small freshwater invertebrates (Pont 2019).

Mycetophilidae Fungus gnats of the family Mycetophilidae first appeared during the Cretaceous, radiating during this period to especially occupy moist temperate forests (Blagoderov and Grimaldi 2004). Approximately 3500 species are now known world-wide (Grimaldi and Engel 2005).

Australian Mycetophilidae (excluding Keroplatidae) include 'old Transanarctic ... pantropical and Indomalayan elements' (Matile 1989b). The genera *Mycomya* and *Epicypta* are widely distributed, including the Afrotropical, Australasian, Nearctic, Neotropical and Oriental regions. *Mycomya* also occurs in New Zealand and New Caledonia (Bugledich 1999). Regional species include *Paraleia fulvescens* (Leiinae), *Epicypta flavipennis* (Mycetophilinae), *Mycomya richmondensis* and *Neoemphera speiseri* (Mycomyinae) (Bugledich 1999; Kurina *et al.* 2019; Kurina and Oliveira 2013; Matile 1989b; Tonnoir 1929).

This nematoceran family is particularly diverse and abundant in wet forest types, especially so at higher elevations, and in North American temperate forests individuals have been shown to be far more abundant at ground level rather than higher within the subcanopy. This is possibly owing to the positioning of their food resources (fungi) there (Preisser *et al.* 1998) and the propensity of adults to undertake mating flights near the ground rather than in the canopy (Peck and Russell 1976). In Australian subtropical rainforests adults can occur in large numbers, and often shelter in low dense vegetation in the understorey along the margins of dissecting streams, or forest margins. Larvae feed on the spores and hyphae of fleshy Basidiomytes, but some species are associated with other fungi (e.g. polypores – bracket fungi). Larvae spin glutinous webs under the loose bark of dead tree trunks (e.g. *Mycomya*), and others are predaceous, though little is known of the larval biology of Australian species (Matile 1989b). Species occur at night on flowering rainforest trees (e.g. on *Alphitonia excelsa*, *Syzygium floribundum* [Williams 1995]), but their nocturnal habits on these are poorly investigated. Mycetophilids (*Mycomya*) are pollinators of non-rewarding *Pterostylis* ground orchids (Bernhardt 1995; Phillips *et al.* 2014). Gnats are attracted by sexual deception and their probing pseudocopulatory behaviour on the orchid's labellum results in its triggering and temporary entrapment of the gnat in the flower. Pollen deposition occurs as the fly escapes via reproductive structures. *Pterostylis* is an endemic and speciose Australasian genus of terrestrial orchids frequently encountered in sunlit openings in littoral rainforest and the margins of rainforests at higher elevations (*Pterostylis ophioglossa*, *P. collina*, *P. concinna*, *P. hildae*, *P. pulchella*), including cool temperate rainforest in the Barrington Tops (Harden 1993).

Mydidae This is a small family of moderate- to large-sized brachyceran flies related to Bombyliidae, Apioceridae, Asilidae and Therevidae (Asiloidea in part), and has a largely austral, southern, distribution (Yeates and Irwin 1996). Adults are characteristically very elongate in form with long clubbed antennae. The Australian genera (*Diochlistus*, *Miltinus*) are endemic. 'The World fauna seems to be an old one … and has suffered much extinction due to climatic changes' (Colless and McAlpine 1991). Larvae and some adults may be predaceous. *Diochlistus gracilis* and *D. hackeri* are recorded from Mt Tamborine (Hardy 1925; Paramonov 1950). Mydidae are not very common and have localised distributions. Adults are active during the day and are usually found in bright sunlit areas flying near, or resting on, the soil surface (where they lay eggs) in open vegetation, rather than rainforest. However, *Miltinus viduatus* will enter dune-associated littoral rainforest when there are large sunlit canopy gaps within. *Miltinus* is mainly an arid-adapted genus. When disturbed, *Miltinus* adults typically fly short distances, before alighting again. In flight they can be confused with similarly sized Asilidae and Apioceridae sharing the same habitat.

Nemestrinidae The Nemestrinidae are a rather primitive family of Brachycera, and probably one of first to feed on the nectar of Early Cretaceous angiosperms (Willemstein 1987). World-wide the family consists of ~500 species and is considered an austral group owing to an extant distribution that includes South America and southern Africa. Nemestrinidae (excluding *Exeretonevra* – Exeretonevridae) are large hairy flies with prominent long probosces that in rainforests are occasionally seen feeding at canopy and edge-flowering trees and shrubs. Larvae are parasitic.

Bernardi (1989) recognised six nemestrinine genera from the Australasian area, of which *Trichophthalma*, *Nycterimyia*, *Nycterimorpha* and *Trichopsea* are recorded from subtropical Queensland and New South Wales (Bernardi 1989; Carter 1933; Daniels 1978; Hardy 1924; I. Mackerras 1925; Paramonov 1953; Williams 2002). *Exeretonevra*, previously in the Nemestrinidae, was subsequently placed in the Exeretonevridae by Nagatomi (1977).

Nycterimorpha (Atriadopinae) is represented in Australia by a single picture-winged species *N. speiseri* (Bernardi 1989; Paramonov 1953), in this region recorded from the Bunya Mountains, Mt Tamborine, Victoria Park Nature Reserve, Red Cedar Flora Reserve and Moonpar State Forest. The genus extends to West Papua and the Bismarck Archipelago. *Trichopsidea oestracea* (Trichopsidinae) occurs in the Bunya Mountains; however, the species also extends south to Victoria and north to Papua New Guinea (Bernardi 1989; I. Mackerras 1925).

Eastern New South Wales appears to be a geologically recent evolutionary centre for the Gondwanan genus *Trichophthalma* (Nemestrinae) (I. Mackerras 1925; Paramonov 1953); it is a genus also with numerous species in Chile and Argentina (Devoto and Medan 2006). *Trichophthalma* is well represented in montane areas of south-eastern mainland Australia (e.g. *Trichophthalma bancrofti*, *T. dubiosa*, *T. harrisoni*, *T. laetilinea*, *T. rufonigra*; see also Williams 2002) with more than half of the known Australian species recorded from this area (see Bernardi 1989; I. Mackerras 1925; Paramonov 1953).

Adult *Trichophthalma* are conspicuous when feeding on blossoms, or hovering over favoured flowering plants, but when disturbed will quickly fly away. Rarely are they seen hovering near the ground, unlike bee flies of the subfamily Bombyliinae. However, they do not fully hover while feeding, instead grasping flowers with their legs but continuing to vibrate their wings. This behaviour very likely places a high energy demand on individuals (see discussion in Dafni 1992). Species also exhibit 'hill-topping' behaviour (e.g. *Trichophthalma novaehollandiae*), commonly in company with large tabanids and tachinids, and sometimes pelecorhynchids, all hovering in concert several metres above vegetation – in particular at the summits of volcanic plugs where these are surrounded by wet sclerophyll forest or rainforest (as in the Manning Valley). Although more usually a genus found on flowering shrubs (e.g. *Leptospermum*, *Kunzea*) adjacent to rainforest, nevertheless some species (e.g. *Trichophthalma nigrovittata* [Plate 208], *T. obscura*, *T. confusa*) do visit flowering rainforest trees (e.g. *Syzygium floribundum*, *S. smithii*, *Ackama paniculata*) (Williams 1995). Owing to their relatively frequent within-plant, and sometimes long-distance inter-plant movements, they are likely important agents of pollen transport and out-crossing (Williams and Adam 2010).

The subtropical rainforest plants that nemestrinids have been recorded visiting possess shallow floral structures, with individually small flowers, that do not require long probosces to reach nectar within. Studies by Devoto and Medan (2006) in Argentina noted similarly that *Trichophthalma* visited small flowers with shallow corollas and with little depth effect, and that species exhibited different degrees of floral specificity.

Neminidae Neminids are known from Australia, Madagascar, South Africa and New Guinea (Freidberg 1994). The genus *Nemo* (previously in Aulacigastridae [Colless and McAlpine 1991]) is restricted to eastern Australia. The minute adults frequent *Alocasia* and smooth-trunked eucalypts. The closely related neminine genus *Ningulus* occurs in South Africa (McAlpine 1983). The poorly known *Nemo kentae* was described from the vicinity of Mt Banda Banda in Werrikimbe National Park, north-west of Port Macquarie (McAlpine 1983).

Neriidae The family is mainly distributed in the tropics and is closely related to Micropezidae. Larvae are thought to be saprophagous and feed on decaying vegetable matter (Pitkin 1989a). It includes *Derocephalus angusticollis* (synonym *Telostylinus angusticollis*), the banana stalk fly, a distinctive long-legged species commonly found on rotting fruit, as well as 'flowering' *Lepidozamia peroffskyana* (Zamiaceae) cones (G. Williams pers. obs.). Adults often posture with one another, followed by quick short-distance movements.

Neurochaetidae Neurochaetidae are small, slender brachyceran flies (Colless and McAlpine 1991; McAlpine 1978, 1988a, 1988b, 1993b; Woodley 1989).The family is known from southern Africa, Madagascar, and the Oriental and Australasian regions (McAlpine 1993b). Conspicuous on 'flowers' of *Alocasia brisbanensis* (Araceae) is the upside-down fly *Neurochaeta inversa* (McAlpine 1978 [as *Alocasia macrorrhizos*]; Williams 1993) (Plate 209). Adults assume a 'head downward' stance and are adept at walking backwards, often in preference to forward motion. Though they feed on pollen, adults do not perform pollination, this being undertaken by small nitidulid and derelomine weevils (Williams and Adam 2010). Larvae develop in the watery medium at the base of the spadix. *Neurochaeta inversa* is usually encountered in subtropical rainforests (e.g. Wilson River Primitive Reserve, Terania Creek–Nightcap National Park, Lansdowne Escarpment), wherever flowering host plants occur. Loss of host plants, as at Wingham Brush Nature Reserve, as canopy restoration removed available sunlit floor habitat for *Alocasia*, can result in localised extinction of *Neurochaeta* populations.

Araceae (*Alocasia*), Pandanaceae (*Pandanus*), Zingiberaceae (*Zingiber*) and Musaceae (*Musa*) are known hosts for the Australo-Oriental neurochaetid fauna. Apparent Afrotropical plant hosts are Pandanaceae (*Pandanus*) and Strelitziaceae (*Strelitzea*, *Ravenala*), with a preference by the world neurochaetid fauna largely for the order Zingiberales (Musaceae, Strelitzaceae, Zingiberaceae) (McAlpine 1993b). Though I have sampled *Pandanus* inflorescences at Port Macquarie on several occasions, no neurochaetids have been collected. McAlpine (1993b) suggests that Zingiberales may represent the ancestral host association for Neurochaetidae with subsequent movement to alternative monocot hosts.

Nycteribiidae and **Streblidae** These two families are obligate, blood-feeding parasites of microchiropteran bats and are often collectively referred to as bat flies (Colless and McAlpine 1991; Maa 1989b, 1989c). Adults are long-legged and spider-like in appearance and move with great speed over the fur of their tiny hosts. Roost and nesting caves facilitate the transfer of adult flies between bats. G. Griffiths (1972) makes the argument that nycteribiids and streblids are highly modified Hippoboscidae. The evolutionary history of both families can be assumed to be intimately tied to that of their geologically recent hosts. In addition to being predators of insects, Grimaldi and Engel (2005) note that bats have been a 'sink' for the evolution of specialised insect ectoparasites.

Nycteribiidae are ectoparasitic, wingless, dorsoventrally compessed and spider-like in appearance. Their distribution is principally Paleotropical (Maa 1989b). Adult emergence may be delayed until 'triggered by warmth or physical contact' of a potential host (Colless and McAlpine 1991). The Australasian-Oceanian fauna is relatively rich, with Australia and New Guinea being centres of diversity (Maa 1989b). At least three species

(*Basilia troughtoni*, *Nycteribia parilis* and *N. allotopa*) are known from the Gondwana rainforests region. *Basilia troughtoni* is widespread in Australia (but apparently absent from Tasmania), and *Nycteribia parilis* and *N. allotopa* have extralimital distributions in the Oriental region, the Indonesian Archipelago and New Guinea.

Streblidae (e.g. *Brachytarsina amboinensis*) are found throughout the tropics and subtropics. Their distribution, however, largely corresponds with the winter isotherm of 10°C. This is the temperature at which bats begin to hibernate (Maa 1989c). Streblids normally have fully developed wings and are largely ectoparasitic. Female Ascodipterinae (two species are known from Queensland) are faculatively endoparasitic and, upon finding a suitable host, shed wings and legs and burrow into the body of the host (Maa 1989c).

Odiniidae This family occurs in all major zoogeographical regions but is very rare (D. K. McAlpine pers. comm.). Cogan (1989) does not list any species as yet described from Australia. Undescribed material from northern New South Wales–southern Queensland includes *Traginops* sp., known only from littoral rainforest at Iluka Nature Reserve, and a possible additional species collected from the vicinity of the Lamington Plateau. Additional undescribed species have been collected from Iron Range and Jimmys Scrub (Queensland), and near Coonabarabran, New South Wales (D. K. McAlpine pers. comm.). An undescribed species oviposits in rotting logs (Colless and McAlpine 1991). Neotropical species are closely associated with trees that are rotting and infested with other insects, fungi or exuding sap. Larvae are usually associated with the galleries of wood-boring insects (Gaimari 2010). Holarctic *Odinia* are known to oviposit eggs near the entrance to cerambycid tunnels, the hatching larvae eating the beetle frass, then sometimes attacking the beetle pupa (Pulkkinen and Yang 1984).

Pelecorhynchidae Pelecorhynchids were originally treated as a subfamily of the Tabanidae, and more recently of Rhagionidae (Stuckenberg 2001). They are moderate- to large-sized, robust and hairy flies that superficially resemble Tabanidae (to which they are related), and when first encountered can be mistaken for flower-feeding *Scaptia* tabanids. The family is known only from Australia and South America, and is considered to be a primitive or Antarctic element (Mackerras and Fuller 1942). This Gondwanan distribution mirrors that of *Trichophthalma* (Nemestrinidae). Mackerras and Mackerras (1953) considered that pelecorhynchids, along with *Trichophthalma* and *Scaptia*, originated in the now frozen Antarctic landmass, from which they then radiated and developed.

Only one Australian genus, *Pelecorhynchus*, is recognised (Colless and McAlpine 1991); additional species (nine) are also recorded from Chile (González 2015; Llanos *et al.* 2015). The bright orange-coloured *Pelecorhynchus fulvus* at a glance resembles large asilid flies (*Chrysopogon*) or pompilid wasps (*Heterodontonyx*), and may be an upland participant in a mimicry complex involving these.

The Australian fauna (and potentially so that of Chile) may be especially vulnerable to global warming for these exclusively flower-feeding flies are principally restricted to high montane zones along the eastern massif between southern Queensland (Fraser Island) and Tasmania (Daniels 1989a). The Australian centre of pelecorhynchid distribution and diversity is in mountainous areas of eastern New South Wales where 27 of the 32 or so described Australian species occur (Mackerras and Fuller 1942; Mackerras and Mackerras 1953; Daniels 1977, 1978, 1989a). A single species occurs in the south-west of Western Australia (Daniels 1977). Williams (2002a) lists 15 species and subspecies from south-east Queensland–northern New South Wales; the fauna is especially richly known from the Dorrigo Plateau and the Barrington Tops. Species include *Pelecorhynchus disinctus*, *P. fulvus*, *P. fusconiger*, *P. interruptus*, *P. mirabilis*, *P. niger* and *P. tillyardi*. However, several species inhabit southern regions in New South Wales (*Pelecorhynchus personatus*, *P. fulvus*, *P. deuqueti*), including alpine and subalpine zones (*Pelecorhynchus kippsi*, *P. nero*, *P. rubidus*).

Adult *Pelecorhynchus* feed exclusively on nectar and usually inhabit open vegetation types, but can be found in association with rainforest vegetation where suitable flowering food plants (e.g. *Leptospermum* – Myrtaceae) occur on margins. They are also recorded from *Eucryphia moorei* (Eucryphiaceae/Cunoniaceae) blossoms (Daniels 1989a). Many species will feed from the same flower cluster for several minutes or more (e.g. *Pelecorhynchus niger*, *P. nigripennis* [Plate 211]), their wings at rest; this behaviour is quite unlike bombyliids and nemestrinids, which move more rapidly from flower to flower. Individuals can be so sluggish that they remain undetected for some time, or bury themselves within flowers and remain completely hidden. However, populations are not necessarily present each year, for those that are abundant in one season may be absent the next. Mackerras and Fuller (1942) noted that a dry spring season may be unfavourable for emergence. Even daily patterns of intermittent sunlight appear to effect activity, with adults suddenly becoming inactive during cloud cover. Adults (e.g. *Pelecorhynchus distinctus*, Plate 210) sometimes 'hill-top' (as at Rowleys Rock, Tapin National Park), hovering several metres in the air above food plants, darting away when disturbed, then returning later to a similar position. Larvae, which when fully grown are elongate and worm-like in appearance (Mackerras and Fuller 1942; Mackerras and Mackerras 1953), can be found in damp areas such as creek banks where they may feed on soft-bodied invertebrates, including earthworms and possibly scarab larvae, but they pupate in drier areas (Daniels 1977).

Perissommatidae This is a very small nematoceran fly family with few Australian representatives. Adults have been found only during winter months, and larvae are associated with rotting fungi and microcarrion (Colless and McAlpine 1991). The austral genus *Perissomma* has four known Australian species and one from Chile. In Australia perissomatids inhabit rainforests and sclerophyll forests of New South Wales and Victoria. Larvae thought to be that of *Perissomma mcalpinei* have been collected in moist leaf litter.

Phoridae Phorids are small humpbacked brachyceran flies commonly referred to as scuttle flies (Borgmeier 1968; Disney 1994). The family is in great need of revision. Preserved phorids are known from Cretaceous amber, the family undergoing a huge radiation in the Tertiary (Grimaldi and Engel 2005). Phoridae are now one of the most ecologically diverse of all Diptera and currently number ~2500 described species, though the actual number is very likely much greater. Genera occurring in New South Wales and southern Queensland include *Diplonevra*, *Dohrniphora*, *Megaselia* and *Aphiura* (Evenhuis 1989b). Phoridae are a major dipteran family within subtropical rainforest, but seem to show a greater species richness and abundance at lower levels (Kitching *et al.* 2004). During the day adults can be seen moving frenetically on the upper surfaces of sunlit understorey foliage. The females of some species are wingless or possess imperfectly formed wings incapable of flight, those of certain species being carried by males during copulation (Colless and McAlpine 1991). Grimshaw *et al.* (2013) suggest phorids may be pollinators of the vine *Pararistolochia praevenosa*, a larval food plant of the threatened Richmond Birdwing Butterfly *Ornithoptera richmondia* (Papilionidae). Phoridae (*Megaselia sakaiae*) are pollinators of the related Neotropical *Aristolochia inflata*, the flies also using fallen *Aristolochia* flowers as breeding sites (Sakai 2002). Disney (1994) suggests phorids are specialised predators and parasitoids, their hosts and prey including gastropod snails (Coupland and Barker 2004) and the eggs of frogs (Davis and Disney 2003). Phoridae have also been recorded as dispersal agents for phoretic mites (Clift and Larsson 1987), associates and parasites of social insects (Disney and Roisin 2000), and saprophages (Yeates and Wiegmann 2005). Phoridae are one of the most significant Diptera families attracted to carcasses (and thus of forensic importance), particularly in the early stages of decomposition. Vasconcelos *et al.* (2013), studying necrophagous fly assemblages visiting an animal corpse in a rainforest remnant in north-eastern Brazil, found that Phoridae comprised more than 24 per cent of attracted individual flies in the first 3 hours of death, including immediately after death, and individuals comprised over 25 per cent of all Diptera that later emerged from the corpse.

Although phorids are seemingly more abundant in lower strata, they are to be found feeding on floral nectar of flowering trees in the canopy (i.e. *Alectryon coriaceus*, *Alphitonia excelsa*, *Cuttsia viburnea*, *Euroschinus falcatus*, *Guioa semiglauca*, *Schizomeria ovata*, *Scolopia braunii*, *Syzygium floribundum*, *S. smithii*, *Tristaniopsis laurina*). On these they are part of the Diptera assemblage, along with Coleoptera and miscellaneous Hymenoptera, which in lowland subtropical rainforest remnants dominate the 'small insect' pollinator guild on mass-flowering trees with open floral structures (Williams 1995).

Platypezidae This is a small family consisting of ~250 species, and although well represented in the Australasian and Oriental regions the family is apparently absent from New Caledonia (P. Chandler 1994). Platypezidae probably originated in the earliest Cretaceous. They are small flies that can be mistaken for Phoridae, owing to their humped thorax and their frenetic jerking movements as they run about on the broad leaves of woody plants. This activity is thought related to feeding on various organic deposits on the leaf surface. Larvae are known to breed in the fruiting bodies of fungi. Those of *Lindneromyia* often develop in *Agaricus* species (P. Chandler 2010).

The Australian fauna is placed in the subfamilies Microsaniinae, Callomyiinae and Platypezinae, of which Platypezinae (*Lindneromyia*) are the most diverse in New South Wales and southern Queensland. Several

Linderomyia species (*Linderomyia fergusoni*, *L. argyrognya*, *L. gressitti*) are recorded from Iluka Nature Reserve, but the genus also occurs at higher elevations in the hinterland (*Linderomyia denticulata* – Cunninghams Gap, *L. pulchra* – Mt Tamborine, *L. austraquilla* – Mt Glorious). *Agathomyia austrocollinella* (Callomyiinae) is recorded from Dorrigo and Werrikimbe national parks. *Microsania boycei* occurs in the western Blue Mountains, and the smoke fly *Microsania australis* (Microsaniinae) is sometimes seen flying in apparent male-dominated mating swarms in wood smoke following bushfires (Bickel 1996b; Klocke *et al.* 2011). However, the environmental cue that causes swarms to aggregate is poorly understood (Bickel 1996b).

Platystomatidae Platystomatids, or signal flies, are a family of acalyptrate flies that has its major centre of diversity in Australasia where more than 400 species have been described (Evenhuis 1989c). Platystomatids are one of the largest families of acalyptrate flies recorded from the Australo-Oceanic region (Evenhuis 1989c). Owing to marked wing patterning some species bear a resemblance to Tephritidae and Pyrgotidae (McAlpine 2001). They are normally encountered during the day in the lower and ground-level strata. Adults of this often picture-winged family feed on vertebrate faeces, especially that of mammals, but notably also of small reptiles and amphibians. Adults may arrive at these food sources within seconds of faecal deposition. Being able to utilise the excrement of small terrestrial vertebrates, such as tiny skinks, allows platystomatids and other necrophagous Diptera to survive in rainforest remnants where large vertebrates are now extinct. Species are also attracted to flowers, rotting fruit and decomposing snails (McAlpine 1972b), with adults commonly displaying distinctive signal wing movements in either courtship display or as a simple behavioural trait (as with *Pogonortalis*), sometimes apparently territorial. Larvae are found on carrion and decaying vegetation, but may also be predaceous. The larvae of *Rivellia melliginis* in Ohio are known to attack and destroy the nitrogen-fixing root nodules of legumes (McMichael *et al.* 1990).

The major southern Australian genera are *Duomyia*, *Euprosopia*, *Lamprogaster* and *Rivellia* (Platystomatinae) (Colless and McAlpine 1991; Evenhuis 1989c). Adult *Euprosopia* are particularly conspicuous at mammal dung; those of *Lamprogaster* often frequent lower tree trunks. *Pterogenia* and *Loxoneuroides* rest on the dark and cool, moister undersurfaces of fallen, but elevated, rainforest tree trunks. *Duomyia* (Plate 212) contains the largest number of Australian species and is confined to Australia and Lord Howe Island. In addition to being attracted to dung, adult *Duomyia*, *Euprosopia*, *Microepicausta*, *Pogonortalis* and *Rivellia* can be encountered on the inflorescences of mass-flowering trees (e.g. *Alphitonia excelsa*, *Guioa semiglauca*, *Syzygium smithii*), particularly in littoral rainforest remnants (Williams 1995). They are infrequent visitors occurring in low abundance (usually singletons) and possibly recruited mainly from adjacent shrubby woodland.

The World Heritage Area-listed littoral rainforest site, Iluka Nature Reserve, is an important type locality for numerous Platystomatidae, particularly in the endemic genus *Duomyia* (*Duomyia cancellata*, *D. curta*, *D. latipes*, *D. latipilus*, *D. octoseta*, *D. picta*, *D. scutellaris* [see McAlpine 1972b; Williams 2002]). Iluka Nature Reserve is also the holotype locality for *Zealandortalis gregi*. *Zealandortalis* is restricted to eastern and south-western Australia and the north and south islands of New Zealand. *Hysma lacteum* (Plate 214) is recorded from the Barrington Tops region, Coorabakh National Park and Dorrigo National Park, but also extends south to Bodalla State Forest and northwards to the Wet Tropics of North Queensland. *Hysma* is known from eastern Australia and Papua New Guinea, occurring principally in rainforest but occasionally in wet sclerophyll forest (McAlpine 2001). The endemic genus *Loxoneuroides* occurs in tropical and subtropical rainforests of the east coast (e.g. Woko National Park [Williams 1993]). *Scotinosoma* is possibly confined to the same areas (McAlpine 1972b). Endemic *Lenophila* (e.g. *Lenophila achilles*, *L. coerulea*) are known from eastern Queensland, New South Wales, Victoria, Tasmania, South Australia and Western Australia (D. K. McAlpine pers. comm.; McAlpine and Kim 1977). The poorly known *Lamprogaster maculipennis* has been collected in the vicinity of Ebor and Cathedral Rocks National Park, northern New South Wales. The species was known from three modern specimens and the original two syntypes at Oxford University collected before the middle of the 19th century (McAlpine 1972b; D. K. McAlpine pers. comm.). *Lamprogaster imperialis* occurs in littoral rainforest and elevated hinterland forests at Lamington National Park. *Rhytidorsalis* (*Rhytidorsalis averni*, *R. perforata*) is predominantly coastal Australian (six species) in distribution, but with one species recorded from Asia (D. McAlpine 1999). *Euprosopia* (Plate 213) is a speciose Oriental-Australian genus characterised by adults with distinctively patterned wings. The genus occurs in lowland rainforests (*Euprosopia anostigma*, *E. armipes*, *E. filicornis*, *E. piperata*), as well as high elevated warm and cool temperate rainforests of the hinterland (*Euprosopia albipila*, *E. celsa*, *E. conjuncta*, *E. megastigma*, *E. subacuta*), with several species ranging in altitudinal distribution from littoral rainforest to cool temperate rainforest (e.g. *Euprosopia holmsi*) and latitudinal distribution from tropical Queensland to subtropical climes (*Euprosopia conjuncta*, *E.tenuicornis*, *E. scatophaga*, *E. sericata*) (McAlpine 1972b, 2007b).

Psychodidae Psychodidae or moth flies (sometimes called sand flies) are abundant and delicate, small, hairy-winged and -bodied nematocerans that prefer moist microhabitats and shaded forest areas (Alexander 1928; Colless and McAlpine 1991; Duckhouse 1965, 1966, 1990; Duckhouse and Lewis 1989; Wagner 2018), including caves (Dyce and Wellings 1991). They are diverse in Cretaceous amber (Grimaldi and Engel 2005).

Moth flies are generally weak fliers. Some are nocturnal, but adults can also be found during the day motionless on sheltered foliage. Most adults do not feed, the exception being blood-feeding female Phlebotominae (*Phlebotomus*, *Sergentomyia*), species of which transmit bacteria, viruses and protozoan parasites (Duckhouse and Lewis 1989). They are infrequently collected on flowers (e.g. *Acradenia euodiiformis*, *Alphitonia excelsa*, *Cuttsia viburnea*) (Williams 1995). Larvae feed on fallen leaves and other decomposing plant material, or are found in dung. The larvae of *Trichomyia* probably bore holes in wood, but many require moist habitats (including temporary water-filled tree holes) to complete their larval cycles (Colless and McAlpine 1991).

Australian Psychodidae include an Antarctic faunal element (e.g. *Trichomyia* – Trichomyiinae) related to the Neotropical fauna and, less so, to the Afrotropical fauna (Duckhouse and Lewis 1989). Genera such as *Paratelmatoscopus* and *Peripsychoda* (= *Telmatoscopus*) (Psychodinae) entered Australia from the Oriental region (Duckhouse and Lewis 1989). Psychodid species recorded from south-eastern Queensland and New South Wales include *Nemapalpus australiensis* (Bruchomyiinae), *Paratelmatoscopus brevistylis*, *Peripsychoda obscurus*, *Brunettia ciliaris*, *B. collessi*, *B. tricorniculata*, *Rotundopteryx sylviae*, *R. notata*, *Theischinger imbricatus*, *Abcharis wilsoni* (Psychodinae), *Trichomyia brachypennis*, *T. crucis*, *T. reducta* and *T. triana* (Trichomyiinae); many of these (e.g. *Nemapalpus australiensis*, *Paratelmatoscopus brevistylis*, *Rotundopteryx sylviae*, *Trichomyia brachypennis*, *Theischinger imbricatus*) apparently have very localised distributions on the north coast of New South Wales (Bugledich 1999; Duckhouse 1990; Wagner 2018).

Pyrgotidae These superficially resemble certain Tephritidae. Pyrgotids are mainly nocturnal, and thus rare in collections. However, species can be attracted to light. Some species are known to be internal parasitoids of Hymenoptera and Coleoptera (Pitkin 1989b). *Frontalia* and *Prodalmannia* may oviposit on Scarabaeidae, such as the small melolonthine *Phyllotocus macleayi* (Paramonov 1958b). Adult *Prodalmannia* are also known from flowers by day (Pitkin 1989b). The regional fauna includes *Epicerella plagiata*, *E. barringtoni* and *Facilina campbelli* (V. Korneyev pers. comm., I. McMillan pers. comm.).

Rhagionidae Rhagionidae (= Leptidae), or snipe flies, are widely distributed, and fossils attributed to the family are diverse and common from the Jurassic and Cretaceous (Grimaldi and Engel 2005). *Atherimorpha* (*Atherimorpha corpulenta*, *A. flavofasciata*, *A. uptoni*) is found throughout Australia, and also occurs in South America and southern Africa (Colless and McAlpine 1991; Nagatomi and Evenhuis 1989b; Paramonov 1962). *Chrysophilus* is very widespread and within Australia belongs to the later northern element of Colless and McAlpine (1991). The genus is speciose in the upland forests of southern Queensland and northern New South Wales (e.g. *Chrysophilus androgynus*, *C. commoni*, *C. hardyi*, *C. mackerrasi*) (Daniels 1978; Paramonov 1962).

Adults of some species may be predaceous, but little is known about the biology of the Australian fauna. In subtropical rainforest adults frequent sunny leaves, particularly foliage near ground level. Larvae may be aquatic, and associated with stream-side vegetation or found in moist soils. The larvae of Neotropical *Chrysophilus* are predators of earthworms and insect larvae (Kerr 2009; M. Roberts 1969). In Australia, adult snipe flies are occasionally collected on flowers in lowland rainforest remnants (*Chrysophilus* sp. on *Alphitonia excelsa*, *Athermorpha* sp. on *Schizomeria ovata*) (Williams 1995).

The Australian genus *Spaniopsis*, sometimes included in a separate family, Spaniidae, are biting flies and can be abundant in the autumn (D. Bickel pers. comm.).

Sarcophagidae The brachyceran sarcophagous flies (e.g. *Sarcophaga*) are very richly represented, with high levels of generic endemism in the Australasia-Oceanian region (de Souza Lopes 1989). Sarcophagids are important agents in the decomposition of dead animals (de Souza

Lopes 2010), participating in a larger guild of necrophagous Diptera (along with Calliphoridae, Muscidae, Phoridae, and others such as micropezids and neriids in smaller numbers) that are attracted to animals, both immediately after death, but sequentially in time as decomposition progresses to a dry state (Vasconcelos *et al.* 2013). Members of this family include food parasites in the nests of solitary wasps and bees, are predators of spider eggs, and breed in insect and snail hosts.

Adult *Sarcophaga* are distinguished by the conspicuous parallel, longitudinal bands on the thorax. Commonly encountered on foliage and carrion, adult *Sarcophaga* also feed on the floral nectar of rainforest trees (*Alphitonia excelsa*, *Euroschinus falcatus*); the attraction to *Alphitonia excelsa* occurs possibly because of the strong urine-like odour of its flowers. Sarcophagidae are also part of the pollinator assemblage that visits the inflorescences of *Archontophoenix cunninghamiana* (Arecaceae) in subtropical rainforest, this native palm attracting a cohort of numerous small nematoceran and brachyceran flies (e.g. Scatopsidae, Drosophilidae, Syrphidae, Tephritidae), in addition to other insects (Williams and Adam 2010).

Scatopsidae Scavenger gnats first appeared in the Jurassic and are known from Cretaceous amber. Adults are usually black in colour. They can occur in abundant numbers near rotting matter, and larvae are found in decaying plant material and animal excreta. Several species inhabit rainforest and wet sclerophyll forest, where adults are occasionally found feeding during daylight on floral nectar (e.g. of *Alphitonia excelsa*, *Archontophoenix cunninghamiana*, *Cuttsia viburnea*, *Rhodomyrtus psidioides*, *Syzygium smithii*) (Williams 1995; G. Williams unpubl. data). Neotropical species have been collected over piles of rotting grass, and in association with rotten wood and ant nests (Amorin 2009).

Only two, Psectrosciarinae and Scatopsinae, of the four known subfamilies occur in Australia. Both are present in New South Wales and Queensland (Bugledich 1999; E. Cook 1971, 1989). Species recorded from subtropical Queensland and New South Wales include *Colobostema commoni*, *C. cyclum*, *C. metarhamphe*, *Rhegmoclema hirtepenne*, *Swammerdamella richmondensis and S. sessionis* (all Scatopsinae). *Colobostema* occurs widely on mainland Australia and is also distributed in Papua New Guinea, and the Oriental, Palaearctic, Nearctic and Afrotropical regions. *Rhegmoclema* is a cosmopolitan genus and *Swammerdamella* occurs widely in Australasia and the south-western Pacific. Species of *Scatopse*, *Reichertella*, *Ferniella*, *Coboldia* and *Rhexoza* are also distributed in the south-east of the mainland (E. Cook 1989). Of the Psectrosciarinae, E. Cook (1989) records species of *Anapausis* (New South Wales) and *Psectrosciara* (Queensland, New South Wales, Victoria).

Sciaridae Sciarids are known as dark-winged fungus gnats and are related to Cecidomyiidae, Mycetophilidae, Perissommatidae, Scatopside and Bibionidae (Colless and McAlpine 1991). The family is very widespread and can be commonly caught in light traps. Most species are nocturnal. They are small, fragile flies often seen hovering in mating swarms near tree trunks. The larvae feed variously on detritus, rotten wood, fungi and decaying plant tissue, while the short-lived adults take only nectar or other liquids. They can occur in immense numbers, and were the most abundant family in the tree trunk sticky trap survey of Werrikimbe National Park and the Carrai Plateau (Bickel and Tasker 2004). The family comprised 48 per cent of all insect specimens and was numerically the dominant taxon at all sites from all seasons. Like the Chironomidae in aquatic ecosystems, the Sciaridae are important in terrestrial food chains, transforming plant matter into animal tissue.

Sepsidae These are small ant-like flies that are attracted to rotting fruit, animal dung and urine, and carcasses (where they are particularly common during 'wet' and 'bloat' decomposition phases). Large numbers of *Sepsis* can be attracted to animal dung within minutes after defecation by the host. Larvae live mainly in mammal dung (Colless and McAlpine 1991). The Australian species (Zuska and Colless 1984) are grouped into three lineages: species derived from ancestors that reached Australia well before the Pleistocene, immigrants originating in or near New Guinea that entered Australia possibly during the Pleistocene, and an Oriental group that reached Australia during or after the Pleistocene (Colless 1980). The genus *Parapalaeosepsis* is endemic to Australasia. All four Australian *Parapalaeosepsis* species are restricted to rainforest. *Parapalaeosepsis compressa* extends to the north coast of New South Wales (Colless 1980).

In addition to their well-known saprophagous and necrophagous habits (Zuska and Colless 1984), adults can be abundant and diverse visitors to mass-flowering rainforest trees, in particular the odorous inflorescences of *Alphitonia excelsa* (*Australosepsis niveipennis*, *Lasionemapoda hirsuta*, *Parapalaeosepsis plebeia*, *Sepsis dissimilis*,

S. nitens) both in littoral rainforest and hinterland west sclerophyll forests. However, adults are also attracted to flowering trees whose flowers produce sweet fragrances (e.g. *Acradenia euodiiformis*, *Elaeocarpus obovatus*, *Elaeodendron australe*, *Euroschinus falcatus*, *Guioa semiglauca*, *Tristaniopsis laurina*, *Syzygium floribundum*, *S. smithii*) or inconspicuous fragrances (*Alectryon coriaceus*) (Williams 1995; Williams and Adam 2010). Although adults of several species (*Sepsis dissimilis*) can be abundant, they generally spend long periods of time feeding on nectar produced by individual florets, with infrequent movements to other inflorescences, and so are probably only minor pollinators.

Simuliidae Known in New Zealand as sand flies and elsewhere as black flies, this is a family of ferocious biters. They are small, hump-backed flies and, in Australia, females of a few species of *Austrosimulium* can be serious pests at times (both of animals and humans). The larvae are aquatic in fast-flowing streams, and attach themselves to the substrate with silk and feed on detritus. They have foldable labial fans surrounding their mouths that expand when feeding, catching passing debris. Compared to other regions of the world, the Australian fauna is rather small (Craig *et al.* 2019).

Stratiomyidae Stratiomyid larvae are preserved in Early Cretaceous limestone, the most primitive fossil species being the tiny *Cretaceogaster pygmaeus* in Canadian amber (Grimaldi and Engel 2005). Stratiomyids, or soldier flies, are a moderately large and speciose group, diverse in form; the world fauna numbers ~2000 species, that of Australia ~100 species. The Australian fauna has older elements that have related taxa in South America, and more specialised elements that have relationships with Oriental lineages (Colless and McAlpine 1991). However, individual species are rarely abundantly encountered in lowland rainforests (Williams 1995; G. Williams pers. observ.), though Lambkin *et al.* (2011) found them to be one of the major families of Diptera sampled in upland rainforest at Lamington National Park. Adults can be found resting on leaves, flowers, or at rotting fruit and decaying plant material. Larvae of some are associated with water bodies, plant matter and soil. Dodson and Daniels (1988) reared *Hermetia illucens* and *Saldubela margaritifera* (not a subtropical species) from the decaying trunk of the tree *Dysoxylum gaudichaudianum* (Meliaceae) in northern Queensland rainforest; this tree is a species that also occurs in subtropical Queensland (Harden *et al.* 2006).

Adults range in form from those that are elongate (e.g. *Xanthoberis* [Plate 216], *Hermetia*), to rotund (*Antissa*, *Antisella*, *Lecomyia* [Plate 215]). A number of Hermetiinae resemble wasps or bees (Lessard *et al.* 2019). Adult *Hermetia* (Hermetiinae) can exhibit territorial behaviour. The largest genus is *Odontomyia* (Stratiomyinae), several of which are more or less metallic in colour and frequent flowers. Adult soldier flies, however, possess no specialised morphological feeding adaptations, such as long probosces (as found in Bombyliidae, Nemestrinidae, Tabanidae); instead their mouthparts are short, and in this respect are akin to flower-frequenting Calliphoridae, Tachinidae, Pelecorhynchidae, and many other unadapted anthophilous Diptera. Stratiomyids can be diverse on flowering *Alphitonia excelsa* growing on margins of littoral rainforest remnants (*Hermetia illucens*, *Odontomyia decipiens*, *O. laterimaculata*, *O. ?sydneyensis* and additional *Odontomyia* species), but the suite of species differs on *Alphitonia excelsa* growing in hinterland wet sclerophyll forest (e.g. *Acanthosargus* sp., *Damaromyia ?clivosa*) (Williams 1995). However, foliage-frequenting species are diverse in hinterland lowland rainforest and wet sclerophyll forest (G. Williams unpubl. data).

Stratiomyidae recorded from the Gondwana rainforests region include *Antisella parvidentata*, *Lecomyia notha*, *Pachygaster nitens* (Antissinae), *Xanthoberis* sp. (Beridinae), *Damaromyia clivosa*, *D. discolor*, *D*, *nitens*, *D. trina*, *D. whitei*, *Dochmiocera aurilineata* (Pachygastrinae), *Sargus darius* (Sarginae) and numerous *Chiromyza* (Chiromyzinae) (Daniels 1978; Hardy 1931, 1932a; Woodley 1989; Williams 2002). A conspicuous species in lowland and upland subtropical and warm temperate rainforests (e.g. at Coorabakh and Tapin Tops national parks) is the large (15 mm) orange and black wasp mimic *Syndipnomyia auricincta* (Clitellariinae), adults of which will fly repeatedly about the base of trees in search of oviposition sites. When disturbed they quickly fly away, often to return to the same tree minutes later. A spectacular species occasionally seen on fallen trees, or at ground level, is the metallic blue-green *Lecomyia notha* (Antissinae).

Syrphidae Syrphidae, or hover flies, are a widespread family (~6000 described species), known from Eocene Baltic amber, but one that probably originated in the Late Cretaceous (Grimaldi and Engel 2005). Regional Syrphidae include relatively large species such as *Orthoprosopis multicolor* (Plate 223), of ~14 mm in length, to diminutive *Graptomyza* (Plate 220) of barely 3 mm. Adult syrphids are predominently flower-frequenting and nectar- and pollen-feeding in habit (Colless and McAlpine 1991; Ferguson 1926a, 1926b; Hardy 1921b, 1933; Paramonov 1955; Riek 1954a; Reemer and Ståhls 2013; Ståhls *et al.* 2003; Thompson and Vockeroth 1989; Vockeroth 1969). Larvae are saprophagous (many Eristalinae) or predators (Microdontinae).

Though Australia is overall poor in species, suggesting the Syrphidae are a geologically recent group (Colless and McAlpine 1991), the fauna recorded from elevated hinterland sites in the Gondwana rainforests region is a rich one. It includes *Allograpta alamacula*, *A. notiale*, *A. terraenovae*, *Archimicrodon barringtonensis*, *A. nicholsoni*, *A. vittatus*, *Ceriana apicalis*, *C.mellivora*, *Chalcosyrphus pleuralis* (Plate 217), *Cyphipelta rufocyanea*, *Deineches hackeri* (Plate 218), *Eristalinus maculata*, *Eristalis copiosa*, *E. tenax*, *Malometasternum rufocaudatum*, *Melangyna damastor*, *Mesembrius bengalensis*, *M. hilaris* (Plate 221), *Microdon nigromarginalis*, *Monoceromyia macleayi*, *M. saundersi*, *M. subarmata*, *Odyneromyia spadix* (Plate 222), *Orthoprosopa grisea*, *O. multicolor*, *Syritta hackeri* and *Xylota flavitarsis* (Plate 224) (Carver and Thompson 2003; Mengual and Thompson 2015; Riek 1954a; Thompson and Vockeroth 1989; Williams 2002). Many of these also frequent lowland rainforest, but lowland rainforests (littoral rainforests in particular) are the preserve of the large stripe-winged *Dideopsis aegrota*, an opportunistic flower-frequenting species found on numerous flowering plants, including the mangrove *Avicennia marina* (Plate 219); the distinctive dark bands on its wings create a blurred image of the whole fly when in flight.

Syrphidae are the iconic anthophilous dipteran family (Williams and Adam 1994, 2010; Willmer 2011), but syrphids are also pollen feeders (Faegri and van der Pijl 1979; Holloway 1976), such that their pollination contribution may be achieved at a reproductive cost to the plant. Species show differential morphological and behavioural adaptations to the floricolous habit (yet lack elongate probosces), and also exhibit forest stratification responses to available flowers, but demonstrate foraging fidelity only in the sense of general constancy to whatever plant happens to be flowering at any one time. Syrphids are not commonly encountered on flowers in shaded rainforest strata, more usually being inhabitants of sunlit canopies, rainforest margins, or large sunny openings below damaged tree crowns. The smaller-sized species of Syrphini (e.g. *Allograpta*, *Betasyrphus*, *Episyrphus*, *Eupeodes*, *Melangyna*) and Eristalinae (*Chalcosyrphus*, *Syritta*, *Triglyphus*) characterise the syrphid fauna visiting flowering herbs (*Pollia*, *Dianella*, *Pratia*, *Lobelia*) and vines (*Cayratia*, *Clematis*, *Hibbertia*) near or at ground level; the tiny flowers of many ground-hugging herbs there are able to supply sufficient floral resources to meet the energy needs of small flies. Elongate and narrow-waisted, *Episyrphus* participate with bees in the pollination of the white-flowered herb *Pollia crispata* (Williams and Walker 2003), and mixed bee–Syrphini pollinator assemblages dominate the pollination strategies of herbaceous plants growing on forest edges.

Syrphidae are also common at flowering rainforest trees and shrubs: *Acradenia euodiiformis* (*Psilota*, *Eristalis*, *Melangyna*, *Archimicrodon*), *Acronychia oblongifolia* (*Graptomyza*), *Archontophoenix cunninghammiana* (*Graptomyza*), *Alectryon coriaceus* (*Episyrphus*, *Graptomyza*, *Microdon*, *Syritta*), *Alphitonia excelsa* (*Dideopsis*, *Eristalinus*, *Mesembrius*, Syrphini), *Ackama paniculata* (*Psilota*), *Cuttsia viburnea* (*Psilota*), *Euroschinus falcatus* (*Mesembrius*, *Psilota*), *Guioa semiglauca* (*Ceriana*, *Mesembrius*), *Tristaniopsis laurina* (*Psilota*), *Syzygium floribundum* (*Eristalinus*) and *S. smithii* (*Psilota*) (Williams 1995; G. Williams unpubl. data). These are all plants with shallow, easily accessible nectaries and pollen. The large wasp mimic *Orthoprosopa multicolor* has been recorded from *Syzygium smithii* in lowland wet sclerophyll forest; and in littoral rainforest *Mesembrius hilaris*, in addition to being a visitor to trees and shrubs, is common on the flowers of the vine *Cissus antarctica* and is also frequently observed on the flowers and foliage of the mangrove *Avicennia marina* (Avicenniaceae) in adjacent estuaries. *Chalcosyrphus pleuralis* feeds on the nectar of *Acronychia oblongifolia* in lowland wet sclerophyll forest and adults have been observed on fallen rotting tree trunks in subtropical and dry rainforest. *Xylota flavitarsis* frequents foliage, and the flowers of *Clematis glycinoides*, on the margins of lowland rainforest and wet sclerophyll forest (G. Williams unpubl. data).

Some genera are often unassociated with flowers. The large pelecorhynchid-like *Deineches hackeri* visits flowers of *Leptospermum* where these grow alongside rainforest (as in Barrington Tops), but commonly adults are encountered patrolling the vegetation near the forest floor or resting on low understorey foliage, individuals moving indistinctly below sheltering ferns and fallen tree branches. These are likely to be females looking for oviposition sites. The equally sized *Odyneromyia spadix* and *Orthoprosopa multicolor* are also species seen near the forest floor. Species of *Archimicrodon* are associated with ant nests, where their larvae are scavengers or predators (Colless and McAlpine 1991), and can be found resting on foliage, fallen tree limbs and rocks. Adults of *Eristalis copiosa* noisily hover above low shrubs, their colour and form suggestive of co-occurring tabanids and calliphorids, and *Eristalinus hervibazini* and *Eristalis tenax* resemble honey bees (*Apis mellifera*), though the possible mimicry of *Apis* by these syrphids is puzzling owing to the historically recent introduction of *A. mellifera* into the Australian landscape, and the cosmopolitan distribution of *Eristalis tenax* whose occurrence in Australia is assumed to pre-date it. Some species, such as the elongate stratiomyid-like *Odyneromyia iridecescens* and large calliphorid-like *Merodon*, are the objects of chance and infrequent encounters unassociated with any particular habit or plant. *Psilota* is a complex of calliphorid-looking syrphids, well represented in Australia, that frequent flowers and foliage. Species of *Ceriana* (*Ceriana ablepta*, *C. ornata*) resemble conopid flies, paper wasps (*Polistes*, *Ropalidia*) and small eumenine wasps (e.g. *Paralastor*), though Rashed and Sherratt (2007) question whether the wasp-like warning signals of hover flies prevent other flower visitors from sharing flower resources.

Tabanidae Tabanidae, commonly called March flies or horse flies, are closely related to the brachyceran families Pelecorhynchidae, Rhagionidae, Stratiomyidae, and Nemestrinidae. They appear to have evolved in the Early–Mid Cretaceous (Grimaldi and Engel 2005). Most are medium- to large-sized, robust flies; all are distinguished by large eyes and a prominent strong proboscis. The tabanid fauna of subtropical Queensland and New South Wales is a rich and ecologically diverse one (Williams 2002), and includes numerous species that are blood feeders (*Scaptia* spp.), those that feed on floral nectar (*Scaptia auriflua*), and those that do both (e.g. *Scaptia quadrimaculata*, Plate 229). This may in part reflect sexually dimorphic feeding habits for blood feeding, as is known to be exhibited by adult females (Lessard 2014). Flower-frequenting species generally have hairy bodies, this facilitating the transport of pollen between plants. All subfamilies known from the Australasian area (i.e. Chrysopinae, Pangoniinae, Tabaninae) occur in the Gondwana rainforests region (Daniels 1978, 1989b). The more primitive genera (e.g. in the tribe Pangoniini) largely have a southern distribution and are related to taxa in New Zealand, South America and South Africa (Colless and McAlpine 1991). A major feature of the Australian tabanid fauna is that 'it contains the largest assemblage of generalised Pangoniini, Scionini, Bouvieromyiini and Diachlorini in the World' (I. Mackerras 1956a): ~90 per cent of our fauna. Examples of Pangoniini occurring in the Gondwana rainforests region are *Ectenopsis* (*Ectenopsis angusta*, *E. vulpecula*), *Caeonoprosopon* (*Caeonoprosopon australe*, *C. nigrovittatum*, *C. trichocerum*) and *Austroplex* (*Austroplex brevipalpis*, *A. chrysophilus*) (I. Mackerras 1956b) (Plates 19, 225). The Australian Tabanini (e.g. *Tabanus davidsoni*, *T. parvicallosus*) generally belong to the Oriental, or Indo-Malayan, element in the tabanid fauna. However, several species exhibit a limited Bassian distribution (I. Mackerras 1971). The Papuan Tabanini are more diverse than the Australian, and are Oriental in origin (I. Mackerras 1971). Diachlorini include several species of *Cydistomyia* (*Cydistomyia nigropicta*, *C. victoriensis*), and the predominantly austral Scionini (Lessard 2014; Lessard and Yeates 2013; I. Mackerras 1960) comprises in part numerous species of *Scaptia* (Plates 226, 227, 229, 230), including the large black and white coloured *S. guttata* (Plate 228), a species that exhibits territorial behaviour. *Scaptia guttata* can be a commonly encountered species in wet sclerophyll forest at high elevations (e.g. >1200 m in Werrikimbe National Park). *Scaptia baylessi* is known only from the Bunya Mountains National Park (Lessard and Yeates 2012a). *Anzomyia* is known only from Australia and New Zealand. *Anzomyia chrysomallis* is recorded from Mount Wilson (Lessard and Yeates 2012b).

Tabanids inhabit forests at all elevations, from coastal littoral rainforests (e.g. *Austroplex brevipalpus*), to intermediate forests such as the Lansdowne Escarpment and Comboyne Plateau (*Cydistomyia casuarinae*, *Scaptia auriflua*, *S. barbara*, *S. limbithorax*, *S. vicina*), and high altitude cool temperate localities such as the Barrington Tops and the Border Ranges (*Scaptia aurinotum*, *S. jacksoniensis*, *S. plana*, *S. pulchra*, *Copidapha auripleura*). Individual species are distributed through a range of elevations from coastal forests to those of the hinterland (e.g. *Caeonoprosopon trichocerum*, *Scaptia violacea*, *S. monticola*). Several are attractively coloured, notably *Scaptia barbara* (Plate 227), a large black and orange species, and *S. subcana* (Plate 230), whose eyes are vividly rainbow-coloured; both species are known from the Lansdowne Escarpment. However, many tabanids are only associated with rainforest – open forest margins and large sunlit openings, rather than being inhabitants of the rainforest core – and these are likely to be species that more characteristically inhabit eucalypt-dominated open forest and that opportunistically exploit rainforest edge environments. Few enter the shaded rainforest environment unless large sunlit understoreys are available. Blood-feeding species are ecologically tied to their hosts, and so their distribution within individual rainforest stands will mirror the availability and abundance of appriate prey. Tabanids recorded from rainforest flowers are few, but include *Scaptia auriflua* (Plate 226) and species of *Tabanus* (at *Alphitonia excelsa*), *Scaptia quadrimaculata* (Plate 229) (at *Cuttsia viburnea*) and *Cydistomyia nigropicta* (at *Syzygium floribundum*, *Tristaniopsis laurina*).

Tachinidae There are over 8000 species of described tachinid flies (Cantrell and Crosskey 1989). The larvae of all Tachinidae are endoparasites of terrestrial arthropods. Adult females either lay their eggs directly on the host or insert them. Some species have free-living first instar planidia that actively seek out a host, and others broadcast minute eggs on foliage of the host's food plant that are ingested by the host as it feeds, and that then hatch in the host's gut (Grimaldi and Engel 2005). Hosts of Phasiinae are almost strictly hemipteran bugs, and those of Dexiinae are mainly beetle larvae; however, Goniinae parasitise several insect orders (Cantrell and Crosskey 1989). Adult *Austrophorocera*, *Blepharella*, *Blepharipa*, *Palexorista*, *Tritaxys* (Goniinae), *Prosena*, *Rutilia*, *Senostoma* (Dexiinae) and *Cylindromyia* (Phasiinae) feed on floral nectar as part of a mixed diet, many being attracted to the odorous fragrance of *Alphitonia excelsa* (Williams 1995; Williams and Adam 2001, 2010). Foetid odours are known to stimulate feeding or oviposition instincts in Tachinidae, and in related sapromyophilous Calliphoridae, Muscidae and Sarcophagidae.

The Australian tachinid fauna, although still poorly known, is highly emdemic (*c.* 41 per cent of genera [Cantrell and Crosskey 1989]). All subfamilies (Dexiinae, Goniinae, Phasiinae, Tachininae) occurring in the Australasian and Oceanic regions are present in south-east mainland Australia. *Cylindromyia* (Phasiinae) is widespread in Australia, but most species are are recorded from the eastern states (Paramonov 1956). *Cylindromyia flavifrons* and *C. nigricosta* occur in the Barrington Tops. In Australia *Macrochloria* (Tachininae) is recorded from New South Wales and Tasmania, and *Winthemia* (Goniinae) is widely distributed in the Australasian region, including Tasmania and New Caledonia, but is apparently absent from New Zealand (Cantrell 1989; Cantrell and Crosskey 1989; Daniels 1978; Williams 2002).

Of the Dexiinae, the tribe Dexiini are almost totally cosmopolitan and form a significant tachinid tribe within the Australian fauna (Barraclough 1992). Although largely confined to the Australian mainland and Tasmania, the Australasian fauna is diverse. However, dexiines are absent from New Zealand, and depauperate in New Guinea, Melanesia and Polynesia. Although there is an extension of the Australian dexiine fauna into Melanesia, there are almost no relatives in the adjacent Oriental region. 'This disjunction may reflect distinctive tachinid-host interactions in Australasia', but the resolution of this puzzle requires further host records (Barraclough 1992). Dexiini occurring in the Gondwana rainforests region include the endemic genera *Trichostylum*, *Heterometopia*, *Geraldia* (which has a superficial resemblance to the little-known Afrotropical genus *Piligena*), *Rutilotrixa* (distributed from eastern New South Wales–Tasmania, and southern Western Australia), *Amphitropesa* (mainly occurs in south-east Queensland, eastern New South Wales) and *Senostoma* (Williams 2002). *Senostoma* is generally restricted to Australia, but with two species in New Caledonia. The genus is notably absent from the Northern Territory and Cape York Peninsula (Barraclough 1992). The distinctive long-legged adults are often encountered resting near the base of tree trunks. *Senostoma basale* is endemic to the Gondwana rainforests region, *S. hirticauda* is known only from Barrington Tops, and *S. simulcercus* is known only from Mt Tamborine.

The dexiine tribe Rutilini is confined to the Indo-Oriental and Australasian regions, but is absent from New Zealand. Rutilini form a dominant and especially abundant group in the Australian tachinid fauna (Crosskey 1973), and are well represented in the subtropical eastern Australia (Cantrell and Crosskey 1989; Crosskey 1973; Daniels 1978; Paramonov 1968; Williams 2002). It includes several species of the genera *Rutilia* (e.g. *Rutilia cryptica*, *R. curpreiventris*, *R. lepida*, *R. dorsomaculata*, *R. vivipara*), *Amphibolia* (*Amphibolia albocincta*, *A. ignorata*, *A. stolida*), *Chetogaster* (*Chetogaster violacea*, *C. viridis*) and *Prodiaphania* (*Prodiaphania deserta*, *P. fullerae*), which inhabit upland forests (e.g. Bunya Mountains, McPherson Range, Dorrigo, Barrington Tops). *Rutilia* is a widespread genus distributed from India, to the Pacific, the large *R. lepida* (Plate 201) and *R. dorsomaculata* often seen on flowers, but particularly conspicuous on sunny warm days displaying 'hill-topping' behaviour hovering high above vegetation on rocky outcrops. This they often do in concert with other large rutiliines, such as the russet-winged *Chetogaster violacea* and the distinctive black and white-coloured *Formosia speciosa*, these two often descending to rest on shaded rock overhangs.

Tanyderidae The primitive crane flies are rare nematoceran flies with a world fauna of ~45 species. Tanyderidae are essentially Australasian in distribution.

Adults resemble Tipuloidea but have a distinctive long-necked form, and are usually found together with them (Colless and McAlpine 1991). The family has been considered the most primitive of living flies, but this has been questioned (Grimaldi and Engel 2005).

Most genera are regional endemics occurring in the south-west Nearctic region, southern South America, southern Africa, Australia and New Zealand (Oosterbroek 1989). *Nothoderus* (restricted to Tasmania) and *Eutanyderus* are Australian endemics, the larvae of the latter boring into the outer layers of submerged rotting logs in alpine streams at high elevations. Two genera, *Protanyderus* (Holarctic and Oriental regions) and *Radinoderus* (West Papua, Australia and western Pacific), have more extensive distributions. *Radinoderus dorrigensis* is known only from northern New South Wales (Bugledich 1999).

Tephritidae The tephritid fauna numbers more than 4500 described species world-wide (Norrbom 2010). Tephritids are known commonly as fruit flies, and are small-sized brachyceran flies <10 mm in body length. A number (e.g. Dacinae) are infamous as serious agricultural pests of fleshy fruit (Anon 2016). The wings of adults are usually highly ornamented, a feature parallelled in certain Lauxaniidae (*Trypetisoma*) and species of the larger Platystomatidae (*Euprosopia*, *Lenophila*), Bombyliidae (*Anthrax*), Syrphidae (*Dideopsis aegrota*), Tabanidae (*Cydistomyia casuarinae*), Therevidae (*Johnmannia*, *Pipinnipons*), Pelecorhynchidae, Otitidae (Ulidiidae) and also in the New World Richardiidae. Adult *Euphranta* resemble Pyrgotidae (*Neotoxura*), a small and mainly nocturnal family (Colless and McAlpine 1991) sometimes attracted to ultraviolet lights placed in subtropical rainforest. Natural enemies of tephritid larvae in rainforest include frugivorous vertebrates such as pigeons and rodents. Fruit predators are capable of eating a high proportion of fruit in which larval flies are present (Drew 1987).

Tephritids may exhibit specialised host preferences to the fruit of particular plant families, and even genera (Novotny *et al.* 2005). The Dacinae (*Anastrepha*, *Bactrocera*, *Ceratitella*, *Dacus*) in particular have a strong biological association with host plants (Drew 2004) and are all fruit breeders (e.g. *Dacus cacuminatus* on *Pouteria australis* [Drew 1987]). The Dacinae-Dacini, largely composed of species of *Dacus* and *Bactrocera*, appear to have co-speciated with rainforest plant species primarily during the Tertiary, this 'influenced by combinations of oscillations in topography, localised climate, and land bridges during glaciation cycles' (Drew 2004).

Some Trypetinae and Phytalmiinae also breed in fleshy fruit; for example, *Dirioxa pornia* has been recorded from *Syzygium smithii* (Permkam and Hancock 1995). However, the majority of the world's tephritid fauna do not feed on fruit (Aluja and Allen 2001; Hardy and Foote 1989). *Dirioxa* has been reared from rotting wood of *Dysoxylum* sp. (Dodson and Daniels 1988) and the cones of *Araucaria cunninghamii* and dry-seeded *Castanospermum australe* (Permkam and Hancock 1995), and the larvae of *Lumirioxa araucariae* develop beneath the loose bark of *A. cunninghamii* trees (Permkam and Hancock 1995). But otherwise tephritids breed in living plant tissue that includes flower heads, seeds, stems and roots (Hardy and Foote 1989; Norrbom 2010). In Australian subtropical rainforests fruit flies are occasional visitors to flowers of trees in lowland rainforest remnants: *Spathulina* (on *Alphitonia excelsa*, *Euroschinus falcatus*, *Guioa semiglauca*, *Syzygium floribundum*, *S. smithii*, *Tristaniopsis laurina*); ?*Austrotephritis* (on *Tristaniopsis laurina*); *Bactrocera*, *Trupanea* (on *Alphitonia excelsa*); *Trupanea* (on *Syzygium floribundum*); and *Philophylla* (on *Alectryon coriaceus*) (Williams 1995). In lowland rainforests at Cooloola in southern Queensland Zalucki *et al.* (1984) found that tephritids exhibited specific patterns of spatial abundance and that this related in part to changing population abundance levels, fly movements, and breeding and feeding behaviour.

The subfamilies Trypetinae and Phytalmiinae include several genera of biogeographical interest (Drew 2004; Hancock 2013; Hancock and Drew 2003a, 2003b; Permkam and Hancock 1995): *Austrorioxa* (endemic genus), *Clusiosoma* (genus confined to Australia, Solomon Islands and Moluccas), *Dirioxa* (occurring from North Queensland to central coastal New South Wales, and New Caledonia) and *Lumirioxa* (restricted to south-eastern Queensland and northern New South Wales). *Micronevrina* has no close allies and is restricted to eastern Australia. Species are well represented in regional rainforests and include *Acanthonevroides mayi*, *A. nigriventris*, *Austrorioxa acidiomorpha* (Plate 231), *Clusiosomina puncticeps*, *Dirioxa pornia*, *Elleipsa distincta*, *Lumirioxa auraucariae*, *Micronevrina apicalis*, *M. gloriosa*, *M. hyalina*, *M. montana*, *Fusciluda unicuneata*, *Euphranta leichhardtiae* and *Philophylla erbia*. *Euphranta numeralis*, originally described from specimens captured in a mixed mangrove–lowland rainforest margin (Lansdowne Nature Reserve) in northern New South Wales (Permkam and Hancock 1995), has also been collected at ultraviolet lights in adjacent escarpment rainforest (G. Williams unpubl. data) and reared from the fruit of *Maclura conchinchinensis* (Moraceae) in south-east Queensland (Hancock and Drew 2003b). *Maclura conchinchinensis* is a robust, woody, armed vine that occurs in rainforest south to Ulladulla, southern New South Wales (Harden *et al.* 2007). *Clusiosomina puncticeps* has been bred from the fruit of *Ficus fraseri* and swept from *F. coronata* (Hancock and Drew 2003a; Permkam and Hancock 1995).

Tephritinae include the endemic genera *Leipana* and *Paraspathulina*. *Leipana* (*Leipana helichrysii*) is restricted to south-eastern Australia (Hardy and Drew 1996). *Trupanea bifida*, *T. prolata*, *Austrotephritis bushi*, *A. furcata* and *A. protrusa* are recorded from upland forests (Hancock and Drew 2003c; Williams 2002).

The agriculturally significant Dacinae (Anon 2016) include the Indo-Australasian genus *Ceratitella*, which comprises two species in New Guinea, one species recorded from Pakistan, Thailand, and Indonesia, one species from China and the Ryukyu Islands, and four species (*Ceratitella amyemae*, *C. bifasciata*, *C. loranthi*, *C. recondita*) whose ranges include the Gondwana rainforests region (Permkam and Hancock 1994). *Ceratitella recondita* is endemic to the region. There are high levels of endemism within the tribe Dacini, with *Bactrocera* in Australia exhibiting extensive speciation (Drew 2004).

Thaumaleidae This is a small family, known as trickle midges, mainly occurring in the Holarctic region (Peterson and Theischinger 1989; B. Sinclair 2008) with a fossil record dated to the Late Jurassic–Early Cretaceous (Grimaldi and Engel 2005). Adults are stoutly built with wingspans of ~5–10 mm. Larvae prefer cool streams and vertical shaded surfaces where water flows very thinly like a film over rocks, so that their backs are always exposed above the surface (Peterson and Theischinger 1989; B. Sinclair 2000b; Wirth and Stone 1956). They feed on detritus and diatoms, actively moving to feed and attach themselves to wet rocks in flowing streams, not necessarily ones that are fully submerged. Pupation occurs among moist leaves, in beds of moss, mud, or the undersurface of stream rocks or between stones (Wirth and Stone 1956). Adults are largely diurnal in habit and are found on rocks and stream-side vegetation. Little is otherwise known about their biology (Colless and McAlpine 1991; Grimaldi and Engel 2005; Peterson and Theischinger 1989).

The only Australian genera are *Austrothaumalea* and *Niphta*, both of which have relatives in New Zealand and South America (Theischinger 1986). *Niphta bickeli* is recorded from Dorrigo and *N. farecta* is recorded from the Blue Mountains National Park (Bugledich 1999; B. Sinclair 2008). *Austrothaumalea* recently described from rainforests and moist microhabitats in eastern New South Wales (e.g. Blue Mountains National Park, Barrington Tops) and southern Queensland (Bunya Mountains) include *A. barrydayi*, *A. bickeli*, *A. capricorni*, *A. concava*, *A.denticulata*, *A. macalpinei*, *A. minnamurrae*, *A. ramosa*, *A. spinosa*, *A. theischingeri* and *A. zentae* (B. Sinclair 2008). Several have localised known distributions.

Therevidae Therevidae, or stiletto flies, are elongate, moderately sized, day-active brachyceran flies. Owing to their form some species can be mistaken for small asilid flies or pompilid wasps. Therevids are mainly semi-arid and open forest adapted (Lambkin *et al.* 2005). However, a number, for example the metallic-coloured members of the genus *Johnmannia* and the putative wasp-mimicking *Collessiama*, inhabit rainforest and wet sclerophyll forest (Lambkin and Turco 2013; Lambkin *et al.* 2005).

The family is known from Mesozoic Burmese amber (Grimaldi and Engel 2005) and the extant fauna has a cosmopolitan distribution (>1200 described species [Lambkin and Turco 2013]). The subfamily Therevinae is especially diverse and may have radiated from an origin in the Southern Hemisphere (Winterton *et al.* 2015). The Australasian region contains the greatest number of stiletto flies of any biogeographical region, amounting to more than 700 described and undescribed species or >40 per cent of the known world therevid fauna (Winterton *et al.* 2009). *Ectinorhynchus* is considered an ancient Antarctic element (Colless and McAlpine 1991); however, the Australian therevid fauna is almost exclusively endemic (except for *Irwiniella*) at the genus level and comprises three endemic genus groups: (1) *Anabarhynchus*, (2) *Agapophytinae* and (3) *Taenogera* (Colless and McAlpine 1991; Irwin and Lyneberg 1989; Lambkin *et al.* 2005; Lyneborg 2001; Mann 1928, 1929; Winterton and Lambkin 2012; Winterton *et al.* 1999a, 2001).

Adult therevids often have patterned wings (e.g. *Johnmannia powerae*, *Collessiama narellae*, *Agapophytes albobasalis*, *Acraspisoides monticola*) (Plates 232, 233). Those of *Eupsilocephala* are bright orange. Larvae are predaceous and are known to be parasitised by bombyllid flies (Irwin and Lyneberg 1989; Irwin and Yeates 1995). Several species have been observed captured by the large bulldog ant *Myrmecia nigrocincta* (G. Williams pers. obs.), and species of *Agapophytus* are often observed at the edges of rainforest streams or on moss covered boulders on the sunlit forest floor. However, little is known about the habits of adults (Irwin and Lyneberg 1989). At times they are seen hovering above tall flowering shrubs (e.g. *Melaleuca linariifolia* – Myrtaceae), or low near the ground, in open forests. The family is infrequently observed on flowering rainforest trees, though individuals have occasionally been collected from flowers of *Alphitonia excelsa*, *Euroschinus falcatus*, *Guioa semiglauca*, *Scolopia braunii* and *Tristaniopsis laurina* in lowland subtropical remnants (Williams 1995). Adults exhibit apparent altitudinal clines in abundance. Lambkin *et al.* (2011) found that therevids were more abundant at lower altitudes (300–500 m) in rainforest at Lamington National Park, and at higher altitudes were uncommon (700 m) or absent (1000 m).

The southern Queensland–northern New South Wales fauna includes many species that inhabit open forest, woodland and heath, but also many that are associated with wet sclerophyll forest, rainforest and sunny rainforest margins (Ferguson *et al.* 2014; Irwin and Lyneberg 1989; Irwin and Winterton 2016; Lambkin and Turco 2013; Lambkin *et al.* 2005; Lyneborg 2001; G. Williams unpubl. data; Winterton 2006, 2007a, 2007b, 2009, 2011; Winterton *et al.* 1999; Winterton and Irwin 2001; Winterton and Lambkin 2012; Winterton *et al.* 1999a, 1999b). *Laxotela* is restricted to southern Australia. *Laxotela hauseri* is recorded from the Barrington Tops (Winterton and Irwin 1999). *Nanexila* is confined to northern, western and eastern Australia, but unknown from Tasmania. Most *Nanexila* species have a southern distribution but are known from few specimens. *Nanexila livea* occurs in rainforest in the Bunya Mountains, as well as Mt Tamborine, Mt Glorious and Mallangenee. *Acatopygia* is restricted to eastern Australia, with *Acatopygia pulchella* recorded from Minnamurra Falls in southern New South Wales. Additional species associated with wet forests include *Acatopygia pulchella*, *Acupalpa imitans*, *Agapophytus albobasalis*, *A. palmulus*, *Anabarhynchus iancommoni*, *A. poona*, *Collessiama narelleae*, *C. damani*, *Eupsilocephala albodorsalis*, *E. sulfurea*, *Johnmannia powerae*, *Manestella obscura*, *Neodialineura stratithorax*, *Taenogera notatithorax* and *Zelothrix warrumbungles*. The genus *Johnmannia* is rarely collected, though the large *J. powerae* is often found in intercept traps set on the margins of lowland subtropical rainforest at Lorien Wildlife Refuge (Lansdowne Escarpment), this being the southern-most known distribution record for the species (Lambkin *et al.* 2005; G. Williams unpubl. data). *Collessiama narelleae* and *C. damani* (both from northern New South Wales–southern Queensland) are apparent wasp mimics, their distinctively banded and coloured wings (particularly of the males) reminiscent of pompilid wasps (e.g. *Phanagenia fasciata*, *Calopompilus raptor*) and also those of the asilid fly *Orthogonis ornatipennis*. Lambkin and Turco (2013) noted that *Collessiama narelleae* is often collected in association with the pompilid wasp *Priocnemus erythrothorax*. *Anabarhynchus* (e.g. *Anabarhynchus montanus*, *A. barrington*, *A. glorious*) is diverse in Australia with close relatives in New Zealand and southern South America; the whole complex represents a Gondwanan faunal element (Lyneborg 2001).

Tipuloidea The Tipuloidea (= Pediciidae, Limoniidae, Cylindrotomidae, Tipulidae) are an extremely large cosmopolitan superfamily of fragile, stalk-legged nematoceran flies commonly referred to as crane flies (though the term 'tipulids' has practical currency) (Oosterbroek 2019). The families cited above are often given subfamily rank within Tipulidae. More than 15 000 species are known world-wide, and it is the largest taxon group of Diptera occurring in Australia (>1000 species) (Colless and McAlpine 1991; Theischinger 2015). Most world species were reputedly described by one person, Charles P. Alexander (1889–1981), professor at the University of Massachusetts (Grimaldi and Engel 2005); this achievement required a prodigious effort.

Historically tipulids have been considered an early branch of Diptera, closely related to Trichoceridae and Tanyderidae, but they may be more highly derived (de Jong *et al.* 2008; Theischinger 2015). The oldest fossils date to the Lower Triassic ~240 mya (Grimaldi and Engel 2005; de Jong *et al.* 2008), and present day patterns of distribution of many higher taxa suggest they have either a Pangaean or Gondwanan origin. Tipulids are associated with humid environments, and moist microhabitats such as moss beds and liverworts; deep, saturated detritus; streams; margins of rills; algal mats; and tree fern gullies.

Some inhabit caves (Darlington and Gelhaus 1994) and others are associated with spider webs (Robinson and Robinson 1976). Larvae are generally aquatic (Plate 235, species under stones from a stream bed in Werrikimbe National Park) although some live in semi-aquatic habitats such as damp moss beds, humus, and cliff faces where filmy algal growth is found. Some aquatic species live completely submerged and lack functional breathing spiracles. Others come to the surface to take oxygen, achieved by spiracles positioned at the tip of the abdomen (de Jong *et al.* 2008). Larvae generally feed on decaying organic matter but some are predaceous, or feed on living fungi and algae. Adults are slim and elongate, are short lived, and most do not appear to actively feed, although those genera with lengthy mouthparts (*e.g. Ptilogyna* – Tipulinae) are capable of imbibing floral nectar (e.g. unidentified genera/spp. on *Acradenia euodiiformis*, *Alphitonia excelsa*, *Guioa semiglauca*, *Syzygium smithii* [Williams 1995]). There is great divergence in size, ranging from large colourful *Leptotarsus* with wingspans over 50 mm, to those <8 mm. Some have ornamented wings (*Gymnoplista bella*, *Dapanoptera richmondiana*), and though most adults are crepuscular or fully nocturnal in habit, individuals are often encountered flying during the day in sheltered locations. During daylight adults normally rest on the upper surfaces of rainforest foliage, on overhanging rocks by streamsides, and at the base of shaded tree trunks. Species may also exhibit distinct seasonal emergence patterns. For example, the ornately winged *Dapanoptera richmondiana* (Plate 234), a species widely distributed, but poorly known, from North Queensland to Sydney (Theischinger 2016), and at Lorien Wildlife Refuge (New South Wales), has only been recorded in early spring (G. Williams pers. obs.).

There are numerous endemic *Molophilus* (Limoniinae) species in the region (e.g. *Molophilus eurygramma*, *M. sinclairi*, *M. smithersi*) (Theischinger 1992, 1996, 1999c; Williams 2002). Several have limited known distributions: *Molophilus nini* (Lower Creek State Forest), *M. mina* (Tuglo Wildlife Refuge north of Singleton), *M. flavocingulata* (Barrington Tops), *M. hylandensis* (Mt Hyland Nature Reserve) and *M. bickeli* (Gloucester Tops, in *Nothofagus* forest). The genus is widespread but the greatest concentration of species is in the Southern Hemisphere (Theischinger 1992). Larvae occur in or beneath wet mosses, liverworts, leaf mould, though generally near streams. *Limonia bickeli* (Limoniinae) is known only from subtropical rainforest at Barrington Tops (Theischinger 1996). Additional limoniine genera from the region include *Austrolimnophila* and *Epiphragma*. Both are widely distributed (Theischinger 1996).

The genus *Dolichopeza* (Tipulinae) is recorded from North America, Europe, Africa, south and eastern Asia, New Guinea, Australia and New Zealand, but includes many species restricted to the Gondwana rainforests region (e.g. *Dolichopeza berrimilla*, *D. oresitropha*, *D. segnis*, *D. geometrica*, *D. zenta*) (Theischinger 1993a). *Gynoplistia* has a Gondwanan distribution, occurring in northern, eastern and south-western Australia, New Zealand, New Caledonia, New Guinea, Argentina and Chile. The Barrington Tops area has a diverse *Gynoplistia* fauna and includes *G. histrionica*, which is known by a single female specimen collected in 1925 (Theischinger 1993b). *Gymnoplista boomerang* inhabits rainforest in New England National Park, and is the only known species in the genus in which the female has brachypterous, non-functional wings (Theischinger 2015). Additional tipuline genera occurring in northern New South Wales and southern Queensland include *Clytocosmus* (endemic), *Leolimnophila* (endemic), *Paralimnophila* (Neotropics and Australia) and *Leptotarsus* (Daniels 1978; Theischinger 1996; Williams 2002). Chioneinae include the endemic genus *Ozeoura* (Theischinger *et al.* 2018a), which primarily occurs along the Great Dividing Range. Regional species include *Ozeoura convoluta* (Brooklana, Upper Hastings River), *O. tonnoiri* (Tapin National Park, Moonee Moonee Creek) and *O. dingo*, the last currently known only from Tapin National Park. In Australia the genus *Styringomyia* is most diverse north of the Tropic of Capricorn; however, one species, *S. williamsi*, extends to coastal New South Wales along the eastern slope of the Great Dividing Range (e.g. Bruxner Park, Sea Acres Nature Reserve, Dooragan National Park, Minnamurra Falls [Theischinger *et al.* 2018b]). *Styringomyia* larvae inhabit decaying organic matter such as rotten wood and leaf litter.

Ulurumyiidae This monotypic and monospecific calyptrate family has been erected for a single species, *Ulurumyia macalpinei*; this species was reared only from cow manure, with no records from manure of marsupials (Michelsen and Pape 2017). *Ulurumyia macalpinei* has been recorded in forest mainly from coastal sites ranging from northern to southern New South Wales (e.g. Gibraltar Range, vicinity of Kempsey, Kiwarrak State Forest, Upper Allyn, Durras Lake), and inland to Kanangra Boyd National Park, and Tidbinbilla near Canberra. Michelsen and Pape (2017) opined that the taxon probably predates the separation of Australia from Antarctica and South America, and were tempted to suggest (given its single known larval food source: cow dung) the possibility that *Ulurumyia macalpinei* may be the last survivor of a group of coprophagous flies that once bred in the dung of now extinct megafauna.

TRICHOPTERA (caddis flies)

Trichoptera are small moth-like insects that are closely related to Lepidoptera, differing most conspicuously in the lifestyles of their respective larvae. The earliest known Trichoptera date from the Early Jurassic (180–185 mya) (Ansorge 2002). The larvae of most Trichoptera are aquatic (though a few live in moist leaf litter) and predominantly occur in freshwater habitats; most adults are crepuscular in behaviour. Adults may be commonly attracted to light. Both larvae and adults are important components of broader rainforest foodwebs, and fall prey to a diverse number of vertebrate and invertebrate predators. They are important macroinvertebrate indicators for determining the health of freshwater ecosystems (Turak and Bickel 1994). Although adults tend to be dull in colour, several diurnal species are colourful and have patterned wings (e.g. Stenopsychidae).

Larval habitats in subtropical rainforests include fast-flowing, highly oxygenated streams that seasonally degenerate into small isolated or ephemeral pools, and larger permanent streams and rivers that are generally slower flowing, though in reality there are graded transitions between these. Rainfall patterns, and associated annual total rainfall, strongly influence larval habitat quality and availability. Shading vegetation, water temperature, water chemistry, in-stream microhabitats (e.g. stones, detritus, submerged wood) and the nature of the stream substrate (for case-making species) also contribute to habitat quality.

Larval feeding behaviour is varied, and comprises herbivores, filter feeders and predators (Neboiss 1991). Numerous groups construct protective enveloping cases from organic matter, coarse sand or silt. The cases can be architecturally outstanding, and allow the delicate larva within both mobility and shelter. Those of certain Calocidae, for example, can be constructed from sand, but also from delicate overlapping plate-like structures formed from plant debris (Plate 236). Their common name, caddis fly, may be derived from the Old English *cadice men*, who pinned pieces of cloth to their clothing so as to advertise their fabrics, portrayed in like fashion by trichopteran larval cases (Grimaldi and Engel 2005, citing Izaak Walton's *The Compleat Angler*, published in 1653).

The Australian Trichoptera are closely related to families and genera occurring in New Zealand and South America, but the northern Australian fauna has strong Oriental-Papuan influences (Neboiss 1991). In Australia, the change from Torresian to Bassian faunas is noticeable in the Townsville–Rockhampton area, rather than further south (Neboiss 1981). This demonstrates the difficulty of assigning faunal provinces across major taxonomic groups.

At the species level endemicity is high. The families Plectrotarsidae and Antipodoeciidae are endemic, and Calocidae, Chathamiidae, Conoesucidae and Oeconesidae also occur in New Zealand. Helicophidae, Kokiriidae and Philortheithridae reflect transantarctic distributions between Australia and South America. The Atriplectididae are constituted by two monotypic genera, one each in the Seychelles Islands and Australia. The endemic Plectrotarsidae (three genera, five species) are restricted to the south-east and south-west of Australia. Tasimiidae exhibit an Australian and Neotropical distribution with Australian species confined to the eastern mainland and Tasmania. Conoesucidae are restricted in distribution to the eastern Australian mainland. The Antipodoeciidae are monogeneric and the single known species is restricted to eastern Australia (Neboiss 1991). The type species, *Antipodoecia turneri*, was described from Ebor adjacent to New England and Dorrigo national parks (in W. Houston 1988). The Hydropsychidae include *Diplectrona spinata*, which is restricted to south-east Queensland (Neboiss 1986).

Families (and examples of species), as larvae, recorded from stream environments in lowland, intermediate and high elevation rainforests of northern New South Wales include Antipodoeciidae (*Antipodoecia turneri*), Calamoceratidae (*Anisocentropus bicoloratus*, *A. valgus*), Calocidae (*Caenota plicata*, *C. simulans*, *Caloca eba*), Conoesucidae (*Coenoria boera*, *Lingora plicata*), Ecnomidae (*Ecnomina* spp.), Glossosomatidae (*Agapetus evansi*), Helicopsychidae (*Helicopsyche cochleaetesta*), Hydrobiosidae (*Apsilochorema spinosum*, *Ethchorema brunneum*, *Ptychobiosis rieki*, *Psyllobetina perkinsi*, *Taschorema asmanum*, *Ulmerochorema seona*), Hydropsychidae (*Diplectroma angusta*, *Sciops spinata*), Hydroptilidae (*Hellyethria allyensis*, *Maydenoptila antennifera*,

Orthotrichia rostrata), Leptoceridae (*Triplectides tambina*, *Oecetis australis*, *Triaenodes forficata*), Odontoceridae (*Barynema australicum*), Philopotamidae (*Hydrobiosella* spp.), Philorheithridae (*Aphorheithrus* spp.), Stenopsychidae (*Stenopsychodes syrdena*) and Tasimiidae (*Tasimia* spp.) (Hose and Turak 2004; W. Houston 1988; Johanson 1995; Neboiss 1962, 1974, 1980, 1983, 1986, 1991; Neboiss and Wells 1998; Wells 1979).

Calamoceratidae This family is widespread, with the greatest number of species occurring in islands between Asia and Australia. In Australia, the Calamoceratidae are better represented in subtropical areas. The family is represented by the genus *Anisocentropus* (*Anisocentropus bicoloratus* [Plate 237], *A. latifascia*, *A. valgus*), which is restricted to Bassian and Torresian provinces (Neboiss 1980). Larvae tend to frequent more sluggish flowing water.

Helicopsychidae The larvae of this family feed on organic matter, commonly inhabit cool mountain streams, and construct protective cases from grains of sand. Helicopsychidae are most diverse in the Oriental (India and China) and Neotropical Antillean, Mexican and Brazilian subregions, with the majority of species having restricted distributions (Johanson 1995). *Helcopsyche* (*Helcopsyche cochleaetesta*, *H. heacota*, *H. murrumba*, *H. tillyardi*) is the only genus known from the Australian region, and also occurs in New Zealand and New Caledonia. In Australia most species are found associated with the wetter, higher relief of the Great Dividing Range (Johanson 1995). Helicopsychids are apparently absent from the south-west of the continent even though suitable habitat occurs there. Johanson (1995) suggests that this absence from the south-west of Western Australia may be due to extinctions resulting from past climatic events such as the drying of rivers ~18 000 years ago.

Hydrobiosidae Hydrobiosidae occur mainly in the Australian and Neotropical regions (in W. Houston 1988). The ancestral Australian stock is thought to have entered from the north with a later evolutionary radiation based on an ancestral nucleus in the southern highlands (Neboiss 1962). Species are moderately sized, and the larvae are free living and are mainly found in cool, fast-flowing streams in mountainous areas.

The Australian fauna is predominantly confined to the east and south-east of the continent, including Tasmania, and the number of species rapidly decreases to the north and west (Neboiss 1962). Most Australian hydrobiosid genera are endemic (Neboiss 1962). Only *Apsilochorema* extends beyond the continent. Numerous caddis fly species have been recorded from the Barrington Tops (*Austrochorema patulum*, *A. spinosum*, *Psyllobetina plutonis*, *Ulmerochorema seaona*) and the Bunya Mountains (*Apsilochorema gisbum*, *A. obliquum*, *Taschorema asmanum*, *T. brunneum*) (Neboiss 1962).

Hydropsychidae Hydropsychids are a large family with a world-wide distribution. The larvae build fixed shelters made from sand and plant material, at the front of which are constructed nets that filter food fragments from the flowing water (Neboiss 1991). Small invertebrates are also captured. Regional species include *Baliomorpha banksi*, *Diplectrona angusta*, *D. bourina*, *D. spinata* and *Sciops spinata* (W. Houston 1988; Neboiss 1983).

Leptoceridae Leptocerid larvae construct tubular cases from plant and mineral material, and inhabit streams and ephemeral pools. Some are predators and others are omnivorous (Neboiss 1991). The family includes the genus *Triaenodes*, which is one of the most diverse Australian leptocerid genera. The Australian fauna comprises approximately one-third of all described species (Neboiss and Wells 1998). Elsewhere the genus is distributed in the Afrotropical, Holarctic and Neotropical regions, and New Guinea, but in Australia *Triaenodes* is confined to the coastal zone of the mainland and Tasmania (within the 500 mm isohyet). *Triaenodes* is not known from New Zealand or New Caledonia (Neboiss and Wells 1998). Neboiss and Wells (1998) consider that *Triaenodes* is possibly a relatively recent element of the Australian fauna owing to the similarities of some species with species occurring in New Guinea and the western Pacific.

LEPIDOPTERA (moths and butterflies)

The earliest fossil lepidopteran comes from the Jurassic but Lepidoptera appear to have radiated more recently than that, and are now the largest group of plant-feeding insects (Grimaldi and Engel 2005). Lepidoptera diversified on flowering plants during the Cretaceous and Tertiary (Grimaldi and Engel 2005; Vane-Wright 2004), participated in a coevolutionary radiation (Edgar *et al.* 1974; Ehrlich and Raven 1964; Ferrer-Paris *et al.* 2013; Grehan 1991; Smiley 1978), and were probably important pollinators at least from the Early Tertiary (Crepet 1984). Most lepidopteran larvae live and feed on angiosperms, and the majority of adult butterflies feed on nectar (Braby 2000; Common 1990; Common and Waterhouse 1981; Zborowski and Edwards 2007). Many of the flowers to which lepidopterans are attracted contain nectar with significant levels of amino acids, and these add to the nutritional value derived from its ingestion (Williams and Adam 2010).

Flowers adapted to be pollinated by butterflies and/or moths generally have long and slender floral tubes (Bertin 1989), and the length of the flower's corolla and that of the lepidopteran proboscis (the long, curled feeding tongue) are interdependent limiting factors with respect to a butterfly's or moth's ability to reach the nectar held within (Williams and Adam 2010). In the majority of Lepidoptera (a few feed on pollen), food uptake can only occur via the proboscis and so nectar viscosity can restrict the ability of moths and butterflies to feed from flowers. Viscosity influences capillarity (Faegri and van der Pijl 1979), and if the nectar is too viscous, then this food resource cannot be imbibed.

Though frequently overlooked, many nocturnal moths in subtropical rainforest communities are able to function as pollinators. In addition to day-active groups (e.g. *Amata* – Erebidae, *Snellenia* – Oecophoridae) and the well-documented hawkmoths (Sphingidae), Noctuidae, Geometridae and Pyralidae may be found at night on flowers of numerous mass-flowering trees. However, behavioural divergence exists within the nectar-feeding moths. Some groups, most markedly the Sphingidae, hover during feeding and others (e.g. Noctuidae, Erebidae – Arctiinae) alight and clamber over individual blossoms. Hovering is energy-expensive, but the difference in behaviour also directs placement of pollen. In moths that hover, pollen tends to be deposited mainly on the proboscis, or the head and upper thorax. With species that land directly on flowers pollen deposition happens more widely on the body, but particularly on the undersurface of the thorax (Williams and Adam 2010).

Most adult rainforest- and wet sclerophyll forest-associated Lepidoptera in the subtropical zone are only active during the warmer months from spring through autumn (September–May). Though even among the butterflies, *Graphium macleayanus* (Papilionidae) (Plate 258) and *Delias nigrina* (Pieridae) (Plate 254) can be seen at higher latitudes in mid-winter flying about the edges of lowland rainforest. However, at this time flowers are almost totally absent (the tree *Synoum glandulosum* being a particular exception). The local distribution of Lepidoptera is strongly determined by the presence of specific food plants, and is also influenced by latitude and altitude (and the climatic factors associated with these), and in some instances (e.g. for certain Lycaenidae) the presence of attendant ants (Hölldobler and Wilson 1990; Schmidt 2002). Indeed, host plant diversity, and that of butterflies and moths, is linked such that the diversification and maintenance of host plants in rainforest stands will probably promote and support an increase in the diversity of butterflies that interact with them. Ultimately, as stated earlier, species distribution is strongly dictated by the presence of larval food plants or hosts (as in Lycaenidae, *Ornithoptera richmondia*) (Braby 2000; Common and Waterhouse 1981), but for many moths their food plants are unknown.

Adult butterflies and some moths (e.g. various Erebidae – Arctiinae and Noctuidae) are ecological generalists with respect to nectar sources and will largely feed on whatever nectar-producing plants are seasonally in flower (as long as these are not toxic and nectar is of an appropriate viscosity), and commonly recruit to feed in plant communities in which their larval food plants are absent as long as appropriate nectar resources are available; their ability to do this is only constrained by the distance over which they are able to fly. But in the absence of larval food plants, the associated nature of which is usually plant, butterfly or moth specific (Braby 2000; Common 1990; Common and Waterhouse 1981), species are unable to persist across seasons or establish founding populations. Equally, some Lepidoptera are able to migrate over long distances during periods of ameniable temperature, but are unable to survive there permanently as harsher seasonal conditions return.

Although angiosperm-feeding Lepidoptera were already present at ~97 mya (Ferrer-Paris *et al.* 2013), there is little fossil evidence to indicate the origin and evolution of the Australian Lepidoptera (Braby 2000). However, the fossil history of food plants, the relationship of lepidopteran behaviour to climate patterns within areas they occupy, and the present-day distribution of extant taxa and their food plants allow some deduction of information (Common 1990; see also Grehan 1991). Of the extant world fauna ~20 000 species of moths and butterflies are recognised from Australia (Braby 2000) and many (though this is likely to be the preserve of moths alone) are estimated to await discovery (Common 1990).

Butterflies are divided into the superfamilies Hesperioidea (skippers, darts) and Papilionoidea (all others), and constitute only ~2 per cent of the Lepidoptera fauna. Nymphalidae and Lycaenidae, together, constitute the bulk of butterfly diversity (Grimaldi and Engel 2005).

Although there are general features (e.g. day-flying activities, clubbed antennae, dorsal closure of wings, lack of a wing-coupling device) that distinguish butterflies from moths, there are intermediate structures and behaviours. For example, the butterfly-like castaniid moths fly during the day and have clubbed antennae, and the males of the skipper genus *Euschemon* are able to couple the fore- and hindwings in moth-like fashion (Braby 2000; Common and Waterhouse 1981). Moths possess a high number of endemic taxa, especially in more primitive families (Common 1990). Approximately 60 per cent of the Australian moth fauna belongs to the microlepidoptera, whereas only 35–40 per cent of the Northern Hemisphere species belongs to this group (Common 1990).

All major families of Lepidoptera occur in Australia. However, the abundance of several taxa is significantly different from other zoogeographical regions (Nielsen and Common 1991). The Castniidae, Cossidae, Heliozelidae, Hepialidae, Incurvariidae and Palaephatidae are diverse but the evolutionary history of individual taxa is uncertain (Nielsen and Common 1991). Anomosetidae, Carthaeidae, Cyclotornidae and Lophocoronidae are endemic to Australia, and Anthelidae and Hypertrophidae are known only from Australia and Papua New Guinea. Families with the largest number of endemic taxa include Carposinidae, Castniidae, Cossidae, Incurvariidae, Nepticulidae and Oecophoridae (Common 1990; Williams 2002).

Glossata (Hesperoidea and Papilionoidea) (butterflies)

These are also grouped as the Rhopalocera (Scoble 1995), though sometimes Hesperoidea have been excluded. Butterflies are largely recently evolved, and could, as Grimaldi and Engel (2005) posited, be viewed simply as a later-evolved and radiated 'lineage of gaudy, day-flying moths, a situation analagous to dinosaurs and birds'. The Australian butterfly fauna commonly is considered to have entered the continent relatively recently from the Northern Hemisphere via South-east Asia and/or via what is now New Guinea (Kondandaramaiah *et al.* 2018). However, an alternative hypothesis is that part or all of the fauna ultimately evolved in remnant Gondwana or Southern Gondwana when Australia was connected to South America through Antarctica (Kondandaramaiah *et al.* 2018).

The males of many butterflies possess very conspicuous, frequently brightly coloured, wing markings. These are putatively considered as sexual ornaments related to mating (Kemp and Macedonia 2006; Silbergleid 1984). The superfamilies Hesperioidea (Hesperiidae only) and Papilionoidea (Papilionidae, Nymphalidae, Pieridae, Lycaenidae) exhibit several distinctive features of form (Kristensen 1976), but historically have been collectively classified as butterflies, the Lycaenidae being the most numerous (>140 species) (Braby 2000). The butterfly fauna largely exhibits a strong affinity with the Oriental-Papuan region, but Satyrinae and especially the skipper subfamily Trapezitinae possess high levels of endemicity. Many are endemic to montane regions in south-eastern Australia; however, their known larval food plants are primarily shrubs, herbs and grasses (Atkins 1975; Atkins *et al.* 1991; Braby and Wurtz 2018). Torresian and Bassian butterfly faunas broadly overlap in south-eastern Queensland and northern New South Wales with no clearly defined line of distinction. Altitude and the local distribution of vegetation strongly influence the presence of species and patterns of distribution (Braby and Wurtz 2018; Williams 2002). *Heteronympha paradelpha deervalensis*, *Oreixenica kershawiella* (Nymphalidae), *Acrodipsas arcana*, *Pseudalmenus chlorinda barringtonensis*, *Candalides heathi heathi*, *Neolucia hobartensis monticola* (Lycaenidae) and *Ornithoptera richmondia* (Papilionidae), as examples within the butterfly fauna, are restricted to the Gondwana rainforests or adjacent areas of southern Queensland and northern New South Wales (Braby 2000; Common and Waterhouse 1981; Sands 1980).

Butterflies, along with bees, are classically considered pollinators of flowering plants, but in lowland rainforest remnants, at least, both their diversity and abundance on flowering trees and large shrubs is generally low and often apparently absent (e.g. *Syzygium floribundum*), even though large nectar volumes suggest an ample and attractive food resource (Williams 1995). Records include *Zizina labradus* (on *Syzygium smithii*, *Tristaniopsis laurina*); *Candalides absimilis*, *Hypocysta metirius*, *Delias nigrina* (on *Acradenia euodiiformis*); *Candalides absimilis*, *Erysichton lineata lineata*, *Nacaduba berenice berenice*, *Catopsilia* sp. (on *Alectryon coriaceus*); *Candalides consimilis*, *Jalmenus* sp., *Paralucia aurifera*, *Hypocysta metirius* (on *Diospyros australis*); and *Hypocysta metirius*, *Delias nigrina* (on *Pittosporum undulatum*, *Tristaniopsis laurina*) (Williams 1995; G. Williams unpubl. data). Only on *Alphitonia excelsa*, in littoral rainforest, has a diverse assemblage of butterflies been noted (*Trapezites* sp., *Candalides absimilis*, *Deudorix epijarbas diovis*, *Erysichton lineata lineata*, *Tirumala hamata hamata*, *Hypolimnas bolina nerina*, *Junonia vallida callibe*, *Graphium eurypylus lycaon*, *Delias nysa nysa*) (Williams 1995). These records were mainly of single individuals, with observations spread over several seasons. Nevertheless, even infrequent visits by butterflies may make significant contributions to outcrossing in plant populations (Williams and Adam 1998).

Hesperiidae These are known as skippers and darts. A high proportion of Hesperiidae are found in temperate latitudes, and though found in all zoogeographical zones they are notably absent from New Zealand. They represent about one-third of the Australian butterfly fauna and comprise over 120 species, more than half of which are endemic. A high proportion of genera are also endemic. All four subfamilies (Coeliadinae, Hesperinae, Pyrginae, Trapezitinae) recorded from Australia occur in the eastern subtropical region (Braby 2000; Common and Waterhouse 1981; Daniels 1976; De Baar 1976, 1977; Miller and Morhaus 1975; Moss and Popple 2000; Smithers 1994b). Many species 'hill-top' and species resident in open forest and woodland readily visit flowering rainforest trees and shrubs, though generally will not enter shaded understoreys.

The subfamily Trapezitinae is almost exclusively Australian and contains the greatest proportion of Australian hesperiid species. It and the subfamily Pyrginae are probably the oldest butterflies in Australia. Elsewhere the Trapezitinae are represented by a few species in New Guinea and the Aru Islands. The Australian temperate zone fauna is particularly rich. Endemic genera in the eastern subtropical region include *Anisynta*, *Hesperilla*, *Mesodinia*, *Motasingha*, *Pasma*, *Signeta* and *Trapezites*. *Trapezites genevievae* is restricted to old growth rainforest from Barrington Tops to southern Queensland and is the only member of the genus restricted to rainforest (Atkins 1987, 1997; A. Atkins pers. comm.). Its larvae feed upon *Lomandra spicata* (Lomandraceae) (A. Atkins and D. Sands pers. comm.). Additional species and subspecies recorded from subtropical rainforest and wet sclerophyll forest (though not necessarily restricted to them) include *Anisyta tillyardi*, *Hesperilla crypsargyra hopsoni*, *H. crypsigramma*, *H. furva*, *H. malindeva*, *H. mastersi mastersi*, *H. ornata ornata*, *H. picta*, *Pasma tasmanica*, *Signeta tymbophora*, *Toxidia andersoni*, *T. rietmanni rietmanni*, *Trapezites maheta* and *T. symmomus* (Williams 2002)

The second largest subfamily, Hesperiinae, is mainly tropical and many species have a wide distribution in the Oriental region, New Guinea and islands of the south-west Pacific (Common and Waterhouse 1981). Hesperinae are diverse in South-east Asia and New Guinea, and the Australian fauna may be derived from these regions (Braby 2000). Unlike the Trapezitinae, the Australian temperate hesperiine fauna is relatively poor. Genera have extralimital distributions, though many are restricted to the western Pacific, New Guinea and islands of eastern Indonesia. Genera occurring in the region include *Arrhenes*, *Cephrenes*, *Ocybadistes*, *Parnara*, *Pelopidas*, *Suniana*, *Taratrocera* and *Telicota*. Species recorded from rainforest include *Cephrenes augiades*, *Ocybadistes ardea*, *Suniana lascivia* and *Telicota anisodesma* (Williams 2002). Larvae of *Ocybadistes ardea* feed on grasses, *Suniana lascivia* in New South Wales feeds on *Imperata cyclindrica*, *Telicota anisodesma* feeds on *Flagellaria indica* (Flagellariaceae), and larvae of *Cephrenes augiades* feed on palms (Arecaceae) (e.g. *Archontophoenix cunninghamiana*, *Livistona australis*) (Braby 2000).

The Pyrginae are most diverse in the tropics. In Australia the subfamily includes the rainforest- and wet sclerophyll forest-inhabiting Regent Skipper *Euschemon rafflesia rafflesia*. This is almost certainly an ancient Gondwanan relic (A. Atkins pers. comm.). *Euschemon rafflesia rafflesia* is of further interest because the method of wing coupling in the male resembles that in the majority of moths; this is effected by means of a frenulum and retinaculum (Common and Waterhouse 1981). In late summer adults are to be seen flying in sunlit gaps in the understorey, the larvae feeding on the coarse-leaved shrubs *Wilkiea huegeliana* and *W. macrophylla* (Monimiaceae) (Plates 255, 256). The larva joins adjacent leaves (usually two) with silk, and shelters within, at night eating irregular sections from the edges.

Additional rainforest- and wet sclerophyll forest-inhabiting Pyrginae include *Chaetocneme beata* and *Netrocoryne repanda repanda* (Williams 2002). The larval food plants of *Chaetocneme beata* include Myrtaceae, Eupomatiaceae, Euphorbiaceae, Lauraceae and Rutaceae. Those of *Netrocoryne repanda repanda* include Myrtaceae, Elaeocarpaceae, Lauraceae, Sapindaceae, Flacourtiaceae, Oleaceae and Podocarpaceae (Braby 2000; Common and Waterhouse 1981). The Coeliadinae (awls) occur mainly in the Afrotropical, Oriental and Australian regions but none of the genera found in Australia is endemic (Braby 2000). Most species are restricted to tropical and subtropical areas and are largely associated with rainforest (Braby 2000). *Hasora discolor* extends south from New Guinea, and in Australia in lowland rainforest in disjunct populations that extend to north-eastern New South Wales. *Hasora chromus* has isolated vagrant records in south-eastern Queensland and New South Wales (Braby 2000). *Hasora khoda haslia* occurs mainly in subtropical, temperate and littoral rainforest from southern Queensland to southern New South Wales. The possibly itinerent *Badamia exclamationis* has coastal records in eastern Australia that extend to Victoria.

Lycaenidae Lycaenids have a world-wide distribution but reach their greatest diversity in tropical regions. The Lycaenidae are the largest butterfly family in Australia, comprising more than 140 species placed in 45 genera, of which six genera and 65 species are endemic (Braby 2000; see also Common and Waterhouse 1981). However, many species are small in size. The subfamilies Liphyrinae, Riodininae, Theclinae and Polyommatinae occur in Australia, but only the latter two are present in the Gondwana rainforests region (Braby 2000; Common and Waterhouse 1981; Daniels 1976; De Baar 1977; Eastwood 1997; Miller and Edwards 1978; Miller and Morhaus 1975; Moss and Popple 2000; Smithers 1994b; Smithers and Peters 1990; Williams 2002; W. Wilson 1984). Theclinae are cosmopolitan in distribution, and in subtropical eastern Australia are found in rainforest, eucalypt forests and woodland, *Melaleuca*-dominated swamps and mangrove communities (Braby 2000). Polyommatinae are also widely distributed but are less frequently encountered in rainforest. Many Australian lycaenid genera are widely distributed in the Oriental-Papuan region; however, the region includes the Australian endemic genera *Acrodipsas, Lucia, Neolucia, Paralucia* and *Pseudalmenus*.

Larvae of numerous lycaenid butterflies are associated with ants (Braby 2000; Common and Waterhouse 1981; Schmidt 2002). Many larvae possess specialised glands that produce compounds that appease ants that otherwise might be aggressive to them. Ants can protect the larvae, but the range of ant species that attend particular lycaenids may reflect the degree of ecological specialisation in the relationship (Eastwood and Fraser 1999).

Several species and subspecies (e.g. *Pseudalmenus chlorinda barringtonensis, Acrodipsas arcana*) occupy very small ranges or have highly localised populations within wider limits of distribution. Species and subspecies recorded from rainforest (including vine thickets) and wet sclerophyll forest, though not necessarily restricted to these, include *Anthene seltuttus affinis, Candalides absimilis, C. consimilis, C. margarita, C. xanthospilos, Catopyrops florinda halys, Erysichton lineata lineata* (Plate 257), *E. palmyra tasmanicus, Nacaduba berenice berenice, N. biocellata biocellata, N. kurava parma, Prosotas felderi, Psychonotis caelius taygetus, Sahulana scintillata, Theclinesthes miskini miskini, T. onycha onycha* (Polyommatinae), *Deudorix epijarbas diovis, Hypochrysops byzos byzos, H. miskinsi, Ogyris abrota, Paralucia aurifera, Philiris innotatus innotatus, Pseudalmenus chlorinda barringtonensis, Pseudodipsas cephenes* and *Rapala varuna* (Williams 2002). Several are localised and occur only at higher elevations (e.g. *Pseudalmenus chlorinda barringtonensis*); however, where species are recorded from markedly differing vegetation communities (e.g. *Sahulana scintillata* from littoral rainforest and woodland), knowledge of their food plants may serve to indicate their restricted or obligate habitat community.

Rainforest food-plant records include *Rapala varuna* on flowers of *Jagera pseudorhus* (Sapindaceae) and *Alphitonia excelsa* (Rhamnaceae); *Deudorix epijarbas diovis* feeding on fruits or seeds of *Arytera foveolata, Cupaniopsis parvifolia, Harpullia hillii* (Sapindaceae), *Macadamia integrifolia* and *M. tetraphylla* (Proteaceae); *Candalides absimilis* on flowers of *H. pendula* (Sapindaceae) and *Stenocarpus sinuatus* (Proteaceae), and foliage of *Cupaniopsis anacardioides* (Sapindaceae); and *Nacaduba berenice* on fruits of *C. anacardioides* and *Elattostachys microcarpa* (Sapindaceae), foliage of *Arytera foveolata, C. parvifolia* and *E. nervosa* (Sapindaceae), and flowers of *Jagera psedorhus*. The larvae of *Erysichton lineata* are recorded feeding on the flowers of *Brachychiton acerifolius* (Sterculiaceae/Malvaceae), *Jagera pseudorhus, Elattostachys microcarpa, Alectryon subcinereus* (Sapindaceae), *Harpullia pendula, Syzygium francisii* (Myrtaceae); *Psychonotis caelius* on flowers and leaves of *Alphitonia excelsa*; and *Prosotas felderi* and *P. dubiosa* on flowers of *H. pendula* (Schmidt and Rice 2002). These records are largely from urban areas in the region of Brisbane, and demonstrate the ability of some species to persist as long as food plants are available, though areas isolated from natural habitat may support only a few common species (Hill and Kitching 1983). Additional food-plant records are given in Braby (2000).

Nymphalidae Nymphalidae have a world-wide distribution. Some are chemically defended, potentially integrating toxic compounds from their host plants, and participate in mimicry complexes (Ackery 1987; Braby 2000). Twelve of the 17 nymphalid subfamilies occur in Australia, seven of which are in the Gondwana rainforests region. However, Braby (2000) emphasises that the distinction between some subfamilies is weak. Most nymphalids are restricted to rainforest margins, often recruiting there from adjacent open forest (see open forest records in Braby 2000; Williams 2002). However, sunlit openings below the canopy can facilitate the occurrence of several species (e.g. *Heteronympha mirifica*, *Hypolimnas misippus*, *Melanitis leda*, *Mycalesis terminus*, *Phaedyma shepherdi*).

Satyrinae are the most diverse (Braby 2000; Kodandaramaiah *et al.* 2018), and this subfamily has a cosmopolitan distribution. Twenty-six species and six genera are endemic (Braby 2000). The non-endemic genera have Oriental-New Guinean affinities. The Australian temperate fauna is particularly rich and may represent a Gondwanan element. The endemic genera *Geitoneura*, *Heteronympha*, *Oreixenica* and *Tisiphone* occur in the Gondwana rainforests region, and it is in this region that *Argynnina* reaches its northern-most distribution. Several subspecies, or forms, within the *Tisiphone abeona* complex occur in south-east Queensland and northern New South Wales. *Tisiphone abeona morrisi* is locally extinct in south-east Queensland (i.e. Mt Tamborine) (Braby 2000) and now ranges from Tweed Heads to Crescent Head in coastal northern New South Wales. *Tisiphone abeona regalis* is distributed in mountainous areas from near Stanthorpe to the Barrington Tops, and *T. a. aurelia* is distributed from Port Macquarie to near Newcastle. The distribution and maintenance of populations are dependent on the occurrence of larval food plants (*Gahnia* spp.). Satyrinae found in rainforest, or wet sclerophyll forest, include *Argynnina cyrila*, *Geitoneura acantha*, *Heteronympha mirifica*, *H. paradelpha*, *Hypocysta irius*, *H. metirius*, *H. pseudirius*, *Melanitis leda* and *Mycalesis terminus*. *Melanitis leda* exhibits crepuscular activity patterns related to seeking mates (Kemp 2002). The dark, broadly white-banded females of *Heteronympha mirifica* (Plate 253) are possibly the most conspicuous butterfly found in the understorey of rainforests at higher latitudes. The species is strongly sexually dimorphic in appearance for the males are visually quite different, being brownish-orange in colour with mottled broken bands, especially on the forewings. The larval food plant is the grass *Oplismenus* (Poaceae), a common low species growing on the margins of subtropical rainforests and in wet sclerophyll forest (e.g. Lansdowne Escarpment). The larvae hide by day at the base of the stems and emerge at night to feed on the leaves. Males and females appear in September and fly until late autumn (G. Williams pers. observ.). Adults actively fly during the day but the females are well camouflaged when at rest on the forest floor, the white bands providing a cryptic pattern when in dappled sunlight. Males often exhibit territorial behaviour and will actively defend sunlit patches, below canopy openings or on forest margins, against conspecific males and other butterfly species. *Heteronympha mirifica* ranges in distribution from south-eastern Queensland to far-eastern Victoria, occurring at elevations up to 1400 m (Braby 2000).

The region also includes a single wide-ranging charaxine, *Polyura sempronius sempronius*. Charaxinae are a predominantly tropical subfamily, which is well represented in the Afrotropical and Oriental zones. It frequents numerous habitats, including the sunny margins of rainforest and wet sclerophyll forest. Adults are sometimes attracted to the juices of rotting fruit. Rainforest larval food plants include *Guilfoylia monostylis* (Surianaceae), *Pararchidendron pruinosum* (Fabaceae/Mimosaceae), *Brachychiton acerifolius*, *B. discolor* (Sterculiaceae/Malvaceae), and *Celtis paniculata* (Ulmaceae/Cannabaceae) (Braby 2000). The Argynninae comprise the Oriental-Australian *Argyreus* (*Argyreus hyperbius*) and *Cupha* (*Cupha prosope*). These genera range from the Indo-Oriental region to north-eastern New South Wales (e.g. Port Macquarie and Clarence River respectively [Common and Waterhouse 1981; Sundholm 1978]). The larvae of *Argyreus hyperbius* feed on small herbaceous species of *Viola* (Violaceae). Those of *Cupha prosope* are recorded feeding on species of Salicaceae/Flacourtiaceae, especially *Scolopia braunii*, and *Breynia oblongifolia* (Phyllanthaceae/Euphorbiaceae). These two plants are not uncommon and extend much further south than the known range of the butterfly; such that it will be of interest to see if the distribution of *Cupha prosope* can track southwards as the climate warms. Limenitinae include *Phaedyma* (*Phaedyma shepherdi shepherdi*), which is distributed from India to eastern Australia and the western Pacific; reaching its distribution limit in New South Wales. The species has been observed in remnant rainforest near Kangaroo Valley, south of Sydney (Braby 2000, citing F. Douglas), and is otherwise common in coastal rainforest. Its larval food plants include *Aphananthe philippinensis*, *Celtis paniculata* (Ulmaceae/Cannabaceae), *Cordia dichotoma*, *Ehretia acuminata* (Boraginaceae), and *Brachychiton* spp. (Sterculiaceae/Malvaceae). Acraeinae are represented solely by the wide-ranging *Acraea andromacha andromacha*; however, the subfamily also occurs in the Oriental, Neotropical and Afrotropical regions (where Acraeinae species are diverse). *Acraea andromacha andromacha* is a frequenter of woodland but occurs as a vagrant on the margins of littoral rainforest associated with dune systems. Food plants are mainly species of Passifloraceae and Violaceae. Heliconiinae are represented only by *Vindula arsinoe*, which ranges from New Guinea south to about Yeppoon in Queensland (Braby 2000, citing K. and L. Dunn). This is a lowland species found in littoral rainforest, gallery forest and coastal vine thickets.

The Nymphalinae have a wide distribution and are well represented in Australia, though none of the genera is endemic. *Doleschallia* (*Doleshallia bisaltidae australis*, Plate 252) ranges from India to New Caledonia and Fiji, occurs in the Iluka Nature Reserve (D. K. McAlpine pers. comm.), is recorded from the Port Macquarie area (Moore and Ginn 2009), and has been recorded further south at Lorien Wildlife Refuge (G. Williams pers. obs.). Adult *Doleshallia bisaltidae* resemble leaves when the wings are vertically closed. The species breeds in lowland rainforest where its larvae feed on the small herb *Pseuderanthemum variabile* (Acanthaceae), also a food plant of the related nymphaline *Hypolimnas alimena* (Braby 2000), which has been observed in subtropical rainforest–wet sclerophyll forest ecotones north of Lansdowne (G. Williams pers. obs.). The pierid-like *Mynes geoffroyi* occurs mainly in rainforest and reaches its southern-most distribition in northern New South Wales (e.g. Victoria Park Nature Reserve). Food plants include the stinging tree *Dendrocnide photinophylla* (Urticaceae).

Rainforest-associated Danainae include *Euploea sylvester, E. tulliolus, E. darcia* and *E. core*. Their larval food plants include species of Moraceae, Apocynaceae and Asclepiadaceae. In addition to *Euploea*, several species of Danainae (e.g. *Tirumala hamata*, *Danaus chrysippus*) enter open areas in disturbed lowland rainforest and wet sclerophyll forest (e.g. via vehicle trails) as vagrants from adjacent open forest, coastal shrubland or mangrove communities. The subfamily is mainly a tropical group with only three genera and 14 species in Australia. These may be derived from the Papuan and Timorese subregions (Braby 2000).

Papilionidae Only the single subfamily Papilioninae, and the included tribes Troidini, Graphini and Papilionini, occurs in Australia. Of these *Protographium* (Graphini) is endemic. Adult Papilionidae, commonly called swallowtails, are frequently conspicuous invertebrates in rainforest, owing to their large size, the bright colour and ornamentation of their wings, and the habit of many to undertake prolonged feeding bouts. Species exhibit 'hilltopping' behaviour (e.g. *Protographium leosthenes*, *Graphium macleayanus*), and are encountered at flowering shrubs (e.g. *Leptospermum*, *Kunzea*) at the summit of karst outcrops and eroded volcanic plugs. *Graphium macleayanus* (Plate 258) extends through a variety of altitudes, from lowland coastal forests to elevations >1100 m (e.g. Werrikimbe National Park).

Papilionids variously associated with subtropical rainforest formations, including vine thickets, are *Protographium leosthenes*, *Graphium aristeus* (this with a highly disjunct coastal distribution), *G. macleayanus*, *G. sarpedon*, *G. euryphylus*, *G. agamemnon*, *Papilio aegeus*, *P. fuscus*, *P. ulysses* and *P. cressida cressida* (Braby 2000; Common and Waterhouse 1981; Williams 2002).With the notable exception of *Ornithoptera richmondia*, these are widespread (particularly *Graphium sarpedon*). Though diverse in tropical rainforests, there are progressively fewer species in Australian subtropical rainforests at higher latitudes and elevations, and the distinctive *Papilio ulysses*, *Graphium agamemnon* and *G. aristeus* extend little past Rockhampton. Adults feed on nectar from a wide variety of flowering trees, shrubs, herbs and even vines, including introduced invasive weeds such as *Lantana camara*. And though the choice of larval food plants is generally more circumspect, several wide-ranging butterflies (*Graphium macleayanus*, *G. sarpedon*) utilise a diversity of plants (Braby 2000; Common and Waterhouse 1981; Sankowsky 1975).

A single species of birdwing butterfly *Ornithoptera richmondia* (tribe Troidini: *Troides*, *Trogonoptera*, *Ornithoptera*) occurs in the northern sector of the Gondwana rainforests, being restricted in its range to localised sites between southern Queensland and far north-east New South Wales (Sands and New 2013); and is dependent on the presence of its primary larval food plant *Pararistilochia praevenosa* (Aristolochiaceae), and a secondary food plant *P. laheyana*. The Richmond Birdwing Butterfly has undergone a contraction of its past known distribution, within its overall range, since European settlement but is currently well represented in conservation reserves (D. P. A. Sands pers. comm.). Current areas where populations survive are the Blackall, Conondale and D'Aguilar Ranges, Glass House Mountains, Mt Tamborine, McPherson Range, Mt Warning, Nightcap and Richmond ranges, and Clarence Peak (Grimshaw *et al.* 2013; Sands and New

2013). The species has also been recorded from the MacKellar Range north-west of Lismore (J. Window pers. comm.). Local birdwing populations are threatened by the introduced Dutchmans Pipe *Aristolochia elegans*, which is toxic to larvae (Sands *et al.* 1997). Populations in high montane sites (e.g. >800 m) may suffer periodic extinction due to climatic stress and successful re-colonisation of these sites is thought to be dependent on immigration from lowland populations (Sands *et al.* 1997). Several regional lowland nature reserves (e.g. Broken Head, Hayters Hill, Victoria Park) possess populations (B. Moffatt pers. comm.). *Ornithoptera richmondia* is the subject of a community-based conservation program to increase availability of food plants and reduce the availability of *A. elegans* to larvae.

Ornithoptera richmondia, and the related *O. priamus* complex from north Queensland and the Papuan region, represent the most primitive lineage of the genus, and *O. richmondia* life stages possess several unique and primitive morphological characters (e.g. monomorphic green pupa, shape of male genitalia) that set it apart from other species (Parsons 1996). Rather than being derived from *Troides*-like ancestors in South-east Asia, *Ornithoptera* is considered to have evolved independently from other Troidini butterflies as the Australian continental plate drifted northwards (Parsons 1996). Consequently, the genus can be considered to represent an ancient Gondwanan or southern taxon. The superficially similar genus *Troides* (with which various evolutionary studies have lumped *Ornithoptera*) evolved separately on the Indian plate with the two genera eventually weakly overlapping in distribution in New Guinea as the separate landmasses later approached each other (Parsons 1996).

Pieridae Pierids, whites and yellows, are largely a tropical family of butterflies, predominantly coloured white and black with contrasting areas of red, orange or yellow (particularly species of *Delias*). These are presumed to be warning colours, and that the butterflies are distasteful or toxic to predators (Orr 1999). Males have more ultraviolet-absorbent pigments in their wings than females, and because butterflies are very sensitive to ultraviolet (Silberglied 1984) this has led to the expression of sexual dimorphism in the intensity of bright coloration and patterns (especially in *Delias*) (Grimaldi and Engel 2005).

Only the subfamilies Coliadinae and Pierinae occur in Australia, with most found in northern Australia. Pieridae become progressively less diverse extending into higher latitudes, with 22 species found in southern Queensland and 10 in Victoria (Braby 2000). Numerous Coliadinae (e.g. *Catopsilia* spp., *Eurema* spp.) inhabit grassy woodlands and are widely distributed on the mainland. Pierinae also inhabit open vegetation, but include several species and subspecies that frequent subtropical rainforest (*Appias paulina ega*, *Delias nigrina*, *D. nysa nysa*, *D. mysis*, *D. harpalyce*, *Elodina angulipennis*, *E. parthia*, *E. queenslandica*, *Cepora perimale*). Larvae of *Elodina angulipennis*, *E. parthia*, *E. queenslandica* and *Cepora perimale* feed on Capparaceae, those of *Appias paulina* on Capparaceae and Euphorbiaceae, *Delias nysa* on Viscaceae/Santalaceae, and *D. mysis* and *D. nigrina* on Loranthaceae (Braby 2000; Common and Waterhouse 1981). *Delias nigrina* occasionally occurs as episodic swarms on understorey trees such as *Pittosporum undulatum* (Pittosporaceae) (Plate 254 and *Macadamia tetraphylla* and *M. integrifolia* (Proteaceae) (G. Williams unpubl. data; Williams and Adam 2010). In some seasons *Cepora perimale* can be seen in large numbers flying in the understorey, and associated canopy gaps, of vine thickets and dry rainforest formations; these are habitats where its larval food plants (e.g. *Capparis arborea* – Capparaceae) are to be commonly found.

Glossata (part – remaining superfamilies), Aglossata and Zeugloptera (moths)

The moths discussed here comprise the primitive Zeugloptera (Micropterigidae only) and Aglossata (Agathiphagidae only) and a heterogeneous selection of families in the suborder Glossata (excluding butterflies). Families are listed alphabetically.

Some species (e.g. of Noctuidae, Pyralidae, Sphingidae) are equipped with sound receptors that allow them to perceive and avoid predators (i.e. bats); however, multiple lineages of moths independently evolved hearing organs well before the origin of bats (Kawahara *et al.* 2019).

Agathiphagidae Commonly referred to as kauri moths, this primitive family solely constitutes the superfamily Agathiphagoidea and suborder Aglossata (Common 1990; Kristensen 1967; Nielsen and Common 1991; Speidel 1978). Only two species are known: *Agathiphaga queenslandensis* from southern Queensland (Cooloola), and *A. vitiensis* from islands in the Pacific (e.g. New Caledonia, Fiji, Vanuatu, Solomon Islands) (Common 1990; Dumbleton 1952). These small moths resemble Trichoptera and Zeugloptera – Micropterigidae. The larvae feed

The Regent Skipper, *Euschemon rafflesia*: a peculiar rainforest butterfly

Almost all butterflies fly with their fore- and hindwings uncoupled. There is one known exception, this being the Australian endemic *Euschemon rafflesia*, or Regent Skipper, a member of the Hesperiidae. Males are unique in that their wings are coupled in flight by means of structures called the frenulum and the retinaculum, devices widely found in the moths but not butterflies. The retinaculum is a membranous hook on the forewing which couples with the frenulum, a stout bristle arising from the base of the hindwing (Common and Waterhouse 1981). Females do not have these structures.

There are two subspecies, the most southern of which, *Euschemon rafflesia rafflesia*, inhabits the dappled understorey of subtropical rainforests and adjacent wet sclerophyll forest. It is a large black- and yellow-banded species, and with its wings outstretched measures ~5 cm in width. Adults are normally active in late afternoon, though they can be encountered throughout the day. Flight movements are rapid and characteristically jerky, but individuals frequently come to rest on exposed leaves, with their wings spread out. Unless disturbed, they will remain motionless in this position for some minutes. Southern populations are usually on the wing from mid-spring to autumn, but are more commonly encountered in the later months of summer through to April. The range of *Euschemon rafflesia rafflesia* extends from southern Queensland to Seal Rocks on the lower north coast of New South Wales (Atkins 1984). Throughout the species' southern range the larval food plants are exclusively *Wilkiea huegeliana* and *W. macrophylla* (Monimiaceae), leathery and serrated-leaved shrubs and small trees of the understorey that are exclusively pollinated by thrips (Williams *et al.* 2001). Larvae have also been reported from the tree *Lophostemon confertus* (Myrtaceae) but this seems unlikely. The northern subspecies, *Euschemon rafflesia alba*, feeds on *Tetrasynandra* spp. (also Monimiaceae) and *Wilkiea huegeliana* (see Braby 2000). The pale green eggs are usually laid singly on the undersurface of the leaves, and the developing larvae construct shelters formed by binding two adjacent leaves with strong silk. At night maturing larvae emerge to feed, though early instars may feed briefly in the morning (Braby 2000). Pupation occurs within the larval shelter, the pupa being fastened in a horizontal position and held in place by a silk girdle. The subspecies is usually uncommon, its presence dependent on the presence of its obligate food plant and, though currently existing through a wide latitudinal range, individual Regent Skipper populations are vulnerable to habitat fragmentation.

exclusively on the seeds of *Agathis* (Araucariaceae), those of *Agathiphaga queenslandensis* being associated with *Agathis robusta*. Larvae feed within the seeds, excavating the interior, and then pupating in the cell that this forms (Common 1990). Pupation may be delayed for several years. Larvae potentionally destroy a high proportion of seeds (10–48 per cent), but the extent of infestation appears related to season (Brimblecombe 1965; Dumbleton 1952). Pupae possess enormous mandibles, which are used by the adult enclosed in the pupal cuticle to grind a circular passage out of the seed cell. Adults have been collected attracted to light, but emergence may be related to rainfall events (Common 1990).

Anomosetidae This small family comprises the single endemic species *Anomoses hylecoetes* from the rainforests of northern New South Wales and southern Queensland (Nielsen and Common 1991). Little is known about the species. Adults have been collected at night by beating foliage and in malaise traps. Common (1990) considered that the larvae breed in rotten logs or in detritus on the floor of the rainforest. Recorded localities include Lamington and the Border Ranges national parks, Tooloom Scrub and Rous. These sites extend over a diverse range of elevation.

Anthelidae The Anthelidae (Plates 238, 239) are nocturnal and crepuscular moths restricted to Australia and New Zealand, and include small to very large species (with wingspans > 160 mm). The family is restricted to Australia and New Guinea; however, the Australian fauna is the larger of the two with more than 60 species, the majority of these (e.g. *Anthela excellans*, *Chelepteryx chalepteryx*) placed in the Anthelinae (Common 1990). Adult males of the large *Chelepteryx chalepteryx* (Plate 239) are attracted to lights (including house lights) in wet sclerophylly forest and the margins of lowland rainforest, and often fly in early autumn, even during light rain. The larvae of several species of *Anthela* feed on *Acacia* (Mimosaceae), which in lowland wet sclerophyll forest commonly includes *Acacia melanoxylon*.

Copromorphidae The Coprosmorphidae are a small, mainly Asian-Australian family. Species also occur in

Madagascar, India, New Zealand, and Central, South and North America (Common 1990; Nielsen and Common 1991). The family is mainly restricted to rainforest along the east coast, but most Australian species are rare in collections. The genera *Osidryas* (*Osidryas phyllodes*), *Copromorpha* (*Copromorpha lichenitis*) and *Phycomorpha* (*Phycomorpha prasinochroa*) occur in southern Queensland and northern New South Wales. The larvae of *Copromorpha* and *Phycomorpha* feed on *Ficus* (Moraceae), attacking fruit, shoots, young twigs, or leaf petioles (Common 1990). *Phycomorpha prasinochroa* feeds on *Ficus coronata* and *F. fraseri*, *F. coronata* being widespread in lowland rainforest and *F. fraseri* mostly found in littoral rainforest.

Cossidae Cossid wood moths are best represented in the Ethiopian, Oriental and Australian regions, some of which are among the largest moths inhabiting the subtropical region of eastern Australia. Wingspans vary considerably, however, ranging from small species to those in excess of 20 cm; gravid females of the largest (e.g. *Xyleutes affinis*) are so encumbered by the weight of their massive abdomens that they are incapable of flight. Two of the four recognised subfamilies (Cossinae, Zeuzerinae) occur in Australia, of which Zeuzerinae are the most diverse (Common 1990; Kallies and Hilton 2012). Adults are attracted to house lights, often during rainy autumn nights.

The fauna is best represented in sclerophyll forest. Larvae bore in the wood of living and dead trees, the Zeuzerinae particularly attacking that of eucalypts (*Xyleutes cinereus*, *X. affinis*) and *Acacia* (*Xyleutes encalypti*, *X. liturata*). Regional Cossinae include *Charmoses dumigani*, *Idioses littleri* and *Archaeoses pentasema*. These are moths of moderate size. Zeuzerinae include the largest species (e.g. *Xyleutes cinereus*, *X. donovani*, *X.epicycla*).

Erebidae Now placed in the Erebidae, but previously placed in the noctuid subfamily Catocalinae, is the endangered endemic Southern Pink Underwing Moth *Phyllodes imperialis smithersi* (Common 1990; Clarke and Spier-Ashcroft 2003; Hunter 1939; Sands 2012). This is a large nocturnally active moth with distinctive pink-coloured bands on the hind wings. The species is recorded from Nambour in south-east Queensland, and from the Border Ranges, Mt Warning, Richmond Range, Richmond River, Billinudgel, Rosewood River, Huonville, Dorrigo and Bellingen Island Reserve in New South Wales. Different subspecies occur in northern Queensland, the New Hebrides, Fiji, New Guinea, Solomons, and New Caledonia. Thus *Phyllodes imperialis smithersi* is an endemic subspecies but a non-endemic species. Its known habitat is restricted to below 600 m in subtropical rainforest in association with the woody climber *Carronia multisepalea* (Menispermaceae), its recorded larval food plant. This plant is restricted to, but widespread in, subtropical rainforest north of Dorrigo in New South Wales to Kin Kin, south-east Queensland (Harden 1990; Harden *et al.* 2007). Larvae utilise collapsed *Carronia* vines but upright vines are apparently not utilised. Adults feed on fermenting fruit and may be attracted to light. Adult dispersal capability is unknown but movements are likely to be restricted to old growth subtropical rainforest, and possibly contiguous eucalypt-dominated forest with a well-developed rainforest understorey (i.e. wet sclerophyll forest). However, there is poor scientific knowledge of its ecology and population dynamics. Threats include loss of habitat, habitat fragmentation, loss of larval food plants, and tourist impacts. The status of individual surviving populations is uncertain.

Members of the subfamily Arctiinae (previously the family Arctiidae [Zahiri *et al.* 2012]), or tiger moths, are distributed in all zoogeographical regions, and comprise a world fauna of ~11 000 species (Grimaldi and Engel 2005). Larvae are phytophagous (see records in Common 1990), some being associated with lichens. Arctiinae are mainly Neotropical in distribution, the adults are often coloured in aposematic (warning) patterns of red or orange and black (Common 1990), and many larvae feed on toxic herbaceous plants from which they integrate substances such as pyrrolizidine alkaloids and cardenolides. In addition to acting as toxic defensive compounds, these may also function as pheromones (Grimaldi and Engel 2005). Species also possess thoracic tymbal organs that produce ultrasonic clicks, which are thought to disrupt the ecolocation search patterns of microchiropteran bats (Common 1990; Nielsen and Common 1991; Corcoran *et al.* 2010).

Species found in subtropical eastern Australia include *Amata aperta*, *Asura cervicalis*, *A. lydia*, *Calamidia hirta*, *Chionaema meyricki*, *Eressa angustipenna*, *Heterallactis euchrysa*, *Manulea replana*, *Nyctemera amica*, *Paralacydes maculifascia*, *Prinasura structa*, *Rhodogastria crokeri*, *Spilosoma canescens*, *S. curvata*, *Termessa gratiosa*, *T. laeta*, *T. conographa*, *Tigrioides alterna* and *Utetheisa pulchelloides* (Common 1990; Moss and Popple 2000; Williams 1995). *Spilosoma*, *Rhodogastria*, *Utetheisa*, *Amata*, *Asura*, *Eressa*, *Nyctemera* and *Tigrioides* are widespread genera, *Calamidia* is found in Australia and New Guinea, and *Termessa* is endemic. *Chionaema* widely ranges in the Oriental region and New Guinea; however, *Chionaema meyricki* is a rainforest species confined to southern Queensland and northern New South Wales. *Manulea* is an Old World genus; the larvae of *M. replana* feed on lichens growing on shaded rocks (Common 1990). Many Arctiinae are nocturnal and are attracted to light, though species of *Amata*, *Asura* and *Eressa* are active during warm sunny days; adults typically spend several minutes feeding from single flowers or individual inflorescence masses, and undertaking only infrequent interplant movements. Records of *Amata* feeding on nectar from flowering subtropical rainforest trees include *Acronychia oblongifolia*, *A. imperforata*, *Alectryon coriaceus*, *Alphitonia excelsa*, *Diospyros australis*, *Elaeocarpus reticulatus*, *Euroschinus falcatus*, *Guioa semiglauca*, *Syzygium floribundum*, *S. smithii* and *Tristaniopsis laurina*. *Nyctemera amica* visits *Tristaniopsis laurina* and *Asura* has been recorded from *Guioa semiglauca* (Williams 1995).

Adults of the widely distributed Aganainae (previously Aganaidae) display aposematic coloration, indicating they are chemically protected against predators or are members of mimic complexes. They are largely nocturnal. The larvae of *Asota iodamia* (southern Queensland to northern New South Wales) (Plate 240) feed on the leaves of *Ficus macrophylla* (Moraceae), as do those of *Asota plagiata*, though the latter has a more northern distribution, its range extending from northern New South Wales to the Northern Territory.

The subfamily Lymantriinae (previously Lymantriidae) is especially numerous in the north and east of Australia (Common 1990). Adults are mostly nocturnal in habit. Two species of *Lymantria* are known from Australia: *L. luneata* ranges throughout coastal Queensland, and *L. nephrographa* occurs in rainforest from southern Queensland to southern New South Wales. Additional genera present in the region include *Orgyia* and the endemic *Leptocneria*. The larvae of *Leptocneria* feed on the leaves of *Melia azedarach* (Meliaceae) (Common 1990).

Ethmiidae Adult Ethmiidae are often brightly coloured and include several species of rainforest-inhabiting *Ethmia* (*Ethmia heliomela*, *E. hemadelpha*, *E. sphaerosticha*, *E. thoraea*) (Common 1990; Moss and Popple 2000). Larvae of *Ethmia sphaerosticha* feed on the leaves of *Ehretia acuminata* (Boraginaceae), at first forming a web on the undersurface, from which each later emerges to feed on younger foliage. The larvae of *Ethmia heliomela*, a species extending from the Bunya Mountains to Mt Dromedary, also feed on *Ehretia acuminata* but at first among the flowers, before later feeding on leaves (Common 1990).

Geometridae The family is distributed throughout the world, and at an estimated 21 000 described species (Scoble 1999), is one of the largest lepidopteran families. Geometrids are biologically distinctive in that most larval hosts are trees (Grimaldi and Engel 2005). The subfamily Ennominae is the largest in Australia (Common 1990) and includes several colourful rainforest and wet sclerophyll-inhabiting species (*Casbia rectaria*, *Cleora repetita*, *Idioides rhacodes*, *Larophylla amimeta*, *Lophosigna catasticta*, *Lyelliana phaeochlora*, *Medasina strixaria*, *Niceteria macrosoma*, *Paralaea porphyrinaria*, *Thalaina* spp., *Tigridoptera matutinata*, *Xenomusa monoda*, *Xylodryas leptoxantha*) (Williams 2002; G. Williams unpubl. records). Larval food plants of *Thalaina selenaea* include *Acacia melanoxylon*, a tree commonly encountered in lowland wet sclerophyll forests. The larvae of *Casbia rectaria* feed on *Alphitonia excelsa* and those of *Cleora repitata* on *Flindersia australis* (Rutaceae). *Paralaea* is a large and widespread genus occurring mainly in eucalypt forests of southern Australia, and reaches its approximate northern limit to distribution in the Gondwana rainforests region (McQuillan *et al.* 2001). Adults of *Thalaina* (Plate 242) are mostly distinctively black-patterned species with silky-white wings. *Tigridoptera matutinata* is a large blue-black and orange-spotted species that is attracted commonly to light. The brightly coloured *Niceteria macrosoma* (Plate 241) is a widely distributed species (North Queensland to Victoria and South Australia) and in subtropical rainforest and wet sclerophyll forest is encountered during late summer and autumn.

Geometrinae associated with moist forest include *Aleochroma viridicata*, *Comostola leucomerata*, *Hypodoxa bryophylla*, *Oenochlora imperialis* and *Pingasa chlora*. Larvae of *Pingasa chlora* feed on *Flindersia schottiana*. Oenochrominae are diverse in Australia, the subtropical rainforest fauna including *Epidesmia tricolor*, the large *Systatica xanthastis*, the green-winged *Agathia prasinaspis*, the strikingly toned *Gastrophora henricaria* and *Onycodes traumataria*; the latter species is restricted to montane sites to 1700 m in altitude and ranges in distribution from northern New South Wales to Tasmania (Common 1990). Larval food plants of *Gastrophora henricaria* include eucalypts and the related *Lophostemon confertus*. The mainly temperate subfamily Larentiinae includes *Chaetolopha leucophragma*, a species found in wet sclerophyll forest from southern Queensland to Tasmania, and *Sauris lichenias*, a species found in both rainforest and sclerophyll forest.

Gracillariidae This is a diverse family of small elegant moths that mine leaves, leaf petioles, twigs and stems of Dicotyledons. However, larvae start their lives as sap feeders before commencing to feed on leaf tissue. Within the subfamily Gracillariinae the larvae of *Cyphosticha bryonoma* are known to feed on leaves of *Nothofagus moorei* (Nothofagaceae); *Lophostemon confertus* (Myrtaceae) is a possible host of *Macarostola polyplaca*; and *M. formosa* mines the leaves of *Syzygium smithii* (Myrtaceae). *Catoptilia xanthopharella* larvae mine the leaves of *Glochidion ferdinandi* (Euphorbiaceae) and those of *C. octopunctata* mine the leaves of *Homalanthus populifolius* (Euphorbiaceae). The known larval food plants of *Acrocercops chionsema*, a species distributed from southern Queensland to southern New South Wales, include *Macadamia tetraphylla*, *M. integrifolia* and *Stenocarpus salignus* (Proteaceae). Those of the related *Acrocercops macaria* mine members of the Rutaceae (*Acronychia laevis*, *Melicope micrococca*, *Halfordia kendack*) (Common 1990).

Hepialidae The primitive Hepialidae, or ghost moths, are widely distributed and are diverse in Australia. Many are attracted to light. There is a rich fauna also represented in South America, New Zealand and New Caledonia (Common 1990), suggesting a Gondwanan origin. *Fraus* is the most primitive genus. Some species are very large (e.g. *Zelotypia stacyi*) with wingspans of ~25 cm. *Onycanus* is the largest Australian genus (*c*. 40 species) is also occurs in New Guinea. The forewings (and sometimes hindwings) of many adult *Aenetus* are a brilliant green. There are 15 described species of *Aenetus* in Australia; the genus also occurs in New Guinea (five species), New Caledonia (one species) and New Zealand (one species). The South-east Asian–Indian genus *Endoclita* replaces *Aenetus* west of the Wallace Line (Wallace 1876).

Hepialid larvae bore either into soil or in living plant material, those concealed in the soil coming to the surface at night to feed. The larvae of the wood-boring *Aenetus* graze at night, below a protective web, on the callus which forms around the tunnel entrance in which the larva is concealed (Beaver and Grehan 2018). Although many hepialids frequent open forest, numerous species inhabit wet sclerophyll forest and rainforest rainforest (e.g. *Abantiades hylalinatus*, *Aenetus eximia*, *A. labyrinthicus*, *A. lewinii*, *A. ligniveren*, *A. montanus*, *A. ramsayi*, *A. scotti*, *Fraus crocea*, *Oncopera brunneata*, *O. epargyra*, *Oxycanus byrsus*, *O. gelidus* and *Zelotypia stacyi*) (Common 1990; Moss and Popple 2000; Nielsen and Common 1991; Nielsen and Kristensen 1989; Williams 2002) (Plates 243, 244, 245). *Abantiades*, *Fraus*, *Oncopera* and *Zelotypia* are endemic, *Aenetus* occurs in New Guinea, New Caledonia and Indonesia, and *Oxycanus* is also found in New Guinea (in addition to Australia). Larval hosts of *Aenetus scotti* include *Dendrocnide excelsa* (Urticaceae), *Diploglottis australis* (Sapindaceae), *Melicope micrococca* (Rutaceae) and *Daphnandra micrantha* (Atherospermataceae). Those of *Aenetus eximia* (Plate 244) include *Syzygium floribundum*, *Eucalyptus grandis* (Myrtaceae), *Doryphora sassafras*, *Daphnandra micrantha* (Atherospermataceae), *Glochidion ferdinandi* (Euphorbiaceae) and *Nothofagus moorei* (Nothofagaceae). *Aenetus lewinii* is recorded from *Alphitonia excelsa* (Rhamnaceae) (Beaver and Grehan 2018).

The Bent-wing Ghost Moth, *Zelotypia stacyi* (Plate 245), is the largest Australian hepialid. Host plants include *Eucalyptus grandis* and *E. salignus*. The species is rarely encountered but has been recorded from Cunninghams Gap in southern Queensland and Cambewarra Mountain south of Sydney, and in wet sclerophyll forest of the Lansdowne Escarpment, northern New South Wales (Common 1990; G. Williams unpubl. data).

Micropterigidae The family Micropterigidae (= Micropterygidae) is known from Early Cretaceous fossils and is the sole member of the basal lepidopteran suborder Zeugloptera. These species represent the most primitive extant moth family, and are possibly living relics surviving the breakup of Gondwana (Grimaldi and Engel 2005).

Micropterigidae are small, mostly diurnal moths that fly in full sunlight. They are widely distributed in most continents, but only number ~100 species. Adults are detritivorous or phytophagous, and possess functional, dentate mandibles that are able to crush pollen grains and fern spores for food (Common 1990). The world fauna can be divided into two groups: the Sabatincta group centred on the Pacific Basin (with outliers in South Africa and North America), and the Micropterix group confined to Eurasia (Grehan 1991). Most species occur in the Southern Hemisphere, and this fauna is likely to have constituted part of the original Gondwanan fauna (Common 1990). The Australasian distribution of the family is restricted to relictual rainforest in eastern Australia (including Tasmania), New Zealand and New Caledonia.

Sabatincta calliplaca occurs as far south as north-eastern New South Wales; however, additional Micropterigidae occur in rainforest along the east coast as far south as Mt Dromedary in southern New South Wales, and Tasmania. Adults have been collected by beating or sweeping vegetation in rainforest, or at mercury vapour lights.

Noctuidae This is the largest family of moths, and world-wide contains more than 25 000 known species placed in more than 4000 genera (Common 1990). The Australian fauna contains ~1500 described species in ~380 genera, and the major proportion of the fauna is closely related to Papuan and Oriental taxa (Common 1990). Noctuids possess a unique pair of tympanal organs positioned either side of the metathorax. These may function in relation to courtship, and may be adaptations against predation by microchiropteran bats (Grimaldi and Engel 2005).

Regionally occurring endemic genera include *Crioa*, *Niguza*, *Pansemna*, *Hecatesia*, *Platagarista* and *Zalissa* (Common 1990). Species associated with rainforest or wet sclerophyll forest include *Parerastria castaneata* (Acontiinae), *Argyrolepidia subaspersa*, *Zalissa catacalina* (Agaristinae), *Calogramma picta*, *Hedymiges aridoxa*, *Pachythrix hampsoni*, *P. trichoma*, *Pansemna beryllodes* (Amphipyrinae), *Calyptra minuticornis*, *Crioa emmelopis*, *Ericeia sobria*, *Eudocima salaminia*, *Ophyx ochroptera*, *Parallelia constricta*, *P. solomonensis*, *Plusiodonta coelonota*, *Serrodes campana*, *Trigonodes hyppasia* (Catocalinae) and *Xanthodes transversa* (Chloephorinae) (Common 1990; Moss and Popple 2000; Williams 2002); however, the actual number of species occurring in the region is much greater. *Fodina* (Plate 246) is an Oriental and Australian genus (Common 1990). Its strikingly patterned species are sometimes attracted to light.

Most adult noctuids are nocturnal, are attracted to light, and feed on nectar or liquid in fermenting fruit (e.g. *Serrodes campana*). Most larvae feed either on foliage or flowers. Larvae of *Argyrolepidia subaspersa* feed on *Cissus antarctica* and *Cayratia clematidea* (Vitaceae). Those of *Calyptra minuticornis* and *Plusiodonta coelonota* feed on *Stephania japonica* (Menispermaceae). The large and colourful adults of *Eudocima salaminia* pierce fruit, feeding on the juices within; however, its larvae feed on the vines *Stephania japonica* and *Sarcopetalum harveyanum* (Menispermaceae). *Othreis fullonia* is one of several large moths in that genus that are pests of fruit crops, but its larval hosts also include numerous native Menispermaceae (e.g. *Legnephora moorei*, *Pleogyne australis*, *Sarcopetalum harveyanum*). *Crioa emmelopis* feeds on *Acacia melanoxylon* (Mimosaceae), *Parallelia constricta* on *Elaeocarpus obovatus* (Elaeocarpaceae) and *P. solomonensis* on the shrub *Breynia oblongifolia* (Phyllanthaceae/Euphorbiaceae). By day the larvae of *Parallelia solomonensis* shelter in detritus. *Xanthodes transversa* larvae feed on *Hibiscus splendens* (Malvaceae), a large shrub growing on the margins of lowland rainforest and wet sclerophyll forest (Common 1990). *Calogramma picta* attacks *Crinum pedunculatum* (Amaryllidaceae), a succulent large herbaceous shrub sometimes found on the margins of lowland rainforests and in upper estuary zones. The gregarious larvae often defoliate plants, such that only their leafless bases remain (G. Williams pers. obs.).

Notodontidae These are medium- to large-sized moths related to Noctuidae. Larvae mostly feed on the leaves of woody trees and shrubs, relying during the day on cryptic coloration to avoid predators. If alarmed they may expose 'eye spots', suddenly display startling postures or deploy defensive chemicals (e.g. as with *Cerura australis*) (Common 1990). The colour and pattern of several species (e.g. of *Destolmia*, Plate 247), when their wings are folded at rest, superficially resemble elaterid beetles in the genus *Paracalais* (Plate 147).

Approximately half of the Australian genera are endemic, with the remaining genera shared with New Guinea or South-east Asia (Common 1990). Examples of the regional fauna include *Gallaba eugraphes*, which ranges from northern New South Wales to western Victoria, and *Cascera muscosa*, which has two isolated populations: one located in North Queensland on the Atherton Tableland, and a second extending from southern Queensland to southern New South Wales. *Cerura* is a widely distributed Old World genus with two species extending to Australia (Common 1990). *Cerura australis* ranges from central coastal Queensland to southern New South Wales. Its larvae are recorded from *Scolopia braunii* (Flacourtiaceae/Salicaceae). *Sorama* is an endemic genus widely distributed in southern Australia. The larvae of *Sorama bicolor* feed on *Eucalyptus* (Common 1990).

Oecophoridae These are very small to medium-sized moths. The Australian fauna exhibits a high level of endemism (Austin *et al.* 2004). The oecophorid fauna inhabiting regional open eucalypt-dominated forest and woodland is diverse, and it is not uncommon for species to be found in both mesic and dry forest types. Many species are nocturnal, though not necessarily attracted to lights (e.g. some Xyloryctinae). However, a number are brightly coloured day-fliers (*Snellenia* – Stathmopodinae, Plate 248). Larvae feed on dead (*Eulechria, Philobota*) and living plant tissue, and some are leaf miners. Larvae conceal themselves in various kinds of shelter, including webbed or otherwise joined leaves (*Agriophora, Calicotis, Plectophila*), crevices (*Lichenaula*), excavated tunnels and galleries in branches (*Barea, Machetis, Crypsicharis*) and stems (*Pilostibes*), or in the soil (*Procometis*). Some make portable cases (*Aristeis*) or carry leaves back to their tunnels to feed upon at night.

The Oecophorinae are the largest subfamily of Australia's moths, comprising ~20 per cent of the moth fauna (Common 1997). Genera are almost entirely endemic with little relationship between the Australian fauna and that of New Zealand and South America. This suggests that the main radiation of taxa in Australia occurred following the breakup of Gondwana (Common 1990). The two largest genera, *Philobota* and *Eulechria*, have their greatest concentration of species centred on the Macpherson–Macleay Overlap of south-east Queensland and northern New South Wales (Common 1990). The monotypic genera *Diplogrypa* and *Ascetoloba*, and numerous species, are endemic to the Gondwana rainforests region (Common 1997; Williams 2002). *Echinocosma, Disscobba, Haplodyta* and *Atheropla* are endemic genera restricted to south-eastern Australia, with south-eastern Queensland forming their northern limit of distribution (Common 1997). The oechophorine fauna recorded from subtropical rainforest and wet sclerophyll forest is a large one. Examples include *Aescetoloba lochmaea, Atheropla barytypa, A. calamaea, Baria consignatella, Callimima lophoptera, Chezala carphodes, C. osteochroa, C. silvestris, C. torva, Diplogrypa microptera, Disscobba caseicolor, Epithymema disparile, E. idiophanes, E. oridroma, E. parile, E. tryodes, Eulechria adoxodes, E. periphanes, Hesperoptila poliochroa, Sclerocris albipalpis, S. styphlodes, S. thetica, Telanepsia astatopis, Thema acratopa* and *Wingia lambertella.*

Stathmopodinae include mimics of lycid beetles and braconid wasps. Some are predators of scale insects. Regional species include *Calicotis crucifera* and *Hieromantis ephodophora*. Xyloryctinae include *Cryptophasa epadelpha, C. albacosta, Echiomima mythica, Pilostibes stigmatias, Plectophila discalis, Thymiatris cephalochroa* and *Lichenaula lichenea*. The Stenomatinae are very diverse in the New World, particularly in the Neotropics. Some Old World genera occur in Madagascar, Sri Lanka, India, the Moluccas, New Guinea, Solomon Islands, New Zealand and Australia. The greatest number of Old World species occurs in Australia (37 species, in three genera), followed by Madagascar (21 species, in four genera).

Larval food plants are known for only a small proportion of oecophorid species recorded from subtropical rainforest. Stenomatine larvae are particularly associated with Myrtaceae (e.g. *Syzygium, Eucalyptus, Syncarpia, Lophostemon*) with many species in the genus *Agriophora* inhabiting rainforest and wet sclerophyll forest (Common 1990). The larvae of *Callimima lophoptera* (Oecophorinae) feed on *Neolitsea australiensis* (Lauraceae), but the Oecophorinae are also associated with Myrtaceae (e.g. *Leptospermum, Eucalyptus, Lophostemon*). *Hieromantis ephodophora* (Stathmopodinae) larvae feed on the catkins of *Araucaria cunninghamii* (Araucariaceae), and *Calicotis crucifera* (Stathmopodinae) feed on the sporangia of *Platycerium* ferns (Common 1997). Of the subfamily Xylorctinae, the larvae of *Lichenaula lichenea*, unsurprisingly as the name suggests, feed on lichens. *Pilostibes stigmatias* is a stem borer of *Elaeocarpus obovatus* (Elaeocarpaceae) and *Plectophila discalis* larvae feed on shelters that have been constructed from leaves of *Cryptocarya glaucescens* (Lauraceae) and *Syncarpia glomulifera* (Myrtaceae). The larval hosts of *Echiomima mythica* include *Elaeocarpus grandis* (Elaeocarpaceae), *Backhousia myrtifolia, Syzygium floribundum* (Myrtaceae) and *Cleistanthus cunninghamii* (Euphorbiaceae). *Cryptophasa epadelpha* feed on *Lophostemon confertus* (Myrtaceae) and recorded rainforest hosts for the common *C. albacosta* include *Callicoma serratifolia* and *Ceratopetalum gummiferum* (Cunoniaceae) (Common 1990).

Palaephatidae Palaephatids are small moths known only from Australia, and central Chile and adjacent areas of Argentina (Nielsen 1987). The Australian fauna comprises *Ptyssoptera* and *Azaleodes*. *Azaleodes* occurs in rainforest and wet sclerophyll forest from northern Queensland to southern New South Wales (Nielsen 1987). Adults are attracted to lights. *Azaleodes brachyceros* is recorded from the vicinity of Barrington Tops, *A. megaceros* is recorded from Dorrigo National Park and adjacent localities, and *A. micronipha* ranges from Mt Tamborine, southern Queensland, to Mt Keira in southern New South Wales. Several species of *Ptyssoptera* also inhabit rainforest (Nielsen and Common 1991).

Pyralidae This is one of the largest moth families (>16 000 species), has a world-wide distribution, is ecologically diverse, and in Australia numbers more than

1000 species (Common 1990; Gaskin 1975; Grimaldi and Engel 2005). Pyralids are concealed feeders, sheltering in folded leaves or in webbing constructed among flowers, seed clusters, and leaves. Larvae feed on dead leaves and detritus, fungi, lichens, algae, the combs of social insects, dried fruit, and several are predators of scale insects. Others are inquilines. The saprophagous habit of some larvae is probably secondarily derived (Grimaldi and Engel 2005).

Phyctinae genera have strong affiliations with Oriental and African regions (Horak 1997). The genus *Faveria* is distributed widely throughout the Australian, Oriental and African regions. *Ptyobathra* is distributed in Australia, Sri Lanka, Sumatra and Japan. *Ptyobathra hades* is restricted to the Gondwana rainforests region. Scopariinae include species (e.g. *Scoparia gethosyna*) that feed on mosses. *Styphlolepis agenor* (Cybalomiinae) produces galleries in the trunks of *Capparis mitchelli* (Capparaceae), a species growing in vine thickets north of Gunnedah, New South Wales (Harden *et al.* 2006). The Pyraustinae species *Glyphodes cosmarcha* and *G. flavizonalis* extend from northern Australia to southern Queensland and northern New South Wales. The larvae of *Glyphodes cosmarcha* feed on the leaves of *Ficus* (Moraceae). Larvae of the pyraustine *Parotis atlitalis* feed on *Glochidion ferdinandi* (Euphorbiaceae) (Common 1990).

Saturniidae The Saturniidae are moderate- to large-sized, mainly nocturnal, moths and reach their greatest diversity in subtropical and tropical areas (Common 1990). Two included species, *Opodiphthera astrophela* and *O. rhythmica*, inhabit mesic forests. Larvae of *Opodiphthera rhythmica* (*Antheraea rhythmica*), a species encountered in rainforest at intermediate elevations (e.g. Mt Tamborine, Lansdowne Escarpment), feed on the leaves of *Litsea reticulata* (Lauraceae), and also are known to accept leaves of *Doryphora sassafras* (Atherospermataceae). Those of *Opodiphthera astrophela* (= *Antheraea astrophela*) feed on *Flindersia schottiana*, *F. australis* (Rutaceae), *Alphitonia excelsa* (Rhamnaceae) and *Callicoma serratifolia* (Cunoniaceae). *Syntherata janetta* feeds on numerous species, including *Geijera salicifolia* (Rutaceae) and *Glochidion ferdinandi* (Euphorbiaceae).

Sphingidae The Sphingidae, or hawkmoths, are best represented in the Wet Tropics. Most are quite large. Australia possesses ~59 species of hawkmoths in 25 genera, and two subfamilies (Sphinginae, Macroglossinae) (Common 1990; Moulds 1985). Most are placed in the Macroglossinae. Adults are long-tongued, and mostly crepuscular or fully nocturnal in behaviour. However, bee-hawkmoths (*Cephonodes*) are diurnal, and hover conspicuously at flowers in bright sunlight when feeding.

The largest species from south-eastern Australia, *Coequosa triangularis* and *C. australasiae* (syn. *Metamimas australasiae*) (Macroglossinae), occur in the Gondwana rainforests region; the range of *C. triangularis* extends from southern Queensland to southern New South Wales, and that of *C. australasiae* (Plate 249) from northern Queensland to southern New South Wales (Common 1990; A. Rose 1975). Both are attracted to house lights, and ultraviolet light, in late summer. *Coequosa* is an endemic Australian genus (Common 1990), but no other genera are endemic to the Gondwana rainforests region (M. S. Moulds pers. comm.). Additional species recorded from subtropical eastern Australia include *Acosmeryx miskini*, *Cephonodes kingii*, *Cizara ardeniae*, *Daphnis placida*, *Gnathothlibus erotus*, *Hippotion scrofa*, *H. celerio*, *H. velox*, *Macroglossum hirundo*, *M. micaceum*, *Psilogramma menephron*, *Theretra queenslandi* and *T. tryoni* (Common 1990; Moss and Popple 2000; Moulds 1985, 1998; Williams 2002). *Psilogramma menephron* (Sphinginae) has an extralimital distribution, and in Australia its larval native food plants include members of the Bignoniaceae, Oleaceae and Verbenaceae. *Coenotes eremophilae* (Sphininae) is an endemic species whose larvae utilise numerous plants from a broad diversity of families, but especially Myoporaceae. Regarding members of the subfamily Macroglossinae, recorded larval food plants for *Coequosa triangularis* are members of the Proteaceae (e.g. *Grevillea robusta*, *Macadamia integrifolia*, *Stenocarpus sinuatus* [Moulds 1981, 1984]), *Coequosa australasiae* larvae feed only on *Eucalyptus* (Myrtaceae) (Common 1990), *Gnathothlibus erotus* feed on Vitaceae (e.g. *Cayratia clematidea*, *Cissus* spp.), and *Cizara ardenia* includes *Coprosma quadrifida* and *Gynochthodes jasminoides* (Rubiaceae). *Daphnis placida* has been recorded feeding on *Alangium villosum* (Alangiaceae), and *Acosmeryx miskini* on *Cayratia clematidea* (Vitaceae). The usual larval food plants of *Cephonodes kingii* are members of the Rubiaceae (e.g. *Canthium* spp.). *Theretra nessus* (Plate 250) feeds on *Dioscorea discolor* (Dioscoreaceae), a yam that grows in wet sclerophyll forest, and those of *T. tryoni* feed on Araceae (Common 1990, Moulds 1998). The larval food plants of *Hippotion celerio* include *Alocasia brisbanensis* (Araceae) (A. Rose 1975). These are not exhaustive records, and in addition the larvae of native hawkmoths are known to utilise exotic plants (Moulds 1998).

Sphingidae are important pollinators in most biodiverse regions of the world (Johnson *et al.* 2016). In subtropical rainforest they include species (e.g. *Theretra nessus*) which are important pollinators of particular plants (e.g. of *Crinum pedunculatum* [Amaryllidaceae]), and also in the long-distance transfer of pollen between dispersed plant populations (Williams and Adam 2010; Willmer 2011). Plants better adapted to hawkmoth visits generally possess deep tubular flowers, and secrete relatively large volumes of nectar. Indeed, the length of visitation may be longer where flowers provide larger nectar rewards (Johnson *et al.* 2016). Sphingidae utilise visual (including flower conspicuousness) and olfactory cues to locate and orientate to flowers, and plants particularly attractive to sphingids tend to release attractant fragrances timed to parallel the activity patterns of the moths. Hawkmoth flowers were reported by Miyake *et al.* (1998) to have strong fruity or flowery scents in the evening (see also Dobson 2006; Willmer 2011). The classic hawkmoth pollination syndrome, sphingophily, includes flowers that open at dusk and night, flowers are white, cream, or pale green in colour, there are no nectar guides, scent is strong/sweet, the flower shape is usually radial with a long tube or spur, the nectar site is concealed, and the nectar is of moderate volume and mid to low concentration (Willmer 2011). However, flower colour can vary (e.g. including red, yellow), and flowers pollinated by hawkmoths may also be bird and bee pollinated. Adult hawkmoths normally need to visit many flowers per minute to provide for their energy needs and, within the constraints of appropriate nectar and flower types, are generalists, visiting whatever flowers are available. Trap-lining foraging patterns are also known (Linhart and Mendenhall 1977), with pollen movement by hawkmoths under favourable foraging conditions (clear and warm nights) extending beyond 350 m.

Thyrididae This is the only family in the superfamily Thyridoidea, and is represented in Australia by ~50 species in 13 genera. Thyrididae have a predominantly subtropical and tropical distribution (Common 1990). Adults are nocturnal and are attracted to lights. Larvae include stem borers and gall formers, as well as those that feed on leaves. The widely distributed genus *Aglaopus* (Striglininae) includes the rainforest-inhabiting *A. stramentaria* from central and southern coastal Queensland. *Addaea subtessellata* (Siculidinae) feeds on *Mallotus philippensis* (Euphorbiaceae), and occurs in southern Queensland and northern New South Wales.

Tineidae At least six subfamilies occur within the Gondwana rainforests region (Regier *et al.* 2014; Robinson and Nielsen 1993). The fauna comprises endemic genera (*Vanna*, *Thomintarra*, *Timaea*), genera that are primarily Australian, but with limited Pacific-Oceanic extensions (*Parochmastis*), genera with essentially Gondwanan distributions (*Amphixystis*), genera with broad but disjunct distributions (*Oenoe*, *Phaeoses*), and those loosely confined to the Indo-Australasian regions (*Ectropoceros*, *Gerontha*). Adults are nocturnal and the larvae consume decaying wood and bark, fungi, lichen, hair and feathers. Others tunnel in the soil or build shelters made by joining adjacent leaves together with silk. The larvae of *Morophaga clonodes* (Scardiinae) tunnel in bracket fungi (Common 1990).

Uraniidae The subfamily Microniinae possesses six Australian species (in four genera) which are mainly rainforest-inhabiting taxa occurring in northern New South Wales (e.g. *Aploschema discata*, *Acropteris nanula*), Queensland and the Northern Territory (Common 1990; Nielsen and Common 1991). Epipleminae include *Balantiucha decorata*, a species widely distributed in northern Australia and south to New South Wales. Larvae excavate sections from the underside of the leaves of *Canthium coprosmoides* (Rubiaceae) (Common 1990). The flower-frequenting Uraniinae (e.g. including the large butterfly-like *Alcides metaurus*) are apparently absent from this region.

Miscellaneous families The larvae of *Batrachedra arenosella* (Batrachedridae) have been recorded as possible predators of scale insects infesting *Macadamia* trees (Common 1990). Depressariidae are small delicate moths that include the rainforest-inhabiting *Barantola pulcherrima*, *Scorpiopsis pyrobola* and *Peritornenta circulatella* (which feeds on *Cupaniopsis anacardioides*). Palaeosetidae contain the endemic genus *Palaeoses*, of which *P. scholastica* occurs in southern Queensland (Nielsen and Common 1991). Cosmopterigidae include several species of *Limnaecia* (e.g. *Limnaecia scoliosema*), the larvae of which are associated with bracket fungi. Zygaenidae are mainly a tropical group with only a few species extending to central New South Wales. Most are nocturnal and are likely to be distasteful to vertebrate predators (Common 1990). They include the endemic genus *Onceropyga* (*O. anelia*) and the more widespread *Lactura* (e.g. *Lactura suffusa*, *L. caminaea*). The larvae of *O. anelia* feed on

Cissus antarctica (Vitaceae). *Lactura suffusa* (Plate 251) larvae feed fully exposed during the day on *Pouteria australis* (Sapotaceae), and *L. caminaea* feeds on *Ficus*, including *F. rubiginosus* (Moraceae) (Common 1990). Adults of Tortricidae (e.g. *Trymalitis optima*, *Cryptaspasma sordida*, *Cryptoptila immersana*) shelter by day in leaf litter, the foliage of their food plants or in bark crevices. A few are diurnal. The larvae are concealed feeders that may form shelters by spinning together shoots or joining living leaves. They bore in twigs and stems, gymnosperm cones, fruits (e.g. *Syzygium* [Juniper and Britton 2010]) and tree bark. *Cryptoptila australiana* feeds on the shrub *Polyscias sambucifolius* (Araliaceae). The vine *Eustrephus latifolius* (Liliaceae) is a larval food plant of *Cryptoptila immersana*. *Cryptaspasma sordida* has been reared from the fruits of *Syzygium ingens* (as *Acmena brachiandra*) in Lamington National Park (Common 1990, citing G. Monteith). Adult *Thymiatris cephalochroa* (Xyloryctidae) resemble small twigs when the wings are held in repose. The species is distributed from Cooktown, North Queensland (Common 1990), to Lansdowne, northern New South Wales (G. Williams pers. obs.), and is recorded from dry sclerophyll forest (I. MacMillan pers. comm.), lowland subtropical rainforest and wet sclerophyll forest (G. Williams pers. obs.).

HYMENOPTERA (wasps, bees, ants, sawflies)

Along with Coleoptera and Diptera, the order Hymenoptera is one of the most numerous of insect groups to be found in the rainforest world of subtropical eastern Australia (Chapman 2009). Hymenoptera include insects variously referred to as sawflies, wasps, hornets, ants, velvet-ants and bees. Many are active predators, capturing prey so as to provision for the developmental needs of their young. Others are parasites (Leucospidae) of other hymenopterans, or like fig wasps (Agaonidae), participate in highly evolved plant mutualisms (Weiblen 2002; Wiebes 1979; Willmer 2011). And among their ranks are groups that are specialised social insects with distinct caste classes (ants, and some bees), or that function as socially cooperative members of nest communities (as with paper wasps). These are derived behaviours from an ancestral plant-feeding state.

Hymenoptera are divided into two suborders: the geologically old Symphyta, or sawflies (e.g. Pergidae), and the far more diverse Apocrita (Kukalová-Peck 1991; Naumann 1991b), which includes ants, wasps and bees. Symphyta does not form not a natural group, but is a collective name for the most basal lineages of Hymenoptera (Vilhelmsen 1997). Apocrita are probably monophyletic (Goulet and Huber 1993), consisting of an ancestor and all its descendants, and are further divided into two groups: parasitic terebrants – the Parasitica or Tenebrantia (those that bore; Ichneumonoidea, Chalcidoidea), and Aculeata or aculeates (e.g. Crabronidae, Vespidae, Pompilidae, Apidae, Megachilidae) in which the ovipositor is usually modified for stinging. Aculeata are further divided into Apoidea–Apiformes (bees), which are also known historically as the Anthophila (Michener 2007), and Apoidea–Spheciformes (wasps). In the Parasitica the ovipositor retains its primitive egg-laying function, but an important phylogenetic distinction between Spheciformes and Apiformes is that, although they are closely related, there is strong morphological and molecular evidence that bees are derived directly from ancestral Spheciformes, such as Crabronidae (Debevec *et al.* 2012; but see Sann *et al.* 2018 for a further consideration of the status of the latter). In addition, while bees essentially are consumers of pollen and nectar, wasps are mainly predators. However, there are morphological and biological exceptions within both groups, such that a clear distinction between Parastica and Aculeata is difficult to achieve.

Additionally, the informal term 'microhymenoptera' is often applied to wasps (normally parasitic) in the smallest size classes (<3–6 mm), while 'macrohymenoptera' are larger insects that include wasps, bees, and ants (though many ants are very small). However, size ranges can vary considerably within families (e.g. Torymidae 0.9–30 mm, Chalcididae 1.5–16 mm, Proctotrupidae 3–15 mm, Pteromalidae 0.6–40 mm. The primitive Mymaridae, or fairy flies, are minute (0.2–1 mm and are some of the smallest of all insects (Lin *et al.* 2007). Many microhymenopterans are recorded from subtropical eastern Australia (e.g. Austin *et al.* 1998; Bouček 1988; Early and Naumann 1989; Naumann 1982, 1991b; Riek 1955a; Williams 2002) and exhibit variation in assemblage diversity with respect to altitude (Hall *et al.* 2014).

Numerous microhymenopteran families visit the flowers of subtropical rainforest trees (e.g. *Alectryon coriaceus*, *Alphitonia excelsa*, *Ackama paniculosa*, *Cryptocarya microneura*, *Syzygium floribundum*, *S. smithii*, *Tristaniopsis laurina*) (Williams 1995; Williams and Adam 2010). Typically, the flower-frequenting families include Aphelinidae, Ceraphronidae, Diapriidae, Encyrtidae, Eulophidae, Figitidae, Mymaridae, Platygastridae, Pteromalidae, and Torymidae; their individual species are all <5 mm in size (Williams 1995). Families of microhymenoptera are particularly diverse at trees with open and shallow floral structures in which nectar is offered as a reward, but poor or absent at species with depth effect in the corolla (e.g. *Diospyros australis*, *Elaeocarpus obovatus*, *E. reticulatus*) (Williams 1995). Numerous families (Diapriidae, Dryinidae, Encyrtidae, Eulophidae, Eurytomidae, Figitidae, Mymaridae, Perilampidae, Platygastridae, Proctotrupidae, Pteromalidae, Torymidae) are attracted to the flowers of *Schizomeria ovata*, a small tree in rainforest but also in the understorey of wet sclerophyll forest. However, most are snared by the viscous nectar and are unable to escape (Williams 1995; Williams and Adam 2010).

Apocrita are mainly predators or parasites, and are usually characterised in form by a distinct waist-like connection (absent is some families, e.g. Mymaridae) between abdominal segments 1 and 2 (e.g. Vespidae – Eumeninae, Evaniidae) and the apparent 'abdomen' (the metasoma or gaster) being only a part of the adbomen; the first segment is fused into the thorax and appears to be, and functions as, part of the thorax The ovipositor is sometimes extremely long and thin (as in Agaonidae, Aulacidae, Braconidae, Gasteruptiidae, Ichneumonidae), or modified as a powerful and venomous sting (this ability has been lost in some females) (see Piek 1986 for a venom review of the Hymenoptera). In some families males or females may be wingless (e.g. Agaonidae, Bethylidae, Mutillidae, Thynnidae). The larvae of Apocrita differ from the caterpillar-like larvae of Symphyta in being grub-like, without legs or eyes, and are generally featureless (Goulet and Huber 1993). Some Apocrita have returned to a specialised plant diet – notably bees (Apoidea/Apiformes), fig wasps (Agaonidae) and gall-making Cynipidae. But during the larval stage most Apocrita (except for most bees) are parasitoids of other insects or spiders.

Though functionally and fundamentally carnivorous (feeding on other insects and spiders), adults of many aculeate wasp families (e.g. Scoliidae, Vespidae, Sphecidae, Crabronidae) will imbibe floral nectar, and incidentally serve as pollinators (Williams and Adam 1994, 1995, 1998, 2001, 2010). The Apocrita underwent a rapid episode of evolutionary radiation in the Early Jurassic (*c.* 195 mya), with the origin of the Aculeata in the Late Jurassic and radiation of many aculeate families in the Early Cretaceous (*c.* 140 mya) (Oberprieler *et al.* 2012 record fossil Evanioidea from the Upper Jurassic of Australia). Ants and bees originated in the Mid Cretaceous, parallelling the emergence and evolution of flowering plants. Eusociality (characterised by transport of larval provisions, social hive and nest development and behaviour [e.g. Johnson *et al.* 2013]) for ants and wasps, and later bees, sequencially developed from the Mid to Late Cretaceous (Grimaldi and Engel 2005). Stow and colleagues (Stow *et al.* 2010) suggest fungal pathogens (against which ants and bees may deploy a range of anti-microbial secretions) may have played a possible role in the early evolution of eusociality, with full sociality in bees generating a better defence against fungal pathogens owing to the prescence of abundant nest mates.

Among the Australian hymenopteran taxa only the wasp families Austroniidae and Peradeniidae, the bee family Stenotritidae and ~8 per cent of subfamilies are endemic and this is likely to be due to the ability of most taxa to fly and disperse (Naumann 1991b). Levels of species and generic endemicity, however, are high. The tropical rainforests share many groups with wider Oriental and Melanesian affinities. Some very primitive families, which otherwise have a relict world-wide distribution, are well represented in Australia (i.e. Aulacidae, Megalyridae, Stephanidae, Trigonalyidae and Xiphydriidae – Xyphdriinae).

Many relatively archaic Australian higher taxa have their closest affinities with the South American fauna (Naumann 1982, 1991b) (e.g. Diapriidae – Ambositrinae, Gasteruptiidae – Hyptiogastrinae, Ichneumonidae – Labeninae, Monomachidae, Orussidae, Pergidae – Philomastiginae, Pompilidae – Epipompilinae, Thynnidae – Thynninae); these Gondwanan groups are usually more common in wet temperate areas (Naumann 1991b). The Aphidiidae, which are parasites of aphids, are associated with *Nothofagus* in Australia and Chile and are probably very archaic. The Monomachidae and Proctotrupidae have representatives in Australia and South America (Naumann 1982). The Gondwana rainforests region is diverse in Thynnidae and numerous new species await description (G. R. Brown pers. comm.).

In the following discussion the various families of Hymenoptera are grouped sequentially as bees, wasps and ants. These three groups are so morphologically (visually) and behaviourally distinctive in the minds of most readers that this departure from the format used for other orders (except moths and butterflies within Lepidoptera) seems a logical treatment.

Symphyta (sawflies)

Symphyta (Plates 279, 280, 281) first appeared in the Triassic ~200 mya, and have preserved the ancestral plant-feeding habits of the order, numerous wing veins, and an abdomen in which each segment is similar in general appearance, unlike the wasp-waist of the Apocrita (Goulet and Huber 1993). In most Symphyta the ovipositor is saw-like, and used to pierce plant tissue. The larvae are caterpillar-like with functioning legs; they can aggregate, may actively translocate in distinct clusters to new feeding resources, and in some instances adult females exhibit quasi brood care. The gregarious larvae of certain species exhibit distinctive defence postures and signalling behaviour (L. Fletcher 2007). Toxic peptides have been isolated from the larvae of Pergidae and Argidae (Boevé *et al.* 2014), and the range of compounds that larvae are known to sequester from their host plants can function in a defensive role against attack by predators (e.g. rodents, birds, other insects) and parasites (e.g. Braconidae, Eulophidae, Ichneumonidae, Tachinidae) (Macdonald and Ohmart 1993; Morrow *et al.* 1976).

The families Argidae, Orussidae, Pergidae, Tenthredinidae and Xiphydriidae are represented in Australia (Benson 1938b, 1939; Naumann 1991b; Schmidt *et al.* 2002). However, Symphyta are poorly represented in rainforest. This may reflect loss of a once more-diverse ancestral rainforest-associated fauna as mesic habitat changed to an increasingly arid-adapted one (Schmidt and Walter 2014). Fossil Xyelidae (*Archexyela ipswichensis*) from south-east Queensland are among the earliest known Hymenoptera (Engel 2005). This family no longer occurs in Australia, extant xyelids now being restricted to the Holarctic (Naumann 1991b).

Sawfly larvae are largely phytophagus and feed either externally on leaves (and are sometimes serious defoliators), or are leaf miners or stem borers. However, the Orussidae are ectoparasitic, the females ovipositing into the galleries formed by tunnelling cerambycid and buprestid larvae, and the orussid larvae feeding externally on the host, but entering the dead host body to pupate (Naumann 1991b).

Pergidae are the most common of the Australian Symphyta and world-wide comprise more than 450 species. Females may deploy pheremones to attract males over long distances (Schmidt *et al.* 2006). Although a small number occur in North America, the family is otherwise restricted to Australia, Central and South America (Benson 1938a, 1938b, 1939; Goulet and Huber 1993; Naumann 1991b). Major pergid lineages likely evolved during the breakup of the Gondwana supercontinent, and then co-radiated with myrtaceous hosts (e.g. *Angophora*, *Callistemon*, *Corymbia*, *Eucalyptus*, *Leptospermum*, *Lophostemon*, *Syncarpia*, *Melaleuca* [arguably now including *Callistemon* – see Craven 2006; Udovic and Spencer 2012]) (Macdonald and Ohmart 1993; Naumann 1991b; Schmidt and Walter 2014). The Perginae (e.g. *Perga*, *Pseudoperga*, *Xyloperga*) are the most speciose of the subfamilies occurring in Australia and appear to have coevolved with their myrtaceous larval hosts (Macdonald and Ohmart 1993). Larvae of *Perga* travel in groups on the ground, individuals signalling with raised abdominal tips as to whether to proceed or halt. Females of some pergid species (e.g. *Philomastix* – Philomastiginae) guard their eggs and young, the adult close in attendance on stems or leaves. The larvae of *Philomastix* feed on *Rubus* (e.g. *Rubus moluccanus*). The distinctive black-orange-banded *Pterygophorus cinctus* is recorded from the flowers of *Syzygium smithii* (Williams 1995). Larval Euryinae mostly feed on decaying leaves and leaf litter. The adults of small-sized *Neoeurys* (Euryinae) feed on the nectar of *Alphitonia excelsa*, *Ackama paniculata*, *Cuttsia viburnea*, *Rhodomyrtus psidioides*, *Syzygium floribundum*, *S. smithii* and *Tristaniopsis laurina* (Williams 1995; Williams and Adam 2010). Species of *Eurys* and *Neoeurys* possess elongate mouthparts adapted to a nectar-feeding habit; adaptations to facilitate nectar and or pollen feeding having arisen independently in several symphytan groups (see Jarvis and Vilhelmsen 2000). The endemic Pteryperginae regionally include *Pteryperga galla* (Tweed River, Dorrigo), associated with a species of *Elaeocarpus*, and *P. hyaloptera* (Bunya Mountains) (Schmidt *et al.* 2002). The Styracotechyinae are represented by a single species, *Styracotechys dicelysma* (Plate 280). The species is recorded from Lamington National Park, Tooloom and Lorien Wildlife Refuge (Schmidt *et al.* 2006; G. Williams and J. Macdonald unpubl. records). Adults are sexually dimorphic with males possessing conspicuous bipectinnate antennae and females having a deeply emarginate sawsheath (Schmidt *et al.* 2006). Female *Styracotechys* lay their eggs in the midvein on the undersurface of the young leaves their larval host, the vine *Austrosteenisia blackii* (Fabaceae), which grows in rainforests of the east coast north from the Hunter River (Harden *et al.* 2007). Males may sometimes be seen in temporary swarms flying about and landing on *Austrosteenisia blackii* leaves. The related *Austrosteenisia glabristyla*, which is distributed from Dorrigo north to Eumundi in southern Queensland, may also be a larval host (Schmidt *et al.* 2006).

World-wide Tenthredinidae are the largest sawfly family with over 5000 species (Naumann 1991b). Larvae mainly feed on the leaves or stems of angiosperms. However, few species are recorded from Australia, and the majority of these are introduced with only one, or possibly two, species being endemic. Of these *Cheilophleps xantha* is recorded only from Ulong east of Dorrigo, and Bulga inland from the central coast of New South Wales (Naumann *et al.* 2002).

Symphyta are also associated with wet sclerophyll forest, where they are often observed resting in isolation on exposed foliage. These include *Pergagrapta bicolor*, *P. latreillii* (e.g. both from Barrington Tops), *Philomastix macleaii* (Werrikimbe National Park) and *Xyloperga leachii* (Tapin National Park).

Apocrita (Aculeata – Apiformes) (bees)

Within the Hymenoptera one wasp superfamily, the Sphecoidea, has recently been subsumed within the superfamily Apoidea, the latter a taxon rank that was previously constituted only by true bees (Melo 1999; Melo and Gonçales 2005; Michener 2007). World-wide, membership of Apoidea now comprises Spheciformes, or sphecoid wasps, (e.g. families Ampulicidae, Sphecidae and Crabronidae) plus Apiformes – bees (families Stenotritidae, Colletidae, Andrenidae, Halictidae, Melittidae, Megachilidae and Apidae) (Michener 2007). Andrenidae and Melittidae do not occur in Australia (T. Houston 2018). Genera previously assigned to Anthophoridae and Ctenoplectridae (Michener and Greenberg 1980) are now placed within Apidae (T. Houston 2018; Michener 2007).

The present world bee fauna numbers more than 17 500 described species (Michener 2007), and this fauna feeds almost exclusively on the floral resources offered by flowering plants. The Australian fauna currently comprises more than 1500 named species, but the actual number is higher – possibly 2000 (T. Houston 2018). Bees represent one of the most important of pollinating insect taxa (e.g. Michener 2007; Nic Lughadha and Proença 1996); however, the origin of bees is uncertain. The fossil *Cretotrigona prisca* dates from the Late Cretaceous but bees may have originated earlier, perhaps in the Early to Mid Cretaceous (Grimaldi and Engel 2005; Michener and Grimaldi 1988; Sann *et al.* 2018). This time frame corresponds roughly with the emergence of angiosperms (Danforth *et al.* 2006). In addition, there are fossil leaves from the Mid Eocene, and earlier, with damage consistent with that caused by leaf-cutter bees (Megachilidae). Megachilidae are a derived bee family, which again indicates a much earlier evolution of true Apiformes (Wappler and Engel 2003). Regardless of their evolutionary emergence, it appears that the earliest bee lineages, originating from crabronid wasp ancestors, included taxa that were host plant specialists (Williams and Adam 2010).

'Melittophily' is the term for pollination by bees, and flowers adapted to be pollinated primarily by bees are termed 'melittophilous'. Many species are highly efficient foragers and exhibit varying degrees of morphological and behavioural specialisation (e.g. having long, as opposed to short tongues, and utilising trap-lining techniques) as potential pollinators (Cunningham *et al.* 2013; Gross 1993; Michener 2007; Roubik 1992; Williams and Adam 2010), and can exert a precise flower selection choice mainly driven by their pollen requirements (Aronne *et al.* 2012). Bees normally restrict their foraging activity to daylight hours, and their foraging behaviour, and specialisation in the diversity of plant taxa visited, varies between opportunistic polytropic (polylectic) bees that visit many species of plants to oligotropic (oligolectic) or monotropic (monolectic) bees that visit a restricted range of flowers (Faegri and van der Pijl 1979). In subtropical rainforests most bees appear to be opportunistic, at most partly restricted in the choice of flowering trees and shrubs upon which they will forage. These generalist foraging strategies are likely to benefit solitary bees (which most are) owing to the reduced resources needed to provision individually small broods. And although the bee fauna of subtropical easternAustralia is diverse, including at regional and local scales (Williams and Adam 1997, and references in Williams 2002), the diversity of species present on any one plant species varies with site and season (Williams 1995; Williams and Adam 1997). Similar temporal and spatial variation in native bee diversity in rainforest stands and other plant communities (including fragmented agricultural landscapes) has been widely reported (e.g. Dainese *et al.* 2018; Oertli *et al.* 2005; Potts *et al.* 2005; Ramírez *et al.* 2016; Roubik 1992). Although the reasons for local scale fluctuations in species diversity are not fully understood, likely influences include resilence restraints to uncertain flowering events, nest site availability, interactions with predators and competition from other anthophilous faunas, storm damage to and localised extinction of specific food plants (loss of resources), and seasonal variation in rainfall patterns and general climate conditions that are also affecting rainforest communities at the landscape scale.

Specialist and generalist bees (with regard to social habit, feeding and nesting behaviour), and rare species occurring at low population levels, may be affected differently by adverse events and conditions.

The Australian bee fauna (Apoidea – Apiformes) includes more than 1500 known species (T. Houston 2018), but the temperate fauna is the least diverse, with only ~18 genera recorded from Tasmania (Cardale 1993). Many of the known native bee species are undescribed and more than half of the named taxa are in need of revision (Batley and Hogendoorn 2009). Australia is unique in possessing a bee fauna dominated by the primitive Colletidae. Approximately half of the species, and the greatest diversity of genera, occur in this one family (Cardale 1993); however, dispersal events linking the Australian and African colletid fauna may have occurred in the Eocene (Kayaalp *et al.* 2017). Historically, the short-tongued Colletidae have been considered the basal family in the genealogical development of bees. This is because colletids possess morphological characters that link them to sphecoid wasps (e.g. Crabronidae, Sphecidae) (Michener 2007), but more recent studies indicate the Melittidae (also short-tongued, though believed to be the progenitors of long-tongued bees) are basal relative to other bees (Danforth *et al.* 2006; Michener 2007), and this position is also suggested by the fossil record.

Although the conservation of bees world-wide has drawn considerable interest, the status of native bee populations within individual subtropical rainforest stands, and especially of lowland remnants, is poorly understood.

Apidae Apidae are a diverse family that includes social bees (*Apis*, *Tetragonula*, *Austroplebeia*), solitary nest builders (*Amegilla*, *Exoneura*, *Xylocopa*), shared nest builders in which several females cooperate (*Braunsapis*), and parasites (e.g. *Thyreus*, *Exoneura* subgenus *Inquilina*). It is one of two families of long-tongued bees, the other being Megachilidae. *Apis mellifera* has been introduced into Australia for commercial honey production, but wild (feral) colonies are widespread throughout the landscape. Feral *Apis* can be found foraging on all subtropical rainforest trees that possess open, readily accessible, flower structures and which produce large quantities of nectar and pollen (Williams 1995; Williams and Adam 1997). *Apis* can occur in large numbers during peak phase flowering events in subtropical rainforest, especially at those trees flowering in the canopy or at lower heights of exposed margins, and in particular trees with massed inflorescences composed of many florets (e.g. *Guioa semiglauca*). The presence of *Apis mellifera*, both in commercial hives and feral occupants of numerous forested areas, is a competitive threat to indigenous flower-dependent

Sonicating bees and buzz pollination

Approximately 15 000–20 000 species of flowering plants possess flower types that can only release pollen via small pores at the tips of their anthers (De Luca and Vallejo-Marín 2013). This number represents ~6–8 per cent of all angiosperms (Willmer 2011). Insects (many bees, but also some wasps, and rarely flies) that visit such poricidal flowers often resort to vibrating the anthers so as to release the pollen; though this behaviour is widespread in bees it is noticeably absent in *Apis mellifera* (Michener 2007). This method of pollen release and collection is called sonication, or buzz pollination (Buchmann 1983), and some floral morphologies appear to have evolved specifically in response to the collection of pollen by sonicating bees. Michener (2007) suggests that the vibration strategy may have developed from the requirement by certain bees during nest construction to push past objects that otherwise could not be moved. Sonicating behaviour can vary in duration and amplitude (Harder and Barclay 1994; Willmer 2011). Usually the bee directs the anther aperture towards its body so as to maximise pollen capture. Bees can be quite vigorous in their attempts to access pollen, and the vibration of the wing muscles, allowing pollen release, results in distinct buzzing sounds; thus the name 'buzz pollination'.

Buzz-pollinated flowers often lack nectar, and the expression of anther structure necessitating sonification to release pollen occurs in many, often unrelated, plant families (in more than 70 families [Willmer 2011]) and represents a remarkable example of convergent evolution (De Luca and Vallejo-Marín 2013). De Luca and Vallejo-Marín (2013) consider that the evolution of buzz pollination was probably triggered by the necessity of nectarless plants to restrict pollen access to optimal dispersal agents and to reduce gamete loss to inefficient pollinators and pollen thieves (e.g. Hylaeinae in Australian subtropical rainforest), these being visitors that remove pollen with little or no contribution to pollen transfer.

taxa (see Pyke and Balzer 1985; Williams 1995; Williams and Adam 1997). Their impact on the reproduction of native plants in subtropical rainforest is poorly known.

Austroplebeia (i.e. *Austroplebeia australis*, *A. cassiae*) extends into southern Queensland and northern New South Wales (Cardale 1993; T. Houston 2018); however, the prominent native social species at higher latitudes in lowland subtropical rainforest is the diminutive stingless *Tetragonula carbonaria* (more correctly, *Tetragonula* has an atrophied and non-functional sting), distributed from southern Queensland to southern New South Wales. *Tetragonula carbonaria* individuals have homing ranges that can exceed 700 m (Smith *et al.* 2017), and forage on a broad range of canopy trees, sunlit shrubs and herbs (Dollin *et al.* 1997; Williams 1995; Williams and Adam 1997; Williams and Walker 2003) (Plate 283). Although *Tetragonula carbonaria* individuals exhibit a degree of temporal foraging constancy (and thus enhancing pollinator efficacy) (White *et al.* 2001), flight activity is strongly influenced by temperature and radiation (Heard and Hendrikz 1993). Both *Apis* and *Tetragonula* generally avoid shaded flowers and may cease foraging during intermittent periods of cool or rainy weather (Williams and Adam 1997); these conditions are generally an anathema to all bee species. Pollen is carried in corbiculae located on the hind legs, though scattered pollen grains are normally lodged elsewhere on the body (Williams and Walker 2003). *Tetragonula carbonaria* arranges its brood cells in combs, though brood cell construction and positioning (e.g. as clusters) can vary in other *Tetragonula* species. *Tetragonula* hives can usually be found in tree cavities, including in decaying fallen tree trunks, often close to or on the ground. Such populations are vulnerable to fire. *Tetragonula* and its close generic relatives are restricted to tropical America, and from India and southern Asia to Australia; this disjunct distribution is possibly due to the reduction of a wider range owing to climatic deterioration during the Tertiary (Michener 2007).

Of the genera previously placed in the Anthophoridae (Cardale 1993), *Xylocopa* (Xylocopinae – Xylocopini) contains medium-sized to very large and conspicuous bees, and occurs world-wide in tropical and subtropical, and occasionally temperate, regions. Two subgenera, *Lestis* and *Koptortosoma*, occur in Australia (Leys 2000). *Lestis* is endemic to the Australo-Papuan region but *Koptortosoma* is more widely distributed from southern Africa to the Middle East, and India to Australia. The ranges of four species, *Xylocopa* (*Lestis*) *aeratus*, *X.* (*L.*) *bombylans*, *X.* (*Koptortosoma*) *aruana* (Plate 287) and *X.* (*K.*) *lieftincki*, span the Gondwana rainforests region (Cardale 1993; Leys 2000; Steen and Schwarz 2000). *Xylocopa aeratus* and *X. bombylans* were known from Victoria and South Australia, but there have been no records of *X. bombylans* from Victoria or South Australia since the 19th century, and *X. aeratus* has not been recorded from either state since 1938. Both species are thought to be extinct in those states (Leys 2000). *Xylocopa aeratus* forages on a wide range of flowering trees, shrubs and vines, but foraging is more commonly restricted to flowers (e.g. *Diospyros australis*, *Hibbertia scandens*) at or near the forest floor. The tribe Allodapini (Xylocopinae) includes the geographically widespread genus *Braunsapis*, and two endemic genera *Exoneura* and *Exoneurella* (T. Houston 2018). All are small in size. Individuals of *Exoneura*, though long-tongued, are restricted as efficient pollinators by their size to readily accessible flowers with little or no depth effect in the corolla, or in the tubular flowers into which they can burrow to access resources. Brood cells may be positioned in stem hollows, including in exotic and infesting *Lantana camara*. *Exoneura lawsoni* has been recorded from flowering *Alphitonia excelsa* in littoral rainforest, and *Exoneura hamulata* from *Rubus rosifolius* in wet sclerophyll forest. Additional species recorded from subtropical eastern Australia include *Exoneura albolineata*, *E. baculifera*, *E. diversipes*, *E. obscuripes* and *E. robusta* (Cardale 1993). Members of the *Exoneura* subgenus *Inquilina* (*E. excavata*) are parasites on other *Exoneura* (Michener 2007). Chenoweth and Schwarz (2011) suggest that the Australian exoneurine allopodine bee fauna reflects a trans-oceanic dispersal event from Africa to Australia (most likely via Antarctica) with an accelerated diversification of temperate-adapted lineages during the Late Miocene.

Strikingly blue-black patterned species of *Thyreus*, of the tribe Melectini, are cleptoparasites of *Amegilla* (Apidae), but have also been recorded parasitising *Megachile* (Megachilidae) in lowland mesic forests (G. Williams unpubl. data). Female Melectini break into the closed brood cells of their host, oviposit on the cell cap or wall, and reclose the cell using earth moistened with a secretion (Michener 2007). Adult *Thyreus lugubris* (Plate 284) and *T. nitidulus* are occasionally observed collecting nectar from canopy flowers (Williams and Adam 1997).

The medium-sized members of the genus *Amegilla* (Apinae – Anthophorini) are distinctive blue-banded, or brown and densely setaceous, bees (e.g. *Amegilla aeruginosa*, *A. adelaidae*, *A. asserta*, *A. distincta*, *A. pulchra*, *A bombiformis*) (Plate 282) commonly found foraging on the sunny margins or large sunlit canopy gaps of rainforest, wet sclerophyll forest, and adjacent cultivated gardens. However, they are infrequently observed at the canopy or subcanopy strata (rarely on *Rhodomyrtus psidioides*, *Tristaniopsis laurina* – Myrtaceae [Williams and Adam 1997]); rather, they are specialist visitors at lower levels to shrubby plants such as *Alpinia* (Zingiberaceae). They either alight directly on flowers to feed, or hover with their long tongues (glossa, and associated structures) extended so as to feed on floral nectar. Individuals exhibit trap-lining behaviour, repeatedly and systematically visiting plants within flowering populations. Though showing constancy in foraging behaviour, *Amegilla* is highly polylectic and forages over a broad range of plants (including exotic species), particularly if there is depth effect in the corolla. *Amegilla* nests are often found in large aggregations that are reused over several generations. Activity in south-east Queensland and New South Wales peaks in October and May, but adults are relatively short lived (*c.* 6 weeks) (Leijs *et al.* 2017).

Colletidae These are short-tongued bees divided into five subfamilies (Michener 2007), of which the Euryglossinae, Colletinae and Hylaeinae occur in Australia (e.g. Batley 2016; Batley and Brandley 2014; Batley and Houston 2012; Cardale 1993; Cockerell 1921; Dollin *et al.* 2000; Exley 1968a, 1968b, 1968c, 1972, 1975, 1976, 1983; Hacker 1921; T. Houston 1975, 1981, 2018; Maynard 1992, 2013; Rayment 1935). Colletids are often pollen thieves, harvesting pollen for their own nutritional needs but without any pollination benefit, or only a small one, for the plant (see discussion in Williams and Adam 2010, 2013). They visit numerous flowering plants in subtropical rainforest and open forest communities, and forage in all forest strata (Williams 1998; Williams and Adam 1994, 1997, 2010; Williams and Walker 2003), but show a level of temporal foraging constancy to individual plants when in flower. Flower types visited range from those with little depth effect (most subtropical rainforest canopy trees), tubular flowers (as in *Diospyros australis*, *Pittosporum undulatum*), and specialised flowers requiring sonication, or buzz pollination (Buchmann 1983), to release pollen (i.e. *Senna acclinis* – Caesalpinioideae); buzz pollination behaviour has been recorded for several Hylaeinae (e.g. *Amphylaeus nubilosellus* [Plate 288], *Hylaeus turgicollaris* [Williams 1998]), but it is probably a behaviour more widespread among Hylaeinae and Colletinae.

The Euryglossinae are endemic (T. Houston 2018), and include minute species as small as ~2 mm in length (Exley 1968c). The subfamily is especially diverse on flowering *Syzygium floribundum* in lowland riparian rainforest stands, the smaller species appearing as blurred images when in flight (Williams 1995). The endemic genus *Heterohesma* is known by only two species: *H. clypeata* described from Jamberoo (in New South Wales) and *H. weiri* described from New England National Park (Cardale 1993; Exley 1983), though each is distributed more widely. Additional genera include *Brachyhesma*, *Euryglossa*, *Euryglossella*, *Euryglossina*, *Hyphesma* and *Pachyprosopis* (Exley 1968a, 1968b, 1968c, 1972, 1975, 1976; Williams 2002). *Euryglossa terminata* is one of the largest regional euryglossines, and is recorded in lowland rainforest foraging on *Tristaniopsis laurina* (Myrtaceae) (Williams 1995). Additional *Euryglossa* species have been recorded from mass-flowering *Syzygium floribundum*, *Tristaniopsis laurina* (Myrtaceae) and *Euroschinus falcatus* (Anacardiaceae), as well as *Trochocarpa laurina* (Ericaceae/Epacridaceae), a large shrub or small tree often found growing on sunny margins in lowland wet sclerophyll forest.

Colletinae are cosmopolitan in distribution but only the pan-austral tribe Paracollini occurs here (Cardale 1993). The most diverse genus is *Leioproctus* (Batley and Brandley 2014; Maynard 1992, 2013), a huge group which also occurs in New Guinea, New Caledonia, New Zealand and temperate South America (Cardale 1993; Michener 2007). The subgenus *Cladocerapis* is an endemic taxon with a Bassian distribution (Maynard 1992). Maynard (1992) lists two species, *Leioproctus* (*Cladocerapis*) *incanescens* and *L.* (*C.*) *speculiferus*, from the Gondwana rainforests. However, the genus as a whole is widespread in subtropical rainforest and adjacent sclerophyll forest, particularly at low and intermediate elevations; though a number have very limited known distributions (e.g. *Leioproctus recusus*, *L. melanoproctus*) (Cardale 1993; Cockerell 1921). Species encountered in lowland rainforest include *Leioproctus alleynae*, *L. amabilis*, *L. carinatus*, *L. cristatus* and *L. irroratus.* Most of these have been recorded from relatively few flowering trees (e.g. *Leioproctus alleynae* on *Scolopia braunii*, *L. amabilis* on *Syzygium floribundum*) (Williams and Adam 1997). Additional genera in the region include *Paracolletes* (*Paracolletes brevicornis*, *P. leptospermi*), *Callomelitta* (*Calomelitta littleri*, *C. fulvicornis*) and *Trichocolletes* (*Trichocolletes burnsi*, *T. hackeri*) (Batley and Houston

2012; Cardale 1993; Hacker 1921). All are endemic (Michener 2007). *Trichocolletes hackeri* has been collected from the Bunya Mountains, Mt Tamborine, Toonumbar State Forest and the Lansdowne Escarpment (on flowering *Syzygium smithii*) (Batley and Houston 2012). *Trichocolletes burnsi* is known from the Blue Mountains, west of Sydney, but is distributed as far as Victoria (Batley and Houston 2012). In general appearance *Paracolletes* resemble *Apis mellifera*, the genus occurring in forests from intermediate to high elevations (e.g. Barrington Tops). In subtropical rainforest *Callomelitta fulvicornis* has been recorded at the flowers of *Abrophyllum ornans* (Escalloniaceae/Rousseaceae), a large understorey shrub or small tree, but individuals will also forage at flowering *Leptospermum* and *Baeckea* (Myrtaceae) growing in adjacent eucalypt forest.

Hylaeinae are also cosmopolitan in distribution but only in Australia do they comprise a major and diverse faunal element (Cardale 1993; T. Houston 1975). Six Australian genera (out of a total of nine) are endemic; 14 of the 16 subgenera of the diverse genus *Hylaeus* are endemic (T. Houston 1975). Of the *Hylaeus* subgenera recorded from Australia, *Euprosopoides* apparently occurs also in the Caroline Islands, *Hylaeorhiza* occurs in New Guinea, and *Prosopisteron* is recorded from New Guinea, New Zealand and the Chatham and Tuamotu Islands (T. Houston 1981). Species are diverse in subtropical rainforest, the fauna comprising *Amphylaeus* (e.g. *Amphylaeus nubilosellus*, *A. obscuriceps*), *Hemirhiza* (*Hemirhiza melliceps*), *Heterapoides* (*Heterapoides digitata*, *H. nigriconcava*), *Hyleoides* (*Hyleoides bivulnerata*, *H. concinna*), *Meroglossa* (*Meroglossa itamuca*, *M. sulcifrons*) and the diverse *Hylaeus* (*Hylaeus amiculinus*, *H. jacksonae*, *H. ofarrelli*) (Cardale 1993; T. Houston 1975, 1981; Williams 2002). All except *Hylaeus* are endemic, and in colour most individuals are black with distinctive yellow markings. Individual species are generally foragers on a wide range of flowering trees and shrubs (Williams 1995, 1998; Williams and Adam 1997, 2010) and also herbs (Williams and Walker 2003).

Halictidae The short-tongued Halictidae are among the most widespread families of bees. In Australia, their greatest diversity occurs in the southern two-thirds of the continent (Walker 1995a). Two of the four subfamilies, Halictinae (*Homalictus*, *Lasioglossum*) and Nomiinae (*Nomia*), occur in Australia (Michener 2007). Halictinae have a cosmopolitan distribution, and the Nomiinae are widely distributed but are absent from South America and Tasmania (Cardale 1993).

The Halictinae are a very large group whose constituents exhibit behaviour ranging from solitary lifestyles to communal (e.g. of single generation related sisters) and fully social structures with queens and castes. In an Australian *Lasioglossum* – subgenus *Chilalictus* species, communal colonies were made up of unrelated sisters (Forbes *et al.* 2002). Halictine nests consist of burrows made in the soil, and short lateral cell chambers may lead off the main burrow (Michener 2007). Pollen is usually carried in specialised scopae on the ventral surface of the abdomen, though pollen readily adheres to other parts of the body (e.g. see Williams and Walker 2003).

Lasioglossum is diverse and occurs on all continents, except Antarctica (Walker 1995a). Several *Lasioglossum* subgenera occurring in Australia are endemic (e.g. *Australictus*, *Callalictus*). The subgenus *Austrevylaeus* is confined to Australia and New Zealand, *Chilalictus* is mainly restricted to Australia except for a single species recorded from New Caledonia, and the subgenus *Parasphecodes* extends to New Guinea (Walker 1995a). All five subgenera occur in the Gondwana rainforests region (K. Walker pers. comm.). *Lasioglossum* is diverse in the rainforests and wet sclerophyll forests of south-eastern Queensland and northern New South Wales (e.g. *Lasioglossum imitans*, *L. demicapillum*, *L. tamburinei*) (Cardale 1993; Walker 1995a, 1995b). Williams (2002) lists 42 species from elevated sites in the Gondwana Rainforests World Heritage Area and adjacent regions (e.g. Bunya Mountains, Lamington, New England, and Barrington Tops national parks); however, the genus is also common in lowland forests (Williams 1995, 1998; Williams and Adam 1997, 2010; Williams and Walker 2003). Adults forage at all levels of the forest, but can be particularly diverse at plants flowering at or near ground level. Williams and Walker (2003) recorded 10 species (in the subgenera *Austrevylaeus*, *Chilalictus* and *Parasphecodes*) visiting the flowering herb *Pollia crispata* (Commelinaceae) in the understorey of regenerating subtropical rainforest (Lansdowne Escarpment), and Williams (1998) recorded three species (*Lasioglossum turgicollaris*, *L. polygoni*, *Lasioglossum* sp.) visiting the understorey, buzz pollinated, threatened shrub *Senna acclinis* (Fabaceae – Caesalpinioideae) in littoral rainforest (Hallidays Point). These carried large *Senna* pollen loads, and *Lasioglossum turgicollaris* and *L. polygoni* both removed pollen by vibrating anthers with their flight muscles. Species such as *Lasioglossum convexum*, *L. atronitens* and *L. callomelittinum* are visitors to flowering herbs (e.g.

Dianella caerulea) and vines (e.g. *Geitonoplesium cymosum*, *Clematis glycinoides*) growing on the margins of wet sclerophyll forest.

Bees in the genus *Homalictus* occur in South-east Asia, Sri Lanka and eastwards to Samoa and the Mariana Islands, but their centre of diversity is in Australia (Walker 1986, 1997). *Homalictus* is almost ubiquitous in subtropical rainforest, especially in rainforests at intermediate and lower elevations. They are generally small in size, the abdomen of most species conspicuously metallic blue or green in colour. Adult *Homalictus*, like those of *Lasioglossum*, forage for nectar and pollen from a wide range of plants with readily accessible floral structures, in all forest strata, but can be conspicous on mass-flowering canopy (e.g. *Alphitonia excelsa*, *Ackama paniculosa*, *Syzygium smithii*) and subcanopy (*Acradenia euodiiformis*, *Alectryon coriaceus*) trees (Williams and Adam 1997). *Homalictus flindersi*, and particularly *H. megastigmus*, are commonly encountered in rainforest and wet sclerophyll forest, but other species include *H. punctatus*, *H. brisbanensis*, *H. niveifrons* and *H. sphecodoides*.

Nomiinae include solitary and communal nesting species. Nests are made in the ground with burrows leading to cell clusters. *Nomia* includes several medium-sized and distinctively red-orange banded bees (*Nomia aurantifer*, *N. swainsoniae*). The genus is distributed in the Palaearctic, Afrotropical and Australian regions (Cardale 1993). In lowland rainforest *N. aurantifer* (Plate 286) has been recorded from herbaceous *Pollia crispata* (Williams and Walker 2003) and the woody vine *Austrosteenisia blackii* (Williams and Adam 2010).

Megachilidae These are long-tongued bees that are commonly referred to as leaf-cutter and resin bees. Species range in size from 5 to 39 mm with three tribes (Anthidiini, Lithurgini, Megachilini) occurring in Australia (Cardale 1993; Gonzalez *et al.* 2013; T. Houston 2018; King 1994). In eastern Australia Lithurgini are represented by *Lithurge* and *Austrothurgus*. However, the tribe does not occur further south than far north-eastern New South Wales (T. Houston 2018). Lithurgini are known to nest in rotten wood (T. Houston 1971). *Megachile* (e.g. *Megachile mackayensis*, *M. lucidiventris*, *M. mundifica*, *M. deanii*, *M. maculariformis*, *M. pictiventris*, *M. punctata*) dominates the Megachilini; the genus *Chalicodoma* is now subsumed by *Megachile*. Pollen is carried in conspicuous ventral scopae (e.g. *Megachile pictiventris*), but individual grains commonly lodge elsewhere on the body and legs. Adult Megachilini nest in existing burrows, or holes and crevices in wood. *Megachile* uses pieces of leaves, resin and mud, and/or masticated leaves, to construct brood nests (King 1984; Cardale 1993; Rayment 1935). They collect pollen from a range of flowering canopy trees (e.g. *Arytera divaricata*, *Diospyros australis*, *Syzygium floribundum*, *S. smithii*, *Tristaniopsis laurina*) (Williams 1995; Williams and Adam 1997), but individual species exhibit marked seasonal differences in foraging patterns and food host choices, and may often be absent one year on plant species they foraged heavily from in a preceding year (Williams 1995). In general, megachilids in lowland subtropical rainforests appear to forage from far fewer tree species than Colletidae, Halictidae and Apidae (Williams 1995; Williams and Adam 1997), and are usually absent from flowering herbs and low shrubs.

In lowland subtropical rainforest the resin-collecting *Megachile mystaceana* is a host for cleptoparasitic *Coelioxys* and *Thyreus* bees, and wasps of the genus *Leucospis* (Leucospidae) (G. Williams pers. observ.) (Plates 268, 284, 285). *Coelioxys* is the usual recorded cleptoparasite of Megachilidae. It is closely related to *Megachile* (Michener 2007; Rocha-Filho 2016), is widely distributed in the Northern and Southern Hemispheres, but is poorly represented in Australia (e.g. *Coelioxys albolineata*, *C. reginae*). Females have a tapering abdominal, metasomal apex, which is used to insert the eggs into the brood cell of the host bee.

Apocrita (Aculeata and Parasitica) (wasps)

The name 'wasp' is applied collectively to a heterogeneous group of families (of varied ecological specialisation and evolutionary derivation) that encompass the Apocrita (Aculeata plus Parasitica), but exclude the Apoidea – Apiformes (true bees) and the Formicidae (ants) (Goulet and Huber 1993). Most Parasitica are predominantly small species that reside and function rather inconspicuously in rainforest, with only the larger-sized members of the Aulacidae, Gasteruptiidae, Braconidae and Ichneumonidae more readily displaying their presence. The aculeate wasps (i.e. Crabronidae, Pompilidae, Scoliidae, Sphecidae, Thynnidae, Vespidae), however, are generally much larger in size, the most notable exception being many of the superfamily Chrysidoidea (e.g. Bethylidae, Chrysididae, Dryinidae). Large black-and-orange-banded *Heterodontonyx* (Pompilidae) and black-coloured *Scolia* (Scoliidae) actively patrol forest margins, and various thynnids (e.g. *Lophocheilus*, *Lestriocothynus*) speed close to the forest floor in search of prey or mates, their flight often so rapid that they appear only as blurred images, occasionally settling to rest on leaves, low down, in sunlit openings below the rainforest canopy. But unlike scoliids and thynnids, *Heterodontonyx* and other pompilids will actively seek for prey by searching in leaf litter on the forest floor.

Wasp mouthparts are adapted to capture prey, and for excision from the pupal cell or host. In 'higher' aculeate groups (e.g. Vespoidea), thought to have originated ~150 mya (Grimaldi and Engel 2005), there are tube-like modifications to the mouthparts (particularly prolongation and development of the glossae, paraglossae and prementum) that allow nectar to be imbibed (Faegri and van der Pijl 1979; Naumann 1991b; Williams and Adam 1995, 2010). Aculeate wasps search for insect and spider prey to sustain the development of their young, but to sustain their own daily energy needs the adults of many species feed on floral nectar (Williams and Adam 1995). Examples include *Pseudotriellus zonata*, *Scolia verticalis* (Scoliidae), *Calopompilus defensor*, *Heterodontonyx bicolor*, *Phanagenia fasciata*, *Ferreola handschinii* (Pompilidae); species of *Acanthostethus*, *Bembix*, *Cerceris*, *Ectemnius*, *Pison*, *Williamsita* (Crabronidae); *Sphex ephippium*, *S. fumipennis* (Sphecidae), *Anthobosca australasiae*, *Diamma bicolor* and numerous Thynninae (Thynnidae); and *Abispa splendida*, *Bidentodynerus bicolor* and *Deuterodiscoelius ephippium* (Vespidae) (Williams and Adam 1995). As with native bees, aculeate wasps show no obligate plant association, but may forage in large numbers at particular species, or flowering individuals, during one season; they may be absent in the following season. Large scoliids (*Radumeris*, *Pseudotrielis*, *Scolia*) and sphecids (*Sphex*) frequently exhibit trap-lining behaviour, repeatedly flying between individual plants in flowering populations and alighting intermittently to feed. Although their movements are partly the seeking of prey, their behaviour also involves feeding on nectar. In lowland subtropical rainforest, flower-visiting activity by aculeate wasps peaks from November to February; this activity tracks peak phenological episodes in mass-flowering trees and shrubs (G. Williams 1995; Williams and Adam 1995). At littoral rainforest remnants the aculeate fauna, and especially that of Thynnidae, can be abundant and diverse (Williams and Adam 1995). However, large aculeate wasps tend to avoid plants in which florets individually offer little nectar reward, and flowers that are shaded, have depth effect in the corolla, or do not provide floral structures strong enough to support wasps while feeding.

Agaonidae Minute agaonid wasps are known from Miocene amber (Grimaldi and Engel 2005), and extant species are generally confined to warmer zones. Agaonids are members of the superfamily Chalcidoidea, a group that has undergone a massive level of speciation, second only to that of the Ichneumonoidea, such that there are now ~20 000 described chalcidoid species, and a vast number of undescribed ones. Chalcidoidea exhibit a greater host range than any other insect parasitoid group; their hosts include insects, mites, spiders, pseudoscorpions and nematodes (Grimaldi and Engel 2005). Many are specialist parasites of particular arthropods, Aphelinidae being associated with eggs and the nymphal stages of Hemiptera – Sternorrhyncha (Polaszek 1991). Others have a broader host range, for example Mymaridae which parasitise the eggs of Hemiptera – Auchenorrhyncha, but also Coleoptera, Psocoptera, Diptera and Orthoptera (Lin *et al.* 2007). A number (e.g. Ormyridae, Torymidae, Pteromalidae) include gall-forming or gall-associated species.

Agaonid wasps, however, differ from most other Chalcidoidea in that they are secondarily phytophagous, with those of the subfamily Agaoninae ('true' fig wasps) being obligate pollinators of figs (*Ficus* – Moraceae) (Bouček 1988; Darwell *et al.* 2014; Elias *et al.* 2018; Williams and Adam 1994, 2010). Fig wasps and figs have apparently coevolved, owing to agaonines being the only pollinators of the fig flowers (these being located internally within the fig syconium). In most cases there is a 1:1 mutualistic relationship between fig wasps and *Ficus* species (Lopez-Vaamonde *et al.* 2002). Volatile chemicals emitted by the figs attract the wasps. The female wasp enters the syconium through the minute apical opening, the ostiole, and inserts its ovipositor into specially adapted 'gall flowers', within which the deposited eggs hatch and develop (Bouček 1988; Goulet and Huber 1993; Willmer 2011). But while some wasps are active pollinators, others are passive. Active pollinators usual coxal corbiculae to move pollen to and from their mesosternal pollen pockets, while passive pollinators lack this behaviour (Lopez-Vaamonde *et al.* 2002). The complexity of the fig wasp–fig mutualism and the complexity of the pollination process is discussed in Willmer (2011).

Agaonid genera recorded from the Gondwana rainforests include *Meselatus* (which is known from Australia and Lord Howe Island), *Watshamiella* (from Africa, southern Asia and Australia) and *Pleistodontes* (from Australia, Lord Howe Island, New Guinea and Solomons)

(Bouček 1988). Known fig host–fig wasp records include *Ficus macrophylla* (*Pleistodontes froggatti*), *F. obliqua* (*Pleistodontes greenwoodi*, *P. xanthocephalus*), *F. rubiginosa* (*Pleistodontes imperialis*) and *F. watkinsiana* (*Pleistodontes nigriventris*) (Lopez-Vaamonde *et al.* 2002).

Aulacidae The parasitic and ancient Aulacidae possess relatively few extant species but formed a major part of the Hymenoptera during the Mesozoic era (Goulet and Huber 1993). The family is grouped with Evaniidae and Gasteruptiidae in the superfamily Evanioidea, being so similar in general appearance to the latter that they have sometimes been placed in the same family (Townes 1949, 1950).

The Aulacidae are widely known from subtropical rainforests and wet sclerophyll forests. Species of *Aulacus* (Plate 267) and *Pristaulacus* can be found commonly on felled eucalypt timber following logging operations, and on the trunks of fallen rainforest trees after storm damage or the collapse of senescent individuals; adults often prefer to frequent the shaded underside of felled trunks. There they oviposit on wood-boring beetle larvae, such as those of Buprestidae and Cerambycidae. Adults tend to emerge before the emergence of surviving host species. One species of *Aulacus* from the Lansdowne Escarpment appears to mimic (in colour) the bulldog ant *Myrmecia nigrocincta* (see Plates 261, 267).

Braconidae The Braconidae (Parasitica) are known from the Lower Cretaceous, and are the second largest family of Hymenoptera, containing at least 40 000 species (Goulet and Huber 1993). Braconids are small to medium-sized wasps. They comprise endoparasitic species that paralyse their prey and then lay eggs inside their host (e.g. larvae of wood-boring Coleoptera), and ectoparasitoids that lay their eggs on or alongside the host; the larvae commencefeeding on the host immediately after hatching (Quicke 2015). However, adult braconids are occasionally encountered on the inflorescences of mass-flowering rainforest trees (Williams 1995).

Cardiochilinae are represented only by *Cardichiles*; however, Australasia and the Neotropics possess a large number of new species which are important for the reinterpretation of generic limits within the subfamily (Dangerfield and Austin 1995). The subfamily Braconinae includes the genus *Chaoilta*, which ranges from India to Australia, and on some Oceanic islands (Quicke 1991). Braconine host range is broad, and includes the larvae of Cerambycidae and Chrysomelidae (Coleoptera), Cossidae (Lepidoptera) and Symphyta. *Iphiaulax* has been collected from the flowers of *Alphitonia excelsa* (Rhamnaceae) and has emerged from dying branches of *Cupaniopsis anacardioides* (Sapindaceae) in littoral rainforest (Williams 1995; G. Williams unpubl. data). The Microgastrinae (e.g. *Microgaster*, *Diolcogaster*) are endoparasitic on the larvae of Lepidoptera. The subfamily has a world-wide distribution and is the largest braconid subfamily (Goulet and Huber 1993; Saeed *et al.* 1999). The distribution of *Microgaster* is centred upon the Holarctic region and is absent or rare in other regions (Austin and Dangerfield 1992). Only three species are recorded from Australasia: one species (*Microgaster nixoni*) is known only from Tooloom Scrub in northern New South Wales and Mt Field National Park, Tasmania. *Diolcogaster* has an almost world-wide occurrence, including Tasmania, New Guinea, New Zealand and New Caledonia (Saeed *et al.* 1999), suggesting radiation predating the separation and breakup of Gondwana. Although there is a rich *Diolcogaster* fauna recorded from the Gondwana rainforests region, all species are found more widely within Australia; and two species have interesting extra-Australian distributions (*Diolcogaster sons* from New Caledonia and Sulawesi; *D. perniciosus* from New Zealand) (Saeed *et al.* 1999). Agathidiinae, Cheloninae and Rogadinae are parasitoids of Lepidoptera (Kittel and Austin 2014; Naumann 1991b; Quicke *et al.* 2014). *Agathis* (Agathidiinae) has been collected from the flowers of *Cissus sterculifolia* (Vitaceae) in littoral rainforest (G. Williams unpubl. data). The Opiinae and Alysiinae parasitise various Diptera, including Tephritidae, Agromyzidae and Platypezidae (Carmichael *et al.* 2005; Naumann 1991b; Wharton 2002). Euphorinae are parasites of Orthoptera, Coleoptera and Lepidoptera (Naumann 1991b), and include the genera *Paramannokeraia* (e.g. *Paramannokeraia gibsoni*, Mt Keira) and *Mannokeraia* (e.g. *Mannokeraia albipalpis*, Mt Glorious, Main Range, Cunninghams Gap) (van Achterberg *et al.* 2017).

Chalcididae Chalcidid wasps are mainly parasites of lepidopteran pupae and dipteran larvae (Bouček 1988; Goulet and Huber 1993), and are known from Tertiary amber. Chalcididae range in size up to 16 mm though most are minute, and are primarily solitary parasites of the larvae and pupae of Lepidoptera, but hosts also include Diptera, Neuroptera, Hymenoptera, Coleoptera and Stresiptera (Naumann 1991b). Chalcidinae include *Brachymeria corneillei*, *B. lugubris*, *B. marmonti* and *B. sidnica*. This is a world-wide genus, also occurring in New Caledonia, with a host range that includes Braconidae, Ichneumonidae, Tachinidae and numerous Lepidoptera.

Chrysididae Chrysididae are commonly metallic green-blue-coloured aculeate wasps that are cleptoparasites in the nests of vespid wasps and bees, and also parasitise the larvae of Symphyta (Kimsey and Bohart 1991; Naumann 1991b). The family is known from Cretaceous and Tertiary amber (Grimaldi and Engel 2005). Lobosscelidiinae are not metallic in colour and apparently parasitise the eggs of Phasmatodea (Goulet and Huber 1993). Species of brightly coloured *Stilbum* and *Praestochrysis* (Chrysidinae) have been collected on the flowers of *Alphitonia excelsa* (Williams 1995). *Praestochrysis* may be a parasite of *Megachile mystaceana* (G. Williams pers. observ.).

Crabronidae The Crabronidae (here including Pemphredoninae, Larrinae, Nyssoninae) are a polyphyletic group of predatory wasps, comprising more than 9000 species. The family now includes most of the species previously placed in the Sphecidae (see Walton 1985a). Definitive crabronids are known from the Mid Cretaceous (Antropov 2000), and it is from within Crabronidae that bees arose (Debevec *et al.* 2012; Melo 1999; see also Sann *et al.* 2018). Extant species range in size from minute *Spilomena* (barely 2 mm) to large *Bembix* (>2 cm) and *Exeirus* (3 cm). Adults nest in existing cavities (e.g. *Spilomena*) or in twigs (Goulet and Huber 1993), though some fossorial forms (e.g. *Bembix*, *Tachytes*, *Tachysphex*) excavate nests in the ground. Species of the cosmopolitan genus *Pison* nest in pre-existing cavities, construct mud nests, or are fossorial. Cooperative and communal nesting behaviour is recorded for some taxa (e.g. *Cerceris* – Philanthinae [Alcock 1980]). Benefits in using pre-existing cavities and burrows include time and energy savings, and where nesting is communal some members can devote time to guarding the nest.

Crabronid prey include numerous arthropods (Evans and Matthews 1973; Naumann 1991b; Walton 1985a). The subfamily Pemphredoninae (sometimes considered a separate family) includes small species (e.g. *Spilomena*, *Paracrabro*) that prey on Hemiptera and Thysanoptera. Larrinae (*Larra*, *Pison*, *Sericophorus*, *Sphodrotes*, *Tachytes*, *Tachysphex*) are predators of various insects (Orthoptera, Diptera, Mantodea, Blattodea) and spiders. Crabroninae (*Ectemnius*, *Notocrabro*, *Williamsita*) are mostly predators of Diptera. Nyssoninae (*Acanthostethus*, *Bembicinus*, *Bembix*, *Clitemnestra*, *Exeirus*) include predators of Hemiptera, Neuroptera, Odonata, but also bees and other wasps. *Acanthostethus* is cleptoparasitic on *Sericophorus*. Philanthinae (*Cerceris*) prey on small beetles (Naumann 1991b). Although predatory, Crabronidae commonly feed on nectar. Species of *Acanthostehus*, *Bembecinus*, *Bembix*, *Cerceris*, *Ectemnius*, *Larra*, *Pison*, *Sericophorus*, *Sphodrotes*, *Tachysphex* and *Williamsita* are recorded visiting mass-flowering trees and shrubs in lowland subtropical rainforests (Williams 1995; Williams and Adam 1995, 2010). Many of these genera have a dense cover of setae on their bodies to which pollen grains are readily captured, or pollen grains aggregate in discrete parts of the body (Williams and Adam 1998), and so individuals can function as pollinators.

The crabronid fauna recorded from subtropical eastern Australia is a rich one. Though not typical of the rainforest interior, species of the bee-like genus *Bembix* may aggregate commonly where broad sunlit trails dissect littoral rainforest growing behind coastal dunes. There they excavate nests, or participate in fierce territorial disputes. Soils with loose and fine sediments are preferred sites for burrow construction. *Bembix* has an almost world-wide distribution, and is best developed in the Southern Hemisphere (Evans and Matthews 1973). However, *Bembix* is absent from South America, where it is largely replaced by other genera. The widely distributed genus *Cerceris* (Evans and Hook 1986) includes several, typically yellow- or orange-banded, species (*Cerceris minusculus*, *C. australis*, *C. froggati*) that inconspicuously patrol the ground below canopy openings in search of prey, often venturing in almost hover-like flight among low shrouding shrubs. *Williamsita* and *Ectemnius* include two large flower-visiting species (*Williamsita smithiensis*, *E. reginellus*) distinctive with their red-banded abdomens, broad head and large eyes. The largest crabronid is *Exeirus lateritus*, or cicada killer, in appearance superficially resembling *Heterodontonyx* (Pompilidae), but distinguished in the field by its less frenetic flight. However, it is a species encountered in open forest, and rarely even in wet sclerophyll forest. The largest crabronid most commonly encountered in lowland rainforest is *Bembix promontorii*, adults often patrolling flowering trees or burrowing in sandy soils. However, this is a habitat generalist, also to be found foraging in open forest, woodland and estuarine mangrove communites.

Diapriidae These are tiny wasps (1–8 mm) that typically parasitise Diptera (Goulet and Huber 1993) and are well represented in subtropical rainforest (see Williams 2002). The subfamily Ambrositrinae has a disjunct southern (Gondwanan) distribution and occurs primarily in Australia, New Zealand and South America (Naumann

1982). Their hosts are mainly Diptera larvae and pupae (e.g. of Nematocera). More than half of the described species exhibit some degree of wing reduction. The genus *Perissodryas* (e.g. *Perissodryas daedalma*) is often associated with *Nothofagus* forest, and also occurs in New Caledonia and New Guinea (Naumann 1982). Additional *Nothofagus*-associated ambrostrines include *Acanthobetyla drymoides*, *A. tooloomensis* and *Diphoropria pecki*. In the subfamily Diapriinae the flightless genera *Austropria* from Australia and *Coecopria* from Brazil appear more closely related to each other than to other Diapriinae (Naumann 1982). The endemic Diapriinae genus *Rostropria* is restricted to eastern Australia, and includes two species known only from south-east Queensland (*Rostropria casta*, *R. garbo*) (Early and Naumann 1989). Most Diapriinae are endoparasites of Diptera (e.g. Tabanidae, Stratiomyidae, Chloropidae, Tephritidae), with host divergence by some species to Coleoptera (Staphylinidae, Psephenidae) and larvae of Formicidae (Goulet and Huber 1993).

Eucharitidae Eucharitidae are widely distributed chalcidoids, but most species inhabit tropical and subtropical regions. Adults are often coloured metallic green or blue and in size range from <2 to 11 mm. They are known parasites of ants. Female wasps lay their eggs into plant tissue, and the mobile larvae, on hatching and if fortunate, attempt to attach themselves to passing worker ants, which then carry them into the nest. Therein, eucharitid larvae feed on ant larvae and pupae (Goulet and Huber 1993; Naumann 1991b). Known rainforest ant hosts include *Camponotus* and *Myrmecia*.

Of the two subfamilies (Oraseminae and Eucharitinae) recognised by Goulet and Huber (1993), the Oraseminae are recorded from Africa, southern Eurasia, South America and Australia. The orasemine fauna of Gondwana rainforests region includes the endemic genus *Orasemorpha*, and also *Orasema* which occurs more widely (the Americas, Africa, Madagascar, southern Asia, New Guinea in addition to Australia). The Eucaritinae include the endemic genus *Substilbula*, and genera with Australasian (*Austeucharis*) or Austro-Asian (*Losbanus*, *Parapsilogastrus*, *Chalcura*) distributions (Bouček 1988).

Eulophidae These are minute to small, frequently metallic-hued, wasps (0.5–6 mm). Eulophidae, including species previously placed in Elasmidae, are one of the largest chalcidoid families, and are particularly diverse in Australia, with many species recorded from southern Queensland and north-eastern New South Wales (Williams 2002). Numerous species have been collected from Tooloom Scrub, Mt Tamborine and Mt Glorious (Bouček 1988). Most genera are widely distributed; however, *Renaniana* (Eulophinae) is endemic and is recorded from Queensland and New South Wales. *Pediocharis*, *Obesulus* and *Derostenoides* (Entedoninae) are endemic to Queensland and New South Wales. The Entedoninae genera *Parahorismenus*, *Zaommomyiella* and *Parzaommomyia* have restricted Indo-Australasian distributions (Bouček 1988).

Most Eulophidae are parasitoids of leaf-mining Lepidoptera, Diptera, Hymenoptera and Coleoptera, though some are phytophagous (Goulet and Huber 1993; Naumann 1991b). Entedoninae (e.g. *Chrysonotomyia pulchrella*, *Closterocerus westwoodi*, *Obesulus ater*, *Pediobius elasmi*, *Zaommomyiella sol*) are mainly endoparasites of Lepidoptera and Coleoptera larvae. *Pediobius* has also been reared from the pupae of parasitic Hymenoptera and Diptera, spider egg sacs and the oothecae of praying mantids (Naumann 1991b). Euderinae (*Allocerastichus versicolor*, *Astichus mirissimus*, *Euderus nestor*, *Parasecodella dickensi*) are endoparasites of the eggs or larvae of Lepidoptera, Coleoptera and Diptera. *Astichus* species have been reared from fungi where they parasitise inhabiting Ciidae beetles (Naumann 1991b). Eulophinae (*Ascotolinx reticoxa*, *Cirrospilus tau*, *Entedomorpha albiclava*, *Euplectromorpha atriflagellum*) include the most primitive genera and are predominantly parasitoids of Diptera, Lepidoptera, Coleoptera and Hymenoptera larvae, and also Hemiptera. Most Tetrastichinae (*Melittobia australica*, *Neotrichoporoides aeneus*, *Sigmophora io*) are endoparasites of the eggs, larvae and pupae of Lepidoptera, Coleoptera and Diptera (Goulet and Huber 1993). *Melittobia* is a parasite in the nests of aculeate Hymenoptera (Naumann 1991b).

Eupelmidae These primitive, minute to small-sized chalcidoid wasps (1–10 mm) are capable of jumping (G. Gibson 1986). The family is most diverse in the tropics and subtropics. Calosotinae are mainly ectoparasities of wood-boring Coleoptera, but some have been reared from the nests of solitary bees and wasps (Goulet and Huber 1993). Neanastatinae parasitise similar host larvae and also cecidomyiid flies. The host associations of the sexually dimorphic Eupelminae include praying mantids and cockroaches, Heteroptera, Coleoptera, Lepidoptera, Diptera and spiders (Bouček 1988; Goulet and Huber 1993; Naumann 1991b). Eupelmines recorded from subtropical eastern Australia include *Anastatus lutheri* (Mt

Tamborine), *Reikosiella marxi* (Tooloom Scrub) and *Calymmochilus marksiae* (Mt Glorious, Barrington Tops) (Austin *et al.* 1998; Bouček 1988).

Eurytomidae Eurytomids are minute to small-sized chalcidoid wasps that are mostly parasites, but also include phytophagous species that are stem miners, gall formers and seed eaters (Goulet and Huber 1993). Species that develop in galls may also feed on the gall-forming insect and the gall itself (Naumann 1991b). Regional species include *Bruchophagus niger*, *Eurytoma nigra* (Eurytominae), *Bootanomyia guttatipennis*, *Malostigmus pergratus* (Megastigminae) and *Torymoidellus reticulatus* (Monodontomerinae) (Bouček 1988). *Bruchophagus* is phytophagous. Most *Eurytoma* are parasitic, though some are also phytophagous in seeds (Naumann 1991b). As with several other Chalcidoidea (see Williams 1995), eurytomids are sometimes associated with flowering trees in subtropical rainforest (on *Alectryon coriaceus*, *Syzygium smithii*), including *Mesamotura* on the flowers of *Alphitonia excelsa* in littoral rainforest (Williams 1995).

Evaniidae Though cosmopolitan in distribution the Evaniidae, commonly called hatchet or ensign wasps, are most abundant in tropical regions. This distinctly narrow-waisted and stout family is placed with Aulacidae and Gasteruptiidae in the superfamily Evanioidea (Townes 1949). Unlike these two families, the ovipositor of evaniid females is concealed. Evaniids are solitary parasites of the egg masses (oothecae) of cockroaches, the larvae feeding on the host eggs inside. Adults often feed on nectar, and when foraging characteristically move their short metasoma up and down; thus the 'hatchet' assignation. Species of *Evania* are often encountered on myrtaceous genera (e.g. *Baeckea*, *Leptospermum*, *Angophora*) in open eucalypt forest and woodland, but species are also known from *Alphitonia excelsa* (Rhamnaceae), *Guioa semiglauca* (Sapindaceae) and *Syzygium smithii* (Myrtaceae) in lowland rainforest (Williams 1995).

Gasteruptiidae These are very slender-bodied parasitic wasps that range in size from ~4 to 85 mm (Naumann 1991b). The ovipositor is concealed or exserted, and may be much longer than the metasoma. Females oviposit in the nests of bees and vespid wasps, and there the developing larvae, after eating the host's egg or larva, may consume the provisions intended for the parasitised host (Goulet and Huber 1993). Gasteruptiidae are frequently observed flying slowly, or hovering (helicopter-like), very close to the ground in search of hosts. Adult *Gasteruption* visit mass-flowering rainforest trees and shrubs (*Alectryon coriaceus*, *Euroschinus falcatus*, *Syzygium floribundum*, *S. smithii*, *Tristaniopsis laurina*), where they will feed on nectar (Williams 1995).

Gasteruptiidae comprise the cosmopolitan Gasteruptiinae, and the Hyptiogastrinae, which have a Gondwanan distribution and occur in Australia, New Zealand and southern South America (Jennings and Austin 1997). Hyptiogastrinae are the most primitive subfamily and are most diverse in Australia. Gasteruptiinae include the large genus *Gasteruption*, which has an almost cosmopolitan distribution. Its hosts include the bee families Halictidae, Megachilidae and Colletidae. The hyptiogastrine genus *Aulacofoenus* is restricted to Australia and South America (Jennings and Austin 1997), and includes *A. whiani* known only from the region of the Gondwana rainforests. *Crassifoenus* and *Hyptiogaster* are Australian endemics, while *Eufoenus* occurs mainly in Australia but is also found in New Guinea and the south-west Pacific (Jennings and Austin 1997).

Ichneumonidae Ichneumonoids are well represented in the fossil record, and particularly the Early Cretaceous. The Ichneumonidae are the largest family in the Hymenoptera, are cosmopolitan in distribution, and include at least 60 000 species (Goulet and Huber 1993). They are closely related to the Braconidae. Species range in size from the minute to very large (120 mm), the ovipositor is often exserted, and females lay between 10 and 1000 eggs (Naumann 1991b). They are parasites and are found in all subtropical rainforest formations, and at all altitudes. The family is best represented in Australia's south-east. Ichneumonids favour moist habits, and adults (e.g. *Enicospilus* – Ophioninae, Plate 265) are often seen flying in the lush understorey, often low among large shading ferns.

The Australian fauna is particularly rich in the moist south-east region. Relatively few species occur in arid zones (Gauld 1977; Naumann 1991b). However, the fauna is small in comparison to that of the Afrotropical, Nearctic, Neotropical and Palaearctic regions (Naumann 1991b). Six species of the genus *Gotra* (Phygadeuontinae) have been recorded from the Gondwana rainforests region. The distribution of *Gotra* includes New Guinea, Indonesia, Japan, Taiwan, the Solomon Islands and Australia (G. Holloway 1986). In Australia *Gotra* is confined to the wetter eastern and south-western regions and primarily occurs in sclerophyll forest. Holloway (1986) considers that the genus had its origins in New Guinea and radiated from there to the Indo-Malayan area (Taiwan, Korea, Japan and the Philippines) and Australia, east to

New Caledonia. *Heteropelma scaposum* (Anomaloninae) (Plate 266) is in an apparent regional mimicry complex (owing to size, form, colour and habitat cooccupation) that includes the cerambycid beetle *Macrones besti* and several asilid genera. *Echthromorpha intricatoria* (Pimplinae) has been collected from the flowers of *Alphitonia excelsa*, and *Eucaphila* spp. (Campopleginae) from flowering *Scolopia braunii*. Additional genera from the region include *Gahus* (Cremastinae), *Certonotus* (Labeninae), *Xanthocrytptus* (Phygadeuontinae/Cryptinae), *Camptotypus* (Pimplinae) and *Xanthopimpla* (Pimplinae) (Townes and Chiu 1970), but this is a tiny fraction of the known diversity. *Gahus* is an endemic genus, and *Certonotus* occurs in Australia and South America.

Ophioninae and Anomaloninae are endoparasitic on the larvae of Lepidoptera, and Phygadeuontinae are ecto- or endoparasites of Lepidoptera, Hymenoptera and Neuroptera. Campopleginae are one of the most common subfamilies of Ichneumonidae, and are endoparasites mainly of Lepidoptera or Symphyta larvae (Goulet and Huber 1993). Cremastinae are endoparasites of Lepidoptera, and less commonly Coleoptera larvae. Hosts of Pimplinae include Lepidoptera and spiders (Goulet and Huber 1993). Labeninae parasitise Coleoptera, solitary bees, Neuroptera and spiders. Species of *Certonotus* (e.g. *Certonotus andrewi*) are parasites of larval Cerambycidae and Symphyta. The subfamily is primarily Australian and South American in distribution, and is probably Gondwanan (Naumann 1991b).

Megalyridae The Megalyridae are small to very large wasps that parasitise wood-boring Coleoptera found under tree bark (Naumann 1991b), but also immature stages of spheciforme Crabronidae – Pemphredoninae (as Pemphredonidae in Goulet and Huber 1993). The exserted ovipositor can be extremely long, much longer than the length of the body. The family has a pantropical distribution (Africa, Australia, South America, South-east Asia), is most diverse in Australia, and is associated with ancient tropical forests (S. Shaw 1990). Cretaceous fossil evidence indicates Megalyridae were once more widely distributed (Grimaldi and Engel 2005).

The Australian megalyrid fauna has adapted to drier environmental conditions, occurs in eucalypt woodlands and is associated with *Acacia* (S. Shaw 1990). *Megalyra caudata*, *M. fasciipennis* and *M. rufipes* (Megalyrinae) occur in the Gondwana rainforests region. *Megalyra* is primarily Australian in distribution but also occurs in Ceram, Philippines, New Guinea and New Caledonia.

Monomachidae This is a small family comprised of three species in eastern and southern Australia (e.g. *Monomachus australicus*), and at least 10 species in South America (Naumann 1985). Adults prefer cool and moist forest habitats. Monomachids are probably Gondwanan in origin and possess several character states, suggesting they are very ancient (Naumann 1985). Species parasitise stratiomyid flies of the subfamily Chiromyzinae. The Chiromyzinae have a similar austral-disjunct distribution to the Monomachidae, and are also most likely to be Gondwanan.

Platygastridae The platygastrid subfamily Scelioninae (previously Scelionidae) comprises minute to small-sized parasitic wasps generally little larger than 7 mm in length (Naumann 1991b). Scelioninae are very diverse in the range of hosts that they attack (Coleoptera, Orthoptera, Heteroptera, Mantodea, and spiders); however, they exhibit a 'high degree of host partitioning' (i.e. little overlap in host groups used by different taxa within the family) (Galloway and Austin 1984). The main hosts are Orthoptera and heteropteran Hemiptera. The Teleasinae apparently restrict their host utilisation to carabid ground beetles (which are diverse in the Gondwana rainforests region) (Galloway and Austin 1984).

Galloway and Austin (1984) recognised 48 Australian genera within the subfamily Scelioninae (e.g. *Neoscelio*, *Crama*, *Duarina*, *Genatropis*, *Lispotela*, *Macroteleia*). South-eastern New South Wales–south-eastern Queensland is the centre of diversity for the endemic scelionine genus *Neoscelio* (e.g. *N. doddi*, *N. lateralis*) (Galloway and Austin 1984, Galloway *et al.*1992). *Neoscelio* occurs in a wide range of plant communities comprising rainforest, dry sclerophyll forest and open scrubland (Galloway *et al.* 1992). The genus *Genatropis* includes two Australian species (*Genatropis curtata* from Victoria, and *G. pretiosa* from south-east Queensland to Tasmania) and is otherwise recorded from Vanuatu, New Zealand and New Caledonia. *Lispotela* is recorded from the east coast of Queensland, New Guinea, the Pacific and South-east Asia. *Lispotela tamborina* is known only from Mt Tamborine in south-east Queensland. *Crama* consists of two species: *C. reticulata* from south-east Queensland to southern New South Wales and *C. albicoxa* from northern Queensland (Galloway and Austin 1984). *Duarina* (*Duarina venustella*) is restricted to southern Queensland and northern New South Wales (Galloway 1978a), but *Macroteleia* (*Macroteleia magna*) and *Trissolcus* (*Trissolcus personatus*) are world-wide genera (Galloway 1978b; Johnson 1991).

Pompilidae The Pompilidae, or spider wasps, are a cosmopolitan family containing more than 4000 described species. They probably originated ~43 mya, with that of most of the extant subfamilies dated from the Late Eocene to the Oligocene (Waichert *et al.* 2015). All are solitary diurnal, frequently wide-ranging species. They are often conspicuously coloured and patterned. *Ctenostegus cinctus* is striped black and white, species of *Heterodontonyx* (Wahis 2008) (Plate 274) are characteristically orange and red, *Ferreola handschinii* (see H. Evans 1982) (Plate 272) is black-bodied with a bright orange-yellow mesoscutum, *Calopompilus defensor* has orange wings, while the forewings of *C. raptor* (Plate 273) are black with a large yellow spot near their apex. *Phanagenia fasciata* is black-grey in colour; however, the antennae are yellow, the appearance of the wasp reminiscent of the clerid beetle *Trogodendron fasciculatum*. Brachypterous, dark red and black coloured *Psoropempula* (H. Evans 1974), as they move about on the rainforest floor, resemble large foraging *Myrmecia* ants.

Pompilids mainly prey on spiders (Plate 275), by either short flights or hunting on the ground. Each larva normally develops in a cell constructed by the female, into which she places the provisions. Prey are normally paralysed, and only rarely are larvae ectoparasitoids of active spiders or cleptoparasites in cells provisioned by other pompilids (Goulet and Huber 1993). Though adults capture prey, so as to provision for the development of their young, they are frequent visitors to flowering trees in subtropical rainforest (Williams and Adam 1995) to imbibe nectar; however, nectar consumption by flower-visiting pompilids (and other solitary wasps) may be an activity relating to maintaining water balance during hot days, rather than fulfilling daily energy needs (see Willmer 2011). Nevertheless, while visiting flowers to feed, wasps snare pollen on the body integument (Williams and Adam 1998) such that they may unwittingly serve as agents of pollen transfer. Flower host records include *Calopompilus defensor* (on *Syzygium floribundum*), *Phanagenia fasciata* (on *Euroschinus falcatus*) and *Ferreola handschinii* (on *Alphitonia excelsa*) (Williams and Adam 1995).

Pteromalidae This family comprises minute to large-sized parasites of the eggs, larvae and pupae of numerous insects (e.g. Coleoptera, Diptera, Hemiptera), and is the most morphologically and biologically diverse of the Chalcidoidea (Naumann 1991b). Some are phytophagous as gall makers or inquilines in galls (e.g. Austrosystasinae, Ormocerinae, some Pteromalinae and Coelocybinae). Species of *Agamerion* (Cleonyminae) parasitise the oothecae of cockroaches. Cleonyminae (e.g. *Mesomatura keatsi*, *Marxiana grandiosa*, *Thaumasura imperialis*) are cosmopolitan but most diverse in the tropics. The Ormocerinae (e.g. *Edgaria blackburni*, *Semiotellus ormyroides*, *Encyrtocephalus mozarti*) occur in Africa, southern Asia and the Americas but are well developed only in the Australian region where ~28 genera occur. The Austrosystasinae (e.g. *Austrosystasis atricorpus*) are restricted to south-east Queensland (Bouček 1988) and are associated with galls on *Elaeocarpus* (Naumann 1991b). Coelocybinae (*Coelocyba varifasciata*, *Ormyromorpha biargentinotata*) have a southern distribution, occurring in Australia (~13 genera), New Zealand (two genera), South America (one genus, associated with *Nothofagus*) and New Guinea (one genus, shared with Australia). Austroterobiinae (e.g. *Austroterobia partibrunnea*) are restricted to Africa, South-east Asia and Australia. Asaphinae (*Ausasaphes atripes*), Colotrechinae (*Hetreulophus bifasciatifrons*), Diaprinae (*Neapterolelaps lodgei*, *Parurios fusca*), Eunotinae (*Ophelosia viridinotata*), Herbertiinae (*Herbertia setosa*), as well as the regionally diverse Pteromalinae (*Inkaka quadridentata*, *Norbanus cyaneus*, *Mirekia planiventris*), are cosmopolitan. Pireninae have a world-wide distribution, but endemic genera (e.g. *Epiterobia*, *Premiscogaster*, *Amuscidea*) occur in the Gondwana rainforests region. Nefoeninae (*Nefoenus* sp.) are restricted to Australia. (Bouček 1988; Goulet and Huber 1993).

However, there is a high proportion of endemic genera, more generally, within the Australian pteromalid fauna (e.g. *Ausasaphes*, *Austrosystasis*, *Coelocyboides*, *Edgaria*, *Longiucha*, *Mesamotura*, *Marxiana*, *Mirekia*, *Nefoenus*, *Hetreulophus*, *Glorimontana*, *Nosodipara*, *Longinucha*, *Hillertia*, *Yrka*) (Bouček 1988). The genus *Enoggera* (Asaphinae) is confined to south-west Western Australia and coastal south-east Australia, including Tasmania. Five species are known, with *E. tanythrix* recorded only from Tooloom Plateau and the Border Ranges National Park (Naumann 1991c). Several genera (e.g. *Eupelmophotismus*, *Halticopterella*, *Neapterolelaps*, *Epanogmus*, *Pseudanogmus*) are known only from the Australasian region (e.g. New Guinea, New Caledonia, New Zealand, Solomon Islands, Lord Howe Island). *Parepistenia* has an essentially Gondwanan distribution but also occurs in South-east Asia (Bouček 1988). Numerous species are recorded from subtropical eastern Australia, with many collected from Mt Tamborine and Mt Glorious (Bouček

1988; Williams 2002). Species associated with rainforest include *Ennoggera nassani* (Asaphinae), *Yrka dahmsi* (Coelocybinae), *Glorimontana flaveola* (Colotrechinae), *Nosodipara monteithorum* (Diparinae), and *Halticopterella nigriflavellum* (Pteromalinae) (Bouček 1988; Naumann 1991a).

Scoliidae The Scoliidae are a cosmopolitan family (*c.* 300 species world-wide) of large (up to 36 mm) day-active solitary aculeate wasps (Elliott 2011; Goulet and Huber 1993; Naumann 1991b; O'Neill 2001; Turner 1909). Fossils attributed to the basal subfamily Archaeoscoliinae are known from the Early Cretaceous (Rasnitsyn 1993). Adult scoliids superfically resemble Anthoboscinae (Thynnidae). Species of *Radumeris* and *Pseudotrielis* (Plate 276) are distinctively colour banded. *Scolia* is mainly or all black in colour. The Australian fauna comprises one subfamily (Scoliinae), two tribes (Campsomerini, Scoliini), 11 genera and 20 species (Elliott 2011). The larvae are ectoparasites of beetle larvae (usually Scaraboidea). *Radumeris tasmaniensis* (= *Campsomeris tasmaniensis*) has been reported attempting copulation with the ground orchid *Calochilus camprestris* (in Elliott 2011). In addition to being observed patrolling close to the ground, or flying about flowering trees, scoliids also feed on floral nectar. Records of flower visitation in lowland subtropical rainforest include *Pseudotrielis zonata* (= *Campsomeris zonata*) and *Scolia verticalis* (at *Tristaniopsis laurina*), and *Radumeris ?tasmaniensis* (at *Diospyros australis*, *Alphitonia excelsa*) (Williams and Adam 1995). Adults are often densely setaceous and are able to carry heavy pollen loads, and the composition of the pollen carried (>95 per cent plant-specific pollen) indicates they exhibit foraging fidelity (Williams and Adam 1998).

Sphecidae Sphecidae are known from the Mid Cretaceous (Antropov 2000), and are commonly called digger wasps, mud-daubers and thread-waisted wasps. Genera remaining in Sphecidae, following the excision of most taxa to Crabronidae, include *Sceliphron*, *Sphex*, *Prionyx* and *Ammonophila* (the Sphecinae of Naumann 1991b). Individuals in these genera are large predatory wasps seen patrolling close to the ground, visiting flowers, or occasionally resting on foliage. *Sceliphron laetum* is a distinctive yellow and black, very narrow-waisted, wasp that visits flowering rainforest trees (e.g. *Tristaniopsis laurina*, *Alphitonia excelsa*), as well as those of the mangrove *Avicennia marina*, to feed on nectar (Williams and Adam 1995; G. Williams unpubl. data). The elongate waist is formed by the long and thin petiole. *Sceliphron* is a spider hunter and builds several cylindrical mud cells that are clustered together. *Sphex* preys on Orthoptera, and digs holes in the ground or reuses existing holes. Nest entrances are often well hidden under low matted vegetation, so that the wasp's entry and exit is obscured from view. Species of *Sphex* are territorial and highly combatitive against other species as well as conspecific individuals. The genus is a cosmopolitan one with numerous, frequently similar looking species, occurring in the region and often at the same site. Recorded species are *Sphex bilobatus*, *S. carbonicolor*, *S. cognatus*, *S. ephippium*, *S. ermineus*, *formosellus*, *S. fumipennis*, *S. gilberti*, *S. luctuosus*, *S. modestus*, *S. resplendens*, *S. sericeus* and *S. vestitus* (Dörfel and Ohl 2015). Many of these are widely distributed in mainland Australia. Individuals usually avoid the shaded rainforest interior but can be conspicuous nectar-feeding visitors to flowering trees in the canopy, particularly at *Alphitonia excelsa* in lowland rainforest (e.g. *Sphex ephippium*, *S. fumipennis*) (Williams 1995; Williams and Adam 1995, 1998, 2010). Individuals undertake frequent flights between individual inflorescences and flowering trees. The similar but slightly smaller *Prionyx globosus* has also been recorded from *Alphitonia excelsa*. Adult *Prionyx* prey on Orthoptera and those of *Ammonophila* are predators of larval Lepidoptera (Naumann 1991b; Walton 1985a).

Thynnidae These are often called flower wasps, and were previously grouped in the Tiphiidae (e.g. Naumann 1991b). However, molecular studies (Johnson *et al.* 2013; Pilgrim *et al.* 2008) have indicated new relationships among taxa historically placed within Vespoidea and Tiphiidae (Goulet and Huber 1993; Kimsey 1991; Naumann 1991b). World-wide, Thynnidae comprise the subfamilies Anthoboscinae, Diamminae, Myzininae, Methochinae and Thynninae. Tiphiidae, as now constituted, contain only the subfamilies Brachycistidinae and Tiphiinae. Brachycistidinae and Methochinae do not occur in Australia, and Tiphiinae are restricted to northern Australia. Thynninae and Diamminae are restricted to Australia and South America (Goulet and Huber 1993; Kimsey 1992; Naumann 1991b). Of the four subfamilies listed from Australia by Naumann (1991b) only the Anthoboscinae (*Anthobosca*) and Thynninae (numerous genera) occur in the subtropical eastern coastal zone.

These visually impressive hymenopterans are an ecologically diverse group of small to large-sized predatory aculeate wasps, first known from the Early Cretaceous (Darling and Sharkey 1990). Females are apterous and are generally smaller than males. Females of *Diamma bicolor* (Diamminae) (Plate 271) are larger than males and consequently are not carried. Thynnids parasitise the larvae of Coleoptera (e.g. ground-dwelling Scarabaeidae, Carabidae), but *Diamma* parasitises mole crickets (Campbell and Brown 1994; Grimaldi and Engel 2005; Naumann 1991b; O'Neill 2001). Thynnid diets have been reviewed by Brown and Phillips (2013).

The fauna recorded from subtropical eastern Australia is a rich one (e.g. G. Brown 1983; 1987, 1996, 1998, 2001; Salter 1953; Williams 1995, 2002), and at individual sites can be diverse, and individual species abundant. For example, Campbell and Brown (1994) recorded 19 genera and ~50 species at two remnant native vegetation sites, within an otherwise agricultural landscape, near Armidale, northern New South Wales.

The large *Anthobosca signata* is occasionally encountered foraging on flowering canopy trees (*Alphitonia excelsa*, *Tristaniopsis laurina*). Pollen loads (>95–100 per cent plant-specific loads or 'home' pollen) recorded by Williams and Adam (1998) indicate individuals may exhibit high levels of foraging fidelity during host plant flowering events. The first tarsomere has a specialised pollen comb. Thynnidae, in subtropical rainforests, range in size from the small *Epactiothynnus tasmaniensis* (<1 cm) to the very large *Hemithynnus rufiventris* (Plate 270) and *H. apterus* (>2.8 cm). *Hemithynnus tillyardi* is a large black and yellow species that frequents elevated woodlands (where it feeds on the nectar of flowering *Leptospermum*) but is also encountered in sunlit gaps below the rainforest canopy (e.g. Barrington Tops, Werrikimbe National Park). Some of the more commonly encountered species in rainforest and wet sclerophyll forest (e.g. *Lophocheilus obscurus*) generally are not flower visitors. However, many are pollinators (via pseudocopulation) of Australian terrestrial orchids (e.g. male wasps of *Neozeleboria tabulata*, *Chilothynnus sandaracus*) (G. Brown 1996) and subtropical rainforest trees (e.g. *Acanthothynnus ater*, *Agriomyia maculata*, *Dimorphothynnus dimidiatus* [Plate 269], *Epactiothynnus tasmaniensis*, *Eirone schizorhina*, *Hemithynnus apterus*, *Tmesothynnus dispersus*, *Zeleboria contigua*) (Williams 1995; Williams and Adam 1994, 1995, 2010). In rainforest, flower-visiting males are often found with small wingless females attached. And though Thynnidae can be abundant on flowering trees and their visits to individual flowers relatively long (up to several minutes per inflorescence), the presence of individual species can be episodic (i.e. only for a single, or limited number of days, during flowering events). Nevertheless, Thynnidae are an important element of the pollinator, and predator, assemblage found in subtropical rainforests and adjacent forests.

Vespidae This is a large cosmopolitan family of predatory wasps that contains ~4000 species (Goulet and Huber 1993). Naumann (1991b) placed the Australian Vespidae in four subfamilies (Eumeninae, Masarinae, Polistinae, and introduced Vespinae). Eumeninae are mainly medium- to large-sized solitary wasps that nest in the soil, in natural cavities and in abandoned nests of other Hymenoptera, or construct mud nests themselves. Adults are often brightly and aposematically coloured in bands of orange, red or yellow. They can be common at water pools, and are frequenters of flowering trees, where they will feed on nectar. In subtropical rainforest, flower-visiting species include *Abispa splendida*, *Pseudabispa confusa*, *Rhynchium mirabile*, *Bidentodynerus bicolor* (Plate 277), *Deuterodiscoelius ephippium* (Plate 278) and species of *Epiodynerus* and *Paralastor* (Williams 1995; Williams and Adam 1994, 1995, 1998, 2010; G. Williams unpubl. data). *Abispa* species, in particular, can be territorial and will aggressively fight with intruding conspecific adults. The distinctive narrow- and elongate-waisted genera *Elimus* and *Delta* are usually restricted to open forest and woodland, though *D. companiformis* is occasionally found in disturbed sites within wet sclerophyll forest. Masarinae are solitary wasps that, like bees, provision their larval cells with nectar and pollen (Naumann 1991b). Regional Masarinae include *Paragia deceptrix* and *P. hirsuta* (Walton 1985a), *Paragia* being an endemic genus. Polistinae are characterised by the social-nesting paper-wasp genera *Polistes* and *Ropalidia*, these being beneficial predators of Lepidoptera caterpillars. *Polistes humilis*, and the somewhat larger *P. tepidus*, are flower-visiting species, although *P. humilis* commonly is observed patrolling foliage on rainforest margins. The inverted paper comb nests of *P. humilis* are hidden below foliage. Colony guards are aggressive. *Ropalidia plebiana* builds very large nest combs that hang from projecting rock ledges. Rarely are nests placed in exposed vegetation. The nests are formed from macerated plant fibre and fluids produced by individual wasps.

Apocrita (ants)

Formicidae Ants are useful bio-indicators to monitor environmental impacts and changes (Majer *et al.* 2004). They are placed in a single family, Formicidae, within the aculeate wasp superfamily Vespoidea, but are treated separately here owing to their distinctive social and physical characters. The Vespoidea otherwise include the families Mutillidae, Pompilidae, Scoliidae, Thynnidae, Tiphiidae and Vespidae (Naumann 1991b), and the non-Australian Bradynobaenidae, Rhopalosomatidae, Sapygidae and Sierolomorphidae (Goulet and Huber 1993); these are all typically wasp-like in appearance. Their origins date from the Cretaceous (Barden *et al.* 2017), from where they subsequently rose to dominance in the Tertiary (~70 mya) (Grimaldi and Engel 2005).

As with many other insect orders there is a tendency for overall ant diversity to reduce with increases in latitude; this is influenced by the progressive loss of taxa with tropical affinities (Andersen and Majer 2000). Variation in altitude and climatic stability also influences species richness and patterns of distribution (Nowrouzi *et al.* 2016). However, Australian ant diversity is especially high (>65 genera) in the Wet Tropics and in the region extending from south-eastern Queensland to northern New South Wales (Shattuck 1999). There is also a high level of endemicity with more than 20 genera restricted to coastal Queensland and north-eastern New South Wales (Shattuck 1999).

Ants commonly exhibit morphologically and behaviourally distinct caste structures (Bellas and Hölldobler 1985; Hölldobler and Wilson 1990; Schmidt and Blum 1978; Traniello 1978), but the division of labour (age polyethism) may progressively change (e.g. from nursing to foraging behaviour) as workers age (Traniello 1978). Many species are chemically defended, and utilise chemical trails and signals to facilitate foraging, food retrieval, and the recruiting of defenders against attack. Ants may exhibit stratification of diversity between terrestrial and arboreal habitats (Bray 2014); however, species may differ in their ability to navigate freely within their foraging landscape and some, more than others, may be particularly reliant on path integration to return to the nest (Bühlmann *et al.* 2011). Many species are prodigous excavators of soil (e.g. *Aphaenogaster*), and in so doing affect soil structure and the distribution of organic matter within the soil profile (Hölldobler and Wilson 1990; Shattuck 1999). They contribute significantly to the structure of ecological communities (Fittkau and Klinge 1973), for their foraging activities and interactions with other organisms (including plants) throughout the rainforest can have profound impacts as to how individual species aggregate or disperse in response to their presence and behaviour (Blüthgen and Stork 2007). Impacts can range from the reduction or exclusion of herbivorous insect populations on plants, disruption to pollination systems (or, conversely, benefits to pollination events), the survival of commensal species, and even the dispersal of plants themselves (through secondary dispersal of small seeds by ants) (Hölldobler and Wilson 1990; Huxley and Cutler 1991). Indeed, seed dispersal by ants may be an important ecological interaction, for their selective burying of seeds has been documented in other plant communities to reduce seed loss by seed-eating rodents (Bond and Breytenbach 1985; Christian 2001). Such ant–plant reproduction mutualisms are poorly understood in subtropical rainforests, but the loss of keystone seed-dispersing ants may have important consequences for the survival of individual plant species in remnants where populations are already small and disjunct.

Ant colonies also interact with one another (e.g. competition for resources such as hemipteran-trophobiont honeydew), and this is reflected in territory patch size, and patterns of nest dispersal and abundance of conspecifics and other species (Bigger 1976; Blüthgen and Fiedler 2002; Blüthgen and Stork 2007; Blüthgen *et al.* 2004; Ellwood *et al.* 2016). Inter- and intraspecific competition can be asymmetrical, the outcome dependent on the size of individuals, population size and colony density, and the level of aggression or tolerance that species exhibit. This can lead to competitive exclusion or aggregation (Ellwood *et al.* 2016).

Ants interact with lycaenid butterflies, and their attendance may act to protect butterfly larvae from predators (Eastwood and Fraser 1999). Examples of ant genera recorded in attendence with Lycaenidae from south-eastern Queensland include *Anonychomyrma*, *Camponotus*, *Crematogaster*, *Dolichoderus*, *Iridomyrmex*, *Meranoplus*, *Paratrachina*, *Pheidole*, *Polyrhachis*, *Prolasius* and *Rhytidoponera* (Schmidt 2002).

Australian ants include many old endemic genera, especially in southern areas. The fauna is particularly rich in genera, and species richness is often high within genera (Naumann 1991b). Rainforest ant faunas trend to being less diverse at higher latitudes and elevations (Majer *et al.* 2001). The subfamilies and genera discussed below include many species that are typical of open forest, rather than rainforest. Nevertheless, the rainforest ant fauna is taxonomically and ecologically diverse and contains species that are restricted to rainforest or wet sclerophyll forest (*Notostigma*), as well as species that occupy a broader range of habitats in addition to that of rainforest (e.g. *Leptomyrmex erythrocephalus*, and species of *Myrmecorhynchus*, *Notoncus*, *Orectognathus*, *Polyrhachis*, *Prolasius*, *Pseudonotoncus* and *Strumigenys*).

Formicidae – Aenictinae This subfamily contains a single genus, *Aenictus* (*Aenictus ceylonica*), which extends south to north-eastern New South Wales (Shattuck 1999). These are army ants that forage day and night, with large numbers of workers attacking other ants, social wasps and other arthropods. During raids the workers may attack and concentrate on a single nest or a small area. Species are nomadic, and also migratory. Most forage on the ground, but some are also arboreal. New colonies are formed by the division of existing nests.

Formicidae – Cerapachyyinae The subfamily Cerapachyyinae occurs throughout Australia, including Tasmania. The genus *Cerapachys* (*Cerapachys senescens*) is widespread and is a specialist predator of other ants. Workers are active during the day, and may forage in clustered groups of fewer than 20 individuals (G. Williams pers. obs.). Captured ant larvae are paralysed and taken back to the nest. *Sphintomyrmex* (*Sphintomyrmex mjobergi*, *S. steinheili*) has a mainly tropical distribution, though in Australia species penetrate to temperate regions. It is also a predator of other ants, but workers generally hunt below the ground rather than coming to the surface (Shattuck 1999).

Formicidae – Dolichoderinae Members of this family have lost the sting (Grimaldi and Engel 2005). Many are arboreal, and are mostly general scavengers and predators. However, a number (e.g. *Dolichoderus*, *Tapinoma*) are associated with Hemiptera, from which they collect honeydew, or tend (*Papyrius*) the caterpillars of butterflies. They will also use captured workers to help with the raising of their young. The distinctive Dolichoderinae genus *Leptomyrmex* (Plate 260), which is the apparent model for several mimicry complexes involving Diptera, Coleoptera and juvenile Phasmatodea, is also found in New Guinea, Aru, Seram and New Caledonia. Fossil evidence (Dominican amber) suggests the genus was more widespread (Shattuck 1999), though Lucky and Ward (2010) consider that the fossil Dominican amber *Leptomyrmex neotropicus* might belong to a stem lineage of *Leptomyrmex*. Most *Leptomyrmex* species are restricted to rainforest and wet sclerophyll forest. Species are most frequently encountered foraging on the ground, though adults occasionally are found feeding on floral nectar (e.g. of *Acronychia oblongifolia*, *Cryptocarya microneura*) and may also be attracted to fallen over-ripe and succulent fruit (Williams 1995; G. Williams pers. observ.). The regional *Leptomyrmex* fauna is a rich one and includes *L. aitchinsoni* (rainforest, wet sclerophyll forest records), *L. burwelli* (rainforest), *L. cnemidatus*, *L. erythrocephalus*, *L. mjobergi*, *L. nigriventris* and *L. rothneyi* (rainforest and dry sclerophyll forest), and *L. rufipes*, *L. rufithorax*, *L. tibialis* and *L. wiburdi* (rainforest, vine forest, wet and dry sclerophyll forest) (Lucky and Ward 2010; Smith and Shattuck 2009). Although a number are widely distributed, *Leptomyrmex ramorniensis* has only been recorded from near Grafton (Smith and Shattuck 2009).

Additional representative dolichoderines from the region include *Doleromyrma darwiniana*, *Dolichoderus scabridus*, *Ochetellus glaber*, and species of *Anonychomyrma*, *Bothriomyrmex*, *Froggattella*, *Iridomyrmex*, *Papyrius*, *Tapinoma*, *Technomyrmex* and *Turneria* (Shattuck 1999; Walton 1985a). The genus *Doleromyrma* is endemic but the others have a more widespread distribution. *Turneria* reaches its southern limit in northern New South Wales (Shattuck 1999). Genera include predominantly ground-nesting species (e.g. *Doleromyrma*, *Iridomyrmex*, *Leptomyrmex*), ground-nesting species that may also nest in trees (*Anonychomyrma*, *Dolichoderus*, *Froggatella*, *Ochetellus*), and tree-nesters (*Turneria*) (Shattuck 1999). *Iridomyrmex* species are often common at the flowers of canopy trees (e.g. *Acradenia euodiiformis*, *Alectryon coriaceus*, *Alphitonia excelsa*, *Cuttsia viburnea*, *Drypetes deplanchei*, *Euroschinus falscatus*, *Guioa semiglauca*, *Syzygium floribundum*, *S. smithii*, *Tristaniopsis laurina*) in lowland rainforest remnants (Williams 1995).

Formicidae – Formicinae Formicinae have also lost the sting but defend themselves by spraying formic acid from the tip of the abdomen. They comprise numerous common ants (Shattuck 1999; R. Taylor 1992), including the weaver ant genus *Oecophylla*, and are both day-active and nocturnal scavengers on the forest floor and in trees and shrubs. Nests range from small, cryptically positioned ones in foliage (e.g. *Acropyga*, some *Polyrhachis*) to those that are large and conspicuous. Some species are aggressive and vigorous defenders of their colonies.

The formicine fauna recorded from the Gondwana rainforests and adjacent regions is a large one (Williams 2002). Representative genera from the world heritage area include *Acropyga*, *Calomyrmex*, *Camponotus*, *Myrmecorhynchus*, *Notoncus*, *Notostigma*, *Opisthopsis*, *Paratrechina*, *Plagiolepis*, *Polyrhachis*, *Prolasius*, *Pseudonotoncus*, *Stigmacros* and *Teratomyrmex*. *Calomyrmex* forage on foliage and flowers and collect nectar, including that from extrafloral nectaries. Species also collect honeydew from tended Hemiptera, but are also general scavengers. *Camponotus* (e.g. *Camponotus innexus, C. consobrinus*) is a widespread and diverse genus, species often being found in rotting logs or standing rotting tree trunks. *Camponotus innexus* has been observed feeding on nectar of *Schizomeria ovata* (Williams 1995), and *C. ?nigriceps* on that of the dioecious understorey rainforest shrub *Tasmannia insipida* (G. Williams unpubl. data). Many species are nocturnal, and in daylight secretively remain in their nests, only to emerge in great numbers at night. Ants of the endemic *Myrmecorhynchus* (*Myrmecorhynchus carteri, M. emeryi*) are small and inconspicuous species that nest in the ground, but also in twigs in foliage. *Notoncus* (*Notoncus spinisquamis*) is restricted to Australia except for one species in New Guinea. They are general predators that forage on the forest floor. *Notostigma* (*Notostigma foreli*) are large nocturnal ground foragers that nest in the soil under stones or wood. *Paratrechina* are habitat generalists that also occur in rainforest and wet sclerophyll forest. They form large colonies on the forest floor, nesting in the soil as well as under stones and fallen wood. *Plagiolepis* are inconspicuous predatory ants that nest under stones and rotting wood. They forage on the ground as well as on tree trunks and branches. (Shattuck 1999; R. Taylor 1992).

Polyrhachis (Plate 264) is found in the southern Palaearctic, Ethiopian and Oriental regions, and is widespread in Australia but absent from New Zealand (Kohout 1988, 2006; Kohout and Taylor 1990; Shattuck 1999; Walton 1985a). The genus is omnivorous and is commonly encountered in rainforest on flowers (e.g. *Polyrhachis ammon, P. clio, P. hookeri, P. pilosa*) (Williams 1995), or generally foraging on foliage, tree trunks and branches. They nest in the soil, in tree holes, or in nests formed from living plant foliage and larval silk. Individual species are either day-active or nocturnal in their foraging patterns. *Polyrhachis australis, P. lydiae* and *P. aurea* are known to attend lycaenid larvae (Schmidt 2002). *Prolasius* (*Prolasius clarki, P. convexus*) most often forage at ground level, and also less commonly on low vegetation. Some species feed on seeds (Ashton 1979). Endemic, but uncommon, *Pseudonotoncus* (*Pseudonotoncus hirsutus, P. turneriseudonotoncus*) principally forage on tree trunks and foliage. *Stigmacros* (*Stigmacros australis*) is an endemic genus whose species are general predators. They nest mainly in the soil and under stones, and habitats range from rainforest to open woodlands and savannahs. *Teratomyrmex* is an endemic genus with the single rare species, *T. greavesi*, restricted to the rainforests of south-east Queensland and north-eastern New South Wales (R. Taylor 1992).

Formicidae – Leptanillinae This subfamily in Australia comprises a single species (*Leptanilla swani*), the recorded habitats of which range from rainforest to woodland. Though known from the extreme south-east of Queensland, and north-eastern New South Wales, the species is rarely encountered (Shattuck 1999). Extralimital species are known to be ground raiders, prey on small arthropods, and form colonies in the soil.

Formicidae – Myrmeciinae The Myrmeciinae (excluding *Nothomyrmecia* – Nothomyrmeciinae) arose in the Mid Cretaceous (Grimaldi and Engel 2005) and comprise the single primitive genus *Myrmecia* (bulldog ants) (J. Clark 1951; Ogata 1991; Ogata and Taylor 1991; Shattuck 1999) (Plates 261, 262). Although several species occur within the Gondwana rainforests region, the genus is principally a Bassian open forest one (Andersen 1991). They are more common in southern regions than in the tropics (Shattuck 1999). *Myrmecia* also occurs naturally in New Caledonia (one rare endemic species, *Myrmecia apicalis*), and introduced to New Zealand (Walton 1985a), and exhibits a Gondwanan or southern distribution. *Myrmecia* all possess very long toothed mandibles, prominent eyes, and most species are large in size: commonly >2 cm, though members of the urens group may be <1 cm. Most species forage during the day, but some are also nocturnal in rainforest. The distinctive orange–black-coloured *Myrmecia nigrocincta* (Plate 261) is an aggressive species that inhabits disturbed rainforest and wet sclerophyll forest, building moderately sized nests (of at least 100 individuals) under stones or in open soil (though nest entrances are usually hidden below leaf litter). Individuals commonly forage on foliage and feed on floral nectar (Williams 1993, 1995; Williams and Adam 2010). Additional species present in wet eucalypt and closed forest types include *Myrmecia analis, M. auriventris, M. brevinoda, M. comata, M. chrysogaster, M. flavicoma, M. gigas, M. gulosa, M. michaelseni, M. midas, M. nigriscapta, M. pyriformis, M. queenslandica, M. simillima, M. tarsata* and *M. tridentata*. These are predominantly ground foragers though several may also frequent tree trunks (e.g. the diminutive *Myrmecia chrysogaster*).

Formicidae – Myrmicinae This is the most diverse subfamily (ecologically and in numbers of species), and is the largest ant subfamily occurring in Australia

(Ettershank 1966; Shattuck 1999; R. Taylor 1973, 1980); it includes species with restricted habitat requirements but also those (e.g. *Pheidole*) that have readily adapted to disturbed or artificial ones such as urban parks and gardens. Myrmicinae range in size from ~1 mm long to ~10 mm (Shattuck 1999). *Strumigenys* and *Orectognathus*, though cryptic and small in size, possess bizarrely shaped heads and long mandibles with distinctive apical teeth. They are found in leaf litter, and nest under loose tree bark and in and under rotten logs. Many (e.g. *Orectognathus mjobergi* [Plate 263], *O. rostratus*, *O. kanangra*) are nocturnal or crepuscular in habit. though some commonly nocturnal species (e.g. *O. phyllobates*) will forage during daylight if the weather conditions are dull, overcaste or showery (Taylor 1977). Many Myrmicinae are predators of small arthropods but others are scavengers or harvest seed. Individual species nest under stones, in rotting wood or in open exposed soil. The rainforest and wet sclerophyll forest-restricted *Anillomyrma* lack eyes, and therefore may be wholly subterranean in habit. However, individual genera and species (e.g. *Colobostruma*, *Crematogaster*, *Epopostruma*, *Pheidole*, *Podomyrma*) commonly inhabit several kinds of vegetation communities, though some (such as *Orectognathus*, *Europalothrix*, *Metapone*, *Oligomyrmex*, *Strumigenys*) may be more common in certain types. Species of *Crematogaster* can be common on the flowers of canopy and subcanopy trees (e.g. *Acradenia euodiiformis*, *Alphitonia excelsa*, *Elaeodendron australe*, *Elaeocarpus obovatus*, *Guioa semiglauca*, *Mischocarpus pyriformis*, *Rhodomyrtus psidioides*, *Syzygium smithii*) and may be pollinators of the small subtropical rainforest tree *Alectryon coriaceus* (Sapindaceae) (G. Williams and P. Adam unpubl. data).

The subfamily is diverse in the Gondwana rainforests region. Included genera are *Adlerzia*, *Aphaenogaster*, *Calyptomyrmex*, *Monomorium*, *Crematogaster*, *Epopostruma*, *Eurhopalothrix*, *Lordomyrma*, *Mayriella*, *Meranoplus*, *Mesostruma*, *Metapone*, *Myrmecina*, *Oligomyrmex*, *Orectognathus*, *Pheidole*, *Podomyrma*, *Pristomyrmex* (*Odontomyrmex*), *Rhophalothrix* and *Strumygenys* (Shattuck 1999; Williams 2002). Many species inhabit rainforest: *Calyptomyrmex beccarii*, *Monomorium ilia*, *M. kiliani*, *M. leae*, *Eurhopalothrix australis*, *Metapone leae*, *M. tillyardi*, *Podomyrma gratiosa*, *Rhophalothrix orbis*, *Strumygenys* sp., and numerous species of *Orectognathus* and *Pristomyrmex* (R. Taylor 1977, 1979; Walton 1985a; Williams 2002). Of the genera, *Orectognathus* is shared with New Guinea, New Caledonia, and the North Island of New Zealand, *Pristomyrmex* also occurs in the Ethiopian, Malagasy and Oriental regions, and east Melanesia, *Rhophalothrix* occurs in New Guinea, Central and South America, and *Adlerzia*, *Epopostruma* and *Mesostruma* are endemic (R. Taylor 1973; W. Brown 1961; Shattuck 1999; Walton 1985a). The remaining genera are widespread.

Formicidae – Ponerinae Ponerinae are considered to be a basal grade of Formicidae (Grimaldi and Engel 2005), and are small to large predaceous ants that include both cryptic (*Amblyopone*, *Platythyrea*) and conspicuous species (Schmidt and Shattuck 2014). Several have very painful stings (e.g. *Leptogenys*, some *Rhytidoponera*) used to immobilise prey and also for defence. Of the genera that inhabit rainforest and wet sclerophyll forest, the primitive *Ambylopone* (Plate 259) normally nest under rotten tree trunks, or under loose bark of fallen dead trees. Nests often contain only a few workers. *Amblyopone* species are carnivorous, showing a strong preference for feeding on soft-bodied arthropods (Shattuck 1999), but worker behaviour and the division of labour tasks have been recorded to be independent of age (Traniello 1978).

Anochetus and *Odontomachus* possess fantastically elongate mandibles and nest in the soil, under fallen wood, and in termite nests (*Anochetus*). *Diacamma* nest in loose debris or in soil. *Cryptopone*, *Discothyrea*, *Heteroponera*, *Hypoponera*, *Leptogenys*, *Myopias*, *Pachycondyla*, *Platythyrea*, *Ponera*, *Prionopelta*, *Proceratium* and *Rhytidoponera* nest either in the soil, under stones or loose bark, or under or in rotten logs on the ground (Shattuck 1999). The uncommon, but endemic, *Onychomyrmex* live a nomadic army-ant lifestyle, forming temporary camps on the forest floor or semi-permanent nests in the soil or in rotten logs. New colonies are founded by the division of established ones (Shattuck 1999). Ponerine species recorded from the rainforests and wet sclerophyll forests of subtropical eastern Australia include *Amblyopone aberrans*, *Cryptopone rotundiceps*, *Heteroponera leae*, *Leptogenys excisa* and *Platythyrea parallela*, the latter an uncommon species occasionally found nesting below the loose bark adhering to the trunks of fallen rainforest trees.

MOLLUSCA (land and freshwater snails, slugs)

Molluscs occupy a diversity of habitats, and are likely best known via the various introduced shelled snails (e.g. *Cornu*

Ants, myrmecophilous Hemiptera and trophobiosis

Some ants harvest honeydew, a metabolism exudate produced by bugs (trophobionts) in the families Cercopidae, Cicadellidae, Coccidae, Eurymelidae, Fulgaridae, Membracidae (Auchenorrhyncha), Aphidae, Pseudococcidae and Psyllidae (Sternorrhyncha) (Hölldobler and Wilson 1990; Naumann 1991b). This relationship is likely an ancient one, for aphids in association with ants are known from Oligocene Baltic amber (W. Wheeler 1914). The majority of ants that participate in this attending symbiotic relationship (termed 'trophobiosis') are members of the Dolichoderinae, Formicinae and Myrmecinae. In Australia the Myrmeciinae, particularly represented by *Myrmecia nigricincta* in lowland subtropical rainforest, have not been recorded tending to honeydew-producing Hemiptera. Honeydew is a carbohydrate-rich solution, secreted as a by-product of the bugs feeding on plant phloem. Phloem sap is a chemically diverse mixture of sugars, proteins, minerals, amino acids, amides and vitamins, some of which are absorbed or converted in the digestion process, the unrequired excess then passing along the gut and finally expelled through the anus. Honeydew contains numerous sugars (e.g. fructose, glucose, trehalose, mannose); its overall chemical composition may be very similar to that in the phloem sap, but the combination of sugars within it can be species-specific. At least in the case of Aphidae (aphids), bugs unattended by ants have been observed to flick away the honeydew droplets by using their hind legs or contractions of the rectum or whole abdomen (Hölldobler and Wilson 1990). The expelled honeydew can accumulate in havestable quantity, and is thought to be the basis for the biblical reference to 'manna' given to the Israelites of the Old Testament.

Myrmecophilous hemipterans often have special structural and behavioural adaptations for their interactive association with ants. These can involve special honeydew-holding hairs, production of waxy coatings (as with mealybugs – Pseudococcidae), easing out of secretions, and leg movements that signal to ants that droplets are to be secreted. Ants may chemically mark the location of trophobiont populations within their foraging territory; however, the relationship may be affected by the life cycle of the trophobiont, since highly synchronus maturing hemipteran populations will mean that honeydew is only available for a certain portion of each year. At higher latitudes subtropical rainforest is subject to substantial episodes of seasonal temperature change, to which invertebrates dramatically respond, and while there are distinct peaks in tropical invertebrate diversity and abundance that relate to temperature and rainfall (Frith and Frith 1985), at southern latitudes adult activity is greatly suppressed during the colder months. For some ants honeydew is important for their nutrition (Degen *et al.* 1986), but it is uncertain whether participating hemipterans and ants gain a mutual benefit. Certainly, the presence of ants potentially deters predator attacks on hemipterans, singly or as aggregations, and removing excess sugary material may facilitate a hygiene service (Strickland 1947). The presence or absence of ants has been shown to influence patterns of brood care behaviour by attendant female membracids, who may even abandon their nymph clusters in the presence of ants; this may not be to their detriment as female membracids can produce new egg batches, and ants may continue to protect the overwise abandoned nymph broods (Bristow 1983). Beyond this, rainforest plants themselves potentially gain protection from herbivorous insects by the presence of ants (Hölldobler and Wilson 1990; Huxley and Cutler 1991), this being incidental to the dynamics of trophobiont hemipteran–ant interraction.

But for ants in subtropical rainforests the harvesting of honeydew may only be an occasional event, simply one small food resource within a suite of many, that is randomly procured during daily foraging patterns.

aspersum – Helicidae, Bradybaenidae, Hygromiidae, Zonitidae) and slugs (Agriolimacidae, Limacidae, Milacidae) that occupy urban and agricultural landscapes. The native fauna includes grazers (Hydrocenidae) and carnivores, and in turn are prey to reptiles, amphibians, birds and mammals. Excluding the division between those that are either terrestrial, or inhabitants of freshwater streams, snails in subtropical eastern Australia are found in moist litter, in and under rotting logs, in soil, under loose bark, beneath rocks and within rock crevices, on tree trunks, and on or underneath leaves. As with most other invertebrate species, snail populations tend to occur centred on favoured microhabitats, rather than uniformly positioned throughout. Though most are nocturnal in habit some species are frequently found active and exposed during daylight, especially during or after rain (e.g. Cystopeltidae, *Triboniophorus graeffei* [Plates 289, 290, 296]). They can also remain inactive for long periods, sealing themselves against leaf surfaces, bark and stones until the return of conditions conducive to renewed activity (Stanisic *et al.* 2010).

As with invertebrate groups previously discussed, the following families are only an indication of the molluscan fauna known to occur in subtropical rainforests and wet sclerophyll forests.

Terrestrial fauna

The fossil record for land snails begins in the Upper Carboniferous of Europe and North America. Many Australian non-marine mollusc families also occur in other continents; however, there is a high level of endemism at the genus and species level (Ponder *et al.* 1998; see Table 5).

More than 40 families of land snails (collectively possessing >1100 species) (J. Stanisic pers. comm.; Stanisic *et al.* 2010) occur along the east coast of Australia. These are a mix of ancient and more recent taxa from landmasses to the north, with speciation events apparently driven by repeated, climate-induced, expansion and contraction of rainforest, resulting in fragmentation of ancestral populations (Stanisic *et al.* 2010). Species diversity is high in east coast rainforests due to their role as refugia over a long geological period (Beesley *et al.* 1998), and the occurrence there of over 90 per cent of the land snail taxa indicates the importance of these refugia to the fauna (Stanisic 1994). Stable moisture regimes and volcanically derived acidic soils have provided an ideal environment for the evolution of slugs (i.e. Athoracophoridae, Cystopeltidae, Rathouisiidae).

The area bounded approximately by the Nambucca and Hastings Rivers (northern New South Wales), and the eastern escarpment of the Great Dividing Range, are particularly rich in land snails. More than 100 species have been recorded in this area, with much of the diversity being centred upon limestone outcrops in the Macleay Valley, the so-termed Macleay Valley refugium. The average rainforest site in New South Wales yields 10–20 species, and there are few areas in the World where diversity at any one site exceeds 30 species (Stanisic 1994). However, the Macleay Valley (supporting rainforest in conjunction with limestone outcrops) possesses a much greater diversity due to the presence of limestone endemics. The number of species recorded from the Macleay Valley compares favourably with the Wet Tropics biogeographical region, yet the Wet Tropics is a much larger area (Stanisic 1997).

Major families occurring in eastern Australia are Charopidae, Camaenidae and Helicarionidae. The Hydrocenidae, Helicinidae, Cyclophoridae, Pupinidae and Diplommatinidae are largely confined to the rainforests of eastern Australia (Stanisic 1994; Stanisic *et al.* 2010). Species occurring in rainforest and associated with limestone outcrops generally have much more restricted ranges than species living in eucalypt woodland.

The association between land snails and rainforest is an ancient one and factors, such as high nutrient soils and moisture, that favour rainforest also favour land snails (Stanisic 1994; Stanisic *et al.* 2010). However, faunal diversity can differ significantly from site to site, and can vary with changes in latitude (Stanisic 1994). Subtropical rainforest (in a floristic sense) is partically rich, and the Border Ranges area of south-east Queensland and northern New South Wales, in particular, possesses complex, species-rich communities (J. Stanisic, cited in Beesley *et al.* 1998). Though the various formations of dry rainforest maintain a less diverse land snail fauna, these sites often include endemic species as well as serving as refugia within open forest and woodland (Stanisic *et al.* 2010). Araucarian vine forests in southern Queensland (e.g. Goodnight Scrub) represent important habitats for land snails, possessing diverse and sympatric land snail communities (Stanisic 1994). Limestone outcrops, such as those in the Macleay Valley of northern New South Wales, also support diverse sympatric island-like land snail faunas in a surrounding landscape matrix of often species-poor habitats (Stanisic 1994, 1997). Limestone outcrops act as fire shadows and moisture reservoirs, and represent important secondary habitats and refugia (Stanisic 1997). Though the fauna associated with temperate rainforest at higher altitudes and latitudes is not as common or as diverse as that found in subtropical rainforest, this rainforest type represents important habitat for Charopidae, Helicarionidae and Rhytididae (Stanisic *et al.* 2010). Sites with large numbers of Charopidae, or with endemic taxa, are significant because they indicate long-term moisture stability (Stanisic 1994).

Land snails are prone and sensitive to desiccation, and may serve as indicators of climatic refugia. As such they are an important group in the biodiversity and conservation debate, and may have a predictive role in the identification of potential reserve areas (Stanisic 1994; Stanisic *et al.* 2010).

Smith and Kershaw (1979) divided Australia into six faunal regions: compare with the freshwater snail fluvifauna provinces of McMichael and Hiscock (1958). The Gondwana rainforests region is enclosed completely within Smith and Kershaw's Oxleyan fauna region. This extends across south-east Queensland and northern New

Table 5: Charopidae genera recorded from wet forest formations in subtropical eastern Australia.

Genus	Status	Distribution	Habitat
Acheronopa	Endemic	North-east NSW	Vine thicket
Barringtonia	Endemic	Northern NSW	Subalpine woodland and forest
Cancellocochlea	Endemic	Northern NSW	Rainforest, open forest
Choridaropa	Endemic	Southern Qld	Rainforest
Coenocharopa	Endemic	Southern Qld–northern NSW	Rainforest, vine thickets
Comboynea	Endemic	Northern NSW	Rainforest
Coricudgia	Endemic	Central NSW	Rainforest
Cralopa	Endemic	South-east Qld–northern NSW	Rainforest, open forest
Cumberlandia	Endemic	NSW	Rainforest, vine thicket, open forest
Decoriropa	Endemic	North-east NSW–south-east NSW	Rainforest, woodland
Dictyoropa	Endemic	South-east Qld–north-east NSW	Rainforest
Diphyoropa	Endemic	South-east Qld– south-east NSW	Rainforest, vine thickets, open forest, woodland
Discocharopa	Extralimital	South-east Qld–north-east NSW, Indonesia, New Hebrides, Fiji, Kermadecs, Samoa, Society Islands	Rainforest, vine thicket
Egilodonta	Endemic	Northern NSW–eastern Vic	Vine thickets, open forest, limestone outcrops
Egilomen	Endemic	Southern Qld–northern NSW	Rainforest
Elsothera	Endemic	South-east Qld–central NSW	Rainforest, open forest
Gyrocochlea	Endemic	North-east Qld– south-east NSW	Rainforest, and drier vine forest
Hedleyoconcha	Extralimital	Northern Qld–northern NSW, Lord Howe Is.	Rainforest
Kooreelahropa	Endemic	Northern Qld–north-east NSW	Rainforest
Lenwebbia	Endemic	North-east Qld–south-east Qld	Rainforest, vine thicket
Letomola	Endemic	Northern NSW	Rainforest, limestone outcrops
Leurocochlea	Endemic	South-east Qld–NSW	Rainforest
Luturopa	Endemic	Central Qld–northern NSW	Araucarian vine forest, vine thicket
Macleayropa	Endemic	Northern NSW	Rainforest
Macphersonea	Endemic	South-east Qld–north-east NSW	Rainforest
Macrophallikoropa	Endemic	South-east Qld–north-east NSW	Rainforest, woodland, limestone outcrops
Mussonola	Endemic	Southern Qld–northern NSW	Rainforest and vine thicket
Nautiliropa	Endemic – monotypic	South-east Qld–north-east NSW	Rainforest
Ngairea	Endemic	South-east Qld–Illawarra region of central NSW	Temperate, subtropical and dry rainforest
Omphaloropa	Endemic	Central Qld–SEQld	Rainforest, vine thicket
Planorbacochlea	Endemic	Central NSW–northern NSW	Rainforest, vine thicket, open forest
Reticularopa	Endemic	Southern Qld	Rainforest, open forest

Rhophodon	Endemic	Southern Qld–Vic	Rainforest, limestone refugia
Richardsoniana	Endemic	Central NSW	Vine thickets
Richmondaropa	Endemic	South-east Qld–north-east NSW	Rainforest, vine thicket
Rotacharopa	Endemic	Southern Qld	Rainforest
Setomedea	Endemic	Northern Qld–northern NSW	Rainforest
Sharniropa	Endemic	Northern NSW–southern NSW	Vine thicket, scrub, woodland
Shearopa	Endemic	Central Qld–south-east Qld	Rainforest, vine thicket
Sinployea	Extralimital	Central Qld–north-east NSW, Micronesia, Melanesia, Polynesia	Rainforest
Spiraliropa	Endemic	South-east Qld	Vine thicket, woodland
Whiteheadia	Endemic	–north-east NSW	Rainforest
Ygernaropa	Endemic	Southern Qld–north-east NSW	Rainforest

South Wales. Dominant terrestrial genera occurring in the Oxleyan fauna region are *Meridolum* and *Pommerhelix* (Camaenidae), and *Hedleyella* and *Pedinogyra* (Caryodidae) (Ponder *et al.* 1998).

Smith and Kershaw's (1979) six Australian faunal regions are not distinct. There are transitional interzones between each region. For example, the Solandrian overlaps with the north-east corner of the Oxleyan and the Peronian overlaps with the south-east corner of the Oxleyan.

Achatinellidae This family is restricted to islands of the Pacific Ocean and some adjacent continents (B. Smith 1992). The family is known from the Upper Carboniferous of North America, indicating a post-Palaeozoic shift in the distribution of the family (Beesley *et al.* 1998). They are small, delicate-shelled snails that graze on leaf and bark films. Most are restricted to rainforest, including vine thickets and limestone outcrops. *Elasmias wakefieldiae* occurs in lowland rainforest (e.g. from Yessabah Caves). *Tornatellinops jacksonensis* inhabits rainforest (Mt Biggenden), including vine thicket formations. Both genera are restricted to the Indo-Pacific regions (Stanisic et al. 2010).

Athoracophoridae The Athoracophoridae represent a Gondwanan element in the Australian pulmonate fauna (Heatwole 1987). A single monotypic genus, *Triboniophorus* (*Triboniophorus graeffei*), occurs in Australia (Beesley et al. 1998) (Plates 289, 290, 291). Related genera are found in New Caledonia, Vanuatu, New Britain, the Admiralty Islands and New Zealand, but the biogeographical history of the family is unresolved. *Triboniophorus graeffei* is mainly arboreal, and occurs in rainforest, and in wet and dry sclerophyll forest (Beesley *et al.* 1998), where it feeds on organic films, fungi and algae growing on rocks, tree trunks and the forest floor (Stanisic *et al.* 2010). It is distributed widely from the south coast of New South Wales, inland to Mount Kaputar in the Nandewar ranges, north to the Kroombit Tops in central coastal Queensland, and Mount Bellenden Ker in North Queensland. Individual populations of *T. graeffei* often exhibit distinct and striking colour forms, including bright red (Mt Kaputar), orange (Mt Bellenden Ker), white with red highlights (Mt Kuringai), and green-grey (Lansdowne Escarpment). Immature stages are almost transparent.

Camaenidae Camaenids range in size from small to very large (5–70 mm) snails, with a diversity of shell forms (Stanisic *et al.* 2010). The genus *Thersites* (Plates 292, 293) includes some of the largest subtropical species. Camaenidae are now likely to be the most diverse and successful family of land snails in Australia (S. Clark 2009; Scott 1996; Stanisic *et al.* 2010). Species are either terrestrial (e.g. *Eurytrachia mucosa*) or partly-arboreal (e.g. *Papuexul bidwilli*), and feed on decaying leaves, fungi, algae and lichens.

Camaenidae are known from South-east Asia, mainland Australia, Central America and the northern tip of South America, and the nature of the Australian fauna suggests a complex history that has been shaped and directed by long-term environmental changes (Hugall and Stanisic 2011; Solem 1997). The family is particularly rich along the mesic chain of forests east of the Great Dividing Range (but also possessing unique species in inland vine thickets [Stanisic 2015]), and though absent from Tasmania they are elsewhere so abundant as to represent the bulk of the mainland terrestrial land snail fauna. However, there is a marked reduction in diversity south of the Hunter Valley in New South Wales (this being a faunal gap for several additional higher taxa), such that in eastern Victoria there is a low diversity of Camaenidae; there, Gondwanan Rhytitidae and Caryodidae are dominant (Hugall and Stanisic 2011).

The Australian Camaenidae are of northern origin, stemming from Miocene immigrants, and are the sister group to the Asian Bradybaenidae (Scott 1996). Although many camaenid taxa are derived from northern Laurasian ancestors via the Malesian South-east Asian archipelago, others represent a divergent endemic element, and it appears that several ancestral lineages were extant before the breakup of the Tertiary mesic forests that had once dominated Gondwanan Australia (Hugall and Stanisic 2011).

Prominent genera are *Austrochloritis*, *Meridolum*, *Pommerhelix* and *Galadistes*, though the latter two genera in particular include numerous species only known from open forest and woodland. Regional rainforest-inhabiting species include *Austrochloritis nambucca*, *A. illawarra*, *A. nundinalis*, *A. paucisetosa*, *A porteri*, *Meridolum marshalli*, *Papuexul bidwilli*, *Posorites consendens*, *P. turneri*, *Ramogenia challengeri*, *Sphaerospira fraseri*, *Thersites mitchellae*, *T. novaehollandiae*, *T. richmondiana* and *Ventopelita leucocheilus* (M. Shea pers. comm.; B. Smith 1992; Stanisic undated; Stanisic *et al.* 2010; Williams 2002). *Thersites mitchellae* (Plate 292) from northern New South Wales is a threatened species (Parkyn *et al.* 2014, 2015). As in other families individual species are sometimes found in rainforest and in dry open woodland or forest (e.g. *Austrochloritis kanangra*, *Meridolum grayi*, *Pommerhelix monacha*, *P. carmelae*, *P. insularis*, *Sphaerospira fraseri*), or in rainforest and wet sclerophyll forest (e.g. *Austrocloritis metuenda*, *Ramogenia challengeri*, *Squamagenia yabba*). Many are associated with vine thickets (e.g. *Calvigenia blackmani*, *Lynfergusonia munduberra*, *Mussonena maxima*, *M. martinlogari*, *Pseudodistes reevesi*, *Trachygenia microlepida*), or vine thickets and woodland (e.g. *Calvigenia cootha*, *Figuladra mattea*, *Galadistes pustulosa*, *Moretonistes mansueta*, *Neveritis misella*, *Trachygenia lanuginosa*, *Ventopelita leucheilus*, *Xanthomelon pachystylum*). Several species have geographically restricted distributions, as with *Pseudodistes reevesi* from the Bunya Mountains, *Meridolum gulosum* and *M. marshalli* from the Illawarra region, and *Thersites darlingtoni* from the Border Ranges (Stanisic *et al.* 2010).

Caryodidae These are endemic to eastern Australia, being distributed largely along the Great Dividing Range south from mid-eastern Queensland to Tasmania (B. Smith 1992; Stanisic *et al.* 2010; Williams 2002). Caryodidae are large, nocturnal snails that feed on decayed plant material and fungi (Stanisic *et al.* 2010), and appear closely related to the family Acavidae from Madagascar, the Seychelles and Sri Lanka (Beesley *et al.* 1998). Individuals are mainly terrestrial, though occasionally arboreal, and generally live under fallen logs, forest debris and rocks. Caryodids inhabit vine thickets, subtropical and warm temperate rainforest, as well as more open vegetation types. Predators include pitta birds (e.g. *Pitta versicolor*). Several are threatened by forestry practices (Beesley *et al.* 1998). Two of Australia's largest land snails, *Hedleyella falconeri* and *H. maconelli* (height to 90 mm and 82 mm respectively [Stanisic *et al.* 2010]), and the monotypic genus *Brazieresta* (*Brazieresta larreyi*), have been described from the region. *Brazieresta larreyi* is known only from moist forests of northern New South Wales. *Hedleyella falconeri* (Plate 294) is restricted to moist forests from southern Queensland to northern New South Wales (Murphy 2002; B. Smith 1992; Stanisic 1990; Stanisic *et al.* 2010; Williams 1993). Adult *Hedleyella falconeri* are generally nocturnal, but they will remain active during the early hours of daylight if heavy mists prevail. *Pygmypanda atomata* (Plate 295) inhabits cool temperate rainforest in the Barrington Tops, warm temperate rainforest in Werrikimbe National Park, and dry rainforest at lower elevations in the Manning Valley. *Pedinogyra rotabilis* is concentrated on the araucarian vine scrubs of the Border Ranges (Stanisic undated), *P. kroombit* is known only from the Kroombit Tops, south-west of Gladstone, and *P. minor* occurs in the Rockhampton area (Stanisic *et al.* 2010).

Charopidae This is a family of small to minute land snails that are very abundant in leaf litter. Charopids are a Gondwanan element of the Australian fauna, and their distribution is centred on Australia, New Zealand, South America, South Africa, and islands of the southern Pacific Ocean (Shea *et al.* 2012; B. Smith 1992; Stanisic 1990). Along with the Rhytididae, the Charopidae are one of only two Gondwanan families shared by Australia and Africa (van Bruggen 1980). The evolutionary development of Gondwanan land snail families in Africa has been relatively poor, and although the charopid faunas are comparable in diversity, the African rhytidid fauna is 'only sparsely represented …' (van Bruggen 1980).

Shell shape of the Australian and New Zealand faunas is more diverse than for taxa occurring in the Pacific Islands, suggesting that the former have undergone a more complex

Saving the Mitchell's Rainforest Snail, *Thersites mitchellae*

Found in the far north-east of New South Wales is the large native land snail *Thersites mitchellae* (Plate 292), a member of the Camaenidae. The species is listed as endangered under the New South Wales *Biodiversity Conservation Act 2016*. *Thersites mitchellae* grows to ~55 mm wide and 50 mm high and its shell is triangular in profile. It is restricted to lowland subtropical rainforest, itself an endangered ecological community, and swamp sclerophyll (*Melaleuca quinquinervia*/paperbark) forest remnants with a rainforest understorey on alluvial soils. A nocturnal snail, *Thersites mitchellae* has low dispersal abilities and spatially limited daily foraging patterns. Though normally found on the forest floor, the species may also be arboreal. It can persist in narrow strips of rainforest, but key habitat components are dense leaf material on the forest floor and an intact canopy to maintain a necessary soil microclimate. Populations survive in generally small isolated and endangered habitat communities at Stotts Island Nature Reserve (the largest population), Byron Bay, Lennox Head, Banora Point and Suffolk Park. These are located in the Tweed-Richmond region of far northern New South Wales. However, the size and density of individual populations are inadequately known. The snail is a generalist feeder, the food source varying with substrate; otherwise its biology and ecology are poorly known. Fungal material comprises a high proportion of diet, the availability of this food source strongly influencing viability and dispersal of populations. *Thersites mitchellae* is nocturnal and shelters in cryptic microhabitats such as palm fronds and forest detritus and bark during the day. Its movement is influenced by humidity preferences for what has been termed 'sedge retreat', and a selection for palm frond and coarse woody debris in all habitats. But individuals vary in selection of retreat sites.

There are two related regionally occurring species, *Thersites novaehollandiae* (Plate 293) and *T. richmondia*. These are associated with subtropical rainforest, and other rainforest formations, at higher elevations.

Threats to the survival of Mitchell's Rainforest Snail (and many other endemic snails) include ongoing habitat clearing, including underscrubbing, removal of forest floor debris, fragmentation of habitat, trampling, grazing by livestock, weed invasion, loss of habitat corridors, habitat edge effects such as desiccation of microhabitat, loss of vegetation buffers, predation by rats, herbicide use, fire, and potentially snail baits and domestic fowl where *Thersites mitchellae* populations may occur in or near residential developments.

The New South Wales Office of Environment and Heritage has developed a species recovery plan to prevent the extinction of this species. The recovery actions include the scientific survey of areas of high habitat potential as well as sites further inland in an effort to gauge the existence of possible additional populations, studies into the species' ecology, life cycle and genetic variation, monitoring of known sites to assess continuing status, as well as monitoring the impact of habitat restoration works on populations (Andrade *et al.* 2011; Druart *et al.* 2011; NSW NPWS 2000; Parkyn *et al.* 2014, 2015).

evolutionary history (Stanisic 1990). In Australia the family has a predominantly east coast distribution with charopid evolution being closely linked to climatic changes which influenced the nature and extent of mesic communities since the Cretaceous (Stanisic 1990). Post-Cretaceous (<70 mya) transitions into drier vegetation types have been made by relatively few species (Stanisic 1990).

Charopidae are a very species-rich Australian family (Beesley *et al.* 1998; Shea *et al.* 2012; Stanisic 2016; Stanisic *et al.* 2010; Williams 2002; Table 5). In northern and mid-coastal Queensland Charopidae are largely confined to mountain tops, but in subtropical latitudes lowland and submontane rainforests have diverse faunas (Shea *et al.* 2012; Stanisic 1990). The minute size of Charopidae may impose dispersal difficulties and drive speciation in isolated populations.

B. Smith (1992) listed 24 species known from south-eastern New South Wales and south-eastern Queensland, but this has greatly increased (Shea *et al.* 2012; Stanisic *et al.* 2010). Stanisic (1990) recorded 15 genera from the Gondwana rainforests region (Williams 2002), however more than 30 genera are now known from rainforest (including vine thickets) in subtropical eastern Australia, most of which are endemic (Shea *et al.* 2012; Stanisic 2016; Stanisic *et al.* 2010; Table 5). Dominant genera include *Coenocharopa* (e.g. *Coenocharopa alata*, *C. multiradiata*), *Gyrocochlea* (*Gyrocochlea convoluta*, *G. vinitincta*) and *Rhophodon* (*Rhophodon colmani*, *R. duplicostatus*). *Ngairea* includes several species recorded from the Gondwana rainforests region, one of which (*N. corticola*) is restricted to the region. *Nautiliropa* is a monotypic genus (*Nautiliropa omicron*) restricted to south-eastern

Queensland and far northern New South Wales. *Hedleyoconcha* is a relict genus containing one species restricted to the summit of Mt Bellenden-Ker in North Queensland, a second species restricted to Lord Howe Island, and a widespread species (*Hedleyoconcha delta*) found in northern New South Wales and southern Queensland – 'This is the first time a land snail group has been identified as having such an unusual and biogeographically complex pattern' (Stanisic 1990). The endemic genus *Rhophodon* includes *R. consobrinus* (which is restricted to the Gondwana rainforests region) and *R. kempseyensis* (known only from limestone outcrops west of Kempsey), as well as *R. duplicostatus* from the single locality of Glenugie Peak south-east of Grafton, New South Wales (Stanisic *et al.* 2010). The endemic genus *Cralopa* is of interest as it has an east–west distribution, rather than the prevailing north–south distribution exhibited by other charopid genera, and the small amount of conchological variation exhibited by taxa may reflect geologically recent fragmentation of populations (Stanisic 1990). *Cralopa stroudensis* ranges from south-east Queensland to northern New South Wales. The two other species in the genus are restricted to Mt Kaputar (*Cralopa kaputarensis*) in north-western New South Wales, and to the southern to northern tablelands of New South Wales (*Cralopa carlessi*) (Stanisic 1990). *Elsothera* (e.g. *Elsothera nautilodea*, *E. grossa*, *E. brazieri*) is one of the few east coast charopid genera that occur in drier sclerophyll forests as well as rainforest. *Reticularopa minjerribah* is restricted to south-eastern Queensland (North Stradbroke Island, Stradbroke Island) where it inhabits dry sclerophyll forest and rainforest on sand (Stanisic 2016).

Numerous endemic and localised genera have been described by Shea *et al.* (2012) from New South Wales and southern Queensland (see Table 5). These include *Cumberlandia* (e.g. *Cumberlandia impressa*, *C. wilsoni*), *Planorbacochlea* (*Planorbacochlea hawkesburyana*, *P. parriwiensis*, *P. graemei*), *Barringtonia* (*Barringtoni polblue*, *B. montana*), *Comboynea* (*Comboynea winghamensis*, *C. boorgannna*), *Cancellocochlea* (*Cancellocochlea coolongolook*), *Dictyoropa* (*Dictyoropa eurythma*), *Richmondaropa* (*Richmondaropa conjuncta*, *R. parva*) and *Macphersonea* (*Macphersonea canalis*). Most of the included species are associated with various rainforest formations, but several currently only have highly localised distributions, for example *Comboynea boorganna* (Comboyne Plateau), *Comboynea winghamensis* (Wingham Brush Nature Reserve), and *Cumberlandia wilsoni* (Mt Wilson). *Planorbacochlea graemei* is only known from littoral rainforest in the Lake Macquarie district.

Many charopid snails inhabit vine thickets (e.g. *Coenocharopa macromphala*, *C. sordidus*, *C. yessabahensis*, *Cumberlandia impressa impressa*, *Discocharopa aperta*, *Hedleyoconcha delta*, *Letomela contortus*, *Luturopa macleayensis*, *L. rockhamptoniana*, *Macleayropa carraiensis*, *Ngairea dorrigoensis*, *Planorbacochlea parriwiensis*, *P. watagan*, *Pedinogyra rotabilis*, *Rhophodon kempseyensis*, *Richmondaropa parva*, *R. conjuncta*, *Rotacharopa densilamellata*, *Setomela seticostata*), though many of these are more widely encountered in other rainforest formations (Shea *et al.* 2012; Stanisic *et al.* 2010; Williams 2002). *Letomola contortus* is a highly localised species known only from a limestone outcrop west of Kempsey (B. Smith 1992; Stanisic 1990). *Macleayropa boonangi*, *M. carraiensis* and *M. kookaburra* are restricted to the Macleay Valley of New South Wales and adjacent areas (Stanisic *et al.* 2010).

Cystopeltidae This small Gondwanan family (Heatwole 1987) of shell-less snails, more correctly slugs in appearance (Plate 296), is endemic to eastern Australia, with species confined to the region between southern Queensland and Tasmania (Beesley *et al.* 1998). They are herbivores and feed on lichen and microalgae, and are found in a diversity of habits that include rainforest, wet sclerophll forest, upland snow gum forest and tussock grassland.

The relationship of the Cystopeltidae to other families is uncertain (Beesley *et al.* 1998). *Cystopelta* is the only genus in the family. *Cystopelta septentrionalis* is recorded from rainforest and wet sclerophyll forest (e.g. Werrikimbe, New England, Gibraltar and Lamington national parks) (Stanisic *et al.* 2010). *Cystopelta astra* is an open forest-inhabiting species distributed further to the south (e.g. Blue Mountains and Deua national parks). These delicate snails resemble bird droppings and are found during rainy days openly moving about on foliage.

Diplommatinidae This is a family of minute terrestrial snails (shell height < 5 mm). All Australian species are essentially found in rainforest, and are herbivores located in detritus on the forest floor, leaf litter and sometimes moss (Stanisic *et al.* 2010). There are two subfamilies; the Australian species are placed in the Diplommatininae, which occurs in South-east Asia, Australia, Micronesia and Melanesia (Beesley *et al.* 1998). The second subfamily, Cochlostomatinae, is known from Europe. *Velepalaina strangei* inhabits rocky talus and is restricted to the Border Ranges. Most mainland species are otherwise restricted to the Wet Tropics.

Helicarionidae These are small to large (*c.* 30 mm) snails in which the shell only covers a portion of the otherwise exposed animal, or is hidden wholly by the mantle. Because of this helicarionids are sometimes aptly termed 'semi-slugs'. They are herbivores that feed on fungi and organic matter occurring as thin films on the surfaces of leaves and rocks (Stanisic *et al.* 2010). The range of the Helicarionidae is centred upon South-east Asia, Australia and the central and western Pacific (Beesley *et al.* 1998). In Australia the family is restricted largely to montane forests of eastern Australia, including Tasmania (B. Smith 1992; Stanisic *et al.* 2010). The fauna recorded from the Gondwana rainforests region is a diverse one and includes the genera *Coneuplecta*, *Fastosarion*, *Helicarion*, *Melocystis*, *Mysticarion*, *Nitor*, *Peloparion* and *Parmavitrina* have been (B. Smith 1982; Stanisic undated; Stanisic *et al.* 2010; Williams 2002) (Plates 297, 298).

Helicarionids include arboreal (e.g. *Dendronitor inscensa*, *Mysticarion porrectus*, *Sheaia glenrockensis*, *Cucullarion parkini*) and terrestrial species (*Parmavitrina megastoma*, *Macularion aquilla*, *Nitor reisi*) that shelter by day under rotten logs and stones. Most are found inhabiting rainforest and wet sclerophyll forest (e.g. *Helicarion dispositus*, *Nitor medioximus*, *N. pudibunda*, *Sigaloeista orara*), but some occupy rainforest as well as open forest or woodland (*Helicarion mastersi*, *Peloparion iridis*). Individual species may inhabit different rainforest formations through a range of altitudes, as with the large species *Desidarion dispositus*, which occurs in subtropical rainforest at intermediate altitudes (Comboyne Plateau) but extends to altitudes above 1000 m in cool temperate *Nothofagus* forest in the Barrington Tops. A number have successfully developed in drier environments (e.g. *Parmavitrina megastoma*, *Fastosarion freycineti*), vine thickets (e.g. *Delinitesta gayndahensis*, *Burnettia subglobosa*, *Eddiella ingrami*), or where relict moist gullies survive in otherwise eucalypt-dominated woodland and forest (*Brevisentis kaputarensis*, *Levidens ponderi*, *Tarocystis responsiva*).

Helicinidae These are mainly small snails and are herbivores that live on the forest floor, on tree trunks and on leaves (Stanisic *et al.* 2010). Many are found associated with limestone outcrops. Their distribution encompasses northern and eastern Australia, New Guinea, South-east Asia, Polynesia, Micronesia (in part), Central America and South America. In eastern Australia they extend to the Hunter Region of coastal New South Wales. Earliest records for the family are from the Upper Cretaceous of Europe, and possibly the Carboniferous of North America (Beesley *et al.* 1998). Australian species are all placed in the genus *Pleuropoma*, which ranges across northern Australia to New South Wales. *Pleuropoma draytonensis* is an arboreal species occurring in northern New South Wales and southern Queensland, and frequents leaves of low shrubs and trees. *Pleuropoma gladstonensis* is known from Gympie to south of Rockhampton (e.g. Goodnight Scrub, Kroombit Tops), inhabiting vine forest and thickets. The common *Pleuropoma talusata* is found in leaf litter and under stones and fallen timber in dry vine thickets (Stanisic *et al.* 2010).

Punctidae This family is widely distributed, occurring in the Northern Hemisphere, Africa, parts of the Pacific, New Zealand and the subantarctic islands. Diversity is high in the New Zealand fauna. Greatest diversity of the Australian fauna apparently occurs in the south-east of the continent (Beesley *et al.* 1998). Species are minute to small in size (*c.* 1–8 mm). They are thought to be herbivores and detritivores, and inhabit rainforest, vine thickets, grassland, limestone outcrops, coastal dunes and semi-arid environments (Stanisic *et al.* 2010). *Pseudiotula eurysiana* is found in rainforest and vine thickets (e.g. Gibralar Range, Lamington National park), *Iotula microcosmos* inhabits rainforest and dry sclerophyll forest (Macleay Valley, Bulburin State Forest), and *Kaputaresta nandewarenis* is found in damp gullies and remnant rainforest in the Nandewar Ranges west of the New South Wales Northern Tablelands.

Pupillidae Pupillids are minute to small-sized snails that feed on vegetable matter, including detritus. Most species are found in dry areas where there is calcareous bedrock (Stanisic *et al.* 2010); however, a number are to be found in rainforest where they inhabit leaf litter, sheltering rocks, branches and twigs. *Omegapilla australis* inhabits littoral rainforest and rock outcrops, and *Pumilicopta bifurcata*, *Pupisoma evada*, *Cylindrovertilla hedleyi* and *Glyptopupoides egregia* inhabit vine thickets, as well as other rainforest formations.

Pupinidae Pupinids are small to medium-sized, ground-dwelling detritus feeders, found among fallen leaves and rock-strewn forest floors (Stanisic *et al.* 2010). The family is distributed in South-east Asia, islands of the western Pacific, New Zealand and eastern Australia (B. Smith 1992). They are restricted to rainforest formations and occur in the coastal zone from northern Queensland, south to northern New South Wales. Individual species have localised distributions (Beesley *et al.* 1998; Stanisic *et al.* 2010). *Signepupina* is richly represented and includes *S. glenugie*, known only from vine thickets at Glenugie Peak south-east of Grafton (New South Wales), and *S. wilcoxi* from Brunswick Heads (northern New South Wales) to Maryborough in south-eastern Queensland. Additional examples of regional pupinid species include *Hildapina subpolita*, *H. kenilworth* and *Necopupina costata*; however, the greatest diversity is in northern Queensland (Stanisic *et al.* 2010).

Rathouisiidae These are cryptic, slow-moving, large slugs that lack a shell of any kind. They are predators that eat the eggs of other snails, but also feed on fungi and vegetable matter. Rathouisiids live in florest litter and among large broken rocks, and possess knife-like radular teeth that allow them to eat other snails; the teeth enable them to bore through the shells of their prey (Stanisic *et al.* 2010). Two rathouisiid genera are recognised, one of which (*Atopos*) occurs in Australia, and is also recorded from India, Indonesia, the Philippines and Papua New Guinea (Beesley *et al.* 1998). The biogeography of the family is presently unresolved. *Atopos australis* is distributed in rainforest from northern Queensland to north-eastern New South Wales.

Rhytididae This carnivorous Gondwanan family is distributed in Australia, New Guinea, New Zealand, New Caledonia, islands of the western Pacific, Indonesia, the Seychelles and South Africa (van Bruggen 1980; B. Smith 1992). Species usually have long necks and range in shell size from the minute (2 mm) to the large (45 mm) (Stanisic *et al.* 2010). Rhytidids prey on other snails and other soft invertebrates such as earthworms, but may also prey on one another. They are found in moist leaf litter, as well as under rocks and rotten logs. In Australia, rhytidids are diverse and are confined to the eastern and south-eastern regions (except for one species in south-west Western Australia) with the greatest radiation of species occurring in the south-east faunal region. Rhytididae reach their greatest diversity in south-eastern Queensland and north-eastern New South Wales (e.g. *Annabellia* spp., *Austrorhytida* spp., *Echotrida strangeoides*, *Griffithsina brisbanica*, *Montidelos* spp., *Protorugosa alpica*, *Pseudechotrida* spp., *Saladelos urarensis*, *Scagacola* spp., *Strangesta ramsayi*, *S. strangei*, *Terrycarlessia bullacea*, *Vitellidelos* spp.) (J. Stanisic pers. comm.; Stanisic *et al.* 2010). *Austrorhytida barringtoniana* is restricted to the Barrington Tops area where it inhabits rainforest, wet sclerophyll forest and snow gum woodland. *Austrorhytida harriettae* is restricted to the Wauchope–Nambucca and adjacent montane forests, inhabiting vine thickets and living under rotten logs and rocks. *Annabellia bingara* is found in vine thickets and dry woodland from the Liverpool Range, north to tenterfield. *Vitellidelos costata* is a small species found living under rotten logs and leaf litter in rainforest and eucalypt forest from Barrington Tops to west of Kempsey.

Miscellaneus terrestrial families The small-sized snails of the widespread gastropod family Hydrocenidae (e.g. *Georissa* spp.) are herbivores that probably feed on microflora growing on leaf and stone surfaces. Some occur in large aggregations, and are often associated with limestone outcrops (e.g. *Georissa laseroni*, *G. beerwah*). Subulinidae are herbivorous snails that have elongate shells, and reproduce by parthenogenesis or self-fertilisation (Stanisic *et al.* 2010). They inhabit leaf litter on the forest floor and are found in habitats ranging from woodland, vine thickets and woodland. Several exotic subulinids (e.g. *Paropeas achatinaceum*, *Subulina octona*) are now established in eastern Australia; however, a representative native species is *Eremopeas tuckeri* (Iluka Nature Reserve, Rockhampton Caves), which occurs under logs and leaves in littoral rainforest, dry vine thickets and woodland. Succineidae have an almost worldwide distribution, but the distribution of the Australian fauna is sporadic and poorly known (Beesley *et al.* 1998). Megaspiridae are very elongate-shelled snails (possibly herbivorous) that are found among rock scree and in litter in vine thickets and dry rainforest. Species include *Coelocion craigeddei*, *C. circumumbilicata* and *C. australis* (Stanisic *et al.* 2010). Arionidae are slugs and include a single exotic species, *Arion intermedius*, which has invaded wet forest areas in south-eastern Australia (e.g. Mount Wilson west of Sydney, Tasmania). The implications of this are unknown (Stanisic *et al.* 2010). The Euconulidae were earlier placed in the Helicarionidae and include small algae-eating herbivores found hidden in leaf litter in moist rainforest and vine thickets (e.g. *Coeneuplecta calculosa*, *Turrisitala normalisi*). Microcystidae are small to medium-sized, ground-dwelling and arboreal snails that include *Liardetia scandens* found in littoral rainforest (Iluka Nature Reserve), dry vine thickets and moist rainforest from New South Wales to northern Queensland (Cairns). Individuals are found on vines and shrubs. *Wilhelminaia mathildae* occurs in leaf litter, and though having a similar Australian range, it also extends to Indonesia, Papua New Guinea and the western Pacific.

Freshwater fauna

McMichael and Hiscock (1958) divided the continent into fluvifaunula provinces to define the zoogeographical affiliations of the Australian freshwater snail fauna (Ponder *et al.* 1998). Two fluvifaunula zones overlap in northern New South Wales–south-east Queensland:

1. Lessonian – this extends along the southern half of the eastern coastal drainage system (i.e. rivers that flow mainly eastward) from northern New South Wales to western Victoria and northern Tasmania. This group is characterised by species of *Hyridella*.
2. Krefftian – this extends along the east coast of southern Queensland. The Krefftian fluvifaunula is similar to the Lessonian but the genera *Alathyria* and *Cucumerunio* are characteristic (Ponder *et al.* 1998).

Within the freshwater snail fauna there are very few geographical instances of sympatry. An exception to this is the Dorrigo–Rockhampton region where a great diversity of taxa occupy wet rainforests and vine thickets (Ponder *et al.* 1998).

Glacidorbidae Northern New South Wales represents the northern-most known distribution for glacidorbid gastropods on the mainland (Ponder and Avern 2000). The majority of species occur in Tasmania, and of the four Australian genera (*Tasmodorbis, Benthodorbis, Striadorbis, Glacidorbis*) only *Glacidorbis* is found in the world heritage area (Ponder and Avern 2000). The swamp-dwelling *Glacidorbis hedleyi* ranges from Victoria to the Barrington and GloucesterTops, Werrikimbe National Park and Dorrigo. *Glacidorbis isolatus* is restricted to the Barrington Tops and the Northern Tablelands area of New South Wales (Ponder 1986; Ponder and Avern 2000). Several subgenera are recognised within *Glacidorbis*: *Glacidorbis* (*sensu strictu*) is confined to south-east and south-west Australia and Tasmania, while the subgenus *Gondwanorbis* is restricted to Chile (Ponder 1986).

Hydrobiidae Hydrobiid gastropods probably arose in the Jurassic (Ponder 1988), and are a diverse family comprising numerous, often cryptic, species. Although the Australian fauna is species-rich, it comprises few genera. Most species are confined to south-eastern Australia, including Tasmania. High hydrobiid species diversity is a result of low dispersal potential, and this has resulted in isolated and endemic populations that are vulnerable to adverse land use impacts and threatening processes generally (W. Ponder pers. comm.; Ponder 1991; Ponder and de Keyzer 1998). Ninety-three Australian species are considered to be vulnerable or endangered (Baillie and Groombridge 1996; Miller *et al.* 1999; Ponder 1994; Ponder and de Keyzer 1998).

Several species with very limited ranges occur in south-east Queensland and northern New South Wales (e.g. *Austropyrgus bunyaensis, Bithynia richmondiana, Fluvidona dorrigoensis, F. petterdi*) (Miller *et al.* 1999; W. Ponder pers. comm.; B. Smith 1992). *Fluvidona* and *Austropygrus* are recorded from southern Queensland to Tasmania, southern Western Australia, Lord Howe Island, New Zealand and New Caledonia (Ponder and de Keyzer 1998). *Posticobia* occurs in south-east Queensland, northern New South Wales and Norfolk Island (Ponder 1991).

The occupation of restricted ranges does not lend the family to habitat-based conservation strategies. Hydrobiidae may provide insight into 'identification of areas where other freshwater invertebrates with poor dispersal powers have speciated' (Ponder 1994).

Hyriidae Freshwater bivalves, or mussels, in the family Hyriidae are widespread in mainland Australia, and also occur in northern Tasmania, New Guinea, New Zealand, and western and southern South America (B. Smith 1998). They include the genera *Alathyria* (*Alathyria pertexta, A. profuga*) and *Velesunio* (*Velesunio ambiguus*), which are confined to Australia and New Guinea (McMichael and Hiscock 1958). *Hyridella* occurs in coastal south-east Australia, northern Tasmania, New Zealand, and south-west New Guinea (B. Smith 1998). Species are hosts to temnocephalid flatworms (Sewell *et al.* 2006).

McMichael and Hiscock (1958) suggest that, in general, the Australasian freshwater mussel fauna arose from a single ancestral stock, and consequently has differentiated into several higher taxa within the region. This fauna is generally distinct from that of the neighbouring Indo-Malayan region.

Appendix 1. Additional ecosystem values – the stygofauna

Most of the eastern Australian stygofauna (fauna inhabiting groundwater environments) is largely concentrated in the upland karst systems of New South Wales and Tasmania, and represents a major element of continental subterranean biodiversity. Some eastern Australian karst systems have acted as refugia for stygofauna, and further investigation of the stygal fauna may contribute significantly to our understanding of the evolutionary history of the Australian continent and individual karst areas. However, many prospective areas have yet to be studied, and most fauna collections are incidental to other surveys (Thurgate *et al.* 2001). The karsts of eastern Australia are not uniformly distributed, being a series of scattered outcrops, and have 'limited surface expression' (areal extent) (Thurgate *et al.* 2001). The 'impounded karsts of New South Wales and Tasmania are complex systems, and many of them have experienced multiple phases of karstification interspersed with periods of marine transgression, burial, infilling and exhumation' (see discussion in Thurgate *et al.* 2001). Karsts of the New South Wales north coast are considered to be impounded formations – mostly relictual and dry – throughout the landscape. Few caves reach groundwater, but some have small stream systems. Representative sites on the New South Wales north coast are Stockyard Creek, Timor, Pigna Barney and Willi Willi (Thurgate *et al.* 2001).

Historically, less research has been paid to the stygofauna of eastern Australia, yet the southern section of the the Gondwana Rainforests of Australia World Heritage Area contains a significant concentration of stygofauna sites (Thurgate *et al.* 2001). Much of the fauna has yet to be determined but there are many highly localised and endemic taxa (e.g. Neoniphargidae – Amphipoda, Anaspidacea – Syncarida) (Table 6). A considerable proportion of the eastern Australian stygofauna exhibit Gondwanan and Pangaea affinities (Thurgate *et al.* 2001; see Table 7); particularly so among the 'higher' crustaceans.

There are ~238 stygofauna taxa present in Australia. The fauna includes species with a range of adaptive dependencies on subterranean habitats:

1. *Stygobites*: obligate and specialised taxa (82 spp.)
2. *Stygophiles*: facultative taxa that seek out and utilise groundwater environments (34 spp.)
3. *Stygoxenes*: epigean taxa that occur accidentally and that exhibit no special adaptation or dependence upon groundwater (Thurgate *et al.* 2001).

Over one-third of the eastern Australian stygofauna are obligate groundwater animals (Thurgate *et al.* 2001). The stygofauna of New South Wales and Tasmania are distinct at the generic and species level, but New South Wales contains fewer taxa (83–98). However, the proportion of obligate and specialised stygobites, relative to included less specialised stigophiles and stygoxenes, is higher.

The stygofauna of Queensland is poorly known, but, in eastern Australia generally, most stygobites are crustaceans. Relative to the composition of the world fauna, crustaceans dominate the Australian taxa, and include polyphyletic taxa and taxa with relictual distribution patterns. These groups are valuable for biogeographic and evolutionary research (Thurgate *et al.* 2001). Gastropod snails (dominated by Hydrobiidae) are an additional important group within the faunal diversity. Insects are only a minor component in individual karst areas. Amphipods and Hydrobiidae dominate the stygophiles, and most insects and Decapoda are stygoxenes. There is a high proportion of stygobites among some families (e.g. the Paramelitidae, Neoniphargidae and Psammaspididae [see Table 6]), which in addition to the high level of local endemism 'suggests strong selection pressures or long periods of isolation in subterranean waters for this fauna' (Thurgate *et al.* 2001). Abundance of stygofauna may relate to available food resources and caves that contain root mats (a substantial and reliable food source) have recorded higher abundances of fauna.

Table 6. Composition of major stygofauna groups in Eastern Australia (after Thurgate *et al.* 2001).
Figures in brackets indicate undetermined taxa that are probably stygobites or stygophiles. Species that have only been collected from a single karst area or single cave are considered endemic (Thurgate *et al.* 2001).

Taxa	Total no. genera	Total no. species	No. stygobites	No. stygophiles	Total no. endemics
Amphipoda					
Paramelitidae	10	35	21	9 (5)	18
Neoniphargidae	10	23	20	3	21
Other families	5	8	1	3	3
Syncarida					
Psammaspididae	2	13	13	–	11
Koonungidae	4	5	3	(2)	3
Aspididae	1	3	1	2	1
other families	2	3	3	–	3
Decapoda					
Parastacidae	7	17	–	3	6
Other families	2	2	–	1	1
Isopoda					
Phreatoicidae	3	5	3	(2)	1
Janiridae	2	5	4	1	1
Amphisopidae	1	1	1	–	1
Other Crustacea	6	10	2 (5)	(1)	–
Gastropoda					
Hydrobiidae	7	25	6 (2)	9	12
Other families	8	9	–	–	6
Coleoptera					
Dysticidae	2	4	–	2 (1)	–
Other families	2	5	–	–	–
Trichoptera					
Philopotamidae	1	2	–	1	–
Other families	2	2	–	–	–
Other Insecta	11	15	–	–	–
Other taxa	16	38	4 (1)	(2)	10
Totals	**104**	**230**	**82–90**	**34–47**	**98**

Table 7. The origins and present global distribution of major stygal crustacean families collected from Eastern Australia (after Thurgate *et al.* 2001).

Order and family	Affinities	Present distribution
Amphipoda		
Paramelitidae	East Pangaea	Australia
Neoniphargidae	East Gondwana	Australia, South Africa, South America
Ceinidae	East Gondwana	Australia, New Zealand, South Africa
Syncarida		
Psammaspididae	Gondwana	Australia
Koonungidae	Gondwana	Australia
Aspididae	Gondwana	Australia
Stygocaridae	Gondwana	Australia, New Zealand, South America
Parabathynellidae	Gondwana	Cosmopolitan
Isopoda		
Phreatoicidae	East Gondwana	Australia, New Zealand, South Africa, India
Janiridae	Gondwana	Australia, South America
Amphisopidae	East Gondwana	Australia, New Zealand, South Africa, India

Appendix 2. The Gondwana Rainforests of Australia (CERRA) World Heritage Area

Also called the 'Gondwana Rainforests', these World Heritage Area reserves span a series of discontinuous rainforest formations stretching from the Barrington Tops National Park in northern New South Wales to Springbrook, Lamington and Main Range national parks, and Goomburra Flora Reserve in south-eastern Queensland. The New South Wales rainforests (in part) were first nominated for inclusion on the list of World Heritage Areas in 1985 (see Adam 1987), and later expanded to include additional sites in New South Wales and Queensland (DASET 1992; Hunter 2003). Details of the biodiversity and geological values of the world heritage area are detailed in Adam (1987), DASET (1992), Hunter (2003), Kitching *et al.* (2010) and Williams (2002, 2003a).

The Gondwana Rainforests are composed of more than 50 individual reserves totalling over 366 000 ha (307 284 in New South Wales, 59 230 in Queensland [see Table 8]). It is listed as a World Heritage Area for the universal significance of its natural heritage: 'representing outstanding examples of the major stages of Earth's evolutionary history, of ongoing geological and biological processes, and of biological diversity' (Hunter 2003). In addition to rainforest, the Gondwana Rainforests reserves include substantial areas of non-rainforest plant communities, particularly eucalypt-dominated wet sclerophyll (tall open) forest and dry sclerophyll (low open) forest.

The Barrington Tops is the most southern-situated of the Gondwana Rainforests. It includes a landscape of great topographic and altitudinal diversity, peaking at over 1500 m. It includes various subtropical, warm temperate and extensive examples of cool temperate rainforest (dominated by *Nothofagus moorei*) mosaics and, in addition, expansive areas of open eucalypt-dominated sclerophyll forest, woodland, and subalpine wetlands. These are subject to periodic snowfall, and have historically drawn the interest of early Australian invertebrate biologists, notably among them the coleopterists Thomas Sloane and Herbert Carter. The Hastings/Werrikimbe and Oxley Wild Rivers/Macleay group extends north of the Barrington Tops along two sections of the Great Escarpment in a series of disjunct reserves, and range in altitude from 140 to 1260 m (at Mount Banda Banda). Except for the Werrikimbe Plateau all sites are situated in rugged and steeply incised gorges. All four rainforest subformations occur in this group, dry rainforest being mostly restricted to the Oxley Wild Rivers National Park. Tall open forest (wet sclerophyll forest) and open low (dry sclerophyll) forest occurs extensively, with subalpine woodland at higher altitudes of the Werrikimbe Plateau.

The New England group consists of parts of Dorrigo, Cunnawarra and New England National Parks and the outlying (to the north-west) Mount Hyland Nature Reserve. All are located on the eastern edge of the New England Tableland. Though disjunct, the separate world heritage reserves are linked by a series of interconnecting national parks and state forests. The topography ranges from the undulating plateau sections of the New England Tableland to rugged, steep slopes and cliff faces. Altitude extends from less than 100 m and peaks at 1434 (Mount Hyland) and 1560 m (Point Lookout, New England National Park). Subtropical and warm temperate rainforest is well developed in the Dorrigo and New England National Parks. Cool temperate rainforest is a major feature of New England National Park, extending from 1500 m there to ~500 m in Dorrigo National Park. This subformation also occurs at Mount Hyland but at this locality *Nothofagus moorei* is absent, the rainforest being dominated instead by *Doryphora sassafras* and *Cryptocarya foveolata* (Lauraceae). Dry rainforest is an insignificant component, though *Araucaria cunninghamii*, which reaches its southern-most distribution just to the south of New England National Park, occurs as an emergent in some areas. Further north, the Washpool- Gibraltar Range group consists of parts of the adjoining Washpool and Gibraltar Range National Parks, and extends from the New England Tablelands in the west to the base of the Great Escarpment in the east. The group ranges in altitude from ~200 to 1175 m. Subtropical rainforest and dry rainforest have a limited distribution, and *Nothofagus moorei* and *Araucaria cunninghamii* are absent. However, there are

Table 8. Land tenure in the Gondwana Rainforests of Australia (CERRA) World Heritage Area (after Williams 2002a; J. Cavanaugh/NSW NPWS pers. comm.)

New South Wales	
Reserve name	**Approximate area (hectares)**
National parks	
Border Ranges (part)	31 508
Nightcap	4945
Mt Warning	2380
Toonumbar (part)	1080
Tooloom (part)	1640
Richmond Range (part)	870
Washpool (part)	27 715
Gibraltar Range (part)	17 273
New England (part)	29 985
Dorrigo	7885
Oxley Wild Rivers (part)	93 220
Werrikimbe (part)	35 288
Willi Willi (part)	1610
Barrington Tops (part)	39 363
Nature reserves	
Limpinwood	2646
Numinbah	858
Iluka	136
Mount Hyland	1636
Mount Seaview	1703
Flora reserves	
Wilsons Peak	184
Mount Clunie	485
Mount Nothofagus	650
Amaroo	36
Mebbin Lagoons	11
Acacia Plateau	585
Captains Creek	380
Tooloom Scrub	25
Bungdoozle	145
Mallanganee	222
Cunnawarra	270
The Castles	2360
Jerusalem Creek	60
Total	307 284
Queensland	
Reserve name	**Approximate area (hectares)**
National parks	
Springbrook (part)	2480
Lamington	20 569
Mount Chinghee	1257
Mount Barney (part)	10 831
Main Range	17 794

State forests	
Goomburra (part)	2067
Spicers Gap (part)	257
Gilbert (part)	84
Emu Vale (part)	268
Gambubal (part)	2260
Teviot (part)	390
Burnett Creek (part)	1076
Rabbit Board paddock reserves	
R475 (Res. 5740)	22
R470 (Res. 11.135)	40
R603 (Res. 3934)	36
R464 (Res. 11.108)	26
R489 (Res. 929)	18
Reserves for prison purposes	
R932 (Res. 12018)	6
R547 (Res. 2678)	42
Road reserves	–
Various road reserves (not listed here) adjacent to the areas listed above	
Total	59 223

significant areas of warm temperate rainforest, including the world's largest remaining example of *Ceratopetalum apetalum* forest (Willowie Scrub).

Iluka Nature Reserve is the single example of littoral (coastal) rainforest in the Gondwana Rainforests of Australia World Heritage Area. Covering just 136 ha in size, this remnant of a once more expansive rainforest is situated north of the entrance to the Clarence River. It is an example of subtropical rainforest on coastal nutrient-enriched sand, a rainforest type otherwise heavily destroyed through mining for mineral sands, in particular rutile. Iluka Nature Reserve protects the largest surving stand of littoral rainforest in New South Wales. This variant of subtropical rainforest occupies a wind-swept and salt-impacted environment that is marginal for the development of rainforest (Adam 1987, 1992; Adam *et al.* 1989). There are two major types of floristic communities: a low seaward community dominated by *Cupaniopsis anacardioides* and *Banksia integrifolia* (Proteaceae), and a taller stratified forest reaching to 30 m in height, dominated by *Syzygium hemilamprum* and *S. luehmannii* (Myrtaceae) in the upper canopy layer (Grantley 2010). Despite its small size Iluka Nature Reserve has high biodiversity significance with several new and unusual invertebrates discovered there (especially through the field studies and research of David McAlpine of the Australian Museum).

Straddling the Queensland–New South Wales border is the Tweed Caldera group, this being the remnants and erosion caldera of the Mount Warning/Tweed shield volcano (active ~23–30 mya). The Tweed Caldera group is composed mainly of Mount Chinghee, Lamington and Springbrook national parks and Numinbah Nature Reserve in Queensland, and Limpinwood Nature Reserve and the Border Ranges (eastern part), Mount Warning, Wollumbin, Mebbin and Nightcap national parks in New South Wales. This group is arguably the central feature of the Gondwana Rainforests of Australia, is a major refugium for Gondwanan ancestral rainforest, and preserves outstanding complex examples of Australia's subtropical rainforest (McDonald 2010; Rossetto *et al.* 2015). It possesses high biodiversity values, with many, endemic, evolutionarily primitive and relictual plants and animals. The altitudinal range extends from 80 to 1150 m, and at the time of European occupation this region preserved one of the most extensive areas of rainforest in Australia. Two rainforest subformations dominate: subtropical rainforest of varying floristic and structural complexity dependent upon altitude, soil fertility and aspect, and dry rainforest with an emergent *Araucaria cunninghamii* overstorey. There are some relatively small stands of cool temperate rainforest in which *Nothofagus moorei* reaches its northern-most distribution (in the Border Ranges).

Warm temperate rainforest is also present, found at sites of lower fertility and where exposure to winds does not allow the establishment of more complex rainforest (McDonald 2010). There are expansive areas of intergrading rainforest and eucalypt-dominated wet sclerophyll forest, as well as woodland and shrub complexes. Stands of *Lophostemon confertus* and *Eucalyptus grandis* (Myrtaceae) are widespread in wet sclerophyll forest, and there are low shrub-dominated communities on exposed rocky knolls and along cliff tops.

The Focal Peak group includes a series of world heritage-listed rainforests interconnected by national parks and state forests. As with the Tweed Caldera group the Focal Peak rainforests include those that straddle the Queensland–New South Wales border as well as isolated sections to the south in Tooloom National Park, Toonumbar National Park and Richmond Range National Park. The eastern extremity joins with the Tweed Caldera group. The rugged landforms and caldera valley are associated with the Focal Peak shield volcano, these ranging in height from 140 to 1359 m on the west peak of Mount Barney. The activity of the Focal Peak volcano, beginning ~23 million years ago, pre-dated that of the larger Mount Warning volcano to the east. Subtropical and dry rainforests are extensively represented, and large stands of subtropical rainforest occur at Koreelah and Tooloom national parks, and in the Richmond Ranges. Dry rainforest, often with emergent *Araucaria cunninghamii*, occurs at sites of lower rainfall than subtropical rainforest, with the two types frequently intergrading. Cool temperate rainforest occurs at Mount Barney and Mount Nothofagus, and includes the largest stand in the Border Ranges region (153 ha).There are small stands of warm temperate rainforest and here this subformation is near its northern-most distribution. *Ceratopetalum apetalum* is absent, though present in the Tweed Caldera group, and tends to be replaced by *Schizomeria ovata*. The largest area of warm temperate rainforest is on the Mount Glennie Plateau (Hunter 2010).

The most northerly of the Gondwanan Rainforest reserves are those of the Main Range. This group extends from Koreelah National Park in New South Wales in the south, and then in Queensland north-west in the Main Range National Park, past Cunninghams Gap, to Mount Mistake. Altitude ranges from 320 to 1375 m on the summit of Mount Superbus. It is an area of spectacular escarpments and erosion resistant volcanic plugs. Subtropical rainforest is the most extensive of the rainforest types represented, in which trees such as *Argyrodendron actinophyllum* (Sterculiaceae/Malvaceae), *Dysoxylum fraserianum* (Meliaceae) and *Sloanea woollsii* (Elaeocarpaceae) are common at different altitudes. In contrast warm temperate rainforest is not well represented. Dry rainforest occurs as scattered stands dominated by *Araucaria cunninghamii* at exposed sites on the Main Range and rocky outcrops of the Acacia Plateau (Duffy Masters 2010).

Appendix 3. Major occurrences of rainforest subforms in subtropical eastern Queensland and New South Wales

Queensland	New South Wales
Subtropical rainforest	
Kroombit Tops	Tweed and Nightcap Ranges
Miriam Vale area (Bulburin and Polmaily)	Basaltic lowlands of Richmond River area
Kilkivan	Dorrigo Plateau, Bellinger Valley
Fraser Island and Cooloola	Comboyne, Hastings and Manning Rivers
Conondale Range	Barrington and Gloucester areas
Upper Bunya Mountains	Illawarra coastal plain and escarpment
Mt Glorious area	
Main Range	
McPherson Range	
Littoral rainforest (shares floristic attributes with subtropical and dry rainforest)	
Tannum Sands, Boyne Island and Eurimbula	Brunswick Heads, Broken Head
Burleigh Heads and Cudgen-Kingscliff area	Bundagen and Iluka
	Sea Acres, Racecourse Head, Point Plomer
	Harrington, Manning Point, Red Head and Blackhead
	Myall Lakes and Seal Rocks
	Jervis Bay, Shellharbour
	Bunya Head and Tuross Head
Dry rainforest	
Mt Larcom and Mt Morgan	Richmond Range
Boyne Valley	Upper Clarence Valley
Kilkivan, Biggenden, Goodnight Scrub, Cania Gorge	Guy Fawkes River National Park
Imbil-Kenilworth area	Macleay River gorge system
Benarkin-Yarraman-Nanango area	Hastings and Manning Valleys
Lower Bunya Mountains	Gloucester area
Mt French area/Boonah	Bega and Shoalhaven Valleys
Vine thickets	
Bajool–Rundle Range	Hills near Gunnedah, Bingara, Yallaroi and Yetman
Monto and Biloela (western Kroombit Tops)	Glenbawn Dam and Upper Hunter Valley

Expedition range, Taroom and isla Gorge	Lower Shoalhaven River gorge
Carnarvon Range	
Western Main Range	
Warm temperate rainforest	
Kroombit Tops	Gibraltar Range–Washpool
McPherson Range	Dorrigo Plateau
	Werrikimbe-Upper Hastings
	Upper Manning Valley
	Gloucester and Barrington Tops, Mt Royal Range
	Liverpool Range
	Blue Mountains
	Robertson Plateau
	Clyde River
Cool temperate rainforest	
McPherson Range	McPherson–Border Ranges
	New England National Park
	Dorrigo Plateau
	Werrikimbe–Mount Boss
	Gloucester and Barrington Tops
	Monga area
	Clyde Mountain and Brown Mountain
	Nadgee and Mt Dromedary
	Clyde River

Note: Adapted from Harden *et al.* 2006, Williams 1993.

Appendix 4. Higher divisions of Coleoptera, Diptera, Hemiptera, Hymenoptera (Insecta) and Mollusca*

This appendix groups families that are alphabetically listed or discussed in the text, and therefore is not an exhaustive treatment of all families occurring in subtropical Australia. For the insect groups the classification generally follows that in *The Insects of Australia* (Naumann 1991a), but placements in that monumental work have increasingly been subject to significant revision (e.g. in Hymenoptera [e.g. in Goulet and Huber 1993]) and additional families (e.g. Ulurumyiidae) and phylogenetic relationships (e.g. that of Tiphiidae/Thynnidae) subsequently recognised or reappraised. That of Coleoptera generally follows Lawrence and Ślipiński (2013). Exceptions are noted in the text. Hemiptera were previously divided into Homoptera and Heteroptera, but Homoptera are now recognised as the Sternorrhyncha and Auchenorrhyncha, with Heteroptera retaining its status. Historically, flies have been classified into two main suborders, the Nematocera and Brachycera. Nematocera are now considered a paraphyletic grouping but are retained here for convenience. Listed families can be further divided (e.g. Brachyceran flies into Orthorrhapha, Cyclorrhapha, Aschiza, Schiophora, and superfamilies etc.). Mollusca follows Beesley *et al.* (1998a) and Stanisic *et al.* (2010).

Insecta
Order Coleoptera

Suborder Archostemata
Cupedidae, Ommatidae

Suborder Myxophaga

Not mentioned

Suborder Adephaga
Carabidae, Dysticidae, Gyrinidae, Hygrobiidae, Noteridae

Suborder Polyphaga
Anthicidae, Anthribidae, Apionidae, Archeocrypticidae, Attelabidae, Belidae, Boganiidae, Boridae, Bothrideridae, Brentidae, Buprestidae, Byrrhidae, Callirhipidae, Cantharidae, Cerambycidae, Chrysomelidae, Clambidae, Cleridae, Coccinellidae, Corylophidae, Cryptophagidae, Cucujidae, Curculionidae, Dermestidae, Elateridae, Elmidae, Endomycidae, Erotylidae, Eucnemidae, Eurhynchidae, Geotrupidae, Histeridae, Hobartiidae, Hybosoridae, Hydrophilidae, Lamingtoniidae, Lampyridae, Latridiidae, Leiodidae, Lucanidae, Lycidae, Lymexylidae, Megalopodidae, Melandryidae, Meloidae, Melyridae, Mordellidae, Nemonychidae, Nitidulidae, Oedemeridae, Passalidae, Passandridae, Phalacridae, Phloeostichidae, Psephenidae, Ptiliidae, Ptilodactylidae, Ptinidae, Pyrochroidae, Rhinorhipidae, Rhynchitidae, Ripiphoridae, Salpingidae, Scarabaeidae, Scirtidae, Scraptiidae, Silphidae, Staphylinidae, Tenebrionidae, Thanerocleridae, Trogidae, Trogossitidae, Throscidae, Ulodidae, Zopheridae

Order Diptera
Suborder Nematocera
Bibionidae, Blephariceridae, Cecidomyiidae, Ceratopogonidae, Chaoboridae, Chironomidae, Corethrellidae, Culicidae, Keroplatidae, Lygistorrhinidae, Mycetophilidae, Perissommatidae, Psychodidae, Scatopsidae, Sciariidae, Simuliidae, Tanyderidae, Thaumaleidae, Tipuloidea/Tipulidae

Suborder Brachycera
Acroceridae, Asilidae, Athericidae, Axiniidae, Bombyliidae, Calliphoridae, Chloropidae, Clusiidae, Conopidae, Dolichopodidae, Drosophilidae, Empididae, Ephydridae, Exeretonevridae, Heleosciomyzidae, Heteromyzidae,

* discussed in the text

Hippoboscidae, Hybotidae, Lauxaniidae, Micropezidae, Muscidae, Mydidae, Nemestrinidae, Neminidae, Neriidae, Neurochaetidae, Nycteribiidae, Odiniidae, Pelecorhynchidae, Phoridae, Platypezidae, Platystomatidae, Pyrgotidae, Rhagionidae, Sarcophagidae, Sepsidae, Stratiomyidae, Streblidae, Syrphidae, Tabanidae, Tachinidae, Tephritidae, Therevidae, Ulurumyiidae

Order Hemiptera

Suborder Sternorrhyncha

Homotomidae, Pseudococcidae, Psyllidae, Triozidae

Suborder Auchenorrhyncha

Achilidae, Aphrophoridae, Cercopidae, Cicadellidae, Cicadidae, Cixiidae, Derbidae, Eurymelidae, Flatidae, Fulgoridae, Membracidae, Myerslopiidae

Suborder Heteroptera

Alydidae, Anthocoridae, Aradidae, Coreidae, Gelastocoridae, Idiostolidae, Largidae, Lygaeidae, Miridae, Ochteridae, Peloridiidae, Pentatomidae, Reduviidae, Rhopalidae, Rhyparochromidae, Schizopteridae, Scutelleridae, Tessaratomidae, Thaumastocoridae, Tingidae, Veliidae

Order Hymenoptera

Suborder Symphyta

Pergidae, Tenthredinidae

Suborder Apocrita (Aculeata) – Apiformes

Apidae, Colletidae, Halictidae, Megachilidae

Suborder Apocrita (Aculeata) – Spheciformes

Chrysididae, Crabronidae, Formicidae, Pompilidae, Scoliidae, Sphecidae, Thynnidae, Vespidae

Suborder Apocrita (Parasitica)

Agaonidae, Aulacidae, Braconidae, Chalcididae, Diapriidae, Eucharitidae, Eulophidae, Eupelmidae, Eurytomidae, Evaniidae, Figitidae, Gasteruptiidae, Ichneumonidae, Megalyridae, Monomachidae, Platygastridae, Proctotrupidae, Pteromalidae

Mollusca
Class Gastropoda

Superorder Neritopsina/superfamily Neritoidea

Helicinidae, Hydrocenidae

Order Architaenioglossa

Diplommatinidae, Pupinidae

Order Basommatophora

Glacidorbidae

Order Eupulmonata

Achatinellidae, Arionidae, Athoracophoridae, Camaenidae, Caryodidae, Charopidae, Cystopeltidae, Euconulidae, Helicarionidae, Megaspiridae, Microcystidae, Punctidae, Pupillidae, Rhytididae, Subulinidae, Succineidae

Order Sorbeoconcha

Hydrobiidae

Order Systellommatophora

Rathouisiidae

Class Bivalvia

Hyriidae

Appendix 5. Divisions of geological time

(Adapted from Hansen 1991 and Barlow 1981; * = elapsed time in millions of years (my) since beginning of the period or epoch)

Era	Period	Epoch	Elapsed time (my)*
Cenozoic	Quaternary	Holocene	11 400 years
		Pleistocene	1.8
	Tertiary	Pliocene	5.3
		Miocene	23.3
		Oligocene	33.9
		Eocene	55.8
		Palaeocene	65.5
Mesozoic	Cretaceous	Late/Upper	99.6
		Early/Lower	145.5
	Jurassic	Late/Upper	161.2
		Middle	175.6
		Early/Lower	199.6
	Triassic	Late/Upper	228
		Middle	245
		Early/Lower	251
Palaeozoic	Permian		299
	Carboniferous	Late/Upper	318
		Early/Lower	359.2
	Devonian		416
	Silurian		443.7
	Ordovician		488
	Cambrian		542

References

Abdullah M (1966) The taxonomic position of the Australian *Anoplopus tuberculatus*, with a proposed new subfamily (Anoplopinae) of the Tenebrionidae, including remarks on the family status of the Merycidae (Coleoptera). *Entomological News* **77**, 143–147.

Abe T, Matsumoto T (1979) Studies in the distribution and ecological role of termites in a lowland rainforest of West Malaysia. 3. Distribution and abundance of termites in Pasoh Forest Reserve. *Japanese Journal of Ecology* **29**, 337–351.

Ackery PR (1987) The danaid genus *Tellervo* (Lepidoptera, Nymphalidae) – a cladistic approach. *Zoological Journal of the Linnean Society* **33**, 203–274. doi:10.1111/j.1096-3642.1987.tb01564.x

Adam P (1987) *New South Wales Rainforests: The Nomination for the World Heritage List.* NSW National Parks and Wildlife Service, Sydney.

Adam P (1992) *Australian Rainforests. Oxford Monographs on Biogeography, Number 6*. Clarendon Press, Oxford.

Adam P (2017) The world heritage list and New South Wales rainforest – reflections on the events of 30 years ago. *The Australian Zoologist* **39**, 228–256. doi:10.7882/AZ.2017.014

Adam P, Stricker P, Wiecek BM, Anderson DJ (1989) The vegetation of seacliffs and headlands in New South Wales, Australia. *Australian Journal of Ecology* **14**, 515–545. doi:10.1111/j.1442-9993.1989.tb01459.x

Adlem LT, Timms BV (2000) Biogeography of the freshwater Pericarida (Crustacea) from Barrington Tops, NSW. *Proceedings of the Linnean Society of New South Wales* **122**, 131–141.

Agnarsson I, Rayor LS (2013) A molecular phylogeny of the Australian huntsman spider (Sparassidae, Deleninae): implications for taxonomy and social behaviour. *Molecular Phylogenetics and Evolution* **69**, 895–905. doi:10.1016/j.ympev.2013.06.015

Aguiar NO, Bührnheim PF (1998) Phoretic pseudoscorpions associated with flying insects in Brazilian Amazônia. *The Journal of Arachnology* **26**, 452–459.

Ahyong ST, Alonso-Zarazaga MA (2017) Anaspidesidae, a new family for syncarid Crustaceans formerly placed in Anaspididae Thomson, 1893. *Records of the Australian Museum* **69**, 257–258. doi:10.3853/j.2201-4349.69.2017.1680

Aide TM, Londoño EC (1989) The effects of rapid leaf expansion of the growth and survivorship of a lepidopteran herbivore. *Oikos* **55**, 66–70. doi:10.2307/3565873

Alcock J (1980) Communal nesting in an Australian solitary wasp, *Cerceris antipodes* (Hymenoptera, Sphecidae). *Journal of the Australian Entomological Society* **19**, 223–228. doi:10.1111/j.1440-6055.1980.tb02093.x

Alexander CP (1928) The Australasian species of the genus *Nemopalpus* (Psychodidae: Diptera). *Proceedings of the Linnean Society of New South Wales* **53**, 291–294.

Ali JR, Aitchison JC (2008) Gondwana to Asia: Plate tectonics, paleogeography and the biological connectivity of the Indian sub-continent from the Middle Jurassic through latest Eocene (166–35 Ma). *Earth-Science Reviews* **88**, 145–166. doi:10.1016/j.earscirev.2008.01.007

Allsopp PG (1984) Checklist of the Hybosorinae (Coleoptera: Scarabaeidae). *Coleopterists Bulletin* **38**, 105–117.

Allsopp PG (1989) *Schizognathus apricagger* sp. nov. (Coleoptera: Scarabaeidae: Rutelinae) from south-east Queensland and new locality records for other *Schizognathus* spp. *Journal of the Australian Entomological Society* **28**, 187–190. doi:10.1111/j.1440-6055.1989.tb00878.x

Allsopp PG (1995) Biogeographhy of Australian Dynastinae, Rutelinae, Scarabaeinae, Melolonthini, Scitalini and

Geotrupidae (Coleoptera: Scarabaeoidea). *Journal of Biogeography* **22**, 31–48. doi:10.2307/2846071

Allsopp PG (2000) Revision of the Australian genus *Anomalomorpha* Arrow (Coleoptera: Scarabaeidae: Dynastinae) with a new species from the wet tropics of Queensland. *Memoirs of the Queensland Museum* **46**, 1–7.

Aluja M, Allen L (2001) *Fruit Flies (Tephritidae): Phylogeny and Evolution of Behavior.* CRC Press, Boca Raton.

Álvarez-Padilla F, Hormiga G (2011) Morphology and phylogenetic atlas of the orb-weaving spider family Tetragnathidae (Araneae: Araneoidea). *Zoological Journal of the Linnean Society* **162**, 713–879. doi:10.1111/j.1096-3642.2011.00692.x

Amhad I, Kamaluddin S (1989) A revision of the Australian genus *Diemenia* Spinola (Hemiptera: Pentatomidae: Pentatominae). *Records of the South Australian Museum* **23**, 21–31.

Amorin D de S (2009) Scatopsidae (minute black scavenger flies). In *Manual of Central American Diptera. Vol. 1.* (Eds BV Brown, A Borkent, JM Cumming, DM Wood, NE Woodley and MA Zumbado) pp. 347–355. NRC Research Press, Ottawa.

Andersen AN (1991) *The Ants of Southern Australia: A Guide to the Bassian Fauna.* CSIRO Publishing, East Melbourne.

Andersen AN, Majer JD (2000) The Australian rain forest ant fauna: A biogeographic overview. In *Sampling Ground-dwelling Ants: Case Studies for the World's Rain Forests.* (Eds D Agosti, J Majer, L Alonso and T Schultz, T) pp. 51–58. Bulletin No. 18. School of Environmental Biology, Curtin University of Technology, Perth.

Andersen NM, Weir TA (2001) New genera of Veliidae (Hemiptera: Heteroptera) from Australia, with notes on the generic classification of the subfamily Microveliinae. *Invertebrate Taxonomy* **15**, 217–258. doi:10.1071/IT00015

Anderson JME (1976) Aquatic Hydrophilidae (Coleoptera), the biology of some Australian species with descriptions of immature stages reared in the laboratory. *Australian Journal of Entomology* **15**, 219–228. doi:10.1111/j.1440-6055.1976.tb01696.x

Anderson RS (1993) Weevils and plants: phylogenetic versus ecological mediation of evolution of host plant associations in Curculioninae (Coleoptera: Curculionidae). *Memoirs of the Entomological Society of Canada* **125**, 197–232. doi:10.4039/entm125165197-1

Anderson RS (1995) An evolutionary perspective on diversity in Curculionoidea. *Memoirs of the Entomological Society of Washington* **14**, 103–114.

Anderson RS, Oberprieler RG, Setliff GF (2018) A review of the *Araucaria*-associated weevils of the tribe Orthorhinini (Coleoptera: Curculionidae: Molytinae), with descriptions of new species of *Ilacuris* Pascoe, 1865 and *Notopissodes* Zimmerman and Oberprieler, 2014 and a new genus, *Kuschelorhinus* Anderson and Setliff. *Diversity (Basel)* **10**, 1–25. doi:10.3390/d10030054

Andrade L, Klootwijk A, Parkyn J, Specht A (2011) Initial observations of a population of Mitchell's Rainforest Snail *Thersites mitchellae. Australian Zoologist* **35**, 590–598. doi:10.7882/AZ.2011.011

Andrew NR, Hughes L (2005) Diversity and assemblage structure of phytophagous Hemiptera along a latitudinal gradient: predicting the potential impacts of climate change. *Global Ecology and Biogeography* **14**, 249–262. doi:10.1111/j.1466-822x.2005.00149.x

Anon (2016) *The Australian Handbook for the Identification of Fruit Flies. Version 2.1.* Plant Health Australia, Deakin.

Ansorge J (2002) Revision of the 'Trichoptera' described by Geinitz and Handlirsch from the Lower Toarcian of Dobbertin (Germany) based on new material. *Nova Supplementum Entomologia, Keltern* **15**, 55–74.

Antropov AV (2000) Digger wasps (Hymenoptera, Sphecidae) in Burmese amber. *Bulletin of the Natural History Museum London (Geology)* **56**, 59–77.

Archer M, Godthelp H, Hand SJ, Megirian D (1989) Fossil mammals of Riversleigh, north-western Queensland: preliminary overview of biostratigraphy, correlation and environmental change. *Australian Zoologist* **25**, 29–66. doi:10.7882/AZ.1989.001

Armstrong JA (1979) Biotic pollination mechanisms in the Australian flora: a review. *New Zealand Journal of Botany* **17**, 467–508. doi:10.1080/0028825X.1979.10432565

Armstrong JE, Irvine AK (1990) Functions of staminodia in the beetle-pollinated flowers of *Eupomatia laurina. Biotropica* **22**, 429–431.

Armstrong JWT (1942) On Australian Dermestidae. Part II: the genus *Trogoderma. Proceedings of the Linnean Society of New South Wales* **67**, 321–330.

Armstrong JWT (1949) On Australian Dermestidae. Part V: notes and the description of four new species. *Proceedings of the Linnean Society of New South Wales* **74**, 107–111.

Aronne G, Giovanetti M, Guarracino MR, Micco V (2012) Foraging rules of flower selection applied by colonies of *Apis mellifera*: ranking and associations of floral resources. *Functional Ecology* **26**, 1186–1196. doi:10.1111/j.1365-2435.2012.02017.x

Ashburner M (1981) Entomophagous and other bizarre Drosophilidae. In *The Genetics and Biology of Drosophila. Vol. 3a.* (Eds M Ashburner, HL Carson and JN Thompson) pp. 395–429. Academic Press, London.

Ashford OS, Foster WA, Turner BL, Sayer EJ, Sutcliffe L, Tanner EVJ (2013) Litter manipulation and the soil arthropod community in lowland tropical rainforest. *Soil Biology & Biochemistry* **62**, 5–12. doi:10.1016/j.soilbio.2013.03.001

Ashlock PD (1967) A generic classification of the Orsillinae of the world (Hemiptera–Heteroptera: Lygaeidae). *University of California Publications in Entomology* **48**, 1–82.

Ashton DH (1979) Seed harvesting by ants in forests of *Eucalyptus regnans* F. Muell. in central Victoria: Effect on natural regeneration. *Australian Journal of Ecology* **4**, 265–277. doi:10.1111/j.1442-9993.1979.tb01218.x

Ashton LA, Nakamura A, Basset Y, Burwell CJ, Cao M, Eastwood R, Odell E, Gama de Oliveira E, Hurley K, *et al.* (2016) Vertical stratification of moths across elevation and latitude. *Journal of Biogeography* **43**, 59–69. doi:10.1111/jbi.12616

Atkins A (1975) The life history of *Anisynta tillyardi* Waterhouse and Lyell (Lepidoptera: Hesperiidae: Trapezitinae). *Australian Entomological Magazine* **2**, 72–75.

Atkins A (1984) Notes and records for some butterflies from eastern Australia. *Victorian Entomologist* **14**, 26–29.

Atkins A (1987) The life history of *Trapezites iacchoides* Waterhouse and *Trapezites phigalioides* Waterhouse (Lepidoptera: Hesperiidae: Trapezitinae). *Australian Entomological Magazine* **13**, 53–58.

Atkins A (1997) Two new species of *Trapezites* Hubner (Lepidoptera: Hesperiidae: Trapezitinae) from eastern Australia. *Australian Entomologist* **24**, 7–26.

Atkins A, Mayo R, Moore M (1991) The life history of *Signeta tymbophora* (Meyrick and Lower) (Lepidoptera: Hesperidae: Trapezitinae). *Australian Entomological Magazine* **18**, 87–90.

Atkinson WD (1985) Coexistance of Australian rainforest Diptera breeding in fallen fruit. *Journal of Animal Ecology* **54**, 507–518. doi:10.2307/4495

Atyeo WT (1992) The pterolichoid feather mites (Acarina, Astigmata) of the Megapodiidae (Aves, Galliformes). *Zoologica Scripta* **21**, 265–305. doi:10.1111/j.1463-6409.1992.tb00331.x

Austin AD, Dangerfield PC (1992) Synopsis of Australasian Microgastrinae (Hymenoptera: Braconidae), with a key to genera and descriptions of new taxa. *Invertebrate Taxonomy* **6**, 1–76. doi:10.1071/IT9920001

Austin AD, Gibson GAP, Harvey MS (1998) Synopsis of Australian *Calymmochilus* Masi (Hymenoptera: Eupelmidae), description of a new Western Australian species associated with a pseudoscorpion, and a review of pseudoscorpion parasites. *Journal of Natural History* **32**, 329–350. doi:10.1080/00222939800770171

Austin AD, Yeates DK, Cassis G, Fletcher WJ, La Salle J, Lawrence JF, McQuillan PB, Mound LA, Bickel DJ, Gullan PJ, Hales FF, Taylor GS (2004) Insects 'Down Under' – Diversity, endemism and evolution of the Australian insect fauna: examples from selected orders. *Australian Journal of Entomology* **43**, 216–234. doi:10.1111/j.1326-6756.2004.00448.x

Bacchus ME (1974) A revision of the Australian species of the genus *Glycyphana* Burmeister (Coleoptera: Scarabaeidae). *Journal of the Australian Entomological Society* **13**, 111–128. doi:10.1111/j.1440-6055.1974.tb02162.x

Baehr B (2003) Revision of the Australian spider genus *Habronestes* (Araneae: Zodariidae). Species of New South Wales and the Australian Capital Territory. *Records of the Australian Museum* **55**, 343–376. doi:10.3853/j.0067-1975.55.2003.1389

Baehr B, Baehr M (1987) The Australian Hersiliidae (Arachnida: Araneae): taxonomy, phylogeny, zoogeography. *Invertebrate Taxonomy* **1**, 351–437. doi:10.1071/IT9870351

Baehr B, Jocqué R (1994) Phylogeny and zoogeography of the Australian genus *Storena*. *Spixiana* **17**, 1–12.

Baehr B, Jocqué R (2017) The new endemic Australian genus *Nosterella* and a review of *Nostera* (Araneae: Zodariidae), including eight new species. *Memoirs of the Queensland Museum – Nature* **60**, 53–76. doi:10.17082/j.2204-1478.60.2016.2016-14

Baehr M (1987) The Australian species of the carabid genus *Perileptus* (Coleoptera: Carabidae: Trechinae). *Invertebrate Taxonomy* **1**, 1–16. doi:10.1071/IT9870001

Baehr M (1989) Revision of the genus *Ochterus* Latreille in the Australian region (Heteroptera Ochteridae). *Entomologica Scandinavica* **20**, 449–477. doi:10.1163/187631289X00438

Baehr M (1995) Revision of *Philipis* (Coleoptera: Carabidae: Bembidinae), a genus of arboreal Tachyine beetles from the rainforests of eastern Australia: taxonomy, phylogeny and biogeography. *Memoirs of the Queensland Museum* **38**, 315–381.

Baehr M (2002a) New species and additional records of the tachyine *Philipis* Erwin (Coleoptera: Carabidae: Bembidiinae). *Memoirs of the Queensland Museum* **48**, 1–7.

Baehr M (2002b) Two new species of *Sloaneana* Csiki from southern Queensland (Coleoptera, Carabidae, Merizodiinae). *Memoirs of the Queensland Museum* **48**, 9–15.

Baehr M (2003) Psydrinae ground beetles (Coleoptera: Carabidae: Psydrinae), excluding Amblytelini, of eastern Queensland rainforests. *Memoirs of the Queensland Museum* **49**, 65–109.

Baehr M (2005) A review of the Australian Odacanthine ground beetles, including checklists for Australia and the Papuan subregion (Insecta: Coleoptera: Carabidae). *Memoirs of the Queensland Museum* **50**, 133–194.

Baehr M (2008) Revision of the genus *Habutarus* Ball and Hilchie from the Australian region (Insecta, Coleoptera, Carabidae, Lebiinae). *Memoirs of the Queensland Museum* **52**, 13–48.

Baehr M (2016) New species of the genera Amblytelus Erichson, 1842 and *Dysrichothorax* Blackburn, 1892 from Queensland and New South Wales, Australia (Insecta: Coleoptera: Carabaeidae: Psydrini: Amblytelina). *Memoirs of the Queensland Museum – Nature* **60**, 13–27. doi:10.17082/j.2204-1478.60.2016.2016-01

Bagchi R, Gallery RE, Gripenberg S, Gurr SJ, Narayan L, Addis CE, Freckleton RP, Lewis OT (2014) Pathogens and insect herbivores drive rainforest plant diversity and composition. *Nature* **506**, 85–88. doi:10.1038/nature12911

Bailey WJ, Rentz DCF (Eds) (1990). *The Tettigoniidae: Biology, Systematics and Evolution*. Crawford House Press, Bathurst.

Baillie J, Groombridge B (Eds) (1996) *IUCN Red List of Threatened Animals*. IUCN, Gland, Switzerland and Cambridge, U.K.

Baker CH (2010) A new subgenus and five new species of Australian glow-worms (Diptera: Keroplatidae: *Arachnocampa*). *Memoirs of the Queensland Museum – Nature* **55**, 11–43.

Baker CH, Graham GC, Scott KD, Cameron SL, Yeates DK, Merritt HDJ (2008) Distribution and phylogenetic relationships of Australia glow-worms Arachnocampa (Diptera: Keroplatidae). *Molecular Phylogenetics and Evolution* **48**, 506–514. doi:10.1016/j.ympev.2008.04.037

Baker CH, Merritt DJ (2003) Life cycle of an Australia glow-worm *Arachnocampa flava* Harrision (Diptera: Keroplatidae: Arachnocampinae). *Australian Entomologist* **30**, 45–55.

Baker GL, Poinar GO, Jr (1994) Studies on the genus *Amphimermis* (Nematoda: Mermithidae): five new species, including four from Orthoptera in southeastern Australia. *Fundamental and Applied Nematology* **17**, 303–321.

Balderson J (1991) Mantodea 'praying mantids'. In *The Insects of Australia*. (Ed. ID Naumann) pp. 348–356. Melbourne University Press, Carlton.

Balderson J, Rentz DCF, Roach AME (1998a) Mantodea. In *Zoological Catalogue of Australia. Volume 23. Archaeognatha, Zygentoma, Blattodea, Isoptera, Mantodea, Dermaptera, Phasmatodea, Embioptera, Zoraptera*. (Eds WWK Houston and A Wells) pp. 347–376. CSIRO Publishing, Melbourne.

Balderson J, Rentz DCF, Roach AME (1998b) Phasmatodea. In *Zoological Catalogue of Australia. Volume 23. Archaeognatha, Zygentoma, Blattodea, Isoptera, Mantodea, Dermaptera, Phasmatodea, Embioptera, Zoraptera*. (Eds WWK Houston and A Wells) pp. 451–456. CSIRO Publishing, Melbourne.

Bale CL, Williams JB (1994) Lost and found: *Nothofagus moorei* at Comboyne. *Cunninghamia* **3**, 529–533.

Balke M (1995) The Hydroporini (Coleoptera: Dytiscidae: Hydroporinae) of New Guinea: systematics, distribution and origin of the fauna. *Invertebrate Taxonomy* **9**, 1009–1019. doi:10.1071/IT9951009

Ballantyne LA, Lambkin C (2000) Lampyridae of Australia (Coleoptera: Lampyridae: Luciolinae: Luciolini). *Memoirs of the Queensland Museum* **46**, 15–93.

Ballantyne LA, Lambkin C (2009) Systematics of the Indo-Pacific fireflies with a redefinition od Australasian *Atyphella* Olliff, Madagascan *Photuroluciola* Pic, and description of seven new genera from the Luciolinae Coleoptera: Lampyridae). *Zootaxa* **1997**, 1–188. doi:10.11646/zootaxa.1997.1.1

Ballerio A (2013) Revision of Australian Ceratocanthinae (Coleoptera, Scarabaeoidea, Hybosoridae). *ZooKeys* **339**, 67–91. doi:10.3897/zookeys.339.6033

Balogh J (1982) New oppioid mites from Australia. *Acta zoologica Acadamiae scientiarum hungaricae* **28**, 3–14.

Balogh J, Balogh P (1983) Data to the oribatid fauna of Australia (Acari) II. *Acta zoologica Acadamiae scientiarum hungaricae* **29**, 283–301.

Barden P, Herhold HW, Grimaldi DA (2017) A new genus of hell ants from the Cretaceous (Hymenoptera: Formicidae: Haidomyrmecini) with a novel head structure. *Systematic Entomology* **42**, 837–846. doi:10.1111/syen.12253

Barker S (1975) Revision of the genus *Astraeus* LaPorte and Gory (Coleoptera: Buprestidae). *Transactions of the Royal Society of South Australia* **99**, 105–142.

Barker S (1979) New species and a catalogue of *Stigmodera* (*Castiarina*) (Coleoptera: Buprestidae). *Transactions of the Royal Society of South Australia* **103**, 1–23.

Barker S (1980) New species and new synonyms of *Stigmodera* (*Castiarina*) (Coleoptera: Buprestidae). *Transactions of the Royal Society of South Australia* **104**, 1–7.

Barker S (1983) New synonyms and new species of *Stigmodera* (*Castiarina*) (Coleoptera: Buprestidae). *Transactions of the Royal Society of South Australia* **107**, 139–169.

Barker S (1986) *Stigmodera* (*Castiarina*) (Coleoptera: Buprestidae): taxonomy, new species and a checklist. *Transactions of the Royal Society of South Australia* **110**, 1–36.

Barker S (1987) Eighteen new species of *Stigmodera* (*Castiarina*) (Coleoptera: Buprestidae). *Transactions of the Royal Society of South Australia* **111**, 133–146.

Barker S (1988) Contributions to the taxonomy of *Stigmodera* (*Castiarina*) (Coleoptera: Buprestidae). *Transactions of the Royal Society of South Australia* **112**, 133–142.

Barker S (1993) Seventeen new species of Australian Buprestidae (Insecta: Coleoptera) and a host plant of *Castiarina uptoni* (Barker). *Transactions of the Royal Society of South Australia* **117**, 15–26.

Barker S (2006a) *Castiarina: Australia's Richest Jewel Beetle Genus*. Australian Biological Resources Study, Canberra.

Barker S (2006b) Twenty five new species of *Cisseis* (*sensu stricto*) and two new synonyms (Coleoptera: Buprestinae: Agrilinae). *Transactions of the Royal Society of South Australia* **130**, 257–284. doi:10.1080/3721426.2006.10887065

Barlow BA (1981) *The Australian Flora: Its Origin and Evolution*. In *Flora of Australia. Vol. 1, Introduction*. Australian Government Publishing Service, Canberra.

Barnes JK (1989) Family Helossciomyzzidae. In *Catalog of the Diptera of the Australasian and Oceanic Regions*. (Ed. NL Evenhuis) pp. 570–571. Bishop Museum Press and E.J. Brill, Honolulu.

Barraclough DA (1992) The systematics of the Australian Dexiini (Diptera: Tachinidae: Dexiinae) with revisions of the endemic genera. *Invertebrate Taxonomy* **6**, 1127–1371. doi:10.1071/IT9921127

Basset Y (1990) The arboreal fauna of the rainforest tree *Argyrodendron actinophyllum* as sampled with restricted canopy fogging: Composition of the fauna. *Entomologist* **109**, 173–183.

Basset Y (1991) The taxonomic composition of the Arthropod fauna associated with an Australian rainforest tree. *Australian Journal of Zoology* **39**, 171–190. doi:10.1071/ZO9910171

Basset Y (1999) Diversity and abundance of insect herbivores foraging on seedlings in a rain forest in Guyana. *Ecological Entomology* **24**, 245–259. doi:10.1046/j.1365-2311.1999.00197.x

Basset Y, Charles E (2000) An annotated list of herbivores foraging on the seedlings of five forest trees in Guyana. *Anais da Sociedade Entomológica do Brasil* **29**, 433–452. doi:10.1590/S0301-80592000000300006

Basset Y, Springate ND (1992) Diel activity of arboreal arthropods associated with a rainforest tree. *Journal of Natural History* **26**, 947–952. doi:10.1080/00222939200770571

Basset Y, Charles E, Hammond DS, Brown VK (2001) Short-term effects of canopy openness on insect herbivores in a rain forest in Guyana. *Journal of Applied Ecology* **38**, 1045–1058. doi:10.1046/j.1365-2664.2001.00660.x

Batley M (2016) New species of *Euryglossula* Michener (Apoidea: Colletidae). *Records of the Australian Museum* **68**, 245–261. doi:10.3853/j.2201-4349.68.2016.1665

Batley M, Brandley B (2014) Phenology of the Australian solitary bee species *Leioproctus plumosus* (Smith) (Hymenoptera: Colletidae). *Australian Entomologist* **41**, 7–14.

Batley M, Hogendoorn K (2009) Diversity and conservation status of native bees. *Apidologie* **40**, 347–354. doi:10.1051/apido/2009018

Batley M, Houston TF (2012) Review of the Australian bee genus *Trichocolletes* Cockerell (Hymenoptera: Colletidae: Paracolletini). *Records of the Australian Museum* **64**, 1–50. doi:10.3853/j.0067-1975.64.2012.1589

Battigelli JP, Berch SM, Marshall VG (1994) Soil fauna communities in two distinct but adjacent forest types on northern Vancouver Island, British Columbia. *Canadian Journal of Forest Research* **24**, 1557–1566. doi:10.1139/x94-203

Baur GN (1957) Nature and distribution of rainforests in New South Wales. *Australian Journal of Botany* **5**, 190–233. doi:10.1071/BT9570190

Baur GN (1965) 'Forest types in New South Wales'. Forestry Commission of NSW Research Note 17. Forestry Commission of NSW, Sydney.

Beadle NCW, Costin AB (1952) Ecological classification and nomenclature. *Proceedings of the Linnean Society of New South Wales* **77**, 61–82.

Beaulieu F, Walter DE, Proctor HC, Kitching RL, Menzel F (2006) Mesostigmatid mite (Acari: Mesostigmata) on rainforest tree trunks: arboreal specialists, but substrate generalists? *Experimental & Applied Acarology* **39**, 25–40. doi:10.1007/s10493-006-0022-2

Beaver EP, Grehan J (2018) A simple rearing technique for larvae and observations on the larval biology of Australian 'splendid ghost moths' *Aenetus* (Herrich-Schäffer, 1855) (Lepidoptera:Hepialidae). *Journal of Insect Biodiversity* **8**, 43–51. doi:10.12976/jib/2018.08.3.1

Beck J, Kitching IJ, Linsenmair KE (2006) Wallace's line revisited: has vicariance or dispersal shaped the distribution of Malesian hawkmoths (Lepidoptera: Sphingidae)? *Biological Journal of the Linnean Society. Linnean Society of London* **89**, 455–468. doi:10.1111/j.1095-8312.2006.00686.x

Beesley PL, Ross GJB, Wells A (Eds) (1998) *Mollusca: The Southern Synthesis. Fauna of Australia. Vol. 5. Parts A, B.* CSIRO Publishing, Melbourne.

Beesley PL, Ross GJB, Glasby CJ (Eds) (2000) *Polychaetes and Allies. The Southern Synthesis. Fauna of Australia Vol. 4A. Polychaeta, Myzostomida, Pogonophora, Echiura, Sipuncula.* CSIRO Publishing, Melbourne.

Beier P, Noss RF (1998) Do habitat corridors provide connectivity? *Conservation Biology* **12**, 1241–1252. doi:10.1111/j.1523-1739.1998.98036.x

Bellamy CL (1987) A revision of the genus *Synechocera* Deyrolle (Coleoptera: Buprestidae: Agrilinae). *Invertebrate Taxonomy* **1**, 17–34. doi:10.1071/IT9870017

Bellamy CL (1991) Further review of the genus *Maoraxia* Obenberger (Coleoptera: Buprestidae). *Invertebrate Taxonomy* **5**, 457–468. doi:10.1071/IT9910457

Bellamy CL (1994) *Balthasarella melandryoides* Obenberger: a relict buprestid becomes less enigmatic (Coleoptera: Buprestidae). *Coleopterists Bulletin* **48**, 300.

Bellamy CL (2002) *Coleoptera: Buprestidae. Zoological Catalogue of Australia. Volume 29.5.* CSIRO Publishing, Melbourne.

Bellamy CL (2013) *Aaaaba* Bellamy, a second replacement name for *Alcinous* Deyrolle, 1864 (Coleoptera: Buprestidae). *The Coleopterists Bulletin* **67**, 32. doi:10.1649/072.067.0107

Bellamy CL, Williams GA (1985) A revision of the genus *Maoraxia* with a new synonym in *Acmaeodera* (Coleoptera: Buprestidae). *International Journal of Entomology* **27**, 147–161.

Bellamy CL, Williams GA (1995) The first Australian *Paratrachys* (Coleoptera: Buprestidae), with comments on the higher classification of the genus. *Invertebrate Taxonomy* **9**, 1265–1276. doi:10.1071/IT9951265

Bellamy CL, Williams GA, Hasenpusch J, Sundholm A (2013) A summary of the published data on host plants and morphology of immature stages of Australian jewel beetles (Coleoptera: Buprestidae) with additional records. *Insecta Mundi* **0293**, 1–172.

Bellas T, Hölldobler B (1985) Constituents of mandibular and Dufour's glands of an Australian *Polyrhachis* weaver ant. *Journal of Chemical Ecology* **11**, 525–538. doi:10.1007/BF00989563

Ben-Dov Y (2012) *ScaleNet, Pseudococcidae*, <http://www.sel.barc.usda.gov/scalenet/scalenet.htm>.

Benson J, Allen C (2007) Vegetation associated with *Wollemia nobilis* (Araucariaceae). *Cunninghamia* **10**, 255–262.

Benson JS, Redpath PA (1997) The nature of pre-European native vegetation in south-eastern Australia: a critique of Ryan, D.G., Ryan, J.R. and Starr, B.J. (1995) *The Australian Landscape – Observations of Explorers and Early Settlers. Cunninghamia* **5**, 285–328.

Benson RB (1938a) On the classification of sawflies (Hymenoptera: Symphyta). *Transactions of the Royal Entomological Society of London* **87**, 353–384. doi:10.1111/j.1365-2311.1938.tb00721.x

Benson RB (1938b) On the Australian Orussidae with a key to the genera of the world (Hymenoptera: Symphyta). *Annals & Magazine of Natural History* **2**, 1–15. doi:10.1080/03745481.1938.9755433

Benson RB (1939) A revision of the Australian sawflies of the genus *Perga* Leach, *sens. lat.* (Hymenoptera: Symphyta). *Australian Zoologist* **9**, 324–357.

Bernardi N (1989) Family Nemestrinidae. In *Catalog of the Diptera of the Australasian and Oceanic Regions.* (Ed. NL Evenhuis) pp. 378–381. Bishop Museum Press and E.J. Brill, Honolulu.

Bernhardt P (1995) Notes on the anthecology of *Pterostylis curta* (Orchidaceae). *Cunninghamia* **4**, 1–8.

Berry A (Ed.) (2002) *Infinite Tropics.* Verso, London.

Bertin RI (1989) Pollination biology. In *Plant–Animal Interactions.* (Ed. WG Abrahamson) pp. 23–86. McGraw-Hill, New York.

Beveridge I, Chilton NB (1999) Revision of the *Rugopharynx australis* (Moennig, 1926) complex (Nematoda: Strongyloidea) from macropodid marsupials. *Invertebrate Taxonomy* **13**, 805–843. doi:10.1071/IT98017

Bickel DJ (1983) Two new Australian *Teuchophorus* Loew (Diptera: Dolichopodidae). *Journal of the Australian Entomological Society* **22**, 39–45. doi:10.1111/j.1440-6055.1983.tb01837.x

Bickel DJ (1986) *Atlatlia*, a new genus of Dolichopodidae (Diptera) from Australia. *Entomologica Scandinavica* **17**, 165–171. doi:10.1163/187631286X00369

Bickel DJ (1987a) A revision of the Oriental and Australasian *Medetera* (Diptera: Dolichopodidae). *Records of the Australian Museum* **39**, 195–259. doi:10.3853/j.0067-1975.39.1987.170

Bickel DJ (1987b) *Kowmungia* (Diptera: Dolichopodidae), a new genus from Australia. *Invertebrate Taxonomy* **1**, 147–154. doi:10.1071/IT9870147

Bickel DJ (1992) The Australian Sympycninae (Diptera: Dolichopodidae): introduction and description of a new genus, *Yumbera*. *Invertebrate Taxonomy* **6**, 1005–1017. doi:10.1071/IT9921005

Bickel DJ (1994) The Australian Sciapodinae (Diptera: Dolichopdidae), with a review of the Oriental and Australian faunas, and a world conspectus of the subfamily. *Records of the Australian Museum Supplement Number* **21**.

Bickel DJ (1996a) Restricted and widespread taxa in the Pacific: biogeographic processes in the fly family Dolichopodidae (Diptera). In *The Origin and Evolution of Pacific Island Biotas, New Guinea to Eastern Polynesia: patterns and processes*. (Eds A Keast and E Miller) pp. 331–346. Academic Publishing, Amsterdam.

Bickel DJ (1996b) Smoke flies (Diptera: Platypezidae) and Sydney bushfires. *Australian Entomologist* **23**, 77–78.

Bickel DJ (1999a) Australian Sympycninae II: *Syntormon* Loew and *Nothorhapium*, gen. nov., with a treatment of the Western Pacific fauna, and notes on the subfamily Rhaphiinae and *Dactylonotus* Parent (Diptera: Dolichopodidae). *Invertebrate Taxonomy* **13**, 179–206. doi:10.1071/IT97028

Bickel DJ (1999b) Australian *Antyx* Meuffels and Grootaert and the New Caledonia Connection (Diptera: Dolichopodidae). *Australian Journal of Entomology* **38**, 168–175. doi:10.1046/j.1440-6055.1999.00102.x

Bickel DJ (2002) *Bandella*, a new hilarine fly genus from Australia (Diptera: Empididae). *Records of the Australian Museum* **54**, 313–324. doi:10.3853/j.0067-1975.54.2002.1369

Bickel DJ (2008) The Dolichopodinae (Diptera: Dolichopodidae) of New Caledonia, with descriptions and records from Australia, New Zealand and Melanesia. In *Zoologica Neocaledonica* 6. Biodiversity studies in New Caledonia. (Ed. P Grandcolas) *Mémoires du Muséum national d'Histoire naturelle* **196**, 13–47.

Bickel DJ (2009) *Dolichopodidae (Long-legged Flies)*. In *Manual of Central American Diptera. Vol. 1.* (Eds BV Brown, A Borkent, JM Cumming, DM Wood, NE Woodley MA Zumbado, MA) pp. 671–694. NRC Research Press, Ottawa.

Bickel DJ (2014) *Pindaia* (Diptera: Dolichopodidae) a new genus from New Caledonia and Australia. In *Zoologia Neocaledonica 8*. (Eds É Guilbert, P Grandcolas, H Jourdan, T Robillard) pp. 209–218. Mémoires du Muséum national d'Histoire naturelle, Paris.

Bickel DJ (2019) *Minjerribah* (Diptera: Dolichopodidae), a new genus from coastal Queensland, Australia. *Zootaxa* **4686**, 593–599.

Bickel DJ, Dyte CE (1989) Family Dolichopodide. In *Catalog of the Diptera of the Australasian and Oceanic Regions*. (Ed. NL Evenhuis) pp. 393–418. Bishop Museum Press and E.J. Brill, Honolulu.

Bickel DJ, Tasker EM (2004) Tree trunk invertebrates in Australian forests: conserving unknown species and complex processes. In *Conservation of Australia's Forest Fauna*. (Ed. D Lunney) pp. 888–899. Royal Zoological Society of New South Wales, Mosman.

Bigger M (1976) Oscillations of tropical insect populations. *Nature* **259**, 207–209. doi:10.1038/259207a0

Bily S (2000) *Euleptodema sainvali*, sp. n. from New Caledonia (Coleoptera: Buprestidae: Polycestinae) and larval morphology of the genus. *Folia Heyrovskyana* **8**, 35–46.

Binns ES (1982) Phoresy as migration – some functional aspects of phoretic mites. *Biological Reviews of the Cambridge Philosophical Society* **57**, 571–620. doi:10.1111/j.1469-185X.1982.tb00374.x

Black DG (1997) Diversity and biogeography of Australian millipedes (Diplopoda). *Memoirs of the Museum of Victoria* **56**, 557–561. doi:10.24199/j.mmv.1997.56.50

Blackburn T (1904) A revision of the Australian Aphodiides, and descriptions of three new species allied to them. *Proceedings of the Royal Society of Victoria* **17**, 145–181.

Blagoderov V (1993) Dipterans (Mesosciophilidae) from the Lower Cretaceous of Transbaikal. *Paleontological Journal* **27**, 123–130.

Blagoderov V, Grimaldi D (2004) Fossil Sciaroidea (Diptera) in Cretaceous ambers, exclusive of Cecidomyiidae, Sciaridae, and Keroplatidae. *American Museum Novitates* **3433**, 1–76. doi:10.1206/0003-0082(2004)433<0001:FSDICA>2.0.CO;2

Blakemore RJ (1994) A review of *Heteroporodrilus* from south-east Queensland (Annelida: Oligochaeta). *Memoirs of the Queensland Museum* **37**, 19–39.

Blakemore RJ (1997) Two new genera and some new species of Australian earthworms (Acanthodrilidae, Megascolecidae: Oligochaeta). *Journal of Natural History* **31**, 1785–1848. doi:10.1080/00222939700770951

Blakemore RJ (2000a) New species of the earthworm genus *Anisochaeta* (Oligochaeta: Megascolecidae) from New South Wales. *Records of the Australian Museum* **52**, 1–40. doi:10.3853/j.0067-1975.52.2000.1307

Blakemore RJ (2000b) Native earthworms (Oligochaeta) from south-western Australia, with descriptions of fifteen new species. *Records of the Australian Museum* **52**, 187–222. doi: 10.3853/j.0067-1975.52.2000.1314

Blanche R, Cunningham SA (2005) Rainforest provides pollinating beetles for *Atemoya* crops. *Journal of Economic Entomology* **98**, 1193–1201. doi:10.1603/0022-0493-98.4.1193

Blest AD, Hardie RC, McIntyre P, Williams DS (1981) The spectral sensitivities of identified receptors and the function of retinal tiering in the principal eyes of a jumping spider. *Journal of Comparative Physiology* **145**, 227–239. doi:10.1007/BF00605035

Blüthgen N, Fiedler K (2002) Interactions between weaver ants *Oecophylla smaragdina*, homopterans, trees and lianas in an Australian rain forest canopy. *Journal of Animal Ecology* **71**, 793–801. doi:10.1046/j.1365-2656.2002.00647.x

Blüthgen N, Stork NE (2007) Ant mosaics in a tropical rainforest in Australia and elsewhere: A critical review. *Austral Ecology* **32**, 93–104. doi:10.1111/j.1442-9993.2007.01744.x

Blüthgen N, Stork NE, Fiedler K (2004) Bottom-up control and co-occurrence in complex communities: honeydew and nectar determine a rainforest ant mosaic. *Oikos* **106**, 344–358. doi:10.1111/j.0030-1299.2004.12687.x

Boag B, Yeates G (1998) Soil nematode biodiversity in terrestrial ecosystems. *Biodiversity and Conservation* **7**, 617–630. doi:10.1023/A:1008852301349

Bock IR (1982) Drosophilidae of Australia. V. Remaining genera and synopsis (Insecta: Diptera). *Australian Journal of Zoology Supplementary Series* **89**.

Bock IR (1987) The Australian species of *Ephydrella* and *Setacera* (Diptera: Ephydridae). *Invertebrate Taxonomy* **1**, 155–166. doi:10.1071/IT9870155

Bock IR (1988) The Australian species of *Paralimna and Notiphila*(Diptera: Ephydridae). *Invertebrate Taxonomy* **2**, 885–902. doi:10.1071/IT9880885

Bock IR (1990) The Australian species of *Hydrellia* Robineau-Desvoidy (Diptera: Ephydridae). *Invertebrate Taxonomy* **3**, 965–993. doi:10.1071/IT9890965

Bock IR, Parsons PA (1978) Australian endemic *Drosophila* V. Queensland rain-forest species associated with fungi, with descriptions of six new species and a redescription of *D. pictipennis* Kertész. *Australian Journal of Zoology* **26**, 331–347. doi:10.1071/ZO9780331

Bock IR, Parsons PA (1979) Australian endemic *Drosophila* VI.* Species collected by sweeping rain forests of Queensland and northern New South Wales. *Australian Journal of Zoology* **27**, 291–301. doi:10.1071/ZO9790291

Boevé J-L, Rozenberg R, Shinohara A, Schmidt S (2014) Toxic peptides occur frequently in pergid and argid sawfly larvae. *PLoS ONE* **9**, e105301. doi:10.1371/journal.pone.0105301

Boland BJ, Brooker MIH, Chippendale GM, Hall N, Hyland BPM, Johnston RD, Kleinig DA, Turner JD (1984) *Forest Trees of Australia*. Thomas Nelson and CSIRO, Melbourne.

Bond WJ, Breytenbach GJ (1985) Ants, rodents and seed predation in Proteaceae. *South African Journal of Zoology* **20**, 150–154. doi:10.1080/02541858.1985.11447928

Borda E, Oceguera-Figueroa A, Siddall ME (2008) On the classification, evolution and biogeography of terrestrial haemadipsoid leeches (Hirudinida: Arhynchobdellida: Hirudiniformes). *Molecular Phylogenetics and Evolution* **46**, 142–154. doi:10.1016/j.ympev.2007.09006

Borda E, Siddall ME (2010) Insights into the evolutionary history of Indo-Pacific bloodfeeding terrstrial leeches (Hirudinida: Arhynchobdellida: Haemadipsidae). *Invertebrate Systematics* **24**, 456–472. doi:10.1071/IS10013

Borgmeier T (1968) A catalogue of the Phoridae of the world (Diptera, Phoridae). *Studies in Entomology* (N.S.) **11**, 1–367.

Borkent A (1989a) Family Corethrellidae. In *Catalog of the Diptera of the Australasian and Oceanic Regions*. (Ed. NL Evenhuis) p. 189. Bishop Museum Press and E.J. Brill, Honolulu.

Borkent A (1989b) Family Chaoboridae. In *Catalog of the Diptera of the Australasian and Oceanic Regions*. (Ed. NL Evenhuis) p. 190. Bishop Museum Press and E.J. Brill, Honolulu.

Borkent A (2008) The frog-biting midges of the world (Corethrellidae: Diptera). *Zootaxa* **1804**, 1–456. doi:10.11646/zootaxa.1804.1.1

Borkent A (2014) World catalog of extant and fossil Corethrellidae (Diptera). *Zootaxa* **3796**, 453–468. doi:10.11646/zootaxa.3796.3.3

Borkent A, Grafe TU (2012) The frog-biting midges of Borneo – from two to eleven species (Corethrellidae: Diptera). *Zootaxa* **3279**, 1–45. doi:10.11646/zootaxa.3279.1.1

Borkent A, Wirth WW (1997) World species of biting midges (Diptera: Ceratopogonidae). *Bulletin of the American Museum of Natural History* **233**, 1–257.

Bornemissza G (1971) Mycetophagous breeding in the Australian dung beetle, *Onthophagus dunningi*. *Pedobiologia* **11**, 133–142.

Borsboom A (1998) Aspects of the biology and ecology of Australian freshwater cratfish, *Euastacus urospinosus* (Decapoda: Parastacidae). *Proceedings of the Linnean Society of New South Wales* **119**, 87–100.

Bouček Z (1988) *Australasian Chalcidoidea (Hymenoptera): A Biosystematic Revision of Genera of Fourteen Families, with a Reclassification of Species*. C.A.B. International, Wallingford.

Bouchard P (2002) Phylogenetic revision of the flightless Australian genus *Apterotheca* Gebien (Coleoptera: Tenebrionidae: Coelmetopinae). *Invertebrate Systematics* **16**, 449–554. doi:10.1071/IT01029

Bouchard P, Yeates DK (2001) Phylogenetic relationships of the Australasian Coelometopini (Coleoptera: Tenebrionidae); a quantitative cladistic analysis with a review of biology. *Organisms, Diversity & Evolution* **1**, 17–43. doi:10.1078/1439-6092-00003

Bouchard P, Lawrence JF, Davies AE, Newton AF (2005) Synoptic classification of the world Tenebrionidae (Insecta: Coleoptera) with a review of family-group names. *Annales Zoologici* **55**, 499–530.

Bourguignon T, Lo N, Cameron SL, Šobotnik J, Hayashi Y, Shigenobu S, Watanbe D, Roisin Y, Miura T, Evans TA (2015) The evolutionary history of termites as inferred from 66 Mitrochondrial Genomes. *Molecular Biology and Evolution* **32**, 406–421. doi:10.1093/molbev/msu308

Bower C, Towle B, Bickel D (2015) Reproductive success and pollination of the Tuncurry midge orchid (*Genoplesium littorale*) (Orchidaceae) by chloropid flies. *Telopea* **18**, 43–55. doi:10.7751/telopea8127

Bowman DMJS (2000) *Rainforests: Islands of Green in a Land of Fire*. Cambridge University Press, Cambridge.

Braby M (2000) *Butterflies of Australia. Their Identification, Biology and Distribution. Vols 1 and 2*. CSIRO Publishing, Melbourne.

Braby MF, Wurtz GE (2018) A new subspecies of *Neolucia hobartensis* (Miskin, 1890) (Lepidoptera: Lycaenidae) from mainland southeastern Australia, witha review of butterfly endemism in montane areas of this region. *Records of the Australian Museum* **70**, 423–433. doi:10.3853/j.2201-4349.70.2018.1715

Brackman AC (1980) *A Delicate Arrangement*. Times Books, New York.

Bradbury JH, Williams WD (1999) Key to and checklist of the inland aquatic amphipods of Australia. *Technical Reports of the Australian Museum* **14**.

Bradshaw CJA (2012) Little left to lose: deforestation and forest fragmentation in Australia since European colonization. *Journal of Plant Ecology* **5**, 109–120. doi:10.1093/jpe/rtr038

Brailovsky H (2002) New genus and species of Mictini (Hemiptera: Coreidae) from Australia. *Entomological News* **113**, 29–34.

Brailovsky H, Monteith GB (1996) A new species of *Pomponatius* Distant from Australia (Hemiptera: Heteroptera: Coreidae: Acanthocorini). *Memoirs of the Queensland Museum* **39**, 205–210.

Braje TJ, Erlandson JM (2013) Human acceleration of animal and plant extinction: a Late Pleistocene, Holocene and Anthropocene continuum. *Anthropocene* **6**, 14–23.

Bray D (2014) Modification of the All Protocol to characterise the overall ant assemblage in temperate eucalypt forest. *Australian Entomologist* **41**, 15–28.

Brimblecombe AR (1965) Microterygidae breeding in kauri pine seed. *Journal of the Entomological Society of Queensland* **4**, 13.

Bristow CM (1983) Treehoppers transfer parental care to ants: a benefit of mutualism. *Science* **220**, 532–533.

Britton EB (1957) *A Revision of the Australian Chafers (Coleoptera: Scarabaeidae: Melolonthinae)*. British Museum (Natural History), London.

Britton EB (1970) A review of *Homolotropus* Macleay (Coleoptera:Scarabaeidae: Melolonthinae). *Journal of the Australian Entomological Society* **9**, 42–48. doi:10.1111/j.1440-6055.1970.tb00768.x

Britton EB (1978) A revision of the Australian chafers (Coleoptera: Scarabaeidae: Melolonthinae). Volume 2. Tribe Melolonthini. *Australian Journal of Zoology Supplementary Series* **60**.

Britton EB (1980) A revision of the Australian chafers (Coleoptera: Scarabaeidae: Melolonthinae). Volume 3. Tribe Liparetrini. Genus *Liparetrus*. *Australian Journal of Zoology Supplementary Series* **76**.

Britton EB (1981) The Australian Hygrobiidae (Coleoptera). *Journal of the Australian Entomological Society* **20**, 83–86. doi:10.1111/j.1440-6055.1981.tb01004.x

Britton EB (1986) A revision of the Australian chafers (Coleoptera: Scarabaeidae: Melolonthinae) Volume 4. Tribe Liparetrini: genus *Colpochila*. *Australian Journal of Zoology Supplementary Series* **118**.

Britton EB (1987) A revision of the Australian chafers (Coleoptera: Scarabaeidae: Melolonthinae). Volume 5. Tribes Scitalini and Comophorinini. *Invertebrate Taxonomy* **1**, 685–799. doi:10.1071/IT9870685

Britton EB (1990) A synopsis of the Australian genera Liparetrini (Coleoptera: Scarabaeidae: Melolonthinae). *Invertebrate Taxonomy* **4**, 159–195. doi:10.1071/IT9900159

Britton EB (1995) A synopsis of the genera Diphucephalini (Coleoptera: Scarabaeidae: Melolonthinae) including *Watkinsia*, gen. et spp. nov. *Invertebrate Taxonomy* **9**, 115–128. doi:10.1071/IT9950115

Britton EB (2000) A review of *Heteronyx* Guérin-Méneville (Coleoptera: Scarabaeidae: Melolonthinae). *Invertebrate Taxonomy* **14**, 465–589.

Britton EB, Stanbury PJ (1981) Type specimens in the Macleay Museum, University of Sydney. VIII. Insects: Beetles (Insecta: Coleoptera). *Proceedings of the Linnean Society of New South Wales* **105**, 241–293.

Broadhead EC (1984) Adaptations for fungal grazing in lauxaniid flies. *Journal of Natural History* **18**, 639–649. doi:10.1080/00222938400770531

Brock PD, Hasenpusch JW (2015) *The Complete Field Guide to Stick and Leaf Insects of Australia*. CSIRO Publishing, Melbourne.

Brock PD, Hasenpusch JW, Petrović S (2019) A spectacular new Australian stick insect genus and species (Phasmida). *Phasmid Studies* **20**, 15–29.

Brock PD, Monteith GB (2018) A striking new species of *Parapodacanthus* Brock (Phasmida: Phasmatidae) from southeastern Queensland. *Australian Entomologist* **45**, 17–26.

Brooks AJ, Haeusler T, Reinfelds I, Williams S (2005) Hydraulic microhabitats and the distribution of macroinvertebrate assemblages in riffles. *Freshwater Biology* **50**, 331–344. doi:10.1111/j.1365-2427.2004.01322.x

Brown GR (1983) *Pentazeleboria*, a new genus of Australian Thynnini (Hymenoptera: Tiphiidae). *Journal of the Australian Entomological Society* **22**, 61–64. doi:10.1111/j.1440-6055.1983.tb01842.x

Brown GR (1987) Revision of the Australian genus *Acanthothynnus* Turner (Hymenoptera: Tiphiidae). *Journal of the Australian Entomological Society* **26**, 181–188. doi:10.1111/j.1440-6055.1987.tb00282.x

Brown GR (1996) *Chilothynnus*, a new genus of Australian Thynninae (Hymenoptera: Tiphiidae) associated with orchids. *The Beagle: Records of the Museums and Art Galleries of the Northern Territory* **13**, 61–72.

Brown GR (1998) Revision of the *Neozeleboria cryptoides* species group of thynnine wasps (Hymenoptera: Tiphiidae): pollinators of native orchids. *Australian Journal of Entomology* **37**, 193–205. doi:10.1111/j.1440-6055.1998.tb01572.x

Brown GR (2001) Status of the *Ariphron* generic group (Hymenoptera: Tiphiidae): a critical review. *Australian Journal of Entomology* **40**, 23–40. doi:10.1046/j.1440-6055.2001.00207.x

Brown GR, Phillips R (2013) A review of the diet of flower wasps (Hymenoptera: Thynnidae). *Northern Territory Naturalist* **25**, 50–63.

Brown WL (1961) A revision of the Dacetine ant genus *Orectognathus. Memoirs of the Queensland Museum* **13**, 84–104.

Brown WV, Doyen JT, Moore BP, Lawrence JF (1992) Chemical composition and taxonomic significance of defensive secretions of some Australian Tenebrionidae (Coleoptera). *Journal of the Australian Entomological Society* **31**, 79–89. doi:10.1111/j.1440-6055.1992.tb00461.x

Bryant LM, Krosch MN (2016) Lines in the sand: a review of the evidence for eastern Australia's 'major' biogeographic barriers to closed forest taxa. *Biological Journal of the Linnean Sociey* **119**, 238–264. doi:10.1111/bij.12821

Buchmann SL (1983) Buzz pollination in angiosperms. In *Handbook of Experimental Pollination Biology.* (Eds CE Jones and JR Little) pp. 73–113. Van Nostrand Reinhold, New York.

Bugledich E-MA (1999) Diptera: Nematocera. *Zoological Catalogue of Australia. Volume 30.1.* (Eds A Wells and WWK Houston). CSIRO Publishing, Melbourne.

Bühlmann C, Cheng K, Wehner R (2011) Vector-based and landmark-guided navigation in desert ants inhabiting landmark-free and landmark-rich environments. *The Journal of Experimental Biology* **214**, 2845–2853. doi:10.1242/jeb.054601

Bullock SH (1994) Wind pollination of Neotropical dioecius trees. *Biotropica* **26**, 172–179. doi:10.2307/2388806

Burbidge NT (1960) The phytogeography of the Australian region. *Australian Journal of Botany* **8**, 75–211. doi:10.1071/BT9600075

Burckhardt D, Ouvrard D (2012) A revised classification of the jumping plant-lice (Hemiptera: Psylloidea). *Zootaxa* **3509**, 1–34. doi:10.11646/zootaxa.3509.1.1

Burrows M (2014) Jumping mechanisms in flatid planthoppers (Hemiptera, Flatidae). *The Journal of Experimental Biology* **217**, 2590–2600. doi:10.1242/jeb.105429

Bush AA, Nipperess DA, Duursma DA, Theischinger G, Turak E, Hughes L (2014) Continental-scale assessment of risk to the Australian Odonata from climate change. *PLoS ONE* **9**, e88958. doi:10.1371/journal.pone.0088958

Byers GW (1991) Mecoptera 'scorpion-flies'. In *The Insects of Australia.* (Ed. ID Naumann) pp. 696–704. Melbourne University Press, Carlton.

Byrne M, Yeates DK, Joseph L, Kearney M, Bowler J, Williams MAJ, Cooper S, Donnellan SC, Keogh JS, Leys R, Melville J, Murphy DJ, Porch N, Wyrwoll KH (2008) Birth of a biome: insights into the assembly and maintenance of the Australian arid zone biota. *Molecular Ecology* **17**, 4398–4417. doi:10.1111/j.1365-294X.2008.03899.x

Cagnolo L, Valladares G, Salvo A, Cabido M, Zak M (2009) Habitat fragmentation and species loss across three interacting trophic levels; effects of life-history and food-web traits. *Conservation Biology* **23**, 1167–1175. doi:10.1111/j.1523-1739.2009.01214.x

Calaby JH, Gay FJ (1959) Aspects of the distribution and ecology of Australian termites. XII. In *Biogeography and Ecology in Australia.* (Eds A Keast, RL Crocker and CS Christian) pp. 211–223. Springer, Dordrecht.

Calaby JH, Murray MD (1991) Phthiraptera. In *The Insects of Australia.* (Ed. ID Naumann) pp. 421–428. Melbourne University Press, Carlton.

Calder A (1986) Classification, relationships and distribution of the Crepidomeninae (Coleoptera: Elateridae). *Australian Journal of Zoology. Supplementary Series Number* **122**.

Calder A (1996) *Click Beetles: Genera of Australian Elateridae. Monographs on Invertebrate Taxonomy. Vol. 2.* CSIRO Publishing, Melbourne.

Calder A (1998) *Zoological Catalogue of Australia. Vol. 29.6. Elateroidea.* Australian Biological Resources Study, Canberra.

Cameron DJ (1987) Temperate rainforests of East Gippsland. In *The Rainforest Legacy. Australian National Forests Study. Vol. 1.* (Eds G Werren and AP Kershaw) pp. 33–46. Australian Government Publishing Service, Canberra.

Campbell AJ, Brown GR (1994) Distribution of parasitoids of scarab larvae in relation to remnant vegetation: a preliminary analysis. *Memoirs of the Queensland Museum* **36**, 27–32.

Camras S (1961) Notes on Australian flies of the family Conopidae. *Proceedings of the United States National Museum* **113**, 61–76. doi:10.5479/si.00963801.113-3453.61

Cantrell BK (1980) Additional harvestmen (Arachnida: Opiliones). *Journal of the Australian Entomological Society* **19**, 241–253. doi:10.1111/j.1440-6055.1980.tb00980.x

Cantrell BK (1989) The Australasian species of *Winthemia* Robineau-Desvoidy (Diptera: Tachinidae) with notes on Oceanian and Oriental species. *Journal of the Australian Entomological Society* **28**, 93–104. doi:10.1111/j.1440-6055.1989.tb01203.x

Cantrell BK, Crosskey RW (1989) Family Tachinidae. In *Catalog of the Diptera of the Australasian and Oceanic Regions.* (Ed. NL Evenhuis) pp. 733–784. Bishop Museum Press and E.J. Brill, Honolulu.

Car CA (2016) The millipede genus *Dicladosomella* (Diplopoda: Polydesmida: Paradoxosomatidae) from New South Wales, Australia, and description of 18 new species. *Austral Entomology* **55**, 63–99. doi:10.1111/aen.12153

Cardale J (1993) Hymenoptera: Apoidea. In *Zoological Catalogue of Australia. Vol. 10.* (Eds WWK Houston and GV Maynard) pp. 1–105. CSIRO Publishing, Melbourne.

Carmichael AC, Wharton RA, Clarke AR (2005) Opiine (Hymenoptera: Braconidae) parasitoids of tropical fruit flies (Diptera: Tephritidae) of the Australian and South Pacific region. *Bulletin of Entomological Research* **95**, 545–569. doi:10.1079/BER2005383

Carne PB (1957a) *Systematic Revision of the Australian Dynastinae.* CSIRO Publishing, Melbourne.

Carne PB (1957b) A revision of the ruteline genus *Anoplognathus* Leach (Coleoptera: Scarabaeidae). *Australian Journal of Zoology* **5**, 88–143. doi:10.1071/ZO9570088

Carne PB (1958) A review of the Australian Rutelinae (Coleoptera: Scarabaeidae). *Australian Journal of Zoology* **6**, 162–240. doi:10.1071/ZO9580162

Carne PB (1965) A revision of the genus *Elaphastomus* Macleay (Coleoptera: Geotrupidae). *Journal of the Entomological*

Society of Queensland **4**, 3–13. doi:10.1111/j.1440-6055.1965.tb00637.x

Carne PB (1974) A review of the *olivaceus* species group of the genus *Paraschizognathus* Ohaus, and description of three new species (Coleoptera: Scarabaeidae). *Journal of the Australian Entomological Society* **13**, 261–266. doi:10.1111/j.1440-6055.1974.tb02205.x

Carne PB (1981) Three new species of *Anoplognathus* Leach, and new distribution records for poorly known species (Coleoptera: Scarabaeidae: Rutelinae). *Journal of the Australian Entomological Society* **20**, 289–294. doi:10.1111/j.1440-6055.1981.tb01049.x

Carne PB, Monteith GB (1971) Three little-known species of Christmas beetle. *News Bulletin of the Australian Entomological Society* **7**, 61–63.

Carnegie AJ, Angel P (2005) *Creiis lituratus* (Froggatt) (Hemiptera: Psyllidae): a new insect pest of *Eucalyptus dunnii* plantations in sub-tropical Australia. *Australian Forestry* **68**, 59–64. doi:10.1080/00049158.2005.10676227

Carnegie AJ, Kathuria A, Pegg GS, Entwistle P, Nagel M, Giblin F (2015) Impact of the invasive rust *Puccinia psidii* (myrtle rust) on native Myrtaceae in natural ecosystems in Australia. *Biological Invasions* **18**, 127–144. doi:10.1007/s10530-015-0996-y-

Carter HJ (1905) Descriptions of new species of Australian Coleoptera. Part I. *Proceedings of the Linnean Society of New South Wales* **30**, 177–189.

Carter HJ (1906) Notes on the genus *Cardiothorax*, with descriptions of new species of Australian Coleoptera. Pt. 2. *Proceedings of the Linnean Society of New South Wales* **31**, 236–260.

Carter HJ (1911) Revision of *Pterhelaeus* (continued) and of *Saragus*; with descriptions of new species of Australian Tenebrionidae. *Proceedings of the Linnean Society of New South Wales* **36**, 179–223.

Carter HJ (1913a) Descriptions of some new species of Coleoptera. *Proceedings of the Linnean Society of New South Wales* **37**, 480–491. doi:10.5962/bhl.part.22353

Carter HJ (1913b) Notes on the genus *Stigmodera*, with descriptions of eleven new species, and of other Buprestidae. *Proceedings of the Linnean Society of New South Wales* **37**, 497–511. doi:10.5962/bhl.part.22355

Carter HJ (1913c) Revision of the Australian species of the subfamilies Cyphaleinae and Cnodaloninae. *Proceedings of the Linnean Society of New South Wales* **38**, 61–105. doi:10.5962/bhl.part.13554

Carter HJ (1914) Revision of the subfamily Tenebrioninae, family Tenebrionidae. *Proceedings of the Linnean Society of New South Wales* **39**, 44–86. doi:10.5962/bhl.part.2275

Carter HJ (1915a) Descriptions of six species of Buprestidae. *Proceedings of the Linnean Society of New South Wales* **40**, 76–82. doi:10.5962/bhl.part.18864

Carter HJ (1915b) The Australian Strongyliinae and other Tenebrionidae, with descriptions of new genera and species (family Tenebrionidae). *Proceedings of the Linnean Society of New South Wales* **40**, 522–539. doi:10.5962/bhl.part.18883

Carter HJ (1916) Descriptions of a new genus and three new species of Australian Tenebrionidae from Barrington Tops, New South Wales. *Proceedings of the Linnean Society of New South Wales* **41**, 209–214. doi:10.5962/bhl.part.15313

Carter HJ (1920) Notes on some Australian Tenebrionidae, with descriptions of new species – also a new genus and species of Buprestidae. *Proceedings of the Linnean Society of New South Wales* **45**, 222–249. doi:10.5962/bhl.part.19543

Carter HJ (1921) Australian Coleoptera: notes and new species. *Proceedings of the Linnean Society of New South Wales* **46**, 301–323. doi:10.5962/bhl.part.14020

Carter HJ (1922) Australian Coleoptera: notes and new species. No. ii. *Proceedings of the Linnean Society of New South Wales* **47**, 65–82.

Carter HJ (1923) A revision of the Australian species of the genus *Melobasis* (Fam. Buprestidae, Order Coleoptera), with notes on allied genera. *The Transactions of the Entomological Society of London* **11**, 63–105.

Carter HJ (1924) Australian Coleoptera – notes and new species. No. ii. *Proceedings of the Linnean Society of New South Wales* **47**, 65–82.

Carter HJ (1925) Revision of the Australian species of *Chrysobothris* (family Buprestidae), together with notes and descriptions of new species of Coleoptera. *Proceedings of the Linnean Society of New South Wales* **50**, 225–244.

Carter HJ (1926a) A checklist of the Australian Tenebrionidae. *Australian Zoologist* **4**, 117–163.

Carter HJ (1926b) Revision of the Australian species of *Anilara* (Fam. Buprestidae) and *Helmis* (Fam. Dryopidae), with notes, and descriptions of other Australian Coleoptera. *Proceedings of the Linnean Society of New South Wales* **51**, 50–71.

Carter HJ (1926c) Revision of *Athemistus* and *Microtragus* (Fam. Cerambycidae) with notes, and descriptions of other Australian Coleoptera. *Proceedings of the Linnean Society of New South Wales* **51**, 492–516.

Carter HJ (1928) Revision of the Australian species of the genera *Curis, Neocuris* and *Trachys*, together with notes and descriptions of new species of other Coleoptera. *Proceedings of the Linnean Society of New South Wales* **53**, 270–290.

Carter HJ (1929a) Australian Coleoptera: notes and new species VI. *Proceedings of the Linnean Society of New South Wales* **54**, 65–79.

Carter HJ (1929b) A checklist of the Australian Buprestidae. *Australian Zoologist* **5**, 265–304.

Carter HJ (1930) Checklist of the Australian Cistelidae. Order Coleoptera. *Australian Zoologist* **6**, 269–276.

Carter HJ (1931) Notes on the genus *Stigmodera* (Family Buprestidae). Together with descriptions of new species of and a retabulation of the subgenus *Castiarina*. *Australian Zoologist* **6**, 337–367.

Carter HJ (1933) *Gulliver in the Bush*. Angus and Robertson, Sydney.

Carter HJ (1936) Australian Coleoptera. Notes and new species. No. 10. *Proceedings of the Linnean Society of New South Wales* **61**, 99–102.

Carter HJ, Zeck EH (1937) A monograph of the Australian Colydiidae. *Proceedings of the Linnean Society of New South Wales* **62**, 181–208.

Carvalho JCM (1978) The distribution in Australia of the grass bugs of the tribe Stenodemini (Heteroptera – Miridae – Mirinae). *Records of the South Australian Museum* **18**, 75–82.

Carvalho JCM, Gross GF (1982) Australian ant-mimetic Miridae (Hemiptera: Heteroptera. I. The *Leucophoroptera* group of the subfamily Phylinae. *Australian Journal of Zoology Supplementary Series* **86**, 1–75.

Carver M (1992) Alloxystinae (Hymenoptera: Cynipoidea: Charipidae) in Australia. *Invertebrate Taxonomy* **6**, 769–785.

Carver M, Gross GF, Woodward TE (1991) Hemiptera 'bugs, leaf-hoppers, cicadas'. In *The Insects of Australia.* (Ed. ID Naumann) pp. 429–509. Melbourne University Press, Carlton.

Carver M, Thompson FC (2003) Two new species of Syrphidae (Diptera) from Australia. *Studia Dipterologica* **10**, 37–41.

Casalla R, Korb J (2019) Termite diversity in Neotropical dry forests of Colombia and the potential role of rainfall in structuring termite diversity. *Biotropica* **51**, 165–177. doi:10.1111/btp.12626

Cassis G (1998) Dermaptera. In *Archaeognatha, Zygentoma, Blattodea, Isoptera, Mantodea, Dermaptera, Phasmatodea, Embioptera, Zoraptera. Zoological Catalogue of Australia. Volume 23.* (Eds WWK Houston and A Wells) pp. 279–345. CSIRO Publishing, Melbourne.

Cassis G, Gross GF (1995) Hemiptera: Heteroptera (Coleorrhyncha to Cimicomorpha). In *Zoological Catalogue of Australia. Vol. 27.3a.* (Eds WWK Houston and GV Maynard) pp. 1–506. CSIRO Publishing, Melbourne.

Cassis G, Gross GF (2002) Hemiptera: Heteroptera (Pentatomomorpha). In *Zoological Catalogue of Australia. Vol. 27.3b.* (Eds WWK Houston and A Wells) pp. 1–737. CSIRO Publishing, Melbourne.

Cassis G, Symonds GF (2012) Systematics and biology of the Australian seed bug genus *Megadrymus* Gross (Insecta: Heteroptera: Rhyparochromidae: Drymini). *Invertebrate Systematics* **26**, 249–273. doi:10.1071/IS11050

Cassis G, Vanags L (2006) Jewel bugs of Australia (Insecta, Heteroptera, Scutelleridae). *Denisia* **19**, 275–398.

Casson DS, Hodkinson ID (1991) The Hemiptera (Insecta) communities of tropical rain forest in Sulawesi. *Zoological Journal of the Linnean Society* **102**, 253–275. doi:10.1111/j.1096-3642.1991.tb00002.x

Caterino MS (2003) New species of *Chlamydopsis* (Histeridae: Chlamydopsinae), with a review and phylogenetic analysis of all known species. *Memoirs of the Queensland Museum* **49**, 159–235.

Cavanaugh J (2010) The conservation of the rainforests: politics and battlelines. In *Remnants of Gondwana.* (Eds R Kitching, R Braithwaite and J Cavanaugh) pp. 259–272. Surrey Beatty and Sons, Baulkham Hills.

Ceballos G, Ehrich PR, Barnosky AD, García A, Pringle RM, Palmer TM (2015) Accelerated modern human-induced species losses: entering the sixth mass extinction. *Science Advances* **1**, e1400253.

Ceccarelli FS (2008) Behavioral mimicry in *Myrmarachne* species (Araneae, Salticidae) from North Queensland, Australia. *The Journal of Arachnology* **36**, 344–351. doi:10.1636/CSt07-114.1

Chadwick CE (1993) The roles of *Tranes lyteroides* and *T. sparsus* Boh (Col., Curculionidae) in the pollination of *Macrozamia communis* (Zamiaceae). In *Proceedings of CYCAD 90, the Second International Conference of Cycad Biology.* (Eds DW Stevenson and NJ Norstog) pp. 77–88. Palm and Cycad Societies of Australia Ltd, Milton, Brisbane.

Chamberlin RV (1920) The Myriopoda of the Australian region. *Bulletin of the Museum of Comparative Zoology* **94**, 3–269.

Chandler DS (2001) *Biology, Morphology, and Systematics of the Ant-like Litter Beetle Genera of Australia (Coleoptera: Staphylinidae: Pselaphinae). Memoirs on Entomology International, Vol. 15.* Associated Publishers, Gainesville, Florida.

Chandler PJ (1994) The Oriental and Australasian species of Platypezidae (Diptera). *Invertebrate Taxonomy* **8**, 351–434. doi:10.1071/IT9940351

Chandler PJ (2010) Platypezidae (flat-footed flies). In *Manual of Central American Diptera. Vol. 2.* (Eds BV Brown, A Borkent, JM Cumming, DM Wood, NE Woodley and MA Zumbado) pp. 715–719. NRC Research Press, Ottawa.

Chapman AD (2009) 'Numbers of living species in Australia and the world'. Report for Australian Biological Resources Study, Canberra.

Charley D, Andren M (2018) The distribution and conservation of the endangered Atlas rainforest ground-beetle *Nurus atlas* (Laporte, 1867) (Coleoptera: Carabidae) in north-east New South Wales. *Australian Entomologist* **45**, 389–402.

Cheesman AW, Cernusak LA, Zanne AE (2018) Relative roles of termites and saprotrophic microbes as drivers of wood decay: a wood block test. *Austral Ecology* **43**, 257–267. doi:10.1111/aec.12561

Chenoweth LB, Schwarz MP (2011) Biogeographic origins and diversification of the exoneurine allodapine bees of Australia (Hymenoptera, Apidae). *Journal of Biogeography* **38**, 1471–1483. doi:10.1111/j.1365-2699.2011.02488.x

Chessman BC, Boulton AJ (1999) Occurrence of the mayfly family Teloganodidae in northern New South Wales. *Australian Journal of Entomology* **38**, 96–98. doi:10.1046/j.1440-6055.1999.00081.x

Chin K, Gill BD (1996) Dinosaurs, dung beetles, and conifers: participants in a Cretaceous food web. *Palaios* **11**, 280–285. doi:10.2307/3515235

Chisholm EC (1925) The Comboyne Plaeau: its general conformation and flora. *Proceedings of the Linnean Society of New South Wales* **50**, 284–298.

Christian CE (2001) Consequences of a biological invasion reveal the importance of mutualism for plant communities. *Nature* **413**, 635–639. doi:10.1038/35098093

Christidas F (2003) Systematics, phylogeny and ecology of Australian Leptophlebiidae (Ephemeroptera). PhD thesis. James Cook University, Australia.

Cipola NG, Wellington de Morais J, Bellini BC (2017) The discovery of *Lepidocyrtoides* Schött, 1917 (Collembola, Entomobryidae, Entombryinae) from the New World, including new species from Brazil and one from Australia. *Zootaxa* **4324**, 201–248. doi:10.11646/zootaxa.4324.2.2

Clark E (1936) The freshwater and land crayfishes of Australia. *Memoirs of the National Museum of Victoria* **10**, 5–58. doi:10.24199/j.mmv.1936.10.01

Clark J (1951) *The Formicidae of Australia. Vol. 1. Subfamily Myrmeciinae.* CSIRO Publishing, Melbourne.

Clark SA (2009) A review of the land snail genus *Meridolum* (Gastropoda: Camaenidae) from central New South Wales, Australia. *Molluscan Research* **29**, 61–120.

Clark WC (1978) Nematoda (Oxyuroidea: Thelastomatidae) from Australian passalid beetles (Coleoptera: Passalidae) with descriptions of new species. *Australian Journal of Zoology* **26**, 603–615. doi:10.1071/ZO9780603

Clarke G, Spier-Ashcroft F (2003) *A Review of the Conservation Status of Selected Australian Non-marine Invertebrates.* Environment Australia, Canberra.

Clarke PJ, Myerscough PJ (1991) Floral biology and reproductive phenology of *Avicennia marina* in South-eastern Australia. *Australian Journal of Botany* **39**, 283–293. doi:10.1071/BT9910283

Clements AN (1985) A taxonomic revision of the tribe Chysopogonini (Diptera: Asilidae). *Australian Journal of Zoology Supplementary Series* **109**.

Clements AN (2000) A revision of *Brachyrhopala* Macquart, an Australian region genus (Diptera: Asilidae). *Invertebrate Taxonomy* **14**, 77–114. doi:10.1071/IT97015

Cleveland LR, Hall SK, Sanders EP, Collier J (1934) The wood-feeding roach *Cryptocercus*, its protozoa, and the symbiosis between protozoa and roach. *Memoirs of the American Academy of Arts and Sciences* **17**, iii–342. doi:10.2307/25058198

Clift AD, Larsson SF (1987) Phoretic dispersal of *Brenandania lambi* (Kcrzal) (Acari: Tarsonemida: Pygmephoridae) by mushroom flies (Diptera: Sciaridae and Phoridae) in New South Wales, Australia. *Experimental & Applied Acarology* **3**, 11–20. doi:10.1007/BF01200409

Cockerell TDA (1921) Australian bees in the Queensland Museum. *Memoirs of the Queensland Museum* **7**, 81–98.

Coddington JA, Kuntner M, Opell BD (2013) Systematics of the spider family Deinopidae with a revision of the genus *Menneus*. *Smithsonian Contributions to Zoology* **636**, 1–61. doi:10.5479/si.00810282.636.1

Cogan BH (1989) Family Odiniidae. In *Catalog of the Diptera of the Australasian and Oceanic Regions.* (Ed. NL Evenhuis) p. 537. Bishop Museum Press and E.J. Brill, Honolulu.

Coley PD, Bryant JP, Chapin FS, III (1985) Resource availability and plant antiherbivore defense. *Science* **230**, 895–899. doi:10.1126/science.230.4728.895

Colledge WR (1911) Notes on a brush-tongued mosquito. *Proceedings of the Royal Society of Queensland* **23**, 121–130.

Colless DH (1970) The Mycetophilidae (Diptera) of Australia Part 1. Introduction, key to subfamilies, and new review of Ditomyiinae. *Journal of the Australian Entomological Society* **9**, 83–99. doi:10.1111/j.1440-6055.1970.tb00777.x

Colless DH (1980) Biogeography of Australian Sepsidae (Diptera). *Australian Journal of Zoology* **28**, 65–78. doi:10.1071/ZO9800065

Colless DH (1986) The Australian Chaoboridae (Diptera). *Australian Journal of Zoology. Supplementary Series Number* **124**.

Colless DH (1994) A new family of muscoid Diptera from Australasia, with sixteen new species in four new genera (Diptera: Axiniidae). *Invertebrate Taxonomy* **8**, 471–534. doi:10.1071/IT9940471

Colless DH (1998) Morphometrics in the genus *Amenia* and revisionary notes on the Australian Ameniinae (Diptera: Calliphoridae), with the description of eight new species. *Records of the Australian Museum* **50**, 85–123. doi:10.3853/j.0067-1975.50.1998.1275

Colless DH, McAlpine DK (1991) Diptera 'flies'. In *The Insects of Australia.* (Ed. ID Naumann) pp. 717–786. Melbourne University Press, Carlton.

Colloff MJ, Halliday RB (1998) *Oribatid Mites: A Catalogue of Australian Genera and Species. Monographs on Invertebrate Taxonomy. Vol. 6.* CSIRO Publishing, Melbourne.

Colwell RK (1995) Effects of nectar consumption by the hummingbird flower mite *Proctolaelaps kirmsei* on nectar availability in *Hamelia patens*. *Biotropica* **27**, 206–217. doi:10.2307/2388996

Common IFB (1990) *Moths of Australia.* Melbourne University Press, Carlton.

Common IFB (1997) *Oecophorine Genera of Australia. II. The* Chezala, Philobota *and* Eulechria *groups (Lepidoptera: Oecophoridae). Monographs on Australian Lepidoptera. Volume 5.* CSIRO Publishing, Melbourne.

Common IFB, Waterhouse DF (1981) *Butterflies of Australia.* Angus and Robertson, London.

Condé B, Pagés J (1991) Diplura. In *The Insects of Australia.* (Ed. ID Naumann) pp. 269–271. Melbourne University Press, Carlton.

Connell JH (1978) Diversity in tropical rain forests and coral reefs. *Science* **199**, 1302–1310. doi:10.1126/science.199.4335.1302

Connell JH, Tracey JG, Webb LJ (1984) Compensatory recruitment, growth, and mortality as factors maintaining rain forest tree diversity. *Ecological Monographs* **54**, 141–164. doi:10.2307/1942659

Connolly NM, Crossland MR, Pearson RG (2004) Effect of low dissolved oxygen on survival, emergence, and drift of tropical stream macroinvertebrates. *Journal of the North American Benthological Society* **23**, 251–270. doi:10.1899/0887-3593(2004)023<0251:EOLDOO>2.0.CO;2

Constant J (2010) Review of the Australian genus *Desudaboides* Musgrave with descriptions of four new species (Hemiptera: Fulgoromorpha: Fulgoridae). *Zootaxa* **2351**, 39–48. doi:10.11646/zootaxa.2351.1.4

Constantine RA, Seeman OD (2014) Three new species of *Eutarsopolipus* (Acari: Podapolipidae) from Australian carabid beetles (Coleoptera: Carabidae). *Systematic and Applied Acarology* **19**, 87–109. doi:10.11158/saa.19.1.8

Constantino R (1992) Abundance and diversity of termites (Insecta: Isoptera) in two sites of primary rain forest in Brazilian Amazonia. *Biotropica* **24**, 420–430. doi:10.2307/2388613

Conway Morris S (1979) Middle Cambrian polychaetes from the Burgess Shale of British Colombia. *Philosophical Transactions of the Royal Society of London. Series B, Biological Sciences* **285**, 227–274. doi:10.1098/rstb.1979.0006

Cook D (1988) Sexual selection in dung beetles. II. Female fecundity as an estimate of male reproductive success in relation to horn size, and alternative behavioural strategies in *Onthophagus binodis* Thunberg (Scarabaeidae: Onthophagini). *Australian Journal of Zoology* **36**, 521–532. doi:10.1071/ZO9880521

Cook EF (1971) The Australian Scatopsidae (Diptera). *Australian Journal of Zoology Supplementary Series* **8**.

Cook EF (1989) Family Scatopsidae. In *Catalog of the Diptera of the Australasian and Oceanic Regions*. (Ed. NL Evenhuis) pp. 182–185. Bishop Museum Press and E.J. Brill, Honolulu.

Cook RM, Parsons PA, Bock IR (1977) Australian endemic *Drosophila* II. A new *Hibiscus*-breeding species with its description. *Australian Journal of Zoology* **25**, 755–763. doi:10.1071/ZO9770755

Corcoran AJ, Conner WE, Barber JR (2010) Anti-bat tiger moth sounds: form and function. *Current Zoology* **56**, 358–369. doi:10.1093/czoolo/56.3.358

Corlett RT (2013) Where are the subtropics? *Biotropica* **45**, 273–275. doi:10.1111/btp.12028

Costa JT (2014) *Wallace, Darwin and the Origin of Species*. Harvard University Press, Cambridge.

Costa M, Allsopp PG (1979) Gamasine mites associated with Australian scarabaeid beetles 1. The genus *Paradoxiphis* Berlese, symbionts of Bolboceratini. *Australian Journal of Zoology* **27**, 825–865. doi:10.1071/ZO9790825

Coughran J (2002) A new species of the freshwater crayfish genus *Euastacus* (Decapoda: Parastacidae) from south-eastern New South Wales, Australia. *Records of the Australian Museum* **54**, 25–30. doi:10.3853/j.0067-1975.54.2002.1362

Coughran J (2005) New crayfishes (Decapoda: Parastacidae: *Euastacus*) from south-eastern New South Wales, Australia. *Records of the Australian Museum* **57**, 361–374. doi:10.3853/j.0067-1975.57.2005.1453

Coughran J (2008) Distinct groups in the genus *Euastacus*? *Freshwater Crayfish* **16**, 125–132.

Coughran J, Leckie S (2007) *Euastacus pilosus* sp. n., a new crayfish from the highland forests of northern New South Wales, Australia. *Fishes of Sahul* **21**, 309–316.

Coughran J, McCormack RB (2011) *Euastacus morgani* sp. n., a new spiny crayfish (Crustacea, Decapoda, Parastacidae) from the highland rainforests of eastern New South Wales, Australia. *ZooKeys* **85**, 17–26. doi:10.3897/zookeys.85.1237

Coupland JB, Barker GM (2004) Diptera as predators and parasitoids of terrestrial gastropods, with emphasis on Phoridae, Calliphoridae, Sarcophagidae, Muscidae and Fanniidae. In *Natural Enemies of Terrestrial Molluscs*. (Ed. GM Barker) pp. 185–158. CABI, Wallingford.

Craig DA, Currie DC, Gil-Azevedo LH, Moulton JK (2019) *Austrocnephia*, new genus, for five species of 'Paracnephia' (Diptera: Simuliidae), with a key to Australian black fly genera. *Zootaxa* **4627**, 1–92.

Crandall KA, Fetzner JW, Kinnersley M, Austin CM (1999) Phylogenetic relationships among the Australian and New Zealand genera of freshwater crayfishes (Decapoda: Parastacidae). *Australian Journal of Zoology* **47**, 199–214. doi:10.1071/ZO99011

Cranston PS (1997) Revision of Australian *Rheotanytarsus* Thienemann and Bause (Diptera: Chironomidae) with emphasis on immature stages. *Invertebrate Taxonomy* **11**, 705–734. doi:10.1071/IT96025

Cranston PS, Edward DHD, Cook LG (2002) New status, species, distribution records and phylogeny for Australian mandibulate Chironomidae (Diptera). *Australian Journal of Entomology* **41**, 357–366. doi:10.1046/j.1440-6055.2002.00304.x

Cranston PS, Gullan PJ, Taylor RW (1991) Principles and practice of systematics. In *The Insects of Australia*. (Ed. ID Naumann) pp. 109–124. Melbourne University Press, Carlton.

Cranston PS, Martin J (1989) Family Chironomidae. In *Catalog of the Diptera of the Australasian and Oceanic Regions*. (Ed. NL Evenhuis) pp. 253–274. Bishop Museum Press and E.J. Brill, Honolulu.

Craven LA (2006) New combinations in *Melaleuca* for Australian species of *Callistemon* (Myrtaceae). *Novon* **16**, 168–175.

Crepet WL (1984) Advanced (constant) insect pollination mechanisms: pattern of evolution and implications *vis-à-vis* angiosperm diversity. *Annals of the Missouri Botanical Garden* **71**, 607–630. doi:10.2307/2399041

Cresson ET (1948) A systematic annotated arrangement of the genera and species of the Indo-Australian Ephydridae (Diptera). II. The subfamily Notophilinae and supplent to Part I on the subfamily Psilopoinae. *Transactions of the American Entomological Society* **74**, 1–28.

Crichton A, Francis N, Doherty S, Tuiwawa C, Hayes S, Stevens MI, Schwarz MP (2019) Low endemic bee diversity and very wide host range in lowland Fiji: support for the pollinator super-generalist hypothesis in island biogeography. *Pacific Conservation Biology* **25**, 135–142. doi:10.1071/PC18037

Crome FHJ, Irvine AK (1986) 'Two bob each way': the pollination and breeding system of the Australian rainforest tree *Syzygium cormiflorum* (Myrtaceae). *Biotropica* **18**, 115–125. doi:10.2307/2388754

Crosskey RW (1965) A systematic revision of the Ameniinae (Diptera: Calliphoridae). *Bulletin of the British Museum of Natural History (Entomology)* **16**, 35–140.

Crosskey RW (1973) A revisionary classification of the Rutilini (Diptera: Tachinidae), with keys to the described species. *Bulletin of the British Museum (Natural History), Entomology Supplement* **19**, 1–167.

Crowson RA (1981) *The Biology of the Coleoptera.* Academic Press, London.

CSIRO (1996) *Murwillumbah Management Area Fauna Survey 1995.* CSIRO Division of Wildlife and Ecology, Sydney.

Culliney TW (2013) Roles of arthropods in maintaining soil fertility. *Agriculture* **3**, 629–659. doi:10.3390/agriculture3040629

Cummins KW, Lauff GF (1969) The influence of substrate particle size on the distribution of stream benthos. *Hydrobiologia* **34**, 145–181. doi:10.1007/BF00141925

Cunningham SA, Schellhorn NA, Marcora A, Batley M (2013) Movement and phenology of bees in a subtropical Australian agricultural landscape. *Austral Ecology* **38**, 456–464. doi:10.1111/j.1442-9993.2012.02432.x

Curletti G (2001) The genus *Agrilus* in Australia (Coleoptera: Buprestidae). In *Jewel Beetles 9.* (Ed. S Endo) pp. 1–5. Kittyokai Society, Tokyo.

Curran TJ, Clarke PJ, Bruhl JJ (2008) A broad typology of dry rainforests on the western slopes of New South Wales. *Cunninghamia* **10**, 381–405.

Curran TJ, Clarke PJ, Warwick NWM (2013) Drought survival of Australian rainforest seedlings is influenced by species evolutionary history and soil type. *Australian Journal of Botany* **61**, 22–28. doi:10.1071/BT12081

Cushing PE (2012) Spider–ant associations: an updated review of myrmecomorphy, myrmecophily, and myrmecophagy in spiders. *Psyche* **2012**, 151–989. doi:10.1155/2012/151989

Rocha-Filho LC (2016) A revision of the cleptoparasitic bee genus *Coelioxys* (Hymenoptera: Megachilidae) from Australia. *European Journal of Entomology* **113**, 9–28. doi:10.14411/eje.2016.002

da Silva FR, de Cerqueira Rossa-Feres D (2017) Fragmentation gradients differentially affect species range distributions of four taxonomic groups in semi-deciduous Atlantic forest. *Biotropica* **49**, 283–292. doi:10.1111/btp.12362

Dafni A (1992) *Pollination Ecology: A Practical Approach.* Oxford University Press, Oxford.

Dainese M, Riedinger V, Holzschuh A, Kleijn D, Scheper J, Steffan-Dewenter I (2018) Managing trap-nesting bees as crop pollinators: spatiotemporal effects of floral resources and antagonists. *Journal of Applied Ecology* **55**, 195–204. doi:10.1111/1365- 2664.12930

Dale M, Kershaw P, Kikkawa J, Parsons P, Webb L (1980) Resolution by the Ecological Society of Australia on Australian rainforest conservation. *Bulletin of the Ecological Society of Australia* **10**, 6–7.

Dalgleish EA, Elgar MA (2005) Breeding ecology of the rainforest dung beetle *Cephalodesmius armiger* (Scarabaeidae) in Tooloom National Park. *Australian Journal of Zoology* **53**, 95–98. doi:10.1071/ZO04067

Danforth BN, Sipes S, Fang J, Brady SG (2006) The history of early bee diversification based on five genes plus morphology. *Proceedings of the National Academy of Sciences of the United States of America* **103**, 15118–15123. doi:10.1073/pnas.0604033103

Dangerfield PC, Austin AD (1995) Revision of the Australasian species of Cardiochilinae (Hymenoptera: Braconidae). *Invertebrate Taxonomy* **9**, 387–445. doi:10.1071/IT9950387

Daniels G (1976) A new locality for *Signeta tymbophora* (Meyrick and Lower) (Lepidoptera: Hesperiidae). *Australian Entomological Magazine* **3**, 29.

Daniels G (1977) Two new species of *Pelecorhynchus* Macquart (Diptera: Pelecorhynchidae), with the first record of the family from Western Australia. *Australian Entomological Magazine* **4**, 73–75.

Daniels G (1978) A catalogue of the type specimens of Diptera in the Australian Museum. *Records of the Australian Museum* **31**, 411–471. doi:10.3853/j.0067-1975.31.1978.222

Daniels G (1987) A revision of *Neoaratus* Ricardo, with the description of six allied new genera from the Australian region (Diptera: Asilidae: Asilini). *Invertebrate Taxonomy* **1**, 473–592. doi:10.1071/IT9870473

Daniels G (1989a) Family Pelecorhynchidae. In *Catalog of the Diptera of the Australasian and Oceanic Regions.* (Ed. NL Evenhuis) pp. 275–276. Bishop Museum Press and E.J. Brill, Honolulu.

Daniels G (1989b) Family Tabanidae. In *Catalog of the Diptera of the Australasian and Oceanic Regions.* (Ed. NL Evenhuis) pp. 277–294. Bishop Museum Press and E.J. Brill, Honolulu.

Daniels G (1989c) Family Exeretonevridae. In *Catalog of the Diptera of the Australasian and Oceanic Regions.* (Ed. NL Evenhuis) p. 321. Bishop Museum Press and E.J. Brill, Honolulu.

Daniels G (1989d) Family Asilidae. In *Catalog of the Diptera of the Australasian and Oceanic Regions.* (Ed. NL Evenhuis) pp. 326–349. Bishop Museum Press and E.J. Brill, Honolulu.

Darling DC, Sharkey MJ (1990) Hymenoptera. *Bulletin of the American Museum of Natural History* **195**, 123–153.

Darlington JPEC, Gelhaus JK (1994) Crane flies (Diptera: Tipulidae) in Trinidad caves. *Living World: Journal of The Trinidad and Tobago Field Naturalists' Club* **1993–1994**, 38–41.

Darlington PJ (1960) Australian carabid beetles IV. List of localities, 1956–1958. *Psyche* **67**, 111–126. doi:10.1155/1960/81583

Darlington PJ (1961a) Australian carabid beetles V. Transition of wet forest faunas from New Guinea - Tasmania. *Psyche* **68**, 1–24. doi:10.1155/1961/64175

Darlington PJ (1961b) Australian carabid beetles VII. *Trichosternus*, especially the tropical species. *Psyche* **68**, 113–130. doi:10.1155/1961/84320

Darlington PJ (1965) *Biogeography of the Southern End of the World. Distribution and History of Far-southern Life and Land, with an Assessment of Continental Drift.* Harvard University Press, Cambridge.

Darlington PJ (1979) Existing trans-tropical and southern pathways of dispersal for Carabidae. In *Carabid Beetles: Their Evolution, Natural History, and Classification.* (Eds TL Erwin, GE Ball, DR Whitehead and AL Halpern) pp. 343–348. Dr W. Junk, Publishers, The Hague.

Darwell CT, al-Beidh S, Cook JM (2014) Molecular species delimitation of a symbiotic fig-pollinating wasp species complex reveals extreme deviation from reciprocal partner specificity. *BMC Evolutionary Biology* **2014**, 189–198. doi:10.11.86/s12862-014-0189-9

Davies VT, Lambkin C (2001) A revision of *Procambridgea* Forster and Wilton, (Araneae: Amaurioidea: Stiphidiidae). *Memoirs of the Queensland Museum* **46**, 443–459.

Davis JA (1986) Revision of the Australian Psephenidae (Coleoptera): systematics, phylogeny and historical biogeography. *Australian Journal of Zoology Supplementary Series* **119**, 1–97.

Davis RA, Disney RHL (2003) Natural history and description of *Aphiura breviceps* Schmitz, a scuttle fly (Diptera: Phoridae) whose larvae prey on the eggs of frogs (Anura: Myobatrachidae) in Western Australia. *Austral Ecology* **42**, 18–21.

Dawes-Gromadzki TZ (2005) The termite (Isoptera) fauna of a monsoonal forest near Darwin, northern Australia. *Austral Entomology* **44**, 152–157. doi:10.1111/j.1440-6055.2005.00452.x

Day MD, Nahrung HF (1998) Records of Australian Cerambycidae (Coleoptera) from *Lantana camara*. *Australian Journal of Entomology* **37**, 285–286. doi:10.1111/j.1440-6055.1998.tb01585.x

Day MF (1999) The genera of Australian Membracidae (Hemiptera: Auchenorrhyncha). *Invertebrate Taxonomy* **13**, 629–747. doi:10.1071/IT98010

Day MF, Fletcher MJ (1994) An annotated catalogue of the Australian Cicadelloidea (Hemiptera: Auchenorrhyncha). *Invertebrate Taxonomy* **8**, 1117–1288. doi:10.1071/IT9941117

Dear JP (1977) A revision of Australian Rhiniinae (Diptera: Calliphoridae). *Australian Journal of Zoology* **25**, 779–826.

De Baar M (1976) Notes on Hesperiidae and Lycaenidae (Lepidoptera) from south-east Queensland. *Australian Entomological Magazine* **2**, 123–124.

De Baar M (1977) Butterflies from an area between the Bunya Mountains and Archookoora State Forest, Queensland. *Australian Entomological Magazine* **3**, 115–119.

De Boer AJ (1997) Phylogeny and biogeography of Australian genera of Chlorocystini (Insecta: Homoptera: Tibicinidae). *Memoirs of the Museum of Victoria* **56**, 91–123. doi:10.24199/j.mmv.1997.56.05

de Jong H, Osterbroek P, Gelhaus J, Reusch H, Young C (2008) Global diversity of craneflies (Insecta, Diptera: Tipulidea or Tipulidae *sensu lato*) in freshwater. *Hydrobiologia* **595**, 457–467. doi:10.1007/s10750-007-9131-0

De Luca PA, Vallejo-Marín M (2013) What's the 'buzz' about? The ecology and evolutionary significance of buzz-pollination. *Current Opinion in Plant Biology* **16**, 429–435. doi:10.1016/j.pbi.2013.05.002

De Sena Oliveira I, Mayer G (2012) A world checklist of Onychophora (velvet worms), with notes on nomenclature and status of names. *ZooKeys* **211**, 1–70. doi:10.3897/zookeys.211.3463

de Souza Lopes H (1989) Family Sarcophagidae. In *Catalog of the Diptera of the Australasian and Oceanic Regions*. (Ed. NL Evenhuis) pp. 721–732. Bishop Museum Press and E.J. Brill, Honolulu.

de Souza Lopes H (2010) *Sarcophagidae*. In *Manual of Central American Diptera. Vol. 2*. (Eds BV Brown, A Borkent, JM Cumming, DM Wood, NE Woodley and MA Zumbado) pp. 721–732. NRC Research Press, Ottawa.

Debenham ML (1972) Australian and New Guinea 'picture wing' species of the genus *Monohelea* Kieffer (Diptera: Ceratopogonidae). *Australian Journal of Zoology Supplementary Series* **12**, 1–40.

Debenham ML (1987a) The biting midge genus *Forcipomyia* (Diptera: Ceratopogonidae) in the Australian region (exclusive of New Zealand). I. Introduction, key to subgenera, and the *Thyridomyia* and *Trichohelea* groups of subgenera. *Invertebrate Taxonomy* **1**, 35–119. doi:10.1071/IT9870035

Debenham ML (1987b) The biting midge genus *Forcipomyia* (Diptera: Ceratopogonidae) in the Australian region (exclusive of New Zealand). IV. The subgenera allied to *Forcipomyia, s.s.*, and *Lepidohelea*, and the interrelationships and biogeography of the subgenera of *Forcipomyia*. *Invertebrate Taxonomy* **1**, 631–684. doi:10.1071/IT9870631

Debenham ML (1989) Family Ceratopogonidae. In *Catalog of the Diptera of the Australasian and Oceanic Regions*. (Ed. NL Evenhuis) pp. 226–251. Bishop Museum Press and E.J. Brill, Honolulu.

Debenham ML (1991) Australian and New Guinea species of the biting midge genus *Brachypogon* (Diptera: Ceratopogonidae). *Invertebrate Taxonomy* **5**, 765–806. doi:10.1071/IT9910765

Debevec AH, Cardinal S, Danforth BD (2012) Identifying the sister group to the bees: a molecular phylogeny of Aculeata with an emphasis on the subfamily Apoidea. *Zoologica Scripta* **41**, 527–535. doi:10.1111/j.1463-6409.2012.00549.x

Degen AA, Gersani M, Avivi Y, Weisbrot N (1986) Honeydew intake of the weaver ant *Polyrhachis simplex* (Hymenoptera: Formicidae) attending the aphid *Chaitophorous populialbae* (Homoptera: Aphididae). *Insectes Sociaux* **33**, 211–215. doi:10.1007/BF02224599

Dem FF (2015) Community parameters of Auchenorrhyncha (Hemiptera: Insecta) along altitudinal gradients in subtropical and tropical rainforests in Australia. PhD thesis. Griffith University, Australia.

Dem FF, Stewart AJA, Gibson A, Weiblen GD (2013) Low host specificity in species-rich assemblages of xylem- and phloem-feeding herbivores (Auchenorrhyncha) in a New Guinea lowland rain forest. *Journal of Tropical Ecology* **29**, 467–476. doi:10.1017/S0266467413000540

Department of the Arts, Sport, the Environment and Territories (DASET) (1992) *Nomination of the Central Eastern Rainforests of Australia for Inclusion on the World Heritage List*. Government of Australia, Canberra.

Department of the Environment, Sport and Territories (DEST) (1994) *Australia's Biodiversity: An Overview of Selected Significant Components*. Commonwealth of Australia, Canberra.

Dettner K (1987) Chemosystematics and evolution of beetle chemical defenses. *Annual Review of Entomology* **32**, 17–48. doi:10.1146/annurev.en.32.010187.000313

Deuquet CM (1938) Description of three new species of *Stigmodera* (Buprestidae). *Proceedings of the Linnean Society of New South Wales* **63**, 305–307.

Deuquet CM (1956) Notes on Australian Buprestidae, with descriptions of three new species and two subspecies of the genus *Stigmodera*, subgenus *Castiarina*. *Proceedings of the Linnean Society of New South Wales* **81**, 153–156.

Deuquet CM (1963) Notes on the genus *Stigmodera* (Buprestidae) and description of new species. *Proceedings of the Linnean Society of New South Wales* **88**, 336–339.

Devetak D, Arnett AE (2015) Preference of antlion and wormlion larvae (Neuroptera: Myrmeleontidae; Diptera: Vermileonidae) for substrates according to substrate particle sizes. *European Journal of Entomology* **112**, 500–509. doi:10.14411/eje.2015.052

Devoto M, Medan D (2006) Diversity, distribution and floral specificity of tangle-veined flies (Diptera: Nemestrinidae) in north west Patagonia, Argentina. *Revista Chilena de Historia Natural* **79**, 29–40. doi:10.4067/S0716-078X2006000100003

Dhileepan K, Treviño M, Snow L (2006) *Application to Release the Leaf-sucking Bug* Carvalhotingis visenda *(Hemiptera: Tingidae), A Potential Biological Control Agent for Cat's Claw Creeper* Macfadyena unguis-cati *(Bignoniaceae)*. Queensland Department of Natural Resources, Mines and Water, Brisbane.

Dhileepan K, Treviño M, Snow L (2007) Specificity of *Carvalhotingis visenda* (Hemiptera: Tingidae) as a biological control agent for cat's claw creeper (Bignoniaceae) in Australia. *Biological Control* **41**, 283–290. doi:10.1016/j.biocontrol.2007.02.006

Dibb JR (1938) Synopsis of Australian Passalideae (Coleoptera). *Transactions of the Royal Entomological Society of London* **87**, 103–124. doi:10.1111/j.1365-2311.1938.tb00092.x

Didham RK (1997) An overview of invertebrate responses to forest fragmentation. In *Forests and Insects*. (Eds AD Watt, NE Stork, and MD Hunter) pp. 303–320. Chapman and Hall, London.

Didham RK, Lawton JH (1999) Edge structure determines the magnitude of changes in microclimate and vegetation structure in tropical forest fragments. *Biotropica* **31**, 17–30.

Dijkstra K-D B, Bechley G, Bybee SM, Dow RA, Dumont HJ, Fleck G, Garrison RW, Hämäläinen M, Kalkman VJ, Karube H, May ML, Orr AG, Paulsen DR, Rehn AC, Theischinger G, Trueman JWH, Tol JV, von Ellenreider N, Ware J (2013) The classification and diversity of dragonflies and damselflies (Odonata). *Zootaxa* **3703**, 036–045.

Dijkstra K-DB, Kalkman VJ, Dow RA, Stokvis FR, van Tol JV (2014) Redefining the damselfly families: a comprehensive phylogeny of Zygoptera (Odonata). *Systematic Entomology* **39**, 68–96. doi:10.1111/syen.12035

Dimitrov D, Benavides LR, Arnedo MA, Giribet G, Griswold CE, Scharff N, Hormiga G (2017) Rounding up the usual suspects: a standard target-gene approach for resolving the interfamilial phylogenetic relationships of ecribellate orb-weaving spiders with a new family-rank classification (Araneae, Araneoidea). *Cladistics* **33**, 221–250. doi:10.1111/cla.12165

Disney RHL (1994) *Scuttle Flies: The Phoridae*. Chapman and Hall, London.

Disney RHL, Roisin Y (2000) The first termitophilous Scuttle fly (Diptera: Phoridae) reported from Papua New Guinea. *Sociobiology* **35**, 17–24.

Dobson HEM (2006) Relationship between floral fragrance composition and type of pollinator. In *Biology of Floral Scent*. (Eds N Dudareva and E Pichersky) pp. 147–198. CRC Press, Boca Raton, FL.

Dodson G, Daniels G (1988) Diptera reared from *Dysoxylum gaudichaudianum* (Juss.) Miq. at Iron Range, northern Queensland. *Australian Entomologist* **15**, 77–79.

Doherty MD, Robertson G, Corcoran D, Wright G (2011) Cool temperate rainforest in the Pilot Wilderness Area, Kosciusko National Park, New South Wales: distribution, composition and impact of the 2003 fires. *Cunninghamia* **12**, 119–127.

Dollin A, Batley M, Robinson M, Faulkner B (2000) *Native Bees of the Sydney Region, A field Guide*. Australian Native Bee Research Centre, Richmond.

Dollin A, Dollin LJ, Sakagami SF (1997) Australian stingless bees of the genus *Trigona* (Hymenoptera: Apidae). *Invertebrate Taxonomy* **11**, 861–896. doi:10.1071/IT96020

Domrow R (1978) New records and species of chiggers from Australasia (Acari: Trombiculidae). *Journal of the Australian Entomological Society* **17**, 75–90. doi:10.1111/j.1440-6055.1978.tb01748.x

Domrow R (1979) Ascid and ameroseiid mites phoretic on Australian mammals and birds. *Records of the Western Australian Museum* **8**, 97–116.

Domrow R (1991) Acari Prostigmata (excluding Trombiculidae) parasitic on Australian vertebrates: an annotated checklist, keys and bibliography. *Invertebrate Taxonomy* **4**, 1283–1376.

Donaldson JF (1983) Revision of the Australian Asiracinae (Homoptera: Fulgoroidea: Delphacidae). *Journal of the Australian Entomological Society* **22**, 277–285. doi:10.1111/j.1440-6055.1983.tb02097.x

Donaldson JF (1991) Revision of the Australian Tropidocephalini (Hemiptera: Delphacidae: Delphaceninae). *Journal of the Australian Entomological Society* **30**, 325–332. doi:10.1111/j.1440-6055.1991.tb00447.x

Dörfel TH, Ohl M (2015) A revision of the Australian digger wasps in the genus *Sphex* (Hymenoptera, Sphecidae). *ZooKeys* **521**, 1–104. doi:10.3897/zookeys.521.5995

Doust JL, Doust LL (Eds) (1990) *Plant Reproductive Ecology: Patterns and Strategies*. Oxford University Press, New York.

Downes JA, Colless DH (1967) Mouthparts of the biting and blood-sucking type in Tanyderidae and Chironomidae (Diptera). *Nature* **214**, 1355–1356. doi:10.1038/2141355a0

Downes MF (2017) Mermithid nematodes hosted by *Polyrhachis* weaver ants (Hymenoptera: Formicidae) in North Queensland, including multiparasitism with an insect parasitoid. *Australian Entomologist* **44**, 147–160.

Drake CJ, Slater JA (1957) The phylogeny and systematics of the family Thaumastocoridae (Hemiptera: Heteroptera). *Annals of the Entomological Society of America* **50**, 353–370. doi:10.1093/aesa/50.4.353

Drew RAI (1987) Reduction in fruit-fly (Tephretidae–Dacinae) populations in their endemic rain-forest habitat by frugivorous vertebrates. *Australian Journal of Zoology* **35**, 283–288. doi:10.1071/ZO9870283

Drew RAI (2004) Biogeography and speciation in the Dacini (Diptera: Tephritidae: Dacinae). In *Contributions to the Systematics and Evolution of Diptera*. (Eds NL Evenhuis and KY Kaneshiro). Elmo Hardy Memorial Volume. *Bishop Museum Bulletin in Entomology* **12**, 165–178.

Druart D, Millet M, Scheifler R, Delhomme O, de Vaufleury A (2011) Glyphosate and glufosinate-based herbicides: fate in soil, transfer to, and effects on land snails. *Journal of Soils and Sediments* **11**, 1373–1384. doi:10.1007/s11368-011-0409-5

Duckhouse DA (1965) Psychodidae (Diptera, Nematocera) of Southern Australia, subfamilies Bruchomyiinae and Trichomyiinae. *Ecological Entomology* **117**, 329–343.

Duckhouse DA (1966) Psychodidae (Diptera, Nematocera) of Southern Australia: subfamily Psychodinae. *Transactions of the Royal Entomological Society of London* **118**, 153–220. doi:10.1111/j.1365-2311.1966.tb00837.x

Duckhouse DA (1990) The Australasian genera of pericomoid Psychodidae (Diptera) and the status of related Enderlein genera in the tropics. *Invertebrate Taxonomy* **3**, 721–746.

Duckhouse DA, Lewis DJ (1989) Family Psychodidae. In *Catalog of the Diptera of the Australasian and Oceanic Regions*. (Ed. NL Evenhuis) pp. 166–179. Bishop Museum Press and E.J. Brill, Honolulu.

Duffy EAJ (1963) *A Monograph of the Immature Stages of Australasian Timber Beetles*. British Museum (Natural History), London.

Duffy Masters C (2010) The Main Range. In *Remnants of Gondwana*. (Eds R Kitching, R Braithwaite and J Cavanaugh) pp. 15–24. Surrey Beatty and Sons, Baulkham Hills.

Dumbleton LJ (1952) A new genus of seed-infesting micro-terygid moths. *Pacific Science* **6**, 17–29.

Dunnet GM, Mardon DK (1973) *Coorilla longictena*, a new genus and species of bat-flea from New South Wales (Siphonaptera: Ischnopsyllidae). *Journal of the Australian Entomological Society* **12**, 3–10. doi:10.1111/j.1440-6055.1973.tb02149.x

Dunnet GM, Mardon DK (1991) Siphonaptera 'fleas'. In *The Insects of Australia*. (Ed. ID Naumann) pp. 705–716. Melbourne University Press, Carlton.

Dyce AL, Wellings G (1991) Phlebotomine sandflies (Diptera: Psychodidae) from caves in Australia. *Parasitology* **33**, 193–198.

Early JW, Naumann ID (1989) *Rostropria*, a new genus of opisthognathus Diapriine wasp from Australia, and notes on the genus *Neurogalesus* (Hymenoptera: Proctotrupoidea: Diapriidae). *Invertebrate Taxonomy* **3**, 523–550. doi:10.1071/IT9890523

Eastwood R (1997) An interesting local form and new larval hostplant of *Hypochrysops byzos* (Boisduval) (Lepidoptera: Lycaenidae). *Australian Entomologist* **24**, 37–38.

Eastwood R, Fraser AM (1999) Associations between lycaenid butterflies and ants in Australia. *Australian Journal of Ecology* **24**, 503–537. doi:10.1046/j.1440-169x.1999.01000.x

Eberhard WG (1980) Horned beetles. *Scientific American* **242**, 166–182. doi:10.1038/scientificamerican0380-166

Edgar JA, Culvenor CCJ, Pliske TE (1974) Coevolution of danaid butterflies with their host plants. *Nature* **250**, 646–648. doi:10.1038/250646a0

Edgecombe GD (2001) Revision of *Paralamcytes* (Chilopoda: Lithobiomorpha: Henicopidae) with six new species from Eastern Australia. *Records of the Australian Museum* **53**, 201–241. doi:10.3853/j.0067-1975.53.2001.1328

Edgerly JS (1987) Maternal behavior of a webspinner (order Embiidina). *Ecological Entomology* **12**, 1–11. doi:10.1111/j.1365-2311.1987.tb00979.x

Edgerly JS (1994) Is group living an antipredator defense in a facultatively communal webspinner? (Embiidina: Clothodidae). *Journal of Insect Behavior* **7**, 135–147. doi:10.1007/BF01990076

Edwards CA, Bohlen PJ (1996) *Biology and Ecology of Earthworms*. Chapman and Hall, London.

Ehrlich PR, Raven P (1964) Butterflies and plants: a study of coevolution. *Evolution* **18**, 586–608. doi:10.1111/j.1558-5646.1964.tb01674.x

Eisner T, Schroeder FC, Snyder N, Grant JB, Aneshansley DJ, Utterback D, Meinwald J, Eisner M (2008) Defensive chemistry of lycid beetles and mimetic cerambycid beetles that feed on them. *Chemoecology* **18**, 109–119. doi:10.1007/s00049-007-0398-4

Eisner T, Smedley SR, Young DK, Eisner M, Roach B, Meinwald J (1996) Chemical basis of courtship in a beetle (*Neopyrochroa flabellata*): cantharadin as precopulatory 'enticing' agent; cantharadin as 'nuptial gift'. *Proceedings of the National Academy of Sciences, U.S.A.* **93**, 6494–6503.

Eldridge DJ, James AI (2009) Soil-disturbance by native animals plays a critical role in maintaining healthy Australian landscapes. *Ecological Management & Restoration* **10**, S27–S34. doi:10.1111/j.1442-8903.2009.00452.x

Elias LC, Lino-Neto J, Pereira RAS (2018) Oogenisis and ovarian morphology in pollinating and non-pollinating fig wasps: evidence from adult and immature stages. *Invertebrate Reproduction & Development* doi:10.1080/07924259.2018.1469549

Elliott MG (2011) Annotated catalogue of the Australian Scoliidae (Hymenoptera). *Technical Reports of the Australian Museum Online (Bergheim)* **22**, 1–17.

Ellwood MDF, Blüthgen N, Fayle TM, Foster WA, Menzel F (2016) Competition can lead to unexpected patterns in tropical ant communities. *Acta Oecologica* **75**, 24–34. doi:10.1016/j.actao.2016.06.001

Emlen DJ (2000) Integrating development with evolution: a case study with beetle horns. *Bioscience* **50**, 403–418. doi:10.1641/0006-3568(2000)050[0403:IDWEAC]2.0.CO;2

Endler JA (1981) An overview of the relationships between mimicry and crypsis. *Biological Journal of the Linnean Society. Linnean Society of London* **16**, 25–31. doi:10.1111/j.1095-8312.1981.tb01840.x

Endrödy-Younga S (1990) A revision of the Australian Clambidae (Coleoptera: Eucinetoidea). *Invertebrate Taxonomy* **4**, 247–280. doi:10.1071/IT9900247

Engel MS (2005) A new sawfly from the Triassic of Queensland (Hymenoptera: Xyelidae). *Memoirs of the Queensland Museum* **51**, 558.

Eriksson ME, Bergman CF, Jeppsson L (2004) Silurian scolecodonts. *Review of Palaeobotany and Palynology* **131**, 269–300. doi:10.1016/j.revpalbo.2004.04.001

Erséus C (2005) Phylogeny of oligochaetous Clitellata. *Hydrobiologia* **535**, 357–372. doi:10.1007/s10750-004-4426-x

Erwin TL (1982) Tropical forests: their richness in Coleoptera and other arthropod species. *Coleopterists Bulletin* **36**, 74–75.

Erwin TL, Ball GE, Whitehead DR, Halpern AL (Eds) (1979) *Carabid Beetles: Their Evolution, Natural History, and Classification.* Dr W. Junk, Publishers, The Hague.

Escalona HE, Ślipiński A (2011) *Australodon*, a remarkable new genus of Australian longhorned beetle (Coleoptera: Cerambycidae: Cerambycinae). *Annales Zoologici* **61**, 731–738. doi:10.3161/000345411X622552

Ettershank G (1966) A generic revision of the world Myrmecinae related to *Solenopsis* and *Pheidologeton* (Hymenoptera: Formicidae). *Australian Journal of Zoology* **14**, 73–171. doi:10.1071/ZO9660073

Evans F (1981) The Tartessinae of Australia, New Guinea and some adjacent islands (Homoptera: Cicadellidae). *Pacific Insects* **23**, 112–188.

Evans H (1974) A review of *Psoropempula*, a new genus of Australian Pompilidae (Hymenoptera). *Transactions of the Royal Entomological Society of London* **126**, 261–278. doi:10.1111/j.1365-2311.1974.tb00854.x

Evans H (1982) The 'Collaris' complex of Australian mimetic Pompilidae (Hymenoptera). *Pacific Insects* **24**, 151–155.

Evans H, Hook AW (1986) Nesting behavior of Austraslian *Cerceris* digger wasps, with special reference to nest reutilization and nest sharing. *Sociobiology* **11**, 275–302.

Evans HE, Matthews RW (1973) *Systematics and Nesting Behavior of Australian* Bembix *Sand Wasps (Hymenoptera, Sphecidae).* American Entomological Institute, Ann Arbor, Mich.

Evans JW (1966) *The Leafhoppers and Froghoppers of Australia and New Zealand (Homoptera: Cicadelloidea and Cercopoidea).* Australian Museum, Sydney.

Evans JW (1988) Some aspects of the biology, morphology, and evolution of leafhoppers (Homoptera: Cicadelloidea and Membracoidea). *Great Basin Naturalist Memoirs* **12**, 61–66.

Evenhuis NL (1979) Studies in Pacific Bombyliidae (Diptera). II. Revision of the genus *Geron* of Australia and the Pacific. *Pacific Insects* **21**, 13–55.

Evenhuis NL (1989a) Family Bombyliidae. In *Catalog of the Diptera of the Australasian and Oceanic Regions.* (Ed. NL Evenhuis) pp. 359–374. Bishop Museum Press and E.J. Brill, Honolulu.

Evenhuis NL (1989b) Family Phoridae. In *Catalog of the Diptera of the Australasian and Oceanic Regions.* (Ed. NL Evenhuis) pp. 423–431. Bishop Museum Press and E.J. Brill, Honolulu.

Evenhuis NL (1989c) Family Platysomatidae. In *Catalog of the Diptera of the Australasian and Oceanic Regions.* (Ed. NL Evenhuis) pp. 482–497. Bishop Museum Press and E.J. Brill, Honolulu.

Evenhuis NL (2002) Review of the Tertiary microbombyliids (Diptera: Mythicomyiidae) in Baltic, Bitterfeld, and Dominican amber. *Zootaxa* **100**, 1–15. doi:10.11646/zootaxa.100.1.1

Evenhuis NL (2018) *Catalog of the Diptera of the Australasian and Oceanian Regions (online version).* <http://hbs.bishopmuseum.org/aocat>.

Evenhuis NL, Gon S, III (1989) Family Culicidae. In *Catalog of the Diptera of the Australasian and Oceanic Regions.* (Ed. NL Evenhuis) pp. 191–218. Bishop Museum Press and E.J. Brill, Honolulu.

Evenhuis NL, Okada T (1989) Family Drosophilidae. In *Catalog of the Diptera of the Australasian and Oceanic Regions.* (Ed. NL Evenhuis) pp. 609–638. Bishop Museum Press and E.J. Brill, Honolulu.

Evenhuis NL, Okadome T (1989) Family Lauxaniidae. In *Catalog of the Diptera of the Australasian and Oceanic Regions.* (Ed. NL Evenhuis) pp. 576–589. Bishop Museum Press and E.J. Brill, Honolulu.

Ewart A (2001) Emergence patterns and densities of cicadas (Hemiptera: Cicadidae). *Australian Entomologist* **28**, 69–84.

Exley EM (1968a) Revision of the genus *Brachyhesma* Michener (Apoidea: Colletidae). *Australian Journal of Zoology* **16**, 167–201. doi:10.1071/ZO9680167

Exley EM (1968b) Revision of the genus *Euryglossella* Cockerell (Apoidea: Colletidae). *Australian Journal of Zoology* **16**, 219–226. doi:10.1071/ZO9680219

Exley EM (1968c) Revision of the genus *Euryglossina* Cockerell (Apoidea: Colletidae). *Australian Journal of Zoology* **16**, 915–1020. doi:10.1071/ZO9680915

Exley EM (1972) Revision of the genus *Pachyprosopis* Perkins (Apoidea: Colletidae). *Australian Journal of Zoology Supplementary Series* **10**, 1–43.

Exley EM (1975) Revision of the genus *Hyphesma* Michener (Apoidea: Colletidae). *Australian Journal of Zoology* **23**, 277–291. doi:10.1071/ZO9750277

Exley EM (1976) Revision of the subgenus *Euryglossa* Smith (Apoidea: Colletidae: Euryglossinae). *Australian Journal of Zoology Supplementary Series* **41**, 1–72.

Exley EM (1983) The genus *Heterohesma* Michener (Hymenoptera: Apoidea: Colletidae). *Journal of the Australian Entomological Society* **22**, 219–221. doi:10.1111/j.1440-6055.1983.tb01880.x

Faegri K, van der Pijl L (1979) *The Principles of Pollination Ecology.* Pergamon Press, Oxford.

Farrow R (2016) *Insects of South-Eastern Australia.* CSIRO Publishing, Melbourne.

Fashing NJ, O'Connor BM, Kitching RL (2000) *Lamingtonocarus*, a new genus of Alogophagidae (Acari: Astigmata) from water-filled tree holes in Queensland, Australia. *Invertebrate Systematics* **14**, 591–606. doi:10.1071/IT99021

Feil JP (1992) Reproductive ecology of dioecious *Siparuna* (Monimiaceae) in Ecuador – a case of gall midge pollination. *Biological Journal of the Linnean Society. Linnean Society of London* **110**, 171–203.

Ferguson DJ, Lambkin CL, Yeates DK (2014) Eight new species of Australian stiletto flies in the genus *Anabarhynchus* Macquart (Diptera: Therevide) from South East Queensland. *Zootaxa* **3802**, 553–582. doi:10.11646/zootaxa.3802.4.7

Ferguson EW (1926a) Revision of Australian Syrphidae (Diptera). Part I. *Proceedings of the Linnean Society of New South Wales* **51**, 136–183.

Ferguson EW (1926b) Revision of Australian Syrphidae (Diptera). Part II, with a supplement to Part I. *Proceedings of the Linnean Society of New South Wales* **51**, 517–544.

Ferguson JE, Metcalf RL (1985) Curcubitacins. Plant-derived defense compounds for Diabroticites (Coleoptera: Chrysomelidae). *Journal of Chemical Ecology* **11**, 311–318. doi:10.1007/BF01411417

Fernandes GW, Santos JC (Eds) (2014) *Neotropical Insect Galls*. Springer, Dordrecht.

Ferrer-Paris JR, Sánchez-Mercado A, Viloria AL, Donaldson J (2013) Congruence and diversity of butterfly-host plant associations at higher taxonomic levels. *PLoS ONE* **8**, e63570. doi:10.1371/journal.pone.0063570

Ferris H (2010) Contribution of nematodes to the structure and function of the soil food web. *Journal of Nematology* **42**, 63–67.

Finlayson GR, Madani G, Dennis G, Harvey M (2015) First reported observation of phoresy of pseudoscorpions on an endemic New Zealand mammal, the lesser short-tailed bat, *Mystacina tuberculata. New Zealand Journal of Zoology* **42**, 298–301. doi:10.1080/03014223.2015.1063517

Fisher EM (2009) Asilidae (robber flies, assassin flies, moscas cazadoras, moscas ladronas). In *Manual of Central American Diptera. Vol. 2.* (Eds BV Brown, A Borkent, JM Cumming, DM Wood, NE Woodley and MA Zumbado) pp. 585–632. NRC Research Press, Ottawa.

Fittkau EJ, Klinge H (1973) On biomass and trophic structure of the Central Amazonian rain forest ecosystem. *Biotropica* **5**, 2–14. doi:10.2307/2989676

Fletcher LE (2007) Vibrational signals in a gregarious sawfly larva (*Perga affinis*): group co-ordination or competitive signaling? *Behavioral Ecology and Sociobiology* **61**, 1809–1821. doi:10.1007/s00265-007-0414-2

Fletcher MJ (1985) Revision of the genus *Siphanta* Stål (Homoptera: Fulgoroidea: Flatidae). *Australian Journal of Zoology* **110**(Supplement), 1–94.

Fletcher MJ (1988) The Australian genera of Flatidae (Homoptera: Fulgoroidea). *General and Applied Entomology* **20**, 9–32.

Fletcher MJ (2005) *Illustrated key to the genera of the family Flatidae found in Australia (Hemiptera: Fulgoroidea)*, <http://www1.dpi.nsw.gov.au/keys/fulgor/flatid/flat00.htm>.

Fletcher MJ (2013) *Illustrated key to the subfamilies and tribes of the family Derbidae found in Australia (Hemiptera: Fulgoroidea)*, <http://www1.dpi.nsw.gov.au/keys/fulgor/derbidae/derbid00.htm>.

Floyd AG (1989) *Rainforest Trees of Mainland South-eastern Australia*. Inkata Press, Melbourne.

Floyd AG (1990) *Australian Rainforests in New South Wales. Vols 1 and 2.* Surrey Beatty and Sons, Chipping Norton.

Floyd AG (2010) The flora. In *Remnants of Gondwana*. (Eds R Kitching, R Braithwaite and J Cavanaugh) pp. 135–146. Surrey Beatty and Sons, Baulkham Hills.

Foelix RF (2010) *Biology of Spiders*. Oxford University Press, Oxford.

Forbes SH, Adams RMM, Bitney C, Kukuk PF (2002) Extended parental care in communal social groups. *Journal of Insect Science* **2**, 1–6. doi:10.1673/031.002.2201

Forster PI (1989) Pollination of *Marsdenia fraseri* (Asclepiadaceae) by *Metrioohynchus lateralis* (Coleoptera: Lycidae). *Coleopterists Bulletin* **43**, 311–312.

Forster PI (1992) Insects associated with the flowers of *Marsdenia cymulosa* Benth. (Asclepiadaceae) and their possible role in pollination. *Australian Entomological Magazine* **19**, 45–47.

Forster RR (1955) Further Australian harvestmen (Arachnida: Opiliones). *Australian Journal of Zoology* **3**, 354–411. doi:10.1071/ZO9550354

Forster RR, Platnick NI (1984) A review of the archaeid spiders and their relatives, with notes on the limits of the superfamily Palpimanoidea (Arachnida, Araneae). *Bulletin of the American Museum of Natural History* **178**, 1–106.

Forster RR, Platnick NI (1985) A review of the austral spider family Orsolobidae (Arachnida, Araneae), with notes on the superfamily Dysderoidea. *Bulletin of the American Museum of Natural History* **181**, 1–229.

Forster RR, Platnick NI, Gray MR (1987) A review of the spider superfamilies Hypochiloidea and Austrochiloidea (Araneae, Araneomorphae). *Bulletin of the American Museum of Natural History* **185**, 1–116.

Forsyth DJ (1972) The structure of the pygidial defence glands of Carabidae (Coleoptera). *Transactions of the Zoological Society of London* **32**, 249–309. doi:10.1111/j.1096-3642.1972.tb00029.x

Fox BJ, Taylor JE, Fox MD, Williams C (1997) Vegetation changes across edges of rainforest remnants. *Biological Conservation* **82**, 1–13. doi:10.1016/S0006-3207(97)00011-6

Framenau VW (2002) Review of the wolf spider genus *Astoria* Thorell (Araneae: Lycosidae). *Invertebrate Systematics* **16**, 209–235. doi:10.1071/IT01028

Framenau VW, Baehr BC, Zborowski P (2014) *A Guide to the Spiders of Australia*. New Holland, London.

Framenau VW, Vink CJ (2001) Revision of the wolf spider genus *Venatrix* Roewer (Araneae: Lycosidae). *Invertebrate Taxonomy* **15**, 927–970. doi:10.1071/IT01008

Frankel S, Berenbaum M (1999) Effects of light regime on antioxidant content of foliage in a tropical forest community. *Biotropica* **31**, 422–429. doi:10.1111/j.1744-7429.1999.tb00384.x

Freeman P (1961) The Chironomidae (Diptera) of Australia. *Australian Journal of Zoology* **9**, 611–737. doi:10.1071/ZO9610611

Freidberg A (1994) *Nemula*, a new genus of Neminidae (Diptera) from Madagascar. *Proceedings of the Entomological Society of Washington* **96**, 471–482.

Freidberg A, Mathis WN (1985) On the feeding habits of *Rhynchopsilopa* (Diptera: Ephydridae). *Entomophaga* **30**, 13–21. doi:10.1007/BF02372278

Freitag R (1979) Reclassification, phylogeny and zoogeography of Australian species of *Cicindela* (Coleoptera: Cicindelidae). *Australian Journal of Zoology Supplementary Series Number* **66**, 1–99.

Fricke FT (1964) A note on *Uracanthus cryptophagus* Oll. (Col., Cerambycidae). *Journal of the Entomological Society of Australia N.S.W.* **1**, 20.

Fricke FT (1965) Notes on Australian Carabidae. *Journal of the Entomological Society of Australia* **2**, 59–60.

Fricke FT (1967) Note on *Cephalodesmius* sp. *Journal of the Entomological Society of Australia* **4**, 40.

Friend JA (1982) New terrestrial amphipods (Amphipoda: Talitridae) from Australian forests. *Australian Journal of Zoology* **30**, 461–491. doi:10.1071/ZO9820461

Friis EM, Crane PR, Pedersen KR (2011) *Early Flowers and Angiosperm Evolution*. Cambridge University Press, Cambridge.

Frith CB, Frith DW (1985) Seasonality of insect abundance in an Australian upland tropical rainforest. *Australian Journal of Ecology* **10**, 237–248. doi:10.1111/j.1442-9993.1985.tb00886.x

Fuller L, Mills K (1985) *Native Trees of Central Illawarra*. Weston and Company, Kiama.

Futuyma DJ, Moreno G (1988) The evolution of ecological specialization. *Annual Review of Ecology and Systematics* **19**, 207–233. doi:10.1146/annurev.es.19.110188.001231

Gagné RJ (1989) *The Plant-feeding Gall Midges of North America*. Cornell University Press, Ithaca, New York.

Gagné RJ (1994) *The Gall Midges of the Neotropical Region*. Cornell University Press, Ithaca, New York.

Gaimari SD (2010) Odiniidae (odiniid flies). In *Manual of Central American Diptera. Vol. 2*. (Eds BV Brown, A Borkent, JM Cumming, DM Wood, D.M Woodley and MA Zumbado) pp. 1049–1055. NRC Research Press, Ottawa.

Galloway ID (1978a) A revision of the Australian genus *Duarina* Dodd (Hymenoptera: Scelionidae: Scelioninae). *Journal of the Australian Entomological Society* **17**, 229–233. doi:10.1111/j.1440-6055.1978.tb00148.x

Galloway ID (1978b) A revision of the Australian species of *Macroteleia* Westwood (Hymenoptera: Scelionidae: Scelioninae). *Journal of the Australian Entomological Society* **17**, 297–310. doi:10.1111/j.1440-6055.1978.tb01494.x

Galloway ID, Austin AD (1984) Revision of the Scelioninae (Hymenoptera: Scelionidae) in Australia. *Australian Journal of Zoology Supplementary Series Number* **99**, 1–138.

Galloway ID, Austin AD, Masner L (1992) Revision of the genus *Neoscelio* Dodd, primitive scelionids (Hymenoptera: Scelionidae) from Australia. with a discussion of the ovipositor system of the tribe Gryonini. *Invertebrate Taxonomy* **6**, 523–545. doi:10.1071/IT9920523

Gardner JA (1989) Revision of the genera of the tribe Stigmoderini (Coleoptera: Buprestidae) with a discussion of phylogenetic relationships. *Invertebrate Taxonomy* **3**, 291–361. doi:10.1071/IT9890291

Garźon-Ordũna IJ, Menchaca-Armenta I, Contreras-Ramos A, Liu X, Winterton SL (2016) The phylogeny of brown lacewings (Neuroptera: Hemerobiidae) reveals multiple reductions in wing venation. *BMC Evolutionary Biology* **16**, 192. doi:10.1186/s12862-016-0746-5

Gaskin DE (1975) A revision of the Australian species of *Pareromene* (Lepidoptera: Pyralidae: Crambinae: Diptychophorini). *Australian Journal of Zoology* **23**, 123–147. doi:10.1071/ZO9750123

Gaston KJ (1991) The magnitude of global insect species richness. *Conservation Biology* **5**, 283–296. doi:10.1111/j.1523-1739.1991.tb00140.x

Gauld ID (1977) A revision of the Ophioninae (Hymenoptera: Ichneumonidae) of Australia. *Australian Journal of Zoology Supplementary Series* **25**, 1–112.

Geoffroy J-J, Ruhberg H (2006) *World Checklist of the Onychophora*, <http://www.mnhn.fr/assoc/myriapoda/ONYLIST.HTM>.

Gerson U, Smith Meyer M (1995) *Neilstigmaeus*, a new Australian genus in the family Stigmaeidae (Acari: Prostigmata). *Acarology* **36**, 219–222.

Gerstmeier R (1990) Revision of the genus Olesterus Spinola, 1841, with description of new species from Australia (Coleoptera: Cleridae). *Mitteilungen der Munchner entomologischen gesellschaft* **80**, 21–38.

Gibbs JP, Stanton EJ (2001) Habitat fragmentation and arthropod community change: carrion beetles, phoretic mites and flies. *Ecological Applications* **11**, 79–85. doi:10.1890/1051-0761(2001)011[0079:HFAACC]2.0.CO;2

Gibson GAP (1986) Mesothoracic skeletonmusculature and mechanics of flight and jumping in Eupelminae (Hymenoptera, Chalcidoidea, Eupelmidae). *Canadian Entomologist* **118**, 691–728. doi:10.4039/Ent118691-7

Gibson R (1995) Nemertean genera and species of the World: an annotated checklist of original names and description citations, synonyms, current taxonomic status, habitats and recorded zoogeographic distribution. *Journal of Natural History* **29**, 271–561. doi:10.1080/00222939500770161

Giffney RA, Kemp DJ (2016) Maternal care behaviour and kin discrimination in the subsocial bug *Tectocoris diophthalmus* Hemiptera: Scutelleridae). *Austral Entomology* **55**, 170–176. doi:10.1111/aen.12164

Gilbert F, Jervis M (1998) Functional, evolutionary and ecological aspects of feeding-related mouthpart specializations in parasitoid flies. *Biological Journal of the Linnean Society. Linnean Society of London* **63**, 495–535. doi:10.1111/j.1095-8312.1998.tb00327.x

Gimmel ML (2013) Genus-level revision of the family Phalacridae (Coleoptera: Cucujoidea). *Zootaxa* **3605**, 1–147. doi:10.11646/zootaxa.3605.1.1

Glover B (1973) The Tanytarsini (Diptera: Chironomidae) of Australia. *Australian Journal of Zoology Supplementary Series* **23**.

Gonźalez CR (2015) Revision of the New World species of the genus *Pelecorhynchus* Macquart, 1850 (Diptera: Pelecorhynchidae). *Zootaxa* **3955**, 188–210.

Gonzalez VH, Engel MS, Griswold TL (2013) The lithurgine bees of Australia (Hymenoptera: Megachilidae), with a note on *Megachile rotundipennis*. *Journal of Melittology* **11**, 1–19.

Goulet H, Huber JT (1993) *Hymenoptera of the World: An Identification Guide to Families*. Agriculture Canada, Ottawa.

Grantley J (2010) The coastal forests. In *Remnants of Gondwana*. (Eds R Kitching, R Braithwaite and J Cavanaugh) pp. 51–58. Surrey Beatty and Sons, Baulkham Hills.

Gray MR (1982) The male of *Progradungulla carraiensis* Forster and Gray (Araneae: Gradungulidae). *Proceedings of the Linnean Society of New South Wales* **107**, 51–58.

Gray MR (1987) Distribution of the funnel web spiders. In *Toxic Plants and Animals: A Guide for Australia*. (Eds J Covacevich, P Davie and J Pearn) pp. 313–321. Queensland Museum, Brisbane.

Gray MR (1992) New desid spiders (Araneae: Desidae) from New Caledonia and eastern Australia. *Records of the Australian Museum* **44**, 253–262. doi:10.3853/j.0067-1975.44.1992.34

Gray MR (1994) A review of the filistatid spiders (Araneae: Filistatidae) of Australia. *Records of the Australian Museum* **46**, 39–69.

Gray MR, Cassis GA (1994) 'Results of ground-dwelling invertebrate fauna surveys of north-east NSW forests'. North East Forests Biodiversity Report No. 3c. Australian Museum Report for the NSW National Parks and Wildlife Service, Sydney.

Gray MR, Smith HM (2002) *Therlinya*, a new genus of spiders from Eastern Australia (Araneae: Amaurobioidea). *Records of the Australian Museum* **54**, 293–312. doi:10.3853/j.0067-1975.54.2002.1368

Gray MR, Smith HM (2004) The 'striped' group of stiphidiid spiders: two new genera from south-eastern New South Wales, Australia (Araneae: Stiphidiidae: Amaurobioidea). *Records of the Australian Museum* **56**, 123–138. doi:10.3853/j.0067-1975.56.2004.1394

Greathead DJ, Evenhuis NL, Lamas CJE (2009) Bombyliidae (bee flies). In *Manual of Central American Diptera. Vol. 1*. (Eds BV Brown, A Borkent, JM Cumming, DM Wood, NE Woodley and MA Zumbado) pp. 565–576. NRC Research Press, Ottawa.

Greenslade PJ (1991) Collembola 'springtails'. In *The Insects of Australia*. (Ed. ID Naumann) pp. 252–264. Melbourne University Press, Carlton.

Greenslade P (1994) Heritage listing of invertebrate sites in south-eastern Australia. *Memoirs of the Queensland Museum* **36**, 67–76.

Greenslade P, Whalley PES (1986) The systematic position of *Rhyniella praecursor* Hirst and Maulik (Collembola). The earliest known hexapod. In *Second International Seminar on Apterygota*. University of Siena, Siena. (Ed. R Dallai) pp. 319–323.

Grehan JR (1991) A panbiogeographic perspective for pre-Cretaceous angiosperm-Lepidoptera coevolution. *Australian Systematic Botany* **4**, 91–110. doi:10.1071/SB9910091

Griffiths GCD (1972) *The Phylogenetic Classification of Diptera Cyclorrhapha, with Special Reference to the Structure of the Male Postabdomen*. Dr W. Junk, Publishers, The Hague.

Griffiths JR (1971) Reconstruction of the south-west Pacific margin of Gondwanaland. *Nature* **234**, 203–207. doi:10.1038/234203a0

Griffiths JR (1974) Revised continental fit of Australia and Antarctica. *Nature* **249**, 336–338. doi:10.1038/249336a0

Grimaldi DA (1999) The co-radiations of pollinating insects and angiosperms in the Cretaceous. *Annals of the Missouri Botanical Garden* **86**, 373–406. doi:10.2307/2666181

Grimaldi DA, Engel MS (2005) *Evolution of the Insects*. Cambridge University Press, New York.

Grimm R (2013) The genus *Cryphaeus* Klug, 1833 in Australia, with description of a new species (Coleoptera: Tenebrionidae: Toxicini). *Australian Entomologist* **40**, 243–247.

Grimshaw P, Sands D, Gynther I (2013) *Birdwing Butterfly Vine* Pararistolochia praevenosa *(F. Muell) M.J. Parsons (Aristolochiaceae)*. Information brochure. Queensland Department of Environment and Heritage Protection, Belbowrie.

Grismado GJ, Goloboff PA (2006) Descripción del macho de *Missulena tussulena* Goloboff 1994 (Areanae, Mygalomorphae, Actinopodidae). *Revista del Museo Argentino de Ciencias Natureles Nueva Serie* **8**, 101–104.

Griswold MW, Berzinis EW, Crisman TL, Golladay SW (2008) Impacts of climate stability on the structure and functional aspects of macroinvertebrate communities after severe drought. *Freshwater Biology* **53**, 2465–2483. doi:10.1111/j.1365-2427.2008.02067.x

Grootaert P, Shamshev I (2011) The genus *Trachydromia* Meigen (Diptera: Hybotidae) from Australia. *Records of the Australian Museum* **63**, 103–112.

Gross GF (1972) A revision of the species of Australian and New Guinea shield bugs formerly placed in the genera *Poecilometis* Dallas and *Eumecopus* Dallas (Heteroptera: Pentatomidae), with descriptions of new species and selection of lectotypes. *Australian Journal of Zoology Supplementary Series* **15**, 1–92.

Gross GL (1993) The breeding system and pollinators of *Melastoma affine* (Melostomataceae): a pioneer shrub in tropical Australia. *Biotropica* **25**, 468–474. doi:10.2307/2388870

Guerra TJ, Romero GQ, Benson WW (2010) Flower mites decrease nectar availability in the rain-forest bromeliad *Neoregelia johannis*. *Journal of Tropical Ecology* **26**, 373–379. doi:10.1017/S0266467410000179

Gullan PJ (1977) A revision of the genus *Anilicus* Candeze, with notes on related genera (Coleoptera: Elateridae). *Memoirs of the National Museum of Victoria* **38**, 209–230. doi:10.24199/j.mmv.1977.38.02

Gullan PJ, Cranston PS (2014) *The Insects: An Outline of Entomology*. John Wiley and Sons, Chichester.

Gunter NL, Weir TA (2017) Two new genera of Australian dung beetles (Coleoptera: Scarabaeidae: Scarabaeinae) with the description of six new species and transfer of six described species. *Zootaxa* **4290**, 201–243. doi:10.11646/zootaxa.4290.2.1

Gunter NL, Weir TA, Ślipiński A, Bocak L, Cameron SL (2016) If dung beetles (Scarabaeidae: Scarabaeinae) arose in association with dinosaurs, did they also suffer a mass co-extinction at the K-Pg Boundary? *PLoS ONE* **11**, e0153570. doi:10.1371/journal.pone.0153570

Gustafson GT, Prokin AA, Bukontaite R, Bergsten J, Miller KB (2017) Tip-dated phylogeny of whirligig beetles reveals ancient lineages surviving on Madagascar. *Scientific Reports* **7**, 8619. doi:10.1038/s41598-017-08403-1

Hacker H (1921) Catalogue of the Australian bees. *Memoirs of the Queensland Museum* **7**, 99–163.

Haddad S, McKenna DD (2016) Phylogeny and evolution of the superfamily Chrysomeloidea (Coleoptera: Cucujiformia). *Systematic Entomology* **41**, 697–716. doi:10.1111/syen.12179

Haddad S, Shin S, Lemmon AR, Lemmon EM, Svacha P, Farrell B, Ślipiński A, Windsor D, McKenna DD (2018) Anchored hybrid enrichment provides new insights into the phylogeny and evolution of longhorned beetles (Cerambycidae). *Systematic Entomology* **43**, 68–89. doi:10.1111/syen.12257

Hadlington P (1965) Variations in diapause and parthenogenesis associated with geographic populations of *Podacanthus wilkinsoni* Macl. (Phasmatodea, Phasmatidae). *Journal of the Entomological Society of Australia N.S.W.* **2**, 45–49.

Halffter G (1977) Evolution of nidification in the Scarabaeinae (Coleoptera, Scarabaeidae). *Quaestiones Entomologicae* **13**, 231–253.

Halffter G, Edmonds WD (1982) *The Nesting Behavior of Dung Beetles (Scarabaeinae).* Instituto de Ecologica, Mexico, D. F.

Halffter G, Matthews EG (1966) The natural history of dung beetles of the subfamily Scarabaeinae. *Folia Entomologica Mexicana* **23–24**, 105–108.

Hall CR, Burwell CJ, Nakamura A, Kitching RL (2014) Altitudinal variation of parasitic Hymenoptera assemblages in Australian subtropical rainforest. *Austral Entomology* **54**(3), 246–258. doi:10.1111/aen.12114

Hall JA, Walter GH, Bergstrom DM, Machin P (2004) Pollination ecology of the Australian cycad *Lepidozamia peroffskyana* (Zamiaceae). *Australian Journal of Botany* **52**, 333–343. doi:10.1071/BT03159

Halliday RB (1997) Revision of the Australian Ameroseiidae (Acarina: Mesostigmata). *Invertebrate Taxonomy* **11**, 179–201. doi:10.1071/IT96010

Halliday RB (2000) The Australian species of *Macrocheles* (Acarina: Macrochelidae). *Invertebrate Taxonomy* **14**, 273–326. doi:10.1071/IT99009

Halliday RB, Walter DE, Lindquist EE (1998) Revision of Australian Ascidae (Acarina: Mesostigmata). *Invertebrate Taxonomy* **12**, 1–54. doi:10.1071/IT96029

Halliday RB, Walter DE, Proctor HC, Norton RA, Colloff MJ (Eds) (2001) Acarology. *Proceedings of the 10th International Congress.* 1998, Canberra. CSIRO Publishing, Melbourne.

Hamilton AG (1897) On the fertilization of *Eupomatia laurina*. *Proceedings of the Linnean Society of New South Wales* **22**, 48–56.

Hamilton KGA (1999) The ground-dwelling leafhoppers Myerslopiidae, new family, and Sagmatiini, new tribe (Homoptera: Membracoidea). *Invertebrate Taxonomy* **13**, 207–235. doi:10.1071/IT96028

Hammond PM, Kitching RL, Stork NE (1997) The composition and richness of the tree-crown Coleoptera assemblage in an Australian subtropical forest. *Ecotropica (Bonn)* **2**, 99–108.

Hancock DL (2013) A revised checklist of Australian fruit flies (Diptera: Tephritidae). *Australian Entomologist* **40**, 219–236.

Hancock DL, Drew RAI (2003a) New species and records of Phytalmiinae (Diptera: Tephritidae) from Australia and the South Pacific. *Australian Entomologist* **30**, 65–78.

Hancock DL, Drew RAI (2003b) New species and records of Trypetinae (Diptera: Tephritidae) from Australia and the South Pacific. *Australian Entomologist* **30**, 93–106.

Hancock DL, Drew RAI (2003c) A new genus and new species, combinations and records of Tephritinae (Diptera: Tephritidae) from Australia, New Zealand and the South Pacific. *Australian Entomologist* **30**, 141–156.

Hangay G, De Keyser R (2017) *A Guide to Stag Beetles of Australia.* CSIRO Publishing, Melbourne.

Hansen WR (Ed.) (1991) *Suggestions to Authors of the Reports of the United States Geological Survey.* 7th edn (STA7). U.S. Geological Survey, Reston, Va.

Hanski I, Cambefort Y (Eds) (1991) *Dung Beetle Ecology.* Princeton University Press, Princeton.

Harbach RE, Kitching IJ (1998) The phylogeny and classification of the Culicidae (Diptera). *Systematic Entomology* **23**, 327–370. doi:10.1046/j.1365-3113.1998.00072.x

Harden G (Ed.) (1990) *Flora of New South Wales. Vol. 1.* New South Wales University Press, Kensington.

Harden G (Ed.) (1991) *Flora of New South Wales. Vol. 2.* New South Wales University Press, Kensington.

Harden G (Ed.) (1993) *Flora of New South Wales. Vol. 4.* New South Wales University Press, Kensington.

Harden G, McDonald B, Williams J (2006) *Rainforest Trees and Shrubs.* Gwen Harden Publishing, Nambucca Heads.

Harden G, McDonald B, Williams J (2007) *Rainforest Climbing Plants.* Gwen Harden Publishing, Nambucca Heads.

Harder LD, Barclay RMR (1994) The functional significance of poricidal anthers and buzz pollination controlled pollen removal from *Dodecantheon. Functional Ecology* **8**, 509–517. doi:10.2307/2390076

Hardy DE, Drew RAI (1996) Revision of the Australian Tephritini (Diptera: Tephritidae). *Invertebrate Taxonomy* **10**, 213–405. doi:10.1071/IT9960213

Hardy DE, Foote RH (1989) Family Tephritidae. In *Catalog of the Diptera of the Australasian and Oceanic Regions.* (Ed. NL Evenhuis) pp. 502–531. Bishop Museum Press and E.J. Brill, Honolulu.

Hardy GHH (1920) Descriptions of Australian flies of the Family Asilidae (Diptera) with synonyms and notes. *Proceedings of the Linnean Society of New South Wales* **45**, 185–202. doi:10.5962/bhl.part.19539

Hardy GHH (1921a) Australian Bombyliidae and Cyrtidae (Diptera). *Papers and Proceedings of the Royal Society of Tasmania* **1921**, 41–83.

Hardy GHH (1921b) Notes on some Australian Syrphidae (Diptera). *Australian Zoologist* **2**, 12–18.

Hardy GHH (1924) Australian Nemestrinidae (Diptera). *Proceedings of the Linnean Society of New South Wales* **49**, 447–460.

Hardy GHH (1925) Australian Mydaidae (Diptera). *Proceedings of the Linnean Society of New South Wales* **50**, 139–144.

Hardy GHH (1926) A reclassification of Australian robberflies of the *Cerdistus-Neoitamus* complex (Diptera-Asilidae). *Proceedings of the Linnean Society of New South Wales* **51**, 643–657.

Hardy GHH (1929a) Revisional notes on the Tribe Brachyrrhopalini (robber flies) with remarks on habits and mimicry. *Proceedings of the Royal Society of Queensland* **41**, 59–72.

Hardy GHH (1929b) Fourth contribution towards a new classification of the Australian Asilidae. *Proceedings of the Linnean Society of New South Wales* **54**, 353–360.

Hardy GHH (1930a) Fourth contribution towards a new classification of Australian Asilidae. *Proceedings of the Linnean Society of New South Wales* **55**, 249–260.

Hardy GHH (1930b) Australian Empididae. *Australian Zoologist* **6**, 237–250.

Hardy GHH (1931) On the genus *Damaromyia* Kertesz (Stratiomyiidae). *Annals & Magazine of Natural History* **8**, 120–128. doi:10.1080/00222933108673366

Hardy GHH (1932a) Notes on Australian Stratiomyiidae. *Proceedings of the Royal Society of Queensland* **44**, 41–49.

Hardy GHH (1932b) Some Australian species of *Calliphora* (subgenus *Neopollenia* and *Proekon*). *Bulletin of Entomological Research* **23**, 549–558. doi:10.1017/S0007485300004326

Hardy GHH (1933) Notes on Australian Syrphinae (Diptera). *Proceedings of the Royal Society of Queensland* **45**, 12–19.

Hardy GHH (1940) Notes on Australian Muscoidea. V. Calliphoridae. *Proceedings of the Royal Society of Queensland* **51**, 133–146.

Harland DP, Jackson RR, Macnab AM (1999) Distance at which jumping spiders (Araneae: Salticidae) distinguish between prey and conspecific rivals. *Journal of Zoology* **247**, 357–364. doi:10.1111/j.1469-7998.1999.tb00998.x

Harms D, Harvey MS (2013) Review of the cave-dwelling species of *Pseudotyrannochthonius* Beir (Arachnida: Pseudoscorpiones: Pseudotyrannochthoniidae) from mainland Australia, with descriptions of two troglobitic species. *Australian Journal of Entomology* **52**, 129–143. doi:10.1111/aen.12009

Harris LD, Silva-Lopez G (1992) Forest fragmentation and the conservation of biological diversity. In *Conservation Biology: The Theory and Practice of Nature Conservation, Preservation, and Management.* (Eds PL Fiedler and SK Jain) pp. 197–237. Chapman and Hall, London.

Harris LF, Johnson SD (2004) The consequences of habitat fragmentation for plant–pollinator mutualisms. *International Journal of Tropical Insect Science* **24**, 29–43.

Harrison RA (1966) Australian glow-worms of the genus *Arachnocampa* Edwards (Diptera: Mycetophilidae). *Pacific Insects* **8**, 877–883.

Harvey MS (1981) A checklist of the Australian Pseudoscorpionida. *Bulletin of the British Arachnological Society* **5**, 237–252.

Harvey MS (1992) The Schizomida (Chelicerata) of Australia. *Invertebrate Taxonomy* **6**, 77–129. doi:10.1071/IT9920077

Harvey MS (1994) Rediscovered pseudoscorpion type material described by Beier from south-eastern Queensland (Arachnida: Pseudoscorpionida). *Memoirs of the Queensland Museum* **37**, 155–156.

Harvey MS (1995) The systematics of the spider family Nicodamidae (Araneae: Amaurobioidea). *Invertebrate Taxonomy* **9**, 279–386. doi:10.1071/IT9950279

Harvey MS (2002) Short-range endemism amongst the Australian fauna: some examples from non-marine environments. *Invertebrate Systematics* **16**, 555–570. doi:10.1071/IS02009

Harvey MS (2003) *Catalogue of the Smaller Arachnid Orders of the World – Amblypygi, Uropygi, Schizomida, Palpigradi, Ricinulei and Solfugae.* CSIRO Publishing, Melbourne.

Hasenpusch JW, Moeseneder CH (2010) *Storeyus*, a new genus and *Storeyus pseudodipterus*, a new species of Cetoniinae (Coleoptera: Scarabaeidae) from Australia, with a redescription of *Lenosoma* Kraatz. *Australian Entomologist* **36**, 221–234.

Hawkeswood TJ (1986) Notes on two new species of Australian fungus beetles (Coleoptera: Erotylidae). *Coleopterists Bulletin* **40**, 27–28.

Hawkeswood TJ (1988) A review of larval host records for twelve Australian Buprestidae (Coleoptera). *Giornale italiano di Entomologia* **4**, 81–88.

Hawkeswood TJ (1990a) Notes on the biology and distribution of *Johannica gemmellata* (Westwood) (Coleoptera: Chrysomelidae) from the rainforests of south-eastern Queensland. *Victorian Entomologist* **20**, 139–140.

Hawkeswood TJ (1990b) Studien zu biologie und verhalten des australischen russelkafers *Enteles vigorsi* Gyllenhal (Coleoptera: Curculionidae: Cryptorhynchinae). *Entomologische Zeitschrift* **100**, 264–267.

Hawkeswood TJ (1990c) Eine ubersicht zu biologie und wirtspflanzen des australischen prachtkafers *Melobasis* Laporte and Gory (Coleoptera: Buprestidae). *Entomologische Zeitschrift* **100**, 340–342.

Hawkeswood TJ (1990d) Neue Wirtsangaban für Imagines einiger pilzfressender Käfer (Coleoptera) von New South Wales und Queensland, Australien. *Entomologische Zeitschrift* **100**, 93–112.

Hawkeswood TJ (1991a) Notes on *Spilopyra sumptuosa* Baly (Coleoptera: Chrysomelidae), a spectacular leaf-beetle from the rainforests of Queensland and New South Wales. *Victorian Entomologist* **21**, 125–126.

Hawkeswood TJ (1991b) Review of the biology and host plants of the Australian weevil *Euthyrinus meditabundus* (Fabricius) (Coleoptera: Curculionidae). *Entomologist* **110**, 58–65.

Hawkeswood TJ (1992) Further notes on *Spilopyra sumptuosa* Baly (Coleoptera: Chrysomelinae). *Victorian Entomologist* **22**, 49–50.

Hawkins CP, MacMahon JA (1989) Guilds: the multiple meanings of a concept. *Annual Review of Entomology* **34**, 423–451. doi:10.1146/annurev.en.34.010189.002231

Hay A, Wise R (1991) The genus *Alocasia* (Araceae) in Australasia. *Blumea* **35**, 499–545.

Heard TA, Hendrikz TK (1993) Factors influencing flight activity of colonies of the stingless bee *Trigona carbonaria* (Hymenoptera: Apidae). *Australian Journal of Zoology* **41**, 343–353. doi:10.1071/ZO9930343

Heatwole H (1987) Major components and distributions of the Terrestrial fauna. In *Fauna of Australia. General Articles. Vol. 1A.* (Eds GR Dyne and DW Walton) pp. 101–135. Australian Government Publishing Service,Canberra.

Heenan PB, Smissen RD (2013) Revised circumscription of *Nothofagus* and recognition of the segregate genera *Fuscospora, Lophozonina* and Trisyngyne (Nothofagaceae). *Phytotaxa* **146**, 1–31. doi:10.11646/phytotaxa.146.1.1

Helman C (1987) Rainforest in southern New South Wales. In *The Rainforest Legacy. Australian National Rainforests Study. Vol. 1.* (Eds G Werren and AP Kershaw) pp. 47–70. Australian Government Publishing Service, Canberra.

Hemmings Z, Andrew NR (2017) Effects of microclimate and species identity on body temperature and thermal tolerance of ants (Hymenoptera: Formicidae). *Austral Entomology* **56**, 104–114. doi:10.1111/aen.12215

Hendrix PF, Baker GH, Callahan MA, Jr, Damoff GA, Fragoso C, González G, James SW, Lachnicht SL, Winsome T, Zou X (2006) Invasion of exotic earthworms into ecosystems inhabited by native earthworms. *Biological Invasions* **8**, 1287–1300. doi:10.1007/s10530-006-9022-8

Hendrix SD (1990) Herbivory and its impact on plant reproduction. In *Plant Reproductive Ecology: Patterns and Strategies.* (Eds JL Doust and LL Doust) pp. 246–263.Oxford University Press, New York.

Henry TJ (1997) Phylogenetic analysis of family groups within the infraorder Pentatomomorpha (Hemiptera: Heteroptera), with emphasis on the Lygaeoidea. *Annals of the Entomological Society of America* **90**, 275–301. doi:10.1093/aesa/90.3.275

Hilbert DW, Graham A, Hopkins MS (2007) Glacial and interglacial refugia within a long-term rainforest tefugium: the Wet Tropics bioregion of NE Queensland, Australia. *Palaeogeography, Palaeoclimatology, Palaeoecology* **251**, 104–118. doi:10.1016/j.palaeo.2007.02.020

Hill CJ (1995) Linear strips of rain forest vegetation as potential dispersal corridors for rain forest insects. *Conservation Biology* **9**, 1559–1566. doi:10.1046/j.1523-1739.1995.09061559.x

Hill CJ (1997) Conservation corridors and rainforest insects. In *Forests and Insects.* (Eds AD Watt, NE Stork and MD Hunter) pp. 381–393. Chapman and Hall, London.

Hill C, Kitching R (1983) Butterflies. In *Wildlife of the Brisbane Area.* (Ed. W Davies) pp. 138–140, 210–212. Jacaranda Press, Brisbane.

Hill RS, Jordan GJ, McPhail MK (2015) Why we should retain '*Nothofagus sensu lato*'. *Australian Systematic Botany* **28**, 190–193. doi:10.1071/SB15026

Hinton HE (1965) A revision of the Australian species of *Austrolimnius* (Coleoptera: Elmidae). *Australian Journal of Zoology* **13**, 97–172. doi:10.1071/ZO9650097

Hirst DB (1991) Revision of Australian species of the genus *Holconia* Thorell (Heteropodidae: Araneae). *Records of the South Australian Museum* **24**, 91–109.

Hirst DB (1992) Revision of the genus *Isopeda* Kock (Heteropodidae: Araneae) in Australia. *Invertebrate Taxonomy* **6**, 337–387. doi:10.1071/IT9920337

Hirst DB (1999) Revision of *Typostola* (Araneae: Heterapodidae) in Australasia. *Memoirs of the Queensland Museum* **43**, 639–648.

Hobbs RJ, Yates CJ (2003) Impacts of ecosystem fragmentation on plant populations: generalising the idiosyncratic. *Australian Journal of Botany* **51**, 471–488. doi:10.1071/BT03037

Hölldobler B, Wilson EO (1990) *The Ants.* Belknap Press, Cambridge.

Holloway BA (1963) Wing development and evolution of New Zealand Lucanidae (Insecta: Coleoptera). *Transactions of the Royal Society of New Zealand (Zoology)* **3**, 99–116.

Holloway BA (1976) Pollen feeding in hover-flies (Diptera: Syrphidae). *New Zealand Journal of Zoology* **3**, 339–350. doi:10.1080/03014223.1976.9517924

Holloway GA (1986) The classification and relationships of the Australian species of *Gotra* (Hymenoptera: Ichneumonidae). MSc thesis. Macquarie University, Australia.

Holt BG, Lessard JP, Borregaard MK, Fritz SA, Araújo MB, Dimitrov D, Fabre PH, Graham CH, Graves GR, Jønsson KA, Nogués-Bravo D, Wang Z, Whittaker RJ, Fjeldså J, Rahbeck C (2013a) An update of Wallace's zoogeographic regions of the world. *Science* **339**, 74–78. doi:10.1126/science.1228282

Holt BG, Lessard JP, Borregaard MK, Fritz SA, Araújo MB, Dimitrov D, Fabre PH, Graham CH, Graves GR, Jønsson KA, Nogués-Bravo D, Wang Z, Whittaker RJ, Fjeldså J, Rahbeck C (2013b) Response to comment on 'An update of Wallace's zoogeographic regions of the world'. *Science* **341**(6144), 343. doi:10.1126/science.1237541

Holynski R (1988) Remarks on the general classification of Buprestidae Leach as applied to Maoraxiina Hol. *Folia Entomologica Hungarica* **49**, 49–54.

Honnay O, Jacquemyn H (2007) Susceptability of common and rare plant species to genetic consequences of habitat fragmentation. *Conservation Biology* **21**, 823–831. doi:10.1111/j.1523-1739.2006.00646.x

Hopkin SP (1997) *Biology of Springtails (Insecta: Collembola).* Oxford University Press, New York.

Horak M (1997) The Phyctine genera *Faveria* Walker, *Morosaphycita* gen. nov., *Epicrocis* Zeller, *Ptyobathra* Turner and *Vinicia* Ragonot in Australia (Pyralidae: Phycitinae). *Invertebrate Taxonomy* **11**, 333–421. doi:10.1071/IT96002

Hormiga G, Eberhard WG, Coddington JA (1995) Web-construction behavior in Australian Phonognatha and the phylogeny of Nephiline and Tetragnathid spiders (Araneae,

Tetragnathidae). *Australian Journal of Zoology* **43**, 313–364. doi:10.1071/ZO9950313

Horton S (Ed.) (1999) *Rainforest Remnants: A Decade of Growth*. NSW National Parks and Wildlife Service, Sydney.

Hose G, Turak E (2004) 'River health in the New South Wales Lower North Coast, Hunter and Central Coast catchments'. Environmental Protection Agency, Canberra.

House SM (1985) Breeding and spatial pattern in dioecious rainforest trees. PhD thesis. Australian National University, Australia.

Houston TF (1971) Notes on the biology of a lithurgine bee (Hymenoptera: Megachilidae) in Queensland. *Journal of the Australian Entomological Society* **10**, 31–36. doi:10.1111/j.1440-6055.1971.tb00007.x

Houston TF (1975) A revision of the Australian hylaeine bees (Hymenoptera: Colletidae. I. Introductory material and the genera *Heterapoides* Sandhouse, *Gephyrohylaeus* Michener, *Hyleoides* Smith, *Pharohylaeus* Michener, *Hemirhiza* Michener, *Amphylaeus* Michener and *Meroglossa* Smith. *Australian Journal of Zoology Supplementary Series Number* **36**.

Houston TF (1981) A revision of the Australian hylaeine bees (Hymenoptera: Colletidae). II.* Genus *Hylaeus* Fabricius, subgenera *Analastoroides* Rayment, *Edriohylaeus* Michener, *Euprosopellus* Michener, *Eprosopis* Perkins, *Euprosopoides* Michener, *Gnathoprosopis* Perkins, *Gnathoprosopoides* Michener, *Hylaeorhiza* Michener, *Hylaeteron* Michener, *Laccohylaeus*, subgen. nov., *Macrohylaeus* Michener, *Meghylaeus* Cockerell, *Planihylaeus*, subgen. nov., *Sphaerhylaeus* Cockerell and *Xenohylaeus* Michener. *Australian Journal of Zoology Supplementary Series Number* **29**, 1–128.

Houston TF (2018) *A Guide to the Native Bees of Australia*. CSIRO Publishing, Melbourne.

Houston WA, Melzer A (2012) Dry rainforests have a distinct and more diverse assemblage of epigaeic invertebrates than eucalypt woodlands: implications for ecosystem health monitoring. *Pacific Conservation Biology* **18**, 133–145. doi:10.1071/PC120133

Houston WWK (Ed.) (1988) *Zoological Catalogue of Australia. Vol. 6. Ephemeroptera, Megaloptera, Odonata, Plecoptera, Trichoptera*. CSIRO Publishing, Melbourne.

Houston WWK (Ed.) (1992) *Zoological Catalogue of Australia. Vol. 9. Coleoptera: Scarabaeoidea*. Australian Government Publishing Service, Canberra.

Houston WWK (Ed.) (1994) *Zoological Catalogue of Australia. Vol. 22. Protura, Collembola, Diplura*. CSIRO Publishing, Melbourne.

Houston WWK, Watson JAL (1988) Odonata. In *Zoological Catalogue of Australia. Vol. 6*. (Ed. WWK Houston) pp. 33–132. Australian Government Publishing Service, Canberra.

Howden HF (1979) A revision of the Australian genus Blackburnium Boucomont (Coleoptera: Scarabaeidae: Geotrupinae). *Australian Journal of Zoology Supplementary Series* **72**, 1–88.

Howden HF (1981) Zoogeography of some Australian Coleoptera as exemplified by the Scarabaeoidea. In *Ecological Biogeography of Australia. Vol. 2*. (Ed. A Keast) pp. 1009–1035. Dr W. Junk, Publishers, The Hague.

Howden HF (1985) A revision of the Australian beetle genera *Bolboleaus* Howden and Cooper, *Blackbolbus* Howden and Cooper, and *Bolborhachium* Boucomont (Scarabaeidae: Geotrupinae). *Australian Journal of Zoology Supplementary Series* **111**, 1–179.

Howden HF (1992) A revision of the Australian beetle genera *Eucanthus* Westwood, *Bolbobaineus* Howden and Cooper, *Australobolbus* Howden and Cooper and *Gilletinus* Boucomont (Scarabaeidae: Geotrupinae). *Invertebrate Taxonomy* **6**, 605–717. doi:10.1071/IT9920605

Howden HF, Cooper JB (1977) The generic classification of the Bolboceratini of the Australian Region, with descriptions of four new genera (Scarabaeidae: Geotrupinae). *Australian Journal of Zoology Supplementary Series* **50**, 1–50.

Howden HF, Howden AT (1991) Nocturnal perching of Scarabaeine dung beetles (Coleoptera, Scarabaeidae) in an Australian tropical rain forest. *Biotropica* **23**, 51–57. doi:10.2307/2388687

Howden H, Howden A, Holloway G (2007) Digging down under: Australian Bolboceratini, their habits and a list of species (Coleoptera: Scarabaeoidea: Geotrupidae). *Zootaxa* **1499**, 47–59. doi:10.11646/zootaxa.1499.1.2

Howden H, Nealis VG (1975) Effects of clearing in a tropical rain forest on the composition of the coprophagous scarab beetle fauna (Coleoptera). *Biotropica* **7**, 77–83. doi:10.2307/2989750

Hsiao Y, Yu Y, Deng C, Pang H (2017) The first fossil wedge-shaped beetle (Coleoptera, Ripiphoridae) from the middle Jurassic of China. *European Journal of Taxonomy* doi:10.5852/ejt.2017.277

Hu SS, Dilcher DL, Jarzen DM, Taylor DW (2008) Early steps of angiosperm pollinator coevolution. *Proceedings of the National Academy of Sciences of the United States of America* **105**, 240–245. doi:10.1073/pnas.0707989105

Hugall AF, Stanisic J (2011) Beyond the prolegomenon: a molecular phylogeny of the Australian camaenid land snail radiation. *Zoological Journal of the Linnean Society* doi:10.1111/j.1096-3642.2010.00644.x

Hughes RD, Greenham PM, Tyndale-Biscoe M, Walker JM (1972) A synopsis of observations on the biology of the Australian bushfly (*Musca vetustissima* Walker). *Journal of the Australian Entomological Society* **11**, 311–331. doi:10.1111/j.1440-6055.1972.tb01635.x

Hull FM (1973) *Bee Flies of the World: The Genera of the Family Bombyliidae*. Smithsonian Institution, Washington.

Humphrey M (2017) A revision and cladistic analysis of the genus *Corasoides* Butler (Araneae: Desidae) with description of nine new species. *Records of the Australian Museum* **69**, 15–64. doi:10.3853/j.2201-4349.69.2017.1671

Hunt GS (1985) Taxonomy and distribution of *Equitius* in Eastern Australia (Opiliones: Laniatores: Triaenonychidae). *Records of the Australian Museum* **36**, 107–125. doi:10.3853/j.0067-1975.36.1985.340

Hunt GS (1994) Orabatids – a mite biodiverse (Acarina). *Memoirs of the Queensland Museum* **36**, 107–114.

Hunt GS (1996a) A review of the genus *Pedrocortessella* Hamer in Australia (Acarina: Cryptostigmata: Pedrocortesellidae). *Records of the Australian Museum* **48**, 223–286. doi:10.3853/j.0067-1975.48.1996.430

Hunt GS (1996b) A review of the genus *Hexachaetoniella* Paschoal in Australia (Acarina: Cyrptostigmata: Pedrocorteesellidae). *Records of the Australian Museum* **48**, 287–302. doi:10.3853/j.0067-1975.48.1996.431

Hunt GS (1996c) Description of predominantly arboreal Plateremaeoid mites from eastern Australia (Acarina: Cryptostigmata: Plateremaeoidea). *Records of the Australian Museum* **48**, 303–324. doi:10.3853/j.0067-1975.48.1996.432

Hunt GS (1996d) A review of the family Pheroliodidae Paschoal in Australia (Acarina: Cryptostigmata: Plateremaeoidea). *Records of the Australian Museum* **48**, 325–358. doi:10.3853/j.0067-1975.48.1996.433

Hunt GS, Cokendolpher JC (1991) Ballarrinae, a new subfamily of phalangioid harvestmen from the Southrn Hemisphere (Arachnida, Opiliones, Neopilionidae). *Records of the Australian Museum* **43**, 131–169. doi:10.3853/j.0067-1975.43.1991.45

Hunt GS, Hickman JL (1993) A revision of the genus *Lomanella* Pocock and its implications for family level classification in the Trakunioidea (Arachnida: Opiliones: Triaenonychidae. *Records of the Australian Museum* **45**, 81–119. doi:10.3853/j.0067-1975.45.1993.131

Hunter RJ (2003) 'World Heritage and associative natural values of the Central Eastern Rainforest Reserves of Australia'. NSW NPWS, Hurstville. Report by the New South Wales National Parks and Wildlife Service for the World Heritage Branch, Environment Australia, Canberra.

Hunter RJ (2010) The Focal Peak group. In *Remnants of Gondwana*. (Eds R Kitching, R Braithwaite and J Cavanaugh) pp. 25–38. Surrey Beatty and Sons, Baulkham Hills.

Hunter RL (1939) The pink underwing moth *Phyllodes meyricki* (Olliff). Family Noctuidae, Sub-family Catocalinae. *North Queensland Naturalist* **7**, 1–2.

Huxley CR, Cutler DF (1991) *Ant-plant Interactions*. Oxford University Press, Oxford.

Hwang WS, Weirauch C (2012) Evolutionary history of assassin bugs (Insecta: Hemiptera: Reduviidae): insights from divergence dating and ancestral state reconstruction. *PLoS ONE* **7**, e45523. doi:10.1371/journal.pone.0045523

Imadaté G (1991) Protura. In *The Insects of Australia*. (Ed. NL Naumann) pp. 265–268. Melbourne University Press, Carlton.

Invasive Species Council (2011) 'Environmental impacts of myrtle rust'. Invasive Species Council, Fairfield.

Irwin ME, Lyneberg L (1989) Family Therevidae. In *Catalog of the Diptera of the Australasian and Oceanic Regions*. (Ed. NL Evenhuis) pp. 353–358. Bishop Museum Press and E.J. Brill, Honolulu.

Irwin ME, Winterton SL (2016) New genera of Australian stiletto flies (Diptera, Therevidae). *ZooKeys* **618**, 97–128. doi:10.3897/zookeys.618.8059

Irwin ME, Yeates DK (1995) An Australian stiletto fly (Diptera: Therevidae) parasitised by a bee fly (Diptera: Bombyliidae). *Journal of Natural History* **29**, 1309–1327. doi:10.1080/00222939500770541

Ishida C, Kono M, Sakai S (2009) A new pollination system: brood-site pollination by flower bugs in *Macaranga* (Euphorbiaceae). *Annals of Botany* **103**, 39–44. doi:10.1093/aob/mcn212

Ismay JW (1993) Revision of *Tricimba* Lioy and *Aprometopis* Becker (Diptera: Chloropidae) from Australia and the Papuan Region. *Invertebrate Taxonomy* **7**, 297–499. doi:10.1071/IT9930297

Jackson RR, Pollard SD, Nelson XJ, Edwards GB, Barrion AT (2001) Jumping spiders (Araneae: Salticidae) that feed on nectar. *Journal of Zoology* **255**, 25–29. doi:10.1017/S095283690100108X

Jamieson BGM (1963) A revision of the earthworm genus *Digaster* (Megascolecidae, Oligochaeta). *Records of the Australian Museum* **26**, 83–111. doi:10.3853/j.0067-1975.26.1963.670

Jamieson BGM (1971) A review of the megascolecoid earthworm genera (Oligochaeta) of Australia. Pt. III – the subfamily Megascolecinae. *Memoirs of the Queensland Museum* **16**, 69–102.

Jamieson BGM (1972) A new species of *Digaster* (Megascolecidae: Oligochaeta) from Queensland. *Memoirs of the Queensland Museum* **16**, 261–264.

Jamieson BGM (1975) The genus *Digaster* (Megascolecidae: Oligochaeta) in Queensland. *Memoirs of the Queensland Museum* **17**, 267–292.

Jamieson BGM (1976) The genus *Diporochaeta* (Oligochaeta: Megascolecidae) in Queensland. *Zoölogische Verhandelingen* **149**, 1–57.

Jamieson BGM (1981) Historical biogeography of Australian Oligochaeta. In *Ecological Biogeography of Australia*. (Ed. A Keast) pp. 885–922. Dr W. Junk, Publishers, The Hague.

Jamieson BGM (1994) Some earthworms from the wet tropics and from the Bunya mountains, Queensland (Megascolecidae: Oligochaeta). *Memoirs of the Queensland Museum* **37**, 157–180.

Jamieson BGM (1995) New species and a new genus of earthworms in the collections of the Queensland Museum (Megascolecidae: Oligochaeta). *Memoirs of the Queensland Museum* **38**, 575–596.

Jamieson BGM, Wampler JE (1979) Bioluminescent Australian earthworms II. Taxonomy and preliminary report of bioluminescence in the genera *Spenceriella*, *Fletcherodrilus* and *Pontodrilus* (Megascolecidae: Oligochaeta). *Australian Journal of Zoology* **27**, 637–669. doi:10.1071/ZO9790637

Jarvis M, Vilhelmsen L (2000) Mouthpart evolution in adults of the basal 'symphytan', hymenopteran lineages. *Biological Journal of the Linnean Society. Linnean Society of London* **70**, 121–146.

Jeekel CAW (1968) *On the Classification and Geographical Distribution of the Family Paradoxosomatidae (Diplopoda, Polydesmida)*. Bonder-offset, Rotterdam.

Jeekel CAW (1974) The taxonomy and geography of the Sphaerotheriida (Diplopoda). *Symposia of the Zoological Society London* **32**, 41–52.

Jeekel CAW (1985) Millipedes from Australia, 9: A new polydesmoid millipede from Queensland (Diplopoda, Polydesmida: Dalodesmesidae). *Entomologische Berichten Deel* **45**, 50–55.

Jeekel CAW (1986) Millipedes from Australia, 10: Three interesting new species and a new genus (Diplopoda: Sphaerotheriida, Spirobolida, Polydesmida). *Beaufortia* **36**, 35–50.

Jeekel CAW (1987) Millipedes from Australia, 11: Australiosomatini from Queensland (Diplopoda, Polydesmida, Paradoxosomatidae). *Beaufortia* **37**, 11–41.

Jell PA (2004) The fossil insects of Australia. *Memoirs of the Queensland Museum* **50**, 1–124.

Jell PA, Duncan PM (1986) Invertebrates, mainly insects, from the freshwater, Lower Cretaceous, Koonwara Fossil Bed (Korumburra Group), South Gippsland, Victoria. *Memoirs of the Association of Australasian Palaeontologists* **3**, 111–205.

Jenkins NL, Hoffmann AA (2001) Distribution of *Drosophila serrata* Malloch (Diptera: Drosophilidae) in Australia with particular reference to the southern border. *Australian Journal of Entomology* **40**, 41–48. doi:10.1046/j.1440-6055.2001.00197.x

Jennings JT, Austin AD (1997) Revision of *Aulacofoenus* Kieffer (Hymenoptera: Gasteruptiidae), Hyptiogastrinine wasps with a restricted Gondwanic distribution. *Invertebrate Taxonomy* **11**, 943–976. doi:10.1071/IT97003

Jocqué R, Baehr B (1992) A revision of the Australian spider genus *Storena* (Araneae: Zodariidae). *Invertebrate Taxonomy* **6**, 953–1004. doi:10.1071/IT9920953

Jocqué R, Baehr B (2001) Revisions of genera in the *Asteron*-complex (Araneae: Zodariidae). *Asteron* Jocqué and the new genus *Pseudasteron*. *Records of the Australian Museum* **53**, 21–36. doi:10.3853/j.0067-1975.53.2001.1321

Johanson KA (1995) Eight new species and a revised key to the Australian *Helicopsyche* (Trichoptera: Helicopsychidae). *Entomologica Scandinavica* **26**, 241–272. doi:10.1163/187631295X00017

Johns PM (1997) The Gondwanaland weta family Anostostomatidae (formerly Stenopelmatidae, Henicidae or Mimmermidae): nomenclatural problems, world checklist, new genera and species. *Journal of Orthoptera Research* **6**, 125–138. doi:10.2307/3503546

Johnson BR, Borowiec ML, Chiu JC, Lee EK, Attalah J, Ward PS (2013) Phylogenomics resolves evolutionary relationships among ants, bees and wasps. *Current Biology* **23**, 2058–2062. doi:10.1016/j.cub.2013.08.050

Johnson LK, Foster RB (1986) Associations of large Homoptera (Fulgoridae and Cicadidae) and trees in a tropical forest. *Journal of the Kansas Entomological Society* **59**, 415–422.

Johnson NF (1991) Revision of Australasian *Trissolcus* species (Hymenoptera: Scelionidae). *Invertebrate Taxonomy* **5**, 211–239. doi:10.1071/IT9910211

Johnson SD, Moré M, Amorin FW, Haber WA, Frankie GW, Stanley DA, Cocucci AA, Raguso RA (2016) The long and the short of it: A global analysis of hawkmoth pollination niches and interaction networks. *Functional Ecology* doi:10.1111/1365.2435.12753

Jokat W, Boebel T, König M, Meyer W (2003) Timing and geometry of early Gondwanan breakup. *Journal of Geophysical Research* **108**(B9), doi:10.1029/2002JB001802

Juniper PA, Britton DR (2010) Insects associated with the fruit of *Syzygium paniculatum* (Magenta Lillypilly) and *Syzygium australe* (Brush Cherry). *Australian Journal of Entomology* **49**, 296–303. doi:10.1111/j.1440-6055.2010.00767.x

Jurado-Rivera JA, Vogler AP, Reid CAM, Petitpierre E, Gómez-Zurita J (2008) DNA barcoding insect-host plant associations. *Proceedings of the Royal Society B* **276**, 639–648.

Kalkman VJ, Dijkstra K-DB, Dow RA, Stokvis FR, van Tol J (2018) Out of Australia: the Argiolestidae reveal the Melanesian Arc System and East Papua Composite Terrane as possible ancient dispersal routes to the Indo-Australian Archipelago (Odonata: Argiolestidae). *International Journal of Odonatology* **21**, 1. doi:10.1080/13887890.2017.1402825

Kalkman VJ, Theischinger G (2013) Generic revision of Argiolestidae (Odonata), with four new genera. *International Journal of Odonatology* **16**, 1–52. doi:10.1080/13887890.2012.749450

Kallies A, Hilton DJ (2012) Revision of Cossinae and small Zeuzerinae from Australia (Lepidoptera: Cossidae). *Zootaxa* **3454**, 1–61. doi:10.11646/zootaxa.3454.1.1

Kaspari M, Clay NA, Lucas J, Yanoviak SP, Kay A (2015) Thermal adaptation generates a diversity of thermal limits in a rainforest ant community. *Global Change Biology* **21**, 1092–1102. doi:10.1111/gcb.12750

Kawahara AY, Plotkin D, Espeland M, Meusemann K, Toussaint EFA, Donath A, Gimnich F, Frandsen PB, Zwick A, dos Reis M, Barber JR, Peters RS, Liu S, Zhou X, Mayer C, Podsiadlowski L, Storer C, Yack JE, Misof B, Breinholt JW (2019) Phylogenetics reveals the evolutionary timing and pattern of butterflies and moths. *Proceedings of the National Academy of Science*. doi:org/10.1073/pnas.1907847116

Kayaalp P, Stevens MI, Schwarz MP (2017) 'Back to Africa': increased taxon sampling confirms a problematic Australia-to-Africa bee dispersal event in the Eocene. *Systematic Entomology* **42**, 724–733. doi:10.1111/syen.12241

Kelley ST, Farrell BD (1998) Is specialization a dead end? The phylogeny of host use in *Dendroctonus* bark beetles (Scolytidae). *Evolution* **52**, 1731–1743. doi:10.1111/j.1558-5646.1998.tb02253.x

Kelton LA (1978) *The Anthocoridae of Canada and Alaska. Heteroptera: Anthocoridae. The Insects of Canada. Part 4*. Agriculture Canada, Ottawa.

Kemp DJ (2002) Visual mate-searching behaviour in the evening brown butterfly, *Melanitis leda* (L.) (Lepidoptera: Nymphalidae). *Australian Journal of Entomology* **41**, 300–305. doi:10.1046/j.1440-6055.2002.00311.x

Kemp DJ, Macedonia JM (2006) *Hypolimnas bolina* L. (Nymphalidae): visual, morphological and ecological properties. *Australian Journal of Zoology* **54**, 235–244. doi:10.1071/ZO06005

Kemp JE, Linder HP, Ellis AG (2017) Beta diversity of herbivorous insects is coupled to high species and phylogenetic turnover of plant communities across short spatial scales in the Cape Floristic Region. *Journal of Biogeography* **44**, 1813–1823. doi:10.1111/jbi.13030

Kerr PH (2009) Rhagionidae (snipe flies). In *Manual of Central American Diptera. Vol. 1.* (Eds BV Brown, A Borkent, JM Cumming, DM Wood, NE Woodley, MA Zumbado) pp. 485–489. NRC Research Press, Ottawa.

Key KHL (1989) Revision of the genus *Praxibulus* (Orthoptera: Acrididae). *Invertebrate Taxonomy* **3**, 1–121. doi:10.1071/IT9890001

Key KHL (1991) Phasmatodea 'stick insects'. In *The Insects of Australia.* Ed. ID Naumann) pp. 394–404. Melbourne University Press, Carlton.

Kim SP (1994) *Australian Lauxaniid Flies: Revision of the Australian Species of* Homoneura *van der Wulp,* Trypetisoma *Malloch, and Allied Genera (Diptera: Lauxaniidae). Monographs on Invertebrate Taxonomy. Volume 1.* CSIRO, Collingwood.

Kimsey LS (1991) Relationships among the tiphiid wasp subfamilies (Hymenoptera). *Systematic Entomology* **16**, 427–438. doi:10.1111/j.1365-3113.1991.tb00677.x

Kimsey LS (1992) Phylogenetic relations among the South American thynnine tiphiid wasps (Hymenoptera). *Systematic Entomology* **17**, 133–144. doi:10.1111/j.1365-3113.1992.tb00326.x

Kimsey LS, Bohart RM (1991) *The Chrysidid Wasps of the World*. Oxford University Press, Oxford.

King J (1984) Immature stages of some Megachilidae (Hymenoptera: Apoidea). *Journal of the Australian Entomological Society* **23**, 51–57. doi:10.1111/j.1440-6055.1984.tb01907.x

King J (1994) The bee family Megachilidae (Hymenoptera: Apoidea) in Australia, I. Morphology of the genus *Chalicodoma* Lepeletier, and a revision of the subgenus *Hackeriapis* Cockerell. *Invertebrate Taxonomy* **8**, 1373–1419. doi:10.1071/IT9941373

King JE, Riegler M, Thomas RG, Spooner-Hart R (2015) Phylogenetic placement of Australian carrion beetles (Coleoptera: Scarabaeidae). *Austral Entomology* **54**, 366–375. doi:10.1111/aen.12138

Kirejtshuk AG, Lawrence JF (1990) Revision of the Australian genus *Idaethina* Reitter (Coleoptera: Nitidulidae). *Journal of the Australian Entomological Society* **29**, 1–9. doi:10.1111/j.1440-6055.1990.tb00305.x

Kirejtshuk AG, Lawrence JF (1992a) Cychramptodini, a new tribe of Nitidulidae (Coleoptera) from Australia. *Journal of the Australian Entomological Society* **31**, 29–46. doi:10.1111/j.1440-6055.1992.tb00456.x

Kirejtshuk AG, Lawrence JF (1992b) Review of the *Thalycrodes* complex of genera (Coleoptera: Nitidulidae) endemic to the Australian region. *Journal of the Australian Entomological Society* **31**, 119–142. doi:10.1111/j.1440-6055.1992.tb00469.x

Kitching RL, Allsopp PG (1987) *Prionocyphon niger* sp. n. (Coleoptera: Scirtidae) from water-filled tree holes in Australia. *Journal of the Australian Entomological Society* **26**, 73–79. doi:10.1111/j.1440-6055.1987.tb00264.x

Kitching RL, Arthur M (1993) The biodiversity of arthropods from Australian rain forest canopies: summary of projects and the impact of drought. *Selbyana* **14**, 29–35.

Kitching RL, Bergelson JM, Lowman MD, McIntyre S, Carruthers G (1993) The biodiversity of arthropods from Australian rainforest canopies: General introduction, methods, sites and ordinal results. *Australian Journal of Ecology* **18**, 181–191. doi:10.1111/j.1442-9993.1993.tb00442.x

Kitching RL, Bickel D, Creagh AC, Hurley K, Symonds C (2004) The biodiversity of Diptera in Old World rain forest surveys: a comparative faunistic analysis. *Journal of Biogeography* **31**, 1185–1200. doi:10.1111/j.1365-2699.2004.01096.x

Kitching RL, Braithwaite R, Cavanaugh J (Eds) (2010) *Remnants of Gondwana: A Natural and Social History of the Gondwana Rainforests of Australia.* Surrey Beatty and Sons, Baulkham Hills.

Kitching RL, Callaghan C (1982) The fauna of water-filled tree holes in box forest in south-east Queensland. *Australian Entomological Magazine* **8**, 61–70.

Kittel RN, Austin AD (2014) Synopsis of Australian cheloninae wasps (Hymenoptera: Braconidae: Cheloninae) with descriptions of two new genera. *Austral Entomology* **53**, 183–202. doi:10.1111/aen.12070

Klein BC (1989) Effects of forest fragmentation on dung and carrion beetle communities in Central Amazonia. *Ecology* **70**, 1715–1725. doi:10.2307/1938106

Klocke D, Schmitz A, Schmitz H (2011) Fire-adaptation in *Hypocerides nearcticus* Borgmeier and *Anabarhynchus hyalipennis hyalipennis* Macquart and new notes about the Australian 'Smoke Fly' *Microsania australis* Collart (Diptera: Phoridae: Therevidae and Platypezidae). *The Open Entomology Journal* **5**, 10–14. doi:10.2174/1874407901105010010

Kneidel KA (1984) Competition and disturbance in communities of carrion-breeding Diptera. *Journal of Animal Ecology* **53**, 849–865. doi:10.2307/4663

Koch LE (1981) The scorpions of Australia: aspects of their ecology and zoogeography. In *Ecological Biogeography of Australia.* (Ed. A Keast) pp. 873–884. Dr W. Junk, Publishers, The Hague.

Koch LE (1983) Revision of the Australian centipedes of the genus *Cormocephalus* Newport (Chilopoda: Scolopendridae: Scolopendrinae). *Australian Journal of Zoology* **31**, 799–833. doi:10.1071/ZO9830799

Kodandaramaiah U, Braby MF, Grund R, Muller CJ, Wahlberg N (2018) Phylogenetic relationships, biogeography and diversification of Coenonymphina butterflies (Nymphalidae: Satyrinae): intercontinental dispersal of a Southern Gondwanan group. *Systematic Entomology* doi:10.1111/syen.12303

Kohen JL, Merrick JR (1998) Limited usage of freshwater crayfishes (genus *Euastacus*) by Aborigines in eastern New

South Wales: records and comments. *Proceedings of the Linnean Society of New South Wales* **119**, 101–105.

Kohout RJ (1988) Nomenclatural changes and new Australian records in the ant genus *Polyrhachis* Fr. Smith (Hymenoptera: Formicidae: Formicinae). *Memoirs of the Queensland Museum* **25**, 429–438.

Kohout R (2006) Review of *Polyrhachis* (*Cyrtomyrma*) Forel (Hymenoptera: Formicidae: Formicinae) of Australia, Borneo, New Guinea and the Solomon Islands with descriptions of new species. *Memoirs of the Queensland Museum* **52**, 87–146.

Kohout RJ, Taylor RW (1990) Notes on ants of the genus *Polyrhachis* Fr. Smith, with a synonymic list of the species (Hymenoptera: Formicidae: Formicinae). *Memoirs of the Queensland Museum* **28**, 509–522.

Kolesik P, Butterill PT (2015) New gall midges (Diptera: Cecidomyiidae) from Papua New Guinea. *Austral Entomology* **54**, 79–86. doi:10.1111/aen.12095

Kolesik P, Butterworth N, Lemmon J, Gibson T, Wallman JF (2019) First gall midge (Diptera: Cecidomyiidae) known to feed on plant family Atherospermataceae: a new species of *Ashondylia* damaging the endangered Australian tree *Daphnandra johnsonii*. *Austral Entomology* doi:10.1111/aen.12387

Kolesik P, Hall CR, Kitching RL, Adair RJ (2018) A new genus of gall midge (Diptera: Cecidomyiidae) inducing leaf galls on *Rubus* (Rosaceae). *Austral Entomology* **57**, 85–91. doi:10.1111/aen.12271

Kolibáč J (1998) New Australian Thanerocleridae with notes on the biogeography of the subtribe Isoclerina Kolibáč (Coleoptera: Cleroidea). *Invertebrate Taxonomy* **12**, 951–975. doi:10.1071/IT96019

Kolibáč J, Huang D (2016) The oldest known clerid fossils from the Middle Jurassic of China, with a review of Cleridae systematics (Coleoptera). *Systematic Entomology* **41**, 808–823. doi:10.1111/syen.12192

Kondandaramaiah U, Braby MF, Grund R, Müller CJ, Wahlberg N (2018) Phylogenetic relationships, biogeography and diversification of Coenonymphina butterflies (Nymphalidae: Satyrinae): intercontinental dispersal of a southern Gondwanan group? *Systematic Entomology* doi:10.111/syen.12303

Kooyman R, Rossetto M, Allen C, Cornwall W (2012) Australian tropical and subtropical rain forest community assemblages: phylogeny, functional biogeography and environmental gradients. *Biotropica* **44**, 668–679. doi:10.1111/j.1744-7429.2012.00861.x

Koteja J (1999) *Eomatsucoccus andrewi* sp. nov. (Hemiptera: Sternorrhyncha: Coccinea) from the Lower Cretaceous of southern England. *Cretaceous Research* **20**, 863–866. doi:10.1006/cres.1999.0187

Krause DW, Prasad GVR, von Koenigswald W, Sahni A, Grine FE (1997) Cosmopolitanism among Late Cretaceous Gondwanan mammals. *Nature* **390**, 504–507. doi:10.1038/37343

Kreft H, Jetz W (2010) A framework for delineating biogeographical regions based on species distributions. *Journal of Biogeography* **37**, 2029–2053. doi:10.1111/j.1365-2699.2010.02375.x

Kreft H, Jetz W (2013) Comment on 'An update of Wallace's zoogeographic regions of the world'. *Science* **341**(6144), doi:10.1126/science.1237471

Kristensen NP (1967) Erection of a new family in the Lepidoptera suborder Dacnonypha. *Entomologiske Meddelelser* **35**, 341–345.

Kristensen NP (1976) Remarks on the family-level phylogeny of butterflies (Insecta, Lepidoptera, Rhopalocera). *Journal of Zoological Sciences and Evolutionary Research* **14**, 25–33.

Kristensen NP (1991) Phylogeny of extant hexapods. In *The Insects of Australia*. (Ed. ID Naumann) pp. 125–140. Melbourne University Press, Carlton.

Krosch MN (2011) Phylogeny of *Echinocladius martini* Cranston (Diptera: Chironomidae) in closed forest streams of Eastern Australia. *Australian Journal of Entomology* **50**, 258–268.

Krzemiński W, Krzemińska E (2003) Triassic Diptera: descriptions, revisions, and phylogenetic relations. *Acta Zoologica Cracoviensia* **46**(Suppl.), 153–184.

Krzemiński W, Soszyńska-Maj A, Bashkuev AS, Kopeć K (2015) Revision of the unique Early Cretaceous Mecoptera from Koonwarra (Australia) with description of a new genus and family. *Cretaceous Research* **52**, 501–506. doi:10.1016/j.cretres.2014.04.004

Kukalová-Peck J (1991) Fossil history and the evolution of Hexapod structures. In *The Insects of Australia*. (Ed. ID Naumann) pp. 141–179. Melbourne University Press, Carlton.

Kumar R (1966) Studies on the biology, immature stages, and relative growth of some Australian bugs of the superfamily Coreoidea (Hemiptera: Heteroptera). *Australian Journal of Zoology* **14**, 895–991.

Kumar R (1969) Morphology and relationships of the Pentatomoidea (Heteroptera). *Australian Journal of Zoology* **17**, 553–606. doi:10.1071/ZO9690553

Kundrata R, Gunter NL, Douglas H, Bocak L (2016) Next step toward a molecular phylogeny of click-beetles (Coleoptera: Elateridae): redefinition of Pityobiinae, with a description of a new subfamily Parablacinae from the Australasian Region. *Austral Entomology* **55**, 291–302. doi:10.1111/aen.12185

Kurahashi H (1989) Family Calliphoridae. In *Catalog of the Diptera of the Australasian and Oceanic Regions*. (Ed. NL Evenhuis) pp. 702–718. Bishop Museum Press and E.J. Brill, Honolulu.

Kurina O, Hippa H, de Souza Amorin D (2019) Notes on *Manota* (Diptera: Mycetophilidae) from Australia and Papua New Guinea, with descriptions of two new species. *Zootaxa* **4555**, 385–395.

Kurina O, Oliveira SS (2013) The first *Cordyla* Meigen species (Diptera, Mycetophylidae) from continental Australia and Tasmania. *ZooKeys* **342**, 29–43.

Kuschel G (1995) A phylogenetic classification of Curculionoidea to family and subfamily level. *Memoirs of the Entomological Society of Washington* **14**, 5–33.

Kuschel G, Leschen RAB (2010) Phylogeny and taxonomy of the Rhinorhynchinae (Coleoptera: Nemonychidae). *Invertebrate Systematics* **24**, 573–615. doi:10.1071/IS09027

Kuschel G, May BM (1989) Palophaginae, a new subfamily for leaf beetles, feeding as adult and larva on Araucarian pollen in Australia (Coleoptera: Megalopodidae). *Invertebrate Taxonomy* **3**, 697–719. doi:10.1071/IT9890697

Kuschel G, May BM (1996a) Palophaginae, their systematic position and biology. In *Chrysomelidae Biology. Volume 3.* (Eds PHA Jolivet and ML Cox) pp. 173–185. SPB Academic Publishing, Amsterdam.

Kuschel G, May BM (1996b) Discovery of Palophaginae (Coleoptera: Megalopodidae) on *Araucaria araucana* in Chile and Argentina. *New Zealand Entomologist* **19**, 1–13. doi:10.1080/00779962.1996.9722015

Lachnicht SL, Hendrix PF, Zou X (2002) Interactive effects of native and exotic earthworms on resource use and nutrient mineralization in a tropical wet forest soil of Puerto Rico. *Biology and Fertility of Soils* **36**, 43–52. doi:10.1007/s00374-002-0501-5

Lackner T, Leschen RAB (2017) A monograph of the AustraloPacific Saprininae (Coleoptera, Histeridae). *ZooKeys* **689**, 1–263. doi:10.3897/zookeys.689.12021

Laidlaw MJ, McDonald WJF, Hunter RJ, Kitching RL (2011) Subtropical rainforest turnover along an altitudinal gradient. *Memoirs of the Queensland Museum - Nature* **55**, 271–290.

Lake DC (1990) Observations on the eastern mouse spider *Missulena bradleyi* Rainbow (Mygalomorphae: Actinopodidae): natural history and envenomation. *Australian Entomological Magazine* **17**, 93–96.

Lambkin CL, Boulter SL, Starick NT, Cantrell BK, Bickel DJ, Wright SG, Power N, Schutze MK, Turco F, Nakamura A, Burwell CJ (2011) Altitudinal and seasonal variation in the family-level assemblages of flies (Diptera) in an Australian subtropical rainforest: one hundred thousand and counting! *Memoirs of the Queensland Museum - Nature* **55**, 315–331.

Lambkin CL, Recsei JM, Yeates DK (2005) Systematic revision of *Johnmannia* Irwin and Lyneborg (Diptera: Therevidae): atypical metallic stiletto flies from Australian mesic habitats. *Zootaxa* **866**, 1–28. doi:10.11646/zootaxa.866.1.1

Lambkin CL, Sinclair BJ, Pape T, Coutenay GW, Skevington JH, Meier R, Yeates DK, Blagoderov V, Wiegmann BM (2013) The phylogenetic relationships among infraorders and superfamilies of Diptera based on morphological evidence. *Systematic Entomology* **38**, 164–179. doi:10.1111/j.1365-3113.2012.00652.x

Lambkin CL, Turco F (2013) *Collessiama* (Diptera: Therevidae: Agapophytinae: *Taenogera* genus group), a new genus from eastern Australia, with a key to the Australian genera of Therevidae. *Zootaxa* **3680**, 96–117. doi:10.11646/zootaxa.3680.1.7

Lambkin KJ (1978) The Australian Achiline genera *Aneipo* Kirkaldy and *Bunduica* Jacobi (Homoloptera: Fulgoroidea: Achilidae). *Journal of the Australian Entomological Society* **17**, 25–40. doi:10.1111/j.1440-6055.1978.tb01742.x

Lambkin KJ (1986) A revision of the Australian Mantispidae (Insecta: Neuroptera) with a contribution to the classification of the subfamily. II. Calomantispinae and Mantispinae. *Australian Journal of Zoology Supplementary Series* **117**, 1–113.

Lambkin KJ (1992) A record of *Austrosialis* Tillyard from the Queensland Paleocene (Insecta: Megaloptera: Sialidae). *Queensland Naturalist* **31**, 84–86.

Lambkin KJ (1994) Revision of the Australian scorpion-fly genus *Harpobittacus* (Mecoptera: Bittacidae). *Invertebrate Taxonomy* **8**, 767–808. doi:10.1071/IT9940767

Lambkin KJ (2014) Psychopsoid Neuroptera (Psychopsidae, Osmylopsycopidae) from the Queensland Triassic. *Australian Entomologist* **41**, 57–76.

Lancaster J, Downes BJ, Glaister A (2009) Interacting environmental gradients, trade-offs and reversals in the abundance - environment relationships of stream insects. *Marine and Freshwater Research* **60**, 259–270. doi:10.1071/MF08226

Lange OL, Nobel PS, Osmond CB, Ziegler H (Eds) (1981) *Physiological Plant Ecology I. Responses to the Physical Environment.* Springer-Verlag, Berlin.

Laurance WF, Bierregaard RO, Jr (Eds) (1997) *Tropical Forest Remnants: Ecology, Management, and Conservation of Fragmented Communities.* The University of Chicago Press, Chicago.

Law BS (1992) Physiological factors affecting pollen use by Queensland blossom bats (*Syconycteris australis*). *Functional Ecology* **6**, 257–264. doi:10.2307/2389515

Law BS (1993) Roosting and foraging ecology of the Queensland blossom bat, *Syconycteris australis*, in north-eastern New South Wales. Flexibility in response to seasonal variation. *Wildlife Research* **20**, 419–431. doi:10.1071/WR9930419

Lawler SH, Crandall KA (1998) The relationship of the Australian freshwater crayfish genera *Euastacus* and *Astacopsis*. *Proceedings of the Linnean Society of New South Wales* **119**, 1–8.

Lawrence JF (1980) A new genus of Indo-Australian Gemplyodini with notes on the constituition of the Colydiidae (Coleoptera). *Journal of the Australian Entomological Society* **19**, 293–310. doi:10.1111/j.1440-6055.1980.tb00989.x

Lawrence JF (1982) Coleoptera. In *Synopsis and classification of living organisms.* (Ed. SP Parker) pp. 482–553. McGraw-Hill, New York.

Lawrence JF (1988) Rhinorphidae, a new beetle family from Australia, with comments on the phylogeny of the Elateriformia. *Invertebrate Taxonomy* **2**, 1–53. doi:10.1071/IT9880001

Lawrence JF (1994a) The larva of *Sirrhas variegates*, sp. nov., with notes on the Perimylopidae, Ulodidae (*stat. nov.*), Zopheridae and Chalcodryidae (Coleoptera: Tenebrionoidea). *Invertebrate Taxonomy* **8**, 329–349. doi:10.1071/IT9940329

Lawrence JF (1994b) Review of the Australian Archeocrypticidae (Coleoptera), with descriptions of a new genus and four new species. *Invertebrate Taxonomy* **8**, 449–470. doi:10.1071/IT9940449

Lawrence JF (1995) Two new species of *Rhopalobrachium* Boheman (Coleoptera: Phloeostichidae: Hymaeinae) from Australia and Chile. In *Biology, Phylogeny and Classification of Coleoptera: Papers Celebrating the 80th Birthday of Roy A. Crowson*. (Eds J Pakaluk and SA Ślipiński) pp. 437–447. Warsaw: Muzeum I Instytut Zoologi PAN.

Lawrence JF, Britton EB (1991) Coleoptera 'beetles'. In *The Insects of Australia*. (Ed. ID Naumann) pp. 543–683. Melbourne University Press, Carlton.

Lawrence JF, Britton EB (1994) *Australian Beetles*. Melbourne University Press, Carlton.

Lawrence JF, Pollock DA (1994) Relationships of the Australian genus *Synerctinus* Newman (Coleoptera: Boridae). *Journal of the Australian Entomological Society* **33**, 35–42. doi:10.1111/j.1440-6055.1994.tb00914.x

Lawrence JF, Ślipiński A (2013) *Australian Beetles: Morphology, Classification and Keys*. CSIRO Publishing, Melbourne.

Le Cesne M, Wilson SW, Soulier-Perkins A (2015) Elevational gradient of Hemiptera (Heteroptera, Auchenorrhyncha) on a tropical mountain in Papua New Guinea. *PeerJ* **3**, e978. doi:10.7717/peerj.978

Lea AM (1895a) Descriptions of new species of Australian Coleoptera. *Proceedings of the Linnean Society of New South Wales* **9**, 589–634. doi:10.5962/bhl.part.18130

Lea AM (1895b) Descriptions of new species of Australian Coleoptera. Part II. *Proceedings of the Linnean Society of New South Wales* **10**, 224–319.

Lea AM (1914) Notes on Australian Cetonides: with a list of species and descriptions of some new ones. *Transactions of the Royal Society of South Australia* **38**, 132–218.

Lea AM (1921) On Coleoptera, mostly from Queensland. *Memoirs of the Queensland Museum* **7**, 182–240.

Lea AM (1923) Australian dung beetles of the subfamily Coprides. *Records of the South Australian Museum* **2**, 353–396.

Lea AM (1926) On some Australian Curculionidae. *Proceedings of the Linnean Society of New South Wales* **51**, 327–362.

Leathers AL (1922) Ecological study of aquatic midges and some related insects with special reference to feeding habits. *Bulletin of the United State Bureau of Fisheries* **38**, 1–62.

Lee TRC, Cameron SL, Evans TA, Ho SYW, Lo N (2015) The origin and radiation of Australian *Coptotermes* termites: from rainforest to desert dwellers. *Molecular Phylogenetics and Evolution* **82**, 234–244. doi:10.1016/j.ympev.2014.09.026

Lee TRC, Evans TA, Cameron SL, Ho SYW, Namyatova A, Lo N (2017) A review of the status of *Coptotermes* (Isoptera: Rhinotermitidae) species in Australia with description of two small termite species from Northern and Eastern Australia. *Invertebrate Systematics* **31**, 180–190. doi:10.1071/IS16031

Leigh C (2013) Dry-season changes in macroinvertebrate assemblages of highly seasonal rivers: responses to low flow, no flow and antecedent hydrology. *Hydrobiologia* **703**, 95–112. doi:10.1007/s10750-012-1347-y

Leijs R, Batley M, Hogendoorn K (2017) The genus *Amegilla* (Hymenoptera, Apoidae, Anthophorini) in Australia: a revision of the subgenera *Notomegilla* and *Zonamegilla*. *ZooKeys* **653**, 79–140. doi:10.3897/zookeys.653.11177

Lessard BD (2014) Revision of the austral horse fly tribe Scionini (Diptera: Tabanidae). *Austral Entomology* **53**, 203–239. doi:10.1111/aen.12076

Lessard BD, Yeates DK (2012a) Seven new species of the Southern Hemisphere horse fly subgenus *Scaptia* (*Psedoscione*) (Diptera: Tabanidae), including descriptions and a revised key. *Journal of Medical Entomology* **49**, 1206–1225.

Lessard BD, Yeates DK (2012b) *Anzyomyia* (Diptera: Tabanidae: Pangoniinae: Scionini): a new genus of Australian and New Zealand horse fly, including the description of three new species. *Insect Systematics and Evolution* **43**, 101–106.

Lessard BD, Yeates DK (2013) New species of the hairy-eyed horse fly subgenera *Scaptia* (*Myioscaptia*) Mackerras, 1955 and *Scaptia* (*Scaptia*) Walker, 1850 (Diptera: Tabanidae) from Australia. *Zootaxa* **3680**, 118–129.

Lessard BD, Yeates DK, Woodley NE (2019) Revision of Hermetiinae (Diptera, Stratiomyidae). *Austral Entomology* **58**, 122–136.

Leston D, Scudder GGE (1957) The taxonomy of the bronze orange-bug and related Australian Oncomerinae (Hemiptera: Tessaratomidae). *Annals & Magazine of Natural History* **10**, 439–448. doi:10.1080/00222935708655982

Levey B (1978a) A new tribe, Epistomentini, of Buprestidae (Coleoptera) with a redefinition of the tribe Chrysochroini. *Systematic Entomology* **3**, 153–158. doi:10.1111/j.1365-3113.1978.tb00111.x

Levey B (1978b) A taxonomic revision of the genus *Prospheres* (Coleoptera: Buprestidae). *Australian Journal of Zoology* **26**, 713–726. doi:10.1071/ZO9780713

Levey B (2012) A revision of the Australian species of the genus *Melobasis* Laporte and Gory 1837 (Coleoptera: Buprestidae), Part 1 (Introductory material, key to species-groups and keys to the species of the *thoracica*, *pusilla*, *formosa*, *propinqua* and *gloriosa* species-groups). *Zootaxa* **3464**, 1–107. doi:10.11646/zootaxa.3464.1.1

Levey B, Bellamy CL (2013) A taxonomic revision of *Neobuprestis* Kerremans (Coleoptera: Buprestidae) with description of a new genus and two new species. *Zootaxa* **3681**, 225–240. doi:10.11646/zootaxa.3681.3.2

Levi HW (1983) The orb-weaver genera *Argiope*, *Gea*, and *Neogea* from the Western Pacific Basin. *Bulletin of the Museum of Comparative Zoology* **150**, 247–338.

Leys R (2000) A revision of the Australian carpenter bees, genus *Xylocopa* Latreille, subgenera *Koptortosoma* Gribodo and *Lestis* Lepeletier and Serville (Hymenoptera: Apidae). *Invertebrate Taxonomy* **14**, 115–136. doi:10.1071/IT98014

Li L, Aguilar R, Berkov A (2017) What shapes cerambycid beetle communities in a tropical forest mosaic? Assessing the effects of host tree identity, forest structure, and vertical stratification. *Biotropica* **49**, 675–684. doi:10.1111/btp.12432

Lin N-Q, Huber JT, La Salle J (2007) The Australian genera of Mymaridae (Hymenoptera: Chalcoidea). *Zootaxa* **1596**, 1–111.

Lindquist EE (1998) Evolution of phytophagy in trombidiform mites. *Experimental & Applied Acarology* **22**, 81–100. doi:10.1023/A:1006041609774

Lingafelter SW, Konstantinov AS (2000) Revision of *Pedethma* Weise (Coleoptera: Chrysomelidae). *Coleopterists Bulletin* **54**, 413–458. doi:10.1649/0010-065X(2000)054[0413:ROPWCC]2.0.CO;2

Linhart YB, Mendenhall JA (1977) Pollen dispersal by hawkmoths in a *Lindenia rivalis* Benth. population in Belize. *Biotropica* **9**, 143. doi:10.2307/2387672

Liu X, Wang Y, Shih C, Ren D, Yang D (2012) Early evolution and historical biogeography of fishflies (Megaloptera: Chauliodinae): implications from a phylogeny combining fossil and extant taxa. *PLoS ONE* **7**, e40345. doi:10.1371/journal.pone.0040345

Llanos L, Gonzalez CR, Saldarriga-Córdoba M (2015) Revision of the New World species of the genus *Pelecorhynchus* Macquart, 1850 (Diptera: Pelecorhynchidae). *Zootaxa* **3955**, 188–210. doi:10.11646/zootaxa.3955.2.2

Lo N, Tokuda G, Watanabe H, Rose H, Slaytor M, Maekawa K, Bandi C, Nada H (2000) Evidence from multiple gene sequences indicates that termites evolved from wood-feeding cockroaches. *Current Biology* **10**, 801–804. doi:10.1016/S0960-9822(00)00561-3

Löcker B (2015) Revision of the Australian species of *Aka* White, 1879 (Fulgoromorpha: Cixiidae) with the description of a new genus. *Zootaxa* doi:10.11646/zootaxa.3956.2.3

Löcker B, Fletcher MJ, Gurr GM (2007) Revision of the genus *Innobindus* Jacobi (Hemiptera: Fulgomorpha: Cixiidae) with description of six new species and comments on other Australian Brixiini genera. *Australian Journal of Entomology* **46**, 45–55. doi:10.1111/j.1440-6055.2007.00586.x

Löcker B, Fletcher MJ, Larivière MC, Gurr GM (2006) The Australian Pentastirini (Hemiptera: Fulgomorpha: Cixiidae). *Zootaxa* **1290**, 1–138.

Lonsdale O (2009) The *Heteromeringia* (Diptera: Clusiidae: Clusiodinae). *Records of the Australian Museum* **61**, 229–262. doi:10.3853/j.0067-1975.61.2009.1531

Lonsdale O, Marshall SA (2010) Clusiidae (clusiid flies). In *Manual of Central American Diptera. Vol. II.* (Eds BV Brown, A Borkent, JM Cumming, DM Wood, NE Woodley and MA Zumbado) pp. 1041–1048. NRC Research Press, Ottawa, Ontario.

Lopez-Vaamonde C, Dixon DJ, Cook JM, Rasplus J-Y (2002) Revision of the Australia species of *Pleistodontes* (Hymenoptera: Agaonidae) fig-pollinating wasps and their host-plant associations. *Zoological Journal of the Linnean Society* **136**, 637–683. doi:10.1046/j.1096-3642.2002.00040.x

Lowman MD (1982) Seasonal variation in insect abundance among three Australian rain forests, with particular reference to phytophagous types. *Australian Journal of Ecology* **7**, 353–361. doi:10.1111/j.1442-9993.1982.tb01310.x

Lowman MD (1985) Temporal and spatial variability in insect grazing of the canopies of five Australian rainforest tree species. *Australian Journal of Ecology* **10**, 7–24. doi:10.1111/j.1442-9993.1985.tb00859.x

Lowman MD (1992) Herbivory in Australian rain forests, with particular reference to the canopies of *Doryphora sassafras* (Monimiaceae). *Biotropica* **24**, 263–272. doi:10.2307/2388521

Luan Y, Mallatt JM, Xie R, Yang Y, Yin W (2005) The phylogenetic positions of three basal-Hexapod groups (Protura, Diplura, and Collembola) based on ribosomal RNA gene sequences. *Molecular Biology and Evolution* **22**, 1579–1592. doi:10.1093/molbev/msi148

Lucky A, Ward PS (2010) Taxonomic revision of the ant genus *Leptomyrmex* Mayr (Hymenoptera: Formicidae). *Zootaxa* **2688**, 1–67. doi:10.11646/zootaxa.2688.1.1

Lughadha E, Proença C (1996) A survey of the reproductive biology of the Myrtoideae (Myrtaceae). *Annals of the Missouri Botanical Garden* **83**, 480–503. doi:10.2307/2399990

Luna-Ramirez L, Miller AD, Rašic G (2017) Genetic and morphological analyses indicate that the Australian endemic scorpion *Urodacus yaschenkoi* (Scorpiones: Urodacidae) is a species complex. *PeerJ* **5**, e2759. doi:10.7717/peerj.2759

Lyneborg L (2001) *The Australian Stiletto-flies of the* Anabarhynchus *Genus-group (Diptera: Therevidae). Entomonograph Vol. 13.* Apollo Books, Stenstrup.

Maa TC (1989a) Family Hippoboscidae. In *Catalog of the Diptera of the Australasian and Oceanic Regions* (Ed. NL Evenhuis) pp. 785–789. Bishop Museum Press and E.J. Brill, Honolulu.

Maa TC (1989b) Family Nycteribiidae. In *Catalog of the Diptera of the Australasian and Oceanic Regions.* (Ed. NL Evenhuis) pp. 790–794. Bishop Museum Press and E.J. Brill, Honolulu.

Maa TC (1989c) Family Streblidae. In *Catalog of the Diptera of the Australasian and Oceanic Regions.* (Ed. NL Evenhuis) pp. 795–796. Bishop Museum Press and E.J. Brill, Honolulu.

Macdonald J, Ohmart CP (1993) Life history strategies of Australian pergid sawflies and their interactions with host plants. In *Sawfly Life History Adaptations to Woody Plants.* (Eds MR Wagner and KF Raffa) pp. 485–502. Academic Press, San Diego.

Mackerras IM (1925) The Nemestrinidae (Diptera) of the Australian region. *Proceedings of the Linnean Society of New South Wales* **50**, 489–561.

Mackerras IM (1956a) The Tabanidae (Diptera) of Australia. I. General review. *Australian Journal of Zoology* **4**, 376–407. doi:10.1071/ZO9560376

Mackerras IM (1956b) The Tabanidae (Diptera) of Australia. II. Subfamily Pangoniinae, tribe Pangoniini. *Australian Journal of Zoology* **4**, 408–443. doi:10.1071/ZO9560408

Mackerras IM (1960) The Tabanidae (Diptera) of Australia III. Subfamily Pangoniinae, tribe Scionini and supplement to Pangoniini. *Australian Journal of Zoology* **8**, 1–152. doi:10.1071/ZO9600001

Mackerras IM (1970) Composition and distribution of the fauna. In *Insects of Australia.* (Ed. ID Naumann) pp. 187–203. Melbourne University Press, Carlton.

Mackerras IM (1971) The Tabanidae (Diptera) of Australia V. Subfamily Tabanidae, tribe Tabanini. *Australian Journal of Zoology Supplementary Series* **4**, 1–54.

Mackerras IM, Fuller ME (1942) The genus *Pelecorhynchus* (Diptera: Tabanidae). *Proceedings of the Linnean Society of New South Wales* **67**, 9–76.

Mackerras IM, Mackerras MJ (1953) A new species of *Pelecorhynchus* (Diptera: Tabanoidea) from the Dorrigo Plateau, New South Wales. *Proceedings of the Linnean Society of New South Wales* **78**, 38–40.

Mackerras MJ (1966) Australian Blattidae (Blattodea). VI. Revision of the genus *Cosmozosteria* Stal. *Australian Journal of Zoology* **15**, 843–882.

Mackerras MJ (1967a) Australian Blattidae (Blattodea). I. General remarks and revision of the genus *Polyzosteria* Burmeister. *Australian Journal of Zoology* **15**, 593–618. doi:10.1071/ZO9670593

Mackerras MJ (1967b) Australian Blattidae (Blattodea). VII. The *Polyzosteria* group; general remarks and revision of the subgenera IPolyzosteria *Brunner* and *Leptozosteria* Tepper. *Australian Journal of Zoology* **15**, 1207–1298. doi:10.1071/ZO9671207

Mackerras MJ (1968a) Australian Blattidae (Blattodea). VIII. The *Polyzosteria* group: subgenus *Melanozosteria* Stal. *Australian Journal of Zoology* **16**, 511–575. doi:10.1071/ZO9680511

Mackerras MJ (1968b) Australian Blattidae (Blattodea). IX. Revision of the Polyzosteriinae tribe Methanini, Tryonicinae and Blattinae. *Australian Journal of Zoology* **16**, 511–575. doi:10.1071/ZO9680511

Mackerras MJ (1968c) *Neolaxta monteithi,* gen. et. sp. n. from eastern Australia (Blattodea: Blaberidae). *Journal of the Australian Entomological Society* **7**, 143–146. doi:10.1111/j.1440-6055.1968.tb00721.x

Mackerras MJ (1968d) Polyphagidae (Blattodea) from eastern Australia. *Journal of the Australian Entomological Society* **7**, 147–154. doi:10.1111/j.1440-6055.1968.tb00722.x

Macleay W (1885) Two new Australian Lucanidae. *Proceedings of the Linnean Society of New South Wales* **10**, 199–202. doi:10.5962/bhl.part.17915

Macleay W (1886) Miscellanea Entomologica. No. 1. The genus *Diphucephala. Proceedings of the Linnean Society of New South Wales* **1**, 381–402.

Main A (1987) Evolution and radiation of the terrestrial fauna. In *Fauna of Australia. General Articles. Vol. 1A*. (Eds DW Walton and GR Dyne) pp. 136–155. Australian Government Publishing Service, Canberra.

Main BY (1981) Eco-evolutionary radiation of mygalomorph spiders in Australia. In *Ecological Biogeography of Australia.* (Ed. A Keast) pp. 853–872. Dr W. Junk, Publishers, The Hague.

Main BY (1983) Systematics of the trapdoor spider genus *Homogona* Rainbow (Mygalomorphae: Ctenizidae: Homogoninae). *Journal of the Australian Entomological Society* **22**, 81–92. doi:10.1111/j.1440-6055.1983.tb01847.x

Majer JD, Kitching RL, Heterick BE, Hurley K, Brennan KEC (2001) North–south patterns within arboreal ant assemblages from rain forests in Eastern Australia. *Biotropica* **33**, 643–661.

Majer JD, Recher HF, Postle AC (1994) Comparison of arthropod species richness in eastern and western Australian canopies: a contribution to the species number debate. *Memoirs of the Queensland Museum* **36**, 121–131.

Majer JD, Shattuck SO, Andersen AN, Beattie AJ (2004) Australian ant research: fabulous fauna, functional groups, pharmaceuticals, and the Fatherhood. *Australian Journal of Entomology* **43**, 235–247. doi:10.1111/j.1326-6756.2004.00435.x

Makinson RO (2018a) Myrtle rust in Australia – a draft action plan. In *National Science Exchange.* 31 May, Melbourne. Plant Biosecurity Cooperative Research Centre, Melbourne.

Makinson RO (2018b) *Myrtle Rust Reviewed: The Impact of the Invasive Plant Pathogen* Austropuccinia psidii *on the Australian environment.* Plant Biosecurity Cooperative Research Centre, Canberra.

Malloch JR (1924) Notes on Australian Diptera. IV. *Proceedings of the Linnean Society of New South Wales* **49**, 348–359.

Malloch JR (1925) Notes on Australian Diptera. V. *Proceedings of the Linnean Society of New South Wales* **50**, 35–97.

Malloch JR (1926) Notes on Australian Diptera. No. viii. *Proceedings of the Linnean Society of New South Wales* **51**, 31–49.

Malloch JR (1927) Notes on Australian Diptera. No. xiii. *Proceedings of the Linnean Society of New South Wales* **52**, 399–446.

Mann JS (1928) Revisional notes on the Australian Therevidae. Part 1. *Australian Zoologist* **5**, 151–194.

Mann JS (1929) Revisional notes on the Australian Therevidae. Part 2. *Australian Zoologist* **6**, 17–49.

Marinho CR, Teixeira SP (2018) Novel reports of laticifers in Moraceae and Urticaceae: revisiting synapomorphies. *Plant Systematics and Evolution* **305**, 13–31. doi:10.1007/s00606-018-1548-6

Marshall AG (1981) *The Ecology of Ectoparasitic Insects.* Academic Press, London.

Marshall SA (1989) Family Sphaeroceridae. In *Catalog of the Diptera of the Australasian and Oceanic Regions.* (Ed. NL Evenhuis) pp. 601–607. Bishop Museum Press and E.J. Brill, Honolulu.

Marshall SA (2000) Agonistic behaviour and generic synonymy in Australian Clusiidae (Diptera). *Studia Dipterologica* **7**, 3–9.

Marshall SA (2012) *Flies – The Natural History and Diversity of Diptera*. Firefly Books, Richmond Hill.

Martin HA (2006) Cenozoic climatic change and the development of arid vegetation in Australia. *Journal of Arid Environments* **66**, 533–563. doi:10.1016/j.jaridenv.2006.01.009

Martínez-Aquino A, Vigliano-relva J, Brusa F, Damborenea C (2017) Historical biogeography of Temnocephalida (Platyhelminthes, Rhabdocoela): testing the Gondwanan hypothesis. *Systematics and Biodiversity* **15**, 327–345. doi:10.1080/14772000.2016.1252441

Marullo R, Mound LA (1993) The *Erythrothrips* complex of tropical Aeolothripidae (Thysanoptera) with new taxa from Australia and South Africa. *Entomologica Scandinavica* **24**, 285–291. doi:10.1163/187631293X00118

Masek M, Motyka M, Kusy D, Bocek M, Li M, Bocak L (2018) Molecular phylogeny, diversity and zoogeography of the net-winged beetles (Coleoptera: Lycidae). *Insects* **9**, 154. doi:10.3390/insects9040154

Mather WB (1955) The genus *Drosophila* (Diptera) in eastern Queensland. I. Taxonomy. *Australian Journal of Zoology* **29**, 907–919.

Mathis WN (1989) Family Ephydridae. In *Catalog of the Diptera of the Australasian and Oceanic Regions.* (Ed. NL Evenhuis) pp. 639–649. Bishop Museum Press and E.J. Brill, Honolulu.

Matile L (1981) A new Australian genus of Keroplatidae with pectinate antennae (Diptera: Mycetophiloidea). *Journal of the Australian Entomological Society* **20**, 207–212. doi:10.1111/j.1440-6055.1981.tb01034.x

Matile L (1989a) Family Keroplatidae. In *Catalog of the Diptera of the Australasian and Oceanic Regions.* (Ed. NL Evenhuis) pp. 128–133. Bishop Museum Press and E.J. Brill, Honolulu.

Matile L (1989b) Family Mycetophilidae. In *Catalog of the Diptera of the Australasian and Oceanic Regions.* (Ed. NL Evenhuis) pp. 135–145. Bishop Museum Press and E.J. Brill, Honolulu.

Matile L (1990) Recherches sur la systamatique et l'evolution des Keroplatidae (Diptera, Mycetophiloidea). *Memoirs du Museum d'Histoire Naturelle de Paris. Serie A (Zoologie)* **148**, 1–682.

Matthews EG (1972) A revision of the scarabaeine dung beetles of Australia. I. Tribe Onthophagini. *Australian Journal of Zoology Supplementary Series* **9**, 1–330.

Matthews EG (1974) A revision of the scarabaeine dung beetles of Australia. II. Tribe Scarabaeini. *Australian Journal of Zoology Supplementary Series* **24** 1–211

Matthews EG (1976a) A revision of the scarabaeine dung beetles of Australia. III. Tribe Coprini. *Australian Journal of Zoology Supplementary Series* **38**, 1–52

Matthews EG (1976b) *Insect Ecology.* University of Queensland Press, St Lucia.

Matthews EG (1984) *A Guide to the Genera of Beetles of South Australia. Part 3. Polyphaga: Eucinetoidea, Dascilloidea and Scarabaeoidea.* South Australian Museum, Adelaide.

Matthews EG (1992) Classification, relationships and distribution of the genera of Cyphaleini (Coleoptera: Tenebrionidae). *Invertebrate Taxonomy* **6**, 437–522. doi:10.1071/IT9920437

Matthews EG (1993) Classification, relationships and distribution of the genera of Heleini (Coleoptera: Tenebrionidae). *Invertebrate Taxonomy* **7**, 1025–1095. doi:10.1071/IT9931025

Matthews EG (1998) Classification, phylogeny and biogeography of the genera of Adelini (Coleoptera: Tenebrionidae). *Invertebrate Taxonomy* **12**, 685–824. doi:10.1071/IT97008

Matthews EG, Bouchard P (2008) *Tenebrionid Beetles of Australia.* Australian Biological Resources Study, Canberra.

Matthews EG, Doyen JT (1989) A reassessment of the Australian species of *Menephilus* Mulsant (Coleoptera: Tenebrionidae) with descriptions of two new genera and a larva and pupa. *Records of the South Australian Museum* **23**, 39–50.

Matthews EG, Reid CAM (2002) *A Guide to the Genera of Beetles of South Australia. Part 8.* South Australian Museum, Adelaide.

Matthews EG, Stebnicka Z (1986) A review of *Demarziella* Balthasar, with a transfer from Aphodiinae to Scarabaeinae (Coleoptera: Scarabaeidae). *Australian Journal of Zoology* **34**, 449–461. doi:10.1071/ZO9860449

Maunsell SC, Burwell CJ, Morris RJ, McDonald WJF, Edward ED, Oberprieler RG, Kitching RL (2017) Host-plants of leaf miners in Australian subtropical rainforest. *Austral Entomology* **56**, 403–411. doi:10.1111/aen.12252

Mavoungou JF, Kohange TL, Acapovi-Yao GL, Gilles J, Kietoh KG, Duvallet G (2013) Vertical distribution of *Stomoxys* spp. (Diptera: Muscidae) in a rainforest area in Gabon. *African Journal of Ecology* **51**, 147–153. doi:10.1111/aje.12018

Maynard GV (1992) Revision of *Leioproctus* (*Cladocerapis*) Cockerell (Hymenoptera: Colletidae). *Journal of the Australian Entomological Society* **31**, 1–11. doi:10.1111/j.1440-6055.1992.tb00451.x

Maynard GV (2013) Revision of *Goniocolletes* and seven Australian subgenera of *Leioproctus* (Hymenoptera: Apoidea: Colletidae), and description of new taxa. *Zootaxa* **3715**, 1–114. doi:10.11646/zootaxa.3715.1.1

Mayr E (1944) Wallace's Line in the light of recent zoogeographic studies. *The Quarterly Review of Biology* **19**, 1–14.

Mazzarolla LA, Amorin DS (2000) *Cratomyia macrorrhyncha*, a Lower Cretaceous brachyceran fossil from the Santana Formation, Brazil, representing a new species, genus and family of the Stratiomyomorpha (Diptera). *Insect Systematics & Evolution* **31**, 91–102. doi:10.1163/187631200X00336

McAlpine DK (1966) Description and biology of an Australian species of Cypselosomatidae (Diptera), with a discussion of family relationships. *Australian Journal of Zoology* **14**, 673–685. doi:10.1071/ZO9660673

McAlpine DK (1972a) A note on the Christmas beetle *Triplognathus griseopilosus* (Coleoptera, Rutelinae). *Australian Entomological Magazine* **1**, 14.

McAlpine DK (1972b) The Australian Platystomatidae (Diptera, Schizophora) with a revision of five genera. *Memoirs of the Australian Museum* **15**.

McAlpine DK (1978) Description and biology of a new genus of flies related to *Anthoclusia* and representing a new family (Diptera, Schizophora, Neurochaetidae). *Annals of the Natal Museum* **23**, 273–295.

McAlpine DK (1983) A new subfamily of Aulacigastridae (Diptera: Schizophora), with a discussion of aulacigastrid classification. *Australian Journal of Zoology* **31**, 55–78. doi:10.1071/ZO9820055

McAlpine DK (1985) The Australian genera of Heleomyzidae (Diptera: Schizophora) and a reclassification of the family into tribes. *Records of the Australian Museum* **36**, 203–251. doi:10.3853/j.0067-1975.36.1985.346

McAlpine DK (1988a) Studies in upside-down flies (Diptera: Neurochaetidae). Part I. Systematics and phylogeny. *Proceedings of the Linnean Society of New South Wales* **110**, 31–59.

McAlpine DK (1988b) Studies in upside-down flies (Diptera: Neurochaetidae). Part II. Biology, adaptations, and specific mating mechanisms. *Proceedings of the Linnean Society of New South Wales* **110**, 59–82.

McAlpine DK (1993a) A new genus of Australian Cypselosomatid flies (Diptera: Nerioidea). *Journal of Applied Entomology* **25**, 2–4.

McAlpine DK (1993b) Review of the upside-down flies (Diptera: Neurochaetidae) of Madagascar and Africa, and evolution of neurochaetid host plant associations. *Records of the Australian Museum* **45**, 221–239. doi:10.3853/j.0067-1975.45.1993.21

McAlpine DK (1998) A review of the Australian stilt flies (Diptera: Micropezidae) with a phylogenetic analysis of the family. *Invertebrate Taxonomy* **12**, 55–134. doi:10.1071/IT96018

McAlpine DK (1999) Australian signal flies of the genus *Rhytidorsalis* (Diptera: Platystomatidae). *Proceedings of the Linnean Society of New South Wales* **121**, 147–174.

McAlpine DK (2001) Review of the Australasian genera of signal flies (Diptera: Platystomatidae). *Records of the Australian Museum* **53**, 113–199. doi:10.3853/j.0067-1975.53.2001.1327

McAlpine DK (2007a) Review of the Borboroidini or wombat flies (Diptera: Heteromyzidae), with reconsideration of the status of the families Heleomyzidae and Sphaeroceridae, and descriptions of femoral gland-baskets. *Records of the Australian Museum* **59**, 143–219. doi:10.3853/j.0067-1975.59.2007.1487

McAlpine DK (2007b) Australian signal flies of the *Euprosopia megastigma* group (Diptera: Platystomatidae). *Tijdschrift voor Entomologie* **150**, 219–235. doi:10.1163/22119434-900000223

McAlpine DK (2012) Notes and descriptions of Australian Helosciomyzidae or comb-winged flies (Diptera: Schizophora). *Records of the Australian Museum* **64**, 51–70. doi:10.3853/j.0067-1975.64.2012.1582

McAlpine DK (2014) Review of the Australian genus *Pentachaeta* (Diptera: Heteromyzidae) with descriptions of new species. *Records of the Australian Museum* **66**, 247–264. doi:10.3853/j.2201-4349.66.2014.1631

McAlpine DK (2019) Review of the species of *Paranomina* (Diptera: Lauxaniidae). *Records of the Australian Museum* **71**, 71–85. doi:10.3853/j.2201-4349.71.2019.1655

McAlpine DK, Kent DS (1982) Systematics of *Tapeigaster* (Diptera: Heleomyzidae) with notes on biology and larval morphology. *Proceedings of the Linnean Society of New South Wales* **106**, 33–58.

McAlpine DK, Kim SP (1977) The genus *Lenophila* (Diptera: Platystomatidae). *Records of the Australian Museum* **30**, 309–336. doi:10.3853/j.0067-1975.30.1977.391

McCafferty WP, Wang TQ (1997) Phylogenetic systematics of the family Teloganodidae (Ephemeroptera: Pannota). *Annals of the Cape Province Museum (Natural History)* **19**, 387–437.

McCairns RF, Freitag R, Rose HA, McDonald FJD (1997) Taxonomic revision of the Australian Cicindelidae (Coleoptera), excluding species of *Cicindela*. *Invertebrate Taxonomy* **11**, 599–687. doi:10.1071/IT94011

McCarthy D, Ebach MC, Morrone JJ, Parenti LR (2007) An alternative Gondwana: Biota links South America, New Zealand and Australia. *Biogeographica* **2**, 2–12.

McCormack RB (2012) *A Guide to Australia's Spiny Freshwater Crayfish*. CSIRO Publishing, Melbourne.

McDermott JJ, Roe P (1985) Food, feeding behaviour and feeding ecology of Nemerteans. *American Zoologist* **25**, 113–125. doi:10.1093/icb/25.1.113

McDonald B (2010) The Tweed Caldera group. In *Remnants of Gondwana*. (Eds R Kitching, R. Braithwaite, J Cavanaugh) pp. 39–50. Surrey Beatty and Sons, Baulkham Hills.

McDonald B, Hunter J (2010) The vegetation. In *Remnants of Gondwana*. (Eds R Kitching, R. Braithwaite, J Cavanaugh) pp. 129–134. Surrey Beatty and Sons, Baulkham Hills.

McEvey S (1992) *North East Forest Fauna Survey: Invertebrate Study – Drosophilidae (Diptera)*. Australian Museum, Sydney.

McEvey S (1994) 'Results of the Drosophilidae (Diptera) invertebrate fauna surveys of north-east NSW forests'. North East Forests Biodiversity Report No. 3d. Unpublished report, NSW National Parks and Wildlife Service.

McEvey S (2009) Taxonomic review of the Australian *Drosophila setifemur* species-group, a new name for the *D. dispar* species-group. *Records of the Australian Museum* **61**, 31–38. doi:10.3853/j.0067-1975.61.2009.1517

McEvey S, Barker JSF (2001) *Scaptodrosophila aclinata*: a new *Hibiscus* flower-breeding species related to *S. hibisci* (Diptera: Drosophilidae). *Records of the Australian Museum* **53**, 255–262. doi:10.3853/j.0067-1975.53.2001.1332

McKay RJ (1979) The wolf spiders of Australia (Araneae: Lycosidae): 13. The genus *Trochosa*. *Memoirs of the Queensland Museum* **19**, 277–298.

McKeown KC (1945) Notes on Australian Cerambycidae VII. *Records of the Australian Museum* **21**, 286–292. doi:10.3853/j.0067-1975.21.1945.546

McKeown KC (1947) Catalogue of the Cerambycidae (Coleoptera) of Australia. *Australian Museum Memoir* **10**, 1–190.

McKie BGL, Cranston PS (1998) Key-stone coleopterans? Colonization by wood-feeding elmids of experimentally immersed wood in south-eastern Australia. *Marine and Freshwater Research* **49**, 79–88. doi:10.1071/MF97086

McKie BGL, Pearson RG, Cranston PS (2005) Does biogeographical history matter? Diversity and distribution of lotic midges (Diptera: Chironomidae) in Australian Wet Tropics. *Austral Ecology* **30**, 1–13. doi:10.1111/j.1442-9993.2005.01408.x

McLaren S, Wallace MW (2010) Plio-Pleistocene climate change and the onset of aridity in Southeastern Australia. *Global and Planetary Change* **71**, 55–72. doi:10.1016/j.gloplacha.2009.12.007

McLeish MJ, González-Orozco CE, Miller JT (2017) Spatial correspondence between specialist galling insect genus *Kladothrips* (Thysanoptera: Tubulifera: Phlaeothripidae) and its *Acacia* host: ecological and evolutionary constraints on host use. *Austral Entomology* **56**, 412–423. doi:10.1111/aen.12256

McLellan ID (1996) A revision of *Stenoperla* (Plecoptera: Eustheniidae) and removal of Australian species to *Cosmioperla* new genus. *New Zealand Journal of Zoology* **23**, 165–182. doi:10.1080/03014223.1996.9518076

McLoughlin S (2001) The breakup history of Gondwana and its impact on pre-Cenozoic floristic provincialism. *Australian Journal of Botany* **49**, 271–300. doi:10.1071/BT00023

McMichael BA, Foote BA, Bowker BD (1990) Biology of *Rivellia melliginis* (Diptera: Platystomatidae), a consumer of the nitrogen-fixing root nodules of black locust (Leguminosae). *Annals of the Entomological Society of America* **83**, 967–974. doi:10.1093/aesa/83.5.967

McMichael DF, Hiscock ID (1958) A monograph of the freshwater mussels (Mollusca: Pelecypoda) of the Australian region. *Australian Journal of Marine and Freshwater Research* **9**, 372–508. doi:10.1071/MF9580372

McQuillan PB (1993) *Nothofagus* (Fagaceae) and its invertebrate fauna – an overview and preliminary synthesis. *Biological Journal of the Linnean Society. Linnean Society of London* **49**, 317–354. doi:10.1111/j.1095-8312.1993.tb00910.x

McQuillan PB, Young CJ, Richardson AMM (2001) A revision of the Australian moth genus *Paralaea* Guest (Lepidoptera: Geometridae: Ennominae). *Invertebrate Taxonomy* **15**, 277–317. doi:10.1071/IT99025

Meinwald J, Meinwald WC, Chalmers AM, Eisner T (1968) Dihidroxymatricaria acid: acetylenic acid secreted by a soldier beetle. *Science* **160**, 890–892. doi:10.1126/science.160.3830.890

Melo GAR (1999) Phylogenetic relationships and classification of the major lineages of Apoidea (Hymenoptera), with emphasis on the crabronid wasps. *Scientific Papers / Natural History Museum, University of Kansas* **14**, 1–55.

Melo GAR, Gonçales RB (2005) Higher level bee classification. *Revista Brasileira de Entomologia* **22**, 153–159.

Mengual X, Thompson FC (2015) Australian *Allograpta* Osten sacken (Diptera, Syrphidae). *ZooKeys* **513**, 65–78. doi:10.3897/zookeys.513.9671

Menzel F, Kitching RL, Boulter SL (2004) Host specificity or habitat structure? The epicortical beetle assemblages in an Australian subtropical rainforest. *European Journal of Entomology* **101**, 251–259. doi:10.14411/eje.2004.035

Merrick JR (1991) *The Biology, Conservation and Management of Australian Freshwater Crayfishes: A Bibliography.* Graduate School of the Environment, Macquarie University, Sydney.

Merrick JR (1995) Diversity, distribution and conservation of freshwater crayfishes in the eastern highlands of New South Wales. *Proceedings of the Linnean Society of New South Wales* **115**, 247–258.

Merrick JR (1997) Conservation and field mangaement of the freshwater crayfish, *Euastacus spinifer* (Decapoda: Parastacidae), in the Sydney region, Australia. *Proceedings of the Linnean Society of New South Wales* **118**, 217–225.

Merritt DJ, Patterson R (2018) Environmental influences on the bioluminescence display of the glow-worm, *Arachnocampa flava* (Diptera: Keroplatidae). *Austral Ecology* **57**, 107–117.

Mesibov R (2018) Three new species of Cladethosoma Chamberlin, 1920 from southeast Queensland (Diplopoda: Polydesmida: Paradoxosomatidae). *Memoirs of the Queensland Museum – Nature* **61**, 53–67.

Michelsen V, Pape T (2017) Ulurumyiidae – a new family of calyptrate flies (Diptera). *Systematic Entomology* **42**, 826–836. doi:10.1111/syen.12252

Michener CD (2007) *The Bees of the World.* 2nd edn. Johns Hopkins University Press, Baltimore.

Michener CD, Greenberg L (1980) Ctenoplectridae and the origin of long-tongued bees. *Zoological Journal of the Linnean Society* **69**, 183–203.

Michener CD, Grimaldi DA (1988) The oldest fossil bee: apoid history, evolutionary stasis, and antiquity of social behaviour. *Proceedings of the National Academy of Sciences of the United States of America* **85**, 6424–6426. doi:10.1073/pnas.85.17.6424

Milledge GA (1997a) Revision of the tribe Archimantini (Mantodea: Mantidae: Mantinae). *Memoirs of the Museum of Victoria* **56**, 1–63. doi:10.24199/j.mmv.1997.56.01

Milledge GA (1997b) A new species of *Progradungla* Forster and Gray (Araneae: Gradungulidae) from Victoria. *Memoirs of the Museum of Victoria* **56**, 65–68. doi:10.24199/j.mmv.1997.56.02

Milledge GA (2005) Revision of the genera *Sphodropoda, Trachymantis* and *Zopheromantis* (Mantodea: Mantidae: Mantinae). *Records of the Australian Museum* **57**, 191–210. doi:10.3853/j.0067-1975.57.2005.1442

Milledge GA (2011) Revision of *Storenosoma* Hogg and description of a new genus *Oztira* (Araneae: Amaurobiidae). *Records of the Australian Museum* **63**, 1–32. doi:10.3853/j.0067-1975.63.2011.1579

Miller AC, Ponder WF, Clark SA (1999) Freshwater snails of the genera *Fluvidona* and *Austropyrgus* (Gastropoda, Hydrobiidae) from northern New South Wales and southern Queensland, Australia. *Invertebrate Taxonomy* **13**, 461–493. doi:10.1071/IT97011

Miller CG, Edwards ED (1978) A new species of *Pseudodipsas* C. and R. Felder (Lepidoptera: Lycaenidae) from northern New South Wales. *Australian Entomological Magazine* **5**, 45–50.

Miller CG, Morhaus IG (1975) Butterfly records of interest from the Northern Rivers district of New South Wales. *Australian Entomological Magazine* **2**, 85–86.

Mills K (1987) The distribution, character and conservation status of the rainforests of the Illawarra district, New South Wales. In *The Rainforest Legacy. Australian National Rainforests Study. Vol. 1.* (Eds G Werren and AP Kershaw) pp. 71–94. Australian Government Publishing Service, Canberra.

Mitchell ML, Adam P (1989) The relationship between mangrove and saltmarsh communities in the Sydney region. *Wetlands Australia* **8**, 37–46. doi:10.31646/wa.116

Miyake T, Yamaoka R, Yahara T (1998) Floral scents of hawk moth-pollinated flowers in Japan. *Journal of Plant Research* **111**, 199–205. doi:10.1007/BF02512170

Mizunuma T, Nagai S (1994) *The Lucanid Beetles of the World.* Mushi-sha, Tokyo.

Monteith GB (1967) A revision and redescription of the genus *Glyptoaptera* Kormilev (Hemiptera: Aradidae). *Proceedings of the Royal Entomological Society of London (B)* **36**, 50–60.

Monteith GB (1969) The relationship of *Kumaressa* Monteith and *Tretocoris* Usinger and Matsuda with a new species of *Kumaressa* (Hemiptera: Aradidae: Chinamyersiinae). *Proceedings of the Royal Society of Queensland* **81**, 75–82.

Monteith GB (1970) Life history of the chrysomelid, *Aproida balyi* Pascoe. *Entomological Society of Queensland News Bulletin* **72**, 9–10.

Monteith GB (1982) Dry season aggregations of insects in Australian monsoon forests. *Memoirs of the Queensland Museum* **20**, 533–543.

Monteith GB (1986a) Some curious insect–plant associations in Queensland. *Queensland Naturalist* **26**, 105–114.

Monteith GB (1986b) Insects from Kroombit Tops, Queensland, with some results of a site survey of Coleoptera. *Queensland Naturalist* **27**, 27–34.

Monteith GB (1993) 'The rainforest insect fauna of the Queensland component of the Border Ranges complex – a brief review'. Unpublished report.

Monteith GB (1997) Revision of the Australian flatbugs of the subfamily Mezerinae (insecta: Hemiptera: Aradidae). *Memoirs of the Queensland Museum* **41**, 1–69.

Monteith GB (2006) Maternal care in Australian oncomerine shield bugs (Insecta, Heteroptera: Tessaratomidae). *Biologiezentrum Linz Neue Serie* **50**, 1135–1152.

Monteith GB (2011) Maternal care, food plants and distribution of Australian Oncomerinae (Hemiptera: Heteroptera: Tessaratomidae). *Australian Entomologist* **38**, 37–48.

Monteith GB, Field LH (2001) Australian king crickets: distribution, habitats and biology (Orthoptera: Anostostomatidae). In *The Biology of Wetas, King Crickets and Their Allies.* (Ed. LH Field) pp. 79–94. CABI Publishing, Wallingford.

Monteith GB, Storey RI (1981) The biology of *Cephalodesmius*, a genus of dung beetles which synthesises 'dung' from plant material (Coleoptera: Scarabaeidae: Scarabaeinae). *Memoirs of the Queensland Museum* **20**, 253–277.

Monteith GB, Turco F (2010) *Mystropomus* carabids from the Mt Etna caves at Rockhampton. *News Bulletin of the Entomological Society of Queensland* **38**, 66–69.

Moore AS, Ginn S (2009) New records of butterflies (Lepidoptera) from the Port Macquarie area, New South Wales. *Australian Entomologist* **36**, 21–22.

Moore BP (1963) Studies on Australian Carabidae (Coleoptera) 3. – the Psydrinae. *Transactions of the Royal Entomological Society of London* **115**, 277–290. doi:10.1111/j.1365-2311.1963.tb00810.x

Moore BP (1964) New cavernicolous Carabidae (Coleoptera) from mainland Australia. *Journal of the Entomological Society of Queensland* **3**, 64–74.

Moore BP (1965) Studies on Australian Carabidae (Coleoptera) 4. – The Pterostichinae. *Transactions of the Royal Entomological Society of London* **117**, 1–32. doi:10.1111/j.1365-2311.1965.tb00042.x

Moore BP (1966) The genus *Perileptus* in Australia (Coleoptera, Trechinae). *Bulletin of the National Science Museum, Tokyo* **9**, 317–320.

Moore BP (1972) A revision of the Australian Trechinae (Coleoptera: Carabidae). *Australian Journal of Zoology Supplementary Series Number* **18**, 1–61.

Moore BP (1973) Larvae of *Zenthicola crassus* (Newman) (Coleoptera: Cleridae) feeding on termites. *Australian Entomological Magazine* **1**, 48–50.

Moore BP (1979) Chemical defense in carabids and its bearing on phylogeny. In *Carabid Beetles: their evolution, natural history, and classification.* (Eds TL Erwin, GE Ball, DR Whitehead and AL Halpern) pp. 193–203. Dr W. Junk, Publishers, The Hague.

Moore BP (1980–96) *A Guide to the Beetles of South-eastern Australia. Fascicles 1–12.* Australian Entomological Press, Greenwich.

Moore BP (1994) A new species of *Eucarteria* Lea (Coleoptera: Lucanidae) and notes on the affinities of the genus. *Australian Entomologist* **21**, 1–6.

Moore BP, Brown WV (1971) Chemical defense in longhorn beetles of the genera *Stenocentrus* and *Syllitus* (Coleoptera: Cerambycidae). *Journal of the Australian Entomological Society* **10**, 142–143. doi:10.1111/j.1440-6055.1971.tb00023.x

Moore BP, Brown WV (1981) Identification of warning odour components, bitter principles and antifeedants in an aposematic beetle: *Metriorrhynchus rhipidius* (Coleoptera: Lycidae). *Insect Biochemistry* **11**, 493–499. doi:10.1016/0020-1790(81)90016-0

Moore BP, Brown WV (1989) Graded levels of chemical defence in mimics of lycid beetles of the genus *Metriorrhynchus* (Coleoptera). *Journal of the Australian Entomological Society* **28**, 229–233. doi:10.1111/j.1440-6055.1989.tb00889.x

Moore BP, Brown WV, Rothschild M (1990) Methylalkylpyrazines in aposematic insects, their host plants and mimics. *Chemoecology* **1**, 43–51. doi:10.1007/BF01325227

Moore BP, Monteith GB (2004) A second species of the Gondwanan stag beetle genus *Sphaenognathus* Buquet (Coleoptera: Lucanidae). *Memoirs of the Queensland Museum* **49**, 693–699.

Moore BP, Wallbank BE (1968) Chemical composition of the defensive secretion in carabid beetles and its importance as a taxonomic character. *Proceedings of the Royal Entomological Society of London* **37**, 62–72.

Moore J (1975) Land nemertines of Australia. *Zoological Journal of the Linnean Society* **56**, 23–43. doi:10.1111/j.1096-3642.1975.tb00809.x

Moore J (1985) The distribution and evolution of terrestrial nemertines. *American Zoologist* **25**, 15–21. doi:10.1093/icb/25.1.15

Moore J, Gibson R (1981) The *Geonemertes* problem (Nemertea). *Journal of the Zoological Society, London* **194**, 175–201. doi:10.1111/j.1469-7998.1981.tb05768.x

Moore J, Gibson R (1988) Further studies on the evolution of land and freshwater nemerteans: generic relationships among paramonostiliferous taxa. *Journal of the Zoological*

Society, London **216**, 1–20. doi:10.1111/j.1469-7998.1988.tb02410.x

Moore KM (1970) Observations on some Australian forest insects. 23. A revision of the genus *Glycaspis* (Homoptera: Psyllidae) with descriptions of seventy-three new species. *Australian Zoologist* **15**, 248–342.

Moradmand M, Schönhofer AL, Jäger P (2014) Molecular phylogeny of the spider family Sparassidae with focus on the genus *Eusparassus* and notes on the RTA-clade + 'Laterigradae'. *Molecular Phylogenetics and Evolution* **74**, 48–65. doi:10.1016/j.ympev.2014.01.021

Moreau CS, Hugall AF, McDonald KR, Jamieson BGM, Moritz C (2015) An ancient divide in a contiguous rainforest: endemic earthworms in the Australian Wet Tropics. *PLoS ONE* doi:10.1371/journal.pone.0136943.

Morgan GJ (1997) Freshwater crayfish of the genus *Euastacus* Clark (Decapoda: Parastacidae) from New South Wales, with a key to all species of the genus. *Records of the Australian Museum* (Supplement), **23**, 1–110.

Moritz C, Joseph L, Cunningham M, Schneider C (1997) Molecular perspectives on historical fragmentation of Australian tropical and subtropical rainforests: implications for conservation. In *Tropical Forest Remnants: Ecology, Management, and Conservation of Fragmented Communities*. (Eds WF Laurance and RO Bierregaard Jr.) pp. 442–454. The University of Chicago Press, Chicago.

Morrow PA, Bellas TE, Eisner T (1976) Eucalyptus oils in the defensive oral discharge of some Australian sawfly larvae (Hymenoptera: Pergidae). *Oecologia* **24**, 193–206. doi:10.1007/BF00345473

Moss JT St L, Moulds MS (2000) A new species of *Psaltoda* Stål, with notes on comparative morphology and song structure (Hemiptera: Cicadidae). *Australian Entomologist* **27**, 47–60.

Moss JT St L, Popple LW (2000) Cicada, butterfly and moth records from the Gibraltar Range, New South Wales. (Hemiptera: Cicadidae: Lepidoptera). *Queensland Naturalist* **38**, 53–60.

Moulds MS (1981) Larval food plants of hawk moths (Lepidoptera: Sphingidae) affecting commercial crops in Australia. *General and Applied Entomology* **13**, 69–80.

Moulds MS (1984) Larval food plants affecting garden ornamentals in Australia. *General and Applied Entomology* **16**, 57–64.

Moulds MS (1985) A review of the Australian hawks of the genus *Macroglossum* Scopoli (Lepidoptera: Sphingidae). *Australian Entomological Magazinee* **12**, 85–105.

Moulds MS (1990) *Australian Cicadas*. New South Wales University Press, Kensington.

Moulds MS (1998) New larval food plants for Australian hawk moths (Lepidoptera: Sphingidae). *Australian Entomologist* **25**, 13–22.

Moulds MS (2005) An appraisal of the higher classification of cicadas (Hemiptera: Cicadoidea) with special reference to the Australian fauna. *Records of the Australian Museum* **57**, 375–446.

Moulds MS (2012) A review of the genera of Australian cicadas (Hemiptera: Cicadoidea). *Zootaxa* **3287**, 1–262. doi:10.11646/zootaxa.3287.1.1

Moulds T, Bannink P (2012) Preliminary notes on the cavernicolous arthropod fauna of Judbarra/Gregory Karst Area, northern Australia. *Helictite* **41**, 75–85.

Moulds T, Cassis G (2006) Review of Australian species of *Peritropis* (Insecta: Heteroptera: Miridae). *Memoirs of the Queensland Museum* **52**, 171–190.

Moulton JK, Wiegmann BM (2007) The phylogenetic relationships of flies in the superfamily Empoidea (Insecta: Diptera). *Molecular Phylogenetics and Evolution* **43**, 701–713.

Mound LA (1972) Further studies on Australian Aeolothripidae (Thysanoptera). *Journal of the Australian Entomological Society* **11**, 37–54. doi:10.1111/j.1440-6055.1972.tb01603.x

Mound LA (1999) Saltatorial leaf-feeding Thysanoptera (Thripidae: Dendrothripinae) from Australia and New Caledonia, with newly recorded pests of ferns, figs and mulberries. *Australian Journal of Entomology* **38**, 257–273. doi:10.1046/j.1440-6055.1999.00112.x

Mound LA (2002) Thrips and their host plants: new Australian records (Thysanoptera: Terebrantia). *Australian Entomologist* **29**, 49–60.

Mound LA (2004) Australian long-tailed gall thrips (Thysanoptera: Phlaeothripinae, Leeuweiini), with comments on related Old World taxa. *Austral Entomology* **43**, 28–37. doi:10.1111/j.1440-6055.2004.00406.x

Mound LA, Heming BS (1991) Thysanoptera. In *The Insects of Australia*. (Ed. ID Naumann) pp. 510–515. Melbourne University Press, Carlton.

Mound LA, Marullo R (1998) Biology and identification of Aeolothripidae (Thysanoptera) from Australia. *Invertebrate Taxonomy* **12**, 929–950. doi:10.1071/IT97014

Mound LA, Williams GA (2003) Host-plant disjunction in a new species of *Neohoodiella* (Insecta, Thysanoptera, Phaeothripinae), with notes on leaf-frequenting thrips in New South Wales subtropical rainforests. *Proceedings of the Linnean Society of New South Wales* **124**, 17–28.

Mroczkowski M (1968) Distribution of the Dermestidae (Coleoptera) of the World with a catalogue of all known species. *Annales zoologici (Warszawa)* **26**, 15–191.

Munasinghe DHN, Burridge CP, Austin CM (2004) The systematic of the freshwater crayfish of the genus *Cherax* Erichson (Decapoda: Parastacidae) in eastern Australia re-examined using nucleotide sequences from 12S rRNA and 16S rRNA genes. *Invertebrate Systematics* **18**, 215–225. doi:10.1071/IS03012

Muona J (1993) Review of the phylogeny, classification and biology of the family Eucnemidae (Coleoptera). *Entomologica Scandinavica Supplement* **44**, 1–133.

Murcia C (1995) Edge effects in fragmented forests: implications for conservation. *Trends in Ecology & Evolution* **10**, 58–62. doi:10.1016/S0169-5347(00)88977-6

Murphy MJ (2002) Observations on the behaviour of the Australian land snail *Hedleyella falconeri* (Gray, 1834) (Pulmata: Caryodidae) using spool-and-line technique. *Molluscan Research* **22**, 149–164. doi:10.1071/MR02007

Murray A (1870) On the geographical relations of the chief Coleopterous faunae. *Zoological Journal of the Linnean Society* **11**, 1–89.

Myers AA, Giller PS (Eds) (1988) *Analytical Biogeography. An Integrated Approach to the Study of Animal and Plant Distributions.* Chapman and Hall, London.

Mynott JH, Suter PJ, Theischinger G (2017) Revision of the genus *Dinotoperla* Tillyard, 1921 (Plecoptera: Gripopterygidae) using morphological characters and molecular data: Establishes two new genera, three new species and updates the larval taxonomy. *Zootaxa* **4224**, 001–076.

Nadolny C (1984) 'Nature and conservation of the invertebrate fauna in New South Wales rainforests – a preliminary report'. (Unpublished report for the NSW National Parks and Wildlife Service.)

Nagamitsu T, Inoue T (1997) Cockroach pollination and breeding system of *Uvaria elmeri* (Annonaceae) in a lowland mixed-dipterocarp forest in Sarawak. *American Journal of Botany* **84**, 208–213. doi:10.2307/2446082

Nagatomi A (1977) Classification of the lower Brachycera (Diptera). *Journal of Natural History* **11**, 321–335. doi:10.1080/00222937700770231

Nagatomi A, Evenhuis NL (1989a) Family Athericeridae. In *Catalog of the Diptera of the Australasian and Oceanic Regions.* (Ed. NL Evenhuis) p. 295. Bishop Museum Press and E.J. Brill, Honolulu.

Nagatomi A, Evenhuis NL (1989b) Family Rhagionidae. In *Catalog of the Diptera of the Australasian and Oceanic Regions.* (Ed. NL Evenhuis) pp. 296–298. Bishop Museum Press and E.J. Brill, Honolulu.

Nageli P, Grant J, Nichols JD, Shell D, Horton B (2016) Bell miner associated dieback: nutrient cycling and herbivore crown damage in *Eucalyptus propinqua*. *Australian Forestry* **79**(1), 74–82. doi:10.1080/00049158.2015.1123565

Naumann ID (1982) Systematics of the Australian Ambositrinae (Hymenoptera: Diapriidae), with a synopsis on non-Australian genera of the subfamily. *Australian Journal of Zoology Supplementary Series* **85**, 1–239.

Naumann ID (1985) The Australian species of Monomachidae (Hymenoptera: Proctotrupoidea), with a revised diagnosis of the family. *Journal of the Australian Entomological Society* **24**, 261–274. doi:10.1111/j.1440-6055.1985.tb00241.x

Naumann ID (Ed.) (1991a) *The Insects of Australia*. Melbourne University Press, Carlton.

Naumann ID (1991b) Hymenoptera 'wasps, bees, ants, sawflies'. In *The Insects of Australia*. (Ed. ID Naumann) pp. 916–1000. Melbourne University Press, Carlton.

Naumann ID (1991c) Revision of the Australian genus *Enoggera* Girault (Hymenoptera: Pteromalidae: Asaphinae). *Journal of the Australian Entomological Society* **30**, 1–17. doi:10.1111/j.1440-6055.1991.tb02184.x

Naumann ID, Williams MA, Schmidt S (2002) Synopsis of the Tenthredinidae (Hymenoptera) in Australia, including two newly recorded introduced sawfly species associated with willows (*Salix* spp.). *Australian Journal of Entomology* **41**, 1–6. doi:10.1046/j.1440-6055.2002.00260.x

Ndueze OU, Okiwelu SN, Oidika CU, Mekeu AEN (2013) Arthropod succession on wildlife carcasses in lowland rainforest, Rivers State, Nigeria. *European Journal of Experimental Biology* **3**, 106–110.

Neall VE, Trewick SA (2008) The age and origin of the Pacific Islands: a geological review. *Philosophical Transactions of the Royal Society of London. Series B, Biological Sciences* **363**, 3293–3308. doi:10.1098/rstb.2008.0119

Neboiss A (1957) The genera *Hapatesus* Candeze and *Toorongus* gen. nov. (Coleoptera: Elateridae). *Australian Journal of Zoology* **5**, 496–520. doi:10.1071/ZO9570496

Neboiss A (1962) The Australian Hydrobiosinae (Trichoptera: Rhyacophilidae). *Pacific Insects* **4**, 521–582.

Neboiss A (1974) Two new species of the genus *Stenopsychodes* Ulmer (Stenopsychidae: Trichoptera). *Australian Entomological Magazine* **1**, 81–86.

Neboiss A (1980) Australian species of the genus *Anisocentropus* McLachlan (Calamoceratidae: Trichoptera). *Australian Journal of Marine and Freshwater Research* **31**, 193–213. doi:10.1071/MF9800193

Neboiss A (1981) Distribution of Trichopteran families in Australia with comments on the composition of fauna in the south-west. In *Proceedings of the 3rd International Symposium on Trichoptera. Series Entomologica Vol. 20.* (Ed. GP Moretti) pp. 1–472. Dr W. Junk, Publishers, The Hague.

Neboiss A (1983) 'Checklist and bibliography of the Australian Caddis-flies (Trichoptera)'. Australian Society for Limnology, Special Publication 5. Chisholm Institute of Technology Printing Service, Caulfield.

Neboiss A (1986) Taxonomic changes in caddis-fly species from the south-west Pacific-Australian region with descriptions of new species. *Memoirs of the Museum of Victoria* **47**, 213–223. doi:10.24199/j.mmv.1986.47.08

Neboiss A (1991) Trichoptera 'caddis flies'. In *The Insects of Australia.* (Ed. ID Naumann) pp. 787–816. Melbourne University Press, Carlton.

Neboiss A, Wells A (1998) Review of Australian species of *Triaenodes* McLachlan (Trichoptera: Leptoceridae). *Memoirs of the Museum of Victoria* **57**, 89–132. doi:10.24199/j.mmv.1998.57.06

Neff Jl, Simpson BB, Evenhuis NL, Dieringer G (2003) Character analysis of adaptations for tarsal pollen collection in the Bombyliidae (Insecta: Diptera): the benefits of putting your foot in your mouth. *Zootaxa* **157**, 1–14. doi:10.11646/zootaxa.157.1.1

Nelson L, Bridger M, Turner S (2018) Endoparasitic flea larvae of *Uropsylla tasmanica* (Siphonaptera: Pygiopsyllidae) infest spotted-tailed quolls in southern Queensland. *Memoirs of the Queensland Museum – Nature* **61**, 68–69.

New TR (1978) Notes on the biology of *Lemidia subaenea* (Coleoptera: Cleridae) of *Acacia* in Victoria. *Australian Entomological Magazine* **5**, 21–22.

New TR (1983) A revision of the Australian Osmylidae: Kempyninae (Insecta: Neuroptera). *Australian Journal of Zoology* **31**, 393–420. doi:10.1071/ZO9830393

New TR (1986) A new species of *Kempynus* Nava (Neuroptera: Osmylidae) from Australia. *Journal of the Australian Entomological Society* **25**, 327–328. doi:10.1111/j.1440-6055.1986.tb01124.x

New TR (1988a) A revision of the Australian Hemerobiidae. *Invertebrate Taxonomy* **2**, 339–411. doi:10.1071/IT9880339

New TR (1988b) The Psychopsidae (Insecta: Neuroptera) of Australia and the Oriental region. *Invertebrate Taxonomy* **2**, 841–883. doi:10.1071/IT9880841

New TR (1989) The genus *Oedosmylus* Kruger (Insecta: Neuroptera: Osmylidae). *Invertebrate Taxonomy* **3**, 135–148. doi:10.1071/IT9890135

New TR (1991) Neuroptera 'lacewings'. In *The Insects of Australia*. (Ed. ID Naumann) pp. 525–542. Melbourne University Press, Carlton.

New TR (2011) *'In Considerable Variety'. Introducing the Diversity of Australia's Insects*. Springer, Dordrecht.

New TR, Smithers CN (1994) Two new species of *Myrmeleon* L. and new records of Myrmeleontini (Insecta, Neuroptera: Myrmeleontidae) from Australia. *Proceedings of the Linnean Society of New South Wales* **114**, 189–194.

Newton AF (1989) Review of *Dactylosternum* Wollaston species of Australia and New Zealand (Coleoptera: Hydrophilidae). *Australian Entomological Magazine* **16**, 49–58.

Nicolai V (1986) The bark of trees: thermal properties, microclimate and fauna. *Oecologica* **69**, 148–160. doi:10.1007/BF00399052

Nielsen ES (1987) The recently discovered primitive moth (non-Ditrysian) family Palaephatidae (Lepidoptera) in Australia. *Invertebrate Taxonomy* **1**, 201–229. doi:10.1071/IT9870201

Nielsen ES, Common IFB (1991) Lepidoptera. In *The Insects of Australia*. (Ed. ID Naumann) pp. 817–915. Melbourne University Press, Carlton.

Nielsen ES, Kristensen NP (1989) *Primitive Ghost Moths: Morphology and Taxonomy of the Australian Genus* Fraus *Walker (Lepidoptera: Hepialidae* s. lat.*). Monographs on Australian Lepidoptera. Volume 1.* CSIRO Publishing, Melbourne.

Nix HA (1982) Environmental determinants of biogeography and evolution in *Terra Australis*. In *Evolution of the Flora and Fauna of Australia*. (Eds WR Barker and PJM Greenslade) pp. 47–66. Peacock Publications, Frewville.

Nobel PS (1981) Wind as an ecological factor. In *Physiological Plant Ecology I. Responses to the Physical Environment.* (Eds OL Lange, PS Nobel, CB Osmond and H Ziegler) pp. 475–500. Springer-Verlag, Berlin.

Noonan GR (1979) The science of biogeography with relation to carabids. In *Carabid Beetles: their evolution, natural history, and classification*. (Eds TL Erwin, GE Ball, DR Whitehead, and AL Halpern) pp. 295–317. Dr W. Junk, Publishers, The Hague.

Nooten SS, Hughes L (2016) Roles of family and architecture in driving insect community structure: a comparison of nine Australian plant species. *Austral Entomology* **55**, 423–432. doi:10.1111/aen.12205

Norrbom AL (2010) *Tephritidae (Fruit Flies, Moscas de Frutas)*. In *Manual of Central American Diptera. Vol. 2.* (Eds BV Brown, A Borkent, JM Cumming, DM Wood, NE Woodley and MA Zumbado) pp. 909–954. NRC Research Press, Ottawa.

Norton RA, Bonamo PM, Grierson JD, Shear WA (1988) Oribatid mite fossils from a terrestrial Devonian deposit near Gilboa, New York State. *Journal of Paleontology* **62**, 259–269. doi:10.1017/S0022336000029905

Norton TW (1994) An overview of modelling fine-scale variation in environmental regimes in complex landscapes with comments on applications to invertebrate survey, monitoring and conservation. *Memoirs of the Queensland Museum* **36**, 173–177.

Novotny V, Clarke AR, Drew RA, Balagawi S, Clifford B (2005) Host specialization and species richness of fruit flies (Diptera: Tephritidae) in a New Guinea rain forest. *Journal of Tropical Ecology* **21**, 67–77. doi:10.1017/S0266467404002044

Nowrouzi S, Andersen AN, Macfadyen S, Staunton KM, VanDerWal J, Robson SKA (2016) Ant diversity and distribution along elevation gradients in the Australian Wet Tropics: the importance of seasonal moisture stability. *PLoS ONE* **11**, e0153420. doi:10.1371/journal.pone.0153420

NSW National Parks and Wildlife Service (2000) 'Mitchell's Rainforest Snail *Thersites mitchellae* (Cox, 1864) recovery plan'. NSW National Parks and Wildlife Service, Hurstville.

Nyffeler M, Pusey BJ (2014) Fish predation by semi-aquatic spiders: a global pattern. *PLoS ONE* **9**, e99459. doi:101371/journal.pone.0099459

Nylander U (2008) A review of the genera *Calodema* and *Metaxymorpha* (Coleoptera: Buprestidae: Stigmoderini). *Folia Heyrovskyana* **13** (Supplementum), 1–84.

O'Brien CW, Askevold SA (1992) Systematics and evolution of weevils of the genus *Bagous* Germar (Coleoptera: Curculionidae). 1. Species of Australia. *Transactions of the American Entomological Society* **118**, 331–452.

O'Neill KM (2001) *Solitary Wasps: Behavior and Natural History*. Cornell University Press, Ithaca.

Obenberger J (1930) *Coleopterorum Catalogus. Buprestidae II.* Dr W. Junk, Publishers, The Hague.

Obenberger J (1959) Sur les especes du genre Agrilus Curtis de l'Australie et Oceanie (Coleoptera, Buprestidae). *Acta entomologica musei nationale Pragae* **33**, 223–240.

Oberprieler RG (1995) The weevils (Coleoptera: Curculionoidea) associated with cycads 1. classification, relationships, and biology. In *Proceedings of the Third International Conference on Cycad Biology*. 5–9 July, South Africa. (Ed. P Voreter) pp. 295–334. Society of South Africa, Stellenbosch.

Oberprieler SK, Rasnitsyn AP, Brothers DJ (2012) The first wasps from the Upper Jurassic of Australia. *Zootaxa* **3503**, 47–54. doi:10.11646/zootaxa.3503.1.3

Ødegaard F, Diserud OH (2011) Taxonomic composition of Coleoptera, Hemiptera (Heteroptera and Coleorrhyncha) and Mutillidae (Hymenoptera) at five different altitudes in

Lamington National Park (Queensland, Australia). *Memoirs of the Queensland Museum – Nature* **55**, 359–374.

Oertli S, Müller A, Dorn S (2005) Ecological and seasonal patterns in a the diversity of a species-rich bee assemblage (Hymenoptera: Apoidea: Apiformes). *European Journal of Entomology* **102**, 53–63. doi:10.14411/eje.2005.008

Ogata K (1991) Ants of the genus *Myrmecia* Fabricius: a review of the species groups and their phylogenetic relationships (Hymenoptera: Formicidae: Myrmeciinae). *Systematic Entomology* **16**, 353–381. doi:10.1111/j.1365-3113.1991.tb00694.x

Ogata K, Taylor RW (1991) Ants of the genus *Myrmecia* Fabricius: a preliminary review and key to the named species (Hymenoptera: Formicidae: Myrmeciinae). *Journal of Natural History* **25**, 1623–1673. doi:10.1080/00222939100771021

Ollerton J (1996) Interactions between gall midges (Diptera: Cecidomyiidae) and inflorescences of *Piper novae-hollandiae* (Piperaceae) in Australia. *Entomologist* **115**, 181–184.

Ollerton J, Cranmer L (2002) Latitudinal trends in plant–pollinator interactions: are tropical plants more specialised? *Oikos* **98**, 340–350. doi:10.1034/j.1600-0706.2002.980215.x

Ollerton J, Liede S (1997) Pollination systems in the Asclepiadaceae: a survey and preliminary analysis. *Biological Journal of the Linnean Society. Linnean Society of London* **62**, 593–610. doi:10.1111/j.1095-8312.1997.tb00324.x

Oosterbroek P (1989) Family Tanyderidae. In *Catalog of the Diptera of the Australasian and Oceanic Regions.* (Ed. NL Evenhuis) pp. 51–52. Bishop Museum Press and E.J. Brill, Honolulu.

Oosterbroek P (2019) *Catalogue of the craneflies of the world*, <https://cw.naturalis.nl>.

Opell BD (1984) Phylogenetic review of the genus *Miagrammopes (sensu lato)* (Araneae: Uloboridae). *The Journal of Arachnology* **12**, 229–240.

Ornduff R (1989) Size distribution and coning behaviour of the Australian cycad *Lepidozamia peroffskyana. Australian Journal of Ecology* **14**, 241–245. doi:10.1111/j.1442-9993.1989.tb01432.x

Orr AG (1999) Evidence for unpalability in the genus *Delias* Hübner (Lepidoptera: Pieridae) and its role in mimetic assemblages. *Australian Entomologist* **26**, 45–52.

Oswald JD (1997) Review of the *sejunctus* species group of the split-footed lacewing genus *Osmylops* Banks (Neuroptera: Nymphalidae), with remarks on the functional morphology of the terminalic coupling. *Australian Journal of Entomology* **36**, 351–358. doi:10.1111/j.1440-6055.1997.tb01485.x

Ozanne CMP (2013) Canopies and climate change. In *Treetops at Risk: Challenges of Global Canopy Ecology and Conservation.* (Eds M Lowman, S Devy and T. Ganesh) pp. 113–118. Springer, New York.

Page TJ, von Rintelen K, Hughes M (2007) Phylogenetic and biogeographic relationships of subterranean and surface genera of Australian Atyidae (Crustacea: Decapoda: Caridea). *Invertebrate Systematics* **21**, 137–145. doi:10.1071/IS06023

Pal TK, Lawrence JF (1986) A new genus and subfamily of mycophagous Bothrideridae (Coleoptera: Cucujoidea) from the Indo-Australian region, with notes on related families. *Journal of the Australian Entomological Society* **25**, 185–210. doi:10.1111/j.1440-6055.1986.tb01101.x

Pape T, Arnaud PH, Jr (2001) *Bezzimyia* – a genus of New World Rhinophoridae (Insecta: Diptera). *Zoologica Scripta* **30**, 257–297. doi:10.1046/j.1463-6409.2001.00064.x

Paramonov SJ (1950) *A Review of the Australian Mydaidae (Diptera). Bulletin 255.* CSIRO Publishing, Melbourne.

Paramonov SJ (1953) A review of the Australian Nemestrinidae (Diptera). *Australian Journal of Zoology* **1**, 242–290. doi:10.1071/ZO9530242

Paramonov SJ (1954) Notes on Australian Diptera (XIII–XV). XIV. Notes on some hippobscids (especially *Ortholfersia*). *Annals and Magazine of Natural History* **7**, 283–292.

Paramonov SJ (1955) Notes on Australian Diptera (XIII–XV). XIV. A review of Australian *Criorrhina* species (Syrphidae). *Annals & Magazine of Natural History* **8**, 135–144.

Paramonov SJ (1956) A review of the Australian species of *Cylindromyia* Meigen and *Saralba* Walker (Tachinidae: Diptera). *Australian Journal of Zoology* **4**, 358–375. doi:10.1071/ZO9560358

Paramonov SJ (1957) A review of the Australian Acroceridae (Diptera). *Australian Journal of Zoology* **5**, 521–546.

Paramonov SJ (1958a) A review of Australian species of *Laphria* (Asilidae, Diptera) with descriptions of three new species from Lord Howe Island. *Pacific Science* **12**, 92–105.

Paramonov SJ (1958b) A review of Australian Pyrgotidae (Diptera). *Australian Journal of Zoology* **6**, 89–138. doi:10.1071/ZO9580089

Paramonov SJ (1962) A review of Australian Leptidae (Diptera). *Australian Journal of Zoology* **10**, 113–169. doi:10.1071/ZO9620113

Paramonov SJ (1967) A review of Australian species of *Ligyra* Newman (*Hyperalonia olim*) (Bombyliidae: Diptera). *Australian Journal of Zoology* **15**, 123–144. doi:10.1071/ZO9670123

Paramonov SJ (1968) A review of the tribe Rutilini (Diptera: Tachinae). I. Genera other than *Rutilia* Robineau-Desvoidy and *Formosia* Guerin-Meneville. *Australian Journal of Zoology* **16**, 349–404. doi:10.1071/ZO9680349

Parker J (2016) Myrmecophily in beetles (Coleoptera): evolutionary patterns and biological mechanisms. *Myrmecological News* **22**, 65–108.

Parkyn J, Brooks SL, Newell D (2014) Habitat use and movement patterns of the endangered land snail *Thersites mitchellae* (Cox, 1864) (Camaenidae). *Malacologia* **57**, 295–307. doi:10.4002/040.057.0203

Parkyn J, Challisthianagara A, Brooks L, Specht A, McMullan-Fisher S, Newell D (2015) The natural diet of the endangered camaenid land snail *Thersites mitchellae* (Cox 1864) in northern New South Wales. *Australian Zoologist* **37**, 343–349. doi:10.7882/AZ.2014.042

Parry L, Tanner A, Vinther J (2014) The origin of annelids. *Palaeontology* **57**, 1091–1103.

Parsons MJ (1996) A phylogenetic reappraisal of the birdwing genus *Ornithoptera* (Lepidoptera: Papilionidae: Troidini) and a new theory of its evolution in relation to Gondwanan vicariance biogeography. *Journal of Natural History* **30**, 1707–1736. doi:10.1080/00222939600771001

Parsons PA, Bock IR (1977) Australian endemic *Drosophila*. I. Tasmania and Victoria, including description of two new species. *Australian Journal of Zoology* **25**, 249–268. doi:10.1071/ZO9770249

Parsons RF, Gill AM (1968) The effects of salt spray on coastal vegetation at Wilson's Promontory, Victoria, Australia. *Royal Society of Victoria* **81**, 1–10.

Paulian R (1977) The Australian Ceratocanthidae (Coleoptera: Scarabaeoidea). *Journal of the Australian Entomological Society* **16**, 261–265. doi:10.1111/j.1440-6055.1977.tb00099.x

Paulian R (1980) Revision taxonomique des Liparochrus Erichson, genre Australo-Melanesien de Coleopteres Hybosoridae, Scarabaeioidea. *Annales de la Société Entomologique de France (N.S.)* **16**, 389–433.

Peakall R, Handel SN, Beattie AJ (1991) The evidence for, and importance of, ant pollination. In *Ant-plant Interactions*. (Eds CR Huxley and DF Cutler) pp. 421–429. Oxford Science Publications. Oxford University Press, Oxford.

Pearson RG (2014) Dynamics of invertebrate diversity in a tropical stream. *Diversity (Basel)* **6**, 771–791. doi:10.3390/d6040771

Peck SB (2001) Review of the carrion beetles of Australia and New Guinea (Coleoptera: Silphidae). *Australian Journal of Entomology* **40**, 93–101. doi:10.1046/j.1440-6055.2001.00216.x

Peck SB, Howden HF (1984) Response of a dung beetle guild to different sizes of dung bait in a Panamanian Rainforest. *Biotropica* **16**, 235–238. doi:10.2307/2388057

Peck SB, Russell DR (1976) Life history of the fungus gnat *Macrocera nobilis* in American caves. (Diptera: Mycetophilidae0. *Canadian Entomologist* **108**, 1235–1241. doi:10.4039/Ent1081235-11

Pederson HA (1995) Anthecological observations on *Dendrochilum longibracteatum* – a species pollinated by facultatively anthophilous insects. *Lindleyana* **10**, 19–28.

Pegg G, Taylor T, Entwhistle P, Guymer G, Giblin F, Carnegie A (2017) Impact of *Austropuccinia psidii* (myrtle rust) on Myrtaceae-rich wet sclerophyll forests in south east Queensland. *PLoS ONE* **12**, e0188058. doi:10.1371/journal.pone.0188058

Permkam S, Hancock DL (1994) Australian Ceratitinae (Diptera: Tephritidae). *Invertebrate Taxonomy* **8**, 1325–1341. doi:10.1071/IT9941325

Permkam S, Hancock DL (1995) Australian Trypetinae (Diptera: Tephritidae). *Invertebrate Taxonomy* **9**, 1047–1209. doi:10.1071/IT9951047

Perotti MA, Braig HR (2009) Phoretic mites associated with animal and human decomposition. *Experimental & Applied Acarology* **49**, 85–124. doi:10.1007/s10493-009-9280-0

Peters WL, Campbell IC (1991) Ephemeroptera 'mayflies'. In *The Insects of Australia*. (Ed. ID Naumann) pp. 279–293. Melbourne University Press, Carlton.

Peterson BV, Theischinger G (1989) Family Thaumaleidae. In *Catalog of the Diptera of the Australasian and Oceanic Regions*. (Ed. NL Evenhuis) pp. 219–220. Bishop Museum Press and E.J. Brill, Honolulu.

Peterson H, Luxton M (1982) A comparative analysis of soil fauna populations and their role in decomposition processes. *Oikos* **39**, 287–388.

Philips TK (2016) Phylogeny of the Oniticellini and Onthophagini dung beetles (Scarabaeidae: Scarabaeinae) from morphological evidence. *ZooKeys* **579**, 9–57. doi:10.3897/zookeys.579.6183

Phillips RD, Scaccabarozzi D, Retter BA, Hayes C, Brown GR, Dixon KW, Peakall R (2014) Caught in the act: pollination of sexually deceptive trap-flowers by fungus gnats in *Pterostylis* (Orchidaceae). *Annals of Botany* **113**, 629–641. doi:10.1093/aob/mct295

Piek T (Ed.) (1986) *Venoms of Hymenoptera. Biochemical, Pharmacological and Behavioural Aspects*. Academic Press, London.

Pilgrim RLC (1972) The aquatic larva and the pupa of *Choristella philpotti* Tillyard, 1917 (Mecoptera: Nannochoristidae). *Pacific Insects* **14**, 151–168.

Pilgrim E, von Dohlen C, Pitts J (2008) Molecular phylogenetics of Vespoidea indicate paraphyly of the superfamily and novel relationships of its component families and subfamilies. *Zoologica Scripta* **37**, 539–560. doi:10.1111/j.1463-6409.2008.00340.x

Pincebourde S, Casas J (2016) Hypoxia and hypercarbia in endophagous insects: larval position in the plant gas exchange network is key. *Journal of Insect Physiology* **84**, 137–153. doi:10.1016/j.jinsphys.2015.07.006

Pincebourde S, Murdock CC, Vickers M, Sears MW (2016) Fine-scale microclimatic variation can shape the responses of organisms to global change in both natural and urban environments. *Integrative and Comparative Biology* **56**, 45–61. doi:10.1093/icb/icw016

Pinder AM, Brinkhurst RO (1997) Review of the Phreodrilidae (Annelida: Oligochaeta: Tubificida) of Australia. *Invertebrate Taxonomy* **11**, 443–523. doi:10.1071/IT95025

Pitkin BR (1973) A revision of the Australian Haplothripini, with descriptions of three new species (Thysanoptera: Phlaeothripidae). *Journal of the Australian Entomological Society* **12**, 315–339. doi:10.1111/j.1440-6055.1973.tb01680.x

Pitkin BR (1989a) Family Neriidae. In *Catalog of the Diptera of the Australasian and Oceanic Regions*. (Ed. NL Evenhuis) p. 468. Bishop Museum Press and E.J. Brill, Honolulu.

Pitkin BR (1989b) Family Pyrgotidae. In *Catalog of the Diptera of the Australasian and Oceanic Regions*. (Ed. NL Evenhuis) pp. 498–501. Bishop Museum Press and E.J. Brill, Honolulu.

Pitkin BR, Evenhuis NL (1989) Family Clusiidae. In *Catalog of the Diptera of the Australasian and Oceanic Regions*. (Ed. NL Evenhuis) pp. 534–536. Bishop Museum Press and E.J. Brill, Honolulu.

Platnick NI (2000) A relimitation and revision of the Australasian ground spider family Lamponidae (Araneae: Gnaphosoidea). *Bulletin of the American Museum of Natural History*

245, 1–328. doi:10.1206/0003-0090(2000)245<0001:ARARO T>2.0.CO;2

Platnick NI, Forster RR (1989) A revision of the temperate South American and Australasian spiders of the family Anapidae (Araneae, Araneoidea). *Bulletin of the American Museum of Natural History* **190**, 1–139.

Poinar GO (1985) Mermithid (Nematoda) parasites of spiders and harvestmen. *The Journal of Arachnology* **13**, 121–128.

Poinar GO, Szadziewski R (2007) *Corethrella andersoni* (Diptera: Corethrellidae), a new species from Lower Cretaceous Burmese amber. *Proceedings of the Entomological Society of Washington* **109**, 155–159.

Polaszek A (1991) Egg parasitism in Aphelinidae (Hymenoptera: Chalcidoidea) with special reference to *Centrodora* and *Encarsia* species. *Bulletin of Entomological Research* **81**, 97–106. doi:10.1017/S0007485300053293

Pollock DA (1995) Classification, reconstructed phylogeny and general history of genera of Pilipalpinae (Coleoptera: Tenebrionoidea: Pyrochroidae). *Invertebrate Taxonomy* **9**, 563–708. doi:10.1071/IT9950563

Pollock D, Lawrence JF (1995) Review of *Anoplopus* Blackburn (Coleoptera: Pythidae), with comments on constituents and systematics of Pythidae. In *Biology, Phylogeny and Classification of Coleoptera: Papers Celebrating the 80th Birthday of Roy A. Crowson*. (Eds J Pakaluk and SA Ślipiński) pp. 449–472. Muzeum I Instytut Zoologi PAN, Warsaw.

Ponder WF (1986) Glacidorbidae (Glacidorbacea: Basommatophora), a new family and superfamily of operculate freshwater gastropods. *Zoological Journal of the Linnean Society* **87**, 53–83. doi:10.1111/j.1096-3642.1986.tb01330.x

Ponder WF (1988) Bioluminescence in *Hinea braziliana* (Lamarck) (Gastropoda: Planaxidae). *The Journal of Molluscan Studies* **54**, 361. doi:10.1093/mollus/54.3.361

Ponder WF (1991) Australian Hydrobiidae – an overview based on current research. In *Proceedings of the Tenth International Malacological Congress.Pt. 2.* 1989, Tübingen. (Ed. C Meier-Brook) pp. 525–528.

Ponder WF (1994) Australian freshwater Mollusca: conservation priorities and indicator species. *Memoirs of the Queensland Museum* **36**, 191–196.

Ponder WF, Avern GJ (2000) The Glacidorbidae (Mollusca: Gastropoda: Heterbranchia) of Australia. *Records of the Australian Museum* **52**, 307–353. doi:10.3853/j.0067-1975.52.2000.1318

Ponder WF, de Keyzer RG (1998) Superfamily Rissooidea. In *Mollusca: The Southern Synthesis. Fauna of Australia. Vol. 5. Part B.* (Eds PL Beesley, GJB Ross and A Wells) pp. 745–766. CSIRO Publishing, Melbourne.

Ponder WF, Wells FE, Solem A (1998) Distribution and affinities of non-marine molluscs. In *Mollusca: The Southern Synthesis. Fauna of Australia. Vol. 5. Part A.* (Eds PL Beesley, GJB Ross and A Wells) pp. 80–88. CSIRO Publishing, Melbourne.

Pont AC (1973) Studies on Australian Muscidae (Diptera) IV. a revision of the subfamilies Muscinae and Stomoxyinae. *Australian Journal of Zoology. Supplementary Series* **21**.

Pont AC (1986) Studies on the Australian Muscidae (Diptera) VII. The genus *Atherigona* Rondani. *Australian Journal of Zoology. Supplementary Series Number* **120**.

Pont AC (1989) Family Muscidae. In *Catalog of the Diptera of the Australasian and Oceanic Regions*. (Ed. NL Evenhuis) pp. 675–699. Bishop Museum Press and E.J. Brill, Honolulu.

Pont AC (2019) Revision of the Australian Muscidae (Diptera). VIII. The genus *Lispe* Latreille, 1797. *Zootaxa* **4557**, 1–232.

Pope RD (1988) A revision of the Australian Coccinellidae (Coleoptera). Pt. 1. subfamily Coccinellinae. *Invertebrate Taxonomy* **2**, 633–735. doi:10.1071/IT9880633

Pople GD, Cowley NB (1982) *The Rainforest Inventory of June 1981*. Forestry Commission of New South Wales, Sydney.

Popple LW (2017) A revision of the *Myopsalta crucifera* (Ashton) species group (Hemiptera: Cicadidae: Cicadettini) with 14 new species from mainland Australia. *Zootaxa* **4340**, 1–98.

Porazinska DL, Giblin-Davis RM, Powers TO, Thomas WK (2012) Nematode spatial and ecological patterns from tropical and temperate rainforests. *PLoS ONE* doi:10.1371/journal.pone.0044641

Postle AC, Steiner MY, Goodwin S (2001) Oriini (Hemiptera: Anthocoridae) new to Australia. *Australian Journal of Entomology* **40**, 231–244. doi:10.1046/j.1440-6055.2001.00074.x

Potts SG, Vulliamy B, Roberts S, O'Toole C, Dafni A, Ne'eman G, Willmer P (2005) Role of nesting resources in organising diverse bee communities in a Mediterranean landscape. *Ecological Entomology* **30**, 78–85. doi:10.1111/j.0307-6946.2005.00662.x

Powers TO, Neher DA, Mullin P, Esquivel A, Giblin-Davis RM, Kanzaki N, Stock SP, Mora MM, Uribe-Lorro L (2009) Tropical nematode diversity: vertical stratification of nematode communities in a Costa Rican humid lowland rainforest. *Molecular Ecology* doi:10.1111/j.1365-294X.2008.04075.x

Preisser E, Smith DC, Lowman MD (1998) Canopy and ground level insect distribution in a temperate forest. *Selbyana* **19**, 141–146.

Price GJ, Sobbe IH (2005) Pleistocene palaeoecology and environmental change on the Darling Downs, southeastern Queensland, Australia. *Memoirs of the Queensland Museum* **51**, 171–201.

Procheş Ş, Ramdhani S (2012) The world's zoogeographical regions confirmed by cross taxon analysis. *Bioscience* **62**, 260–270. doi:10.1525/bio.2012.62.3.7

Proctor HC (1999) *Gallilichus jonesi* sp. n. (Acari: Ascouracaridae): a new species of feather mite from the quills of the Australian brush-turkey (Aves: Megapodidae). *Australian Journal of Entomology* **38**, 77–84. doi:10.1046/j.1440-6055.1999.00088.x

Pulkkinen M, Yang Z-Q (1984) The parasitoids and predators of *Saperda populnea* (Linnaeus) (Coleoptera, Cerambycidae) in Finland. *Annales Entomologici Fennici* **50**, 7–12.

Pullen K, Jennings D, Oberprieler RG (2014) Annotated catalogue of Australian weevils (Coleoptera: Curculionidae). *Zootaxa* **3896**, 1–481.

Purschke G, Bleidorn C, Struck T (2014) Systematics, evolution and phylogeny of Annelida – a morphological perspective. *Memoirs of the Museum of Victoria* **71**, 247–269. doi:10.24199/j.mmv.2014.71.19

Pyke GH, Balzer L (1985) 'The effects of the introduced honeybee (*Apis mellifera*) on Australian native bees'. A report prepared for NSW National Parks and Wildlife Service. Occasional Papers of the New South Wales National Parks and Wildlife Service, Sydney.

Qin TK, Gullan PJ (1989) The Australian mealybugs (Homoptera: Pseudococcidae) of *Xanthorrhoea* (Xanthorrhoeaceae). *Invertebrate Taxonomy* **3**, 759–769. doi:10.1071/IT9890759

Qin TK, Halliday RB (1997) Eriorhynchidae, a new family of Prostigmata (Acarina), with a cladistic analysis of enpodoid species of Australia and New Zealand. *Systematic Entomology* **22**, 151–171. doi:10.1046/j.1365-3113.1997.d01-27.x

Quicke DLJ (1991) A revision of the Australian species of *Iphiaulax* Foerster and *Chaolita* Cameron (Insecta: Hymenoptera: Braconidae). *Records of the Australian Museum* **43**, 63–84. doi:10.3853/j.0067-1975.43.1991.41

Quicke DLJ (2015) *The Braconid and Ichneumonid Parasitoid Wasps*. Wiley Blackwell, Oxford.

Quicke DL, Shaw MR, van Achterberg C, Bland KF, Butcher BA, Lyszkowski R, Zhang YM (2014) A new Australian genus and five new species of Rogadinae (Hymenoptera: Braconidae), one reared as a gregarious endoparasitoid of an unidentified limacodid (Lepidoptera). *Zootaxa* **3881**, 237–257. doi:10.11646/zootaxa.3881.3.3

Rakitov R (2015) Observations on the biology and anatomy of Myerslopiidae (Hemiptera, Membracoidea). *Psyche* doi:10.1155/2015/898063

Ramírez VM, Ayala R, Delfín H (2016) Temporal variation in native bee diversity in the tropical sub-deciduous forest of the Yucatan Peninsula, Mexico. *Tropical Conservation Science* **9**, 718–734. doi:10.1177/194008291600900210

Rashed A, Sherratt TN (2007) Mimicry in hoverflies (Diptera: Syrphidae): a field test of the competitve mimicry complex. *Behavioral Ecology* **18**, 337–344. doi:10.1093/beheco/arl089

Rasnitsyn AP (1993) Archaeoscoliinae, an extinct subfamily of scoliid wasps (Insecta: Vespida = Hymenoptera: Scoliidae). *Journal of Hymenoptera Research* **2**, 85–95.

Rasnitsyn AP, Quicke DLJ (Eds) (2002). *History of Insects*. Kluwer Academic Publishers, Dordrecht.

Rasnitsyn AP, Ross AJ (2000) A preliminary list of arthropod families in the Burmese amber collection of The Natural History Museum, London. *Bulletin of the Natural History Museum, London* **56**, 21–24.

Rasnitsyn AP, Zherikhin VV (1999) First fossil chewing louse from the Lower Cretaceous of Baissa, Transbaikalia (Insecta, Pediculida = Phthiraptera, Saurodectidae *fam. n.*). *Russian Entomological Journal* **8**, 253–255.

Raven PH, Axelrod DI (1972) Plate tectonics and Australasian paleobiogeography. *Science* **176**, 1379–1386. doi:10.1126/science.176.4042.1379

Raven RJ (1978) Systematics of the spider subfamily Hexathelinae (Dipluridae: Mygalomorphae: Arachinida). *Australian Journal of Zoology Supplementary Series* **65**, 1–75.

Raven RJ (1981) A review of the Australian genera of the mygalomorph spider subfamily Diplurinae (Dipluridae: Chelicerata). *Australian Journal of Zoology* **29**, 321–363. doi:10.1071/ZO9810321

Raven RJ (1982) On the mygalomorph spider genus *Xamiatus* Raven (Diplurinae: Dipluridae) with the description of a new species. *Memoirs of the Queensland Museum* **20**, 473–478.

Raven RJ (1984a) Systematics and biogeography of the mygalomorph spider family Migidae (Araneae) in Australia. *Australian Journal of Zoology* **32**, 379–390. doi:10.1071/ZO9840379

Raven RJ (1984b) A new diplurid genus from eastern Australia and a related *Aname* species (Diplurinae: Dipluridae: Araneae). *Australian Journal of Zoology* **96**, 1–151.

Raven RJ (1984c) Systematics of the Australian curtain-web spiders (Ischnothelinae: Dipluridae: Chelicerata). *Australian Journal of Zoology Supplementary Series* **93**, 1–102.

Raven RJ (1994) Mygalomorph spiders of the Barychelidae in Australia and the Western Pacific. *Memoirs of the Queensland Museum* **35**, 291–706.

Raven RJ (1995) 'Border Ranges preliminary proposals, 1995'. Unpublished project report.

Raven RJ (2014) A revision of the ant-mimicking spiders of the family Corinnidae (Araneae) in the Western Pacific. *Zootaxa* **3958**, 1–258. doi:10.11646/zootaxa.3958.1.1

Raven RJ, Hebron W (2018) A review of the water spider family Pisauridae in Australia and New Caledonia with descriptions of four new genera and 23 new species. *Memoirs of the Queensland Museum, Nature* **60**, 233–381. doi:10.17082/j.2204-1478.60.2018.2017-06

Raven RJ, Wishart G (2005) The trapdoor spider *Arbanitis* L. Koch (Idiopidae: Mygalomorphae) in Australia. *Memoirs of the Queensland Museum* **51**, 531–557.

Raven PH, Yeates DK (2007) Australian biodiversity: threats for the present, opportunities for the future. *Australian Journal of Entomology* **46**, 177–187. doi:10.1111/j.1440-6055.2007.00601.x

Rayment T (1935) *A Cluster of Bees*. Endeavour, Sydney.

Raymond A, Cutlip P, Sweet M (2000) Rates and processes of terrestrial nutrient cycling in the Palaeozoic: The world before beetles, termites and flies. In *Evolutionary Palaeoecology, the Ecological Context of Macroevolutionary Change*. (Eds W Allmon and DJ Bottjer) pp. 235–283. Cambridge University Press, Cambridge.

Razzak MA, Ali T, Ali SI (1992) The pollination biology of *Aristolochia bracteolata* Lamk. (Aristolochiaceae). *Pakistan Journal of Botany* **24**, 79–87.

Reay F (1991) A new genus and two new species of plant nematode (Tylenchidae) from Australia. *Invertebrate Taxonomy* **5**, 855–867. doi:10.1071/IT9910855

Rech AR, Dalsgaard B, Sandel B, Sonne J, Svenning J-C, Holmes N (2016) The macroecology of animal versus wind pollination: ecological factors are more important than historical climate stability. *Plant Ecology & Diversity* **9**, 253–262. doi:10.1080/17550874.2016.1207722

Recher HF, Shields J, Kavanaugh R, Webb G (1987) Retaining remnant mature forest for nature conservation at Eden, New

South Wales: a review of theory and practice. In *Nature Conservation: The Role of Remnants of Native Vegetation.* (Eds DA Saunders, GW Arnold, AA Burbidge and JM Hopkins) pp. 177–194. Surrey Beatty and Sons, Chipping Norton, in association with CSIRO and CALM.

Reef R, Lovelock CE (2015) Regulation of water balance in mangroves. *Annals of Botany* **115**, 385–395. doi:10.1093/aob/mcu174

Reemer M, Ståhls G (2013) Generic revision and species classification of the Microdontinae (Diptera, Syrphidae). *ZooKeys* **288**, 1–213. doi:10.3897/zookeys.288.4095

Regier JC, Mitter C, Davis DR, Harrison L, Sohn J-C, Cummings MP, Zwick A, Mitter KT (2014) A molecular phylogeny and revised classification for the oldest ditrysia moth lineages (Lepidoptera: Tineoidea), with implications for ancestral feeding habits of the mega-diverse Ditrysia. *Systematic Entomology* **39**, 1–24. doi:101111/syen.12110

Rehn AC (2003) Phylogenetic analysis of higher-level relationships of Odonata. *Systematic Entomology* **28**, 181–240. doi:10.1046/j.1365-3113.2003.00210.x

Reid AL (1996) Review of the Peripatopsidae (Onychophora) in Australia, with comments on peripatopsid relationships. *Invertebrate Taxonomy* **10**, 663–936. doi:10.1071/IT9960663

Reid AL (2000) Eight new *Planipapillus* (Onychophora: Peripatopsidae) from southeastern Australia. *Proceedings of the Linnean Society of New South Wales* **122**, 1–32.

Reid CAM (1989a) The Australian species of the tribe Zeugophorini (Coleoptera: Chrysomelidae: Megalopodinae). *General and Applied Entomology* **21**, 39–47.

Reid CAM (1989b) A new species of *Calomela* Hope (Coleoptera: Chrysomelidae) from New South Wales, with habitat and distribution notes on other species in the genus. *Australian Entomological Magazine* **16**, 69–73.

Reid CAM (1991) A new genus of Cryptocephalinae from Australia (Coleoptera: Chrysomelidae). *Entomologica Scandinavica* **22**, 139–157. doi:10.1163/187631291X00020

Reid CAM (1992a) Revision of the genus *Cheiloxena* Baly (Coleoptera: Chrysomelidae: Eumolpinae). *Memoirs of the Museum of Victoria* **53**, 101–114. doi:10.24199/j.mmv.1992.53.04

Reid CAM (1992b) The leaf-beetle genus *Microdonacia* Blackburn (Coleoptera: Chrysomelidae: Galerucinae): revision and systematic placement. *Systematic Entomology* **17**, 359–387. doi:10.1111/j.1365-3113.1992.tb00557.x

Reid CAM (1994) Revision of the genus *Platycolaspis* Jacoby (Coleoptera: Chrysomelidae: Cryptocephalinae). *Memoirs of the Museum of Victoria* **54**, 207–220. doi:10.24199/j.mmv.1994.54.10

Reid CAM (1995) A cladistic analysis of subfamilial relationships in the Chrysomelidae *sensu lato* (Chrysomeloidea). In *Biology, Phylogeny, and Classification of Coleoptera. Papers Celebrating the 80th Birthday of Roy A. Crowson.* (Eds J Pakaluk and SA Slipinski) pp. 559–631. Muzeum i Instytut Zoologii PAN, Warszawa.

Reid CAM (1997) New records of the genus *Stenus* Latreille (Coleoptera: Staphylinidae) in Australia. *Australian Entomologist* **24**, 119–126.

Reid CAM (1999a) A new generic synonym in the Australian Lucanidae (Coleoptera). *Coleopterists Bulletin* **53**, 175–177.

Reid CAM (1999b) Revision of leaf beetles of the genus *Cadmus* Erichson, subgenus *Lachnabothra* Saunders (Coleoptera: Chrysomelidae: Cryptocephalinae). *Invertebrate Taxonomy* **13**, 1–66. doi:10.1071/IT97026

Reid CAM (2000) Spilopyrinae Chapuis: a new subfamily in the Chrysomelidae and its systematic placement (Coleoptera). *Invertebrate Taxonomy* **14**, 837–862. doi:10.1071/IT00042

Reid CAM (2006) A taxonomic revision of the Australian Chrysomelinae, with a key to the genera (Coleoptera: Chrysomelidae). *Zootaxa* **1292**, 1–119. doi:10.11646/zootaxa.1292.1.1

Reid CAM (2016) Australopapuan leaf beetle diversity: the contributions of host plants and geography. *Austral Entomology* **56**, 123–137. doi:10.1111/aen.12251

Reid CAM, Beatson M (2010) Revision of the Australo-Papuan genus *Spilopyra* Baly (Coleoptera: Chrysomelidae: Spilopyrinae). *Zootaxa* **2692**, 1–32. doi:10.11646/zootaxa.2692.1.1

Reid CAM, Beatson M (2016) Revision of the stag beetle genus *Ryssonotus* Macleay (Coleoptera: Lucanidae), with descriptions of a new genus and three new species. *Zootaxa* **4250**, 1–39.

Reid CAM, Nally S (2008) Revision of the genus *Menippus* Clark in Australia (Coleoptera: Chrysomelidae: Galerucinae). *Australian Journal of Entomology* **47**, 87–101.

Reid CAM, Storey RI (2000) Revision of the dung beetle genus *Temnoplectron* Westwood (Coleoptera: Scarabaeidae: Scarabaeini). *Memoirs of the Queensland Museum* **46**, 253–297.

Ren D (1998a) Flower-associated brachycera flies as fossil evidence for Jurassic angiosperm origins. *Science* **280**, 85–88. doi:10.1126/science.280.5360.85

Ren D (1998b) Late Jurassic Brachycera from northeastern China (Insecta: Diptera). *Dong Wu Fen Lei Xue Bao* **23**, 65–82.

Ren D, Labandeira CC, Santiago-Blay JA, Rasnitsyn A, Shih CK, Bashkuev A, Logan MA, Hotton CL, Dilcher D (2009) Probable pollination mode before angiosperms: Eurasian, long-proboscid scorpionflies. *Science* **326**, 840–847. doi:10.1126/science.1178338

Rentz DCF (1985) *A Monograph of the Tettigoniidae of Australia. Vol. 1. The Tettigoniinae.* CSIRO Publishing, Melbourne.

Rentz DCF (1991) Orthoptera 'grasshoppers, katydids, crickets'. In *The Insects of Australia.* (Ed. ID Naumann) pp. 369–393. Melbourne University Press, Carlton.

Rentz DCF (1993) *A Monograph of the Tettigoniidae of Australia. Vol. 2. The Austrosaginae, Zaprochilinae and Phasmodinae.* CSIRO Publishing, Melbourne.

Rentz DCF (1996) *Grasshopper Country.* UNSW Press, Sydney.

Rentz DCF (2010) *A Guide to the Katydids of Australia.* CSIRO Publishing, Melbourne.

Rentz DCF (2014) *A Guide to the Cockroaches of Australia.* CSIRO Publishing, Melbourne.

Rentz DCF, Balderson J (1979) 'Catalogue of Australian Tettigoniidae'. Division of Entomology Technical Paper Number 15. CSIRO Publishing, Melbourne.

Rentz DCF, Clyne D (1983) A new genus and species of pollen- and nectar-feeding katydids from eastern Australia (Orthoptera: Tettigoniidae: Zaprochilinae). *Journal of the Australian Entomological Society* **22**, 155–160. doi:10.1111/j.1440-6055.1983.tb01863.x

Rentz DCF, John B (1989) Studies in Australian Gryllacrididae: taxonomy, biology, ecology and cytology. *Invertebrate Taxonomy* **3**, 1053–1210. doi:10.1071/IT9891053

Rentz DCF, Kevan McE DK (1991) Dermaptera 'earwigs'. In *The Insects of Australia*. (Ed. ID Naumann) pp. 360–368. Melbourne University Press, Carlton.

Ribeiro SP, Basset Y, Kitching R (2014) Density of insect galls in the forest understorey and canopy: Neotropical, Gondwana or Global patterns? In *Neotropical Insect Galls*. (Eds GW Fernandes and JC Santos) pp. 129–139. Springer, Dordrecht.

Richards AM (1987) Distribution and relationships of the Australian Rhaphidophoridae (Orthoptera): pointers to past climatic changes. In *Evolutionary Biology of Orthopteroid Insects*. (Ed. B Baccetti) pp. 438–439. Ellis Horwood, Chichester.

Richards OW (1973) The Sphaeroceridae (=Borboridae or Cypselidae; Diptera Cyclorrhapha) of the Australian region. *Australian Journal of Zoology Supplementary Series* **22**.

Richards P (1952) *The Tropical Rain Forest*. Cambridge University Press, Cambridge.

Richardson BJ, Žabka M (2007) A revision of the Australian jumping spider genus *Prostheclina* Keyserling, 1892 (Araneae: Salticidae). *Records of the Australian Museum* **59**, 79–96. doi:10.3853/j.0067-1975.59.2007.1471

Richardson LR (1967) An annotated list of Australian leeches. *Proceedings of the Linnean Society of New South Wales* **92**, 227–245.

Richardson LR (1974) *Amicibdella* and *Microbdella* gen. nov. of Eastern Australia (Hirudinoidea: Haemadipsidae). *Memoirs of the Queensland Museum* **17**, 125–149.

Richardson LR (1975) A contribution to the general zoology of the land-leeches (Hirudinea: Haemadipsoidea superfam. nov.). *Acta Zoologica Academiae Scientiarum Hungaricae* **21**, 119–152.

Riedel A (2001) Revision of the *Euops quadrifasciculatus*-group (Coleoptera: Curculionoidea: Attelabidae) from the Australian region, with a discussion of shifts between *Nothofagus* and *Eucalyptus* host plants. *Invertebrate Taxonomy* **15**, 551–587. doi:10.1071/IT00039

Riek EF (1951) The Australian freshwater prawns of the family Palaemonidae. *Records of the Australian Museum* **22**, 358–367. doi:10.3853/j.0067-1975.22.1951.614

Riek EF (1954a) The Australian syrphid flies of the subfamily Cerioidinae (Diptera: Syrphidae): Systematics and wing folding. *Australian Journal of Zoology* **2**, 100–130. doi:10.1071/ZO9540100

Riek EF (1954b) The Australian Megaloptera of alderflies. *Australian Journal of Zoology* **2**, 131–140. doi:10.1071/ZO9540131

Riek EF (1954c) The Australian Mecoptera or scorpion-flies. *Australian Journal of Zoology* **2**, 143–168. doi:10.1071/ZO9540143

Riek EF (1955a) Australian wasps of the family Proctotrupidae (Hymenoptera: Proctotrupoidea). *Australian Journal of Zoology* **3**, 106–117. doi:10.1071/ZO9550106

Riek EF (1955b) Australian cleptid (Hymenoptera: Chrysidoidea) egg parasites of Cresmododea (Phasmodea). *Australian Journal of Zoology* **3**, 118–130. doi:10.1071/ZO9550118

Riek EF (1969) The Australian freshwater crayfish (Crustacea: Decapoda: Parastacidae), with descriptions of new species. *Australian Journal of Zoology* **17**, 855–918. doi:10.1071/ZO9690855

Riek EF (1973a) A revision of Australian scorpion flies of the family Choristidae (Mecoptera). *Journal of the Australian Entomological Society* **12**, 103–112. doi:10.1111/j.1440-6055.1973.tb01644.x

Riek EF (1973b) The genera of Australian Austroperlidae (Insecta: Plecoptera). *Journal of the Australian Entomological Society* **12**, 289–295. doi:10.1111/j.1440-6055.1973.tb01675.x

Riek EF (1974) The Australian moth-lacewings (Neuroptera: Ithonidae). *Journal of the Australian Entomological Society* **13**, 37–54. doi:10.1111/j.1440-6055.1974.tb02289.x

Roach AME (2000) Review of the Australian species of the dermestid genus *Anthrenocerus* Arrow (Coleoptera: Dermestidae). *Invertebrate Taxonomy* **14**, 175–224. doi:10.1071/IT97031

Roach AME, Rentz DCF (1998a) Blattodea. In *Zoological Catalogue of Australia. Volume 23. Archaeognatha, Zygentoma, Blattodea, Isoptera, Mantodea, Dermaptera, Phasmatodea, Embioptera, Zoraptera*. (Eds WWK Houston and A Wells) pp. 21–162. CSIRO Publishing, Melbourne.

Roach AME, Rentz DCF (1998b) Isoptera. In *Zoological Catalogue of Australia. Volume 23. Archaeognatha, Zygentoma, Blattodea, Isoptera, Mantodea, Dermaptera, Phasmatodea, Embioptera, Zoraptera*. (Eds WWK Houston and A Wells) pp. 163–250. CSIRO Publishing, Melbourne.

Roberts FHS (1928a) A revision of the Australian Bombyliidae (Diptera). Part i. *Proceedings of the Linnean Society of New South Wales* **53**, 90–144.

Roberts FHS (1928b) A revision of the Australian Bombyliidae (Diptera). Part ii. *Proceedings of the Linnean Society of New South Wales* **53**, 411–455.

Roberts FHS (1929) A revision of the Australian Bombyliidae (Diptera). Part iii. *Proceedings of the Linnean Society of New South Wales* **54**, 553–583.

Roberts FHS (1970) *Australian Ticks*. CSIRO, Melbourne.

Roberts MJ (1969) Structure of the mouthparts of the larvae of the flies *Rhagio* and *Sargus* in relation to feeding habits. *Journal of Zoology* **159**, 381–398. doi:10.1111/j.1469-7998.1969.tb08453.x

Robinson GS, Nielsen ES (1993) *Tineid Genera of Australia (Lepidoptera). Monographs on Australian Lepidoptera. Volume 2*. CSIRO Publishing, Melbourne, British Museum (Natural History), London.

Robinson MH, Robinson B (1976) A tipulid associated with spider webs in Papua New Guinea. *Entomologist's Monthly Magazine* **112**, 1–3.

Rodgers DJ, Kitching RL (1998) Vertical stratification of rainforest collembolan (Collembola: Insecta) assemblages: description of ecological patterns and hypotheses concerning their generation. *Ecography* **21**, 392–400. doi:10.1111/j.1600-0587.1998.tb00404.x

Rodríguez-Castañeda G, Dyer LA, Brehm G, Connahs H, Forkner RE, Walla TR (2010) Tropical forests are not flat: how mountains affect herbivore diversity. *Ecology Letters* doi:10.1111/j.1461-0248.2010.01525.x

Rose AB (1975) Moths of the families Sphingidae, Notodontidae and Agaristidae observed in Ku-ring-gai Chase National Park, N.S.W. *Australian Entomological Magazine* **2**, 93–96.

Rose HA (1965) Two new species of Thaumastocoridae (Hemiptera: Heteroptera) from Australia. *Proceedings of the Royal Entomological Society, London (B)* **34**, 141–144.

Rose P (2014) A conceptual model of the species composition of the original riparian rainforests of the Clarence River Floodplain, New South Wales. *Cunninghamia* **14**, 153–178. doi:10.7751/cunninghamia.2014.14.008

Rosenthal GA, Janzen DH (Eds) (1979) *Herbivores: Their Interaction with Secondary Plant Metabolites*. Academic Press, New York.

Ross ES (1991) Embioptera 'wed-spinners', 'embiids'. In *The Insects of Australia*. (Ed. ID Naumann) pp. 405–409. Melbourne University Press, Carlton.

Rossetto M, McPherson H, Siow J, Kooyman R, van der Merwe M, Wilson PD (2015) Where did all the trees come from? A novel multispecies approach reveals the impacts of biogeographical history and functional diversity on rain forest assembly. *Journal of Biogeography* doi:10.1111/jbi.12571

Roth LM (1991a) Blattodea 'cockroaches'. In *The Insects of Australia*. (Ed. ID Naumann) pp. 320–329. Melbourne University Press, Carlton.

Roth LM (1991b) The cockroach genera *Beybienkoa*, gen. nov., *Escala* Shelford, *Eowilsonia*, gen. nov., *Hensaussurea* Princis, *Parasigmoidella* Hanitsch and *Robshelfordia* Princis, (Dictyoptera: Blattaria: Blattellidae). *Invertebrate Taxonomy* **5**, 553–716. doi:10.1071/IT9910553

Roth LM (1992) The Australian cockroach genus *Choristima* Tepper (Blattaria, Blattellidae: Ectobiinae). *Entomologica Scandinavica* **23**, 121–151. doi:10.1163/187631292X00263

Roubik DW (1992) *Ecology and Natural History of Tropical Bees*. Cambridge University Press, New York.

Rousset V, Pleijel F, Rouse GW, Erséus C, Siddall ME (2007) A molecular phylogeny of annelids. *Cladistics* **23**, 41–63. doi:10.1111/j.1096-0031.2006.00128.x

Rugg D, Rose HA (1984) Reproductive biology of some Australian cockroaches (Blattodea: Blaberidae). *Journal of the Australian Entomological Society* **23**, 113–117. doi:10.1111/j.1440-6055.1984.tb01922.x

Rule S, Brook BW, Haberle SG, Turney CSM, Kershaw A, Pand Johnson CN (2012) The aftermath of megafaunal extinction: ecosystem transformation in Pleistocene Australia. *Science* **335**, 1483–1486. doi:10.1126/science.1214261

Russell-Smith A, Stork NE (1995) Composition of spider communities in the canopies of rainforest trees in Borneo. *Journal of Tropical Ecology* **11**, 223–235. doi:10.1017/S0266467400008695

Sabrosky CW (1989) Family Chloropidae. In *Catalog of the Diptera of the Australasian and Oceanic Regions*. (Ed. NL Evenhuis) pp. 650–665. Bishop Museum Press and E.J. Brill, Honolulu.

Saeed A, Austin AD, Dangerfield PC (1999) Systematics and host relationships of Australasian *Diolcogaster* (Hymenoptera: Braconidae: Microgastrinae). *Invertebrate Taxonomy* **13**, 117–178. doi:10.1071/IT97033

Saenger P, Moverley J (1985) Vegetative phenology of mangroves along the Queensland coast. *Proceedings of the Ecological Society of Australia* **13**, 257–265.

Sainval T Neef de, Lander T (1994) Iconographie des especes descrites des genres *Calodema* et *Metaxymorpha* (suite) Coleoptera: Buprestidae). *Bulletin de la Société Sciences Naturelles* **81**, 23–24.

Saito Y (2010) *Plant Mites and Sociality: Diversity and Evolution*. Springer, Tokyo.

Sakai S (1993) *Dermapterum Catalogus XXV: Iconographia IX. A Basic Survey of Integrated Taxonomy of the Dermaptera of the World*. Ikegami, Tokyo.

Sakai S (2002) *Aristolochia* spp. (Aristolochiaceae) pollinated by flies breeding on decomposing flowers in Panama. *American Journal of Botany* **89**, 527–534. doi:10.3732/ajb.89.3.527

Sakai S, Inoue T (1999) A new pollination system: dung-beetle pollination discovered in *Orchidantha inouei* (Lowiaceae, Zingiberales) in Sarawak, Malayasia. *American Journal of Botany* **86**, 56–61. doi:10.2307/2656954

Sakai S, Kato M, Nagamasu H (2000) *Artocarpus* (Moraceae) – gall midge pollination mutualism mediated by a male-flower parasitic fungus. *American Journal of Botany* **87**, 440–445. doi:10.2307/2656640

Sakai S, Momose K, Yumoto T, Kato M, Inoue T (1999) Beetle pollination of *Shorea parvifolia* (section *Mutica*, Dipterocarpaceae) in a general flowering period in Sarawak, Malaysia. *American Journal of Botany* **86**, 62–69. doi:10.2307/2656955

Salter KEW (1953) Studies on Australian Thynnidae I. A checklist of the Australian and Indo-Malayan Thynnidae. *Proceedings of the Linnean Society of New South Wales* **78**, 276–315.

Samuelson GA (1994) Pollen consumption and digestion by leaf beetles. In *Novel Aspects of the Biology of Chrysomelidae*. (Eds PH Jolivet, ML Cox and E Petitpierre) pp. 179–183. Kluwer Academic Publishers, Netherlands.

Sanborn AF, Villet MH (2014) *Catalogue of the Cicadoidea (Hemiptera: Auchenorrhyncha)*. Academic Press, London.

Sandeman DC, Wilkens LA (1982) Sound production by abdominal stridulation in the Australian Murray River crayfish *Euastacus armatus*. *The Journal of Experimental Biology* **99**, 469–472.

Sands DPA (1980) A new genus, *Acrodipsas*, for a group of Lycaenidae Lepidoptera) previously referred to *Pseudodipsas* C. and R. Felder, with descriptions of two new species from northern

Queensland. *Journal of the Australian Entomological Society* **18**, 251–265. doi:10.1111/j.1440-6055.1979.tb00849.x
Sands DPA (2012) Review of Australian *Phyllodes imperialis* Druce (Lepidoptera: Erebidae) with description of a new sub-specicies from subtropical Australia. *Australian Entomologist* **39**, 281–292.
Sands DPA, New TR (2013) *Conservation of the Richmond Birdwing Butterfly in Australia.* Springer, Dordrecht.
Sands DPA, Scott SE, Moffatt R (1997) The threatened richmond birdwing butterfly (*Ornithoptera richmondia* [Gray]: a community conservation project. *Memoirs of the Museum of Victoria* **56**, 449–453. doi:10.24199/j.mmv.1997.56.37
Sanger JC, Kirkpatrick JB (2017) Patterns of morphological and physiological traits of epiphytes within trees and between elevations in subtropical Australian rainforest. *Cunninghamia* **17**, 15–25.
Sankowsky G (1975) Some new food plants for various Queensland butterflies. *Australian Entomological Magazine* **2**, 55–56.
Sanmartín I, Ronquist F (2004) Southern Hemisphere biogeography inferred by event-based models: plant versus animal patterns. *Systematic Biology* **53**, 216–243. doi:10.1080/10635150490423430
Sann M, Niehuis O, Peters RS, Mayer C, Kozlov A, Podsiadlowski L, Blanks S, Meusemann K, Misof B, Bleidorn C, Ohl M (2018) Phylogenomic analysis of Apoidea sheds new light on the sister group of bees. *BMC Evolutionary Biology* **18**, 71–85 doi:10.1186/s12862-018-1155-8
Saunders DA, Hobbs RJ, Margules CR (1991) Biological consequences of ecosystem fragmentation: a review. *Conservation Biology* **5**, 18–32. doi:10.1111/j.1523-1739.1991.tb00384.x
Saunders DS (1977) *An Introduction to Biological Rhythms.* Blackie, Glasgow.
Savory T (1977) *Arachnida*. Academic Press, London.
Sawyer RT (1986) *Leech Biology and Behavior.* Clarendon Press, Oxford.
Scambler DJ (1989) A revision of the genus *Psilomorpha* Saunders (Coleopter: Cerambycidae: Cerambycinae). *Invertebrate Taxonomy* **3**, 163–173. doi:10.1071/IT9890163
Scanlan L, McDonald WJF, Shapcott A (2018) Phylogenetic diversity and conservation of rainforests in the Sunshine Coast region, Queensland, Australia. *Australian Journal of Botany* **66**, 518–530. doi:10.1071/BT18118
Schaefer CW, Mitchell PL (1983) Food plants of Coreoidea (Hemiptera: Heteroptera). *Annals of the Entomological Society of America* **76**, 591–615. doi:10.1093/aesa/76.4.591
Schimper AFW (1903) *Plant-geography upon a Physiological Basis.* (Translator WR Fisher, eds P Groom and IB Balfour). Clarendon Press, Oxford.
Schlinger EI, Jefferies MG (1989) Family Acroceridae. In *Catalog of the Diptera of the Australasian and Oceanic Regions.* (Ed. NL Evenhuis) pp. 375–377. Bishop Museum Press and E.J. Brill, Honolulu.
Schmidt CA, Shattuck SO (2014) The higher classification of the ant subfamily Ponerinae (Hymenoptera: Formicidae), with a review of ponerine ecology and behavior. *Zootaxa* **3817**, 1–242. doi:10.11646/zootaxa.3817.1.1
Schmidt D (2002) Notes on ant-lycaenid associations (Hymenoptera: Formicidae and Lepidoptera: Lycaenidae) in south-east Queensland. *Australian Entomologist* **29**, 61–68.
Schmidt D, Rice S (2002) Lycaenid butterflies (Lepidoptera: Lycaenidae) of Brisbane: new food plant records and life history notes. *Australian Entomologist* **29**, 37–46.
Schmidt JO, Blum MS (1978) A harvester ant venom: chemistry and pharmacology. *Science* **200**, 1064–1066. doi:10.1126/science.653354
Schmidt S, Walter GH (2014) Young clades in an old family: major evolutionary transitions and diversification of eucalypt-feeding pergid sawflies in Australia (Insecta, Hymenoptera, Pergidae). *Molecular Phylogenetics and Evolution* **74**, 111–121. doi:10.1016/j.ympev.2014.02.002
Schmidt S, Smith DR, Macdonald J (2002) Review of the Australian subfamily Pteryperginae (Hymenoptera: Symphyta: Pergidae). *Journal of Hymenoptera Research* **11**, 134–141.
Schmidt S, Walter GH, Grigg J, Moore CJ (2006) Sexual communication and host plant associations of Australian pergid sawflies (Hymenoptera: Symphyta: Pergidae). In *Recent Sawfly Research: Synthesis and Prospects.* (Eds SM Blank, S Schmidt and A Taeger) pp. 173–193. Goecke and Evers, Keltern.
Schneider MA (1991) Revision of the Australasian genus *Poecilohetaerus* Hendel (Diptera: Lauxaniidae). *Journal of the Australian Entomological Society* **30**, 143–168. doi:10.1111/j.1440-6055.1991.tb00406.x
Schneider MA (2010) A taxonomic revision of Australian Conopidae (Insecta: Diptera). *Zootaxa* **2581**, 1–246. doi:10.11646/zootaxa.2581.1.1
Scholtz CH (1986) Revision of the genus *Trox* Fabricius (Coleoptera: Trogidae) of the Australasian Region. *Australian Journal of Zoology Supplementary Series* **125**, 1–99.
Schuh RT, Slater JA (1995) *True Bugs of the World (Hemiptera: Heteroptera).* Cornell University Press, Ithaca.
Schuldt A, Wubet T, Buscot F, Staab M, Assman T, Böhnke-Kammerlander M, Both S, Erfmeier A, Klein A-M, Ma K, Pietsch K, Schultze S, Wirth C, Zhang J, Zumstein P, Bruelheide H (2015) Multitrophic diversity in a biodiverse forest is highly non-linear across spatial scales. *Nature Communications* **6**, 1–8. doi:10.1038/ncomms10169
Schuster JC, Schuster LB (1985) Social behavior in passalid beetles (Coleoptera: Passalidae): Cooperative brood care. *The Florida Entomologist* **68**, 266–272. doi:10.2307/3494359
Scoble MJ (1995) *The Lepidoptera: Form, Function, and Diversity.* Oxford University Press, Oxford.
Scoble MJ (Ed.) (1999) *Geometrid Moths of the World: A Catalogue (Lepidoptera: Geometridae). Volumes 1 and 2.* The Natural History Museum, London.
Scott B (1996) Phylogenetic relationships of the Camaenidae (Pulmonata: Stylommatophora: Helicoidea). *The Journal of Molluscan Studies* **62**, 65–73. doi:10.1093/mollus/62.1.65
Scudder GGE (1962) Results of the Royal Society expedition to southern Chile, 1958–1959: Lygaeidae (Hemiptera), with

description of a new subfamily. *Canadian Entomologist* **94**, 1064–1075. doi:10.4039/Ent941064-10

Seeman OD (2007) A new species of *Paradiplogynium* (Acari: Diplogyniidae) from *Titanolabis colossea* (Dohrn) (Dermaptera: Anisolabididae), Australia's largest earwig. *Zootaxa* **1386**, 31–38. doi:10.11646/zootaxa.1386.1.3

Seiverling EV, Khalsa A, Ahrns HT (2014) Pruritis and palpable purpurea from leeches in the Australian rainforest. *IDCases* **1**, 9–11. doi:10.1016/j.idcr.2014.02.001

Selman BJ, Lowman MD (1983) The biology and herbivory rates of *Novacastria nothofagi* Selman (Coleoptera: Chrysomelidae), a new genus and species on *Nothofagus moorei* in Australian temperate rainforests. *Australian Journal of Zoology* **31**, 179–191. doi:10.1071/ZO9830179

Sen Gupta T, Crowson RA (1969) On a new family of Clavicornia (Coleoptera) and a new genus of Languriidae. *Proceedings of the Royal Entomological Society of London* **38**, 125–131.

Sewell KB, Cannon LRG (1998) New Temnocephalans from the branchial chamber of Australian *Euastacus* and *Cherax* crayfish host. *Proceedings of the Linnean Society of New South Wales* **119**, 21–36.

Sewell KB, Cannon LRG, Blair D (2006) A review of *Temnohaswellia* and *Temnosewellia* (Platyhelminthes: Temnocephalida: Temnocephalidae), ectosymbionts from Australian crayfish *Euastacus* (Parastacidae). *Memoirs of the Queensland Museum* **52**, 199–280.

Shanahan M, Harrison RD, Yamuna R, Boen W, Thornton IWB (2001) Colonization of an island volcano, Long Island, Papua New Guinea, and an emergent island, Motmot, in its caldera lake. V. Colonization by figs (*Ficus* spp.), their dispersal and pollinators. *Journal of Biogeography* **28**, 1365–1377. doi:10.1046/j.1365-2699.2001.00638.x

Shapcott A, Forster PI, Guymer GP, McDonald WJF, Faith DP, Erickson D, Kress WJ (2015) Mapping biodiversity and setting conservation priorities for SE Queensland's rainforests using DNA barcoding. *PLoS ONE* **10**, e0122164. doi:10.1371/journal.pone.0122164

Sharma A, Raman A, Taylor GS, Nicol HI (2015) Feeding and oviposition behaviour of a gall inducing species of *Glycaspis* (*Synglycaspis*) (Hemiptera: Psylloidea: Aphalaridae) and development of galls on leaves of *Eucalyptus macrorhyncha* (Myrtaceae) in central wetsern New South Wales, Australia. *European Journal of Entomology* **112**, 75–90. doi:10.14411/eje.2015.021

Shattuck SO (1999) *Australian Ants: Their Biology and Identification*. CSIRO Publishing, Melbourne.

Shaw E (1925) New genera and species (mostly Australian) of Blattidae, with notes, and some remarks on Tepper's types. *Proceedings of the Linnean Society of New South Wales* **50**, 171–213.

Shaw SR (1990) A taxonomic revision of the long-tailed wasps of the genus *Megalyra* Westwood (Hymenoptera: Megalyridae). *Invertebrate Taxonomy* **4**, 1005–1052.

Shea M, Colgan D, Stanisic J (2012) Systematics of the landsnail genus *Gyrocochlea* and relatives (Mollusca: Charopidae). *Zootaxa* **3585**, 1–109. doi:10.11646/zoootaxa.3585.1.1

Shear WA (2015) The chemical defenses of millipedes (Diplopoda): biochemistry, physiology and ecology. *Biochemical Systematics and Ecology* **61**, 78–117. doi:101016/j.bse.2015.04.033

Shin S, Bayless KM, Winterton SL, Dikow T, Lessard BD, Yeates DK, Wiegmann BM, Trautwein MD (2018) Taxon sampling to address an ancient rapid radiation: a supermatrix phylogeny of early brachyceran flies (Diptera). *Systematic Entomology* **43**, 277–289. doi:10.1111/syen.12275

Shine R (2010) The ecological impact of invasive cane toads (*Bufo marinus*) in Australia. *The Quarterly Review of Biology* **85**, 253–291. doi:10.1086/655116

Short AEZ (2018) Systematics of aquatic beetles (Coleoptera): current state and future directions. *Systematic Entomology* **43**, 1–18. doi:10.1111/syen.12270

Shuh RT (1995) *Plant Bugs of the World (Heteroptera: Miridae). Systematic Catalog, Distributions, Host List and Bibliography.* New York Entomological Society, New York.

Shull HC, Perez-Losada M, Blair D, Sewell K, Sinclair EA, Lawler S, Ponniah M, Crandall KA (2005) Phylogeny and biogeography of the freshwater crayfish *Euastacus* (Decapoda: Parastacidae) based on nuclear and mitochondrial DNA. *Molecular Phylogenetics and Evolution* **37**, 249–263. doi:10.1016/j.ympev.2005.04.034

Siddall ME (2002) Phylogeny of the leech family Erpobdellidae (Hirudinia: Oligochaeta). *Invertebrate Systematics* **16**, 1–16. doi:10.1071/IT01011

Siddall ME, Bely AE, Borda E (2006) Hirudinida. In *Reproductive Biology and Phylogeny of Annelida*. (Eds G Rouse and F Pleijel) pp. 393–429. CRC Press, Enfield, NH.

Silberglied RE (1984) Visual communication and sexual selection among butterflies. In *The Biology of Butterflies*. (Eds RLVane-Wright and PR Ackery) pp. 207–224. Academic Press, London.

Simberloff DS, Wilson EO (1969) Experimental zoogeography of islands: the colonization of empty islands. *Ecology* **50**, 278–296. doi:10.2307/1934856

Sinclair BJ (2000a) Revision of the genus *Clinocera* Meigen from Australia and New Zealand (Diptera: Empididae: Clinocerinae). *Invertebrate Taxonomy* **14**, 347–440. doi:10.1071/IT99019

Sinclair BJ (2000b) Immature stages of Australian *Austrothaumalea* Tonnoir and *Niphta* Theischinger (Diptera: Thaumaleidae). *Australian Journal of Entomology* **39**, 171–176. doi:10.1046/j.1440-6055.2000.00162.x

Sinclair BJ (2003) Taxonomy, phylogeny and zoogeography of the subfamily Ceratomerinae of Australia (Diptera: Empidoidea). *Records of the Australian Museum* **55**, 1–44. doi:10.3853/j.0067-1975.55.2003.1373

Sinclair BJ (2008) New species of *Austrothaumalea* Tonnoir from Australia (Diptera: Thaumaleidae). *Records of the Australian Museum* **60**, 1–12. doi:10.3853/j.0067-1975.60.2008.1490

Sinclair BJ, Cumming JM (2006) The morphology, higher-level phylogeny and classification of the Empoidea (Diptera). *Zootaxa* **1180**, 1–172. doi:10.11646/zootaxa.1180.1.1

Sinclair BJ, Cumming JM (2007) *Leptopezella*, a new southern Hemisphere genus of Ocydromiinae (Diptera: Empidoidea: Hybotidae). *Zootaxa* **1629**, 27–37.

Sinclair DP (2002a) Two new genera of the Tassaratomidae (Hemiptera: Heteroptera: Pentatomoidea). *Memoirs of the Queensland Museum* **46**, 299–305.

Sinclair DP (2002b) A generic revision of the Oncomerinae (Hemiptera: Heteroptera: Tessaratomidae). *Memoirs of the Queensland Museum* **46**, 307–329.

Skuse FAA (1888–1890) Series 'Diptera of Australia', Parts I–III and Supplement I. *Proceedings of the Linnean Society of New South Wales (2nd Ser.)*: 1888a, Part 1, **3**, 657–726: 1890, Suppl. I, **5**, 373–412.

Slater JA (1975) On the biology and zoogeography of Australian Laygaeidae (Hemiptera: Heteroptera) with special reference to the southwest fauna. *Australian Journal of Entomology* **14**, 47–64. doi:10.1111/j.1440-6055.1975.tb02002.x

Ślipiński S (2007) *Australian Ladybird Beetles (Coleoptera: Coccinellidae)*. CSIRO Publishing, Melbourne.

Ślipiński SA, Escalona HE (2013) *Australian Longhorn Beetles (Coleoptera: Cerambycidae). Volume 1. Subfamily Lamiinae.* CSIRO Publishing, Melbourne.

Ślipiński SA, Escalona HE (2016) *Australian Longhorn Beetles (Coleoptera: Cerambycidae). Volume 2. Subfamily Cerambycinae.* CSIRO Publishing, Melbourne.

Ślipiński SA, Lawrence JF (1997) Genera of Colydiinae (Coleoptera: Zopheridae) of the Australo-Pacific region. *Annales zoologici (Warszawa)* **47**, 341–440.

Ślipiński SA, Tomaszewska W, Lawrence JF (2009) Phylogeny and classification of Corylophidae (Coleoptera: Cucujoidea) with descriptions of new genera and larvae. *Systematic Entomology* **34**, 409–433. doi:10.1111/j.1365-3113.2009.00471.x

Ślipiński A, Leschen RAB, Lawrence JF (2011) Order Coleoptera Linnaeus, 1758. In *Animal Biodiversity: An Outline of Higher-level Classification and Survey of Taxonomic Richness.* (Ed. Z-Q Zang) *Zootaxa* **3148**, 203–208.

Sloane TG (1896) On the Australian Clivinides (fam. Carabidae). Revision of the Australian species of the genus *Clivina* with the description of a new genus, *Clivinarchus*. *Proceedings of the Linnean Society of New South Wales* **21**, 143–257. doi:10.5962/bhl.part.8469

Sloane TG (1902) A revision of the genus *Notonomus* (family Carabidae: subfamily Feronini). *Proceedings of the Linnean Society of New South Wales* **27**, 252–325.

Sloane TG (1905) Revisional notes on Australian Carabidae. Part 1. Tribes Carabini, Pamborini, Pseudozaenini, Clivini; and the genus *Nebriosoma*. *Proceedings of the Linnean Society of New South Wales* **29**, 699–733. doi:10.5962/bhl.part.20177

Sloane TG (1911) Carabidae from Dorrigo, N.S.W. *Proceedings of the Linnean Society of New South Wales* **35**, 823–848.

Sloane TG (1913) Revisional notes on Australian Carabidae. Part 4. The genus *Notonomus*. *Proceedings of the Linnean Society of New South Wales* **38**, 404–449. doi:10.5962/bhl.part.13564

Sloane TG (1915) Studies in Australian Entomology. No. 17. New genera and species of Carabidae (Pamborini, Migapodini, Broscini, Cuneipectini, Nomiini, Pterostichini, Platynini, Oodini, Harpalini and Lebiini). *Proceedings of the Linnean Society of New South Wales* **40**, 438–473. doi:10.5962/bhl.part.18880

Sloane TG (1916) Carabidae of the Upper Williams River. *Proceedings of the Linnean Society of New South Wales* **41**, 196–208. doi:10.5962/bhl.part.15312

Sloane TG (1921) Revisional notes on Australian Carabidae. Part 6. *Proceedings of the Linnean Society of New South Wales* **46**, 192–208. doi:10.5962/bhl.part.14008

Sloane TG (1923) Studies in Australian entomology. No. 18. New genera and species of Carabidae. *Proceedings of the Linnean Society of New South Wales* **48**, 17–39.

Sluys R, Kawakatsu M, Riutot M, Baguña J (2009) A higher classification of planarian flatworms (Platyhelminthes, Tricladida). *Journal of Natural History* **43**, 1763–1777. doi:10.1080/00222930902741669

Smales LR (1994) Parasite extinctions – why care? *Memoirs of the Queensland Museum* **36**, 203–206.

Smiley J (1978) Plant chemistry and the evolution of host specificity: new evidence from *Heliconius* and *Passiflora*. *Science* **201**, 745–747. doi:10.1126/science.201.4357.745

Smit H (2007) New records of hyporheic water mites from Australia, with a description of two new genera and ten new species (Acari: Hydrachnidia). *Records of the Australian Museum* **59**, 97–116. doi:10.3853/j.0067-1975.59.2007.1486

Smith BJ (1992) Non-marine Mollusca. In *Zoological Catalogue of Australia. Canberra. Vol. 8.* (Ed. WWK Houston) pp. 1–399. Australian Government Publishing Service, Canberra.

Smith BJ (1998) Unionoidea. In *Mollusca: The Southern Synthesis. Fauna of Australia. Vol. 5. Part A.* (Eds PL Beesley, GJB Ross and A Wells) pp. 296–298. CSIRO Publishing, Melbourne.

Smith BJ, Kershaw RC (1979) *Field Guide to the Non-marine Molluscs of South Eastern Australia.* Australian National University Press, Canberra.

Smith DJ, Shattuck S (2009) Six new, unusually small ants of the genus *Leptomyrmex* (Hymenoptera: Formicidae). *Zootaxa* **2142**, 57–68. doi:10.11646/zootaxa.2142.1.5

Smith G (1998) Archaeognatha. In *Zoological Catalogue of Australia. Canberra. Vol. 23. Archaeognatha, Zygentoma, Blattodea, Isoptera, Mantodea, Dermaptera, Phasmatodea, Embioptera, Zoraptera.* (Eds WWK Houston and A Wells) pp. 1–5. CSIRO Publishing, Melbourne.

Smith IM, Harvey MS (1989) Descriptions of adults of *Arrenurus (?Micruracarus) kitchingi* sp. nov. (Acarina: Arrenuridae) from water-filled tree holes in Australia. *Canadian Entomologist* **121**, 283–289. doi:10.4039/Ent121283-3

Smith JP, Heard TA, Beckman M, Gloag R (2017) Flight range of the Australian stingless bee *Tetragonula carbonaria* (Hymenoptera: Apidae). *Austral Entomology* **56**, 50–53. doi:10.1111/aen.12206

Smith KGV (1989) Family Empididae. In *Catalog of the Diptera of the Australasian and Oceanic Regions.* (Ed. NL Evenhuis) pp. 382–392. Bishop Museum Press and E.J. Brill, Honolulu.

Smith TJ, Duke NC (1987) Physical determinants of inter-estuary variation in mangrove species richness around the tropical coastline of Australia. *Journal of Biogeography* **14**, 9–19. doi:10.2307/2844783

Smithers CN (1973) New species and records of Australian Bittacidae (Mecoptera). *Journal of the Australian Entomological Society* **12**, 296–300. doi:10.1111/j.1440-6055.1973.tb01676.x

Smithers CN (1974) New records of Choristidae (Mecoptera). *Australian Entomological Magazine* **2**, 1.

Smithers CN (1975) Additions to Australian Myopsocidae (Psocoptera). *Australian Entomological Magazine* **2**, 51–54.

Smithers CN (1977) The Psocoptera of Muogamarra Nature Reserve. *Records of the Australian Museum* **31**, 251–306. doi:10.3853/j.0067-1975.31.1977.215

Smithers CN (1981) A preliminary note on the Papilionoidea (Lepidoptera) of Tuglo Wildlife Refuge, New South Wales. *Australian Entomological Magazine* **7**, 91–96.

Smithers CN (1988) New distribution records for Australian Chrysopidae (Neuroptera). *Australian Entomological Magazine* **15**, 35–38.

Smithers CN (1990) New records of Australian Osmylidae. *Australian Entomological Magazine* **17**, 53–56.

Smithers CN (1991a) New records of Australian Hemerobiidae (Neuroptera). *Australian Entomological Magazine* **18**, 139–141.

Smithers CN (1991b) Psocoptera 'psocids, booklice'. In *The Insects of Australia*. (Ed. ID Naumann) pp. 412–420. Melbourne University Press, Carlton.

Smithers CN (1993) A note on the Megaloptera, Neuroptera and Mecoptera of Tuglo Wildlife Refuge, New South Wales. *Australian Entomologist* **20**, 67–71.

Smithers CN (1994a) A note on the Peripsocidae (Psocoptera) of Tuglo Wildlife Refuge, Hunter Valley, New South Wales. *Australian Entomologist* **21**, 7–10.

Smithers CN (1994b) A note on the Hesperiidae (Lepidoptera) (skippers) of Tuglo Wildlife Refuge, New South Wales. *Australian Entomologist* **21**, 103–109.

Smithers CN (1995) Psocoptera from the cycad *Lepidozamia peroffskyana* Regel (Cycadaceae) in New South Wales. *Australian Entomologist* **22**, 107–108.

Smithers CN (1996) New species and new records of Pseudocaeciliidae, Philotarsidae and Elipsocidae (Insecta: Psocoptera) from the Mount Royal area, Hunter Valley, New South Wales. *Proceedings of the Linnean Society of New South Wales* **116**, 233–243.

Smithers CN (1997a) Lepidopsocidae, Trogiidae, Myopsocidae and Psocidae (Insecta: Psocoptera) from the Mount Royal area, New South Wales. *Proceedings of the Linnean Society of New South Wales* **118**, 111–121.

Smithers CN (1997b) Psocoptera from Coocumbac Island Nature Reserve, Taree, New South Wales. *Australian Entomologist* **24**, 49–54.

Smithers CN, Peters JV (1990) New locality records for some butterflies in New South Wales. *Australian Entomological Magazine* **17**, 89–90.

Solem A (1997) Camaenid land snails from Western and Central Australia (Mollusca: Pulmonata: Camaenidae) VII taxa from Dampierland and through the Nullabor. *Records of the Western Australian Museum* **50**(*Supplement*), 5–142.

Solodovnikov A, Newton AF (2010) Review of the rove beetle genus *Antimerus* (Coleoptera, Staphylinidae, Staphylininae), a puzzling endemic Australian lineage of the tribe Staphylinini. *ZooKeys* **67**, 21–63. doi:10.3897/zookeys.67.704

Sousa JRP de, Silva Carvalho-Filho F de, Juen L, Esposito MC (2016) Evaluating the effects of different vegetation types on necrophagous fly communities (Diptera: Calliphoridae: Sarcophagidae): Implications for conservation. *PLoS ONE* **11**(10), e0164826. doi:10.1371/journal.pone.0164826

Specht RL (1970) Vegetation. In *The Australian Environment*. (Ed. GW Leeper) pp. 44–67. CSIRO/Melbourne University Press, Melbourne.

Specht RL (1981) Foliage projective cover and standing biomass. In *Vegetation Classification in Australia*. (Eds AN Gillison and DJ Anderson) pp. 10–21. CSIRO/ANU Press, Canberra.

Specht RL, Specht A, Whelan MB, Hegarty EE (1995) *Conservation Atlas of Plant Communities in Australia*. Centre for Coastal Management, Lismore, in association with Southern Cross University Press.

Speidel W (1978) Ein Versuch zur Unterteilung der Lepidopteren in Unterordnungen. *Atalanta* **8**, 119–121.

Spencer KA (1986) The Australian Chloropinae (Diptera: Chloropidae). *Journal of Natural History* **20**, 503–615. doi:10.1080/00222938600770401

Spencer KA (1989) Family Agromyzidae. In *Catalog of the Diptera of the Australasian and Oceanic Regions*. (Ed. NL Evenhuis) pp. 538–547. Bishop Museum Press and E.J. Brill, Honolulu.

Spratt DM (2006) Description of capillariid nematodes (Trichinelloidea: Capillariidae) parasitic in Australian marsupials and rodents. *Zootaxa* **1348**, 1–82. doi:10.11646/zootaxa.1348.1.1

Ståhls G, Hippa H, Rotheray R, Muona J, Gilbert F (2003) Phylogeny of the Syrphidae (Diptera) inferred from combined analysis of molecular and morphological characters. *Systematic Entomology* **28**, 433–450. doi:10.1046/j.1365-3113.2003.00225.x

Stanisic J (1990) Systematics and biogeography of eastern Australian Charopidae (Mollusca, Pulmonata) from subtropical rainforests. *Memoirs of the Queensland Museum* **30**, 1–241.

Stanisic J (1994) The distribution and patterns of species diversity of land snails in Eastern Australia. *Memoirs of the Queensland Museum* **36**, 207–214.

Stanisic J (1997) An area of exceptional land snail diversity: the Macleay Valley, north-eastern New South Wales. *Memoirs of the Museum of Victoria* **56**, 441–448. doi:10.24199/j.mmv.1997.56.36

Stanisic J (2015) *Pallidelix simonhudsoni* sp. n.: a new land snail from the central highlands of inland southern Queensland, Australia (Gastropoda: Eupulmonata: Camaenidae). *Memoirs of the Queensland Museum – Nature* **59**, 55–60.

Stanisic J (2016) Two new species of Pinwheel snails from contrasting Queensland habitats and a redescription and generic re-assignment of *Gyrocochlea myora* Stanisic, 2010 (Gastropoda: Eupulmonata: Charopidae). *Memoirs of the Queensland Museum* **60**, 1–12. doi:10.17082/j.2204-1478.60.2016.2016-07

Stanisic J, Shea M, Potter D, Griffiths O (2010) *Australian Land Snails, Vol. 1. A Field Guide to Eastern Australian Species.* Bioculture Press, Mauritius.

Stanković S, Dimkić I, Vujisić L, Pavković-Lučić S, Jovanović Z, Stević T, Sofrenić I, Mitić B, Tomić V (2016) Chemical defence in a millipede: evaluation and charcaterization of antimicrobial activity of the defensive secretion from *Pachyliulus hungaricus* (Karsch, 1881) (Diplopoda, Julida, Julidae). *PLoS ONE* **11**(12), e0167249. doi:10.1371/journal.pone.0167249

Staunton KM, Robson SKA, Burwell CJ, Reside AE, Williams SE (2014) Projected distributions and diversity of flightless ground beetles within Australia's Wet Tropics and their environmental correlates. *PLoS ONE* **9**, e88635. doi:10.1371/journal.pone.0088635

Stebnicka ZT, Howden HF (1994) A revision of the Australian genus *Podotenus* A. Schmidt (Coleoptera: Scarabaeoidea: Aphodiini). *Invertebrate Taxonomy* **8**, 17–62. doi:10.1071/IT9940017

Stebnicka ZT, Howden HF (1995) Revision of the Australian genera in the tribes Aphodiini, Aegialini and Proctophanini (Coleoptera: Scarabaeidae: Aphodiinae). *Invertebrate Taxonomy* **9**, 709–766. doi:10.1071/IT9950709

Stebnicka ZT, Howden HF (1996) Australian genera and species in the tribes Odontolochini, Psammodiini, Rhyparini, Stereomerini and part of the Eupariini (Coleoptera: Scarabaeoidea: Aphodiinae). *Invertebrate Taxonomy* **10**, 97–170. doi:10.1071/IT9960097

Steel T (1897) Australian land planarians: descriptions of new species and notes on collecting and preserving. *Proceedings of the Linnean Society of New South Wales* **22**, 104–119. doi:10.5962/bhl.part.12706

Steen Z, Schwarz MP (2000) Nesting and life cycle of the Australian green carpenter bees *Xylocopa* (*Lestis*) *aeratus* Smith and *Xylocopa* (*Lestis*) *bombylans* (Fabricius) (Hymenoptera: Apidae: Xylocopinae). *Australian Journal of Entomology* **39**, 291–300. doi:10.1046/j.1440-6055.2000.00195.x

Stevens MM (1994) Taxonomy, cladistics and biogeography of the Australian genus *Putoniessa* Kirkaldy (Hemiptera: Cicadelloidea: Cicadellidae). *Invertebrate Taxonomy* **8**, 1037–1115. doi:10.1071/IT9941037

Stireman JO (2005) The evolution of generalization? Parasitoid flies and the perils of inferring host range evolution from phylogenies. *Journal of Evolutionary Biology* **18**, 325–336. doi:10.1111/j.1420-9101.2004.00850.x

Stone C (1996) The role of psyllids (Hemiptera: Psyllidae) and bell miners (*Manorina melanophyrs*) in canopy dieback of Sydney blue gum (*Eucalyptus saligna* Sm.). *Australian Journal of Zoology* **21**, 450–458.

Storey RI (1977) Six new species of *Onthophagus* Latreille (Coleoptera: Scarabaeidae) from Australia. *Journal of the Australian Entomological Society* **16**, 313–320. doi:10.1111/j.1440-6055.1977.tb00111.x

Storey RI (1984) A new species of *Aptenocanthon* Matthews from North Queensland (Coleoptera: Scarabaeidae: Scarabaeini). *Memoirs of the Queensland Museum* **21**, 387–390.

Storey RI (1986) A new flightless species of *Aulacopris* White from North Queensland (Coleoptera: Scarabaeidae: Scarabaeini). *Memoirs of the Queensland Museum* **22**, 197–203.

Storey RI, Monteith GB (2000) Five new species of *Aptenocanthon* Matthews (Coleoptera: Scarabaeidae: Scarabaeinae) from tropical Australia, with notes on distribution. *Memoirs of the Queensland Museum* **46**, 349–358.

Storey RI, Weir TA (1988) New localities and biological notes for the genus *Onthophagus* Latreille (Coleoptera: Scarabaeidae) in Australia. *Australian Entomological Magazine* **15**, 17–24.

Storey RI, Weir TA (1989) New species of *Onthophagus* Latreille (Coleoptera: Scarabaeidae) from Australia. *Invertebrate Taxonomy* **3**, 783–815. doi:10.1071/IT9890783

Stork NE (1991) The composition of the arthropod fauna of Bornean lowland rain forest trees. *Journal of Tropical Ecology* **7**, 161–180. doi:10.1017/S0266467400005319

Stork NE, Eggleton P (1992) Invertebrates as determinants and indicators of soil quality. *American Journal of Alternative Agriculture* **7**, 38–47. doi:10.1017/S0889189300004446

Stork NE, Grimbacher PS (2006) Beetle assemblages from an Australian tropical rainforest show that the canopy and the ground strata contribute equally to biodiversity. *Proceedings of the Royal Society (B)* **273**(1596), 1969–1975. doi:10.1098/rspb.2006.3521

Stork NE, Srivastava DS, Eggleton P, Hodda M, Lawson G, Leakey RRB, Watt AD (2017) Consistency of effects of tropical-forest disturbance on species composition and richness relative to use of indicator taxa. *Conservation Biology* **31**, 924–933. doi:10.1111/cobi.12883

Stow A, Turnbull C, Gillings M, Smith S, Holley M, Silberbauer L, Wilson PD, Briscoe D, Beattie A (2010) Differential antimicrobial activity in response to entomopathogenic fungus *Cordyceps* in six Australian bee species. *Australian Journal of Entomology* **49**, 145–149. doi:10.1111/j.1440-6055.2010.00749.x

Strickland AH (1947) Coccids attacking cacao (*Theobroma cacao*, L.), in West Africa, with descriptions of five new species. *Bulletin of Entomological Research* **38**, 497–523. doi:10.1017/S0007485300022318

Struck TH, Bleidom C, McHugh D, Halanych KM (2007) Annelid phylogeny and the status of Sipuncula and Echiura. *BMC Evolutionary Biology* **7**, 57 doi:10.1186/1471-2148-7-57.

Stuckenberg BR (1973) The Athericidae, a new family in the lower Brachycera (Diptera). *Annals of the Natal Museum* **21**, 649–673.

Stuckenberg BR (2000) A new genus and species of Athericidae (Diptera: Tabanoidea) from Cape York Peninsula. *Records of the Australian Museum* **52**, 151–159. doi:10.3853/j.0067-1975.52.2000.1312

Stuckenberg BR (2001) Pruning the tree: a critical review of classification of the Homeodactyla (Diptera, Brachycera), with new perspectives and an alternative classification. *Studia Dipterologica* **8**, 3–41.

Sturm H (1980) Redescription of *Nesomachilis* (Archeognatha: Meinertellidae) with descriptions of new species from the Australian region. *New Zealand Journal of Zoology* **7**, 533–550. doi:10.1080/03014223.1980.11760685

Sturm H, Smith GB (1993) New bristle tails (Archaeognatha: Meinertellidae) from Australia. *Journal of the Australian Entomological Society* **32**, 233–240. doi:10.1111/j.1440-6055.1993.tb00580.x

Suetsugu K, Tetsu S, Hiraiwa MK, Tsutsumi T (2018) Thrips as a supplementary pollinator in an orchid with granular pollinia: is this mutualism? *Ecology* **100**, 1–14.

Sundholm A (1978) Extension of the known range of the Australian Fritillary *Argynnis hyperbius inconstans* Butler (Lepidoptera: Nymphalidae). *Australian Entomological Magazine* **5**, 34.

Suter PJ (1999) *Irpacaenis*, a new genus of Caenidae (Ephemeroptera) from Australia. *Australian Journal of Zoology* **38**, 159–167.

Suter RB, Benson K (2014) Nocturnal, diurnal, crepuscular: activity assessments of Pisauridae and Lycosidae. *The Journal of Arachnology* **42**, 178–191. doi:10.1636/J13-66.1

Svenson GJ, Whiting MF (2009) Reconstructing the origins of praying mantises (Dictyoptera, Mantodea): the roles of Gondwana vicariance and morphological convergence. *Cladistics* **25**, 468–514. doi:10.1111/j.1096-0031.2009.00263.x

Sweet M (2000) *Seed and Chinch Bugs (Lygaeoidea), Heteropoda of Economic Importance*. CRC Press, Boca Raton, Fl.

Szwedo J (2004) An annotated checklist of Myersolpiidae with notes on the distribution and origin of the group (Hemiptera: Cicadomorpha). *Zootaxa* **425**, 1–15. doi:10.11646/zootaxa.425.1.1

Szwedo J (2005) Notes on Otiocerini with a second record of Derbidae in Eocene Baltic amber (Hemiptera: Fulgoromorpha: Derbidae). *Insect Systematics & Evolution* **36**, 161–172. doi:10.1163/187631205788838483

Tagami K, Halliday B (2013) Mites of the family Histiostomatidae associated with the Australian giant earwig *Titanolabis colossea* (Dohrn) (Dermaptera: Anisolabididae). *Australian Journal of Entomology* **52**, 218–226. doi:10.1111/aen.12029

Takami Y, Sota T (2006) Four new species of the Australian *Pamborus* Latreille (Coleoptera, Carabidae) carabid beetles. *Australian Journal of Entomology* **45**, 44–54. doi:10.1111/j.1440-6055.2006.00495.x

Tarasov S, Dimitrov D (2016) Multigene phylogenetic analysis redefines dung beetles relationships and classification (Coleoptera: Scarabaeidae: Scarabaeinae). *BMC Evolutionary Biology* **16**, 257. doi:10.1186/s12862-016-0822-x

Taylor CK (2013) Further revision of the genus *Megalopsalis* (Opiliones, Neopilionidae), with the description of seven new species. *ZooKeys* **328**, 59–117. doi:10.3897/zookeys.328.5439

Taylor DL, Leung LK-P, Gordon IJ (2011) The impact of feral pigs (*Sus scrofa*) on an Australian lowland tropical rainforest. *Wildlife Research* **38**, 437–445. doi:10.1071/WR08138

Taylor RW (1973) Ants of the Australian genus *Mesostruma* Brown (Hymenoptera: Formicidae). *Journal of the Australian Entomological Society* **12**, 24–38. doi:10.1111/j.1440-6055.1973.tb02151.x

Taylor RW (1977) New ants of the Australasian genus *Orectognathus*, with a key to the known species (Hymenoptera: Formicidae). *Australian Journal of Zoology* **25**, 581–612. doi:10.1071/ZO9770581

Taylor RW (1979) New Australian ants of the genus *Orectognathus*, with summary description of the twenty-nine known species (Hymenoptera: Formicidae). *Australian Journal of Zoology* **27**, 773–788. doi:10.1071/ZO9790773

Taylor RW (1980) Australian and Melanesian ants of the genus *Eurhopalothrix* Brown and Kempf –notes and new species (Hymenoptera: Formicidae). *Journal of the Australian Entomological Society* **19**, 229–239. doi:10.1111/j.1440-6055.1980.tb02094.x

Taylor RW (1992) Nomenclature and distribution of some Australian and New Guinea ants of the subfamily Formicinae (Hymenoptera: Formicidae). *Journal of the Australian Entomological Society* **31**, 57–69. doi:10.1111/j.1440-6055.1992.tb00458.x

Taylor GS (1999) New species of *Acizzia* Heslop-Harrison (Hemiptera: Psyllidae) from Australian mistletoe (Loranthaceae). *Australian Journal of Entomology* **38**, 66–71. doi:10.1046/j.1440-6055.1999.00083.x

Taylor TN, Taylor EL, Krings M (2009) *Paleobotany: The Evolution of Fossil Plants*. Academic Press, Amsterdam.

Terry I (2001) Thrips and weevils as dual specialist pollinators of the Australian cycad *Macrozamia communis* (Zamiaceae). *International Journal of Plant Sciences* **162**, 1293–1305. doi:10.1086/321929

Theischinger G (1977) The male of *Antipodophlebia asthenes* (Tillyard, 1916) (Anisoptera:Aeshnidae). *Odonatologica* **6**, 205–209.

Theischinger G (1982) New and little known Dinotoperline stoneflies from Australia (Insecta: Plecoptera: Gripopterygidae). *Memoirs of the Queensland Museum* **20**, 489–525.

Theischinger G (1983) The genus *Stenoperla* McLachlan in Australia (Insecta: Plecoptera: Eusteniidae). *Australian Journal of Zoology* **31**, 541–556. doi:10.1071/ZO9830541

Theischinger G (1985) A revision of the Australian genus *Telephlebia* Selys (Odonata: Aeshnidae: Brachytroninae). *Australian Journal of Zoology* **33**, 245–261. doi:10.1071/ZO9850245

Theischinger G (1986) Australian Thaumaleidae (Insecta: Diptera). *Records of the Australian Museum* **38**, 291–317. doi :10.3853/j.0067-1975.38.1986.184

Theischinger G (1991a) Plecoptera 'stoneflies'. In *The Insects of Australia*. (Ed. ID Naumann) pp. 311–319. Melbourne University Press, Carlton.

Theischinger G (1991b) Megaloptera 'alderflies'. In *The Insects of Australia*. (Ed. ID Naumann) pp. 516–520. Melbourne University Press, Carlton.

Theischinger G (1992) The Limoniinae (Diptera: Tipulidae) of Australia. I. Introduction, Methods, Identification. II. The genus *Molophilus* Curtis. *Stapfia* **18**, 1–150.

Theischinger G (1993a) The Australian species *Dolichopeza* Curtis (Diptera: Tipulidae). *Linzer biologische Beiträge* **25**, 83–911.

Theischinger G (1993b) The Limoniinae (Diptera: Tipulidae) of Australia. III. The genus *Gynoplistia* Macquart. *Stapfia* **29**, 1–106.

Theischinger G (1996) The Limoniinae (Diptera: Tipulidae) of Australia. VI. New and insufficiently known species of *Toxorhina* Loew, *Limonia* Meigem, *Atarba* Osten Sacken, *Amphineurus* Skuse, *Gonomyia* Meigen and *Molophilus* Curtis. VII. The genera *Austrolimnophila* gen. nov., *Tipulimnoea* gen. nov. and *Paralimnophila* Alexander (Limnophilini). VIII. Preliminary key to the genus-group taxa. *Stapfia* **44**, 1–144.

Theischinger G (1998a) A new species of *Eusynthemis* Forster from Australia (Odonata: Synthemistidae). *Stapfia* **30**, 143–146.

Theischinger G (1998b) The *Eusynthemis guttata* (Selys) group of species from Australia (Odonata, Synthemistidae) (Odonata: Synthemistidae) – Part 2. *Stapfia* **30**, 147–153.

Theischinger G (1998c) A new species of *Griseargiolestes* Theischinger from Australia (Odonata: Zygoptera: Megapodagrionidae). *Stapfia* **55**, 623–627.

Theischinger G (1998d) *Tonyosynthemis*, a new dragonfly genus from Australia (Insecta: Odonata: Synthemistidae). *Linzer biologische Beiträge* **30**, 139–142.

Theischinger G (1999a) A new species of *Petalura* Leach from south-eastern-Queensland (Odonata: Petaluridae). *Linzer biologische Beiträge* **31**, 159–166.

Theischinger G (1999b) New and little-known species of Synthemistidae (Insecta: Odonata). *Linzer biologische Beiträge* **31**, 373–379.

Theischinger G (1999c) The Limoniinae (Diptera: Tipulidae) of Australia. IX. New taxa of *Gynoplistia* Macquart, *Gonomyia* Meigen and *Molophilus* Curtis. *Linzer biologische Beiträge* **31**, 493–502.

Theischinger G (2000) *Australian Alderfly Larvae and Adults (Insecta: Megaloptera). A Preliminary Guide to the Identification of Larvae and Survey of Adults of Australian Alderflies. Identification Guide No. 29.* Cooperative Research Centre for Freshwater Ecology, Thurgoona.

Theischinger G (2015) *Minipteryx robusta* – a new genus and species of micropterous Limnophilinae (Insecta: Diptera: Tipuloidea: Limoniidae) from Australia. *Records of the Australian Museum* **67**, 201–206. doi:10.3853/j.2201-4349.67.2015.1650

Theischinger G (2016) 'Bird of Paradise Crane fly' *Dapanoptera richmondiana* in suburban Sydney. *Victorian Entomologist* **46**, 83–86.

Theischinger G, Billingham ZD, Growns I (2018a) *Ozeoura* – a new genus of Chioneinae (Insecta: Diptera: Tipuloidea: Limoniideae) from Australia. *Records of the Australian Museum* **70**, 447–468. doi:10.3853/j.2201-4349.70.2018.1714

Theischinger G, Billingham ZD, Martin J (2018b) The genus *Styringomyia* Loew in Australia (Diptera: Tipuloidea: Limoniidae). *Linzer biologische Beiträge* **50**, 1587–1633.

Theischinger G, Cardale JC (1987) An illustrated guide to the adults of the Australian stoneflies (Plecoptera). *CSIRO Australian Division of Entomology Technical Papers* **26**, 1–83.

Theischinger G, Endersby I (2009) *Identification Guide to the Australian Odonata.* NSW Department of Environment, Climate Change and Water, Sydney South.

Theischinger G, Hawking JH (2000) The larva of *Eusynthemis ursula* Theischinger (Odonata: Synthemistidae). *Linzer biologische Beiträge* **32**, 247–251.

Theischinger G, Hawking JH (2006) *The Complete Field Guide to Dragonflies of Australia.* CSIRO Publishing, Melbourne.

Theischinger G, Houston WWK (1988) Megaloptera. In *Zoological Catalogue of Australia. Vol. 6. Ephemeroptera, Megaloptera, Odonata, Plecoptera, Trichoptera.* (Eds DW Walton and WWK Houston) pp. 23–32. Australian Government Publishing Service, Canberra.

Theischinger G, O'Farrell AF (1986) The genus *Austroargiolestes* Kennedy (Zygoptera: Megapodagrionidae). *Odonatologica* **15**, 387–428.

Theischinger G, Watson JAL (1985) The genus *Episynlestes* Kennedy (Odonata: Synlestidae). *Journal of the Australian Entomological Society* **24**, 143–148. doi:10.1111/j.1440-6055.1985.tb00212.x

Thiele H-U (1977) *Carabid Beetles in their Environments.* Springer-Verlag, Berlin.

Thien LB (1980) Patterns of pollination in primitive angiosperms. *Biotropica* **12**, 1–13. doi:10.2307/2387768

Thomas CD (1994) Extinction, colonization, and metapopulations: environmental tracking of rare species. *Conservation Biology* **8**, 373–378. doi:10.1046/j.1523-1739.1994.08020373.x

Thompson FC (1989) Family Lygistorrhinidae. In *Catalog of the Diptera of the Australasian and Oceanic Regions.* (Ed. NL Evenhuis). p. 134. Bishop Museum Press and E.J. Brill, Honolulu.

Thompson FC, Vockeroth JR (1989) Family Syrphidae. In *Catalog of the Diptera of the Australasian and Oceanic Regions.* (Ed. NL Evenhuis) pp. 437–458. Bishop Museum Press and E.J. Brill, Honolulu.

Thompson RT (1996) The species of *Phaenomerus* Schonherr (Coleoptera: Curculionidae: Zygopinae). *Invertebrate Taxonomy* **10**, 937–993. doi:10.1071/IT9960937

Thurgate ME, Gough JS, Clarke AK, Serov P, Spate A (2001) Stygofauna diversity and distribution in Eastern Australian cave and karst areas. *Records of the Western Australian Museum Supplement No.* **64**, 49–62.

Tishechkin DYu (2016) Newe data on vibrational calling signals of Fulgoroidea (Hemiptera: Auchenorrhyncha) from the Asia part of Palaearctic with new records of three new species of Cixiidae. *Russian Entomological Journal* **25**, 307–322. doi:10.15298/rusentj.25.4.01

Tng DYP, John GJ, Bowman DMJS (2013) Plant traits demonstrate that temperate and tropical giant eucalypt forests are convergent with rainforest and not savanna. *PLoS ONE* **12**, e84378. doi:10.1371/journal.pone.0084378

Todd Davies V (1993) A new spider genus (Araneae: Amaurobioidea) from rainforests of Queensland, Australia. *Memoirs of the Queensland Museum* **33**, 483–489.

Todd Davies V (1998) A revision of the Australian metaltellines (Araneae: Amaurobioidea: Amphinectidae: Metaltellinae). *Invertebrate Taxonomy* **12**, 211–243. doi:10.1071/IT96008

Todd Davies V, Raven RJ (1980) *Megadolomedes* nov. gen. (Araneae: Pisauridae) with adescription of the male of the type species, *Dolomedes australianus* Koch. 1865. *Memoirs of the Queensland Museum* **20**, 135–141.

Tomaszewska W, Ślipiński A (2013) Revision of the genera *Erithionyx* Blackburn and *Wioletta* Ślipiński (Coleoptera: Coccinellidae: Coccidulini). *Australian Journal of Entomology* **52**, 87–100. doi:10.1111/aen.12002

Tonnoir AL (1929) Australian Mycetophilidae. Synopsis of genera. *Proceedings of the Linnean Society of New South Wales* **56**, 584–614.

Townes HK (1949) The Nearctic species of the Evaniidae (Hymenoptera). *Proceedings of the United States National Museum* **99**, 525–539. doi:10.5479/si.00963801.99-3253.525

Townes HK (1950) The Nearctic species of Gasteruptiidae (Hymenoptera). *Proceedings of the United States National Museum* **100**, 85–145. doi:10.5479/si.00963801.100-3259.85

Townes HK, Chiu S-C (1970) The Indo-Australian species of *Xanthopimpla* (Ichneumonidae). *Memoirs of the American Entomological Institute* **14**, 1–372.

Traniello JFA (1978) Caste in a primitive ant: absence of age polyethism in *Amblyopone*. *Science* **202**, 770–772. doi:10.1126/science.202.4369.770

Tree DJ, Mound LA (2011) Thysanoptera of Lamington National Park, Australia, collected during the IBISCA – Queensland Project. *Memoirs of the Queensland Museum – Nature* **55**, 349–358.

Truswell E (1993) Vegetation changes in the Australian Teriary in response to climate and phytogeographic forcing factors. *Australian Systematic Botany* **6**, 533–557. doi:10.1071/SB9930533

Turak E, Bickel D (1994) 'Using biological indicators to monitor the health of freshwater ecosystems'. New South Wales Environment Protection Authority Technical Report 94/69. New South Wales Environment Protection Authority, Sydney.

Turbeville JM (2002) Progress in Nemertean biology: development and hylogeny. *Integrative and Comparative Biology* **42**, 692–703. doi:10.1093/icb/42.3.692

Turner RE (1909) A revision of the Australian species of the genus *Scolia*. *Annals & Magazine of Natural History* **4**, 165–176. doi:10.1080/00222930908692655

Tuxen SL (1967) Australian Protura, their phylogeny and zoogeography. *Journal of Zoological Systematics and Evolutionary Research* **5**, 1–53. doi:10.1111/j.1439-0469.1967.tb00499.x

Udovic F, Spencer RD (2012) New combinations in *Callistemon* (Myrtaceae). *Muelleria* **30**, 23–25.

Valentine JW, Erwin DH, Jablonski D (1996) Developmental evolution of Metazoan bodyplans: the fossil evidence. *Developmental Biology* **173**, 373–381. doi:10.1006/dbio.1996.0033

van Achterberg C, Quicke DLJ, Boring CA (2017) A revision of the tribe Planitorini van Achterberg (Hymenoptera, Braconidae, Euphorinae), with description of a new genus from Australia. *ZooKeys* **718**, 65–94. doi:10.3897/zookeys.718.21151

Van Bocxlaer I, Roelants K, Biju SD, Nagaraju J (2006) Late Cretaceous vicariance in Gondwanan amphibians. *PLoS ONE* doi:10.1371/journal.pone0000074

van Bruggen AC (1980) Gondwanaland connections in the terrestrial molluscs of Africa and Australia. *Journal of the Malacological Society of Australia* **4**, 215–222. doi:10.1080/00852988.1980.10673930

van Klinken RD (1997) Taxonomy and distribution of the *coracina* group of *Scaptodrosophila* Duda (Diptera: Drosophilidae) in Australia. *Invertebrate Taxonomy* **11**, 423–442. doi:10.1071/IT95039

van Klinken RD, Walter GH (2001) Subtropical drosophilids in Australia can be characterized by adult distribution across vegetation type and by height above forest floor. *Journal of Tropical Ecology* **17**, 705–718. doi:10.1017/S0266467401001523

van Oosterzee P (1997) *Where Worlds Collide: The Wallace Line*. Reed, Kew.

VanDerWal J, Shoo LP, Williams SE (2009) New approaches to understanding late Quaternary climate fluctuations and refugial dynamics in Australian wet tropical rain forests. *Journal of Biogeography* **36**, 291–301. doi:10.1111/j.1365-2699.2008.01993.x

Vane-Wright D (2004) Butterflies at that awkward age. *Nature* **428**, 477–480. doi:10.1038/428477a

Vasconcelos SD, Cruz TM, Salgado RL, Thyssen PJ (2013) Dipterans associated with a decomposing animal carcass in a rainforest fragment in Brazil: Notes on the early arrival and colonization by necrophagous species. *Journal of Insect Science* **13**, 145. doi:10.1673/031.013.14501

Verne NC (Ed.) (2007) *Forest Ecology Research Horizons*. NovaScience Publishers, New York.

Vilhelmsen L (1997) The phylogeny of the lower Hymenoptera (Insecta) with a summary of the early evolutionary history of the order. *Journal of Zoological Systematics and Evolutionary Research* **35**, 49–70. doi:10.1111/j.1439-0469.1997.tb00404.x

Vockeroth JR (1969) A revision of the genera of the Syrphini (Diptera: Syrphidae). *Memoirs of the Entomological Society of Canada* **62**, 1–176.

von Keler S (1971) A revision of the Australian Boopiidae (Insecta: Phthiraptera), with notes on the Trimenoponidae. *Australian Journal of Zoology Supplementary Series* **6**.

Wagner R (2018) Remarkable new Australian mothflies (Diptera, Psychodidae). *Zootaxa* **4457**, 285–295. doi:10.11646/zootaxa.4457.2.4

Wahis R (2008) Contributien à la connaissance des Pompilides d'Australie (Hymenoptera: Pompilidae). *Faunistic Entomology –. Entomologie Faunistique* **61**, 23–31.

Waichert C, Rodriguez J, Wasbauer MS, von Dohlen CD, Pitt JP (2015) Molecular phylogeny and systematics of spider wasps (Hymenoptera: Pompilidae): redefing subfamily boundaries and the origin of the family. *Zoological Journal of the Linnean Society* **175**, 271–287. doi:10.1111/zoj.12272

Walker JA, Rose HA (1998) Oothecal structure and male genitalia of the Geoscapheinae and some Australian *Panaesthia* Serville (Blattodea: Blaberidae). *Australian Journal of Entomology* **37**, 23–26. doi:10.1111/j.1440-6055.1998.tb01538.x

Walker KL (1986) Revision of the Australian species of the genus *Homalictus* Cockerell (Hymenoptera: Halictidae). *Memoirs of the Museum of Victoria* **47**, 105–200. doi:10.24199/j.mmv.1986.47.05

Walker K (1995a) Revision of the Australian native bee subgenus *Lasioglossum* (*Chilalictus*) (Hymenoptera: Halictidae). *Memoirs of the Museum of Victoria* **55**, 1–214. doi:10.24199/j.mmv.1995.55.01

Walker K (1995b) Revision of the Australian native bee subgenus *Lasioglossum* (*Chilalictus*) (Hymenoptera: Halictidae) (continued). *Memoirs of the Museum of Victoria* **55**, 215–423. doi:10.24199/j.mmv.1995.55.02

Walker K (1997) Supplement to a revision of the Australian members of the bee enus *Homalictus* (Cockerell) (Hymenoptera: Halictidae). *Memoirs of the Museum of Victoria* **56**, 69–82. doi:10.24199/j.mmv.1997.56.03

Wallace AR (1876) *The Geographical Distribution of Animals: With a Study of the Relations of Living and Extinct Faunas as Elucidating Past Changes of the Earth's Surface.* Macmillan, London.

Wallace MS, Deitz LL (2006) Australian treehoppers (Hemiptera: Membracidae: Centrotinae: Terentiini): phylogeny and biogeography. *Invertebrate Systematics* **20**, 163–183. doi:10.1071/IS05040

Walter DE (1998) *Hoploseius australianus* sp. nov. (Acari: Megostigmata: Ascidae), a unique element in the Australian acarofauna. *Australian Entomologist* **25**, 69–74.

Walter DE (1999) Review of the Australian *Asperoseius* Chant, *Euseius* Wainstein, *Okiseius* Ehara and *Phytoscutus* Mum (Acari: Mesostigmata: Phytoseiidae) with a key to the genera of Australian Ambyseiinae and descriptions of two new species. *Australian Journal of Entomology* **38**, 85–95. doi:10.1046/j.1440-6055.1999.00095.x

Walter DE (2001) Endemism and cryptogenesis in 'segmented' mites: a review of Australian Alicorhagidae, Terpnacaridae, Oehserchestidae and Grandjeanicidae (Acari: Sarcoptiformes). *Australian Journal of Entomology* **40**, 207–218. doi:10.1046/j.1440-6055.2001.00226.x

Walter DE (2004) Hidden in plain sight: mites in the canopy. In *Forest Canopies.* (Eds M Lowman and HB Rinker) pp. 224–241. Elsevier Academic Press, Amsterdam.

Walter DE, Beard JJ (1997) A review of the Australian Phytoseiinae (Acari: Mesostigmata: Phytoseiidae). *Invertebrate Taxonomy* **11**, 823–860. doi:10.1071/IT97010

Walter DE, Behan-Pelletier V (1999) Mites in forest canopies: filling the size distribution shortfall? *Annual Review of Entomology* **44**, 1–19. doi:10.1146/annurev.ento.44.1.1

Walter DE, Halliday RB, Lindquist EE (1993) A review of the genus *Asca* (Acraina: Ascidae) in Australia, with descriptions of three new leaf-inhabiting species. *Invertebrate Taxonomy* **7**, 1327–1347. doi:10.1071/IT9931327

Walter DE, Lindquist EE (1997) Australian species of *Lasioseius* (Acari: Mesostigmata: Ascidae): the *porulosus* group and other species from rainforest canopies. *Invertebrate Taxonomy* **11**, 525–547. doi:10.1071/IT96003

Walter DE, O'Dowd DJ (1992) Leaves with domatia have more mites. *Ecology* **73**, 1514–1518. doi:10.2307/1940694

Walter DE, O'Dowd DJ (1995) Life on the forest phylloplane: hairs, little houses, and myriad mites. In *Forest Canopies.* (Eds MD Lowman and N Nadkarni) pp. 325–351. Academic Press, New York.

Walter DE, O'Dowd DJ, Barnes V (1994) The forgotten arthropods: foliar mites in the forest canopy. *Memoirs of the Queensland Museum* **36**, 221–226.

Walter DE, Proctor HC (2013) *Mites: Ecology, Evolution and Behaviour.* Springer, Dordrecht.

Walter DE, Seeman O, Rodgers D, Kitching RL (1998) Mites in the mist: how unique is a rainforest canopy knockdown fauna? *Australian Journal of Ecology* **23**, 501–508. doi:10.1111/j.1442-9993.1998.tb00760.x

Walton DW (Ed.) (1985a) *Zoological Catalogue of Australia. Vol. 2. Hymenoptera: Formicoidea, Vespoidea and Sphecoidea.* Australian Government Publishing Service, Canberra.

Walton DW (Ed.) (1985b) *Zoological Catalogue of Australia. Vol. 3. Arachnida: Mygalomorphae, Araneomorphae in Part, Pseudoscorpionida, Amblypygi and Palpigradi.* Australian Government Publishing Service, Canberra.

Walton DW (Ed.) (1987) *Zoological Catalogue of Australia. Vol. 4. Coleoptera: Archostemata, Myxophaga and Adephaga.* Australian Government Publishing Service, Canberra.

Wang Y, Dietrich CH, Zhang Y (2017) Phylogeny and biogeography of leafhopper subfamily Evacanthinae (Hemiptera: Cicadellidae) based on morphological and molecular data. *Scientific Reports* **7**(45387), 1–13.

Wanless FR (1988) A revision of the spider group Astieae (Araneae: Salticidae) in the Australian region. *New Zealand Journal of Zoology* **15**, 81–172. doi:10.1080/03014223.1988.10422611

Wappler T, Engel MS (2003) The middle Eocene bee faunas of Eckfeld and Messel, Germany (Hymenoptera: Apoidea). *Journal of Paleontology* **77**, 908–921. doi:10.1017/S0022336000044760

Wardhaugh CW, Stork NE, Edwards W (2013a) Specialization of rainforest canopy beetles to host trees and microhabitats: not all specialists are leaf-feeding herbivores. *Biological Journal of the Linnean Society* **109**, 215–228. doi:10.1111/bij.12029

Wardhaugh CW, Stork NE, Edwards W, Grimbacher PS (2013b) Insects on flowers. The unexpectedly high biodiversity of flower-visiting beetles in a tropical rainforest canopy. *Communicative & Integrative Biology* doi:10.4161/cib.22509

Warman L, Moles AT (2009) Alternative stable states in Australia's wet tropics: a theoretical framework for the field data and a field-case for the theory. *Landscape Ecology* **24**, 1–13. doi:10.1007/s10980-008-9285-9

Warman L, Bradford MG, Moles AT (2013) A broad approach to abrupt boundaries: looking beyond the boundary at soil attributes within and across tropical vegetation types. *PLoS ONE* **8**(4), e60789. doi:10.1371/journal.pone.0060789

Waters JM, Craw D (2006) Goodbye Gondwana? New Zealand biogeography, geology, and the problem of circularity. *Systematic Biology* **55**, 351–356. doi:10.1080/10635150600681659

Watson JAL, Gay FJ (1991) Isoptera 'termites. In *The Insects of Australia*. (Ed. ED Naumann) pp. 330–347. CSIRO, Division of Entomology, Canberra.

Watson JAL, Miller LR, Abbey HM (1998) Isoptera. In *Zoological Catalogue of Australia. Volume 23. Archaeognatha, Zygentoma, Blattodea, Isoptera, Mantodea, Dermaptera, Phasmatodea, Embioptera, Zoraptera*. (Eds WWK Houston and A Wells) pp. 163–250. CSIRO Publishing, Melbourne.

Watson JAL, Moulds MS (1979) New species of Australian Lestidae (Odonata). *Journal of the Australian Entomological Society* **18**, 143–155. doi:10.1111/j.1440-6055.1979.tb00828.x

Watson JAL, O'Farrell AF (1991) Odonata 'dragonflies'. In *The Insects of Australia*. (Ed. ID Naumann) pp. 294–310. Melbourne University Press, Carlton.

Watson JAL, Smith GB (1991) Archaeognatha 'bristletails'. In *The Insects of Australia*. (Ed. ID Naumann) pp. 272–278. Melbourne University Press, Carlton.

Watson JAL, Theischinger G (1980) The larva of *Antipodophlebia asthenes* (Tillyard): a terrestrial dragonfly? (Anisoptera: Aesthnidae). *Odonatologica* **9**, 253–258.

Watson JAL, Theischinger G (1984) Regions of taxonomic disjunction in Australian Odonata and other freshwater insects. *Odonatologica* **13**, 147–157.

Watt JC (1974) A revised classification of Tenebrionidae (Coleoptera). *New Zealand Journal of Zoology* **1**, 381–452. doi:10.1080/03014223.1974.9517846

Watts CHS (1978) A revision of the Australian Dytiscidae (Coleoptera). *Australian Journal of Zoology* **57**, 1–66.

Watts CHS (1985) A faunal assessment of Australian Hydradephaga. *Proceedings. Academy of Natural Sciences of Philadelphia* **137**, 22–28.

Watts CHS (1989) Revision of the Australasian *Sternolophus* Solier (Coleoptera: Hydrophilidae). *Records of the South Australian Museum* **23**, 89–95.

Watts CHS (1995) Revision of the Australasian genera *Agraphydrus* Rogimbart, *Chasmogenus* Sharp and *Helochares* Mulsant (Coleoptera: Hydrophilidae). *Records of the South Australian Museum* **28**, 113–130.

Watts CHS (1997) Four new species of *Antiporus* Sharp (Coleoptera, Dytiscidae) from Australia, with notes on *A. femoralis* (Boh.) and *A. interrogationis* (Clark). *Records of the South Australian Museum* **30**, 35–42.

Watts CHS (1998) Revision of Australian *Enochrus* Thomson (Coleoptera: Hydrophilidae). *Records of the South Australian Museum* **30**, 137–156.

Webb C, Doggett S, Russell R (2016) *A Guide to Mosquitoes of Australia*. CSIRO Publishing, Melbourne.

Webb GA (1987) 'Larval host plants of Cerambycidae (Coleoptera) held in some Australian insect collections'. Technical Paper No. 38. Forestry Commission of New South Wales, Beecroft.

Webb GA, Simpson JA (1991) Notes on some Australian fungus beetles and their hosts and parasites (Coleoptera). *Coleopterists Bulletin* **45**, 42–44.

Webb GA, Williams GA, de Keyzer R (1988) Some new and additional larval host records for Australian Cerambycidae (Coleoptera). *Australian Entomological Magazine* **15**, 95–104.

Webb LJ (1959) A physiognomic classification of rainforests. *Journal of Ecology* **47**, 551–570. doi:10.2307/2257290

Webb LJ (1978) A general classification of Australian rainforests. *Australian Plants* **9**, 349–363.

Webb LJ, Tracey JG (1981) Rainforests: patterns and change. In *Ecological Beogeography of Australia*. (Ed. A Keast) pp. 606–694. Dr W. Junk, Publishers, The Hague.

Webb LJ, Tracey JG, Williams WT (1984) A floristic framework of Australian rainforests. *Australian Journal of Ecology* **9**, 169–198. doi:10.1111/j.1442-9993.1984.tb01356.x

Webb MD (1983) Revision of the Australian Idiocerinae (Hemiptera: Homoptera: Cicadellidae). *Australian Journal of Zoology Supplementary Series* **92**, 1–147.

Webber BL, Curtis ASO, Cassis G, Woodrow IE (2008) Flowering morphology, phenology and flower visitors of the Australian rainforest tree *Ryparosa kurrangii* (Achariaceae). *Australian Entomologist* **35**, 7–17.

Wedmann S, Yeates DK (2008) Eocene records of bee flies (Insecta, Diptera, Bombyliidae, *Comptosia*); their palaeobiogeographic implications and remarks on the evolutionary history of bombyliids. *Palaeontology* **51**, 231–240. doi:10.1111/j.1475-4983.2007.00745.x

Weiblen GD (2002) How to be a fig wasp. *Annual Review of Entomology* **47**, 299–330. doi:10.1146/annurev.ento.47.091201.145213

Weirauch C, Cassis G (2006) Attracting ants: the trichome and novel glandular areas on the sternum of *Ptilocnemus lemur* (Heteroptera: Reduviidae: Holoptilinae). *Journal of the New York Entomological Society* **114**, 28–37. doi:10.1664/0028-7199(2006)114[28:AATTAN]2.0.CO;2

Wellman P (2017) Temperate-climate myxomycetes in Australia and the Southern Hemisphere. *Cunninghamia* **17**, 1–13.

Wells A (1979) A review of the Australian genera *Xuthotrichia* Mosely and *Hellyethria* Neboiss (Trichoptera: Hydroptilidae), with descriptions of new species. *Australian Journal of Zoology* **27**, 311–329. doi:10.1071/ZO9790311

Wells A (Ed.) (1996a) *Zoological Catalogue of Australia. Vol. 26. Psocoptera, Phthiraptera, Thysanoptera*. CSIRO Publishing, Australia.

Wells A (Ed.) (1996b) *Zoological Catalogue of Australia. Vol. 28. Neuroptera, Strepsiptera, Mecoptera, Siphonaptera*. CSIRO Publishing, Australia.

Werren G, Kershaw P (Eds) (1991) *The Rainforest Legacy: Australian Rainforests Study, Volume 2 – Flora and Fauna of the Rainforests*. Australian Government Publishing Service, Canberra.

Wharton RA (2002) Revision of the Australian Alysiini (Hymenoptera: Braconidae). *Invertebrate Systematics* **16**, 7–105. doi:10.1071/IT01012

Wheeler QD (1986) Revision of the genera of Lymexylidae (Coleoptera: Cucujiformia). *Bulletin of the American Museum of Natural History* **183**, 113–210.

Wheeler WM (1914) The ants of the Baltic amber. *Schriften der Physikalisch-Ökonomischen Gesellschaft zu Königsberg* **55**, 1–142.

White IM, Hodkinson ID (1985) Nymphal taxonomy and systematics of Psylloidea (Homoptera). *Bulletin of the British Museum (Natural History) Entomology* **50**, 153–301.

White D, Cribb BW, Heard TA (2001) Flower constancy of the stingless bee *Trigona carbonaria* Smith (Hymenoptera: Apidae: Meliponini). *Austral Entomology* **40**, 61–64. doi:10.1046/j.1440-6055.2001.00201.x

Whitmore TC (1984) *Tropical Rain Forests of the Far East.* Oxford University Press, Oxford.

Whyte R, Anderson G (2017) *A Field Guide to Spiders of Australia.* CSIRO Publishing, Melbourne.

Whyte G, Howard K, Hardy GEStJ, Burgess TI (2011) Foliar pests and pathogens of *Eucalyptus dunnii* plantations in southern Queensland. *Australian Forestry* **74**, 161–169. doi:10.1080/00049158.2011.10676359

Wiebes JT (1979) Co-evolution of figs and their insect pollinators. *Annual Review of Ecology and Systematics* **10**, 1–12. doi:10.1146/annurev.es.10.110179.000245

Will K, Monteith GB (2018) A multi-locus phylogeny, taxonomic review and description of new species of Australian *Nurus* (*sensu stricto*) Motschulsky, 1865 (Coleoptera: Carabidae: Pterostichini). *Australian Entomologist* **45**, 353–388.

Willemstein SC (1987) *An Evolutionary Basis for Pollination Ecology.* Leiden University Press, Leiden.

Williams DJ (1985) *Australian Mealybugs.* British Museum (Natural History), London.

Williams GA (1981) Records of the carrion beetle *Diamesus osculans* Vigor (Silphidae: Coleoptera) from New South Wales. *Australian Entomological Magazine* **8**, 47–48.

Williams GA (1982) A note on Buprestidae (Coleoptera) observed at lights. *Australian Entomological Magazine* **8**, 81.

Williams GA (1983) Observations on the genus *Alcinous* Deyrolle (Coleoptera: Buprestidae). *Victorian Naturalist* **100**, 37–39.

Williams GA (1985) New larval food plants for some Australian Buprestidae and Cerambycidae (Coleoptera). *Australian Entomological Magazine* **12**, 41–46.

Williams GA (1987) A revision of the genus *Nascioides* Kerremans (Coleoptera: Buprestidae). *Invertebrate Taxonomy* **1**, 121–145. doi:10.1071/IT9870121

Williams GA (1993) *Hidden Rainforests: Subtropical Rainforests and their Invertebrate Biodiversity.* UNSW Press/Australian Museum, Sydney.

Williams GA (1995) Pollination ecology of lowland subtropical rainforests in New South Wales. PhD thesis. University New South Wales, Australia.

Williams GA (1998) Bee pollination in the threatened Australian shrub *Senna acclinis* (Caesalpinioideae). *Cunninghamia* **5**, 767–772.

Williams GA (2002) 'A taxonomic and biogeographic review of the invertebrates of the Central Eastern Rainforest Reserves of Australia (CERRA) World Heritage Area, and adjacent regions'. *Technical Reports of the Australian Museum* **16**, 1–208. doi:10.3853/j.1031-8062.16.2002.1353

Williams GA (2003a) Appendix 12, Additional Invertebrate Values. In *World Heritage and Associative Natural Values of the Central Eastern Rainforest Reserves of Australia.* (Ed J Hunter) pp. 1–111. NSW National Parks and Wildlife Service, Hurstville.

Williams GA (2003b) New distribution and biological records for native dung beetles, in the tribe Scarabaeini, from northern New South Wales. *Proceedings of the Linnean Society of New South Wales* **124**, 13–16.

Williams GA (2018) Insects associated with flowering *Rhodomyrtus psidioides* (Myrtaceae): is this a Myrtle Rust (*Austropuccinia psidii*)-induced plant-pollinator interaction extinction event? *Cunninghamia* **18**, 023–027. doi:7751/cunninghamia.2018.002

Williams GA, Adam P (1993) Ballistic pollen release in Australian members of the Moraceae. *Biotropica* **25**, 478–480. doi:10.2307/2388872

Williams GA, Adam P (1994) A review of rainforest pollination and plant-pollinator interactions, with particular reference to Australian subtropical rainforests. *Australian Zoologist* **29**, 177–212. doi:10.7882/AZ.1994.006

Williams GA, Adam P (1995) Records of aculeate wasps from flowering subtropical rainforest trees. *Australian Entomologist* **22**, 51–58.

Williams GA, Adam P (1997) The composition of the bee (Apoidea: Hymenoptera) fauna visiting flowering trees in New South Wales lowland subtropical rainforest remnants. *Proceedings of the Linnean Society of New South Wales* **118**, 69–95.

Williams GA, Adam P (1998) Pollen loads collected from large insects in Australian subtropical rainforests. *Proceedings of the Linnean Society of New South Wales* **120**, 49–67.

Williams GA, Adam P (1999a) Pollen sculpture in subtropical rain forest plants: Is wind pollination more common than previously suspected? *Biotropica* **31**, 520–524. doi:10.1111/j.1744-7429.1999.tb00396.x

Williams GA, Adam P (1999b) Pollination ecology of Australian subtropical rainforests – implications for the conservation of remnant lowland communities. In *Rainforest Remnants: A Decade of Growth.* (Ed. S Horton) pp. 122–147. NSW National Parks and Wildlife Service, Sydney.

Williams GA, Adam P (2001) The insect assemblage visiting the flowers of the subtropical rainforest pioneer tree *Alphitonia excelsa* (Fenzl) Reiss. ex Benth. (Rhamnaceae). *Proceedings of the Linnean Society of New South Wales* **123**, 235–259.

Williams GA, Adam P (2010) *The Flowering of Australia's Rainforests: A Plant and Pollination Miscellany.* CSIRO Publishing, Melbourne.

Williams GA, Adam P (2013) Plants and pollinators – an essential partnership in subtropical rainforest. *Australasian Plant Conservation* **22**, 5–7.
Williams GA, Adam P (2019) A preliminary checklist of flower-visiting insects from *Syzygium floribundum*, *Syzygium smithii* and *Tristaniopsis laurina*: three members of the Myrtle rust-vulnerable plant family Myrtaceae. *Cunninghamia* **19**, 057–074.
Williams GA, Adam P, Mound LA (2001) Thrips (Thysanoptera) pollination in Australian subtropical rainforests, with particular reference to pollination of *Wilkiea huegeliana* (Monimiaceae). *Journal of Natural History* **35**, 1–21. doi:10.1080/002229301447853
Williams GA, Bellamy CL (2002) New species and new records for the genus *Nascioides* Kerremans, 1903. (Insecta: Coleoptera: Buprestidae). *Zootaxa* **58**, 1–9. doi:10.11646/zootaxa.58.1.1
Williams GA, Bickel D (2010) The Invertebrates: insects, crustaceans, arachnids, snails and worms. In *Remnants of Gondwana*. (Eds R Kitching, R Braithwaite and J Cavanaugh) pp. 197–216. Surrey Beatty and Sons, Baulkham Hills.
Williams GA, Gerrand D (1998a) 'Coastline survey of Asparagaceae and other environmental weeds in the Manning Valley, northern New South Wales'. Report for Greater Taree City Council and Coastcare.
Williams GA, Gerrand D (1998b) '"Manning Point" vegetation restoration action plan (1998)'. Report for Environmental Trusts of New South Wales.
Williams GA, Gerrand D (1998c) '"Saltwater Reserve" vegetation restoration action plan (1998)'. Report for Environmental Trusts of New South Wales.
Williams GA, Walker K (2003) Pollination of the wet forest herb *Pollia crispata* (Commelinaceae). *Cunninghamia* **8**, 141–146.
Williams GA, Weir TA (1987) Four new species and new records of Australian Mastogeniinae (Coleoptera: Buprestidae). *Journal of the Australian Entomological Society* **26**, 153–159. doi:10.1111/j.1440-6055.1987.tb00277.x
Williams GA, Williams T (1982) A survey of the Aphodiinae, Hybosorinae and Scarabaeinae (Coleoptera: Scarabaeidae) from small wet forests of coastal New South Wales, Part 1.: Nowra to Newcastle. *Australian Entomological Magazine* **9**, 42–48.
Williams GA, Williams T (1983a) A survey of the Aphodiinae, Hybosorinae and Scarabaeinae (Coleoptera: Scarabaeidae) from small wet forests of coastal New South Wales, Part 2.: Barrington Tops to the Comboyne Plateau. *Victorian Naturalist* **100**, 25–30.
Williams GA, Williams T (1983b) A survey of the Aphodiinae, Hybosorinae and Scarabaeinae (Coleoptera: Scarabaeidae) from small wet forests of coastal New South Wales, Part 3.: Bulahdelah to Taree. *Victorian Naturalist* **100**, 98–105.
Williams GA, Williams T (1983c) A survey of the Aphodiinae, Hybosorinae and Scarabaeinae (Coleoptera: Scarabaeidae) from small wet forests of coastal New South Wales, Part 4.: Lansdowne State Forest. *Victorian Naturalist* **100**, 146–154.
Williams GA, Williams T (1984) A survey of the Aphodiinae, Hybosorinae and Scarabaeinae (Coleoptera: Scarabaeidae) from small wet forests of coastal New South Wales, Part 5.: Littoral rainforests from Myall Lakes to Crowdy Bay National Park. *Victorian Naturalist* **101**, 127–135.
Williams JE, Woinarski JCZ (1997) *Eucalypt Ecology: Individuals to Ecosystems*. Cambridge University Press, Cambridge.
Williams JG, Harden GJ, McDonald WJF (1984) *Trees and Shrubs in Rainforests of New South Wales and Southern Queensland*. Botany Department, University of New England, Armidale.
Willmann R (1978) Mecoptera (Insecta, Holometabola). *Fossilium Catalogus: Animalia* **124**, 1–139.
Willmann R (2003) Die phylogenetischen Beziehungen der Insecta: Offene Fragen und Probleme. *Verhandlungen Westdeutscher Entomologentag* **2001**, 1–64.
Willmer P (2011) *Pollination and Floral Ecology*. Princeton University Press, Princeton.
Wilson FE (1923) New Australian Coleoptera, with notes on some previously described species. Part 2. *Royal Society of Victoria* **35**, 117–133.
Wilson WJ (1984) New distribution records for some Queensland and New South Wales butterflies. *Australian Entomological Magazine* **11**, 59.
Windholz M, Budavari S, Stroumtos L, Fertig MN (Eds) (1978) *The Merck Index; and Encyclopedia of Chemicals and Drugs*. Merck and Company, Rahway, NJ.
Winsor L (1985) The land nemertine *Argonemertes australiensis* (Dendy) in south eastern Australia. *Victorian Naturalist* **102**, 28–36.
Winsor L (1997) The biodiversity of terrestrial flatworms (Tricladida: Terricola) in Queensland: a preliminary report. *Memoirs of the Museum of Victoria* **56**, 575–579. doi:10.24199/j.mmv.1997.56.52
Winsor L (1998) The Australian terrestrial flatworm species (Tricladida: Terricola). *Pedobiologia* **42**, 457–463.
Winter JW, Bell FC, Atherton RG, Pahl LI (1987) An introduction to Australian rainforests. In *The Rainforest Legacy. Australian National Rainforests Study. Vol. 1*. (Eds G Werren and AP Kershaw) pp. 1–7. Australian Government Publishing Service, Canberra.
Winterbourn MJ (1980) The freshwater insects of Australasia and their affinities. *Palaeogeography, Palaeoclimatology, Palaeoecology* **31**, 235–249. doi:10.1016/0031-0182(80)90020-6
Winterton SL (1995) A new genus and species of Apochrysinae (Neuroptera: Chrysopidae) from Australia, with a checklist of Australian Chrysopidae. *Journal of the Australian Entomological Society* **34**, 139–145. doi:10.1111/j.1440-6055.1995.tb01306.x
Winterton SL (2006) New species of *Euposilocephala* Kröber from Australia (Diptera: Therevidae). *Zootaxa* **1372**, 17–25. doi:10.11646/zootaxa.1372.1.2
Winterton SL (2007a) Revision of the Australian stiletto fly genus *Acatopygia* Kröber (Diptera: Therevidae, Agapophytinae). *Zootaxa* **1405**, 51–62.
Winterton SL (2007b) New species of *Laxotela* Winterton and Irwin from Australia (Diptera: Therevidae: Agapophytinae). *Zootaxa* **1407**, 57–64.

Winterton SL (2009) Revision of the stiletto fly genus *Neodialineura* Mann (Diptera: Therevidae): an empirical example of cybertaxonomy. *Zootaxa* **2157**, 1–33. doi:10.11646/zootaxa.2157.1.1

Winterton SL (2011) Revision of the stiletto fly genera *Acupalpa* and *Pinnipons* (Diptera, Therevidae, Agapophytinae) using cybertaxonomic methods, with a key to Australasian genera. *ZooKeys* **95**, 29–79. doi:10.3897/zookeys.95.1461

Winterton SL, Hardy NB, Gaimari SD, Hauser M, Hill HN, Holston KC, Irwin ME, Lambkin CL, Metz MA, Turco F, Webb D, Yang L, Yeates DK, Wiegmann BM (2015) The phylogeny of stiletto flies (Diptera: Therevidae). *Systematic Entomology* **41**, 144–161.

Winterton SL, Irwin ME (1999) *Laxotela* – a new genus of Therevidae (Diptera) from Australia. *Entomologica Scandinavica* **30**, 299–310. doi:10.1163/187631200X00138

Winterton SL, Irwin ME (2001) Phylogenetic revision of *Agapophytus* Guérin (Diptera: Therevidae: Agapophytinae). *Invertebrate Taxonomy* **15**, 467–526. doi:10.1071/IT00027

Winterton SL, Irwin ME, Yeates DK (1999a) Phylogenetic revision of the *Taenogera* Krober genus-group (Diptera: Therevidae), with descriptions of two new genera. *Australian Journal of Entomology* **38**, 274–290. doi:10.1046/j.1440-6055.1999.00126.x

Winterton SL, Irwin ME, Yeates DK (1999b) Systematics on *Nanexila* Winterton and Irwin, gen. nov. (Diptera: Therevidae) from Australia. *Invertebrate Taxonomy* **13**, 237–308. doi:10.1071/IT97029

Winterton SL, Lambkin CL (2012) New Australian stiletto flies: revision of *Manestella* Metz and description of *Medomega* gen. n. (Diptera, Therevidae, Agapophytinae). *ZooKeys* **240**, 1–119. doi:10.3897/zookeys.240.2967

Winterton SL, Yang L, Wiegmann BM, Yeates DK (2001) Phylogenetic revision of Agapophytinae subf. n. (Diptera: Therevidae). *Systematic Entomology* **26**, 173–211. doi:10.1046/j.1365-3113.2001.00142.x

Wirth WW, Stone A (1956) Aquatic Diptera. In *Aquatic Insects of California*. (Ed. RL Usinger) pp. 372–482. University of California Press, Berkeley.

Wishart G (1992) New species of the trapdoor spider genus *Misgolas* Karsch (Mygalomorphae: Idiopidae) with a review of the tube-building species. *Records of the Australian Museum* **44**, 263–278. doi:10.3853/j.0067-1975.44.1992.35

Wishart G, Rowell DM (2008) Trapdoor spiders of the genus *Misgolas* (Mygalomorphae: Idiopidae) from eastern New South Wales, with notes on genetic variation. *Records of the Australian Museum* **60**, 45–86. doi:10.3853/j.0067-1975.60.2008.1495

Wolda H (1978) Seasonal fluctuations in rainfall, food and abundance of tropical insects. *Journal of Animal Ecology* **47**, 369–381. doi:10.2307/3789

Wood GA, Hasenpusch J, Storey RI (1996) The life history of *Phalacrognathus muelleri* (Macleay) (Coleoptera: Lucanidae). *Australian Entomologist* **23**, 37–48.

Wood LM (1926) On some land planarians from Barrington Tops, N.S.W. with descriptions of new species. *Proceedings of the Linnean Society of New South Wales* **51**, 608–613.

Woodley NE (1989) Family Neurochaetidae. In *Catalog of the Diptera of the Australasian and Oceanic Regions*. (Ed. NL Evenhuis) p. 671. Bishop Museum Press and E.J. Brill, Honolulu.

Woodward TE (1956) *On Australian and New Zealand Peloridiidae (Homoptera: Coleorrhyncha)*. University of Queensland Press, Brisbane.

Woodward TE, Postle AC (1986) The Australian species of *Orius* Wolff (Heteroptera: Anthocoridae). *Journal of the Australian Entomological Society* **25**, 245–254. doi:10.1111/j.1440-6055.1986.tb01111.x

Wunderlich J (1976) Spinnen aus Australien. 1. Uloboridae, Theridiosomatidae und Symphytognathidae (Arachnida: Araneida). *Senckenbergiana Biologica* **57**, 113–124.

Xuankan L, Rodrigues PFM, Lamas CJE, Yeates DK (2018) A review of Australian species of *Thevenetimyia* Bigot, 1875 (Bombyliidae, Bombyliinae, Eclimini), with descriptions of four new species and the pupal case of *T. longipalpis*. *Records of the Australian Museum* **70**, 331–375.

Yan Z, Bai M, Ren D (2013) A new genus and species of fossil Hybosoridae (Coleoptera: Scarabaeoidea) from the Early Cretaceous Yixian formation of Liaoning, China. *Alcheringa: An Australasian Journal of Palaeontology* **37**, 139–145. doi:10.1080/03115518.2012.716231

Yeates DK (1985) Altitudinal transect studies at Cape Tribulation, north Queensland IV. The marchflies (Diptera: Tabanidae). *Queensland Naturalist* **26**, 58–61.

Yeates DK (1991a) Revision of the Australian beefly genus *Aleucosia* Edwards (Diptera: Bombyliidae). *Invertebrate Taxonomy* **5**, 133–209. doi:10.1071/IT9910133

Yeates DK (1991b) Revision of the Australian beefly genus *Comptosia* (Diptera: Bombyliidae). *Invertebrate Taxonomy* **5**, 1023–1178. doi:10.1071/IT9911023

Yeates DK (1994) The cladistics and classification of the Bombyliidae (Diptera: Asiloidea). *Bulletin of the American Journal of Natural History* **219**, 1–191.

Yeates DK (1996) Revision of the Australian bee fly genus *Neosardus* Roberts (Diptera: Bombyliidae). *Invertebrate Taxonomy* **10**, 47–75. doi:10.1071/IT9960047

Yeates DK, Harvey MS, Austin HI (2003) New estimates for terrestrial arthropod species-richness in Australia. *Records of the South Australian Museum. Monograph Series* **7**, 231–241.

Yeates DK, Irwin ME (1996) Apioceridae (Insecta: Diptera): Cladistic reappraisal and biogeography. *Zoological Journal of the Linnean Society* **116**, 247–301. doi:10.1111/j.1096-3642.1996.tb00124.x

Yeates DK, Lambkin CL (1998) Cryptic species diversity and character congruence: review of the tribe Anthracini (Diptera: Bombyliidae) in Australia. *Invertebrate Taxonomy* **12**, 977–1078. doi:10.1071/IT97019

Yeates DK, Wiegmann BM (Eds) (2005) *The Evolutionary Biology of Flies*. Columbia University Press, New York.

Yen AL (2002) Short-range endemism and Australian Psylloidea (Insecta: Hemiptera) in the genera *Glycaspis* and

Acizzia (Psyllidae). *Invertebrate Systematics* **16**, 631–639. doi:10.1071/IT01038

Young GR (2003) Life history, biology, host plants and natural enemies of the lilly pilly psyllid, *Trioza eugeniae* Froggatt (Hemiptera: Triozidae). *Australian Entomologist* **30**, 31–38.

Young PAR, McDonald WJF (1987) The distribution, composition and status of the rainforests of southern Queensland. In *The Rainforest Legacy. Australian National Rainforests Study. Vol. 1.* (Eds G Werren and AP Kershaw) pp. 119–141. Australian Government Publishing Service, Canberra.

Yu YL, Ślipiński A, Reid C, Shih CK, Pang H, Ren D (2015) A new longhorn beetle (Coleoptera) from the Early Cretaceous Jehol biota of western Lianing in China. *Cretaceous Research* **52**, 453–460. doi:10.1016/j.cretres.2014.02.019

Žabka M (1990) Salticidae (Araneae) of Oriental, Australian and Pacific regions, IV. Genus *Ocrisiona* Simon, 1901. *Records of the Australian Museum* **42**, 27–43. doi:10.3853/j.0067-1975.42.1990.105

Žabka M (1991) Salticidae (Arachnida: Araneae) of Oriental, Australian and Pacific Regions, V. Genus *Holoplatys* Simon, 1885. *Records of the Australian Museum* **43**, 171–240. doi:10.3853/j.0067-1975.43.1991.46

Žabka M (1992) Salticidae (Arachnida: Araneae) from Oriental, Australian and Pacific regions. VII. *Paralatoides* and *Grayenulla* – new genera from Australia and New Caledonia. *Records of the Australian Museum* **44**, 165–183. doi:10.3853/j.0067-1975.44.1992.31

Žabka M (2000) Salticidae (Arachnida: Araneae) from Oriental, Australian and Pacific regions, XIII: the genus *Sandalodes* Keyserling. *Invertebrate Taxonomy* **14**, 695–704. doi:10.1071/IT99006

Žabka M (2002) Salticidae (Arachnida: Araneae) from the Oriental, Australian and Pacific regions, XV. New species of Astieae from Australia. *Records of the Australian Museum* **54**, 257–268. doi:10.3853/j.0067-1975.54.2002.1365

Žabka M, Gray MR (2004) Salticidae (Arachnida: Araneae) from the Oriental, Australian and pacific regions, XVIII. *Huntglennia* – a new genus from Australia. *Annales Zoologici* **54**, 587–590[Warszawa].

Zahiri R, Holloway JD, Kitching IJ, Lafontaine JD, Mutanen M, Wahlberg N (2012) Molecular phylogenetics of Erebidae (Lepidoptera, Noctuoidea). *Systematic Entomology* **37**, 102–124. doi:10.1111/j.1365-3113.2011.00607.x

Zalucki MP, Drew RA, Hooper GH (1984) Ecological studies of Eastern Australian fruit flies (Diptera: Tephritidae) in their endemic habitat. II. The spatial pattern of abundance. *Oecologia* **64**, 273–279. doi:10.1007/BF00376882

Zborowski P, Edwards T (2007) *A Guide to Australian Moths.* CSIRO Publishing, Melbourne.

Zimmerman EC (1993) *Australian Weevils. Volume III. (Coleoptera: Curculionidae).* CSIRO Publishing, Melbourne.

Zimmerman EC (1994a) *Australian Weevils. Volume I. Orthoceri, Anthribidae to Attelabidae. The Primitive Weevils.* CSIRO Publishing, Melbourne.

Zimmerman EC (1994b) *Australian Weevils. Volume II. Brentidae, Eurhynchidae, Apionidae.* CSIRO Publishing, Melbourne.

Zipperer WC (1993) Deforestation patterns and their effects on forest patches. *Landscape Ecology* **8**, 177–184. doi:10.1007/BF00125349

Zuska J, Colless DH (1984) Australian Sepsidae (Diptera). *Journal of the Australian Entomological Society* **23**, 59–67. doi:10.1111/j.1440-6055.1984.tb01909.x

Zwick P (1981) Australian *Edwardsina* (Diptera: Blephariceridae), new and rediscovered species. *Aquatic Insects* **3**, 75–78. doi:10.1080/01650428109361049

Zwick P (1989) Family Blephariceridae. In *Catalog of the Diptera of the Australasian and Oceanic Regions.* (Ed. NL Evenhuis) pp. 119–121. Bishop Museum Press and E.J. Brill, Honolulu.

Zwick P (1998) Australian net-winged midges of the tribe Apistomyiini (Diptera: Blephariceridae). *Australian Journal of Entomology* **37**, 289–311. doi:10.1111/j.1440-6055.1998.tb01587.x

Zwick P (2000) Phylogenetic system and zoogeography of the Plecoptera. *Annual Review of Entomology* **45**, 709–746. doi:10.1146/annurev.ento.45.1.709

Index

www.ingramcontent.com/pod-product-compliance
Ingram Content Group UK Ltd.
Pitfield, Milton Keynes, MK11 3LW, UK
UKHW061542051026
15625UKWH00011B/574
* 9 7 8 1 4 8 6 3 1 2 9 1 7 *